HANDBOOK OF BEHAVIORAL NEUROSCIENCE

HANDBOOK OF BEHAVIORAL NEUROSCIENCE

Translational Medicine in CNS Drug Development

VOLUME 29

Edited by

GEORGE G. NOMIKOS
Biogen, Clinical Development, Cambridge, MA, United States

DOUGLAS E. FELTNER
Clinical Development, AveXis, a Novartis company, United States

ELSEVIER

ACADEMIC PRESS
An imprint of Elsevier

Academic Press is an imprint of Elsevier
125 London Wall, London EC2Y 5AS, United Kingdom
525 B Street, Suite 1650, San Diego, CA 92101, United States
50 Hampshire Street, 5th Floor, Cambridge, MA 02139, United States
The Boulevard, Langford Lane, Kidlington, Oxford OX5 1GB, United Kingdom

Library of Congress Cataloging-in-Publication Data
A catalog record for this book is available from the Library of Congress

British Library Cataloguing-in-Publication Data
A catalogue record for this book is available from the British Library

ISBN 978-0-12-803161-2
ISSN 1569-7339

For information on all Academic Press publications
visit our website at https://www.elsevier.com/books-and-journals

Publisher: Nikki Levy
Acquisition Editor: Joslyn Chaiprasert-Paguio
Editorial Project Manager: Timothy Bennett
Production Project Manager: Paul Prasad Chandramohan
Cover Designer: Miles Hitchen

Typeset by SPi Global, India

Working together
to grow libraries in
developing countries

www.elsevier.com • www.bookaid.org

Contents

Contributors

Ayman Abou-Aisha Mood and Anxiety Disorders Service, Southern Health NHS Foundation Trust, Southampton, United Kingdom

Stephen P. Arnerić Critical Path for Alzheimer's Disease, Critical Path Institute, Tucson, AZ, United States

Rajendra D. Badgaiyan Department of Psychiatry, South Texas Veteran Health Care System, Audie L. Murphy Memorial VA Hospital; Long School of Medicine, University of Texas Medical Center, San Antonio, TX, United States

David S. Baldwin Clinical and Experimental Sciences, Faculty of Medicine, University of Southampton; Mood and Anxiety Disorders Service, Southern Health NHS Foundation Trust, Southampton, United Kingdom; University Department of Psychiatry and Mental Health, University of Cape Town, Cape Town, South Africa

David Baron Western University Health Sciences, Graduate School of Biomedical Sciences, Pomona, CA; Division of Neuroscience and Addiction Research, Pathway Healthcare, Birmingham, AL, United States

Nicolas Barthelemy Department of Neurology, Washington University School of Medicine, St. Louis, MO, United States

Randall J. Bateman Department of Neurology, Washington University School of Medicine, St. Louis, MO, United States

Kenneth Blum Western University Health Sciences, Graduate School of Biomedical Sciences, Pomona, CA; Division of Addiction Services, Dominion Diagnostics LLC, North Kingstown, RI; Division of Neuroscience and Addiction Research, Pathway Healthcare, Birmingham, AL; Division of Neuroscience Research and Addiction Therapy, Shores Treatment and Recovery Center, Port Saint Lucie, FL; Human Integrated Services Unit, University of Vermont Centre for Clinical & Translational Science, College of Medicine, Burlington, VT; Division of Clinical Neurology, Path Foundation NY, New York, NY; Division of Addiction Research & Therapy, Nupathways, Innsbrook, MO; Division of Precision Medicine, Geneus Health LLC, San Antonio, TX; Department of Psychiatry, Wright State University, Boonshoft School of Medicine, Dayton, OH, United States; Eötvös Loránd University, Institute of Psychology, Budapest, Hungary

James G. Bollinger Department of Neurology, Washington University School of Medicine, St. Louis, MO, United States

Peter Bonate Astellas Pharma, Northbrook, IL, United States

Anjana Bose Synchrogenix, A Certara Company, Delaware Corporate Center, Wilmington, DE, United States

Abdalla Bowirrat Division of Anatomy, Biochemistry and Genetics, Faculty of Medicine and Health Sciences, An-Najah National University, Nablus, Palestine

John R. Cirrito Department of Neurology, Washington University School of Medicine, St. Louis, MO, United States

Emil F. Coccaro Clinical Neuroscience and Psychopharmacology Research Unit, Department of Psychiatry, Pritzker School of Medicine, University of Chicago, Chicago, IL, United States

Patricia E. Cole Cole Imaging and Biomarker Consulting, LLC, Glenview, IL, United States

Joyce Colussi-Mas Victoria University of Wellington, School of Psychology, Behavioural Neurogenetics Group, Wellington, New Zealand

Paul Cumming Department of Nuclear Medicine, Inselspital, University of Bern, Bern, Switzerland; School of Psychology and Counselling and IHBI, Queensland University of Technology, and QIMR Berghofer Medical Research Institute, Brisbane, QLD, Australia

Willie Daniels Department of Physiology, University of the Witwatersrand, Johannesburg, South Africa

Elizabeth C. de Lange Leiden Academic Centre for Drug Research, Gorlaeus Laboratories, Leiden, The Netherlands

Dominique J.-F. de Quervain Transfaculty Research Platform Molecular and Cognitive Neurosciences; Psychiatric University Clinics; Department of Psychology, Division of Cognitive Neuroscience, University of Basel, Basel, Switzerland

Aldemar Degroot Boehringer Ingelheim Pharma GmbH & Co. KG., Biberach an der Riss, Germany

Wayne C. Drevets Janssen Research & Development, LLC, of Johnson & Johnson, Titusville, NJ, United States

Steven G. Einstein Janssen Research and Development, Titusville, NJ, United States

Bart A. Ellenbroek Victoria University of Wellington, School of Psychology, Behavioural Neurogenetics Group, Wellington, New Zealand

Igor Elman Department of Psychiatry, Wright State University, Boonshoft School of Medicine, Dayton, OH, United States

Audrey Gabelle Memory Research and Resources Center, Gui de Chauliac Hospital, Université de Montpellier, Montpellier, France

B.M. Geiger Department of Biomedical and Nutritional Sciences, Zuckerberg College of Health Sciences, University of Massachusetts, Lowell, Lowell, MA, United States

Jogarao V.S. Gobburu Center for Translational Medicine, School of Pharmacy, University of Maryland, Baltimore, Baltimore, MD, United States

Mark S. Gold Division of Neuroscience and Addiction Research, Pathway Healthcare, Birmingham, AL; Department of Psychiatry, Washington University School of Medicine, Saint Louis, MO, United States

Mathangi Gopalakrishnan Center for Translational Medicine, School of Pharmacy, University of Maryland, Baltimore, Baltimore, MD, United States

Radu C. Grovu Mood Disorders Psychopharmacology Unit (MDPU), Toronto Western Hospital, University Health Network, Toronto, ON, Canada

Gerhard Gründer Department of Molecular Neuroimaging, Central Institute of Mental Health (CIMH), Mannheim, Germany

Adam Hanina AiCure, New York, NY, United States

John E. Harrison Metis Cognition Ltd., Kilmington Common, Kilmington; Institute of Psychiatry, Psychology and Neuroscience, King's College London, London, United Kingdom; Alzheimer's Center, VUmc, Amsterdam, The Netherlands

Brian H. Harvey Division of Pharmacology, School of Pharmacy and Centre of Excellence for Pharmaceutical Sciences, North-West University, Potchefstroom, South Africa

Suzanne Hendrix Pentara Corporation, Salt Lake City, UT, United States

Christophe Hirtz Laboratoire de Biochimie Protéomique Clinique, CHU de Montpellier, Université de Montpellier, INSERM U1183, Montpellier, France

Yash B. Joshi Department of Psychiatry, University of California, San Diego, CA, United States

Amir Kalali San Diego, CA, United States

Meyrick Kidwell Victoria University of Wellington, School of Psychology, Behavioural Neurogenetics Group, Wellington, New Zealand

Bruce J. Kinon Lundbeck North America, Deerfield, IL, United States

Paul Kotzbauer Department of Neurology, Washington University School of Medicine, St. Louis, MO, United States

Michael Krams Janssen Research & Development, LLC, of Johnson & Johnson, Titusville, NJ, United States

Royce Lee Clinical Neuroscience and Psychopharmacology Research Unit, Department of Psychiatry, Pritzker School of Medicine, University of Chicago, Chicago, IL, United States

Sylvain Lehmann Laboratoire de Biochimie Protéomique Clinique, CHU de Montpellier, Université de Montpellier, INSERM U1183, Montpellier, France

Gregory A. Light Department of Psychiatry, University of California, San Diego, CA, United States

Enchi Liu E-Scape Bio, South San Francisco, CA, United States

Veronika Logovinsky Regeneron Pharmaceuticals Inc., Tarrytown, NY, United States

Brendan Lucey Department of Neurology, Washington University School of Medicine, St. Louis, MO, United States

Matthew Macaluso Department of Psychiatry and Behavioral Sciences, University of Kansas School of Medicine-Wichita and KU-Wichita Clinical Trials Unit, Wichita, KS, United States

Gerard J. Marek Astellas Pharma Global Development, Inc., Development Medical Science, CNS and Pain, Northbrook, IL, United States

Paul Maruff Florey Institute for Neuroscience and Mental Health, University of Melbourne, Parkville, VIC, Australia; Cogstate, New Haven, CT, United States

Roger S. McIntyre Mood Disorders Psychopharmacology Unit (MDPU), Toronto Western Hospital, University Health Network; Institute of Medical Science; Department of Pharmacology; Department of Psychiatry, University of Toronto; Brain and Cognition Discovery Foundation, Toronto, ON, Canada

Bradley Miller Eli Lilly and Company, Indianapolis, IN, United States

Timothy Miller Department of Neurology, Washington University School of Medicine, St. Louis, MO, United States

Edward J. Modestino Department of Psychology, Curry College, Milton, MA, United States

George G. Nomikos Biogen, Clinical Development, Cambridge, MA, United States

Eamon O'Loinsigh Synchrogenix, A Certara Company, Delaware Corporate Center, Wilmington, DE, United States

Emilia Ouzunova Neuroscience Center of Excellence, IQVIA, Durham, NC, United States

Zihang Pan Mood Disorders Psychopharmacology Unit (MDPU), Toronto Western Hospital, University Health Network; Institute of Medical Science, University of Toronto, Toronto, ON, Canada

Andreas Papassotiropoulos Transfaculty Research Platform Molecular and Cognitive Neurosciences; Department of Psychology, Division of Molecular Neuroscience; Psychiatric University Clinics; Department Biozentrum, Life Sciences Training Facility, University of Basel, Basel, Switzerland

Bruce W. Patterson Department of Internal Medicine, Washington University School of Medicine, St. Louis, MO, United States

Jeffrey Paul Drexel University, Graduate School of Biomedical Sciences, College of Medicine, Philadelphia, PA, United States

Katrina Paumier Department of Neurology, Washington University School of Medicine, St. Louis, MO, United States

E.N. Pothos Program in Pharmacology and Experimental Therapeutics and Pharmacology and Drug Development, Sackler School of Graduate Biomedical Sciences and Department of Immunology, Tufts University School of Medicine, Boston, MA, United States

Sheldon H. Preskorn Department of Psychiatry and Behavioral Sciences, University of Kansas School of Medicine-Wichita and KU-Wichita Clinical Trials Unit, Wichita, KS, United States

Lisa Prock Translational Neuroscience Center; Developmental Medicine Center, Division of Developmental Medicine, Department of Medicine, Boston Children's Hospital, Harvard Medical School, Boston, MA, United States

Sarah Richerson Neuroscience Center of Excellence, IQVIA, Durham, NC, United States

Mustafa Sahin Department of Neurology; Translational Neuroscience Center, Boston Children's Hospital, Harvard Medical School, Boston, MA, United States

Chihiro Sato Department of Neurology, Washington University School of Medicine, St. Louis, MO, United States

Jonathan Savitz Laureate Institute for Brain Research, Tulsa, OK, United States

Laura Shafner AiCure, New York, NY, United States

Neal G. Simon Department of Biological Sciences, Lehigh University, Bethlehem, PA, United States

A. Benjamin Srivastava Department of Psychiatry, Washington University School of Medicine, Saint Louis, MO; Department of Psychiatry, Columbia University Medical Center/New York State Psychiatric Institute, New York, NY, United States

Siddharth Srivastava Department of Neurology, Boston Children's Hospital, Harvard Medical School, Boston, MA, United States

Dan J. Stein Department of Psychiatry, University of Cape Town, Cape Town, South Africa

Diane T. Stephenson Critical Path for Parkinson's, Critical Path Institute, Tucson, AZ, United States

Melissa A. Tarasenko Department of Psychiatry, University of California, San Diego, CA, United States

Joop van Gerven Centre for Human Drug Research, Leiden, The Netherlands

Tim West C2N Diagnostics, LLC, St. Louis, MO, United States

Ryan Westphal Eli Lilly and Company, Indianapolis, IN, United States

Kevin Yarasheski C2N Diagnostics, LLC, St. Louis, MO, United States

Jiun Youn Victoria University of Wellington, School of Psychology, Behavioural Neurogenetics Group, Wellington, New Zealand

Preface

Development of new drugs for the treatment of CNS disorders has been hampered by inefficiencies in advancing compounds from nonclinical discovery to the clinic. This may be due to the lack of appropriate mechanisms for testing the pharmacology of compounds as they transition from bench to bedside. Until recently, emphasis was placed on testing the new molecules in various in vitro and in vivo models and then promoting the best candidates for further evaluation in man. As a result the number of approvals of new medications has remained relatively low, while the unmet medical needs in the treatment of CNS disorders remain high. Over the last decade, game-changing strides have been made in our understanding of the pathophysiology of CNS disorders and the relation of drug exposure in plasma and CNS to pharmacodynamic measures in both animals and humans. In addition, considerable progress has been made in the use of neuroimaging, neurochemical, neurophysiological, and pharmacogenetic/pharmacogenomic biomarker tools in monitoring disease and drug responses. Thus biomarkers that play multiple roles in the diagnosis and progression of the disease and as tools in showing central exposure, target engagement, functional responses, and patient enrichment in clinical trials have been developed with a well-defined context of use. Experimental medicine trials in healthy volunteers or patients also provide confidence in advancing drug development to exploratory clinical trials. The utility of these approaches will ultimately be tested, as drugs progress successfully to confirmatory clinical trials, registration, and launch. Furthermore, biomarker-driven personalized medicine offers the opportunity to treat patients based on their individual characteristics and achieve better efficacy and safety outcomes. Given this background information, it appears timely to attempt at capturing in a systematic manner the latest developments in translational medicine and the biomarker tool bag in use for the development of new drugs for the treatment of CNS disorders. This volume is intended for clinicians; drug developers in industry and academia; and graduate students in behavioral neuroscience, neurobiology, translational neuroscience, clinical psychiatry, and pharmaceutical science. The chapters in this volume cover several aspects of applications of translational medicine in CNS drug development by providing specific examples of (i) biomarker modalities and experimental medicine in early phases of clinical development; (ii) the uptake of these approaches by public-private partnerships, consortia, and regulators and lessons learned from such interactions; and (iii) the use of translational medicine tools in clinical development efforts in selected indications and therapeutic areas.

The first chapter provides examples of translation of nonclinical models of disease and drug action to the clinic (Geiger and Pothos), while the second chapter covers the role of biomarkers in defining early patient studies and clinical trial efficiency (Degroot). The next two chapters discuss the applications of modeling and simulation in the translational pharmacology of CNS drugs (Bonate and de Lange) and functional measurements of CNS drug effects in early human drug development (van Gerven). We continue with a chapter on experimental medicine approaches in CNS drug development (Paul), a chapter on Phase I trials from traditional to newer approaches (Macaluso et al.), a chapter on translational and experimental medicine approaches for antidepressant drug development (Marek), and a chapter on biomarker opportunities for enrichment of clinical trial populations for drug development in schizophrenia and depression (Kinon). The next few chapters focus on describing various biomarker modalities in CNS drug development; these include applications of translational neuroimaging research (Cole and Einstein), PET occupancy and competition in translational medicine (Cumming and Gründer), uses of stable isotope labeling kinetics in CNS translational medicine (Bateman et al.), applications of neurophysiological biomarkers with focus on psychoses (Joshi et al.), heart rate variability as a translational biomarker for emotional and cognitive deficits (Ellenbroek et al.), applications of pharmacogenetic and pharmacogenomic biomarkers (Papassotiropoulos and De Quervain), the use of cognition to guide decisions about the safety and efficacy of drugs in early phase clinical trials (Maruff), and digital health-based biomarkers for use in clinical drug development (Kalali et al.). These are followed by chapters on lessons learned from public-private partnerships and consortia based on the ADNI paradigm (Liu), regulatory perspectives on the use of biomarkers and personalized medicine from the FDA and EMA viewpoint (Gopalakrishnan and Gobburu; O'Loinsigh and Bose), and learnings from the Critical Path Institute efforts in regulatory qualification of biomarkers (Stephenson and Arneric). The next chapter addresses the assessment of cognition in translational medicine and the contrast between the approaches used in Alzheimer's disease and major depressive disorder (Harrison and Hendrix) followed by a number of chapters on translational medicine strategies and experimental medicine approaches in drug development for neurodevelopmental disorders (Sahin and Srivastava); for mood disorders with focus on anhedonia and cognitive dysfunction and the role of inflammatory processes (Pan et al.); in Alzheimer's disease (Logovinsky); in generalized anxiety and social anxiety disorders (Baldwin and Ayman); in PTSD (Stein et al.); in the clinical development of a differentiated antidepressant (Nomikos);

in tobacco, alcohol, drug, and behavioral addictions (Srivastava and Gold): for impulse aggression (Coccaro et al.): and for hypothesizing major depression as a subset of reward deficiency syndrome linked to polymorphic reward genes (Blum et al.). The final chapter presents practical applications of virtual patient monitoring for increased signal detection in CNS drug development (Hanina and Shafner).

We thank all the contributors of the chapters in this volume for sharing their experience and knowledge with us and our readers, who we hope that they find the present data and information useful and applicable to their translational medicine approaches in CNS drug development.

George Nomikos and Douglas Feltner

1

Translating Animal Models of Obesity and Diabetes to the Clinic

B.M. Geiger*, E.N. Pothos[†]

*Department of Biomedical and Nutritional Sciences, Zuckerberg College of Health Sciences, University of Massachusetts, Lowell, Lowell, MA, United States [†]Program in Pharmacology and Experimental Therapeutics and Pharmacology and Drug Development, Sackler School of Graduate Biomedical Sciences and Department of Immunology, Tufts University School of Medicine, Boston, MA, United States

I INTRODUCTION

Preclinical models of disease, specifically animal models, are a necessary tool for a robust and innovative drug development process. The use of animal models to identify and validate pharmacological targets has been part of drug development for decades. Experiments in cell culture systems can provide valuable molecular mechanistic information, but they cannot typically model the more complex interactions of multiple cell types that often contribute to the pathophysiology of a disease. Animal models that contribute to our understanding of the CNS are especially important, as acquiring tissue to study from the human brain is not generally feasible or permissible. In this chapter, we discuss the various attributes of valid animal models and illustrate specific examples of animals that we use for translational research in obesity and diabetes.

II ATTRIBUTES OF VALID ANIMAL MODELS

Animal models can be extremely valuable in the translation of our understanding of various diseases to treatments for those diseases in the clinic. Good disease models have some, if not all, of the following attributes: (1) pathophysiological similarities to human disease, (2) a phenotypic match to the disease state, (3) replicability and/or reproducibility, and (4) cost-efficiency (Blass, 2015). The first two criteria directly address the validity of the model (i.e., whether and how it relates to the disease it is supposed to model).

(1) A Pathophysiological Similarities to Human Systems

When considering which animal models to use, one of the first criteria to consider should be how closely does the onset and development of the disease in the animal mimic the corresponding human disease. For example, in humans, an increase in an individual's body weight and excess blood glucose over time will contribute to decreased sensitivity of their insulin receptors and the eventual development of type 2 diabetes (Reinehr, 2013; Weiss et al., 2004). Therefore, a good model of this disease would also develop insulin resistance over time as body weight, adiposity, and blood glucose levels increase. As an example, C57Bl/6 mouse models of diet-induced obesity have been shown to follow the same metabolic progression of the disease as seen in humans and develop hyperglycemia and insulin resistance over time, similarly to the human disease (Collins, Martin, Surwit, & Robidoux, 2004; Petro et al., 2004; Speakman, Hambly, Mitchell, & Krol, 2007; Wang & Liao, 2012). In studying these models, researchers have been able to elucidate molecular mechanisms involved in the development of insulin resistance over time. We discuss the utility of diet-induced obesity models in greater detail later in this chapter.

(2) B Phenotypic Match to Disease State

Other times, models may be used, where the exact pathophysiology of a disease is not a match to the human pathophysiology, but the presentation of the disease in the model is similar to the human condition.

These models are referred to as phenotypic models of disease. The phenotype of a model is the set of characteristics, usually disease-related, that can be observed. If a model is a phenotypic match to a disease, then it has similar observable traits as a human who has that disease. For example, the *ob/ob* mouse exhibits extreme weight gain in the form of excess adipose tissue, similarly to human obesity. However, the development of obesity in the *ob/ob* mouse arises from leptin deficiency, while the vast majority of obese individuals have high levels of leptin. Only a very small number of human obese patients are actually leptin-deficient (Farooqi, 2008; Farooqi et al., 2007; Farooqi & O'Rahilly, 2008; Wang, Chandrasekera, & Pippin, 2014). Therefore, this model matches the phenotype of obesity, while the pathophysiology presents differences. Investigators should use caution when using models that are phenotypic models of disease to elucidate the molecular mechanisms of the disease. The altered underlying physiology may not match the human pathophysiology of the disease.

C Replicability and Reproducibility

Replicability of a model is the ability to use the same model within a research group and achieve similar results. When conducting research using animal models, a study will often span several months of data collection and use animals that have been born to different dams or in different litters. In this case, replicability of the model is extremely important. If the model is developed using an environmental insult, like 1-methyl-4-phenyl-1,2,3,6-tetrahydropyridine (MPTP) injections for the generation of a Parkinson's disease rodent model, consistent use of the same stereotaxic coordinates is necessary to ensure that the insult is in the same portion of the brain. The results will also replicate better if the same lot and batch of MPTP is used throughout the duration of the study. Numerous other details of the experiment must also be controlled in order to produce results that are replicable (Jackson-Lewis & Przedborski, 2007; Meredith & Rademacher, 2011). In the case of transgenic animal models, the best design to ensure replicability is to use the wild-type littermates of the knockout animals, supplemented by the appropriate backcrossing to ensure genes neighboring to the targeted ones are not affected, as opposed to a more general wild-type animal like the C57Bl6 mouse (Masca et al., 2015; Wolfer, Crusio, & Lipp, 2002).

Reproducibility is the ability of other researchers to reproduce the model in their lab and achieve similar results to those of the original research. Investigators can most easily reproduce results when the original description of the development and use of the model is clear (Masca et al., 2015). Using the same examples as in the preceding text, if a group is reproducing the results from an MPTP-induced Parkinson's disease model, but does not have the coordinates or dosage of MPTP used in the original model, reproduction of the original results will be extremely difficult, as different parameters could easily result in ablation of a different set of neurons than the original research. Additionally, many times a study will limit the model to a specific gender. If the reproducing group uses the opposite gender, they may find that gender differences complicate their ability to reproduce the original results. Furthermore, with all animal models, reproducibility is difficult when different litters of animals are used. This is particularly problematic if the animals are from different suppliers as this introduces more variability in the animals' genetic background and microbiome, which could influence the study results (Masca et al., 2015). Sometimes, animals of the same species, gender, and phenotype may manifest important differences in behavior, physiology, disease state, or microbiome even if they come from different batches of the same or different suppliers housed in different locations and fed different diets (Ussar et al., 2015).

D Cost Efficiency

Finally, the cost of husbandry of the various animal models varies widely and should be considered when determining which animal model is best suited for a particular project. The physiology of nonhuman primates is very similar to human physiology. However, the regulations pertinent to a nonhuman primate facility are much more stringent than with other research models, and therefore the cost associated with the establishment, housing and maintenance of these facilities may not be feasible for many researchers. On the other hand, zebra fish are being used more and more as preclinical models of disease because the construction and maintenance of the facilities to house them are much less expensive (Kim, Carlson, Zafreen, Rajpurohit, & Jagadeeswaran, 2009; Paige, Hill, Canterbury, Sweitzer, & Romero-Sandoval, 2014). The cost of development, housing, and maintenance of rodent models of disease is somewhere in between the cost of a nonhuman primate colony and a zebra fish facility. For most research organizations, both academic and industrial, protocols and resources for conducting research in rodents are well established.

III TYPES OF ANIMAL MODELS USED FOR TRANSLATIONAL RESEARCH

While the vast majority of animal models are rodents, many different animal species are used for translational research in drug discovery and development, ranging from flies and worms to monkeys and other large animals. In some cases, the species is chosen based on its relative simplicity, while other times, it may be chosen based on the similarity of a particular organ system to the human system. In addition to different animal species that may be used, recent advances in genetics technology now allow us better control of DNA modifications in live animal models. In this section, we discuss some of the more common animal models and the inducible knockout and CRISPR-Cas9 DNA editing technologies that are used in research, including CNS drug development.

A Rodent Models

The most common species of animal model used for translational research in the CNS is the rodent model, usually rats or mice. These models have several advantages over other types of models. First, these models are mammalian and have similar physiology to humans for many organ systems, including the brain. While not the least expensive option for this type of research, they are usually more affordable than the other mammalian models that may be used. Finally, researchers are often interested in understanding how learning and memory are affected when developing CNS-related drugs. Robust experimental protocols have been developed to assess changes in these areas in rats and mice that can then be translated to the development of new therapeutics for diseases like Alzheimer's or Parkinson's disease. In a later section of the chapter, we apply these principles and discuss the mouse and rat models used for the study of obesity in more detail. We also provide examples of how the information from these models can then be translated to the clinic.

B Zebrafish

The use of zebra fish as a model for various diseases, particularly in the CNS, is becoming more common (Kalueff, Echevarria, & Stewart, 2014; Kalueff, Stewart, & Gerlai, 2014; Stewart, Braubach, Spitsbergen, Gerlai, & Kalueff, 2014). In addition to the cost-effectiveness of zebra fish, they can be useful for many other reasons. Many of the same cell types, tissues, and organ systems are similar in the zebra fish and humans, including complex systems like the stress endocrine axis. Additionally, when studying developmental disorders, the transparency of the zebra fish larva allows monitoring and modification of organ systems as they develop in vivo. However, like with any model, there are limitations to the use of zebra fish. In adult zebra fish, drug treatment is often done by solubilizing the drug in tank water, but many drugs are not highly water-soluble. With CNS-specific studies, some behaviors are learned and cannot be studied in larval fish, and other brain areas are not as developed in fish as they are in mammals making it difficult to extend the results of the study in zebra fish to mammals (Kalueff, Echevarria, et al., 2014; Kalueff, Stewart, et al., 2014; Stewart et al., 2014). Despite these difficulties, useful zebra fish models related to energy balance and obesity have been developed. We have shown that intestinal melanin-concentrating hormone, which is primarily known for its appetite regulating effects, is upregulated in zebra fish with TNBS-induced enterocolitis constituting a striking molecular similarity to the mouse and human form of the disease (Geiger et al., 2013). A diet-induced obesity model in zebra fish has also been described, which will be useful in understanding more of the pathophysiology associated with the development of obesity (Oka et al., 2010).

C Other Small Animal Models

Besides rodents, other small mammalian models may be used in research. These species are typically used very sparingly, but there are instances when they are necessary due to their similarities to specific human organ systems. These species could include dogs, cats, pigs, gerbils, rabbits, and hamsters. For example, if a study protocol is attempting to address administration of a transdermal nociceptive agent, a pig model may be used as the skin of the pig has striking similarities to human skin and delivery of the agent through the skin is an important consideration in the development of transdermal formulations. Cats have been found to be particularly useful in the study of Alzheimer's disease as they exhibit disease progression very similar to that in humans (Chambers et al., 2015), and dogs are particularly useful in kidney and renal disease research (Norrdin, 1981).

D Nonhuman Primate Models

Nonhuman primates are expensive animal models that still have great utility not only in drug pharmacokinetic analysis but also in the drug discovery and development process due to their similarities to human physiology, particularly in the CNS. Their high level of intellectual ability allows researchers to ask and answer questions related to complex social and cognitive behaviors for which other types of animal models cannot be used. For example, the social interactions that surround food intake in humans are extremely complex and difficult to study in the uncontrolled setting of normal everyday life. However, nonhuman primates, including the macaque monkey, exhibit eating behaviors that are extremely similar to humans. Researchers are able to watch and learn from primates in a more controlled social environment than the complex set of parameters under which humans operate. Furthermore, many of the same gut and neuro-peptide systems control feeding in different situations in these animals as they do in humans, so we can place them in a specific context and evaluate how the eating behavior is affected by the circumstances and which hormones are altered. These correlations can then be used to identify and test specific drug molecules before they are tested in humans (Wilson, Moore, Ethun, & Johnson, 2014).

E Targeted Genetic Modifications of Animal Models

Some of the more recent advances in genetic modification have allowed researchers to alter the DNA of traditional animal models with unprecedented precision. Two of these technologies, inducible knockouts using Cre-LoxP systems and DNA editing using CRISPR-Cas9, are discussed in the succeeding text.

1 Inducible Knockout Animal Models

Traditional knockout animal models eliminate the functioning gene from the DNA of the organism at the embryo stage. This approach leads to the elimination of the gene throughout the whole organism for the duration of their life and has been possible to a large extent only in mice. Over the last two decades, new technology has emerged that can alter gene expression in both time- and tissue-specific manner through the use of Cre recombinase fused to a tissue-specific gene and the use of LoxP of the target gene for deletion (Gierut, Jacks, & Haigis, 2014; Lakso et al., 1992; Lobe & Nagy, 1998). The Cre-LoxP system was used to produce a tamoxifen-induced ablation of G protein-coupled receptor kinase 2 (GRK2) in mice with diet-induced obesity. The use of this model validated GRK2 as a potential drug target for the treatment of obesity (Vila-Bedmar et al., 2015).

2 CRISPR-Cas9 DNA Editing

One of the most recent major advances in model development was the advent of the CRISPR-Cas9 gene-editing technology that can precisely target and edit specific sequences in genes. By introducing a mutation to study or correct a genetic deficit, this method allows for the study of this editing as a therapy (Barrangou et al., 2007; Cong et al., 2013; Savell & Day, 2017). While research related to the CRISPR-Cas9 technology is still relatively new, this technology provides a unique opportunity for investigators to further identify the effects of genetic manipulations on disease phenotypes and genotypes that have not been previously possible in several species.

IV EXAMPLES OF ANIMALS MODELS USED FOR TRANSLATIONAL OBESITY RESEARCH

Because the obesity phenotype is easily identifiable and quantifiable, unlike the phenotype of diseases like schizophrenia, autism, and clinical depression, researchers involved in the discovery and development of obesity therapeutics have used numerous valid and reliable animal models that focus on the CNS pathophysiology of food intake, energy expenditure, and food reward in the translation of disease targets to the clinic. The following section gives an overview of some of these obesity-related models as examples of the use of preclinical models in CNS translational efforts.

A Behavioral and Environmental Models

When some of the contributing factors to the pathogenesis of a disease are environmental and/or behavioral, the preclinical models that best represent the onset and pathophysiology of that disease will often include a simulation of that environment. The rapid increase in obesity in developed countries around the world is in part due to the environment in which we live with increased availability of highly palatable foods and a more sedentary lifestyle. As a result, the most commonly prescribed treatments for obesity are diet and exercise, and some of the most important preclinical models of obesity incorporate diet and/or exercise to investigate the mechanisms by which these factors contribute to the obesity phenotype. In this section, we discuss models that are developed based on behavioral attributes of the animal and specific environmental cues, including diet and exercise. We also include a discussion of the use of changes in the environment within the gut via the gut microbiota and its role in translational obesity research.

1 Animal Models of Dietary Obesity

One particularly useful model in obesity research is the diet-induced obesity (DIO) model as it mimics the most common cause of obesity in humans, dietary obesity. In this model, animals, usually rats and mice, although zebra fish DIO models have also been developed, are exposed to a specific diet that has macronutrient and/or caloric differences than their normal chow diet. There are a variety of types of diets that are used that range from high-sucrose and high-fat diets, high-fat only diets, high-saturated fat only diets, to food variety (cafeteria-style) diets (Geiger et al., 2009; Hariri, Gougeon, & Thibault, 2010; Hariri & Thibault, 2010; Hryhorczuk et al., 2016; Oka et al., 2010). Other variations of diets may also be used depending on the research question being addressed. All of these diets are effective at increasing body weight more rapidly in the animals exposed to the diets than control animals in the experiment that eat a normal lab chow diet. As a result, you can monitor how the increasing adiposity is affecting the various neurochemical pathways that are associated with food intake and energy expenditure. For example, animals that ate a cafeteria-style diet increased weight very rapidly and had excess weight gain over the control animals as early as 2 weeks after the start of the cafeteria diet (Geiger et al., 2009). These animals also showed a preference for a sweetened condensed milk and high-sucrose solutions over other choices that included regular lab chow, chocolate, peanut butter, salami, marshmallow, cookies, and bananas.

One major development in obesity preclinical research is the accumulating evidence on the role of central synaptic plasticity on food intake and weight gain. Since the discovery that lesions in the hypothalamus induce obesity (Stricker, 2012; Teitelbaum & Stellar, 1954), there have been numerous studies that focus on the CNS as the regulator of homeostatic mechanisms of energy balance through the activation of serotonin receptors and orexigenic and anorexigenic circuits in the arcuate nucleus and through the melanocortin-4 (MC4) receptor in the paraventricular nucleus of the hypothalamus (PVH) (Garfield & Heisler, 2009). DIO animal models have been extensively employed in this line of research. However, the rapid rise of common dietary obesity in industrialized societies indicates that nonhomeostatic signaling pathways that allow for chronic positive energy intake may be responsible. A crucial question is why laboratory animals and humans keep on eating energy-rich, palatable food to the degree that they become obese. From an evolutionary perspective, it is to be expected that the brain developed a system to respond to natural rewards, such as food. These central mechanisms are conserved across species in order to ensure survival (Kelley & Berridge, 2002) and could interact with or modulate the circuitry that regulates body weight. Therefore, availability of rewarding palatable food may lead to increased caloric intake and weight gain that homeostasis-driven mechanisms, originating primarily in the hypothalamus, may not overcome. This possibility may explain, at least in part, the epidemic proportions of dietary obesity.

Prominent among neural systems known to code for the perception of reward are the mesolimbic dopamine pathways, where the release of the neurotransmitter dopamine, particularly in the nucleus accumbens and medial prefrontal cortex terminals originating in the ventral tegmental area (VTA), along with the dorsal striatum (that receives dopaminergic projections from the substantia nigra pars compacta), is known to mediate reward. The activation of these pathways includes elevation of dopamine levels and changes in dopamine turnover after natural rewarding behaviors like feeding (Hernandez & Hoebel, 1988). It is, therefore, reasonable to expect that dietary obesity may be linked to the mesolimbic dopamine-releasing ability of palatable high-energy food.

The first evidence of linking mesolimbic dopamine release to dietary obesity came through our investigation of chronic exposure (15 weeks) of rats to a high-energy, palatable cafeteria diet along the diet model developed by Anthony Sclafani (Pothos, Sulzer, & Hoebel, 1998). We found that Sprague-Dawley rats took the majority of their daily caloric intake from high-carbohydrate sources and developed diet-induced obesity (Geiger et al., 2009; Pothos et al., 1998). As shown in Fig. 1.1, DIO rats demonstrated depressed basal dopamine release in the nucleus accumbens and an attenuated dopamine response to a standard chow meal or systemic administration of d-amphetamine, a drug that releases the neurotransmitter dopamine into the extracellular space by being a substrate of the plasma membrane and vesicular transporters for dopamine (DAT and VMAT2, respectively) (Pothos & Sulzer, 1998; Sulzer & Rayport, 1990). In further studies on strains of rodents that were selectively bred to develop dietary obesity, we established that the attenuation of basal and stimulated dopamine release in the brain characterized these animal models since early postnatal life (Fig. 1.2) (Geiger et al., 2008).

Our findings of decreased dopamine availability in DIO rats were replicated in other studies with obesity animal models (Fulton et al., 2006; Geiger et al., 2008; Geiger et al., 2009; Hryhorczuk et al., 2016; Pothos et al., 1998; Wang et al., 2001) and imaging studies in obese humans (Stice, Spoor, Bohon, & Small, 2008; Stice, Spoor, Bohon, Veldhuizen, & Small, 2008; Wang et al., 2001) and led to the theory of dopamine deficiency in obesity that expanded to the reward deficiency hypothesis. It postulates that hyperphagia, or hedonic food intake in excess of energy needs, is a drive to compensate for the compromised dopamine neurotransmission in the brain of obese animals and patients and their blunting of food reward (Allen et al., 2012).

Proof of principle for the importance of reversing dopamine deficits in the brain in order to reduce hyperphagia is the efficacy of the dopamine stimulant D-amphetamine in inducing weight loss (Harris, Ivy, & Searle, 1947). Amphetamine was manufactured as an ephedrine analog in the late 19th century and used as an over-the-counter medication until the early 1970s for weight loss. Its withdrawal from the market following its Schedule II classification ended the tenure of amphetamine as weight-loss medication and highlighted concerns over dopamine-releasing drugs, namely, their addictive and neurotoxic potential that can result in severe extrapyramidal symptoms and, in some cases, suicidality, psychosis, and other adverse psychoactive symptoms (Power et al., 2014). The case of amphetamine further strengthens the statement that stimulation of dopamine release in the brain results in inhibition of feeding. Partial dopaminergic

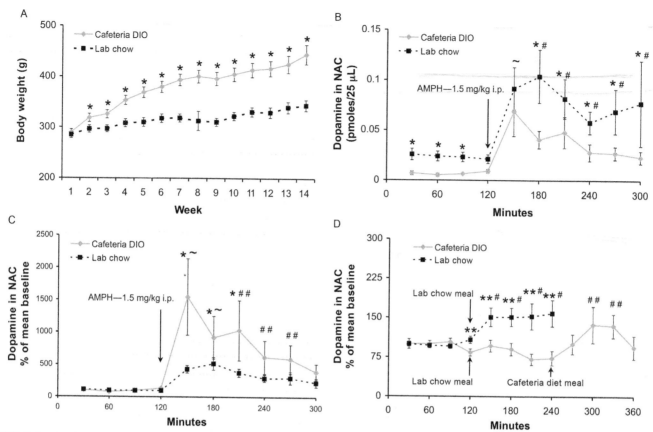

FIG. 1.1   Basal, amphetamine- and laboratory chow meal-challenged nucleus accumbens dopamine levels are decreased in freely moving dietary obese rats in vivo. (A) Body weight of cafeteria DIO rats during a 14-week period was significantly higher than that of the laboratory chow-fed group beginning at week 2 of the dietary regimen (*$P < 0.01$ by one-way ANOVA). (B) Basal and amphetamine-challenged extracellular dopamine levels during week 14 in the nucleus accumbens of cafeteria DIO rats ($n = 9$) was significantly lower than in chow-fed rats ($n = 13$). (C) The percent increase from baseline after amphetamine was higher in the cafeteria DIO rats than in the chow-fed rats (*$P < 0.01$ between groups, ~$P < 0.05$ within both groups, #$P < 0.05$ within laboratory chow-fed group only, and ##$P < 0.05$ within DIO group only relative to baseline prior to the amphetamine injection). (D) A plain chow meal was presented to cafeteria DIO ($n = 18$) and chow-fed groups ($n = 22$) after four baseline samples. Only the chow-fed group showed significant increases in dopamine after the meal. The cafeteria meal that was presented to a subset ($n = 8$) of the cafeteria DIO group 2.5 h after the regular chow meal resulted in a significant increase in dopamine release (**$P < 0.05$ between groups, #$P < 0.05$ within the laboratory chow-fed group only, and ##$P < 0.05$ within the cafeteria DIO group only relative to baseline prior to the laboratory chow or cafeteria meal). *From Geiger, B. M., Haburcak, M., Avena, N. M., Moyer, M. C., Hoebel, B. G., & Pothos, E. N. (2009). Deficits of mesolimbic dopamine neurotransmission in rat dietary obesity. Neuroscience, 159(4), 1193–1199.*

agonists and reuptake inhibitors with much more moderate effects on body weight continue to be introduced in the market throughout the last decades (Padwal, 2009), although their long-term efficacy and adverse effects remain a concern. Overall, central dopamine deficiency (or inefficiency) in obesity seems to be a consistent foundation of our understanding of the neurobiology of obesity in both animal models and humans, and it remains to be established how this deficit can be translated to any deficiencies in the perception of food reward. More importantly, it is potentially significant to demonstrate novel ways through which dopamine neurotransmission in the brain can be corrected in order to reduce hyperphagia without potential for addiction and other serious adverse effects of weight loss-inducing dopaminergic and other drugs.

A number of DIO studies have shown further links between neurochemistry and diet, including showing that macronutrient composition can directly affect central synaptic plasticity. Recently, a total western diet has been introduced that has not only the high fat content of a typical western diet but also has a modified micronutrient content to better match the dietary pattern of the western diet. Initial studies indicate that animals fed this diet are less susceptible to the excessive weight gain typically seen with high-fat diets. This model provides an opportunity for further investigation into the role of micronutrients in the alteration of CNS pathways involved in feeding behavior and their effect on the development of obesity (Monsanto et al., 2016). As a result of these diet studies, health-care providers may make better dietary and lifestyle recommendations for effective weight loss. Remarkably, weight management interventions in one member of a married couple can have crossover benefits for the other partner even if the latter is not treated (Gorin et al., 2018).

caution (handwritten margin note)

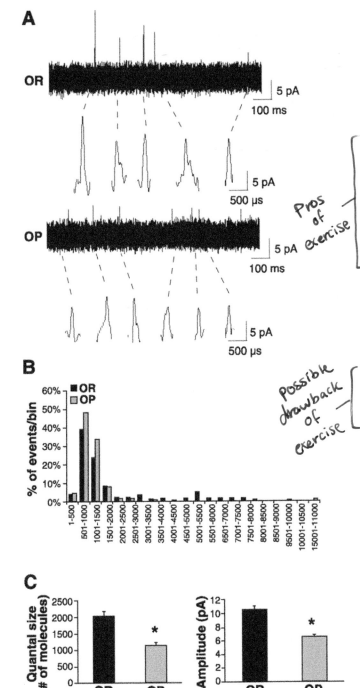

FIG. 1.2   Reduced dopamine quantal size in VTA-derived neurons from P0-P1 obesity-prone pups. (A) Representative amperometric traces from VTA cultures of obesity-resistant (OR) (top) and obesity-prone (OP) (bottom) neonates. Individual events are shown at higher resolution. (B) Quantal size distribution in cultures from neonatal OP and OR animals. Note that the OP distribution is skewed to the left due to the lack of events in the higher quantal size bins. (C) Quantal size (left panel) and amplitude (right panel) of stimulated dopamine release from VTA-derived neuronal cultures from OP pups (*gray bars*, n =110 events) is lower than that in cultures derived from OR pups (*black bars*, n =182 events). *P < 0.01. *From Geiger, B. M., Behr, G. G., Frank, L. E., Caldera-Siu, A. D., Beinfeld, M. C., Kokkotou, E. G., et al. (2008). Evidence for defective mesolimbic dopamine exocytosis in obesity-prone rats. FASEB Journal, 22(8), 2740–2746.*

2 Exercise Models in Obesity Research

Along with diet recommendations, exercise is the most commonly recommended treatment for obesity by physicians, as increased exercise will tip the energy balance scales in favor of energy expenditure and weight loss. To study the effects of exercise, rodents are given access to a treadmill, and the amount of time they spend on the treadmill is tracked. In mouse models of diet-induced obesity, exercise has been linked to not only lower body weight but also improvements in other markers of the metabolic syndrome like hyperglycemia and inflammation (Evans et al., 2014). Exercise also influences neurotransmitter release and metabolism that are associated with both energy homeostasis and hedonic feeding behavior, including that of orexin, leptin, norepinephrine, dopamine, and serotonin. These effects are seen in both adult animals and animals immediately post weaning (Greenwood et al., 2011; Meeusen & De Meirleir, 1995; Novak, Kotz, & Levine, 2006; Obici et al., 2015; Patterson, Bouret, Dunn-Meynell, & Levin, 2009; Patterson, Dunn-Meynell, & Levin, 2008; Patterson & Levin, 2008; Thanos et al., 2010). However, taken to the extreme, excessive exercise can cause addiction-related changes to the opiate pathways in the brain and cause rats to exhibit withdrawal like symptoms (Kanarek, D'Anci, Jurdak, & Mathes, 2009). Taken together, these findings provide more support of the effects of exercise on central synaptic plasticity and the use of exercise, particularly in children, as a safe alternative to pharmacotherapies to combat obesity. In fact, a recent metaanalysis of exercise studies in obese and overweight children indicates that aerobic exercise or a combination of aerobic and strength exercises can decrease body weight and adiposity in this population (Kelley, Kelley, & Pate, 2017).

3 Gut Microbiota Models in Obesity Research

Recent studies have revealed the gut microbiome as a major contributor to the obesity phenotype in both mice and humans (Boursi et al., 2018; Hildebrandt et al., 2009). However, whether alterations in the microbiome cause obesity or are the result of obesity is still not clear (Moran-Ramos, Lopez-Contreras, & Canizales-Quinteros, 2017; Sanmiguel, Gupta, & Mayer, 2015). In general, obesity is linked to lower diversity of microbes in the gut, particularly in animal models. There are also differences in the by-products produced by these microbes between lean and obese humans and animals. However, the exact significance of these changes is not completely understood (Moran-Ramos, Lopez-Contreras, et al., 2017). Others have also shown that following laparoscopic sleeve gastrectomy, the gut microbiome is altered in patients and these alterations are associated with weight loss, reduced appetite, and decreased hedonic eating (Sanmiguel et al., 2017).

The molecular mechanisms that contribute to these changes are not well understood. In animal models, changes in the microbiota are induced in a variety of ways. Diet has significant effects on the composition of the gut microbiota, and studies like the one recently published by Moran-Ramos, He, et al. (2017) indicate that even small changes in the diet of an animal can have significant effects on the gut microbiota that lead to improvements in the obesity status of the animal. Another approach that has shown promise as a therapeutic as well is the fecal transfer model. In animals, fecal matter from lean animals is transferred to obese animals, and changes in the gut microbiota and the overall phenotype of the animal have been shown (Moran-Ramos, Lopez-Contreras, et al., 2017). Based on these results, similar studies are being done in humans that so far have shown similar improvements in obesity (Marotz & Zarrinpar, 2016; Moran-Ramos, Lopez-Contreras, et al., 2017; Vrieze et al., 2012). More work is still necessary in both the preclinical and clinical areas to identify the molecular mechanisms that are altered in these studies, particularly as they relate to the gut-brain axis and changes seen in feeding behavior (Kaczmarek, Musaad, & Holscher, 2017). This work can potentially lead to identification of novel therapeutic targets for obesity.

B Spontaneous and Selectively Bred Animal Models

In obesity research, some of the most useful animal models have been the spontaneous and selectively bred models of obesity and diabetes. These types of models exhibit the phenotype of the disease without specific manipulation of a single gene, in contrast to the genetic models discussed in the next section. For obesity and many other complex diseases of the CNS, these spontaneous models are particularly useful, as the disease will develop over time as the result of a complex interaction of multiple genetic factors. The animal models of obesity discussed in this section are summarized in Table 1.1.

A spontaneous disease model exhibits the phenotype of the disease without any specific intervention by the breeder. The Zucker rat is a spontaneous model of obesity and hypertension that was first discovered in the 1960s. In contrast, selectively bred models typically arise from exposure to a specific environmental influence, and then the animals are grouped based on their response to the influence. The obesity-prone and obesity-resistant rat model is a good example of a selectively bred model of obesity (Levin et al., 1997). In this model, a colony of Sprague-Dawley rats ate a high-fat diet. The rats that gained the highest and lowest amounts of weight were separated into two groups and selectively bred within

their groups. After several generations of offering the high-fat diet to the animals and selective breeding, the offspring of the high weight gainers would spontaneously weigh about 20% more than the offspring of the low-weight gaining rats, even when all of the rats were fed a normal chow diet (Levin et al., 1997). The rats that gained the excess weight are referred to as diet-induced obese (DIO) or obesity-prone, while the rats that remained leaner are referred to as diet-resistant (DR) or obesity-resistant animals. In this chapter, we use the terms obesity-prone (OP) and obesity-resistant (OR) for selectively bred strains to avoid confusion with our discussion of models that are developed by exposing the animals to specific diets as part of the study, which were discussed earlier in this chapter. In this section the Zucker rat and OP/OR models of obesity as well as others like them are discussed in more detail.

1 Obesity-Prone and Obesity-Resistant Rats in Translational Research

The selectively bred OP/OR strains constitute a model with many pathophysiological similarities to the human obese state, which is typically viewed as a polygenic disease. A weight distribution exists between the OP animals and the OR animals, similarly to what is seen in humans. In the presence of a high-fat diet, these weight differences are exacerbated with the OP animals gaining much more weight than the OR animals (Levin et al., 1997). Furthermore, numerous studies have shown that a number of different neuropeptides and other CNS proteins and neurotransmitters are altered in these animals, which also corresponds well to the human condition where multiple neural systems, both homeostatic and hedonic, have been shown to be altered in obese individuals (Berthoud & Morrison, 2008; Levin et al., 1997; Levin & Dunn-Meynell, 2002; Levin & Keesey, 1998; Novak et al., 2006). An analysis of the food reward system in these animals showed an overall depression of dopamine signaling in both males and females (Fig. 1.3). Most importantly, the parent phenotype affected its offspring similarly to human obesity. Specifically, attenuation of dopamine release was evident in neurons that were cultured from the dopamine neurons in newborn pups, indicating that the reduction in dopamine was present from birth and not induced by differences in the animals' diets once they were born (Fig. 1.2; Geiger et al., 2008).

2 Zucker Rats as a Spontaneous Model of Obesity

The Zucker fatty rat is a widely used model of obesity that was discovered in the 1960s when a spontaneous mutation in the recessive fa/fa genotype resulted in obesity and hypertension in these animals (Bray, 1977; Zucker & Zucker, 1967). The fa genotype is a missense mutation on the leptin receptor gene that renders the gene

TABLE 1.1   Selected Genetic Rodent Models Used in Obesity and Diabetes Research

Model	Genotype	Obesity/diabetes phenotype	Reference
Zucker fatty rat (ZFR)	Recessive *fa/fa*	Obese, hypertension, hyperinsulinemia, hyperlipidemia	Bray (1977), Zucker and Zucker (1967)
Zucker diabetic fatty rat (ZDF)	Recessive *fa/fa* selectively bred for hyperglycemia	Obese, hypertension, hyperglycemia	Bray (1977)
Obesity-prone rat	Selectively bred polygenic	Obese on lab chow, hyperinsulinemia, hyperglycemia, reduced leptin sensitivity	Levin and Dunn-Meynell (2002), Levin, Dunn-Meynell, Balkan, and Keesey (1997)
Obesity-resistant rat	Selectively bred polygenic	Lean, resistant to diet-induced obesity from high-fat diet	Levin et al. (1997)
Otsuka Long-Evans Tokushima Fatty (OLETF) rats	Cholecystokinin 1 receptor-deficient	Mildly obese, late-onset hyperglycemia, islet hyperplasia, renal disease	Kawano, Hirashima, Mori, and Natori (1994)
ob/ob mouse	Leptin-deficient	Obese, mild hyperglycemia, hyperinsulinemia, altered feeding behavior	Bray and York (1971), Halaas et al. (1995), Pelleymounter et al. (1995)
db/db mouse	Leptin receptor-deficient	Obese, hyperphagia, hyperglycemia, hyperinsulinemia	Coleman and Hummel (1967, 1974)
POMC −/− mouse	Proopiomelanocortin peptide deficient	Obese, defective adrenal glands, light coat color	Yaswen, Diehl, Brennan, and Hochgeschwender (1999)
Mc4R −/− mouse	Melanocortin-4 receptor-deficient	Obese, hyperphagic, reduced metabolism, hyperinsulinemia, type 2 diabetes	Huszar et al. (1997), Ste Marie, Miura, Marsh, Yagaloff, and Palmiter (2000)
Agrp overexpression mouse	Agouti-related protein overexpression	Obese, hyperinsulinemia, late-onset hyperglycemia, pancreatic-islet hyperplasia	Graham, Shutter, Sarmiento, Sarosi, and Stark (1997)
NMU −/− mouse	Neuromedin U-deficient	Obese, hyperphagia, reduced activity, hyperinsulinemia, hyperglycemia, hyperleptinemia	Hanada et al. (2004)
MCH −/− mouse	Melanin-concentrating hormone-deficient	Lean, hyperactive, hypophagic	Shimada, Tritos, Lowell, Flier, and Maratos-Flier (1998)
CB1R −/− mouse	Cannabinoid 1 receptor-deficient	Lean, resistant to diet-induced obesity	Ravinet Trillou, Delgorge, Menet, Arnone, and Soubrie (2004)

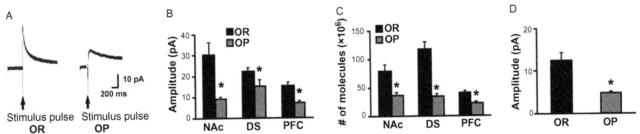

FIG. 1.3   Stimulated dopamine release is attenuated in the nucleus accumbens, dorsal striatum, and prefrontal cortex of adult obesity-prone rats. (A) Representative amperometric traces of stimulated dopamine release from acute accumbens coronal slices. (B and C) Average peak amplitude (pA) (B) and number of molecules released after stimulation of dopamine release (C) were significantly lower in obesity-prone (OP) rats than in obesity-resistant (OR) rats in the nucleus accumbens (NAc) (OP [$n=45$] stimulations in 9 slices; OR [$n=53$] stimulations in 11 slices), dorsal striatum (DS) (OP [$n=40$] stimulations in 8 slices; OR [$n=40$] stimulations in 8 slices), and prefrontal cortex (PFC) (OP [$n=32$] stimulations in 7 slices; OR [$n=35$] stimulations in 9 slices). (D) In 4–5-week-old rats, the average peak amplitude after stimulation of dopamine release was also significantly lower in OP rats than in OR rats (OP [$n=37$] stimulations in 10 slices; OR [$n=56$] stimulations in 13 slices). *$P < 0.01$. *From Geiger, B. M., Behr, G. G., Frank, L. E., Caldera-Siu, A. D., Beinfeld, M. C., Kokkotou, E. G., et al. (2008). Evidence for defective mesolimbic dopamine exocytosis in obesity-prone rats. FASEB Journal, 22(8), 2740–2746.*

nonfunctional. In addition to the traditional Zucker fatty rat, it has been further selectively bred for hyperglycemia to produce a diabetic rat model referred to as the Zucker diabetic fatty (ZDF) rat (Bray, 1977; Wang et al., 2014). When investigating the links between obesity, impaired leptin signaling, and dopamine, the Zucker rat was used to show that bromocriptine, a dopamine D2 receptor agonist, could be used to decrease food intake and body fat and increase locomotor activity in these animals (Davis et al., 2009).

3 Otsuka Long Evans Tokushima Fatty (OLETF) Rat Obesity Model

This model of obesity was first discovered in the 1990s by the spontaneous development of obesity in some of the animals in a colony of Long-Evans rats. These obese animals were then selectively bred to preserve the phenotype (Bi & Moran, 2016; Kawano et al., 1994). The lean control for this model is the Long-Evans Tokushima Otsuka (LETO) rat. The primary deficit in the OLETF rat is the lack of cholecystokinin receptor 1 (CCK1R) in both the GI tract and the brain. CCK is involved in the control of food intake, and disruptions in this system have been shown to cause hyperphagia that leads to increased body weight, reduced insulin sensitivity, and many other characteristics that are consistent with obesity in humans (Anderzhanova, Covasa, & Hajnal, 2007; Hajnal, Margas, & Covasa, 2008). As a result, the neural pathways that are affected by the lack of CCK have been extensively investigated, and the role of brain and gut neuropeptides in the homeostasis of energy balance has been identified (Bi & Moran, 2016; Moran & Bi, 2006). The increased understanding of the neural control of eating that was determined using this model has led to the identification of targets for obesity therapeutics in the clinic. In fact, both CCK1 agonists and antagonists and some of the NPY receptors have been investigated as potential anti-obesity treatments due to their interactions with the homeostatic systems of energy balance. However, in both systems, the single therapy did not result in clinically significant results (Colon-Gonzalez, Kim, Lin, Valentino, & Waldman, 2013; Merlino, Blomain, Aing, & Waldman, 2014).

C The Use of Genetic Models in Translational Obesity Research

As our understanding of not only the human genome but also the genome of other species, like mice and rats, has advanced, we have been able to target genes in various animal models for overexpression or deletion to create a modified animal that exhibits a desired disease phenotype (Table 1.1).

In addition to complete deletion or overexpression of specific genes, we are able to target the altered gene expression to specific tissues or during specific times of development. More recently, these techniques have advanced using CRISPR/Cas9 technology so that we now are able to edit a specific gene and introduce specific disease-related mutations to genes of interest. In this section, we describe some of these genetic-based models that relate to our understanding of that pathophysiology of obesity and its translation to the clinic.

1 Genetic Models of Obesity and Diabetes

Genetically modified animal models are some of the most commonly used models to help understand the obesity and diabetes phenotypes and then translate that knowledge into treatments for these diseases. For example, the leptin-deficient (ob/ob) and leptin receptor-deficient (db/db) mouse models show a distinct obese phenotype, mild hyperglycemia, and altered feeding behavior (Bray & York, 1971; Coleman & Hummel, 1967, 1974; Halaas et al., 1995; Pelleymounter et al., 1995). By providing a disease state that so similarly mimics the human phenotype, we have been able to gain valuable insights into the altered physiology of the obese individual. Such information has in turn helped provide strategies for treatment or prevention of this disease that in this case could be leptin-independent. For example, we found in the ob/ob mouse that there was a significant reduction in central dopamine signaling (Fulton et al., 2006). Because dopamine is a key component of the reward response to feeding, the lack of a proper dopamine response to food and the resulting uncontrolled eating behavior of these animals have contributed to the reward deficiency hypothesis of obesity (Blum et al., 2000). This hypothesis provides a specific target, the dopamine system, for the development of obesity treatments.

Another anorexigenic neuropeptide whose role has been elucidated through the use of genetically modified mice is neuromedin U (NMU), a neuropeptide that is associated with a variety of stress-related and feeding behaviors (Hanada et al., 2004; Kaisho et al., 2017). Mice that are lacking either the gene coding for NMU or its receptor that is more common in the CNS, NMUR2, are obese and exhibit hyperinsulinemia and hyperglycemia along with other metabolic disturbances. Kaisho et al. (2017) have recently shown that treatment of dietary obese mice with a specific NMUR2 agonist, NMU-7005, can reduce body weight in these animals, while it has no effect on NMUR2 knockout mice. This study validates the selectivity of NMU-7005 for NMUR2 and provides another possible target for the treatment of dietary obesity (Kaisho et al., 2017). An additional and potentially important target is alpha–melanocyte-stimulating hormone (α-MSH) that shares proopiomelanocortin as a precursor with other MSHs, corticotropin, and endogenous opiates (Millington & Levell, 2007). Interestingly, aging does not significantly alter hypothalamic mRNA levels of anorexigenic signals such as cocaine- and amphetamine-related transcript (CART) and α-MSH in mice (Kmiec, 2006), which could allow for interventions to accelerate metabolism in older age. Many other genetically modified models of obesity models have been developed for a wide variety of targets, some of which are proopiomelanocortin-deficient, melanocortin-4 receptor–deficient, and agouti-related peptide overexpressing mice, and have led to several pharmacological interventions under clinical development (Graham et al., 1997; Huszar et al., 1997; Ste Marie et al., 2000; Yaswen et al., 1999).

2 Genetic Models of Leanness

Along with the genetically obese animal models, several animal models that exhibit a lean phenotype have also been used to determine what neurobiological pathways and networks may protect from the development of obesity. For example, the melanin-concentrating hormone (MCH −/−) knockout mouse is lean compared with its wild-type littermates (Shimada et al., 1998). MCH is an appetite-promoting hypothalamic neuropeptide, and the reduced food intake and dysregulation of dopamine neurotransmission in the MCH −/− mouse contribute to its lean phenotype (Pissios et al., 2008). Based on this phenotype, antagonism of the MCH-1 receptor, which should have a similar effect on feeding behavior as seen in the MCH −/− mouse, is one approach that researchers are pursuing in the development of obesity therapeutics (Macneil, 2013). Because the MCH system is one of many overlapping hypothalamic systems and because it is well known that obesity is a polygenic disorder, combination therapies including MCH-1 receptor antagonists are also being investigated (Colon-Gonzalez et al., 2013; Valsamakis, Konstantakou, & Mastorakos, 2017).

Interestingly, other knockout mouse models of hypothalamic orexigenic peptides have been less successful in showing the expected lean phenotype. For example, Agouti-related peptide (AgRP) is an orexigenic molecule that has been shown to cause increased food intake and is inhibited by elevated leptin and insulin levels. However, the phenotype of the knockout mouse did not feature significant differences in body weight, response to high-fat diet, or insulin sensitivity compared with its wild-type littermates (Qian et al., 2002). In all likelihood, the use of an AgRP inhibitor alone as a suppressant of food intake in the clinic would not show significant body weight reduction in humans. vTv Therapeutics had an AgRP inhibitor program in which they produced several inhibitors, one of which was evaluated in a phase II clinical trial for safety (Valsamakis et al., 2017). In mice, it was shown that one of the inhibitors, TPP2515, could blunt the effects of AgRP, particularly in situations where AgRP would be expected to be elevated. However, there were several off target effects of this compound and limited efficacy in situations where AgRP would not be artificially elevated, and these compounds no longer appear to be in development for the treatment of obesity (Dutia et al., 2013).

The endocannabinoid system has also emerged as a potential target for obesity therapeutics due to its actions in both the homeostatic pathways of energy balance and the hedonic system of food reward (Di Marzo, Ligresti, & Cristino, 2009; Melis et al., 2007). Mice lacking the cannabinoid 1 receptor exhibit a lean phenotype and resistance to diet-induced obesity when exposed to a high-fat diet (Ravinet Trillou et al., 2004). Rimonabant is a CB1 receptor antagonist that has been developed for the treatment of obesity. However, it was withdrawn from the market due to its significant effects on patients' mood that outweighed the clinical benefits of the drug in the treatment of obesity (Richey & Woolcott, 2017). This clinical failure highlights the limitations of murine models in CNS drug development. Specifically, while we have some indications of a mouse's affective state, it is difficult to assess the extent of mood-related side effects that might occur in humans in a mouse model.

3 Inducible KO Models With Phenotypes Applicable to Food Intake

The most effective treatments of disease will target the protein or gene of interest at the target tissue while leaving it unaltered in other areas of the body. Traditional inducible models of disease use a Cre/LoxP systems where gene expression is altered either in specific tissues or following the use of a trigger stimulus that initiates the introduction of the altered DNA into the host's genome (Gu, Marth, Orban, Mossmann, & Rajewsky, 1994; Heldt & Ressler, 2009). These models have provided invaluable insights into disease mechanisms. Recent advances in Cre/LoxP systems can alter gene expression in the hypothalamus at specific ages or loci in mice and have provided additional insights to the development of obesity. For example, studies have showed that food intake was reduced and energy expenditure increased dramatically by disrupting the synthesis of GABA specifically in agouti-related peptide expressing neurons of the hypothalamus in young adult mice (Meng et al., 2016). These effects were not as significant in older mice, indicating that hypothalamic control of feeding behavior in these neurons is more sensitive to GABA signaling in younger animals. These results demonstrate that targeting GABA signaling might be an effective treatment of obesity in younger populations. Such signaling modulation could be envisioned in human patients, particularly the morbidly obese, through, for example, patterns of localized mild brain stimulation in the hypothalamic region (Palmiter, 2017).

4 CRISPR-Cas9 Technology in Translational Research

The CRISPR-Cas9 gene-editing technology is proving to be one of the most powerful tools available to alter gene sequences with a high level of accuracy and precision, providing researchers in search of a model to understand the role of specific mutations a way to develop a corresponding animal model more easily than ever before (Cong et al., 2013; Savell & Day, 2017). With respect to obesity and diabetes research, this technology has already been used to alter leptin and ghrelin signaling in both rats and mice. As a result of these studies, hundreds of altered genes have been identified both in the peripheral tissues as well as the CNS that could provide information for new potential targets for obesity pharmacotherapies (Guan et al., 2017; Zallar et al., 2019). As further proof of concept, in mice embryos, both leptin and the leptin receptor have been deleted using CRISPR-Cas9 systems, resulting in phenotypes very similar to the ob/ob and db/db mouse models (Roh et al., 2018). This illustrates the possibility to use the CRISPR-Cas9 technology to produce animal models of obesity in other species. As the difficulties with introducing this technology to neurons specifically are overcome, numerous other applications of CRISPR-Cas9 gene editing may also become critical to biomedical research in the CNS (Savell & Day, 2017).

D Surgical/Pharmacological Models

Additional effective approaches to develop a reproducible model of obesity and test novel treatments are through surgical or pharmacological interventions. Bariatric surgery in rats has given us extremely useful insight into the CNS and gut hormone alterations that occur when this surgery is done (Berthoud, Shin, & Zheng, 2011; Hajnal et al., 2010; Tam et al., 2011). In other cases, the disease would be initiated by the surgical or pharmacological treatment. The streptozotocin-induced diabetic rats and mice, which were first used in the 1960s, are an example of this type of model (Szkudelski, 2001). In this section, we discuss examples of each of these types of models and their use in translational research in more detail.

1 Bariatric Surgery

In some cases, translational research begins in the clinic, returns to the lab, and then goes back to the clinic. Bariatric surgery is widely considered one of the most successful weight-loss treatments with beneficial effects extending beyond weight reduction. Roux-en-Y gastric bypass surgical techniques were developed in rat and mouse models to investigate what other molecular mechanisms might be altered by this procedure. Interestingly, which type of model is used for the surgery influences the outcomes of the surgery. For example, DIO mice respond differently to the surgery than ob/ob mice (Hao et al., 2015). It has also been shown that a number of neuropeptides are altered in these animals including CCK, glucagon-like peptide-1, serotonin, and neurotensin (Mumphrey, Patterson, Zheng, & Berthoud, 2013; Shin, Zheng, Townsend, Sigalet, & Berthoud, 2010). Rat models of bariatric surgery have shown alterations to behaviors associated with the hedonic pathways including altered sweet taste and changes in central dopamine (Hajnal et al., 2010; Thanos et al., 2015). Those observations have been confirmed in patients subjected to bariatric surgery through, for instance, functional MRI pilot studies (Wang, Yang, Hajnal, & Rogers, 2016) and may help illustrate changes in alcohol intake observed in human bariatric patients (Smith et al., 2017). More recently, it has been found that the fat-satiety molecule oleoylethanolamide (OEA) is increased in rats that have had Roux-en-Y surgery, while central dopamine signaling through the D1 receptor was also upregulated (Hankir et al., 2017). Taken together, these results indicate that surgery-induced changes in satiety-related gut hormones and a number of contributing neurochemicals may improve weight loss. In the future, as an alternative to surgery, the targeting of these systems or a specific combination of these systems could elicit weight loss in obese individuals.

2 Streptozotocin-Induced Diabetic Rats and Mice

The use of pharmacological agents to damage β-cells and produce a diabetic rodent has been utilized for several decades. Streptozotocin is one agent that is often used to create this model. At high doses, it will cause significant damage to the β-cells, similarly to type 1 diabetes. At lower doses, it will cause the damage over time, mimicking the onset of type 2 diabetes (Szkudelski, 2001). In a variation of this model, the animals are also fed a high-fat diet to become obese. This may be one of the most accurate models to link the pathogenesis of diabetes and obesity (Gheibi, Kashfi, & Ghasemi, 2017; Reed et al., 2000). This model has been used to provide valuable insights on the effects of diabetes on central dopamine signaling and reward response. Furthermore, it has been shown that targeting the D1 and D2 receptors can attenuate the hyperphagia that is seen in these animals (Kuo, 2006). Finally, this is a good model of diabetic neuropathy and as the number of diabetic patients increases, the need for safe and effective treatments for pain in diabetics also increases. Tramadol is a commonly prescribed pain medication for diabetic neuropathy. A recent study in the streptozotocin-induced diabetic rat model showed that while tramadol may work in this population, the pathways that it targets are altered when diabetes is present and therefore the patient's response will also likely be altered (Ezzeldin, Souror, El-Nahhas, Soudi, & Shahat, 2014).

3 Models of Central Insulin Deficiency

Although the role of insulin secretion from pancreatic beta cells is well established in the periphery, an increasingly active area of investigation focuses on the role and function of central insulin receptors in diabetes, obesity, and cognitive function (Arnold et al., 2018). Surprisingly, the targeted deletion of brain insulin receptors does not consistently produce a diabetic phenotype or a phenotype related to metabolic syndrome in mice. Instead, it can produce behavioral indexes of depression and a remarkable decrease in mesolimbic dopamine neurotransmission (Kleinridders, 2016). This is a testament of how animal models can at times offer unexpected contributions to the understanding of disease mechanisms other than the ones originally considered. In this case, the brain insulin animal models may decisively aid in the understanding of the connection between diabetes and neurodegenerative disorders and whether their underlying pathologies are partially overlapping or coincidental and complementary.

V CONCLUSION

The use of animal models of disease is crucial to the successful development of new therapeutics to treat diseases of the central nervous system. However, a number of factors should be considered when choosing an appropriate disease model including pathophysiological and phenotypic similarities to the disease in humans, the replicability and reproducibility of the data generated by the model, and the cost of the development and maintenance of the model. With these considerations in mind, models generated using a wide variety of methods ranging from spontaneous development of the disease to complex genetic manipulations that give rise to the disease phenotype have proved to be an invaluable resource in the drug discovery and development process for the treatment of obesity and diabetes.

References

Allen, P. J., Batra, P., Geiger, B. M., Wommack, T., Gilhooly, C., & Pothos, E. N. (2012). Rationale and consequences of reclassifying obesity as an addictive disorder: neurobiology, food environment and social policy perspectives. *Physiology & Behavior, 107*(1), 126–137.

Anderzhanova, E., Covasa, M., & Hajnal, A. (2007). Altered basal and stimulated accumbens dopamine release in obese OLETF rats as a function of age and diabetic status. *American Journal of Physiology. Regulatory, Integrative and Comparative Physiology, 293*(2), R603–R611.

Arnold, S. E., Arvanitakis, Z., Macauley-Rambach, S. L., Koenig, A. M., Wang, H. Y., Ahima, R. S., et al. (2018). Brain insulin resistance in type 2 diabetes and Alzheimer disease: concepts and conundrums. *Nature Reviews. Neurology, 14*(3), 168–181.

Barrangou, R., Fremaux, C., Deveau, H., Richards, M., Boyaval, P., Moineau, S., et al. (2007). CRISPR provides acquired resistance against viruses in prokaryotes. *Science, 315*(5819), 1709–1712.

Berthoud, H. R., & Morrison, C. (2008). The brain, appetite, and obesity. *Annual Review of Psychology, 59*, 55–92.

Berthoud, H. R., Shin, A. C., & Zheng, H. (2011). Obesity surgery and gut-brain communication. *Physiology & Behavior, 105*(1), 106–119.

Bi, S., & Moran, T. H. (2016). Obesity in the otsuka long evans tokushima fatty rat: mechanisms and discoveries. *Frontiers in Nutrition, 3*, 21.

Blass, B. E. (2015). *Basic principles of drug discovery and development.* Amsterdam; Boston: Elsevier/AP, Academic Press is an imprint of Elsevier.

Blum, K., Braverman, E. R., Holder, J. M., Lubar, J. F., Monastra, V. J., Miller, D., et al. (2000). Reward deficiency syndrome: a biogenetic model for the diagnosis and treatment of impulsive, addictive, and compulsive behaviors. *Journal of Psychoactive Drugs, 32*(Suppl, i–iv), 1–112.

Boursi, B., Werner, T. J., Gholami, S., Houshmand, S., Mamtani, R., Lewis, J. D., et al. (2018). Functional imaging of the interaction between gut microbiota and the human host: a proof-of-concept clinical study evaluating novel use for 18F-FDG PET-CT. *PLoS ONE, 13*(2), e0192747.

Bray, G. A. (1977). The Zucker-fatty rat: a review. *Federation Proceedings, 36*(2), 148–153.

Bray, G. A., & York, D. A. (1971). Genetically transmitted obesity in rodents. *Physiological Reviews, 51*(3), 598–646.

Chambers, J. K., Tokuda, T., Uchida, K., Ishii, R., Tatebe, H., Takahashi, E., et al. (2015). The domestic cat as a natural animal model of Alzheimer's disease. *Acta Neuropathologica Communications, 3*, 78.

Coleman, D. L., & Hummel, K. P. (1967). Studies with the mutation, diabetes, in the mouse. *Diabetologia, 3*(2), 238–248.

Coleman, D. L., & Hummel, K. P. (1974). Hyperinsulinemia in pre-weaning diabetes (db) mice. *Diabetologia, 10*(Suppl), 607–610.

Collins, S., Martin, T. L., Surwit, R. S., & Robidoux, J. (2004). Genetic vulnerability to diet-induced obesity in the C57BL/6J mouse: physiological and molecular characteristics. *Physiology & Behavior, 81*(2), 243–248.

Colon-Gonzalez, F., Kim, G. W., Lin, J. E., Valentino, M. A., & Waldman, S. A. (2013). Obesity pharmacotherapy: what is next? *Molecular Aspects of Medicine, 34*(1), 71–83.

Cong, L., Ran, F. A., Cox, D., Lin, S., Barretto, R., Habib, N., et al. (2013). Multiplex genome engineering using CRISPR/Cas systems. *Science, 339*(6121), 819–823.

Davis, L. M., Michaelides, M., Cheskin, L. J., Moran, T. H., Aja, S., Watkins, P. A., et al. (2009). Bromocriptine administration reduces hyperphagia and adiposity and differentially affects dopamine D2 receptor and transporter binding in leptin-receptor-deficient Zucker rats and rats with diet-induced obesity. *Neuroendocrinology, 89*(2), 152–162.

Di Marzo, V., Ligresti, A., & Cristino, L. (2009). The endocannabinoid system as a link between homoeostatic and hedonic pathways involved in energy balance regulation. *International Journal of Obesity, 33*(Suppl 2), S18–S24.

Dutia, R., Kim, A. J., Modes, M., Rothlein, R., Shen, J. M., Tian, Y. E., et al. (2013). Effects of AgRP inhibition on energy balance and metabolism in rodent models. *PLoS ONE, 8*(6), e65317.

Evans, C. C., LePard, K. J., Kwak, J. W., Stancukas, M. C., Laskowski, S., Dougherty, J., et al. (2014). Exercise prevents weight gain and alters the gut microbiota in a mouse model of high fat diet-induced obesity. *PLoS ONE, 9*(3), e92193.

Ezzeldin, E., Souror, W. A., El-Nahhas, T., Soudi, A. N., & Shahat, A. A. (2014). Biochemical and neurotransmitters changes associated with tramadol in streptozotocin-induced diabetes in rats. *BioMed Research International, 2014*, 238780.

Farooqi, I. S. (2008). Monogenic human obesity. *Frontiers of Hormone Research, 36*, 1–11.

Farooqi, I. S., Bullmore, E., Keogh, J., Gillard, J., O'Rahilly, S., & Fletcher, P. C. (2007). Leptin regulates striatal regions and human eating behavior. *Science, 317*(5843), 1355.

Farooqi, I. S., & O'Rahilly, S. (2008). Mutations in ligands and receptors of the leptin-melanocortin pathway that lead to obesity. *Nature Clinical Practice. Endocrinology & Metabolism*, 4(10), 569–577.

Fulton, S., Pissios, P., Manchon, R. P., Stiles, L., Frank, L., Pothos, E. N., et al. (2006). Leptin regulation of the mesoaccumbens dopamine pathway. *Neuron*, 51(6), 811–822.

Garfield, A. S., & Heisler, L. K. (2009). Pharmacological targeting of the serotonergic system for the treatment of obesity. *The Journal of Physiology*, 587(Pt 1), 49–60.

Geiger, B. M., Behr, G. G., Frank, L. E., Caldera-Siu, A. D., Beinfeld, M. C., Kokkotou, E. G., et al. (2008). Evidence for defective mesolimbic dopamine exocytosis in obesity-prone rats. *The FASEB Journal*, 22(8), 2740–2746.

Geiger, B. M., Gras-Miralles, B., Ziogas, D. C., Karagiannis, A. K., Zhen, A., Fraenkel, P., et al. (2013). Intestinal upregulation of melanin-concentrating hormone in TNBS-induced enterocolitis in adult zebrafish. *PLoS ONE*, 8(12), e83194.

Geiger, B. M., Haburcak, M., Avena, N. M., Moyer, M. C., Hoebel, B. G., & Pothos, E. N. (2009). Deficits of mesolimbic dopamine neurotransmission in rat dietary obesity. *Neuroscience*, 159(4), 1193–1199.

Gheibi, S., Kashfi, K., & Ghasemi, A. (2017). A practical guide for induction of type-2 diabetes in rat: incorporating a high-fat diet and streptozotocin. *Biomedicine & Pharmacotherapy*, 95, 605–613.

Gierut, J. J., Jacks, T. E., & Haigis, K. M. (2014). Strategies to achieve conditional gene mutation in mice. *Cold Spring Harbor Protocols*, 2014(4), 339–349.

Gorin, A. A., Lenz, E. M., Cornelius, T., Huedo-Medina, T., Wojtanowski, A. C., & Foster, G. D. (2018). Randomized controlled trial examining the ripple effect of a nationally available weight management program on untreated spouses. *Obesity (Silver Spring)*, 26(3), 499–504.

Graham, M., Shutter, J. R., Sarmiento, U., Sarosi, I., & Stark, K. L. (1997). Overexpression of Agrt leads to obesity in transgenic mice. *Nature Genetics*, 17(3), 273–274.

Greenwood, B. N., Foley, T. E., Le, T. V., Strong, P. V., Loughridge, A. B., Day, H. E., et al. (2011). Long-term voluntary wheel running is rewarding and produces plasticity in the mesolimbic reward pathway. *Behavioural Brain Research*, 217(2), 354–362.

Gu, H., Marth, J. D., Orban, P. C., Mossmann, H., & Rajewsky, K. (1994). Deletion of a DNA polymerase beta gene segment in T cells using cell type-specific gene targeting. *Science*, 265(5168), 103–106.

Guan, L. J., Xu, K. X., Xu, S. Y., Li, N. N., Wang, X. R., Xia, Y. K., et al. (2017). Profiles of metabolic gene expression in the white adipose tissue, liver and hypothalamus in leptin knockout (Lep(Delta14/Delta14)) rats. *Journal of Biomedical Research*, 31, 528–540.

Hajnal, A., Kovacs, P., Ahmed, T., Meirelles, K., Lynch, C. J., & Cooney, R. N. (2010). Gastric bypass surgery alters behavioral and neural taste functions for sweet taste in obese rats. *American Journal of Physiology. Gastrointestinal and Liver Physiology*, 299(4), G967–G979.

Hajnal, A., Margas, W. M., & Covasa, M. (2008). Altered dopamine D2 receptor function and binding in obese OLETF rat. *Brain Research Bulletin*, 75(1), 70–76.

Halaas, J. L., Gajiwala, K. S., Maffei, M., Cohen, S. L., Chait, B. T., Rabinowitz, D., et al. (1995). Weight-reducing effects of the plasma protein encoded by the obese gene. *Science*, 269(5223), 543–546.

Hanada, T., Teranishi, H., Pearson, J. T., Kurokawa, M., Hosoda, H., Fukushima, N., et al. (2004). Neuromedin U has a novel anorexigenic effect independent of the leptin signaling pathway. *Nature Medicine*, 10(10), 1067–1073.

Hankir, M. K., Seyfried, F., Hintschich, C. A., Diep, T. A., Kleberg, K., Kranz, M., et al. (2017). Gastric bypass surgery recruits a gut PPAR-alpha-striatal D1R pathway to reduce fat appetite in obese rats. *Cell Metabolism*, 25(2), 335–344.

Hao, Z., Munzberg, H., Rezai-Zadeh, K., Keenan, M., Coulon, D., Lu, H., et al. (2015). Leptin deficient ob/ob mice and diet-induced obese mice responded differently to Roux-en-Y bypass surgery. *International Journal of Obesity*, 39(5), 798–805.

Hariri, N., Gougeon, R., & Thibault, L. (2010). A highly saturated fat-rich diet is more obesogenic than diets with lower saturated fat content. *Nutrition Research*, 30(9), 632–643.

Hariri, N., & Thibault, L. (2010). High-fat diet-induced obesity in animal models. *Nutrition Research Reviews*, 23(2), 270–299.

Harris, S. C., Ivy, A. C., & Searle, L. M. (1947). The mechanism of amphetamine-induced loss of weight; a consideration of the theory of hunger and appetite. *Journal of the American Medical Association*, 134(17), 1468–1475.

Heldt, S. A., & Ressler, K. J. (2009). The use of lentiviral vectors and Cre/loxP to investigate the function of genes in complex behaviors. *Frontiers in Molecular Neuroscience*, 2, 22.

Hernandez, L., & Hoebel, B. G. (1988). Feeding and hypothalamic stimulation increase dopamine turnover in the accumbens. *Physiology & Behavior*, 44(4–5), 599–606.

Hildebrandt, M. A., Hoffmann, C., Sherrill-Mix, S. A., Keilbaugh, S. A., Hamady, M., Chen, Y. Y., et al. (2009). High-fat diet determines the composition of the murine gut microbiome independently of obesity. *Gastroenterology*, 137(5), 1716–1724. e1711–e1712.

Hryhorczuk, C., Florea, M., Rodaros, D., Poirier, I., Daneault, C., Des Rosiers, C., et al. (2016). Dampened mesolimbic dopamine function and signaling by saturated but not monounsaturated dietary lipids. *Neuropsychopharmacology*, 41(3), 811–821.

Huszar, D., Lynch, C. A., Fairchild-Huntress, V., Dunmore, J. H., Fang, Q., Berkemeier, L. R., et al. (1997). Targeted disruption of the melanocortin-4 receptor results in obesity in mice. *Cell*, 88(1), 131–141.

Jackson-Lewis, V., & Przedborski, S. (2007). Protocol for the MPTP mouse model of Parkinson's disease. *Nature Protocols*, 2(1), 141–151.

Kaczmarek, J. L., Musaad, S. M., & Holscher, H. D. (2017). Time of day and eating behaviors are associated with the composition and function of the human gastrointestinal microbiota. *The American Journal of Clinical Nutrition*, 106(5), 1220–1231.

Kaisho, T., Nagai, H., Asakawa, T., Suzuki, N., Fujita, H., Matsumiya, K., et al. (2017). Effects of peripheral administration of a neuromedin U receptor 2-selective agonist on food intake and body weight in obese mice. *International Journal of Obesity*, 41(12), 1790–1797.

Kalueff, A. V., Echevarria, D. J., & Stewart, A. M. (2014). Gaining translational momentum: more zebrafish models for neuroscience research. *Progress in Neuro-Psychopharmacology & Biological Psychiatry*, 55, 1–6.

Kalueff, A. V., Stewart, A. M., & Gerlai, R. (2014). Zebrafish as an emerging model for studying complex brain disorders. *Trends in Pharmacological Sciences*, 35(2), 63–75.

Kanarek, R. B., D'Anci, K. E., Jurdak, N., & Mathes, W. F. (2009). Running and addiction: precipitated withdrawal in a rat model of activity-based anorexia. *Behavioral Neuroscience*, 123(4), 905–912.

Kawano, K., Hirashima, T., Mori, S., & Natori, T. (1994). OLETF (otsuka long-evans tokushima fatty) rat: a new NIDDM rat strain. *Diabetes Research and Clinical Practice*, 24(Suppl), S317–S320.

Kelley, A. E., & Berridge, K. C. (2002). The neuroscience of natural rewards: relevance to addictive drugs. *The Journal of Neuroscience*, 22(9), 3306–3311.

Kelley, G. A., Kelley, K. S., & Pate, R. R. (2017). Exercise and BMI z-score in overweight and obese children and adolescents: a systematic review and network meta-analysis of randomized trials. *Journal of Evidence-Based Medicine*, 10(2), 108–128.

Kim, S., Carlson, R., Zafreen, L., Rajpurohit, S. K., & Jagadeeswaran, P. (2009). Modular, easy-to-assemble, low-cost zebrafish facility. *Zebrafish*, 6(3), 269–274.

Kleinridders, A. (2016). Deciphering brain insulin receptor and insulin-like growth factor 1 receptor signalling. *Journal of Neuroendocrinology*, *28*(11).

Kmiec, Z. (2006). Central regulation of food intake in ageing. *Journal of Physiology and Pharmacology*, *57*(Suppl 6), 7–16.

Kuo, D. Y. (2006). Hypothalamic neuropeptide Y (NPY) and the attenuation of hyperphagia in streptozotocin diabetic rats treated with dopamine D1/D2 agonists. *British Journal of Pharmacology*, *148*(5), 640–647.

Lakso, M., Sauer, B., Mosinger, B., Jr., Lee, E. J., Manning, R. W., Yu, S. H., et al. (1992). Targeted oncogene activation by site-specific recombination in transgenic mice. *Proceedings of the National Academy of Sciences of the United States of America*, *89*(14), 6232–6236.

Levin, B. E., & Dunn-Meynell, A. A. (2002). Reduced central leptin sensitivity in rats with diet-induced obesity. *American Journal of Physiology. Regulatory, Integrative and Comparative Physiology*, *283*(4), R941–R948.

Levin, B. E., Dunn-Meynell, A. A., Balkan, B., & Keesey, R. E. (1997). Selective breeding for diet-induced obesity and resistance in sprague-dawley rats. *The American Journal of Physiology*, *273*(2 Pt 2), R725–R730.

Levin, B. E., & Keesey, R. E. (1998). Defense of differing body weight set points in diet-induced obese and resistant rats. *The American Journal of Physiology*, *274*(2 Pt 2), R412–R419.

Lobe, C. G., & Nagy, A. (1998). Conditional genome alteration in mice. *Bioessays*, *20*(3), 200–208.

Macneil, D. J. (2013). The role of melanin-concentrating hormone and its receptors in energy homeostasis. *Frontiers in Endocrinology (Lausanne)*, *4*, 49.

Marotz, C. A., & Zarrinpar, A. (2016). Treating obesity and metabolic syndrome with fecal microbiota transplantation. *The Yale Journal of Biology and Medicine*, *89*(3), 383–388.

Masca, N. G., Hensor, E. M., Cornelius, V. R., Buffa, F. M., Marriott, H. M., Eales, J. M., et al. (2015). RIPOSTE: a framework for improving the design and analysis of laboratory-based research. *eLife*, *4*.

Meeusen, R., & De Meirleir, K. (1995). Exercise and brain neurotransmission. *Sports Medicine*, *20*(3), 160–188.

Melis, T., Succu, S., Sanna, F., Boi, A., Argiolas, A., & Melis, M. R. (2007). The cannabinoid antagonist SR 141716A (rimonabant) reduces the increase of extra-cellular dopamine release in the rat nucleus accumbens induced by a novel high palatable food. *Neuroscience Letters*, *419*(3), 231–235.

Meng, F., Han, Y., Srisai, D., Belakhov, V., Farias, M., Xu, Y., et al. (2016). New inducible genetic method reveals critical roles of GABA in the control of feeding and metabolism. *Proceedings of the National Academy of Sciences of the United States of America*, *113*(13), 3645–3650.

Meredith, G. E., & Rademacher, D. J. (2011). MPTP mouse models of Parkinson's disease: an update. *Journal of Parkinson's Disease*, *1*(1), 19–33.

Merlino, D. J., Blomain, E. S., Aing, A. S., & Waldman, S. A. (2014). Gut-brain endocrine axes in weight regulation and obesity pharmacotherapy. *Journal of Clinical Medicine*, *3*(3), 763–794.

Millington, G. W., & Levell, N. J. (2007). From genesis to gene sequencing: historical progress in the understanding of skin color. *International Journal of Dermatology*, *46*(1), 103–105.

Monsanto, S. P., Hintze, K. J., Ward, R. E., Larson, D. P., Lefevre, M., & Benninghoff, A. D. (2016). The new total Western diet for rodents does not induce an overweight phenotype or alter parameters of metabolic syndrome in mice. *Nutrition Research*, *36*(9), 1031–1044.

Moran, T. H., & Bi, S. (2006). Hyperphagia and obesity in OLETF rats lacking CCK-1 receptors. *Philosophical Transactions of the Royal Society of London. Series B, Biological Sciences*, *361*(1471), 1211–1218.

Moran-Ramos, S., He, X., Chin, E. L., Tovar, A. R., Torres, N., Slupsky, C. M., et al. (2017). Nopal feeding reduces adiposity, intestinal inflammation and shifts the cecal microbiota and metabolism in high-fat fed rats. *PLoS ONE*, *12*(2), e0171672.

Moran-Ramos, S., Lopez-Contreras, B. E., & Canizales-Quinteros, S. (2017). Gut microbiota in obesity and metabolic abnormalities: a matter of composition or functionality? *Archives of Medical Research*, *48*, 735–753.

Mumphrey, M. B., Patterson, L. M., Zheng, H., & Berthoud, H. R. (2013). Roux-en-Y gastric bypass surgery increases number but not density of CCK-, GLP-1-, 5-HT-, and neurotensin-expressing enteroendocrine cells in rats. *Neurogastroenterology and Motility*, *25*(1), e70–e79.

Norrdin, R. W. (1981). Animal model of human disease. Renal osteodystrophy in dogs with radiation nephropathy. *The American Journal of Pathology*, *103*(3), 466–469.

Novak, C. M., Kotz, C. M., & Levine, J. A. (2006). Central orexin sensitivity, physical activity, and obesity in diet-induced obese and diet-resistant rats. *American Journal of Physiology Endocrinology and Metabolism*, *290*(2), E396–E403.

Obici, S., Magrisso, I. J., Ghazarian, A. S., Shirazian, A., Miller, J. R., Loyd, C. M., et al. (2015). Moderate voluntary exercise attenuates the metabolic syndrome in melanocortin-4 receptor-deficient rats showing central dopaminergic dysregulation. *Molecular Metabolism*, *4*(10), 692–705.

Oka, T., Nishimura, Y., Zang, L., Hirano, M., Shimada, Y., Wang, Z., et al. (2010). Diet-induced obesity in zebrafish shares common pathophysiological pathways with mammalian obesity. *BMC Physiology*, *10*, 21.

Padwal, R. (2009). Contrave, a bupropion and naltrexone combination therapy for the potential treatment of obesity. *Current Opinion in Investigational Drugs*, *10*(10), 1117–1125.

Paige, C., Hill, B., Canterbury, J., Sweitzer, S., & Romero-Sandoval, E. A. (2014). Construction of an affordable and easy-to-build zebrafish facility. *Journal of Visualized Experiments*, *93*, , e51989.

Palmiter, R. D. (2017). Neural circuits that suppress appetite: targets for treating obesity? *Obesity (Silver Spring)*, *25*(8), 1299–1301.

Patterson, C. M., Bouret, S. G., Dunn-Meynell, A. A., & Levin, B. E. (2009). Three weeks of postweaning exercise in DIO rats produces prolonged increases in central leptin sensitivity and signaling. *American Journal of Physiology. Regulatory, Integrative and Comparative Physiology*, *296*(3), R537–R548.

Patterson, C. M., Dunn-Meynell, A. A., & Levin, B. E. (2008). Three weeks of early-onset exercise prolongs obesity resistance in DIO rats after exercise cessation. *American Journal of Physiology. Regulatory, Integrative and Comparative Physiology*, *294*(2), R290–R301.

Patterson, C. M., & Levin, B. E. (2008). Role of exercise in the central regulation of energy homeostasis and in the prevention of obesity. *Neuroendocrinology*, *87*(2), 65–70.

Pelleymounter, M. A., Cullen, M. J., Baker, M. B., Hecht, R., Winters, D., Boone, T., et al. (1995). Effects of the obese gene product on body weight regulation in ob/ob mice. *Science*, *269*(5223), 540–543.

Petro, A. E., Cotter, J., Cooper, D. A., Peters, J. C., Surwit, S. J., & Surwit, R. S. (2004). Fat, carbohydrate, and calories in the development of diabetes and obesity in the C57BL/6J mouse. *Metabolism*, *53*(4), 454–457.

Pissios, P., Frank, L., Kennedy, A. R., Porter, D. R., Marino, F. E., Liu, F. F., et al. (2008). Dysregulation of the mesolimbic dopamine system and reward in MCH-/- mice. *Biological Psychiatry*, *64*(3), 184–191.

Pothos, E. N., & Sulzer, D. (1998). Modulation of quantal dopamine release by psychostimulants. *Advances in Pharmacology*, *42*, 198–202.

Pothos, E. N., Sulzer, D., & Hoebel, B. G. (1998). Plasticity of quantal size in ventral midbrain dopamine neurons: possible implications for the neurochemistry of feeding and reward. *Appetite*, *31*(3), 405.

Power, B. D., Stefanis, N. C., Dragovic, M., Jablensky, A., Castle, D., & Morgan, V. (2014). Age at initiation of amphetamine use and age at onset of psychosis: the australian survey of high impact psychosis. *Schizophrenia Research, 152*(1), 300–302.

Qian, S., Chen, H., Weingarth, D., Trumbauer, M. E., Novi, D. E., Guan, X., et al. (2002). Neither agouti-related protein nor neuropeptide Y is critically required for the regulation of energy homeostasis in mice. *Molecular and Cellular Biology, 22*(14), 5027–5035.

Ravinet Trillou, C., Delgorge, C., Menet, C., Arnone, M., & Soubrie, P. (2004). CB1 cannabinoid receptor knockout in mice leads to leanness, resistance to diet-induced obesity and enhanced leptin sensitivity. *International Journal of Obesity and Related Metabolic Disorders, 28*(4), 640–648.

Reed, M. J., Meszaros, K., Entes, L. J., Claypool, M. D., Pinkett, J. G., Gadbois, T. M., et al. (2000). A new rat model of type 2 diabetes: the fat-fed, streptozotocin-treated rat. *Metabolism, 49*(11), 1390–1394.

Reinehr, T. (2013). Type 2 diabetes mellitus in children and adolescents. *World Journal of Diabetes, 4*(6), 270–281.

Richey, J. M., & Woolcott, O. (2017). Re-visiting the endocannabinoid system and its therapeutic potential in obesity and associated diseases. *Current Diabetes Reports, 17*(10), 99.

Roh, J. I., Lee, J., Park, S. U., Kang, Y. S., Lee, J., Oh, A. R., et al. (2018). CRISPR-Cas9-mediated generation of obese and diabetic mouse models. *Experimental Animals, 67*, 229–237.

Sanmiguel, C., Gupta, A., & Mayer, E. A. (2015). Gut microbiome and obesity: a plausible explanation for obesity. *Current Obesity Reports, 4*(2), 250–261.

Sanmiguel, C. P., Jacobs, J., Gupta, A., Ju, T., Stains, J., Coveleskie, K., et al. (2017). Surgically induced changes in gut microbiome and hedonic eating as related to weight loss: preliminary findings in obese women undergoing bariatric surgery. *Psychosomatic Medicine, 79*(8), 880–887.

Savell, K. E., & Day, J. J. (2017). Applications of CRISPR/Cas9 in the mammalian central nervous system. *The Yale Journal of Biology and Medicine, 90*(4), 567–581.

Shimada, M., Tritos, N. A., Lowell, B. B., Flier, J. S., & Maratos-Flier, E. (1998). Mice lacking melanin-concentrating hormone are hypophagic and lean. *Nature, 396*(6712), 670–674.

Shin, A. C., Zheng, H., Townsend, R. L., Sigalet, D. L., & Berthoud, H. R. (2010). Meal-induced hormone responses in a rat model of Roux-en-Y gastric bypass surgery. *Endocrinology, 151*(4), 1588–1597.

Smith, K. E., Engel, S. G., Steffen, K. J., Garcia, L., Grothe, K., Koball, A., et al. (2017). Problematic alcohol use and associated characteristics following bariatric surgery. *Obesity Surgery, 28*, 1248–1254.

Speakman, J., Hambly, C., Mitchell, S., & Krol, E. (2007). Animal models of obesity. *Obesity Reviews, 8*(Suppl 1), 55–61.

Ste Marie, L., Miura, G. I., Marsh, D. J., Yagaloff, K., & Palmiter, R. D. (2000). A metabolic defect promotes obesity in mice lacking melanocortin-4 receptors. *Proceedings of the National Academy of Sciences of the United States of America, 97*(22), 12339–12344.

Stewart, A. M., Braubach, O., Spitsbergen, J., Gerlai, R., & Kalueff, A. V. (2014). Zebrafish models for translational neuroscience research: from tank to bedside. *Trends in Neurosciences, 37*(5), 264–278.

Stice, E., Spoor, S., Bohon, C., & Small, D. M. (2008). Relation between obesity and blunted striatal response to food is moderated by TaqIA A1 allele. *Science, 322*(5900), 449–452.

Stice, E., Spoor, S., Bohon, C., Veldhuizen, M. G., & Small, D. M. (2008). Relation of reward from food intake and anticipated food intake to obesity: a functional magnetic resonance imaging study. *Journal of Abnormal Psychology, 117*(4), 924–935.

Stricker, E. M. (2012). Neurochemical and behavioral analyses of the lateral hypothalamic syndrome: a look back. *Behavioural Brain Research, 231*(2), 286–288.

Sulzer, D., & Rayport, S. (1990). Amphetamine and other psychostimulants reduce pH gradients in midbrain dopaminergic neurons and chromaffin granules: a mechanism of action. *Neuron, 5*(6), 797–808.

Szkudelski, T. (2001). The mechanism of alloxan and streptozotocin action in B cells of the rat pancreas. *Physiological Research, 50*(6), 537–546.

Tam, C. S., Berthoud, H. R., Bueter, M., Chakravarthy, M. V., Geliebter, A., Hajnal, A., et al. (2011). Could the mechanisms of bariatric surgery hold the key for novel therapies? report from a pennington scientific symposium. *Obesity Reviews, 12*(11), 984–994.

Teitelbaum, P., & Stellar, E. (1954). Recovery from the failure to eat produced by hypothalamic lesions. *Science, 120*(3126), 894–895.

Thanos, P. K., Michaelides, M., Subrize, M., Miller, M. L., Bellezza, R., Cooney, R. N., et al. (2015). Roux-en-Y gastric bypass alters brain activity in regions that underlie reward and taste perception. *PLoS ONE, 10*(6), e0125570.

Thanos, P. K., Tucci, A., Stamos, J., Robison, L., Wang, G. J., Anderson, B. J., et al. (2010). Chronic forced exercise during adolescence decreases cocaine conditioned place preference in Lewis rats. *Behavioural Brain Research, 215*(1), 77–82.

Ussar, S., Griffin, N. W., Bezy, O., Fujisaka, S., Vienberg, S., Softic, S., et al. (2015). Interactions between gut microbiota, host genetics and diet modulate the predisposition to obesity and metabolic syndrome. *Cell Metabolism, 22*(3), 516–530.

Valsamakis, G., Konstantakou, P., & Mastorakos, G. (2017). New targets for drug treatment of obesity. *Annual Review of Pharmacology and Toxicology, 57*, 585–605.

Vila-Bedmar, R., Cruces-Sande, M., Lucas, E., Willemen, H. L., Heijnen, C. J., Kavelaars, A., et al. (2015). Reversal of diet-induced obesity and insulin resistance by inducible genetic ablation of GRK2. *Science Signaling, 8*(386), ra73.

Vrieze, A., Van Nood, E., Holleman, F., Salojarvi, J., Kootte, R. S., Bartelsman, J. F., et al. (2012). Transfer of intestinal microbiota from lean donors increases insulin sensitivity in individuals with metabolic syndrome. *Gastroenterology, 143*(4), 913–916. e917.

Wang, B., Chandrasekera, P. C., & Pippin, J. J. (2014). Leptin- and leptin receptor-deficient rodent models: relevance for human type 2 diabetes. *Current Diabetes Reviews, 10*(2), 131–145.

Wang, C. Y., & Liao, J. K. (2012). A mouse model of diet-induced obesity and insulin resistance. *Methods in Molecular Biology, 821*, 421–433.

Wang, G. J., Volkow, N. D., Logan, J., Pappas, N. R., Wong, C. T., Zhu, W., et al. (2001). Brain dopamine and obesity. *Lancet, 357*(9253), 354–357.

Wang, J. L., Yang, Q., Hajnal, A., & Rogers, A. M. (2016). A pilot functional MRI study in Roux-en-Y gastric bypass patients to study alteration in taste functions after surgery. *Surgical Endoscopy, 30*(3), 892–898.

Weiss, R., Dziura, J., Burgert, T. S., Tamborlane, W. V., Taksali, S. E., Yeckel, C. W., et al. (2004). Obesity and the metabolic syndrome in children and adolescents. *The New England Journal of Medicine, 350*(23), 2362–2374.

Wilson, M. E., Moore, C. J., Ethun, K. F., & Johnson, Z. P. (2014). Understanding the control of ingestive behavior in primates. *Hormones and Behavior, 66*(1), 86–94.

Wolfer, D. P., Crusio, W. E., & Lipp, H. P. (2002). Knockout mice: simple solutions to the problems of genetic background and flanking genes. *Trends in Neurosciences, 25*(7), 336–340.

Yaswen, L., Diehl, N., Brennan, M. B., & Hochgeschwender, U. (1999). Obesity in the mouse model of pro-opiomelanocortin deficiency responds to peripheral melanocortin. *Nature Medicine, 5*(9), 1066–1070.

Zallar, L. J., Tunstall, B. J., Richie, C. T., Zhang, Y. J., You, Z. B., Gardner, E. L., et al. (2019). Development and initial characterization of a novel ghrelin receptor CRISPR/Cas9 knockout wistar rat model. *International Journal of Obesity, 43*, 344–354.

Zucker, L. M., & Zucker, T. F. (1967). Fatty, a new mutation in the rat. *The Journal of Heredity, 52*, 275–278.

2

Biomarker-Guided Drug Development for Better Defined Early Patient Studies and Clinical Trial Efficiency

Aldemar Degroot

Boehringer Ingelheim Pharma GmbH & Co. KG., Biberach an der Riss, Germany

I INTRODUCTION

Recent advances in biomarkers and translational medicine have provided a better understanding of the cellular and molecular processes and the underlying neurocircuitry involved in neuropsychiatric and neurological disorders. Improvements have been made through enhanced imaging techniques, patient selection tools, the advent of digital technology, liquid biomarkers, and electrophysiology. In addition, novel diagnostic and enrichment biomarkers allow for better patient selection and improved clinical trial design. This comes coupled with an increased understanding of gene expression, which may particularly benefit monogenetic diseases such as Huntington's disease (Asscher & Koops, 2010). An assessment of genetic differences may further suggest a higher degree of responders in a subpopulation, thus allowing for a precision medicine approach for maximized therapeutic benefit (Espadaler, Tuson, Lopez-Ibor, Lopez-Ibor, & Lopez-Ibor, 2017; Sweatt & Tamminga, 2016). Furthermore, computational modeling is increasingly used to predict outcomes, and the result from modeling can, in turn, stimulate further research (for a review, see Geerts, Gieschke, & Peck, 2018). These developments are all in addition to a changing regulatory environment and an increased focus on the use of decision-making biomarkers in clinical trials.

Biomarker-guided drug development for CNS compounds ensures the confirmation of target engagement, provides early evidence for efficacy, and helps guide future dosing (Kielbasa & Stratford, 2012). Specifically, mechanistic biomarkers can confirm target engagement through, e.g., pharmacodynamic data such as receptor occupancy, enzyme binding, and enzyme activity (Eketjäll et al., 2016). On the other hand, functional biomarkers can measure a functional changes in the brain, such as glucose metabolism, quantitative pharmacoelectroencephalography (EEG), and event-related potential (ERP) (P50/mismatch negativity (MMN)/gamma activity) that occur as a consequence of target engagement (Chen & Zhong, 2013; Danjou et al., 2019). Ideally, receptor occupancy in preclinical models can be correlated with a pharmacodynamic effect, and an analogous measure can be used in healthy volunteers or early patient studies to select doses for proof-of-concept trials. Plasma samples can be banked for future genomic analysis to maximize pharmacogenetics (PGx) and precision medicine opportunities (Pisanu, Heilbronner, & Squassina, 2018). In patient studies, efficacy can then be measured as a symptomatic, behavioral, or functional effect. This biomarker-guided drug development approach can result in an increase in research and development (R&D) productivity, which is needed to sustain innovation and to minimize loss of revenue (Paul et al., 2010). This helps shift attrition rates to earlier stages of drug development for a quick win, fast fail approach.

Digital tools are becoming increasingly relevant in CNS clinical trials with both advances of digital technology and the wide adoption of smartphones in recent years. The advent of digital tools allows for increasing self-assessment, may act to further boost a drug's efficacy, and may serve as a monitoring tool to track disease progression (Hidalgo-Mazzei, Young, Vieta, & Colom, 2018). For instance, voice analysis can now be used for diagnosis and monitoring of a number of neurological and psychiatric conditions (e.g., Martínez-Sánchez, Meilán, Carro, & Ivanova, 2018). The prescription digital medicine company Akili has taken this one step further and has built a broad pipeline of programs to treat cognitive deficiency and improve symptoms associated with medical conditions across neurology and psychiatry using prescription digital medicine. In addition, Abilify MyCite is the first sensor-embedded medicine that monitors adherence to the atypical antipsychotic Abilify by patients suffering from schizophrenia. Abilify MyCite originated from a collaboration between Otsuka, the drug's manufacturer, and Proteus Digital Health, a California-based company that developed the sensor.

All these developments are starting to bear fruit as we enter a new frontier in CNS drug development.

II USING BIOMARKERS AS A STEPWISE APPROACH TOWARDS EVIDENCE GATHERING IN EARLY CLINICAL STUDIES

Before patient recruitment and enrollment into clinical trials, it's imperative to demonstrate that the drug penetrates the CNS, binds to the target, and has a functional effect that is reflective of efficacy. In some cases, this information may then also be used to stratify future patient populations, allowing for smaller patient studies and earlier decision-making. Mechanistic markers used to demonstrate target engagement are usually linked to the mechanism of action of the drug, and the tools needed to directly demonstrate target engagement may not always be available in clinical trials. Whereas in preclinical studies that one can rely on tissue level measurements and in vivo receptor occupancy assays using, for example, liquid chromatography-tandem mass spectrometry (LC/MS/MS) technology (Need, McKinzie, Mitch, Statnick, & Phebus, 2007) and pharmacokinetics (PK)/pharmacodynamics (PD) modeling and simulation (Amore, Gibbs, & Emery, 2010) for dosing, fewer tools are available in the clinic. One solution comes from PET imaging, and if a PET ligand is available, then this becomes a powerful tool, which also translates from preclinical to clinical studies (e.g., Sihver et al., 2000; Smith et al., 2015). If a ligand is not available, then the technique becomes more limited but can still be used to demonstrate CNS penetration. Other tools to demonstrate CNS penetration include bold functional magnetic resonance imaging (fMRI) and cerebral spinal fluid (CSF) measurements (Borsook, Becerra, & Fava, 2013; Nau, Sörgel, & Eiffert, 2010).

Alternatively, functional markers can be used to demonstrate downstream functional effects in response to target engagement, and this can serve as an indirect measure of engagement at the target. An example comes from qEEG, which can be used to demonstrate a physiological effect that corresponds to target binding and that back translates to preclinical studies. This provides reasonable downstream evidence of CNS penetration and target engagement but fails to provide a complete picture since neuronal activity is not limited to electric activity and also involves neurotransmitter and neurohormonal changes (Richard, 2018). To that end, advances in neuroimaging allow for unprecedented spatiotemporal resolution, which enables neurotransmitter measurements in real time, and this results in a much deeper understanding of neurotransmitter dynamics during disease (Richard, 2018). This alternative to qEEG measurements may provide enhanced real-time insight into functioning of brain networks, but this also requires the use of specialty imaging centers, which may not be practical and may prolong timelines and increase costs. It's also possible to image synaptic density, which may be particularly relevant to those compounds in development that are thought to mediate their therapeutic effect through enhanced synaptic signaling (Mercier, Provins, & Valade, 2017). All these markers provide additional insight into the neurocircuitry that is being modulated by the compound in question.

Demonstrating target engagement, either directly or through a downstream functional marker, can result in earlier proof-of-concept studies. A combination of more extensive target validation and earlier POC studies can reduce Phase 2 attrition by up to 50%, which, in turn, may lower the cost of a new molecular entity by approximately 30% (Paul et al., 2010).

III PHYSIOLOGICAL AND FUNCTIONAL CHANGES MAY CORRESPOND TO A MEANINGFUL CLINICAL EFFECT

Observed biomarker changes may correlate to a meaningful clinical outcome, even if those biomarkers are just exploratory and are not surrogate markers of the disease. For instance, an increase in P3A and evoked gamma and a decrease in MMN in a qEEG study may provide evidence that the compound may enhance synchronized activity involved in learning and memory and qEEG can be used as a physiological biomarker to predict future treatment response (Light & Swerdlow, 2015). Thomas et al., 2017, conducted an elaborate analysis of cross-sectional data from 1415 schizophrenia patients in an effort to determine to what extent an enhancement of basic information processing as measured with qEEG might improve cognition and psychosocial functioning in schizophrenia. They found that an intervention that induces a 1 μV change in the mean amplitude of MMN, P3A or RON would produce Cohen's d improvements of 0.78 for cognition and 0.28 for psychosocial functioning in schizophrenia patients. This provides a compelling example to the utility of functional markers to predict clinical response in the target patient population, thus allowing for earlier decision-making.

IV REDEFINING STANDARDIZED TEST BATTERIES AND SELECTING THE MOST RESPONSIVE MEASURES

Standardized test batteries such as MATRICS and ADAS-cog often involve a composite score, and this may result in poor translation from preclinical models to clinical efficacy. The failure rate for the development of compounds used to treat cognitive impairment in Alzheimer's disease is 99.6% even though compounds tested in the clinic demonstrated efficacy in preclinical behavioral models (Cummings, Morstorf, & Zhong, 2014). As such, animal behavioral models do not model the disease, but may reflect efficacy for particular symptoms of the disease. An intermediate step can be used in clinical development where one attempts to mirror those cognitive domains that were found to be affected in animal models and that can be ascribed to the mechanism of action of the drug. This can then further be combined with biomarker assessment as described, and this effect on subdomains of standardized tests in combination with an effect on biomarkers can then be used to potentially stratify future patient populations that will be tested on a composite score (see Table 2.1). Thus a deeper phenotyping in combination with biomarker assessment and clinical endpoint selection can be used to enrich future patient studies and increase the odds of CNS drug development.

TABLE 2.1   Tools of Drug Development and Their Correlation to Efficacy in the Primary Clinical Endpoint

Preclinical models	Biomarkers	Endpoint that is a subset of primary endpoint	Primary endpoint
Do not model the disease, but model the symptoms of the disease	Can be used to demonstrate target engagement, functional response in relation to target engagement and response related to efficacy	Can target those aspects of standard battery that are most responsive to the MOA based on preclinical data and other information related to the target	Needed for registration, but not set in stone
Very sensitive to treatment	Sensitive to treatment and can count as go/no go criteria	Can be used to identify responders for future clinical trials	Extremely high failure rate

V DIAGNOSTIC AND ENRICHMENT MARKERS FOR BETTER CLINICAL TRIALS

In addition to using mechanistic and functional markers to demonstrate CNS penetration, target engagement, and CNS activity reflective of the drug's mechanism of action and its potential efficacy, new diagnostic tools are being developed based on molecular-based techniques, and this may result in blood-based diagnostic markers. This could greatly facilitate earlier diagnosis and make things easier for the patient. For instance, Alzheimer's disease may be diagnosed through PET imaging or spinal tab for CSF neurochemical assessments following the completion of a cognitive test (Nield, 2018). Using PET imaging and performing a spinal tab rule out cognitive deficits through, e.g., stroke, Parkinson's disease, and Huntington's disease. However, spinal tab is uncomfortable to the patient and has a high patient burden (but is good for enrichment based on levels of CSF amyloid beta peptide and neuronal tau protein), and PET takes almost an hour to complete during which the patient has to remain immobile. Also, reimbursement can be an issue, and patients may have to pay thousands out of pocket for expensive imaging techniques. Moreover, often diagnosis occurs only by looking for visible symptoms of the disease at which point there is already considerable disease progression (Nield, 2018). On the other hand a blood-based diagnostic test would be much easier and less expensive to administer and may result in earlier diagnosis. This may allow patient to enroll at earlier stages of the disease, and this may result in better treatment. This may be of particular relevance for neurodegenerative disorders, where earlier treatment may increase the odds of efficacy with a therapeutic intervention. To that end, several research teams are working on blood tests for early Alzheimer's disease diagnosis, which may disrupt the expensive and invasive scanning and spinal fluid technologies (Nield, 2018).

Enrichment markers are also under development. For instance, it was recently determined that intravenous injection of a small amount of radioactive tracer can be used as an imaging agent that very specifically binds to dopamine transporter sites on neurons that are lost in Parkinson's disease, and this can be used to stratify Parkinson's patients in clinical trials using the dopamine transporter (DAT) single-photon emission computerized tomography (SPECT) imaging technique or DaTScan (Critical Path Institute, 2018). Another example comes from schizophrenia where the underlying circuitry related to the disease changes over time. Whereas initially we may see hyperconnectivity, later in the disease, this may revert to a state of hypoactivity and volume loss. Therefore a drug that inhibits cortical activity may be beneficial early in the disease, but not during later stages.

Consequently, it's not only important to test the right patient but also to test the right patient at the right time. This can be achieved through the use of biomarkers combined with an improved understanding of the underlying neurocircuitry and taxonomy of the disease. Also, going back to the example of Alzheimer's disease earlier, regulators have now approved the use of amyloid PET not only for diagnosis but also for the enrichment of Alzheimer's trials.

VI EXAMPLES OF USING BIOMARKER-GUIDED DRUG DEVELOPMENT

Example 1: NMDA antagonists as fast acting antidepressants (e.g., ketamine)

Although depression is commonly treated with compounds that target the monoamine system, these compounds take weeks to take effect and are ineffective in a large percentage of patients. Another possibility is to treat patients with compounds that target the glutamatergic system. To that end the role of NMDA antagonists in depression has received a lot of attention since the nonselective NMDA antagonist ketamine was found to be highly efficacious in treatment-resistant depression almost two decades ago (Berman et al., 2000). What's more, the effect of ketamine was immediate and lasted for days following a single intravenous (i.v.) infusion. However, ketamine also induces dissociative and psychotomimetic effects and is prone to abuse, which resulted in it being listed as a Schedule III drug under the US Controlled Substances Act. Consequently the use of ketamine is limited to treatment-resistant depression (Molero et al., 2018). Therefore there has been an increasing effort over recent years to develop compounds that target the NMDA receptor and that mimic the ketamine effect on depression but with a more favorable side effect profile. To that end, there are a number of biomarkers associated with both the mechanism and the disease, allowing for a biomarker-guided drug development approach for this mode of action (see Fig. 2.1).

Ketamine binds to NMDA receptors on GABAergic neurons that project to glutamatergic neurons. Ketamine binding causes an inhibition of the GABAergic neurons, and this results in increased glutamate release as has been observed using microdialysis in animal models (Lorrain, Baccei, Bristow, Anderson, & Varney, 2003). Increased glutamate release results in an increased activation of AMPA, which increases the activation of mTORC1, and this, in turn, activates brain-derived neurotrophic factor (BDNF). The activation of BDNF results in increased synaptic signaling, and this increases the expression of the synaptic proteins GluR1 and synapsin (Duman, Aghajanian, Sanacora, & Krystal, 2016).

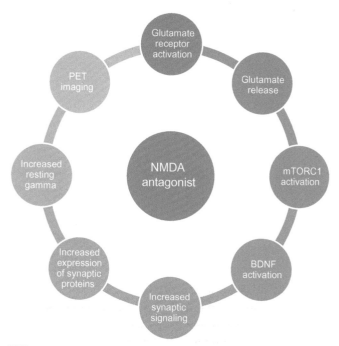

FIG. 2.1   Mechanistic, functional, and disease biomarkers that can be used in the biomarker-guided drug development of compounds that target NMDA receptors to treat depression.

There are thus a number of functional markers to help guide development of compounds that target NMDA receptors to treat depression. Some of these, such as mTOR1 activation and synaptic proteins in the prefrontal cortex, can only be measured in preclinical studies. Others, such as glutamate release and synaptic signaling, can also be measured in the clinic (Cai, Li, Matuskey, Nabulsi, & Huang, 2019; Finnema et al., 2016), although their assessment would involve considerable clinical complexity, requiring specialized centers, which may not be feasible. Nevertheless, even if it may take too much effort to include these markers in clinical trials, they can be used in preclinical models to build clinical confidence and as a more reliable translational tool compared with some of the preclinical behavioral models. In addition, it can be considered to include some of these markers in clinical substudies to get a better sense of target engagement and future dosing.

Another biomarker that can be used to guide drug development of compounds that target the NMDA receptor is qEEG. Antagonists at the NMDA receptor are well known to increase resting gamma oscillations (Sanacora et al., 2014), and qEEG can be assessed in both preclinical and clinical studies (English et al., 2014). Also, it has been known for years that changes in qEEG may be a characteristic of depression (Fitzgerald & Watson, 2018). Measurements of qEEG may, therefore, be related not only to the mechanism of the drug but also to the nature of the disease.

In conclusion, there are a number of biomarkers available for the development and derisking of NMDA antagonists for depression. Although including some of these markers may add considerable clinical complexity, others can be efficiently used to demonstrate CNS penetration and target engagement and to improve the translatability of preclinical efficacy to clinical outcome.

Example 2: PDE4 inhibitors and cognition (e.g., roflumilast)

Another example comes from the phosphodiesterase (PDE) 4 inhibitors, a class of compounds that is being developed to treat a variety of disorders, including cognitive disorders. A number of tools are available to help facilitate and derisk development of the PDE4 inhibitors (see Fig. 2.2). For instance, brain PDE4 receptor occupancy can be measured using a PET ligand (Takano et al., 2018). In addition, PK/PD modeling has been used as an early translational approach for human dose prediction in the development of PDE4 inhibitors (Plock, Vollert, Mayer, Hanauer, & Lahu, 2017). Moreover, PDE4 inhibition modulates a number of biomarkers that are reflective of target engagement and that are correlated to the disease. Specifically, PDE4 inhibition modulates cAMP, protein kinase A (PKA) and cAMP response element binding (CREB) protein signaling (Bolger, 2017). Thus in addition to using PET to demonstrate CNS penetration and target engagement, one can also measure cAMP in the CSF as an indirect demonstration of target engagement in the CNS.

Initial data indicate that the preclinical efficacy data translate to a model of efficacy in healthy volunteers. For instance, the PDE4 inhibitor roflumilast improves memory in rodents (Vanmierlo et al., 2016), including in a mouse model of Alzheimer's disease (Xu et al., 2018). In addition, acute administration of roflumilast enhances immediate verbal word memory in healthy volunteers (Van Duinen et al., 2018). Interestingly, qEEG was recorded during the cognitive tasks on the first day. Different ERPs were considered with special emphasis on P600, as this peak has been related to word learning. Memory performance was significantly improved after acute administration of 100 µg roflumilast, and this improvement was accompanied by an enhanced P600 peak during word presentation (Van Duinen et al., 2018).

Therefore, there are a number of mechanistic and functional markers to help facilitate the development of PDE4 inhibitors for cognitive disorders.

VII CONCLUSION

Advances in the identification and development of biomarkers in CNS drug development provide researchers and clinicians with the tools and techniques to better understand and treat CNS disorders. Moreover, biomarkers improve our understanding of the underlying cellular and molecular processes. In addition, this has provided insight into the neurocircuitry that needs to be targeted to treat some of these conditions. An improved understanding of how neurocircuits relate to specific symptoms provides an opportunity for using brain-based taxonomy to classify CNS conditions. This, in turn, allows for the identification of novel targets along with markers that can be used to guide the development of those compounds that modulate those targets. The techniques used to demonstrate engagement of the neurocircuitry may add considerable clinical complexity, and at times, it may be feasible to include simple behavioral measures in clinical studies that are known to be correlated to modulation of specific circuits. Using biomarker-guided drug development in CNS may result in a new frontier in CNS drug development, with enhanced patient selection, optimized endpoints, and improved target identification.

References

Amore, B. M., Gibbs, J. P., & Emery, M. G. (2010). Application of in vivo animal models to characterize the pharmacokinetic and pharmacodynamic properties of drug candidates in discovery settings. *Combinatorial Chemistry & High Throughput Screening*, 13(2), 207–218.

Asscher, E., & Koops, B. J. (2010). The right not to know and preimplantation genetic diagnosis for Huntington's disease. *Journal of Medical Ethics*, 36(1), 30–33.

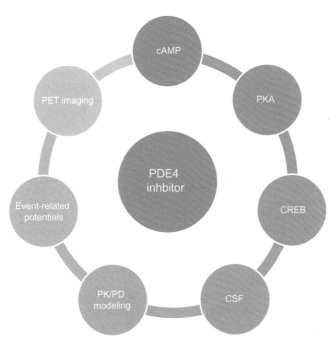

FIG. 2.2 Mechanistic, functional, and disease biomarkers that can be used in the biomarker-guided drug development of PDE4 inhibitors to treat cognitive disorders.

Berman, R. M., Cappiello, A., Anand, A., Oren, D. A., Heninger, G. R., Charney, D. S., et al. (2000). Antidepressant effects of ketamine in depressed patients. *Biological Psychiatry, 47*(4), 351–354.

Bolger, G. B. (2017). The PDE4 cAMP-specific phosphodiesterases: targets for drugs with antidepressant and memory-enhancing action. *Advances in Neurobiology, 17,* 63–102.

Borsook, D., Becerra, L., & Fava, M. (2013). Use of functional imaging across clinical phases in CNS drug development. *Translational Psychiatry, 16*(3), e282.

Cai, Z., Li, S., Matuskey, D., Nabulsi, N., & Huang, Y. (2019). PET imaging of synaptic density: a new tool for investigation of neuropsychiatric diseases. *Neuroscience Letters, 691,* 44–50.

Chen, Z., & Zhong, C. (2013). Decoding Alzheimer's disease from perturbed cerebral glucose metabolism: implications for diagnostic and therapeutic strategies. *Progress in Neurobiology, 108,* 21–43.

Critical Path Institute (2018). *First-ever biomarker qualified for Parkinson's is a vital step toward improved clinical trials.* https://c-path.org/first-ever-biomarker-qualified-for-parkinsons-is-a-vital-step-toward-improved-clinical-trials/.

Cummings, J. L., Morstorf, T., & Zhong, K. (2014). Alzheimer's disease drug-development pipeline: few candidates, frequent failures. *Alzheimer's Research & Therapy, 6*(4), 37.

Danjou, P., Viardot, G., Maurice, D., Garcés, P., Wams, E. J., Phillips, K. G., et al. (2019). Electrophysiological assessment methodology of sensory processing dysfunction in schizophrenia and dementia of the Alzheimer type. *Neuroscience and Biobehavioral Reviews, 97,* 70–84.

Duman, R. S., Aghajanian, G. K., Sanacora, G., & Krystal, J. H. (2016). Synaptic plasticity and depression: new insights from stress and rapid-acting antidepressants. *Nature Medicine, 22*(3), 238–249.

Eketjäll, S., Janson, J., Kaspersson, K., Bogstedt, A., Jeppsson, F., Fälting, J., et al. (2016). AZD3293: a novel, orally active BACE1 inhibitor with high potency and permeability and markedly slow off-rate kinetics. *Journal of Alzheimer's Disease, 50*(4), 1109–1123.

English, B. A., Thomas, K., Johnstone, J., Bazih, A., Gertsik, L., & Ereshefsky, L. (2014). Use of translational pharmacodynamic biomarkers in early-phase clinical studies for schizophrenia. *Biomarkers in Medicine, 8*(1), 29–49.

Espadaler, J., Tuson, M., Lopez-Ibor, J. M., Lopez-Ibor, F., & Lopez-Ibor, M. I. (2017). Pharmacogenetic testing for the guidance of psychiatric treatment: a multicenter retrospective analysis. *CNS Spectrums, 22*(4), 315–324.

Finnema, S. J., Nabulsi, N. B., Eid, T., Detyniecki, K., Lin, S. F., Chen, M. K., et al. (2016). Imaging synaptic density in the living human brain. *Science Translational Medicine, 8*(348), 1–10.

Fitzgerald, P. J., & Watson, B. O. (2018). Gamma oscillations as a biomarker for major depression: an emerging topic. *Translational Psychiatry, 8*(1), 177.

Geerts, H., Gieschke, R., & Peck, R. (2018). Use of quantitative clinical pharmacology to improve early clinical development success in neurodegenerative diseases. *Expert Review of Clinical Pharmacology, 11*(8), 789–795.

Hidalgo-Mazzei, D., Young, A. H., Vieta, E., & Colom, F. (2018). Behavioural biomarkers and mobile mental health: a new paradigm. *International Journal of Bipolar Disorders, 6*(1), 9.

Kielbasa, W., & Stratford, R. E., Jr. (2012). Exploratory translational modeling approach in drug development to predict human brain pharmacokinetics and pharmacologically relevant clinical doses. *Drug Metabolism and Disposition, 40*(5), 877–883.

Light, G. A., & Swerdlow, N. R. (2015). Future clinical uses of neurophysiological biomarkers to predict and monitor treatment response for schizophrenia. *Annals of the New York Academy of Sciences, 1344,* 105–119.

Lorrain, D. S., Baccei, C. S., Bristow, L. J., Anderson, J. J., & Varney, M. A. (2003). Effects of ketamine and N-methyl-D-aspartate on glutamate and dopamine release in the rat prefrontal cortex: modulation by a group II selective metabotropic glutamate receptor agonist LY379268. *Neuroscience, 117*(3), 697–706.

Martínez-Sánchez, F., Meilán, J. J. G., Carro, J., & Ivanova, O. (2018). A prototype for the voice analysis diagnosis of Alzheimer's disease. *Journal of Alzheimer's Disease, 64*(2), 473–481.

Mercier, J., Provins, L., & Valade, A. (2017). Discovery and development of SV2A PET tracers: potential for imaging synaptic density and clinical applications. *Drug Discovery Today: Technologies, 25,* 45–52.

Molero, P., Ramos-Quiroga, J. A., Martin-Santos, R., Calvo-Sánchez, E., Gutiérrez-Rojas, L., & Meana, J. J. (2018). Antidepressant efficacy and tolerability of ketamine and esketamine: a critical review. *CNS Drugs, 32,* 411–420. Epub ahead of print.

Nau, R., Sörgel, F., & Eiffert, H. (2010). Penetration of drugs through the blood-cerebrospinal fluid/blood-brain barrier for treatment of central nervous system infections. *Clinical Microbiology Reviews, 23*(4), 858–883.

Need, A. B., McKinzie, J. H., Mitch, C. H., Statnick, M. A., & Phebus, L. A. (2007). In vivo rat brain opioid receptor binding of LY255582 assessed with a novel method using LC/MS/MS and the administration of three tracers simultaneously. *Life Sciences, 81*(17–18), 1389–1396.

Nield, D. (2018). *New blood test for Alzheimer's is so precise it could predict it 30 years ahead.* https://www.sciencealert.com/new-blood-test-could-be-vital-alzheimers-early-warning-system.

Paul, S. M., Mytelka, D. S., Dunwiddie, C. T., Persinger, C. C., Munos, B. H., Lindborg, S. R., et al. (2010). How to improve R&D productivity: the pharmaceutical industry's grand challenge. *Nature Reviews. Drug Discovery, 9*(3), 203–214.

Pisanu, C., Heilbronner, U., & Squassina, A. (2018). The role of pharmacogenomics in bipolar disorder: moving towards precision medicine. *Molecular Diagnosis & Therapy, 22*(4), 409–420.

Plock, N., Vollert, S., Mayer, M., Hanauer, G., & Lahu, G. (2017). Pharmacokinetic/pharmacodynamic modeling of the PDE4 inhibitor TAK-648 in type 2 diabetes: early translational approaches for human dose prediction. *Clinical and Translational Science, 10*(3), 185–193.

Richard, R. (2018). At the bench-dopamine signaling: new imaging technique offers cell-level insights into dopamine signaling. *Neurology Today, 18*(14), 30–31.

Sanacora, G., Smith, M. A., Pathak, S., Su, H. L., Boeijinga, P. H., McCarthy, D. J., et al. (2014). Lanicemine: a low-trapping NMDA channel blocker produces sustained antidepressant efficacy with minimal psychotomimetic adverse effects. *Molecular Psychiatry, 19*(9), 978–985.

Sihver, W., Nordberg, A., Långström, B., Mukhin, A. G., Koren, A. O., Kimes, A. S., et al. (2000). Development of ligands for in vivo imaging of cerebral nicotinic receptors. *Behavioural Brain Research, 113*(1–2), 143–157.

Smith, J. A., Bourdet, D. L., Daniels, O. T., Ding, Y. S., Gallezot, J. D., Henry, S., et al. (2015). Preclinical to clinical translation of CNS transporter occupancy of TD-9855, a novel norepinephrine and serotonin reuptake inhibitor. *The International Journal of Neuropsychopharmacology, 18*(2), 1–11.

Sweatt, J. D., & Tamminga, C. A. (2016). An epigenomics approach to individual differences and its translation to neuropsychiatric conditions. *Dialogues in Clinical Neuroscience, 18*(3), 289–298.

Takano, A., Uz, T., Garcia-Segovia, J., Tsai, M., Lahu, G., Amini, N., et al. (2018). A nonhuman primate PET study: measurement of brain PDE4 occupancy by roflumilast using (R)-[11C]rolipram. *Molecular Imaging and Biology, 20*(4), 615–622.

Thomas, M. L., Green, M. F., Hellemann, G., Sugar, C. A., Tarasenko, M., Calkins, M. E., et al. (2017). Modeling deficits from early auditory information processing to psychosocial functioning in schizophrenia. *JAMA Psychiatry, 74*(1), 37–46.

Van Duinen, M. A., Sambeth, A., Heckman, P. R. A., Smit, S., Tsai, M., Lahu, G., et al. (2018). Acute administration of roflumilast enhances immediate recall of verbal word memory in healthy young adults. *Neuropharmacology, 131*, 31–38.

Vanmierlo, T., Creemers, P., Akkerman, S., van Duinen, M., Sambeth, A., De Vry, J., et al. (2016). The PDE4 inhibitor roflumilast improves memory in rodents at non-emetic doses. *Behavioural Brain Research, 303*, 26–33.

Xu, Y., Zhu, N., Xu, W., Ye, H., Liu, K., Wu, F., et al. (2018). Inhibition of phosphodiesterase-4 reverses Aβ-induced memory impairment by regulation of HPA axis related cAMP signaling. *Frontiers in Aging Neuroscience, 10*(204), 1–11.

3

Modeling and Simulation in the Translational Pharmacology of CNS Drugs

Elizabeth C. de Lange*, Peter Bonate†

*Leiden Academic Centre for Drug Research, Gorlaeus Laboratories, Leiden, The Netherlands †Astellas Pharma, Northbrook, IL, United States

I INTRODUCTION

We've all seen the numbers. And they are truly staggering. It takes more than $1 billion dollars and more than 10 years of development for a single drug to be approved. The attrition rates of new chemical entities are dismal. For every 5000 compounds screened on average, approximately 15 will be tested in animals, 9 will be tested in man, and only 1 of those will make it to market, for an overall success rate of 4% (Calcoen, Elias, & Yu, 2015). For CNS drugs, the odds are even more staggering. In Alzheimer's disease, for every 119 compounds that are tested preclinically, only one will make it to market leading to an overall success rate of less than 1% (Calcoen et al., 2015).

The attrition rates for drugs that affect the central nervous system (CNS) are lower than many other therapeutic areas for several reasons. The blood-brain barrier (BBB) and blood-cerebrospinal fluid barrier (BCB) limit access of drugs to the brain and CNS, thereby limiting the drug space for successful new chemical entities. The lack of biomarkers in the area, compared with say cardiovascular, make choosing a dose particularly challenging because early clinical studies must be done in patients and hard clinical endpoints used to assess a drug's efficacy are difficult and expensive to obtain. Further, linkage of drug concentrations to clinical endpoints is complicated because of the inability to measure brain concentrations directly. And the translation of animal pharmacology models of CNS drug activity to humans, such as the results of a water maze test for antianxiety compounds, has been particularly challenging. Pharmacokinetic-pharmacodynamic modeling and simulation is being used to help bridge the gap between animal pharmacology models and humans and to better understand the relationship between plasma concentrations, brain concentrations, and drug effects. More recently, efforts to fully model brain activity using computational models and to use these models to understand how drugs might affect brain processes are being developed and utilized. The purpose of this chapter will be to review modeling and simulation and how it is currently being used to help to facilitate translational pharmacology of CNS compounds.

II WHAT IS MODELING AND SIMULATION?

Mathematical models characterize the relationship between a set of inputs and a set of outputs. These inputs might be the dose of the drug administered to animals or the concentration of drug in a test tube, while the outputs could be a drug's effect observed in an animal or the amount of drug remaining in the test tube. Famous mathematical equations in biology include the Hodgkin–Huxley equation for axon membrane potential, the Lotka-Volterra equation for competition among animals, and the Michaelis-Menten equation for enzyme kinetics (Jungck, 1997). Pharmacokinetic-pharmacodynamic modeling is a special class of modeling that relates drug concentrations in biological fluids, like plasma or cerebrospinal fluid (CSF), to a drug's response (Bonate, 2011), which could be something as simple as locomotor response after administration of cocaine or change in clinical global impression (CGI) scores after long-term administration of antipsychotics to schizophrenic patients.

Modeling is useful because it allows a scientist to interpret results from many different experiments in a harmonious manner. For example, a preclinical scientist may have results from pharmacokinetic studies in mice, in vitro receptor binding studies with brain parenchyma, behavioral pharmacology studies in mice, and drug metabolism studies. Using an integrated model, the studies can be tied together to link pharmacokinetics to brain concentrations to receptor binding to drug effect (Fig. 3.1). These types of analyses are useful in that they require scientists think logically about how the systems tie together and interact and how the effects in one experiment might affect another experiment. In a team science environment where many different scientists from different backgrounds work together, such a type of integrative modeling is a useful tool to encourage interaction among the scientists and have frank discussions about their beliefs regarding how a drug works. There are other advantages as well. Such models may lead to the discovery of new questions, are useful to illustrate and explain concepts to other scientists, and can help identify those variables that are of particular importance to the system or identify variables that need to be identified to make the system work properly.

Once a model is identified, simulation may be used to make predictions about the system. Modeling and simulation have different points of view. Modeling looks back in time and uses data to generate the models. Simulation looks forward in time and uses the model to predict data. There are two different types of simulation, deterministic and Monte Carlo. In a deterministic simulation, the variability in the data is ignored. Simulations are based on mean values, and the results are what would be expected in the long term. Monte Carlo simulation, which gets its name from the gambling capital of Monaco, considers stochastic variability in the data, both random in nature and systematic variability varying from individual to individual, and the results capture both the long-term expectation and expected variability around that expectation (Bonate, 2001). As might be expected, deterministic simulation is easier to perform than Monte Carlo expectation, but the trade-off for Monte Carlo expectation is a greater understanding of the results and variability in the results.

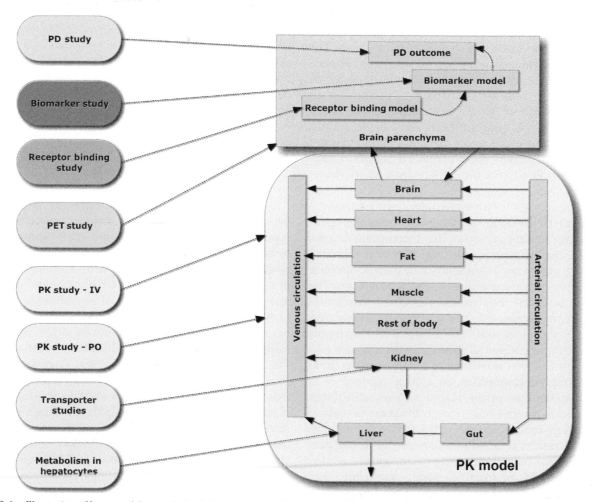

FIG. 3.1   Illustration of how modeling and simulation can be used to integrate different experiments into a harmonious system.

III TOWARD PRECLINICAL TO CLINICAL CNS PHARMACOLOGY

[handwritten: modeling vs. simulation] *[handwritten: Goal]*

The ultimate goal in the preclinical translation of nonclinical data is to identify a human dose that will achieve similar exposures at the site of action that resulted in pharmacological activity in animals. This translation of animal data to humans is complicated and involves many assumptions. One general assumption, one that applies to any translation from animals, is that in the absence of other information, the results in animals apply to humans. For instance, if an IC50 in rat brain is 10 nM, in the absence of any other information, it is assumed that the IC50 in humans will be the same. If this is the case, then a back of the envelope calculation can be made using a body surface area conversion (U.S. Department of Health and Human Services, 2005), which assumes that the difference between animals and humans is simply a matter of size of the animal. First proposed in the 1960s as a way to compare doses of oncology drugs administered to animals with doses of drugs that were clinically efficacious in humans (Freireich, Gehan, Rall, Schmidt, & Skipper, 1966), this method is now standard practice in drug development. The method is to multiply the efficacious dose in animals expressed as milligram/kilogram by a constant factor that is the ratio of the body surface area in humans to the body surface area of the animal in question. For mouse, rat, monkey, and dog, those multipliers are 3, 6, 12, and 20, respectively. Therefore, if the efficacious dose in rats was 10 mg/kg, the equivalent dose and humans would be 60 mg/kg, and an approximate total dose in humans, assuming a body surface area of 1.8 m^2, would be 108 mg. But this method, because it makes lots of assumptions, should be used with caution. Nevertheless, it's a useful tool to make early dose predictions in humans in the preclinical candidate stage where there is little information available.

[handwritten: Assumption #1]

A first important assumption (that holds for all drugs) is that it is the unbound drug concentration at the site of action that is the driver of effect, that is, drug bound to plasma proteins or macromolecules is unavailable to bind to receptors. A corollary of this assumption is that it is the unbound drug concentration in the brain at the site of action that is the driver of efficacy. For example, Watson et al. (2009) determined the receptor occupancy at D2 receptors for six marketed antipsychotic drugs based on total brain concentration, CSF concentration, unbound brain concentration, and unbound blood concentration. They found that unbound brain concentration was a better predictor of receptor occupancy than other measurements. Liu, Vilenski, Kwan, Apparsundaram, and Weikert (2009) compared receptor occupancy with 18 dopamine and serotonin receptor uptake transporter inhibitors and came to a similar conclusion that it is the unbound drug concentration in the brain that is a better

determinant of receptor occupancy than total brain concentration. Hence this is a reasonable assumption and has been accepted as dogma by pharmacologists.

[handwritten: Assumption #2]

A second assumption is that the transfer from blood to the site of action in the brain is the same in animals and humans. For drugs that are not transported mediated into the brain, this may be a reasonable assumption because the important factors that affect a drug's distribution into the brain are related more to its ionic charge and lipophilicity (Wager, Hou, Verhoest, & Villalobos, 2010). For drugs that are subjected to active transport across the BBB, this may not be a reasonable assumption, and there may be large differences in the brain-to-plasma unbound drug concentration ratio between animals and humans. *[handwritten: caveat]* As an example, rat and mouse brain uptake is relatively low, whereas in monkey and humans, brain uptake is relatively high, for the 5HT-1a receptor antagonist RWAY (Liow et al., 2007). Syvanen et al. (2009) showed large species differences in the brain uptake of verapamil, GR205171, and altanserin, compounds that are strong substrates for the P-glycoprotein (PGP) transporter. Even in the presence of complete PGP transporter inhibition, large species differences were still present indicating that differences in PGP activity do not completely explain the results. Nevertheless, this information, whether there is a species-related difference in brain uptake, is not known until the clinical program in human starts, and since there is no a priori way to predict whether a drug will show species differences in brain uptake, the assumption of similar brain uptake between animals and humans is often made. *[handwritten: so, why is the assumption made?]*

Because drug concentrations are not a constant in vivo, neuroscientists use an integrated measure of the extent of the drug distribution into the brain defined under equilibrium conditions as the ratio of unbound brain concentrations to unbound blood concentrations based on either a single time point or integrated over the entire concentration-time profile of a drug. This parameter is referred to as Kpuu, and the higher the Kpuu, the higher is the extent of brain distribution. Mathematically, this is calculated as

$$Kpuu = \begin{cases} \dfrac{C(t)_{brain} \times fu_{brain}}{C(t)_{plasma} \times fu_{plasma}} & \text{for a single time point} \\[2em] \dfrac{AUC(0-\infty)_{brain} \times fu_{brain}}{AUC(0-\infty)_{plasma} \times fu_{plasma}} & \begin{array}{l}\text{integrated over the}\\ \text{entire concentration-time profile}\end{array} \end{cases}$$

(3.1)

Historically, CSF concentrations have been viewed as reflections of brain extracellular fluid (ECF) concentrations, and lots of data are available on CSF concentrations in animals. Moreover, it was assumed that CSF concentrations in animals would reflect those in human. With the introduction of brain microdialysis, however, brain ECF concentrations could be measured, and the assumption of CSF as a good reflection of brain ECF concentrations could be challenged. This was discussed in a review of De Lange (2013). Generally, it can be said that CSF Kpuu values provide a rather good indication, but not a reliable measure, for predicting brain ECF Kpuu values. Furthermore, comparing rat with human CSF Kpuu values, human CSF Kpuu values tend to be higher and display much more variability in which the latter might be influenced by disease conditions. Therefore, it has been concluded that to be able to more accurately predict human brain ECF concentrations, understanding of the complexity of the CNS in terms of intrabrain pharmacokinetic relationships and the influence of CNS disorders on brain pharmacokinetics needs to be increased.

The last assumption is that the pharmacokinetics in animals can be scaled to humans using allometry. First proposed by Adolph (Wainer, 1991), he proposed that physiological processes (denoted Y) can be related to total body weight (TBW) using a power model:

$$Y = \alpha(TBW)^{\beta} \tag{3.2}$$

where α and β are scaling constants. Boxenbaum (1982a, 1982b) proposed that since many physiological processes are related to total body weight, then it makes sense that pharmacokinetic parameters would also be related to total body weight. He showed that clearance and volume of distribution, the two primary parameters necessary to make those predictions, for many drugs could also be described by Eq. (3.2) and that for clearance, β was often very close to 0.75, whereas volume of distribution was very close to 1.0. An example of allometric scaling of tramadol total systemic clearance and volume of distribution is shown in Fig. 3.2. The models to explain tramadol clearance and volume of distribution were

$$CL\,(L/h) = 2.04(TBW)^{0.89}$$
$$Vd\,(L) = 2.53(TBW)^{1.01} \tag{3.3}$$

From the results of allometric scaling, a quick estimate of the required dose can be made using a fundamental formula in pharmacokinetics (Box 3.1). Chien, Friedrich, Heathman, de Alwis, and Sinha (2005) used a similar approach to estimate the EC50 for a drug having differences in protein binding, receptor binding, and differences between a competitor drug and drug of interest.

Going back to the problem at hand, based on the aforementioned assumptions, a basic translational model can be developed (Fig. 3.3) by matching the unbound plasma concentrations in animals to the unbound plasma concentrations in humans. But this basic model, while useful, ignores certain aspects related to CNS pharmacology. First, it fails to consider brain distribution. The model treats concentrations throughout the brain as homogenous, which often is not. For example, Bonate, Swann, and Silverman (1997) showed that the distribution of cocaine 10 min after intraperitoneal administration is

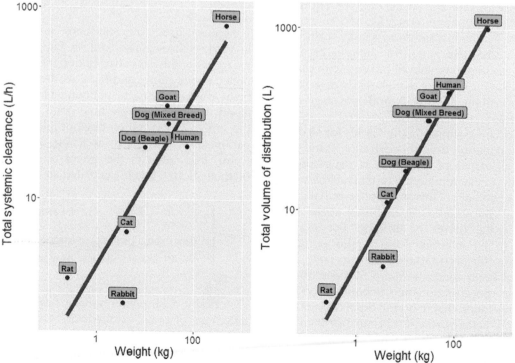

FIG. 3.2   Allometric scaling of the pharmacokinetic parameters for tramadol after intravenous administration. *Srinivas, N. R. (2011). Bridging pharmacokinetics between laboratory/veterinary animal species and man by allometry: a case study of intravenous tramadol.* Journal of Pharmaceutical Sciences and Research, 3, 1368–1372.

BOX 3.1

ESTIMATION OF CLINICALLY EFFICACIOUS DOSE

It is known that

$$Css,u = \frac{F \times D}{CL \times \tau} \times fu$$

where Css is the unbound steady-state drug concentration, F is the oral bioavailability, D is dose, CL is total systemic clearance, τ is the dosing interval, and fu is the unbound fraction of drug in the plasma. An alternative way to express this equation is

$$Css,u = \frac{D}{CL/F \times \tau} \times fu$$

where CL/F is the apparent oral clearance of the drug in humans estimated using allometric scaling. By rearrangement, the required dose is then

$$\frac{Css,u \times CL \times \tau}{F \times fu} = D$$

The ECx can be calculated from the EC50 using

$$ECx = \left(\frac{x}{100-x}\right)EC50$$

where x is the percent of interest. If it is assumed that a steady-state concentration must be greater than the EC80, the concentration that produces 80% receptor occupancy, and that the fu in rats and humans is 0.2 and 0.1, respectively, then a corrected estimate of the required dose is

$$\left(\frac{80}{100-80}\right)\left(\frac{CL}{F}\right)\left(\frac{fu_{human}}{fu_{rat}}\right)EC50 \times \tau = D$$

Note: ensure that the appropriate units are used.

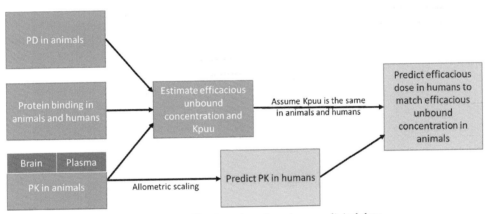

FIG. 3.3   General workflow for the prediction of human efficacious doses based on nonclinical data.

highest in the striatum and hippocampus and lowest in the cortex and hindbrain. Refinements of the model break the brain down into its components through measurement of drug at the site of action (and elsewhere), usually by using microdialysis. For example, Kielbasa and Stratford (2012), in an analysis of duloxetine pharmacokinetics in the rat, modeled the brain as three separate compartments: CSF, brain ECF, and brain cells. In their analysis, they used microdialysis to simultaneously sample plasma, CSF, brain ECF, and end-of-experiment total brain concentrations. Their model contains just enough compartments to explain the data. Johnson et al. (2016) modeled the brain as vascular, brain extravascular, and striatum (bound and free) in an analysis of six D2 receptor antagonists, and Li et al. (2012) modeled in the extracellular fluid of the prefrontal cortex the unbound concentration of PF-04269339 and PF-03529936, two mixed serotonin reuptake inhibitors and 5HT$_{1a}$ partial agonists.

IV EXTENSIONS OF THE MODEL USING PBPK

Typically, plasma concentrations are seen more as a driver of brain concentrations with basic pharmacokinetic-pharmacodynamic models of CNS drugs, rather than as the concentration of interest, because it is of more interest to determine a model of brain pharmacokinetics and pharmacodynamics. As such, plasma concentrations are modeled using a forcing function, like a polyexponential equation, or minimalist compartmental model, like a 1- or 2-compartmental model with absorption. In these models, plasma concentrations and usually brain concentrations are modeled using empiric functions that do not have a basis in reality—they simply fit the data.

Physiologically based pharmacokinetic (PBPK) models take the opposite approach. Each organ of interest in the body is represented by a compartment, and each compartment is linked in an anatomically correct manner with other compartments. Compartment volumes and blood flows are defined by physiological constraints. Generally, these models are used to model drug concentrations in the blood or plasma (Upton, Foster, & Abuhelwa, 2016; Zhuang & Lu, 2016). A representative PBPK model is shown in the lower right of Fig. 3.1, where there are compartments for organs like the liver, kidney, gut, heart, and brain. The beauty of these models is that once developed, they can be used to make predictions in compartments that were not sampled. Suppose it was of interest to know what is the drug concentration in the heart 30 min after dosing. With a PBPK model, it is possible to make such a prediction. They can also be used to make cross species predictions by scaling the compartment volumes and blood flows to the relevant species. PBPK models are also finding increasing use as a tool to estimate the pharmacokinetics in children by scaling down the compartment volumes to the appropriate aged child (Barrett, Della Casa Alberighi, Laer, & Meibohm, 2012). Generally speaking, PBPK models are gaining increasing popularity because of their face validity and their utility.

PBPK modeling that includes the brain has been largely fitted for purpose. In other words, the model should be good enough to do the job it was meant to do. Complicated models add additional layers of complexity that the data may be unable to support and have been regarded as an unnecessary waste of time developing, although complex models in the end will have a better general predictive power. Ball, Bouzom, Scherrmann, Walter, and Decleves (2014) present an excellent review of this topic, and the reader is referred there for more details. For example, the model of Kielbasa and Stratford (Ball et al., 2014), previously described, contains just enough compartments to explain the data. Ball et al., in their review, reported the use of a similar model to explain the pharmacokinetics of atomoxetine; acetaminophen; and S18986, a potential nootropic drug. Westerhout, Danhof, and De Lange (2011), Westerhout, Ploeger, Smeets, Danhof, and De Lange (2012), Westerhout, Smeets, Danhof, and De Lange (2013), Westerhout, van den Berg, Hartman, Danhof, and De Lange (2014) modeled in the rat, using a PBPK approach (see Fig. 3.4) that includes active transport components, unbound acetaminophen, quinidine, and methotrexate concentrations in plasma, brain ECF, and multiple CSF compartments. For acetaminophen and methotrexate, extrapolation to other species, including humans, could be made providing greater trust in the validity and reliability of these models.

Recently, a generic PBPK-based model has been published (Yamamoto et al., 2016) for which the structure (Fig. 3.5) enabled fitting with high precision extensive

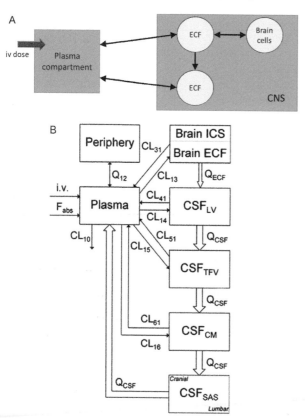

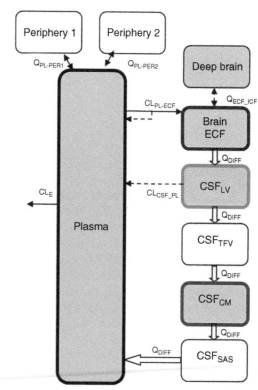

FIG. 3.4 Examples of PBPK models of the CNS used in the literature. *Top* (A): Minimalist PBPK model reported by Kielbasa and Stratford (2012). *Bottom* (B): Expanded model reported by Westerhout et al. (2011, 2012, 2013, 2014).

FIG. 3.5 Structure of the generic CNS drug distribution model. *Yamamoto, Y., Valitalo, P. A., van den Berg, B. D. J., Hartman, R., van den Brink, W., Wong, Y.C., et al. (2016). A generic multi-compartmental CNS distribution model structure for 9 drugs allows prediction of human brain target site concentrations. Pharmaceutical Research, 34, 333–351, 2016.*

plasma, brain ECF, and CSF data obtained in the rat for nine compounds with distinctively different physico-chemical properties, including compounds that are PGP or other active BBB transporter substrates, and for which the data were obtained with and without appropriate blockers (Fig. 3.6). Translation of this generic rat model to the human situation could be validated for acetaminophen and morphine as human data were available for these two compounds. For a direct comparison of rat and human pharmacokinetics in the different CNS compartments in response to plasma pharmacokinetics, simulations for individual compounds with the same plasma exposure in rat and human were used (Fig. 3.7). Generally, human CNS pharmacokinetics, especially in the CSF in the subarachnoid space (like lumbar CSF), was slower than that in the rat. This simulation provides important information on the relationship between brain ECF (which often is the target site) pharmacokinetics and lumbar CSF concentrations (which are often used as biomarker of brain ECF). Larger differences in the pharmacokinetics between brain ECF and lumbar CSF occur at the earlier time points of the different CNS compartments in human than in rat. Over time, these differences dissipate. The consequences for drug-target interaction kinetics (de Witte, Danhof, van der Graaf, & de Lange, 2016) and further processes toward CNS drug effects remain to be determined.

V EXTENSION OF THE MODEL INCORPORATING RECEPTOR OCCUPANCY AND BINDING KINETICS

The basic model also fails to account for effects at the receptor level. Ultimately the effect of any drug is what happens at the receptor, not necessarily with some unbound concentration near the receptor, and by modeling the relationship between unbound drug concentration at the receptor and receptor occupancy, a better understanding of the relationship between dose, concentration, and effect can be made. The modified flow chart for this approach is shown in Fig. 3.8. For example, Wellstein, Palm, Matthews, and Belz (1985) showed that receptor occupancy was missing in modeling the relationship between propranolol plasma concentrations and inhibition of tachycardia. In an experiment with human subjects challenged to ride an exercise bicycle, a direct, linear relationship was shown between beta-adrenergic receptor binding and inhibition of exercise tachycardia. It is generally recognized that for G-protein coupled receptors, neurotransmitter transporters, and ligand-gated ion channels, 60%–90% receptor occupancy is required for clinical effect in the case of antagonists (Grimwood & Hartig, 2009). For agonists, the situation is more complicated and is dependent on the intrinsic

activity of the agonist, the receptor or ion channel reserve at the site of action, and the response that is measured. But generally, for low intrinsic activity agonists, a higher degree of occupancy is required than for high intrinsic activity agonists.

Inclusion of receptor occupancy into the model can lead to a class of models called operational models of agonism. With this class of models, a pharmacodynamic effect is modeled using a semimechanistic model that includes terms normally seen in classical pharmacology, terms like the agonist dissociation rate constant from the receptor and intrinsic activity. To date, all the models developed using this approach have used plasma concentration or unbound plasma concentration as the driver of efficacy, not brain concentration or unbound brain concentration (see, e.g., Cox, Kerbusch, van der Graaf, & Danhof, 1998; Zuideveld et al., 2004), but that is not to say they can't. These models were in vogue in the late 1980s and 1990s but for unclear reasons have fallen out of favor, which is surprising because these models would be useful for CNS drugs and could be easily adapted using concentrations more relevant to the site of action in the brain.

interesting…

VI EXTENSION OF THE PK AND RECEPTOR OCCUPANCY MODEL USING HUMAN DATA

Neils Bohr, the Nobel Prize-winning physicist, once said "Prediction is very difficult, especially if it's about the future." As might be expected, predictions of clinically efficacious doses based entirely on animal data, in the absence of any human data, will have a wide prediction interval because of all the uncertainties and assumptions involved. As drug development progresses through the clinical phase and more information is obtained, identification of the clinically efficacious dose will become more and more refined.

The first clinical studies that are performed for new drug candidates are the single-ascending dose (SAD) and multiple-ascending dose (MAD) studies. In these studies, cohorts of subjects are administered single or multiple doses of the drug to understand the safe dose and exposure range of the drug. From these first studies, the drug's pharmacokinetics can begin to be characterized. Once these data are available, the allometric scaling step in the workflow can be removed because the human data are known. The actual human pharmacokinetics can be substituted in the workflow, thereby improving the accuracy of the dose estimation.

The next study that is often done or is sometimes combined within the SAD or MAD study is a PET study to confirm distribution into the brain and to determine the degree of receptor occupancy in the brain region of interest. It is beyond the scope of this chapter to discuss the

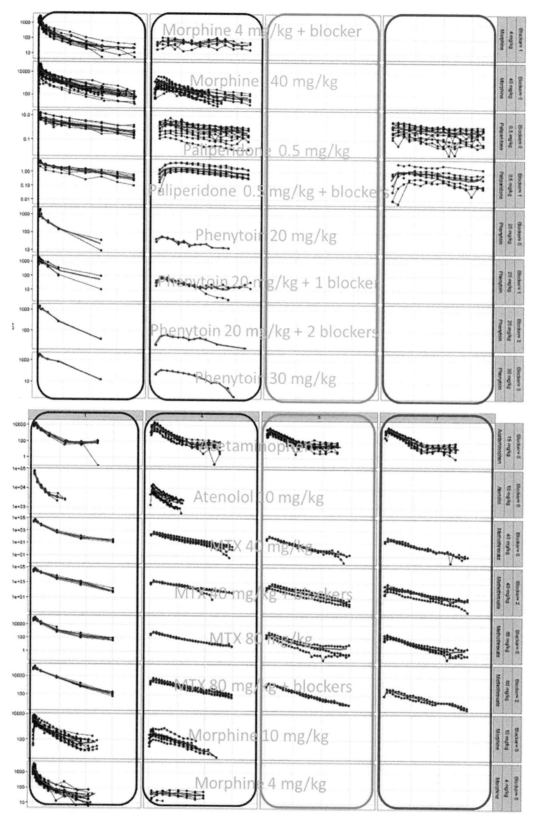

FIG. 3.6 Scatter plots of observed and predicted concentrations for the nine compounds used to develop the generic PBPK model in Fig. 3.5. Observations *(black lines and circles)* and model prediction *(red solid lines)* of the nine compounds in rat for each dose and without and with coadministration of active transport blockers. The *x*-axis represents the time in minutes, and the *y*-axis represents the concentration of the nine compounds in nanogram/milliliter. The panel is stratified by brain compartments *(black=plasma, blue=brain ECF, green=CSF in lateral ventricle (CSF_LV), and red=CSF in cisterna magna (CSF_CM))* and by active transport blockers (colors).

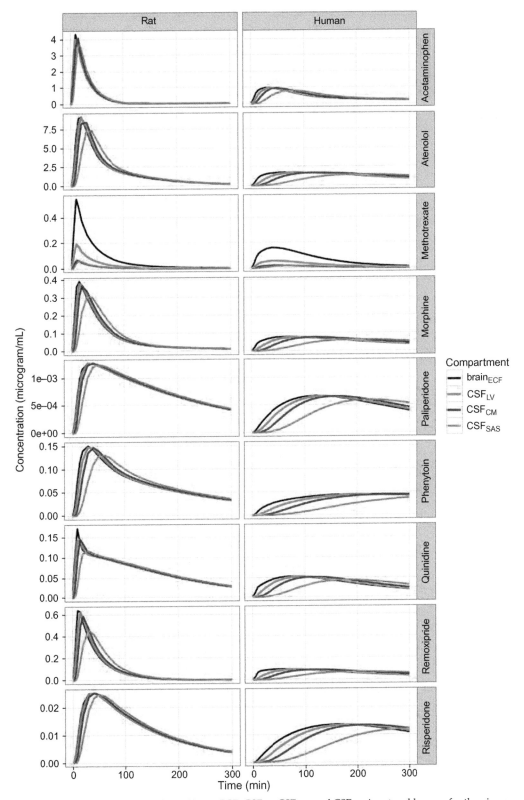

FIG. 3.7   Simulations of time-concentration profiles of brain ECF, CSF$_{LV}$, CSF$_{CM}$, and CSF$_{SAS}$ in rat and human for the nine compounds. Differences in the time concentration between rat and human reflected the differences of the rate and extent of drug distribution in brain between in rat and human since identical plasma exposure was used as an input.

analytic and modeling methods related to PET. If the reader is interested, they can refer to Varnas, Varrone, and Farde (2013), and for a more general introduction, they can refer to Lee and Farde (2006). Once the receptor occupancy data are obtained by PET, external estimation of receptor occupancy using ex vivo or animal data can be discarded and the human data used in its place.

As an example, consider the study reported by Smith et al. (2014). TD-9855 is a dual norepinephrine (NE) and serotonin (5HT) reuptake inhibitor, with greater selectivity for NE, which may be useful for the treatment of pain. In a phase 1 SAD study, healthy volunteers were administered oral doses of TD-9855 and the selective radiotracers [11C]-DASB for 5HT or [11C]-MRB for NE

by intravenous infusion. Displacement of the radiotracers from their binding sites allowed for the estimation of TD-9855 receptor occupancy. Using a pharmacokinetic model built from the data, NE occupancy was estimated by simulating multiple once-daily doses of 4- or 20-mg TD-9855 (Fig. 3.9). From these simulations, doses of 4 mg once daily are expected to produce steady-state NE receptor occupancy values of approximately 80%, and doses of 20 mg once daily are expected to produce receptor occupancy values of 90% or more. At the higher doses, however, 5-HT receptor occupancy starts to occur, and at 20 mg once daily, occupancy values in the range of 15%–30% are expected. Using once-daily doses of 5 and 20 mg, TD-9855 recently completed a phase 2 trial in

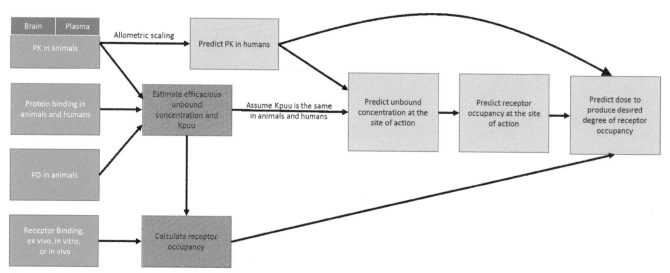

FIG. 3.8   Modified model that incorporates receptor occupancy as a measure of drug effect.

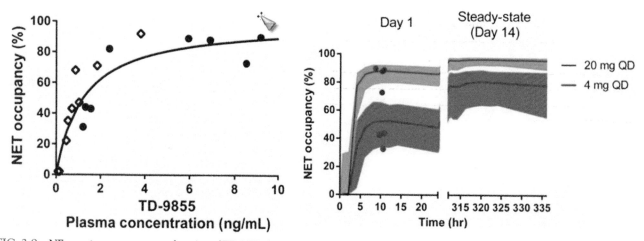

FIG. 3.9   NE receptor occupancy as a function of TD-9855 plasma concentration (*left plot*) and simulated NE receptor occupancy after a single dose (day 1) and repeated administration of 4- and 20-mg TD-9855 at steady state (*right plot*). *Left plot* legend: the circles are the data from day 1; the diamonds are the data on days 3–7. The *solid line* is the model fit predictions from a two-parameter occupancy model. Maximal occupancy was estimated at 92%. The concentration that produces 50% occupancy was 11.7 ng/mL. *Right plot* legend: the symbols represent the observed NE occupancy. The *solid line* represents the mean simulated occupancy in the shaded bands that represent the 95% confidence intervals. *Reprinted with permission from Smith, J. A., Patil, D. L., Daniels, O. T., Ding, Y. S., Gallezot, J. D., Henry, S., et al. (2014). Preclinical to clinical translation of CNS transborder occupancy of TD-9855, a novel norepinephrine and serotonin reuptake inhibitor.* International Journal of Neuropsychopharmacology, 18, 1–11.

patients with fibromyalgia. Although the 5-mg dose did not reach statistical significance, 20-mg dose demonstrated a significant reduction in average pain scores compared with placebo.

Another example is the study reported by Raddad et al. (2016). LY2940094 is a selective nociception/orphanin antagonist that may be useful for a variety of psychiatric conditions. Starting with doses of 40 mg, healthy volunteers received subsequent doses to fully characterize the receptor occupancy profile. Consequently, subjects received doses of 4-, 10-, 20-, and 40-mg LY2940094 and an intravenous administration of [11C]-nociceptin, a 17 amino acid neuropeptide that acts as an endogenous ligand for the opioid-like one receptor. Analysis of the data revealed similar estimates of maximal occupancy (near 100%) and EC50 in the prefrontal cortex, occipital cortex, putamen, and thalamus (values ranged from 2.64 to 3.46 ng/mL). Based on the relationship between plasma concentration and NOP receptor occupancy, the average occupancy at 4-, 10-, 20-, and 40-mg LY2940094 was 73%, 80%, 93%, and 97%, respectively. Similarly, given the long half-life of the compound, receptor occupancy 24 hours after dose administration was estimated to be 28%, 49%, 65%, and 82%, respectively. In this study the authors did not develop a pharmacokinetic model to estimate the degree of receptor occupancy with multiple-dose administration. However, using a back of the envelope calculation, they expected plasma concentrations to be about twofold higher at steady state compared with single dose in the receptor occupancy at the 40-mg dose would be in the 80% range, which is in the range of occupancy associated with clinical efficacy for other G-protein coupled receptors. Based on the results of this study, LY2940094 recently completed trials in major depressive disorder and patients with alcohol dependency. In the 8-week, placebo-controlled, double-blind study in patients with major depressive disorder, once-daily dosing of 40-mg LY2940094 resulted in some minor changes in the GRID Hamilton Rating Scale, but these results did not reach statistical significance (Post et al., 2014).

VII RECENT ADVANCES IN QUANTITATIVE SYSTEMS PHARMACOLOGY OF CNS DRUGS

When considering the human body and all its interactions and interdependent processes, we, as scientists, need to start thinking beyond the single-target approach and start considering drug action more on a system level, as part of modifying one or more networks. In system biology, an area that has been around for some time but has become a very hot area in the last few years, qualitative network identification has been performed by

analysis of large datasets using advanced statistical techniques or pattern recognition approaches followed by extensive annotation. This has been a very useful approach, so far, for mainly oncology and immunology. However, the application to CNS diseases is lagging. Of course, such application on the CNS is a huge challenge as the CNS can be considered as "a universe inside the body." The combination of PKPD modeling and simulation (Hurko & Ryan, 2005) with a CNS systems approach, referred to as "quantitative CNS systems pharmacology," would be a strong way to move forward in understanding drug action on the CNS (Geerts, Spiros, Roberts, & Carr, 2013).

From the system side, for CNS diseases, there is a rich expertise of computational neurosciences that can be useful. In a recent big science initiative, the Blue Brain Project, time-dependent membrane potential changes and action potentials have been simulated using a detailed computer model of a human microcolumn that contains more than 200 different cell types and a wide variety of voltage-gated ion channels (Markram, 2012). Such a model may provide increased understanding of the human neurobiology. However, it currently lacks information on human pathology parameters from imaging and postmortem studies that may reveal neuropharmacological targets of CNS-active drugs or provide a proper description of target engagement of these drugs and a calibration using retrospective and historical clinical data. That would be needed to make it a real translational tool.

Important steps have been made into that direction, however, as studies have been performed on how this approach can make a difference for CNS R&D projects. An interesting example is the development of the humanized computer-based quantitative system pharmacology (QSP) platform that simulates the neurophysiology of the parts of the cortico-striatal-pallido-thalamocortical pathway (Spiros, Roberts, & Carr, 2012). This platform made it possible to blindly predict the clinical outcome of a phase 2 study for a novel investigative drug, purely based on its human pharmacology (Geerts et al., 2012). It was shown that it can be possible to prospectively identify a substantial extrapyramidal side effect (EPS) liability, which, per the authors, would have been completely missed in preclinical animal models. Another study involves the use of a humanized QSP platform as an example of a well-validated phenotypic assay in Parkinson's disease (PD) tremor for filtering out possible interesting molecules that can then be tested in preclinical animal models. The pharmacological profiles of serotonergic drugs in the Prestwick library were screened, and five interesting multipharmacology agents could be identified, including trazodone that in a previously reported study improved clinical PD scales as augmentation strategy.

The QSP platform is a powerful modeling and simulation tool with a relevant human clinical scale as output where multitarget drugs effect can be simulated and promising candidates for further study in pharmacological profiling or animal models can be identified. The platform contains 30 CNS physiologically implemented targets and simulates biophysically realistic neuronal network interactions between supplemental motor cortex and motor striatum based on preclinical neurophysiology and human electrophysiology data. Importantly, the platform is further calibrated using retrospective clinical data on Parkinsonian side effects with antipsychotics (Geerts, Roberts, & Spiros, 2015; Roberts, Spiros, & Geerts, 2016; Spiros, Roberts, & Geerts, 2013). Thus QSP is a recent addition to the modeling and simulation toolbox for (CNS) drug discovery and development and is based upon mathematical modeling of biophysical realistic biological processes in the disease area of interest. The combination of preclinical neurophysiology information with clinical data on pathology, imaging, and clinical scales makes it a real translational tool (Roberts et al., 2016).

VIII SUMMARY

Pharmacokinetic and pharmacodynamic modeling of drugs and their effects continue to evolve. As computers become faster, as more scientists become modelers themselves, and as more classical pharmacologists recognize the value of modeling and simulation as a tool to improve the understanding of experiments and systems, as a means to predict future outcomes, and as a way to better design future experiments, the application of modeling and simulation to translational CNS pharmacology will only continue to increase. Moving from simple, empiric pharmacokinetic models in the 1960s and 1970s, today's models are more mechanistic in nature, incorporating anatomy, physiology, biochemistry, and pathology into their structure. Computational models of the brain are starting to revolutionize our understanding of how the brain works. It is only logical that the future coupling of these mechanistic pharmacokinetic models to computational models of brain activity will lead to better understanding of future drugs, possibly reducing attrition and leading to more successful products for patients.

References

Ball, K., Bouzom, F., Scherrmann, J. -M., Walter, B., & Decleves, X. (2014). A physiologically based modeling strategy during preclinical CNS drug development. *Molecular Pharmaceutics, 11,* 836–848.

Barrett, J. S., Della Casa Alberighi, O., Laer, S., & Meibohm, B. (2012). Physiologically based pharmacokinetic (PBPK) modeling in children. *Clinical Pharmacology and Therapeutics, 92,* 40–49.

Bonate, P. L. (2001). A brief introduction to Monte Carlo simulation. *Clinical Pharmacokinetics, 40,* 15–22.

Bonate, P. L. (2011). *Pharmacokinetic—Pharmacodynamic modeling and simulation.* New York: Springer.

Bonate, P. L., Swann, A., & Silverman, P. B. (1997). Behavioral sensitization to cocaine in the absence of altered brain cocaine levels. *Pharmacology Biochemistry and Behavior, 57,* 665–669.

Boxenbaum, H. (1982a). Comparative pharmacokinetics of benzodiazepines in dog and man. *Journal of Pharmacokinetics and Biopharmaceutics, 10,* 411–426.

Boxenbaum, H. (1982b). Interspecies scaling, allometry, physiological time, and the ground plan of pharmacokinetics. *Journal of Pharmacokinetics and Biopharmaceutics, 10,* 201–227.

Calcoen, D., Elias, L., & Yu, X. (2015). What does it take to produce a breakthrough drug? *Nature Reviews Drug Discovery, 14,* 161–162.

Chien, J. Y., Friedrich, S., Heathman, M., de Alwis, D. P., & Sinha, V. (2005). Pharmacokinetics/pharmacodynamics and the stages of development: role of modeling and simulation. *The AAPS Journal, 7,* E544–E559. Article 55.

Cox, E. H., Kerbusch, T., van der Graaf, P. H., & Danhof, M. (1998). Pharmacokinetic-pharmacodynamic modeling of the electroencephalogram effect of synthetic opioids in the rat: correlation with the interaction at the mu-opioid receptor. *Journal of Pharmacology and Experimental Therapeutics, 284,* 1095–1103.

De Lange, E. C. M. (2013). Utility of CSF in translational neuroscience. *Journal of Pharmacokinetics and Pharmacodynamics, 40,* 315–326.

de Witte, W. E. A., Danhof, M., van der Graaf, P., & de Lange, E. C. M. (2016). In vivo target residence time and kinetic selectivity: the association rate constant as a determinant. *Trends in Pharmacological Sciences, 37,* 831–842.

Freireich, E. J., Gehan, E. A., Rall, D. P., Schmidt, L. H., & Skipper, H. E. (1966). Quantitative comparison of toxicity of anticancer agents in mouse, rat, hamster, dog, monkey, and man. *Cancer Chemotherapy Reports, 50,* 219–244.

Geerts, H., Roberts, P., & Spiros, A. (2015). Assessing the synergy between cholinomimetics and memantine as augmentation therapy in cognitive impairment in schizophrenia. A virtual human patient trial using quantitative systems pharmacology. *Frontiers in Pharmacology, 6,* 22–198.

Geerts, H., Spiros, A., Roberts, P., & Carr, R. (2013). Quantitative systems pharmacology as an extension of PK/PD modeling in CNS research and development. *Journal of Pharmacokinetics and Pharmacodynamics, 40,* 257–265.

Geerts, H., Spiros, A., Roberts, P., Twyman, R., Alphs, L., & Grace, A. A. (2012). Blinded prospective evaluation of computer-based mechanistic schizophrenia disease modeling for predicting drug response. *PLoS One, 7,* 732e49.

Grimwood, S., & Hartig, P. R. (2009). Target site occupancy: emerging generalizations from clinical and preclinical studies. *Pharmacology & Therapeutics, 122,* 281–301.

Hurko, O., & Ryan, J. L. (2005). Translational research in central nervous system discovery. *NeuroRx, 2,* 671–682.

Johnson, M., Kozielska, M., Pilla Reddy, V., Vermeulen, A., Barton, H. A., Grimwood, S., et al. (2016). Translational modeling in schizophrenia: predicting human dopamine D2 receptor occupancy. *Pharmaceutical Research, 33,* 1003–1017.

Jungck, J. R. (1997). Ten equations that changed biology: mathematics in a problem-solving biology curricula. *Bioscene, 23,* 11–36.

Kielbasa, W., & Stratford, R. E. (2012). Exploratory translational modeling approach to drug development to predict human brain pharmacokinetics and pharmacologically relevant clinical doses. *Drug Metabolism and Disposition, 40,* 877–883.

Lee, C. M., & Farde, L. (2006). Using positron emission tomography to facilitate CNS drug development. *Trends in Pharmacological Sciences, 27,* 310–316.

Li, C. S. -W. L., Zheng, L., Haske, T., Dounay, A., Gray, D., Barta, N., et al. (2012). Mechanism-based pharmacokinetic/pharmacodynamic modeling of rat prefrontal cortical dopamine response to dual acting norepinephrine reuptake inhibitor and 5HT-1a partial agonist. *The AAPS Journal, 14,* 365–376.

Liow, J. S., Lu, S., McCarron, J. A., Hong, J., Musachio, J. L., Pike, V. W., et al. (2007). Effect of a P-glycoprotein inhibitor, cyclosporin A, on the disposition in rodent brain and blood of the 5-HT1A receptor radioligand, [11C](R)-(-)-RWAY. *Synapse, 61,* 96–105.

Liu, X., Vilenski, O., Kwan, J., Apparsundaram, S., & Weikert, R. (2009). Unbound brain concentration determines receptor occupancy: a correlation of drug concentration and brain serotonin and dopamine reuptake transporter occupancy for eighteen compounds in rats. *Drug Metabolism and Disposition, 37,* 1548–1556.

Markram, H. (2012). The human brain project. *Scientific American, 306,* 50–55.

Post, A., Smart, T., Krikke-Workel, J., Witkin, J., Statnick, M., Harmer, C., et al. (2014). *The efficacy and safety of LY2940094, a selective nociceptin receptor antagonist, in patients with major depressive disorder: a randomized, double-blind, placebo-controlled study.* .

Raddad, E., Chappell, A., Meyer, J., Wilson, A., Ruegg, C. E., Tauscher, J., et al. (2016). Occupancy of nociceptin/orphanin FQ peptide receptors by the antagonist LY2940094 in rats and healthy human subjects. *Drug Metabolism and Disposition, 44,* 1536–1542.

Roberts, P., Spiros, A., & Geerts, H. (2016). Humanized clinically calibrated quantitative systems pharmacology model for hypokinetic motor symptoms in Parkinson's disease. *Frontiers in Pharmacology, 7,* 6.

Smith, J. A., Patil, D. L., Daniels, O. T., Ding, Y. -S., Gallezot, J. -D., Henry, S., et al. (2014). Preclinical to clinical translation of CNS transborder occupancy of TD-9855, a novel norepinephrine and serotonin reuptake inhibitor. *International Journal of Neuropsychopharmacology, 18,* 1–11.

Spiros, A., Roberts, P., & Carr, R. (2012). A quantitative systems pharmacology computer model for schizophrenia. *Drug Development Research, 73,* 1098–1109.

Spiros, A., Roberts, P., & Geerts, H. (2013). Phenotypic screening of the Prestwick library for treatment of Parkinson's tremor symptoms using a humanized quantitative systems pharmacology platform. *Journal of Parkinsons Disease, 3,* 569–580.

Srinivas, N. R. (2011). Bridging pharmacokinetics between laboratory/ veterinary animal species and man by allometry: a case study of intravenous tramadol. *Journal of Pharmaceutical Sciences and Research, 3,* 1368–1372.

Syvanen, S., Lindhe, O., Palner, M., Kornum, B. R., Rahman, O., Langstrom, B., et al. (2009). Species differences in blood-brain barrier transport of three positron emission tomography radioligands with emphasis on P-glycoprotein transport. *Drug Metabolism and Disposition, 37,* 635–643.

U.S. Department of Health and Human Services (2005). Food and drug administration, and center for drug evaluation and research. In *Guidance to industry: Estimating the maximum safe starting dose in initial clinical trials for therapeutics in adult healthy volunteers.*

Upton, R. N., Foster, D. J., & Abuhelwa, A. Y. (2016). An introduction to physiologically-based pharmacokinetic models. *Paediatric Anaesthesia, 26,* 1036–1046.

Varnas, K., Varrone, A., & Farde, L. (2013). Modeling of PET data in CNS drug discovery and development. *Journal of Pharmacokinetics and Pharmacodynamics, 40,* 267–279.

Wager, T. T., Hou, X., Verhoest, P. R., & Villalobos, A. (2010). Moving beyond rules: the development of a central nervous system multiparameter optimization (CNS MPO) approach to enable alignment of druglike properties. *ACS Chemical Neuroscience, 1,* 435–439.

Wainer, H. (1991). Adjusting for differential rates: Lord's paradox again. *Psychological Bulletin, 109,* 147–151.

Watson, J., Wright, S., Lucas, A., Clarke, K. L., Viggers, J., Cheetham, S., et al. (2009). Receptor occupancy and brain free fraction. *Drug Metabolism and Disposition, 37,* 753–760.

Wellstein, A., Palm, D., Matthews, J. H., & Belz, G. G. (1985). In vitro receptor occupancy allows to establish equieffective doses of beta-blockers with different pharmacodynamic profiles in man. Investigations with propranolol and bufuralol. *Methods and Findings in Experimental Clinical Pharmacology, 7,* 645–651.

Westerhout, J., Danhof, M., & De Lange, E. C. M. (2011). Preclinical prediction of human brain target site concentrations: considerations in extrapolating to the clinical setting. *Journal of Pharmaceutical Sciences, 100,* 3577–3593.

Westerhout, J., Ploeger, B., Smeets, J., Danhof, M., & De Lange, E. C. M. (2012). Physiologically based pharmacokinetic modeling to investigate regional brain distribution kinetics in rats. *The AAPS Journal, 14,* 543–553.

Westerhout, J., Smeets, J., Danhof, M., & De Lange, E. C. M. (2013). The impact of P-gp functionality on non-steady-state relationships between CSF and brain extracellular fluid. *Journal of Pharmacokinetics and Pharmacodynamics, 40,* 327–342.

Westerhout, J., van den Berg, D. -J., Hartman, R., Danhof, M., & De Lange, E. C. M. (2014). Prediction of methotrexate CNS distribution in different species—influence of disease conditions. *European Journal of Pharmaceutical Sciences, 57,* 11–24.

Yamamoto, Y., Valitalo, P. A., van den Berg, B. D. J., Hartman, R., van den Brink, W., Wong, Y. C., et al. (2016). A generic multi-compartmental CNS distribution model structure for 9 drugs allows prediction of human brain target site concentrations. *Pharmaceutical Research, 34,* 333–351.

Zhuang, X., & Lu, C. (2016). PBPK modeling and simulation in drug research and development. *Acta Pharmaceutica Sinica B, 6,* 430–440.

Zuideveld, K. P., van der Graaf, P. H., Newgreen, D., Thurlow, R., Petty, N., Jordan, P., et al. (2004). Mechanism-based pharmacokinetic-pharmacodynamic modeling of 5-HT1A receptor agonists: estimation of in vivo affinity and intrinsic efficacy on body temperature in rats. *Journal of Pharmacology and Experimental Therapeutics, 308,* 1012–1020.

4

Functional Measurements of Central Nervous System Drug Effects in Early Human Drug Development

Joop van Gerven

Centre for Human Drug Research, Leiden, The Netherlands

I INTRODUCTION

Functional measurements of central nervous system (CNS) activity play an essential role in the evaluation of drug effects for neurobehavioral disorders in humans. On the one hand, functional CNS measurements are used in patients to characterize psychiatric or neurological diseases for research and diagnosis and to determine the response to therapy in clinical practice and trials. On the other hand, functional measurements are used in earlier phases of drug development, to learn enough about the effects of the new drug to increase the success of further clinical development. These two objectives are often combined in drug development, although the emphasis shifts during the progress of the program. It is important to realize, however, that the requirements for the measurement of clinical endpoints of psychiatric or neurobehavioral diseases are different from the characteristics of pharmacodynamic measures of neuropsychiatric drug effects in early human development. In the latter the research focus is on the properties of the drug rather than on the disease, and the measurements are therefore more often performed in healthy subjects than patients. This chapter is devoted to measurements of drug actions in these early development phases, rather than on clinical endpoints in patient trials.

Since the brain constitutes the essence of human existence, the scope of CNS functions is unfathomable. Complex cellular processes underlie neurophysiological, synaptic, and humeral interactions between close to 10^{11} neurons, supported by microglia, specialized endothelia, immune cells, and other cell types. Neurons form functional networks with thousands of connections per cell, over distances that range from micrometers between neighboring neurons to more than a meter between cortical and spinal neurons or for peripheral nerves. The networks are connected to the rest of the body and the environment, subserving physical processes and behavioral and motor expressions, steered by perceptions, emotions, and cognitions that are influenced by memories of the past and expectations of the future. This enormous complexity of expressions of the CNS provides for an almost unlimited variety of CNS measurements, and hundreds of different neuropsychological, subjective, neuroendocrine, neurophysiological, and imaging methodologies have been used to study the human brain. This chapter presents a systematic overview of the CNS tests that have been applied to early drug development. Because the number of studies and methods in this area are very large, selections and choices were unavoidable. The first choice was to largely exclude neurophysiological and neuroimaging methods. A few words will be said here and there, but the role of these methods in the demonstration of CNS drug effects will not be discussed in any detail. The second choice was to avoid descriptions of tests. This is inherent to the huge variability of the methods and to the system that was adopted for the integration of study outcomes. A third choice was to limit the exploration of the drug responsiveness of functional CNS tests to a certain number of drug classes. The responsiveness of tests to drugs with a shared mechanism of action was examined rather than the other way around: the sensitivity of a specific test to a range of different drugs. The reason is that early drug development is aimed at learning more about the compounds, not about the tests. The next paragraph will provide more insight into the primary objectives of early drug research and on how functional CNS measurements can be used to address some important questions in CNS drug development.

II CNS MEASUREMENTS IN EARLY DRUG DEVELOPMENT

In early drug development the drug rather than the disease plays a central role. The main objectives of studies in humans are aimed at how a drug acts on the body (pharmacodynamics) and how the body acts on the drug (pharmacokinetics), in concentrations that cover the therapeutic window. One way to think about this is to follow the course that drug takes from the moment of its administration to its effects on cellular and physiological processes in the target organ and other parts of the body. This trajectory is depicted in Fig. 4.1 (Cohen, 2010).

A Steps 1 and 2: From Administration to Site of Action

The administration step (number 1 in the figure) has an important impact on the effect profile of the drug. The dose of the drug determines the concentrations, and the formulation and the route of administration determine the rate at which the drug reaches the brain. Except for intrathecally administered compounds, all drugs reach the brain though the bloodstream and the blood-brain barrier (BBB). For compounds that are poorly absorbed or strongly metabolized, oral administration will also affect the concentrations of the active moiety of the drug. The pharmacokinetics of a compound will determine the

time course of the exposure of the brain and hence of the CNS effects of the drug. The concentrations of a drug in the brain are primarily driven by the free unbound fraction of the drug. For many CNS-active drugs, particularly lipophilic compounds, the time course of CNS effects is delayed and prolonged by accumulation in CNS tissue. In time the activity can be modified by pharmacological or functional adaptations that lead to the development of tolerance or other long-term changes. All these factors have an impact on the drug's time of onset, the maximum peak (T_{max}), and the minimum trough of the effect and how these change during prolonged treatment. As a result the time profile can also determine the therapeutic indication. All benzodiazepines, for instance, have the same pharmacological mechanism of action (they are all GABA-A receptor allosteric agonists), but they have a wide range of registered therapeutic indications, including induction of anesthesia or treatment of status epilepticus (fast onset), sleep disorders (exposure only during sleep phase), anxiety, and muscle pain (continuous low exposure). This diversity is entirely based on the pharmacokinetic properties of the different compounds (including dose), which govern the intensity and the onset and duration of the CNS depressant effects.

Pharmacokinetic variability also contributes to the statistical variability of drug effects between or within studies. It is essential for drug development (and at least important for neuropharmacological research in general) that the concentration profile of a drug is considered in the interpretation of its functional effects. In most development programs the plasma concentrations are used to get an impression of the concentration profile in the brain. This basically ignores the BBB, which can interfere with the distribution of the compound from blood to CNS (step 2 in Fig. 4.1). Moreover the physicochemical properties of nerve tissue differ substantially from those of plasma, and the CNS also has its own clearance mechanisms. This means that the time course of drug concentrations in the brain may be lower and shorter than in plasma, if a drug doesn't readily penetrate the brain or is rapidly eliminated, or that CNS levels remain elevated for a longer period than in plasma, because the drug has a higher affinity for lipophilic brain structures than for the hydrophilic milieu of blood. The problem that plasma concentrations do not accurately represent brain levels can partly be solved by determination of concentrations in cerebrospinal fluid (CSF) as a proxy of the drug's levels in brain tissue. This requires a lumbar puncture or cannula, which is obviously more invasive and cumbersome than plasma sampling. The advantage over plasma is that the presence of drug in CSF almost certainly indicates that the drug has reached the brain. However, the fluid shifts and barriers between brain interstitial fluid and CSF are complicated, and the quantitative and temporal relationships between the two compartments are not

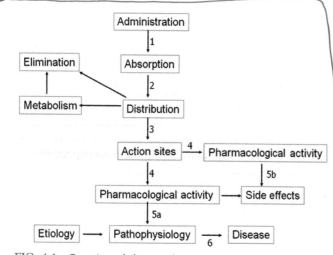

FIG. 4.1 Overview of pharmacokinetic and pharmacodynamic processes involved in drug action. The primary aim of early drug development is to determine the different steps in the routes that leads from drug administration to desired or adverse effects. Numbers refer to steps that are discussed in the text.

always straightforward. Therefore functional CNS measurements are often used as an indirect indication of the presence of a drug in the brain. ↘ *Takeaway from steps 1 + 2*

B Step 3: Drug-Target Interaction

After the drug has penetrated the brain (or other parts of the CNS), it needs to reach its pharmacological or biochemical target (step 3 in Fig. 4.1), often a receptor or an enzyme. Binding to the target is an essential prerequisite for all drug effects, and it is an important goal of early CNS drug development to demonstrate and characterize the drug-target interaction. The most direct way to quantify receptor binding of drugs is with PET and SPECT imaging, by measuring displacement of radioactive tracers that bind to the free fraction of pharmacological targets (usually receptors). Compounds that bind to the same receptor can differ considerably in their affinity for the target, which influences the intensity and the duration of their effects, as well as the reversibility and the development of tolerance. For instance, all antipsychotics bind to D2 receptors, but classic neuroleptics like haloperidol have a much longer residence time on the receptor than atypical drugs like clozapine and quetiapine. This may be one of the mechanisms that determine the propensity for side effects of antipsychotic drugs (Kapur & Seeman, 2001) (others are related to the intrinsic activity of the drug [antagonism or partial agonism] and the selectivity for other receptors with undesirable effects). PET imaging (and to a lesser degree SPECT) are increasingly used in early drug development, but for drugs with a novel mechanism of action or an alternative (allosteric) binding site, new validated tracers need to be developed at the same time as the drug itself, which is often impossible. Both CSF sampling and PET imaging address important questions about the drug, but both approaches have practical limitations that restrict their use in most drug development programs. Moreover the presence of drug in the brain and the demonstration of target binding provide no information on the functional effects, for instance, if the compound is an agonist or an antagonist or not active at all.

C Step 4: Recognizing Pharmacological Activity

For neuropsychiatric drugs the distribution of a drug to the brain and the binding to its pharmacological target (steps 2 and 3 in Fig. 4.1) are often (or additionally) established by inference, from measurements of plasma concentrations (step 1) and drug-induced changes of CNS functions (step 4). However, if a functional CNS measurement shows a statistically significant drug effect, this is not always directly caused by the compound's pharmacological activity. First, there may be statistical restrictions.

Many early drug development studies include a substantial range of CNS tests, each often with a number of readout parameters (like the number of correct responses, reaction time, and sometimes many secondary derivatives). Statistical correction for multiple testing is quite uncommon, so a significant effect may be a false-positive chance finding (type I error). Second the CNS effect may be indirect. Drugs can lead to systemic adverse effects like nausea or hypotension that are strong enough to interfere with the CNS tests, by unblinding of subjects or investigators, reduction of performance, or sometimes secondary changes that affect neurophysiological signals. Third, group effects may mimic as drug effects. Drug studies are usually performed concomitantly in cohorts of subjects, who may be affected by pronounced effects in one of their fellows or other group interactions. In first-in-human studies with a fixed dose escalation and a crossover design, it can sometimes be difficult to distinguish dose relatedness from learning or practice effects.

All these conditions may imitate a drug effect, but the chances of a false-positive result are highest when several different CNS tests (with different outcome parameters) are performed once at a single-dose level. A single measurement around the expected maximum plasma (or brain) concentration may also miss the maximum effect in subjects with a more rapid or slower time profile. Unfortunately, not all CNS methods are suitable for repeated measurements, and such designs are not as common in the literature as would be desired. These methodological considerations should apply to the interpretation of changes in functional CNS measurements that seem to be drug-induced. However, chance findings are quite rare in early development studies, where different doses are applied and plasma concentrations and effects are measured repeatedly, in line with the drug's pharmacokinetic profile. False-positive drug effects don't show a clear dose-response or follow the plasma concentration profile of the drug. Repeated measures also reduce the variability of the time course of concentrations and effects among participants.

The best evidence that a functional CNS effect is directly caused by the drug is derived from a close relationship to the individual plasma concentration profiles of the compound. This provides good evidence for pharmacological activity in the brain and hence BBB penetration and target engagement. In many drug development programs, the time courses of the plasma concentrations and the observed effect are related by pharmacokinetic/pharmacodynamic (PK/PD) modeling, which is essentially based on the mathematics of drug-receptor interactions (Cawello & Antonucci, 1997; Schoemaker, van Gerven, & Cohen, 1998). Distribution and target binding can then be estimated from the model. The model also allows the quantification of factors that contribute to the variability of drug levels and responses, such as weight, age, sex, and renal or hepatic function. PK/PD models can improve the predictions of dosing regimens for subsequent studies (e.g., multiple-dose studies or exploratory clinical trials).

D Step 5a and b: Interpretation of Pharmacological CNS Effects

PK/PD relationships provide important support for the fact that the observed changes in CNS functions after drug administration are caused by the pharmacological activity of the drug. However, PK/PD relations by themselves do not distinguish between therapeutically relevant effects (step 5a) or nonspecific undesirable effects (5b). The interpretation of the effect relies on both the characteristics of the drug and the nature of the CNS effects. The pharmacological properties of the drug provide an important background to interpret the effects. If the compound is highly selective, it is quite likely that CNS effects that are found to be concentration-related will represent the primary pharmacological activity that is designed to cause the intended therapeutic effect (5a). If on the other hand, the drug binds to a number of different receptors with comparable affinity, concentration-related effects may also be caused by activity at any one of the other targets (5b). Binding affinities for different receptors in vitro may be reflected by the relative position of PK/PD relationships for the various CNS effects in humans. In addition, interpretations are facilitated by the nature of the effects, in relation to the known effects of pharmacological mechanisms. The tricyclic antidepressant amitriptyline, for instance, binds not only to serotonin and noradrenalin reuptake sites (which is responsible for the therapeutic effects on depression and neuropathic pain) but also to muscarinic receptors (responsible for parasympatholytic and memory effects), histamine receptors (causing sedation), and others. In healthy subjects, amitriptyline causes dose-related changes in all these domains. The prediction of CNS effects is more uncertain for drugs with new mechanisms of action, although animal models can provide important clues. However, many innovative compounds are also quite selective, so any concentration-related effect is likely to reflect the intended pharmacological activity. The distinction between undesirable (5b) and desirable (5a) pharmacological effects is not necessarily related to the adversity of the effects in study participants. Any pharmacological effect may become adverse if drug levels are too high. Moreover, CNS functions that are deranged in neurobehavioral disorders (and corrected if the drug is effective) operate well in the motivated healthy young students who participate in early drug studies. In this optimally functioning population, CNS tests may show ceiling

effects or even impairments, with a drug that is meant to improve neurobehavioral disorder. Functional changes in healthy subjects that are caused by highly selective compounds will usually affect the same CNS systems, which in patients are impaired by the disease and improved if the drug is efficacious.

III SYSTEMATIC EVALUATION OF FUNCTIONAL CNS MEASUREMENTS

CNS functionality can be measured in numerous ways, but the selection is restricted by the objectives of early drug development. The previous paragraph has outlined how functional CNS measurements can be used to demonstrate the impact of a drug on relevant neurobehavioral functions and by inference BBB penetration and target engagement (Cohen, 2010; Cohen, Burggraaf, van Gerven, Moerland, & Groeneveld, 2015). Questions in early drug development involve distribution to the brain (step 4), targeted (5a) and additional pharmacological activity (5b), and efficacy (step 6) in later clinical phases of development. Such information can provide indications for the therapeutic window between desired effects and side effects (and hence about tolerability and safety) and about their time course and relationships with doses and plasma concentrations. The influence of age, sex, food and drug interactions, and other population variants can also be determined. The utility of measurements that provide these types of information in early drug development should comply with a number of requirements. First the measurements should be responsive to drugs. Not all functional domains of the CNS are equally likely to be affected by drugs. The effects on complex functions like intelligence and language are more likely to be secondary to drug effects on underlying functionalities, like attention or memory. Word-finding difficulties, for instance, are a well-known side effect of carbonic anhydrase inhibiting antiepileptics like topiramate and zonisamide, but this is a manifestation of a more widespread functional impairment (Yasuda et al., 2013). Second the measurement must respond to doses that cover the therapeutic range. Effects that only occur at supratherapeutic doses may provide indications for potential toxic effects, but they are not very informative for dose selection. Third the measurement should demonstrate dose- or concentration-response relationships. This means that the measurement should be repeatable and that responses are graded. Ideally the test should have enough versions, a short duration, and limited carryover and learning effects, to cover an entire time profile of a drug (baseline and increasing and maximum and decreasing concentrations) in a crossover design. Fourth the measurements should also show consistent effects between drugs from the same pharmacological class and across studies (preferably from different research groups). Functional CNS measurements that comply with these criteria can be considered as biomarkers of pharmacological activity.

A Studies of Functional Effects of CNS Active Drugs in Healthy Subjects

Although there are many studies that examine functional CNS effects of drugs in healthy subjects, few are actually part of an early drug development program, and even fewer use validated functional pharmacological biomarkers. Most studies are performed with the purpose of identifying the effects of well-known drugs on brain functions, not to address questions about brain penetration or target engagement or to make predictions of an effective dose range of a novel drug. Studies from actual early development programs are often not published, or only at a later stage. This limits the evaluation of the practical significance and the predictive value of specific CNS tests for the drug development. Some of the information presented in this chapter is based on unpublished data from the authors' institute, which obviously constitutes a personal selection. For the most part, however, the usefulness of functional CNS measurements in drug development has to be derived from studies with a different focus. An advantage is that these studies generally use registered drugs with established mechanisms of action and accepted efficacy and therapeutic windows. In theory, this provides an opportunity to validate the responsiveness of CNS tests to clinically and pharmacologically relevant drug effects. On the basis of this consideration, several systematic reviews have been published, which evaluate the sensitivity of functional CNS tests to various drug classes in healthy subjects. The reviews included the CNS effects in healthy subjects of benzodiazepines (GABA-A receptor agonists) (De Visser et al., 2003); antipsychotics (D2 receptor antagonists) (De Visser, Van der Post, Pieters, Cohen, & Van Gerven, 2001); antidepressants (serotonin reuptake inhibitors (Dumont, De Visser, Cohen, & Van Gerven, 2005) and amitriptyline for the evaluation of REM sleep effects (Rijnbeek, de Visser, Franson, Cohen, & Van Gerven, 2003); and

prototypical compounds like ethanol (Zoethout, Delgado, Ippel, Dahan, & Van Gerven, 2011), methylphenidate (Schrier, Brill, Burggraaf, Cohen, & Van Gerven, 2015), 3,4-methylenedioxymethamphetamine (MDMA, "ecstasy" (Dumont & Verkes, 2006), and cannabis (Zuurman, Ippel, Moin, & Van Gerven, 2009)). These reviews were comprehensive for a number of classes; they only include a limited number of major neurotransmitter systems. In this chapter, tests that are potentially sensitive to some other mechanisms will be discussed occasionally, but the value of these tests has not been determined systematically, and the responses are essentially based on incidental reports. The sensitivity of tests to new pharmacological mechanisms is generally unknown. In these cases, there is rarely enough time for method validation, and measurements are often included for exploratory purposes. The best chance to obtain interpretable effects with a new compound is by using (a combination of) tests with proved sensitivity to a wide range of different pharmacological mechanisms (Cohen et al., 2015; Groeneveld, Hay, & Van Gerven, 2016). The purpose of the next paragraphs in this section is to identify the most responsive functional CNS measurements to different CNS depressants and stimulants.

B Heterogeneity of Investigations of CNS Drug Effects

Some main properties of the studies included in the reviews of drug-sensitive functional CNS measurements are presented in Table 4.1. All studies were limited to the effects of a single dose. This is typical for many studies in early human drug development, but it limits the assessment of cumulative effects of drugs, which may lead to the development of tolerance and in some cases to gradual development of therapeutic or adverse effects. Specific drug development questions related to time-dependent effects are typically addressed in multiple-dose studies in healthy volunteers and patients. Overall, 687 studies were evaluated. In all reviews the numbers of subjects in each study varied substantially between 4 and 161, with average group sizes of 30 for ethanol studies

TABLE 4.1 Properties of Published Systematic Reviews of Test Sensitivities for Various Drug Classes

Drug class	Antipsychotics (De Visser et al., 2001)	Benzodiazepines (De Visser et al., 2003)	Cannabis (Zuurman et al., 2009)	Ethanol (Zoethout et al., 2011)	Methylphenidate (Schrier et al., 2015)	MDMA (Dumont & Verkes, 2006)	SSRIs (Dumont et al., 2005)
Studies	63	56	165	218	78	29	78
Compounds	23	16	1	1	1	1	13
Tests	101	73	318	342	151	150	171
<5 studies/test	84%	75%	85%	89%	93%	93%	80%

and 13–20 for the other compounds. This is similar to the number of healthy volunteers in most single-dose protocols in early drug development. In most individual studies, no more than one or two doses were used. This limits conclusions about the dose relatedness of most CNS measurements and also of their applicability as pharmacological measurements, as explained in Section III.C. A striking finding was the large variability of methods and test variants. As shown in Table 4.1, the number of different tests that were employed in each drug class varied between 73 and 342. In total, about 400 tests and test variants were used. In addition, many tests provided different outcome parameters and often more than one or two. This huge variability reflects the paucity of knowledge about relationships between molecular and cellular biology (where drugs act) and normal or abnormal CNS functionality (which patients suffer from and which can be measured). This limited understanding also contributes to the poor (if not absent) standardization of functional CNS measurements. This variability, in turn, complicates the integration of drug effects and the determination of test sensitivities.

Several companies have tackled this "catch-22" position by offering computerized batteries of CNS tests to the pharmaceutical industry. These batteries were usually spun off from academic research groups, including Cantab (Cambridge, the United Kingdom), Cognitive Drug Research (Reading, the United Kingdom), Cogstate (Northern Australia), and NeuroCart (Centre for Human Drug Research, Leiden, the Netherlands). The Vienna Test System (Wiener Gerät, Schuhfried, Vienna) has also been applied in drug research, but this system is now mainly used for psychological diagnostics in traffic, work, and sports. Other parties offer specialized measurement platforms, like polysomnography or sophisticated systems or analyses for electroencephalography (EEG) and evoked potentials, but neurophysiological measurements are largely beyond the scope of this chapter. These standardized systems present improved logistics and quality control for data procurement and processing, which is obviously useful for drug development programs. The tests in these batteries are still quite diverse, however, and scientific integration across the different platforms is uncommon. Moreover, none of these batteries have been specifically developed for the purpose of demonstrating pharmacological activity. Most tests were derived from neuropsychological or neurophysiological research, for the elucidation of CNS functions (often as safety or tolerability assessments) rather than drug mechanisms. In this respect, components from commercial batteries do not differ from other methods, and all the individual functional tests were treated as separate measurements in the systematic evaluation of their responsiveness to drugs.

C Requirements for Functional CNS Measurements that Reflect Drug Effect

The purpose of the reviews (and of this chapter) was to identify tests or functions that reliably reflect pharmacological effects of certain CNS-active drugs. The integration of results from roughly 400 different tests, each with several different effect parameters, is not straightforward. A stepwise approach was therefore developed, which is described in more detail in the earlier reviews (e.g., De Visser et al., 2001, 2003). In brief, for each review, all publications that reported the effects of drugs in healthy subjects were evaluated. Doses were normalized to the established therapeutic dose range (within, below, or above this range). All results from all tests were selected, including the different effect parameters. If one of the neurobehavioral parameters showed a statistically significant drug effect, this was marked as a positive test result if the outcome was a functional improvement or a negative one in case of a deterioration. Each result was treated as a separate outcome, without corrections for multiple testing. Different parameters from the same test were considered as one single outcome, except if they were clearly related to different CNS functions. For instance, for some cognitive tasks that involve a trade-off between speed and accuracy, reaction time was generally considered separately (as a measure of attention) from the number of correct responses (reflecting the primary cognitive function). The results were represented as the proportion of tests that showed a statistically significant drug effect. In essence, this analysis reflects the chance that a certain test will demonstrate a significant effect of a drug (dose) in an average-sized population of healthy subjects. This is comparable with the situation in most early drug development studies, where rigorous testing of a concrete hypothesis is usually less important than an exploratory identification of plausible drug effects.

The rather coarse reduction of the methodological heterogeneity in the literature, to statistical significance of test outcomes, did not solve all methodological problems. Grouping of data was also thwarted by the huge variety of tests. Since the main purpose of the reviews was to identify individual tests that are sensitive to a therapeutic drug dose, the analysis was first performed on all tests that were similar or highly comparable and that were used at least five times by different research groups. As shown in Table 4.1, this did not apply to 75%–93% of the tests. The small proportion of tests that were used frequently and relatively uniformly were evaluated first for drug sensitivity (as shown by the percentage of statistically significant effects) and dose-response relationships. Subsequently, tests that showed a clear drug response were considered in more detailed, to link the physiology that underlies the method to the pharmacology of the drug. This was impossible for the majority of tests, which were heterogeneous and only used sparsely. These tests were clustered on the basis of catalogs of psychological tests (Lezak, 2004; Strauss, Sherman, & Spreen, 1998). Within each cluster and for each dose level, the proportion of significant drug effects were scored. Each test cluster corresponds to a primary functional CNS domain (for instance, immediate recall within the memory domain), although secondary parameters (which if available were treated separately in the analyses) can reflect a contribution of additional functions (like attention or motor speed). These results will be presented in the next sections of this paragraph, in reverse order of data condensation: first, functional domains that are responsive to drugs; then functional clusters; and finally some individual drug-sensitive tests.

D Drug Sensitive Central Nervous System Domains

The expressions of brain activity are highly complex, and any classification will fail to capture every aspect of CNS function. The reviews of drug-sensitive tests adopted a practical approach, by categorizing functions, on the basis of the methods that were used to study the drug effects in healthy subjects. This led to a primary distinction between neurophysiological functions (EEG, evoked potentials, and eye movements), neuroendocrine functions (hypothalamic and pituitary hormones), subjective effects (visual analog scales (VAS) and questionnaires of emotions, sensations, impressions, etc.), and cognitive effects. Neurocognitive functions are divided into different but interacting domains, which according to the psychiatric Diagnostic and Statistical Manual of Mental Disorders (DSM) 5 (Sachdev et al., 2014) and neuropsychological compendiums (Lezak, 2004; Strauss et al., 1998) distinguish perception motor, executive function, learning memory, complex attention, social cognition, language, and achievement, each with subordinate functional clusters. The first two columns in Tables 4.2 and 4.3 provide an overview of the functional domain and test clusters that were regularly addressed in drug studies. Social cognition, language, and achievement were hardly ever measured in the studies included in the reviews. In recent years, however, there seems to be an increasing interest in aspects of social cognition, like emotion recognition and theory of mind. This growing interest is related (and contributes) to developing hypotheses about the pathogenesis of certain psychiatric conditions like depression and disruptive behavioral or autism spectrum disorders. The number of drug studies is still low, however, and many of these use magnetic resonance

imaging (MRI) to determine the responses of relevant brain areas (Baker, Clanton, Rogers, & De Brito, 2015; George, 2016; Outhred et al., 2013). Functional MRI "challenges" specific neuronal responses with a task that can elicit context-specific drug effects. Functional MRI using emotional face recognition, for instance, can demonstrate acute effects on mood from single doses of antidepressants in healthy subjects (Harmer, Duman, & Cowen, 2017; Pringle & Harmer, 2015), which are not detected in an equable state of mind. However, the focus of this chapter is on direct drug-responsive CNS functions and tasks and not on challenge tests. Resting-state MRI imaging does not employ tasks or stimuli to elicit responses, but measures the direct effects of drugs on voxel-wise changes of BOLD signals (pharmaco-MRI), perfusion (arterial spin labeling), or connectivity (network analysis). These technologies are promising but too complicated to be properly addressed in this chapter. The interested reader is referred to some recent publications in which different MRI technologies in drug research are reviewed (Duff et al., 2015; Khalili-Mahani et al., 2017; Wandschneider & Koepp, 2016).

1 Effects of CNS Depressants on Functional Domains

The most frequently investigated functional domains in early development-type studies in healthy subjects are (visuomotor) perception and (sensori) motor function, executive function, learning memory, and attention. Subjective assessments partly address the same domains, but these measurements were considered as a separate entity in the reviews. The specific neuroendocrine and neurophysiological measurements are discussed in Section III.F. As shown in Table 4.2 (where the first column indicates the neurocognitive domains), CNS depressants like benzodiazepines, neuroleptics, ethanol, and cannabis affected a wide range of domains. Except cannabis, all CNS depressants caused fairly consistent reductions of attention, which reached statistical significance in 63%–70% of all studies with one or more of the tests in this domain. Cannabis mainly impaired the executive domain, which was also strongly affected by ethanol. Subjective assessments were very sensitive to CNS depressants, with significant effects in 79%–96% of studies. The effects of a single dose of an antipsychotic, however, were only detected by healthy subjects with statistical significance in about half of the publications.

2 Influence of Cognitive State on Responsiveness to CNS Depressants

In the single-dose healthy volunteer studies of prototypical CNS depressants, memory was not very sensitive, and hardly any test in this domain showed statistically significant drug effects in more than 50%

TABLE 4.2 Proportions of Statistically Significant Outcomes (%) of Various Test Clusters and Functional Domains, for Different CNS Depressant Drugs and Classes

Domain	Cluster	Drug class			
		Neuroleptics	Benzodiazepines	Cannabinoids	Ethanol
Executive	Inhibition		0%	−52%	−50%
	Information processing	−46%	−43%	−67%	−100[a]
	Time estimation			−17%	−43%
Attention	Search	−70%	<10%	−35%	−50%
	Digit-symbol substitution	−48%	−66%	−43%	−63%
	Divided attention			−36%	−68%
	Flicker discrimination	−45%	−30%	−33%	−50%
	Vigilance			−13%	−42%
	Saccadic eye movements		−80%		−90%
Memory	Memory span	−53%	−50%	−40%	−55%
	Learning			−38%	
	Immediate recall			−58%	−60%
	Delayed recall			−52%	−38%
Visuomotor	Reaction time	−46%	−56%	−47%	−53%
	Eye-hand coordination	−41%	−50%	−56%	−78%
Sensorimotor	Manipulation	−48%	−36%	−33%	−61%
	Motor control			−58%	−86%
Subjective	Alertness	−53%	−79%	−39%	−36%
	Mood	−50%		∼20%	−8%
	Calmness	−19%		−24%	−36%
	Anxiety	0%		−28%	
	"Drug effect"	−57%		−96%	−98%
Most sensitive biomarkers		Prolactin (+98%)	Saccadic peak velocity (−80%)	VAS "high" (+96%)	VAS "drug effect" (98%), body sway (−86%), driving (−100%[b])

Changes indicated by (−) signify functional deteriorations, and equivalent results are indicated by (∼). See Table 4.1 and text for references.

[a] *The sensitivity of "information processing" tests to ethanol is mostly indicated by the consistent impact of ethanol on driving performance (which is a largely on complex information processing task).*

[b] *Effects on complex information processing mostly reflect consistent effect on driving performance.*

of the studies. A systematic review of a variety of benzodiazepines, antihistamines, anticholinergics, and opioids also showed inconsistent anamnestic effects, although benzodiazepine-induced memory impairment was quite consistent in elderly subjects (Tannenbaum, Paquette, Hilmer, Holroyd-Leduc, & Carnahan, 2012). This suggests that the sensitivity of functional domains could be partly related to age or cognitive state. Early drug development-type studies of CNS-active drugs are usually performed in young healthy subjects, typically with a good education. This gives them ample capacity to overcome the effects of CNS-active compounds, at least in a lower dose range. Although most functional domains exhibit an age-related decline, there is less direct evidence for an increased sensitivity to detrimental drug effects. Different factors may be involved in apparent enhanced drug sensitivity in elderly subjects. In the general population, there is the impact of disease and comedication, which may interact with CNS-active drugs. This does not play a major role in healthy elderly subjects who participate in drug development-like studies. In this group, pharmacokinetic differences may play an important role. This can be illustrated with the antimuscarinic scopolamine, which produces many dose-dependent CNS depressant effects in healthy young (Liem-Moolenaar et al., 2011) and elderly subjects (Alvarez-Jimenez et al., 2016). As expected, memory function was lower at baseline, in the older people than in the young. Scopolamine also caused a relatively large impairment of memory in the elderly. However, a comparison of the PK/PD relationships in these two groups provided no indications for a difference

TABLE 4.3   Test Sensitivities for Different CNS Stimulant Drugs

| Drug class | | | | |
Domain	Cluster	Methylphenidate[a]	MDMA	SSRIs
Executive	Inhibition	+63%		
	Info processing	+39%		+33%
Attention	Search		0%	−50%
	Divided attention	+71%		
	Flicker discrimination			~46%
	Vigilance	+76%		
Memory	Memory span			+24%
Visuomotor	Reaction time	+50%		−41%
	Eye-hand coordination			<10%
Sensorimotor	Manipulation	+72%		~22%
	Motor control	+50%		
Subjective	Alertness (observed)	+83%		+18%
	Mood			+33%
	Anxiety		+67%	
	"Drug effect"		+88%	
Other sensitive biomarkers		P300 (+70%)	Oxytocin (+33pmol/l)	REM sleep (−79%)

Proportions of statistically significant outcomes (%) of various test clusters and functional domains are presented for different compounds. Most changes are enhancements or increases (+). Some deteriorations or decreases are indicated (−) and equivalent or biphasic results by (~). References are presented in Table 4.1 and in the text.
[a] Results pertain to the effects of single doses in healthy young volunteers, except methylphenidate that apply to ADHD patients off-treatment.

in drug sensitivity. The higher response in elderly could be largely explained by a reduction of clearance, which caused an elevation and longer persistence of scopolamine exposure (Alvarez-Jimenez et al., 2016). Pharmacokinetic differences are an underestimated source of variability in CNS drug research, although less so in formal early drug development programs. The factors that govern age dependence may differ between drugs. The larger sensitivity to lorazepam in elderly subjects (Tannenbaum et al., 2012) is only partly explained by the limited reductions in clearance (Greenblatt, Allen, Locniskar, Harmatz, & Shader, 1979).

3 Effects of CNS Stimulants on Functional Domains

Three reviews were performed to identify drug-sensitive tests for drugs with stimulant effects on CNS functions: (selective) serotonin reuptake inhibitors, methylphenidate, and MDMA ("ecstasy"). Table 4.3 shows the percentages of statistically significant effects of these three drug categories,

with functional domains in the first column. The effects on cognitive domains and subjective assessments seemed more variable than for the CNS depressants. This was partly caused by the selection of tests, which was often adapted to the application of these agents in certain groups of users or indications. The large majority of single-dose MDMA studies were devoted to subjective effects, and neuropsychological functions were rarely examined. Therefore not all domains or clusters could be evaluated as well as for the CNS depressants. Methylphenidate caused consistent improvements in functional and subjective domains of attention and alertness. Statistically significant effects of a therapeutic dose were reached in 71%–83% of the studies. There were also some enhancements of executive and motor functions, which could in part be secondary to improvement of concentration.

4 Influence of Cognitive State on Responsiveness to CNS Stimulants

As for CNS depressants the cognitive state of study participants can also affect their sensitivity to CNS stimulants. The consistent effects of methylphenidate are related to the fact that this review was performed in patients with attention deficit and hyperactivity disorders (ADHD), who were either untreated or withdrawn from therapy. These subjects were otherwise healthy, and there is probably a significant overlap with the average student population that often participates in early drug studies. It should be realized, however, that the test sensitivity for methylphenidate that is indicated in Table 4.3 may not fully apply to healthy young subjects without ADHD. The effects of SSRIs and MDMA on cognition were much more limited than for methylphenidate and also smaller than for the CNS depressants. MDMA and SSRIs both specifically enhance serotonergic activity, which involves a dozen different receptor subtypes that are only partly stimulatory. From these reviews the impression arise that neurocognitive functions are less sensitive to mild CNS stimulant agents in healthy subjects who function optimally, than in subjects with an impairment in these areas. This could be an indication of the ceiling effects discussed previously, as a limitation of functional CNS measurements in healthy young volunteers. However, the findings with MDMA and SSRIs cannot be generalized to other mild CNS stimulants. There is no systematic information that functional domains are most sensitive to other CNS stimulants like histamine agonists, adenosine antagonists (caffeine), and noradrenergic drugs. There is little doubt that these compounds can cause stimulatory effects, but there are only anecdotal indications that tests might qualify as biomarkers for pharmacological activity, for instance, by exhibiting clear concentration-related effects. Some potential examples from the literature and some specific neurophysiological and neuroendocrine effects of serotonergic compounds from the reviews will be discussed in Section III.F on individual tests.

E Drug Sensitive Functional Test Clusters

The close to 700 drug studies that were included in the reviews all showed statistically significant drug effects in one or more domains. This suggests that healthy people are quite susceptible to CNS drug effects (if the possibility is ignored that negative studies without any effect may never have been published). The effects of CNS-active drugs seem to span different functional domains, and for all classes, tests were identified that already show an effect after a single clinically relevant dose. In principle, this makes healthy volunteers a suitable group for the exploration of new compounds in early drug development, with tests that cover a range of functional CNS domains. However, there are some limitations to this conclusion. First the reviews were restricted to seven well-known different drugs and classes. The responsiveness of CNS tests may differ for drugs with other mechanisms. Second, some cognitive domains do not seem to be very sensitive to most drugs, except secondary to changes in other domains. This is the case for intellectual achievement and language, which so far haven't attracted much attention from drug developers, except perhaps regarding potential side effects. Social cognition may be an interesting target, but drug-sensitive methods in this domain cannot be identified as long as there are no published drug effects. There are also differences in test sensitivity between compounds. Within the group of CNS depressants, benzodiazepines have more pronounced effects than antipsychotics. It may be even more difficult to demonstrate the effects of clinically relevant doses of CNS stimulants, due to ceiling effects in healthy subjects (Section III.D.1). Finally, not nearly all studies produce statistically significant effects, even with drugs that have well-known CNS effects and proved clinical efficacy. Not all functional CNS measurements are equally drug-sensitive or suitable to demonstrate pharmacological activity. Only a small proportion of tests (7%–20% in the reviews) are used frequently enough to determine their value as biomarkers of pharmacological activity. These methods will be discussed in the next section of this paragraph (Section III.F). The majority of tests (80%–93%, Table 4.1) are used too infrequently for a reliable evaluation of the accumulated experience. These roughly 300 heterogeneous tests were grouped into functional clusters, which essentially measure similar brain activities. About 20 neurocognitive clusters contained more than 10 individual items, which was considered a minimum for generalization. These are presented in column 2 of Tables 4.2 and 4.3.

1 Executive Test Cluster

Within the executive functional domain, tasks that involve complex information processing were particularly sensitive to ethanol and cannabis. Significant impairments were found in more than two-thirds of the studies with these compounds, mostly with driving studies. Impaired response inhibition was improved by a single dose of methylphenidate in 63% of the studies, which all included patients with ADHD who were off medication. Memory was the least sensitive functional domain in healthy subjects. Different processes are involved in learning, memory consolidation, and retrieval, and most of these steps can only be tested once after the administration of a drug. Since drug concentrations vary strongly after single-dose administration, memory is less suitable as a pharmacological effect measure. An additional limitation is that the number of versions of memory tests and their equivalence (words, figures, etc.) are limited.

2 Other Test Clusters

Memory span is a working memory test that is strongly influenced by attention. Span test are relatively simple and repeatable and fairly sensitive to CNS depressants (about 40%–50% of results were significant). Overall, SSRIs had little impact on memory span, but significant improvements were found in two-thirds of studies with a low dose. Low doses of SSRIs also improved flicker fusion recognition, another attention test cluster, in 75% of the studies. Increased attention was detected less frequently in higher dose ranges. The number of tests that evaluated attention was large. Divided attention, search, and digit-symbol substitution-like tests were the most responsive test clusters, which showed significant deteriorations with most CNS depressants and improvements with methylphenidate (in ADHD) in about two-thirds of the studies. Tests within the visuomotor domain only showed significant CNS drug effects in about 50% of the cases, both for depressants and stimulants. This domain contains reaction time, which is one of the most frequently used measures of CNS function. Reaction time is also one of the most variable test clusters, partly because it is a secondary parameter of many tests. This may explain the lack of sensitivity. Alcohol had a significant effect on eye-hand coordination in 78% and on sensorimotor control in 85% of the studies.

F Drug Sensitivity of Selected CNS Tasks

The clustering of tests that address similar functions within a domain provides information on the drug responsiveness of certain types of CNS measurements. The sensitivity of this analysis is reduced by the variability and the nonresponsive tests that belong to the same cluster. Individual tests may perform better, but these are rarely used frequently enough to give reliable indications of drug responsiveness. To give an impression the most frequently used tests within a cluster are presented in Table 4.4, next to the maximum sensitivities within

TABLE 4.4 Sensitivities of Most Frequently Used Clusters, to Detect Drug-Induced Deteriorations/Decreases (↓) or Improvements/Increases (↑)

Domain	Cluster	Sensitivity ↓	↑	Test
Executive	Inhibition	50%	62%	Go-no go, Stroop conflict, prepulse inhibition
	Info processing	43%/67%	39%	Simulated driving, card sorting, arithmetic
	Time estimation	17%/43%		Time estimation
Memory	Memory span	40%/55%		Sternberg, digit span, n-back
	Learning	38%/60%		Word, spatial, picture
	Immediate recall			
	Delayed recall			
Attention	Search	35%/70%		Letter cancellation, symbol copying, trail making
	DSST-like	43%/66%		Symbol-digit substitution
	Divided attention	36%/68%	71%	Divided attention
	Flicker discrimination	30%/50%		Critical flicker fusion
	Vigilance	13%/42%	76%	Continuous performance task
	Saccadic eye movements	80%		Saccadic peak velocity
Visuomotor	Reaction time	46%/56%	50%	Simple/complex reaction time
	Eye-hand coordination	41%/78%		Pursuit rotor, spiral maze tracing, (adaptive) tracking
Sensorimotor	Manipulation	33%/61%	72%	Finger tapping, pegboard
	Motor control	58%/86%	50%	Force platform (Wright's ataxiameter)
Subjective	Alertness	36%/79%	83%	VAS Bond and Lader, Von Zerssen Befindlichkeitsskala
	Mood	8%/50%		Profile of mood scales
	Anxiety	28%		Spielberger trait anxiety index
	Drug effect	57%	96%/98%	VAS Bowdle (Bowdle et al., 1998), VAS "high," "drug effect"
				Addiction Research Center Inventory (ARCI)

very common in rodent studies

subjective, self-reported

Response percentages reflect proportions of statistically significant result from literature. Ranges refer to extremes for different drugs or classes (e.g., within "Search" cluster, up to 35% of studies with cannabis showed significant deterioration and up to 70% with antipsychotics). Last column shows most frequently used tests (>5 times for any drug class by different research groups; tests between brackets are used in the author's research institute and not used for response calculations).

these clusters to CNS depressants or stimulants. For a description of these tests, the reader is referred to neuropsychological reference works (e.g., Lezak, 2004; Strauss et al., 1998) and the pertaining references. The most responsive tests, which were able to demonstrate significant drug effects in 70%–98% of the studies, were found in the domains of attention, visuomotor function, subjective assessments, and neuroendocrine hormones.

1 Executive Tasks

Executive tests are not among the most sensitive methods to show CNS-active drug actions. Executive functions are complex integrated cognitive processes that include reasoning, planning, inhibition, and flexibility, which depend on underlying functions like attention, vigilance, memory, and visuomotor control. The dependence of executive cognition on different drug-sensitive functions increases its vulnerability to many drugs that have an impact on the brain or on the physical condition. However, this impact can be influenced (and partly compensated) by adapting strategies, practice, and motivation. Because of this complexity the mechanisms that underlie executive drug effects are difficult to interpret, and cognitive tasks are less suitable to demonstrate the pharmacological origins of drug-induced CNS effects. This is better achieved with simpler methods that are reproducible, consistent, and repeatable enough to follow the time course of changing drug levels.

DRIVING TASKS

Executive tasks can be quite useful however to reflect the impact of drugs on aspects of daily life or for clinical conditions that are characterized by derangements of everyday activities. Driving is one of the most widely used complex executive tasks. It requires an integration of information processing and visuomotor coordination, and it is frequently used to investigate effects of CNS-active drugs (Verster & Mets, 2009). The effects of medications on driving are often compared with ethanol. Alcohol affects a wide range of CNS functions that are relevant to driving (Jongen, Vuurman, Ramaekers, & Vermeeren, 2016), but it has a particularly large impact on eye-hand coordination (Table 4.2). Most driving tests use a measure of a visuomotor stability as the primary parameter for the impact of drugs and medication on driving: the standard deviation of the lateral position (SDLP) of the vehicle. This reflects the sustained ability of the driver to maintain driving lane position in a monotonous driving environment. Most CNS depressants with a warning label ("potentially driver-impairing medications") show in an increase of SDLP. Conversely, SDLP can improve with CNS stimulants like armodafinil, particularly in case of sleep deprivation (Drake, Gumenyuk, Roth, & Howard, 2014), or with methylphenidate in subjects with ADHD (Barkley and Cox, 2007).

SDLP does not address other important driving skills like risk taking and reaction speed (Jongen et al., 2016; Laude & Fillmore, 2015), which may also be affected by drugs and disorders (Food and Drug Administration, 2015). Driving performance can also be impaired by increased distraction and impulsivity, which are core symptoms of ADHD. A single dose of methylphenidate in subjects with ADHD enhances behavioral inhibition in 63% of studies, mainly shown on go/no go tasks (Schrier et al., 2015). Methylphenidate also consistently improves divided attention (71%) and vigilance (76%) in this population. The wide range of cognitive and noncognitive effects of methylphenidate in ADHD illustrates the dependence of "higher" neuropsychological processes on such "supportive" CNS functions.

2 Attention Tasks

Attention is fundamental for cognition and performance, and it is a prerequisite for most neuropsychological function tests that are used in drug research. Most tests require comprehension of instructions, perception of stimuli, and active participation, although there may be some exceptions like EEG and MRI. The effects of drugs cannot be interpreted without information about the influence on attention, and many complicated tests include reaction time as a measure of attention (and of sensorimotor function, which is another frequent underlying aspect of test performance). Attention is particularly sensitive to CNS depressants, as shown by the high frequencies of 63%–70% statistically significant outcomes with most of the test clusters in this domain (Table 4.2). There seem to be differences in test sensitivity between different drug classes. Reduced search performance was one of the most consistent effects of antipsychotics

(in 70% of the studies), whereas letter cancellation tests and similar search tasks were much more variably affected by benzodiazepines (<10% reached significance). Benzodiazepines and ethanol on the other hand caused an impairment of digit-symbol substitution in about two-thirds of the studies but in less than half of the publications with antipsychotics. It remains to be determined whether these apparent differences are related to the impact on dopamine systems by antipsychotics, versus GABA-A systems by benzodiazepines (or to the methodological limitations of integrating highly diverse results from the literature).

Attention tests are generally not very sensitive to CNS stimulants. SSRIs only caused a small significant improvement in the lowest dose range, in 75% of studies that used flicker fusion and 50% for the attention modules of the Vienna Test System. This may partly reflect ceiling effects on this domain in healthy (student) volunteers, and more demanding methods or circumstances may be required to show improvements in subjects who already perform at high levels. Most studies show improvements of methylphenidate on attention and vigilance in untreated subjects with ADHD, who have some impairment of these functions. Relatively simple attention tests like selective attention and simple reaction times show significant drug effects in 43%–50% of studies, but this increases with more demanding tasks like divided (71%) and sustained attention (76%) (Schrier et al., 2015).

SACCADIC EYE MOVEMENTS

Saccadic eye movements are rapid autonomic responses that play a role in the scanning of objects and the detection of stimuli that appear in the peripheral field of view. Saccadic eye movements are strongly influenced by attention and therefore often used as a measure of this functional domain. The velocity of saccadic eye movements is significantly diminished by sleep deprivation (Van Steveninck et al., 1999). Benzodiazepines also caused a strong reduction of peak saccadic velocity in 80% of the studies (De Visser et al., 2001). The effects of ethanol are less prominent but were quite consistent in 90% of the (small number of) studies where this method was used (Zoethout et al., 2011). Antipsychotics caused a reduction of saccadic peak velocity in 83% of the studies (De Visser et al., 2001). Interestingly, almost no study in the cannabis review had a significant impact on eye movements (Zuurman et al., 2009), and this is corroborated by the lack of effects in the large number of cannabis studies that were performed at the authors' institute (which were not included in the review). This suggests that saccadic eye movement only partly reflect the multitude of pharmacological mechanisms that govern alertness. Saccadic peak velocity is particularly influenced by the GABA-A system, as will be discussed in more detail in Section IV.B on pharmacological mechanisms of functional CNS measurements. Saccadic eye movements are also susceptible to the effects of CNS stimulants, although their sensitivity is difficult to quantify. The systematic reviews offer limited information on the effects of CNS stimulant on the speed of saccades. Saccadic eye movements were used only once in the methylphenidate review, which was limited to single-dose studies in ADHD patients after withdrawal of treatment (Schrier et al., 2015). Another smaller but less restrictive review of methylphenidate identified more saccadic eye movement studies in ADHD, which showed improvements in 83% of the cases (Pietrzak, Mollica, Maruff, & Snyder, 2006). About half of the publications with SSRIs that include saccadic eye movement report an increase of peak velocity (Dumont & Verkes, 2006). These studies used a single dose in the therapeutic range. A significant increase of saccadic peak velocity is observed with supratherapeutic levels of serotonergic stimulation (Gijsman et al., 2002). Saccadic peak velocity also increased with other CNS stimulants, including low-dose caffeine (Wilhelmus et al., 2017), modafinil (Baakman et al., 2019) and dexamphetamine in nondrug-using healthy subjects (unpublished data) and with an experimental H3 receptor antagonist (Baakman et al., 2019).

3 Eye-Hand Coordination Tasks

Eye-hand coordination includes a number of closely regulated neurophysiological processing loops, which involve different neurotransmitter systems. Consequently, it is one of the most sensitive functions for the detection of a wide range drug effects. The reviews indicate that eye-hand coordination tests show a decrease in up to 78% of ethanol studies (Zoethout et al., 2011) (Table 4.2). The reviews provide little evidence for sensitivity to CNS stimulants of the many types of eye-hand coordination tests, which include pursuit rotor, wire maze tracing, and several tracking methods. This apparent lack of sensitivity for improvement seems to be largely attributable to the heterogeneity of eye-hand coordination methods. The cluster of eye-hand coordination tests only showed significant effects in half of all studies with benzodiazepines, but tracking tasks provided significant results in eight of nine cases (De Visser et al., 2003). The authors have similar experience with their own adaptive tracking method (Groeneveld et al., 2016. Most CNS depressants including all those in Table 4.2 showed clear effects on adaptive tracking, in several studies, and with different drugs in each class. The reviews of CNS stimulants included too few studies of eye-hand coordination to allow conclusions about test sensitivities, and there were no published studies with adaptive tracking. However, an unpublished study in the authors' institute with methylphenidate in children and adolescents with ADHD showed clear improvements of adaptive tracking. Significant enhancements were also observed with

a range of other CNS stimulants, like low-dose caffeine (Wilhelmus et al., 2017, modafinil and an experimental H3 receptor antagonist (Baakman et al., 2019), and dexamphetamine in nondrug-using healthy subjects (unpublished data). The fact that adaptive tracking is not identified as one of the most responsive measures of CNS effects is related to a selection criterion in the reviews, which was restricted to tests that were applied in the literature by different research groups, for generalizability. Some tests in other clusters were also excluded for this reason, although these sometimes showed quite consistent drug effects (like the "Von Zerssen Befindlichkeitsskala" questionnaire for drug effects and the alphabetic cross-out search test, both from the antipsychotic review (De Visser et al., 2001)).

4 Sensorimotor Tasks

The sensorimotor effects of CNS-active drugs can be divided into two major groups: manipulation and postural stability. There is no behavioral expression without movement, but the sensorimotor methods that are most frequently used in drug research are relatively simple compared with the large variability of tests in other domains. Manual dexterity is often tested with tapping tests, which measure rapid alternating movements of the hand or fingers. Pegboard-like tests examine the positioning of small geometric objects. This sensorimotor task is also visually guided, and there is some overlap with visuomotor coordination. Fine motor performance does not seem very sensitive to CNS-active drugs. The most consistent effects were observed with ethanol, which showed sensorimotor effects in less than two-thirds of the studies. Methylphenidate caused sensorimotor improvements in 72% of the studies with ADHD. In both cases, tapping was the most frequently used test. For other drugs the sensitivity of sensorimotor tests was less than 50%.

Postural stability can be measured in different dimensions. The most complicated systems evaluate all aspects of bodily movement and sensory input in sophisticated "balance rooms" (Pasma et al., 2016). Static force platforms determine changes in the center of gravity in two dimensions. String ataxiameters only measure displacement in the anterior-posterior plane. Balance rooms are mainly designed to provide insights into the complex biophysical processes that govern gait and posture. A recent comparative study suggests that such complex measurements are not necessarily more sensitive to drugs than easier tests of body sway, although they offer more insight into the underlying derangements (Pasma et al., 2016). Sway platforms and string ataxiameters are used most frequently in drug research. There is little doubt about the negative impact of benzodiazepines and other CNS depressants on postural stability. Psychoactive drugs clearly increase the risk of falls in elderly subjects (Thapa, Gideon, Fought, & Ray, 1995; Van der Hooft et al., 2008). However, the effects in healthy volunteers are more difficult to identify from the literature, largely because dedicated studies are infrequent. Still, consistent effects are found with the "body sway," an adapted version of the Wright ataxiameter (which was developed at the authors' research institute and for this reason not included in the literature reviews). Strong dose-related impairments of body sway are found with basically all CNS depressants that were examined, including several compounds from Table 4.2. Body sway improvements were also found with CNS stimulants, like low-dose caffeine (Wilhelmus et al., 2017), modafinil (Baakman et al., 2019), dexamphetamine, and methylphenidate (unpublished data).

5 Subjective Assessments

After the administration of a CNS-active compound, many healthy subjects notice the effects before this starts to affect their test performance. Subjective effects belonged to the more sensitive measures of CNS-active drugs. For most reviewed compounds, awareness of a drug effect reached significance in 79%–96% of the studies. The subjective effect of benzodiazepines and other sedative agents are most reliably found with instruments that quantify different aspects of alertness, concentration, sleepiness, and tiredness, such as the VAS of "alertness" (Bond & Lader, 1976; Norris, 1971). These scales show significant effects of benzodiazepines in 79% of the studies. In the remaining 21%, drug-induced reduction of alertness could not be distinguished from placebo, probably because sleep deprivation, fatigue, and boredom are common during study days. Healthy subjects have more difficulty identifying subtle effects of compounds that they have not experienced before. Antipsychotics are only detected in 57% of the studies in healthy subjects and most accurately with poorly circumscribed terms like "feeling drug effect" or "feeling spacey." Unusual or unexpected sensations are often quantified with visual analog scales of psychomimetic effects, subscales of the Addiction Research Center Inventory (ARCI), and sometimes the Profile of Mood States (POMS) or the Short Form (36) Health Survey (SF-36). Such scales readily demonstrate the effects of drugs with which subjects have at least some prior experience. The subjective effects of cannabis, alcohol, and MDMA ("ecstasy") were consistent detected in most studies with these compounds (88%–98%).

6 Neuroendocrine Responses

Neuroendocrine responses are not measured nearly as frequently as neuropsychological or neurophysiological effects in studies of CNS-active drugs. It is difficult therefore to derive the responsiveness of neuroendocrine hormones to other drug classes from the systematic literature reviews. The value of most neuroendocrine hormones as indicators of pharmacological activity of CNS-active drugs is still largely undetermined. A notable exception is the increase of prolactin, which is a well-known effect of antipsychotics. Hyperprolactinemia can be excessive and clinically manifest, particularly in patients who are treated with high doses of classic neuroleptics. More limited elevations of plasma prolactin levels were observed in almost all healthy volunteer studies with antipsychotic agents, even after a single dose (Table 4.2). Prolactin release is closely associated with dopaminergic activity, as will be discusses in paragraph 4.2. Prolactin is also increased by high levels of serotonergic stimulation (MacIndoe & Turkington, 1973). This may partly be an indirect (dopaminergic) stress response associated with nausea and vomiting (Jacobs et al., 2008). MDMA, which is a strong serotonin releasing agent that doesn't cause much nausea in experienced users, only shows prolactin elevation in 56% of the studies (Dumont & Verkes, 2006).

The literature is equivocal regarding the effects of serotonergic stimulation on cortisol. There is however strong evidence that cortisol is a useful marker of serotonergic activity. The authors of this chapter have found consistent elevations after single-dose administration of SSRIs (Klaassens et al., 2015, 2017) and dose-related increases with several other (partly experimental) serotonergic compounds (Smarius et al., 2008; unpublished data). MDMA, which is serotonin releasing agent, showed significant cortisol increases in 92% of studies (Dumont & Verkes, 2006). MDMA also caused a significant elevation of oxytocin (Dumont et al., 2009).

An important question is whether a neuroendocrine drug effect actually signifies pharmacological activity in the brain. The release of neuroendocrine hormones is largely governed outside the BBB by the pituitary gland (prolactin and releasing hormones like adrenocorticotropin-releasing hormone) and peripheral endocrine glands like the adrenals (cortisol). The central regulation of all neuroendocrine axes relies on specialized neurons in the hypothalamus that excrete releasing proteins into the portal system to the pituitary gland, which cannot be measured directly in peripheral blood. Neuroendocrine responses to CNS-active drugs often involve several hormones, which is difficult to attribute to a simultaneous effect on different cell populations in the pituitary, which are highly specialized and pharmacologically diverse. A hypothalamic site of action is functionally more plausible if several hormones respond concomitantly to a CNS-active compound, because of the integrative role of the hypothalamus: this important autonomic command center governs the concerted activity of many neurobehavioral and peripheral physiological processes.

IV PHARMACOLOGICAL RELEVANCE OF FUNCTIONAL CNS MEASUREMENTS

This chapter started with an outline of the main objectives of early drug development, which is primarily devoted to learning about the drug's pharmacological properties in humans (Section II). This information is necessary for the design of subsequent clinical trials and the prediction of therapeutic efficacy, dose ranges, potential interactions with comedication and organ failure, and expected adverse effects. The essentials of human pharmacology are represented in Fig. 4.1, as the route that is followed by the compound from administration to (patho)physiological impact. Measurements of functional CNS effects play an important role in understanding these processes. However, the interpretation of drug effects and their clinical consequences are complicated by the large diversity of CNS tests. The previous paragraph (Section III) integrated the results of a series of systematic reviews, designed to establish which neuropsychological, subjective, and neuroendocrine domains, functions, and methods are most sensitive to CNS-active drugs. This yielded a number of tests with a reasonable (>60%–70%) chance to show significant effects of a therapeutic dose of a CNS-active drug (Table 4.4). As outlined in Section II, however, the mere demonstration of a drug effect leaves several important questions about the pharmacological properties of the drug unanswered, and not all drug-sensitive methods identified in Section III provide suitable information about drug-target interactions and relevant impact. This paragraph will try to identify some functional CNS measurements that are able to give more detailed and quantitative information about pharmacological activity, which can be used for the design of clinical trials and other decisions in drug development. Methods that provide evidence that the drug has its intended pharmacological effect ("pharmacological proof of concept") should satisfy certain criteria that are based on the mechanisms of action of CNS-active drugs. First, if the effect is caused by the pharmacological activity, it should demonstrate meaningful relationships with the concentrations of the drug. This implies dose dependence and predictable time courses and can ultimately be proved by pharmacokinetic/pharmacodynamic models. Second the method should be able to reliably differentiate between compounds that (only) differ in the pertaining pharmacological activity. Third the method should be able to show quantitative relations between the pharmacological mechanisms of a drug (e.g., receptor binding) and the associated effect. CNS tests that comply with these criteria can be considered as measures of pharmacological activity. The following sections will discuss how these pharmacological effect measurements can be validated and how they can contribute to early drug development.

A Time-, Dose- and Concentration-Dependence of CNS Drug Effects

One of the most important questions in early development is how the effects of a drug change over time, in relation with the concentrations. Fluctuations and variations of drug concentrations are strong sources of variability of drug effects. Compared with pharmacodynamic differences in drug sensitivity between and within patients, pharmacokinetic variations of plasma concentrations are relatively easy to predict and control, by the selection and adaptation of dosing regimens and formulations. The resolution of concentration-effect relationships relies on the assessment of the dose dependence and the time dependence of the effects of the drug. Fig. 4.2 provides an example of the integration of time curves for concentrations (of benzodiazepines) and effects (on saccadic peak velocity). As argued in Section II, useful tests for drug development should be able to capture these two aspects of drug activity. Although these are relatively simple requirements, the literature is sparse on information about CNS tests that are able to show dose- and time-dependent drug effects. Most drug studies in the literature were not part of a drug development program, so the time course was not considered in much detail: tests were often only performed at a single fixed time point after drug administration (often compared with another measurement at baseline or during placebo). Many CNS drug studies also use a single dose and most disregard the pharmacokinetic profiles that underlie the effects. Time dependence could therefore not be investigated systematically in the reviews. Dose dependence was identified if the proportion of statistically significant studies increased with dose. This is a crude measure of

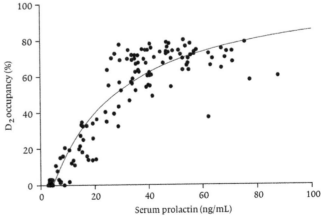

FIG. 4.2   Correlation between serum concentration of prolactin and striatal dopamine D2 receptor occupancy of fast-dissociating D2 antagonist JNJ-37822681, limited to time interval between 2 and 3 h after administration of doses between 0.5 and 20 mg. *Copied with permission from Te Beek, E. T. (2014). Chapter 8. Discussion. In: E. T. Te Beek (Ed.), Neuropharmacology of novel dopamine modulators (pp. 145–158). Leiden University, Thesis.*

TABLE 4.5   Sensitivities of Different CNS Tests to Various Nonselective and Subtype-Selective GABA-A Agonists

GABA-A agonist	Subtype (subunit-containing	α1/α2-ratio	Body sway	VAS alertness	Adaptive tracking
SL65.1498 25 mg	α2,3 partial agonist	0.39	42%	33%	NA
TPA023 1.5 mg	α2,3 partial agonist	0	16%	26%	NA
MK-0343 0.75 mg	α2,3 partial agonist	0.78	9%	78%	NA
AZD7325 10 mg	α2,3 partial agonist	0	17%	23%	33%
AZD6280 40 mg	α2,3 partial agonist	0.25	51%	NA	48%
NS11821 300 mg	α2,3 partial agonist	0.24	11%	12%	−32%
Zolpidem 10 mg	α1 affinity	0.96	108%	121%	85%
Alprazolam 1 mg	Nonselective	0.78	67%	58%	131%

Results are derived from studies with individual drugs. To cope with dose differences, each effect was related to the reduction of saccadic peak velocity (SPV), as slopes of the regression lines. In this overview, these results are presented as percentages of the lorazepam effect/SPV slopes (see Section IV.C for further explanation). *Results are adapted from Chen, X., De Haas, S., De Kam, M., & Van Gerven, J. (2012). An overview of the CNS-pharmacodynamic profiles of non-selective and selective GABA agonists.* Advances in Pharmacological Sciences, 2012, 134523; *Sanna, E., Busonero, F., Talani, G., Carta, M., Massa, F., Peis, M., et al. (2002). Comparison of the effects of zaleplon, zolpidem, and triazolam at various GABA(A) receptor subtypes.* European Journal of Pharmacology, 451, 103–110; *Crestani, F., Martin, J. R., Möhler, H., Rudolph, U. (2000). Mechanism of action of the hypnotic zolpidem in vivo.* British Journal of Pharmacology, 131, 1251–1254.

dose relatedness, which yielded only a small number of dose responsive tests. The most consistent indications of dose dependence were found for tests that showed at least 80% overall response rates within a drug class (Table 4.5). This included saccadic peak velocity for benzodiazepines (De Visser et al., 2003) and prolactin for antipsychotics (De Visser et al., 2001). These parameters closely reflect the pharmacological activity of these drug classes, as will be discussed in more detail in the next sections of this paragraph. It seems to make sense that a fairly basic neurophysiological or neuroendocrine system is more closely linked to a specific pharmacological mechanism than a complex neuropsychological process. However, Table 4.4 indicates that even quite composite CNS tests like VAS of subjective effects can be very reliable measures of pharmacological effects. Visual analog scales (VAS) can be repeatedly recorded during the course of an experiment, and numerous studies have shown the time and concentration dependence of these subjective effects. VAS of "alertness," "high feeling," or "drug effect" all demonstrate dose and concentration dependence. VAS for the subjective effects of alcohol ("feeling drunk") produced significant effects in 59% of the studies at ethanol levels below 0.5 g/L and over 90% at higher concentrations (Zoethout et al., 2011). Previous cannabis users identified the "high" effects of low doses of cannabis in 94% of placebo-controlled studies, and this increased to close to 100% with higher "recreational" doses (Zuurman et al., 2009). The "high" effects of cannabis show very complicated relationships with plasma concentrations of tetrahydrocannabinol (THC, the most important psychoactive constituent of cannabis). These relationships can however be accurately described with a pharmacokinetic/pharmacodynamic (PK/PD) model, which takes the rapid disappearance of THC into the brain (and other lipophilic tissues) into

account (Strougo et al., 2008). PK/PD models for the "high" effects of THC played an essential role in the development of a number of cannabinoid-1 (CB1) receptor antagonists (Guan et al., 2016). This entire class of compounds was prematurely abandoned after the demise of the first registered representative, rimonabant, which caused an increase in depression and suicidality. Nonetheless, early development studies with surinabant showed that doses that reduced weight gain after smoking cessation could be accurately predicted from the PK/PD models that were based on VAS "high" measurements (Guan et al., 2016; Klumpers et al., 2013; Tonstad & Aubin, 2012). These examples clearly illustrate that a functional CNS measurement can be a very reliable indicator of a pharmacological effect, even if the causative mechanisms are not entirely understood—or entirely not, like for the "high" effects of THC. If a highly selective compound shows good concentration-effect relationships that can be accurately described with PK/PD modeling, this provides compelling evidence that the effects represent a pharmacological activity of the drug (Section II). Even if the functional CNS effects are highly diverse and show no direct plausible link with the therapeutic indication, a PK/PD model for the effect can still be very informative for predictions of pharmacologically active dosing regimens.

PK/PD relationships are difficult to establish if the effect does not provide a relatively simple parameter that can be measured repeatedly and related with plasma concentrations. This is the case for many neurophysiological and neuroendocrine methods, which belong to some of the most drug-responsive measures in the literature (shown in Tables 4.2 and 4.3). REM sleep reduction of on average about 30% is reported in 79%–85% of the studies after single- or multiple-dose administration of SSRIs (Rijnbeek et al., 2003), and P300-evoked potential increases are found in 70% of studies with methylphenidate in

ADHD (Schrier et al., 2015). Polysomnography (PSG), evoked potentials (EP), and EEG often demonstrate drug effects, although it is technically not always easy to identify concentration-effect relationships for these effects. Modern pharmacological EEGs consist of a large number of leads, which generate dozens of signals with different frequencies, time series, and networks, from which numerous parameters can be extracted. PSGs produce long recordings of sleep phases, with complex and shifting patterns during normal sleep and placebo conditions. Sometimes a relatively straightforward pharmacological parameter can be extracted from these complex neurophysiological registrations, such as REM sleep reduction for serotonergic activity (Rijnbeek et al., 2003) or EEG beta power for GABA-A receptor activation (Danhof & Mandema, 1992). Increasingly, however, "big data sets" are a challenge for PK/PD modeling, and innovative analytic strategies are required, to fully understand the complicated pharmacological interactions of drugs with sleep stages (Karlsson et al., 2000) and neuronal networks (Bewernitz & Derendorf, 2012) and to use these models for dose predictions. Such limitations also apply to the use of functional resting state MRI in drug development (Khalili-Mahani et al., 2017).

Neuroendocrine systems are highly integrated with neurobehavioral functions that are related to stress, feeding, sleep, sex, and many other physiological and homeostatic processes. Perhaps, this explains why neuroendocrine hormones generally respond quite sensitively to many current CNS-active drugs, which often affect neurotransmitter systems that also play a dual role in the central and autonomic nervous system. In the literature reviews the neuroendocrine sensitivity to drugs is demonstrated by the reliable responses of prolactin (for antipsychotics, dopamine) and oxytocin (for MDMA, serotonin). Cortisol was not measured very frequently in these reviews, but this parameter is also very sensitive to serotonergic compounds, as argued in Section III.F.6. Other hormones have not been investigated as frequently. However, the time and concentration dependence of hormonal drug effects are difficult to establish. Endocrine systems are characterized by pulsatile release patterns, diurnal variations, and feedback regulations, and the interactions of drugs with these regulatory systems are highly complex (e.g., Stevens et al., 2012). Nonetheless, neuroendocrine hormones can provide important insights into the specific pharmacological effects of CNS-active drugs, as will be illustrated for the relations between prolactin and D2 receptor antagonists in Section IV.C.

B Differentiation of Pharmacological Characteristics by Profiling of CNS Effect

Many older CNS-active drugs have an established mechanism of action, but they are relatively nonselective.

Nonselective drugs bind not only to a primary pharmacological target that is responsible for the therapeutic effect (for instance, serotonin reuptake receptors for antidepressants) but also to secondary targets that are responsible for many of the side effects (e.g., memory impairment caused by older tricyclic antidepressants that bind to cholinergic receptors). An important way of improving the efficacy and tolerability of a drug class is by pharmacochemical modification of their binding affinities and intrinsic efficacies on receptors (or other targets). This has led to the introduction of many new drugs in most areas of medicine. Many neurotransmitter receptors consist of families of different subtypes, and these may be linked with distinct CNS functions and pathologies. Based on new insights into the three-dimensional structure of receptors, compounds can now be designed that specifically affect or bind to a specific subtype of a receptor and hence selectively affect the associated function. This could result in a higher therapeutic potency and an improved safety profile. Promising experimental pipelines include subtype-selective muscarinic receptor agonists (M1, M4, and others) for different cognitive disorders (Kruse et al., 2014), nicotinic agonists (α4 and α7) for pain and a variety of other CNS disorders (Hurst, Rollema, & Bertrand, 2013), and dopamine antagonists (D3) for addiction and other compulsive disorders (Merlo Pich & Collo, 2015). There is an ongoing search for functional CNS measurements that reliably distinguish between different subtype-selective compounds. Currently, there is more support for the pharmacological relevance of tests that are associated with the activity of subtypes of GABA-A receptors. Clinical examples of drugs that act on GABA-A receptor subtypes are nonbenzodiazepine z-hypnotics, which bind with high affinity to the α1-subtype (α1-subunit containing) of the GABA-A receptor family. This is intended to offer improved sleep induction with reduced side effects, compared with benzodiazepine hypnotics, that are nonselective full agonists (positive allosteric modulators) at most GABA-A receptors. More recently a number of α2,3 subtype-selective GABA-A partial agonists have been developed for the treatment of anxiety or pain. These compounds have a limited α1-efficacy and are hence expected to be less sedative than benzodiazepines. As shown in Table 4.2, benzodiazepines reliably affect a number of different functional domains and test clusters. This reflects their nonselective impact on widely distributed GABA-A receptor subtypes. As discussed in Section III.F, the most sensitive methods for nonselective benzodiazepines include saccadic peak velocity, VAS of alertness, body sway, and adaptive tracking. This raises the possibility that these anatomically and functionally diverse effects are related to distinct GABA-A receptor subtypes. A series of studies with different subtype-selective and nonselective GABA-A-ergic compounds were performed, to examine their effects on these distinct functional CNS measurements

(Chen, De Haas, De Kam, and Van Gerven, 2012; Chen et al., 2014, 2015; De Haas et al., 2007, 2008, 2009, 2010; Zuiker et al., 2016). The nonselective full GABA-A agonist lorazepam, a benzodiazepine that acts on the same receptor site as the subtype-selective partial agonists, was used as a positive control in most of these experiments. Because the effects of the drugs were influenced by different doses and varying degrees of receptor agonism, all CNS effects were related to saccadic peak velocity (SPV). This parameter is closely associated with anxiolytic effects and GABA-A receptor activity, as will be discussed in more detail in Section IV.C. Table 4.5 provides an overview of the effect profiles of different nonselective and subtype-selective GABA-A-ergic compounds. To allow a simple comparison, all results are expressed as percentages of the effects of lorazepam. The effect profiles of zolpidem and alprazolam did not differ significantly from that of lorazepam (which by definition had effects of 100% for all CNS tests). Zolpidem may be slightly more sedative (121% alertness reduction) and alprazolam somewhat less (58%), which may be in line with their therapeutic indications (hypnotic and anxiolytic). Compared with lorazepam, most α2,3 subtype-selective GABA-A partial agonists had limited impacts on alertness, postural stability, and eye-hand coordination (relative to their effects on SPV). Some of the variability may be related to pharmacological differences, as will be argued in the next section. These results illustrate how drug-sensitive functional CNS measurements can be used to characterize the pharmacological properties of novel CNS-active compounds in early drug development. However, they provide only limited information about the underlying pharmacological mechanisms of drug-responsive CNS measurements. Particularly for novel compounds, it may be important to provide evidence that the purported new mechanism of the drug also occurs in

humans. This requires a method with an established link to the specific pharmacological mode of action. This has only been established for only a few functional CNS measurements, which will be presented in Section IV.C.

C Associations Between Drug Effects and Receptor Binding

The validity of a specific CNS test as a pharmacological effect measure can be corroborated, by comparing the test responses with the potencies of drugs from the same pharmacological class. This approach was followed for saccadic peak velocity, because this was closely associated with GABA-A activation. SPV showed consistent reductions in 80% of the benzodiazepine studies in the literature (De Visser et al., 2003). For all benzodiazepines where this has been studied, SPV reductions can be readily related to plasma concentrations by PK/PD modeling. PK/PD models for SPV have been successfully used to predict therapeutic doses for novel benzodiazepines (Dingemanse et al., 1997; Van Gerven et al., 1997), including a rapid onset of action and the short effect duration that limited their clinical utility (Tang, Wang, White, Gold, & Gold, 1999). Fig. 4.3 provides an illustration of the time profiles and concentration-effect relationships for these one of these studies. The relationships between saccadic peak velocity reductions and intrinsic receptor efficacies for some of the GABA-A receptor agonists are presented in Table 4.5. To account for dose differences, all pharmacodynamic effects were expressed relative to the SPV effects of lorazepam (as described in Chen et al., 2012). Although in vitro subtype receptor efficacies were determined with similar laboratory methods, they all came from various laboratory sources, and direct quantitative comparisons were difficult. Therefore α2 and α3 efficacies were normalized to the results for α1

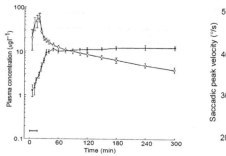 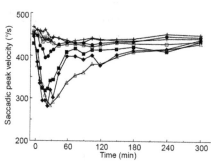 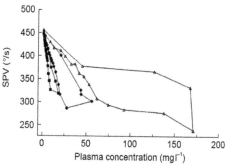

FIG. 4.3   Integration of time profiles of plasma concentrations (A) and functional CNS effects (B) into concentration-effect relationship (C) for ascending doses of new benzodiazepine Ro48-8684 (*closed symbols*) and midazolam (*open triangles*). The loops in panel C represent the prolongation of the effect, by longer persistence of the benzodiazepines in the central nervous system than in plasma. (A) Plasma concentration time profile of Ro48-8684 10-mg dose with metabolite. (B). Saccadic peak velocity over time for Ro48-8684 and midazolam. (C) Concentration-effect relationships for Ro48-8684 3 and 10 mg and midazolam. *Copied with permission from Van Gerven, J. M. A., Roncari, G., Schoemaker, R. C., Massarella, J., Keesmaat, P., Kooyman, H., et al. (1997). Integrated pharmacokinetics and pharmacodynamics of Ro 48-8684, a new benzodiazepine, in comparison to midazolam during first administration to healthy male subjects.* British Journal of Clinical Pharmacology, 44, *487–493.*

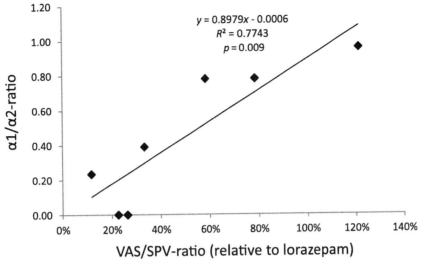

FIG. 4.4   Correlation between VAS effects and α1 selectivity. Results are normalized to account for differences in experimental conditions. For GABA-A efficacy, α1 selectivity was expressed as the ratio to α2 efficacy from the same publication. VAS effects were corrected for dose, by the ratio to the SPV effects. All results are relative to those of benzodiazepines.

efficacy from the same lab. These relative efficacy data were only available for eight GABA-A-ergic compounds (Table 4.5), but Fig. 4.4 shows that a higher selectivity for α2 or α3 (relative to α1) is associated with less effects on VAS alertness ($R^2 = 0.77$, $P = 0.009$ for α1/α2; $R^2 = 0.72$, $P = 0.016$ for α1/α3) and body sway ($R^2 = 0.41$, $P = 0.083$ for α1/α2; $R^2 = 0.71$, $P = 0.008$ for α1/α3). Further support for the applicability of saccadic peak velocity as a pharmacological effect for GABA-A receptor agonists is provided by the significant correlation of the SPV effects with the binding affinities of several benzodiazepines ($R^2 = 0.89$, and $P < 0.01$). SPV reduction was also significantly related to the therapeutic doses for treatment of anxiety (De Visser et al., 2003). These studies clearly support the close relationships between saccadic peak velocity and the activation of GABA-A receptors, in particular α2 and α3 subtypes. They also confirm the relevance of α1 activity for sedation (and the applicability of VAS as pharmacological effect measurements).

The experience with saccadic peak velocity makes it clear that extensive investigations are required for a thorough validation of pharmacological effect measures in humans. It may seem almost impossible to perform such intensive method development programs for new pharmacological targets. However, the validations were largely based on "routine" drug studies, and some form of ongoing method evaluation can be incorporated into most drug development programs. In this way a pharmacological effect measurement can be gradually validated by experience. The predictive value of a functional CNS measurement for drug development decisions often becomes apparent only after enough clinical experience has been gathered with a certain drug class. This is the case for the effect of dopamine D2 receptor antagonists on prolactin release. One of the first physiological roles

that were identified for dopamine was that it functions as a prolactin-inhibiting factor acting on the D2 receptors (Ben-Jonathan & Hnasko, 2001). There is also a well-known association between D2 binding affinity of antipsychotics and their therapeutic dose levels (Seeman & Tallerico, 1998). In healthy subjects, prolactin release is by far the most predictable effect of antipsychotic agents, with a statistically significant increase in 98% of the studies, irrespective of whether the antipsychotics are atypical or classic (although the extent of prolactin elevation differed) (De Visser et al., 2001). This strongly corroborates the use of prolactin as a measure of D2 receptor inhibition. However, the question remains whether prolactin release represents a hypothalamic or a pituitary effect. This is essential for the use of prolactin release as an indicator of activity at central D2 receptors, which is obviously more relevant for neuropsychiatric indications. The literature review identified a highly significant relationship between the therapeutic maintenance dose of registered antipsychotics and their prolactin releasing propensities (De Visser et al., 2001). This strongly suggests that prolactin release represents a central antidopaminergic effect—at least for compounds that enter the BBB. Further evidence came from the early development program of JNJ-37822681, a novel highly selective D2 antagonist that was founded on the "fast dissociation" hypothesis for atypical antipsychotics, which was mentioned in Section II (Kapur & Seeman, 2001). The compound caused a significant dose-related prolactin release in healthy subjects (Te Beek et al., 2012), and a concentration-dependent D2 receptor occupancy in a PET study (Te Beek et al., 2012). As shown in Fig. 4.2, prolactin release was closely associated with striatal D2 receptor binding in these studies (Te Beek, 2014). The pharmacodynamic results suggested a low therapeutic dose of around 5 mg, which was confirmed by significant clinical improvements in schizophrenia, with all doses in a range of 10–30 mg (Schmidt et al., 2012). Prolactin can also be used to exclude D2 inhibition for other subtype-selective dopaminergic compounds such as D3 receptor antagonists (Te Beek et al., 2012) or investigate indirect modulators of dopaminergic systems (Fitzgerald & Dinan, 2008; Te Beek et al., 2015).

V CONCLUSIONS

The development of new drugs for neurobehavioral disorders is complicated and often disappointing. To a large extent, this is related to the complexity of the brain and the poor understanding of the pathophysiology of CNS disorders. Even if a compound is effective in animal models, there is still a large chance that this will not be the case in a clinical trial in patients, particularly if the drug has a new mode of action. These uncertainties can only be reduced by advancements in pathogenic research and basic sciences. However, the efficacy of all drugs relies on its impact on a molecular, pharmacological target, which for neuropsychiatric disorders is in the CNS. The demonstration and optimization of drug-target interactions can largely be established in early drug development. Development programs that disregard these important steps have a high chance of failure in clinical trials (Morgan et al., 2012). One of the reasons why many early drug development programs ignore the demonstration of CNS effects in healthy subjects is the uncertainty about the interpretation of results. There is a huge variety of functional CNS measurements, many of which have shown irreproducible or incidental effects. Therefore it is important to select useful pharmacological effect measures, based on a number of criteria that are related to the principles of drug-target interaction. The most important requirement is that the test should be able to follow the changing drug effects over time. This means that a functional CNS measurement should be short and repeatable and have enough versions without unpredictable learning or carryover effects. In this way the associations can be established with increasing and decreasing drug concentrations in plasma and brain tissue. A good understanding of concentration-effect relationship is important for dose predictions during development. A dose- and concentration-related CNS effect provides clear proof that the effect is a consequence of interaction with the pharmacological target. As shown in the chapter, this can even be the case for CNS measurements and subjective assessments where the functional relationship between the target and the effect is poorly understood. By itself a drug-related effect does not indicate which target is involved. For compounds that are pharmacologically highly selective, all concentration-related effects

usually represent the intended mode of action, which should also be relevant for the drug's putative therapeutic efficacy. For less specific drugs the effect can also represent a secondary pharmacological target that is involved in adverse effects. A second important characteristic of a pharmacological effect measure in drug development is that it can be related to a certain target. Full validation of a pharmacological parameter generally requires extensive basic scientific and experimental evidence, which is rarely obtained for neurobehavioral effects. In this chapter, this was only identified for saccadic peak velocity related to GABA-A receptor agonism and for prolactin release with D2 activity. For new compounds the test that best represents the mode of action is often still uncertain. In such cases the drug can often be examined with a battery of different drug-sensitive tests, in comparison with the effect profile of a well-known positive control (Cohen et al., 2015; Groeneveld et al., 2016). Functional CNS measurements are essential tools to answer some of the most important questions regarding new drugs, and their rational selection and proper implementation must be part of every early development program for CNS-active drugs.

References

Alvarez-Jimenez, R., Groeneveld, G. J., van Gerven, J. M., Goulooze, S. C., Baakman, A. C., Hay, J. L., et al. (2016). Model-based exposure-response analysis to quantify age related differences in the response to scopolamine in healthy subjects. *British Journal of Clinical Pharmacology, 82*(4), 1011–1021.

Baakman, A. C., Zuiker, R., van Gerven, J. M. A., Gross, N., Yang, R., Fetell, M., et al. (2019). Central nervous system effects of the histamine-3 receptor antagonist CEP-26401, in comparison with modafinil and donepezil, after a single dose in a cross-over study in healthy volunteers. *British Journal of Clinical Pharmacology, 85*(5), 970–985.

Baker, R. H., Clanton, R. L., Rogers, J. C., & De Brito, S. A. (2015). Neuroimaging findings in disruptive behavior disorders. *CNS Spectrums, 20*(4), 369–381.

Barkley, R. A., & Cox, D. (2007). A review of driving risks and impairments associated with attention-deficit/hyperactivity disorder and the effects of stimulant medication on driving performance. *Journal of Safety Research, 38*(1), 113–128.

Ben-Jonathan, N., & Hnasko, R. (2001). Dopamine as a prolactin (PRL) inhibitor. *Endocrine Reviews, 22*(6), 724–763.

Bewernitz, M., & Derendorf, H. (2012). Electroencephalogram-based pharmacodynamic measures: a review. *International Journal of Clinical Pharmacology and Therapeutics, 50*(3), 162–184.

Bond, A., & Lader, M. (1976). Self-concepts in anxiety states. *The British Journal of Medical Psychology, 49*, 275–279.

Bowdle, T. A., Radant, A. D., Cowley, D. S., Kharasch, E. D., Strassman, R. J., & Roy-Byrne, P. P. (1998). Psychedelic effects of ketamine in healthy volunteers: relationship to steady-state plasma concentrations. *Anesthesiology, 88*(1), 82–88.

Cawello, W., & Antonucci, T. (1997). The correlation between pharmacodynamics and pharmacokinetics: basics of pharmacokinetics-pharmacodynamics modeling. *Journal of Clinical Pharmacology, 37*(1 Suppl), 65S–69S.

Chen, X., De Haas, S., De Kam, M., & Van Gerven, J. (2012). An overview of the CNS-pharmacodynamic profiles of non-selective and selective GABA agonists. *Advances in Pharmacological Sciences*, *2012*, 134523.

Chen, X., Jacobs, G., de Kam, M., Jaeger, J., Lappalainen, J., Maruff, P., et al. (2014). The central nervous system effects of the partial GABA-Aα2,3-selective receptor modulator AZD7325 in comparison with lorazepam in healthy males. *British Journal of Clinical Pharmacology*, *78*(6), 1298–1314.

Chen, X., Jacobs, G., de Kam, M. L., Jaeger, J., Lappalainen, J., Maruff, P., et al. (2015). AZD6280, a novel partial γ-aminobutyric acid A receptor modulator, demonstrates a pharmacodynamically selective effect profile in healthy male volunteers. *Journal of Clinical Psychopharmacology*, *35*(1), 22–33.

Cohen, A. F. (2010). Developing drug prototypes: pharmacology replaces safety and tolerability? *Nature Reviews. Drug Discovery*, *9*, 856–865.

Cohen, A. F., Burggraaf, J., van Gerven, J. M., Moerland, M., & Groeneveld, G. J. (2015). The use of biomarkers in human pharmacology (Phase I) studies. *Annual Review of Pharmacology and Toxicology*, *55*, 55–74.

Danhof, M., & Mandema, J. W. (1992). Modelling of the pharmacodynamics and pharmacodynamic interactions of CNS active drugs. *International Journal of Clinical Pharmacology, Therapy, and Toxicology*, *30*(11), 516–519.

De Haas, S. L., De Visser, S. J., Van der Post, J. P., De Smet, M., Schoemaker, R. C., Rijnbeek, B., et al. (2007). Pharmacodynamic and pharmacokinetic effects of TPA023, a GABAA α2,3 subtype-selective agonist, compared to lorazepam and placebo in healthy volunteers. *Journal of Psychopharmacology*, *21*, 374–383.

De Haas, S. L., De Visser, S. J., Van der Post, J. P., Schoemaker, R. C., Van Dyck, K., Murphy, M. G., et al. (2008). Pharmacodynamic and pharmacokinetic effects of MK-0343, a GABA-A α2,3 subtype selective agonist, compared to lorazepam and placebo in healthy volunteers. *Journal of Psychopharmacology*, *22*, 24–32.

De Haas, S. L., Franson, K. L., Schmitt, J. A. J., Cohen, A. F., Fau, J. -B., Dubruc, C., et al. (2009). The pharmacokinetic and pharmacodynamic effects of SL65.1498, a GABA-A α2,3 selective agonist, in comparison to with lorazepam in healthy volunteers. *Journal of Psychopharmacology*, *23*, 625–632.

De Haas, S. L., Schoemaker, R. C., van Gerven, J. M., Hoever, P., Cohen, A. F., & Dingemanse, J. (2010). Pharmacokinetics, pharmacodynamics and the pharmacokinetic/pharmacodynamic relationship of zolpidem in healthy subjects. *Journal of Psychopharmacology*, *24*(11), 1619–1629.

De Visser, S. J., Van der Post, J., Pieters, M. S. M., Cohen, A. F., & Van Gerven, J. M. A. (2001). Biomarkers for the effects of antipsychotic drugs in healthy volunteers. *British Journal of Clinical Pharmacology*, *51*, 119–132.

De Visser, S. J., Van der Post, J. P., De Waal, P. P., Cornet, F., Cohen, A. F., & Van Gerven, J. M. A. (2003). Biomarkers for the effects of benzodiazepines in healthy volunteers. *British Journal of Clinical Pharmacology*, *55*, 39–50.

Dingemanse, J., Van Gerven, J. M. A., Schoemaker, H. C., Roncari, G., Oberyé, J. J., Van Oostenbruggen, M. F., et al. (1997). Integrated pharmacokinetics and pharmacodynamics of Ro 48-6791, a new benzodiazepine, in comparison to midazolam during first administration to healthy male subjects. *British Journal of Clinical Pharmacology*, *44*, 477–486.

Drake, C., Gumenyuk, V., Roth, T., & Howard, R. (2014). Effects of armodafinil on simulated driving and alertness in shift work disorder. *Sleep*, *37*(12), 1987–1994.

Duff, E. P., Vennart, W., Wise, R. G., Howard, M. A., Harris, R. E., Lee, M., et al. (2015). Learning to identify CNS drug action and efficacy using multistudy fMRI data using multistudy fMRI data. *Science Translational Medicine*, *7*(274), 274ra16.

Dumont, G. J. H., De Visser, S. J., Cohen, A. F., & Van Gerven, J. M. A. (2005). Biomarkers for the effects of selective serotonin reuptake inhibitors (SSRIs) in healthy volunteers. *British Journal of Clinical Pharmacology*, *59*, 495–510.

Dumont, G. J. M., & Verkes, R. J. (2006). A review of acute effects of 3,4-methylenedioxymethamphetamine in healthy volunteers. *Journal of Psychopharmacology*, *20*, 176–187.

Dumont, G. J. H., Sweep, F. C. G. J., Van der Steen, R., Hermsen, R., Donders, A. R. T., Touw, D. J., et al. (2009). Increased oxytocin concentrations and prosocial feelings in humans after ecstasy (3,4-methylenedioxymethamphetamine) administration. *Social Neuroscience*, *4*, 366.

Fitzgerald, P., & Dinan, T. G. (2008). Prolactin and dopamine: what is the connection? A review article. *Journal of Psychopharmacology*, *22*(2 Suppl), 12–19.

Food and Drug Administration. (2015). *Evaluating drug effects on the ability to operate a motor vehicle*. Draft Guidance for the Industry.

George, N. (2016). A neuroimaging point of view on the diversity of social cognition: evidence for extended influence of experience- and emotion-related factors on face processing. *Cult Brain*, *4*(2), 147–158.

Gijsman, H. J., Van Gerven, J. M. A., Verkes, R. J., Schoemaker, R. C., Pieters, M. S. M., Pennings, E. J. M., et al. (2002). Further evidence of saccadic eye movements as end-point for serotonergic challenge tests. *Human Psychopharmacology: Clinical and Experimental*, *17*, 83–89.

Greenblatt, D. J., Allen, M. D., Locniskar, A., Harmatz, J. S., & Shader, R. I. (1979). Lorazepam kinetics in the elderly. *Clinical Pharmacology and Therapeutics*, *26*(1), 103–113.

Groeneveld, G. J., Hay, J. L., & Van Gerven, J. M. (2016). Measuring blood-brain barrier penetration using the NeuroCart, a CNS test battery. *Drug Discovery Today: Technologies*, *20*, 27–34.

Guan, Z., Klumpers, L. E., Oyetayo, O. O., Heuberger, J., van Gerven, J. M., & Stevens, J. (2016). Pharmacokinetic/pharmaco-dynamic modelling and simulation of the effects of different cannabinoid receptor type 1 antagonists on Δ(9)-tetrahydrocannabinol challenge tests. *British Journal of Clinical Pharmacology*, *81*(4), 713–723.

Harmer, C. J., Duman, R. S., & Cowen, P. J. (2017). How do antidepressants work? New perspectives for refining future treatment approaches. *Lancet Psychiatry*, *S2215-0366*(17), 30015–30019.

Hurst, R., Rollema, H., & Bertrand, D. (2013). Nicotinic acetylcholine receptors: from basic science to therapeutics. *Pharmacology & Therapeutics*, *137*(1), 22–54.

Jacobs, G. E., Kamerling, I. M., de Kam, M. L., Derijk, R. H., van Pelt, J., Zitman, F. G., et al. (2008). Enhanced tolerability of the 5-hydroxytryptophane challenge test combined with granisetron. *Journal of Psychopharmacology*, *24*(1), 65–72.

Jongen, S., Vuurman, E. F., Ramaekers, J. G., & Vermeeren, A. (2016). The sensitivity of laboratory tests assessing driving related skills to dose-related impairment of alcohol: a literature review. *Accident; Analysis and Prevention*, *89*, 31–48.

Kapur, S., & Seeman, P. (2001). Does fast dissociation from the dopamine D2 receptor explain the action of atypical antipsychotics? A new hypothesis. *The American Journal of Psychiatry*, *158*, 360–369.

Karlsson, M. O., Schoemaker, R. C., Kemp, B., Cohen, A. F., Van Gerven, J. M. A., Tuk, B., et al. (2000). A pharmacodynamic Markov mixed effect model for temazepam's effect on sleep. *Clinical Pharmacology and Therapeutics*, *68*, 175–188.

Khalili-Mahani, N., Rombouts, S. A., van Osch, M. J., Duff, E. P., Carbonell, F., Nickerson, L. D., et al. (2017). Biomarkers, designs, and interpretations of resting-state fMRI in translational pharmacological research: a review of state-of-the-Art, challenges, and opportunities for studying brain chemistry. *Human Brain Mapping*, *38*, 2276–2325.

Klaassens, B. L., Van Gorsel, H. C., Khalili-Mahani, N., Van der Grond, J., Wyman, B. T., Whitcher, B., et al. (2015). Single-dose serotonergic stimulation shows widespread effects on functional brain connectivity. *NeuroImage*, *122*, 440–450.

Klaassens, B. L., Rombouts, S. A., Winkler, A. M., van Gorsel, H. C., van der Grond, J., & van Gerven, J. M. (2017). Time related effects on functional brain connectivity after serotonergic and cholinergic neuromodulation. *Human Brain Mapping*, 38(1), 308–325.

Klumpers, L. E., Roy, C., Ferron, G., Turpault, S., Poitiers, F., Pinquier, J. -L., et al. (2013). Surinabant, a selective CB1 antagonist, inhibits THC-induced central nervous system and heart rate effects in humans. *British Journal of Clinical Pharmacology*, 76, 65–77.

Kruse, A. C., Kobilka, B. K., Gautam, D., Sexton, P. M., Christopoulos, A., & Wess, J. (2014). Muscarinic acetylcholine receptors: novel opportunities for drug development. *Nature Reviews Drug Discovery*, 13(7), 549–560.

Laude, J. R., & Fillmore, M. T. (2015). Simulated driving performance under alcohol: effects on driver-risk versus driver-skill. *Drug and Alcohol Dependence*, 154, 271–277.

Lezak, H. D. L. D. (2004). *Neuropsychological assessment* (4th ed.). New York: Oxford University Press.

Liem-Moolenaar, M., de Boer, P., Timmers, M., Schoemaker, R. C., van Hasselt, J. G., Schmidt, S., et al. (2011). Pharmacokinetic-pharmacodynamic relationships of central nervous system effects of scopolamine in healthy subjects. *British Journal of Clinical Pharmacology*, 71(6), 886–898.

MacIndoe, J. H., & Turkington, R. W. (1973). Stimulation of human prolactin secretion by intravenous infusion of l-tryptophan. *The Journal of Clinical Investigation*, 52(8), 1972–1978.

Merlo Pich, E., & Collo, G. (2015). Pharmacological targeting of dopamine D3 receptors: possible clinical applications of selective drugs. *European Neuropsychopharmacology*, 25(9), 1437–1447.

Morgan, P., Van Der Graaf, P. H., Arrowsmith, J., Feltner, D. E., Drummond, K. S., Wegner, C. D., et al. (2012). Can the flow of medicines be improved? Fundamental pharmacokinetic and pharmacological principles toward improving phase II survival. *Drug Discovery Today*, 17(9–10), 419–424.

Norris, H. (1971). The action of sedatives on brain stem oculomotor systems in man. *Neuropharmacology*, 10, 181–191.

Outhred, T., Hawkshead, B. E., Wager, T. D., Das, P., Malhi, G. S., & Kemp, A. H. (2013). Acute neural effects of selective serotonin reuptake inhibitors versus noradrenaline reuptake inhibitors on emotion processing: implications for differential treatment efficacy. *Neuroscience and Biobehavioral Reviews*, 37(8), 1786–1800.

Pasma, J. H., Engelhart, D., Maier, A. B., Aarts, R. G., Van Gerven, J. M., Arendzen, J. H., et al. (2016). Reliability of system identification techniques to assess standing balance in healthy elderly. *PLoS One*, 11(3), e0151012.

Pietrzak, R. H., Mollica, C. M., Maruff, P., & Snyder, P. J. (2006). Cognitive effects of immediate-release methylphenidate in children with attention-deficit/hyperactivity disorder. *Neuroscience and Biobehavioral Reviews*, 30, 1225–1245.

Pringle, A., & Harmer, C. J. (2015). The effects of drugs on human models of emotional processing: an account of antidepressant drug treatment. *Dialogues in Clinical Neuroscience*, 17(4), 477–487.

Rijnbeek, B., de Visser, S. J., Franson, K. L., Cohen, A. F., & Van Gerven, J. M. A. (2003). REM sleep reduction as a biomarker for the effects of antidepressants in healthy volunteers. *Journal of Psychopharmacology*, 17, 196–203.

Sachdev, P. S., Blacker, D., Blazer, D. G., Ganguli, M., Jeste, D. V., Paulsen, J. S., et al. (2014). Classifying neurocognitive disorders: the DSM-5 approach. *Nature Reviews. Neurology*, 10, 634–642.

Schoemaker, R. C., van Gerven, J. M., & Cohen, A. F. (1998). Estimating potency for the Emax-model without attaining maximal effects. *Journal of Pharmacokinetics and Biopharmaceutics*, 26(5), 581–593.

Schrier, L., Brill, M., Burggraaf, J., Cohen, A. F., & Van Gerven, J. M. A. (2015). Chapter 3. Biomarkers of acute methylphenidate effect in children and adolescents with attention-deficit/hyperactivity disorder. In L. Schrier (Ed.), *Non-invasive monitoring of pharmacokinetics and pharmacodynamics for pharmacological drug profiling in children and adolescents* (pp. 45–74): Leiden University. Thesis.

Seeman, P., & Tallerico, T. (1998). Antipsychotic drugs which elicit little or no parkinsonism bind more loosely than dopamine to brain D2 receptors, yet occupy high levels of these receptors. *Molecular Psychiatry*, 3(2), 123–134.

Schmidt, M. E., Kent, J. M., Daly, E., Janssens, L., Van Osselaer, N., Hüsken, G., et al. (2012). A double-blind, randomized, placebo-controlled study with JNJ-37822681, a novel, highly selective, fast dissociating D2 receptor antagonist in the treatment of acute exacerbation of schizophrenia. *European Neuropsychopharmacology*, 22(10), 721–733.

Smarius, L. J. C. A., Jacobs, G. E., Hoeberechts-Lefrandt, D. H. M., Kemme, M., De Kam, M. L., Van der Post, J. P., et al. (2008). Pharmacology of rising oral doses of 5-hydroxytryptophan with carbidopa. *Journal of Psychopharmacology*, 22, 426–433.

Stevens, J., Ploeger, B. A., Hammarlund-Udenaes, M., Osswald, G., van der Graaf, P. H., Danhof, M., et al. (2012). Mechanism-based PK-PD model for the prolactin biological system response following an acute dopamine inhibition challenge: quantitative extrapolation to humans. *Journal of Pharmacokinetics and Pharmacodynamics*, 39(5), 463–477.

Strauss, E., Sherman, E. M. S., & Spreen, O. (1998). *A compendium of neuropsychological tests; administration, norms and commentary* (3rd ed). New York: Oxford University Press, Inc.

Strougo, A., Zuurman, L., Roy, C., Pinquier, J. -L., Cohen, A. F., Van Gerven, J. M. A., et al. (2008). Modelling of the concentration-effect relationship of THC on central nervous system parameters and heart rate—insight into its mechanisms of action and a tool for clinical research and development of cannabinoids. *Journal of Psychopharmacology*, 22, 717–726.

Tang, J., Wang, B., White, P. F., Gold, M., & Gold, J. (1999). Comparison of the sedation and recovery profiles of Ro 48-6791, a new benzodiazepine, and midazolam in combination with meperidine for outpatient endoscopic procedures. *Anesthesia and Analgesia*, 89(4), 893–898.

Tannenbaum, C., Paquette, A., Hilmer, S., Holroyd-Leduc, J., & Carnahan, R. (2012). A systematic review of amnestic and non-amnestic mild cognitive impairment induced by anticholinergic, antihistamine, GABAergic and opioid drugs. *Drugs & Aging*, 29(8), 639–658.

Te Beek, E. T., Moerland, M., De Boer, P., Van Nueten, L., De Kam, M. L., Burggraaf, J., et al. (2012). Pharmacokinetics and central nervous system effects of the novel dopamine D2 receptor antagonist JNJ-37822681. *Journal of Psychopharmacology*, 26, 1119–1127.

Te Beek, E. T., De Boer, P., Moerland, M., Schmidt, M. E., Hoetjes, N. J., Windhorst, A. D., et al. (2012). In vivo quantification of striatal dopamine D2 receptor occupancy by JNJ-37822681 using [11C] raclopride and positron emission tomography. *Journal of Psychopharmacology*, 26, 1128–1135.

Te Beek, E. T., Zoethout, R. W. M., Bani, M. S. G., Andorn, A., Iavarone, L., Klaassen, E. S., et al. (2012). Pharmacokinetics and central nervous system effects of the novel dopamine D3 receptor antagonist GSK598809 and intravenous alcohol infusion at pseudo-steady state. *Journal of Psychopharmacology*, 26, 304–315.

Te Beek, E. T. (2014). Chapter 8. Discussion. In E. T. Te Beek (Ed.), *Neuropharmacology of novel dopamine modulators* (pp. 145–158): Leiden University. Thesis.

Te Beek, E. T., Chen, X., Jacobs, G. E., Nahon, K. J., De Kam, M. L., Lappalainen, J., et al. (2015). The effects of the nonselective benzodiazepine lorazepam and the α2/α3 subunit-selective GABAA receptor modulators AZD7325 and AZD6280 on plasma prolactin levels. *Clinical Pharmacology Drug Development*, 4, 49–154.

Tonstad, S., & Aubin, H. J. (2012). Efficacy of a dose range of surinabant, a cannabinoid receptor blocker, for smoking cessation: a randomized controlled clinical trial. *Journal of Psychopharmacology*, 26(7), 1003–1009.

Thapa, P. B., Gideon, P., Fought, R. L., & Ray, W. A. (1995). Psychotropic drugs and risk of recurrent falls in ambulatory nursing home residents. *American Journal of Epidemiology, 142*, 202–211.

Van der Hooft, C. S., Schoofs, M. W., Ziere, G., Hofman, A., Pols, H. A., Sturkenboom, M. C., et al. (2008). Inappropriate benzodiazepine use in older adults and the risk of fracture. *British Journal of Clinical Pharmacology, 66*(2), 276–282.

Van Gerven, J. M. A., Roncari, G., Schoemaker, R. C., Massarella, J., Keesmaat, P., Kooyman, H., et al. (1997). Integrated pharmacokinetics and pharmacodynamics of Ro 48-8684, a new benzodiazepine, in comparison to midazolam during first administration to healthy male subjects. *British Journal of Clinical Pharmacology, 44*, 487–493.

Van Steveninck, A. L., Van Berckel, B. N. M., Schoemaker, R. C., Breimer, D. D., Van Gerven, J. M. A., & Cohen, A. F. (1999). The sensitivity of pharmacodynamic tests for central nervous system effects of drugs on the effects of sleep deprivation. *Journal of Psychopharmacology, 3*, 10–17.

Verster, J. C., & Mets, M. A. (2009). Psychoactive medication and traffic safety. *International Journal of Environmental Research and Public Health, 6*(3), 1041–1054.

Wandschneider, B., & Koepp, M. J. (2016). Pharmaco fMRI: determining the functional anatomy of the effects of medication. *Neuroimage Clinical, 12*, 691–697.

Wilhelmus, M. M., Hay, J. L., Zuiker, R. G., Okkerse, P., Perdrieu, C., Sauser, J., et al. (2017). Effects of a single, oral 60 mg caffeine dose on attention in healthy adult subjects. *Journal of Psychopharmacology, 31*(2), 222–232.

Yasuda, C. L., Centeno, M., Vollmar, C., Stretton, J., Symms, M., Cendes, F., et al. (2013). The effect of topiramate on cognitive fMRI. *Epilepsy Research, 105*(1–2), 250–255.

Zoethout, R. W. M., Delgado, W. L., Ippel, A. E., Dahan, A., & Van Gerven, J. M. A. (2011). Functional biomarkers for the acute effects of alcohol on the central nervous system in healthy volunteers. *British Journal of Clinical Pharmacology, 71*, 331–350.

Zuiker, R., Chen, X., Østerberg, O., Mirza, N., Muglia, P., De Kam, M., et al. (2016). NS11821, an α2,3 subtype-specific GABAA agonist, elicits selective effects on the central nervous system in randomized controlled trial with healthy subjects. *Journal of Psychopharmacology, 30*(3), 253–262.

Zuurman, L., Ippel, A. E., Moin, E., & Van Gerven, J. M. A. (2009). Biomarkers for the effects of cannabis and THC in healthy volunteers. *British Journal of Clinical Pharmacology, 67*, 5–21.

5

Experimental Medicine Approaches in CNS Drug Development

Jeffrey Paul

Drexel University, Graduate School of Biomedical Sciences, College of Medicine, Philadelphia, PA, United States

I INTRODUCTION AND DEFINITION OF EXPERIMENTAL MEDICINE

The definition of experimental medicine (EM), according to the Medical Research Council of the United Kingdom (MRC) (http://www.mrc.ac.uk/research/initiatives/experimental-medicine/), is, "Investigation undertaken in humans, relating where appropriate to model systems, to identify mechanisms of pathophysiology or disease, or to demonstrate proof-of-concept evidence of the validity and importance of new discoveries or treatments." The key term is "model systems." This approach is to differentiate from clinical trials where the purpose is to determine the safety and efficacy of a new drug. There are a number of characteristics of EM:

1. Model system: EM uses experimental human models to gain knowledge about certain attributes of the drug. Experimental models do not substitute for clinical trials whose objective is to determine safety and efficacy in patients with disease. In the context of drug development, the outcome of the experimental model can be then used to make decisions on dose, segments of patients who may benefit, and the disease indications where the drug may have utility. The experimental medicine study can be used to make decisions on whether the drug project or the target remains valid. To use EM studies for decision-making, statistical criteria can be employed. As in any model, one must have knowledge of its validity and positive and negative predictive values for how the drug will perform in the intended patient population to use the results for decision-making. Human EM models can be used as a translational tool to inform on the biological relevance of new drug targets, biological processes, and biomarkers.

2. EM is a subset of translation medicine. The translation medicine concept is formed around the concept that a laboratory discovery can be directly brought to patients for their benefit. In the translation medicine concept, the term "bench-to-bedside" refers to the direct transfer of a drug target modulation in an in vitro or in vivo experimental model to improving the disease in patients. Thus, the term "bench-to-bedside" assumes that the finding in the drug experiment in the laboratory will "translate" to patients, including the dose/regimen and choice of patient segments (Wehling, 2009). EM human studies impose a middle step in the drug development process. Instead of immediately testing the drug's efficacy in a Phase 2 proof-of-concept study (POC) following the laboratory discovery, a human experimental model study is conducted to optimize the conditions of the future POC study and improve the probability of a successful trial (Littman & Williams, 2005).

3. An important difference between an EM study and safety/efficacy clinical trial is the endpoints. In the former the endpoint is a biomarker, usually pharmacological response(s), whereas, in the latter, the endpoint is a validated measure showing how a patient feels, functions, or survives or a validated outcome surrogate (e.g., LDL-cholesterol lowering). Additionally, there are differences in the underlying approach in the design of an experimental study versus a clinical trial.

Furthermore, there are notable differences in the selection of subjects for EM studies. EM studies are generally

Translational Medicine in CNS Drug Development, Volume 29
ISSN: 1569-7339
https://doi.org/10.1016/B978-0-12-803161-2.00005-9

nontherapeutic; that is, the subjects do not derive benefit from the treatment intervention. Therefore the selection of subjects in EM studies is not based on potential-benefits; rather, it is based on pathway sensitivity to the experimental drug and expression of response parameters. One design approach is to study a subset of patients with a known high sensitivity to the research drug, based on its mechanism of action. Alternatively, healthy volunteers can be used but only if the drug responses can be detected. Often the EM model in healthy subjects requires manipulating, stressing, or challenging the system that permits the evaluation drug response in the perturbed system. Indeed the challenged or perturbed system is designed to model specific attributes of the target disease. For example, scopolamine is administered to induce amnesia, ketamine to induce psychotic-like symptoms, and sleep deprivation to induce physiological stress/depression/mood changes. See Table 5.1 for comparison of attributes between EM study and standard clinical trial.

Why use EM approach instead of directly assessing the drug's therapeutic potential in a patient trial?

1. EM study is typically less expensive than a traditional clinical trial.
2. EM study can be conducted faster than a clinical trial.
3. EM study can be used to optimize the design of a future safety/efficacy clinical trial in the patient disease indication.
4. EM studies have smaller sample size and are conducted in very select centers; therefore exposure of drug to patients is limited and closely monitored, resulting in low risk to the patient. Also, EM studies

are conducted at a limited number of clinical sites, which increases model reliability and reduces measurement error due intersite and interrater error.
5. EM study can be used to derisk a project and increase the probability of a successful future Phase 2 clinical trial.

As the cost of drug development increases and the success rate decreases, EM provides a useful informative step to derisk a novel drug or can be used to compare a new drug with a validated target using a modeled endpoint (Littman & Williams, 2005).

For CNS drug development the EM approach is not only attractive because it informs but also controversial because the models may have limited predictive validation. Therefore human EM models should be used to answer specific scientific questions about mechanism of action and biological processes. If the EM approach is to be used for project decision-making for a novel drug target, the EM model result will carry a level of risk for a false-positive or false-negative outcome, unless the model has good predictive validation across multiple drugs of varying mechanisms of action.

CNS drug development is particularly difficult and has a lower success rate compared with other therapeutic areas. Many large pharmaceutical companies report a high attrition rate with CNS drugs leading to a downsizing or removal of CNS as an active therapeutic area for drug development. From 1999 to 2013 the average approval time for each clinical phase for CNS compounds approved for marketing in the United States was 19.3 months, 31% longer than the 14.7 months for non-CNS. During this period the average time required for review and

TABLE 5.1   Comparison Between EM and Standard Clinical Trial

	Experimental medicine	Clinical trial
Purpose	Understand underlying mechanism of action or pathophysiology. Typically to demonstrate proof of pharmacology, proof of mechanism, disease mechanism, or exposure response relationships	Answer a specific question about a drug's safety and efficacy. In Phase 2: to be used as proof of concept
Approach	Perturb or challenge a model system to assess a pharmacological response to drug; use a sensitive surrogate patient population or subset	Assess patient outcome following treatment
Size of study	Small, typically less than 50 subjects	Large, typically over 100 subjects
Endpoints	Biomarkers that have high measurement precision and accuracy due to the small sample size. Biomarkers do not require full validation, only enough data to meet the study objectives	Validated and qualified patient endpoints showing how a patient feels, functions, or survives
Number of sites	Very few trial centers, only highly qualified experts who are familiar with the experimental model, response measurements, and patient groups	Multicenter study, experts in patient management and disease assessment
Statistical hypothesis	Mostly, exploratory, hypothesis-generating. However, for a well-characterized or validated model system, inferential hypothesis testing can be used for decision-making with prospective effect-size criteria	Definitive inferential hypothesis for confirmation; statistical hypothesis clearly stated
Subjects	Can vary based on the model objectives: healthy volunteers, specific patient disease groups, phenotypes or genotypes. Subjects are chosen to optimize detection of drug response	Target patient population and may have enrichment of highly sensitive responders

approval of marketing applications (e.g., NDAs) was 12.8 months, or 18% longer than for non-CNS compounds. A closer look revealed only one in six CNS compounds received a priority review designation from the FDA, compared with nearly half of all non-CNS compounds (Kaitin, 2014).

Using the EM approach, one can derisk the inherent high risk of novel CNS drug development by taking full advantage of multiple biomarker response platforms such as omics, imaging, and electrophysiology. Additionally, new drug discovery approaches such as small-molecule GPCRs, ion channel selectivity, antibodies/immunology, and epigenetics benefit from some experimentation and optimization prior to initiating multicenter Phase 2 POC clinical studies. Additionally, new hopeful approaches in drug target identification including in disease genetics (i.e., GWAS); epigenetic disease mapping; and the use of mechanistic, diagnostic, and prognostic biomarkers all require further characterization prior to Phase 2. Although these exciting new approaches bring hope for novel drugs, the lack of understanding of disease pathophysiology and the use of these technologies contributes to the low success rate and a high unmet medical need of CNS drugs. In the succeeding texts are the challenges of CNS development, which provide a basis for using translational/EM approaches (Cook et al., 2014; Pangalos, Schechter, & Hurko, 2007; Potter, 2012).

1. There is a lack of predictive validity from preclinical models and very few validated CNS targets. Very few pathophysiological mechanisms are known for psychiatric disease; therefore drugs are generally developed to treat symptoms (McGonigle & Ruggeri, 2014). Typically, modulation of certain neurotransmitters is known to ameliorate disease symptoms. For example, it is well known that increasing synaptic serotonergic tone by serotonergic reuptake inhibition lowers the symptomatology for patients with major depression. Another example is lowering dopaminergic tone that is associated with an antipsychotic drug action in schizophrenia. For neurological diseases, where more mechanisms are understood, it allows for more opportunities to test various pathway regulatory points in a cascade.

2. The blood-brain barrier presents a high hurdle for developing drugs with an acceptable therapeutic window. Because most drug molecules are excluded by the blood-brain barrier, high doses must be administered to provide a pharmacologically active concentration in the CNS. However, a high administered dose has a higher probability of peripheral adverse events and an unacceptable or problematic safety profile. There are some examples of approved drugs where a

nasal or intrathecal administration is efficacious, which avoids peripheral side effects.

3. Assessments of patient status are typically done using subjective tests that are susceptible to a high placebo response rate. Therefore, for a CNS drug to be therapeutic, it must have a very robust efficacy. In fact the placebo effect is real and has a basis in neuroanatomy and reward behavior (Vallance, 2007). In other therapeutic areas, validated objective laboratory tests are acceptable as surrogates for a patient's disease status.

The challenges to CNS drug development are improving with better central rating technologies and more objective measures, better diagnostics to identify subpopulations, and understanding of pathophysiology. EM studies, where appropriate, can provide an effective derisking plan allowing more rational development of future therapeutics.

II EXPERIMENTAL MEDICINE IN THE CLINICAL DEVELOPMENT PLAN

Conducting EM studies in a clinical development plan is optional. There is no regulatory requirement to conduct EM studies. As part of the normal exercise of developing a clinical plan, one identifies the list of studies required, prior to initiating the Phase 2 and 3 studies. However, for a new chemical entity or a new drug target, key questions should be answered to optimize the chances of a successful program and to reduce uncertainty, called "derisking." EM can provide a path forward of stepwise derisking the project by filling knowledge gaps on the pharmacologically active dose range, exposure response relationships, and proof of pharmacology in humans. Regarding translation the purpose of EM is to provide the demonstration that the intended pharmacological target modulated in an animal model is also "carried forward" and modulated in humans in a predicted manner. Morgan et al. (2011) describes the "three pillars" required to effectively derisk a compound entering into an expensive POC study (i.e., Phase 2A) in the target population. The three pillars form the basic pharmacokinetic/pharmacodynamic understanding of the new compound, and they include proof of adequate exposure at the site of action, proof of binding to the intended target, and expression of pharmacological activity. EM studies are conducted during early clinical development to provide demonstration of fundamental PK/PD relationships, to optimize the design of the POC study, and to derisk the project, prior to initiating expensive Phase 2 studies in patients (Littman & Williams, 2005). See Fig. 5.1. An example is the use of human PET imaging to demonstrate that novel drug enters the CNS and specifically binds to the intended receptor to give a reasonable probability of

FIG. 5.1 The three assessments comprising the derisking exercise, as part of a translational plan:

1. Demonstrate that the dose/regimen in humans result in adequate exposure at the target receptor. Typically, this information is provided by pharmacokinetics and modeling. Also, be knowledgeable of blood-brain drug concentrations to use systemic exposure as a surrogate for CNS exposure.

2. Demonstrate that the drug is bound to its intended target.

3. Demonstrate that following binding, the signal is transduced, resulting in a pharmacological response. There can be a variety of response biomarkers. In this figure, Response 1, bioimaging; Response 2, serum or CSF biomarker; and Response 3, electrophysiological.

The strength of the derisking exercise relies on the ability to provide convincing information for all three assessment points. Legend: *Cp*, plasma concentration of drug; *Ce*, tissue concentration of drug; *BBB*, blood-brain barrier; *Resp*, response. Based on Morgan, P., Graaf, P., Arrowsmith, J., Feltner, D., Drummond, K., Wegner, C., et al. (2011). Can the flow of medicines be improved? Fundamental pharmacokinetic and pharmacological principles toward improving Phase II survival. Drug Discovery Today, 17(9-10), 419–24.

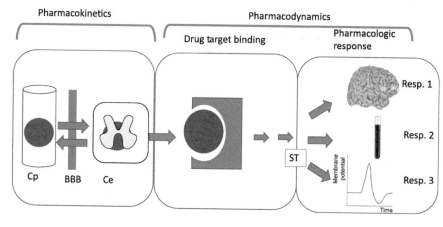

FIG. 5.2 Placement of EM studies in clinical development plan. The sequence of SAD, MAD, POC, and Phase 2B clinical studies is critical path to starting Phase 3. Methodology studies can be conducted prior to or in parallel with the SAD study. These are human studies where biomarker and procedures are developed and optimized for the EM study. The EM study is best conducted in parallel with Phase 1 and should be completed prior to the MAD study so as not to delay the decision to initiate Phase 2. Following completion of the MAD study, a decision to initiate Phase 2 POC study is based on safety, PK, and EM study results.

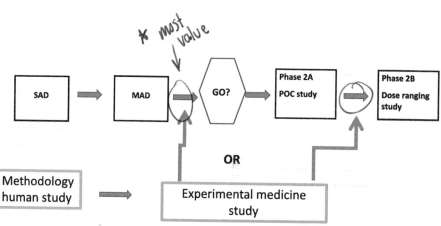

a downstream pharmacological response. PET imaging studies are typically conducted in less than 10 subjects requiring just several months, compared with the standard Phase 2 study requiring much more time and resources.

Louis Shreiner (1997) identified two distinct phases of clinical drug development, learn and confirm, in his classic paper on drug development. The Learn phase provides information on dose and regimen, potential patients who may potentially benefit (i.e., active pharmacology), potential for tolerance developing, etc. It is where EM studies provide most value as these studies will give information on drug attributes, mechanism, and potential patient segment responders (Fig. 5.2). The Confirm phase is where definitive trials are conducted

for regulatory approval. In a typical clinical development plan, the EM studies are best conducted up to and in parallel with the Phase 2 studies to derisk the project and to provide information to optimize the next study.

Alternatively, EM studies can be conducted during early Phase 2, in parallel with POC patient studies. Although this approach diminishes the value, the EM study would provide scientific support for dose selection for Phase 2B or give support to an equivocal outcome of a Phase 2A study. Additionally an EM study could be used to identify alternative disease indications. Fig. 5.2 describes how EM studies are placed within the early clinical plan.

Fig. 5.3 shows a simple exercise used for determining response endpoints to use when designing EM studies,

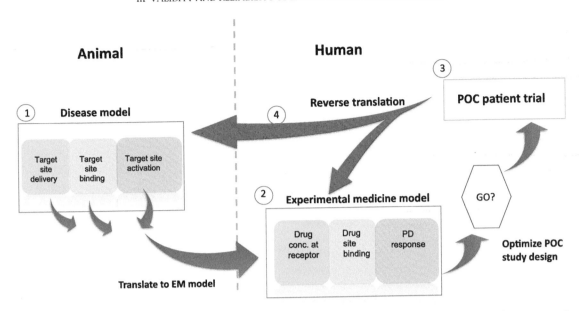

FIG. 5.3   For a schematic of translational exercise in designing an EM study. Animal models provide quantitative information on drug concentrations at target receptor, binding to receptor, and PD/biomarker responses, which may be utilized in future human studies. Human EM studies provide translation information ensuring that the human drug target is present and pharmacological responses are conserved. Derisking occurs by demonstration of adequate drug exposure at target site, receptor binding, and downstream PD responses. Human EM study results are integrated into a project decision (i.e., safety, PK, and PD) to initiate Phase 2 studies in patients. Reverse translation is when patient observations are used to modify or optimize animal and human EM models.

based on the goal for providing evidence that the modulation of the target in animal models are carried forward (i.e., translated) to humans. Alternatively, one can map a validated response in patients "backward" (i.e., reverse translation) into human EM or animal models.

III VALIDITY AND RELIABILITY OF EXPERIMENTAL MEDICINE MODELS

EM, as with animal testing of drugs, uses the model approach to gain knowledge about human disease pathology, to predict responses to new drugs, and to aid in derisking the process of drug development. As with any model, there are criteria that should be used to judge the model's value based on its reliability and validity.

The reliability of a model refers to how well the model produces stable and consistent results. Knowing and identifying the acceptable criteria for the model can greatly help reduce error and increase the reliability of the experimental model. In the case of all biomedical models, evaluation of test-retest reliability is important because it captures how closely the results are consistent over time and defines the acceptable experimental time window. The stability refers to the model's usable time window for which test compounds can be evaluated. If the experimental model protocol is to collect measurements of repeated responses before and periodically after drug administration, it is important to understand the variation in the results that is caused by the model's

reliability, quantified by test-retest reliability (Guttman, 1945). For example, in the scopolamine challenge model, scopolamine is administered to healthy volunteers to induce amnesia, and the available time window can be maximized up to 9 h, depending on the dose and route of administration of scopolamine. Additionally, instrumentation drift and noise as well as variations in technical procedures add to measurement error. CNS experimental models may have cognition or behavioral testing as a PD response. Where the PD response requires human raters, the interrater and intrarater consistency becomes relevant, especially when using neuropsychological testing (McDowell, 2006). An example of a test-retest reliability relevance is the concept of practice of the trial subjects administered cognitive tests. In a recent evaluation of a computerized cognitive testing for brain concussions in athletes, the practice effect with repeat measures was large enough to give a false result of clinical improvement. The authors concluded that the test should have a better reliability by changing the schedule of testing and incorporate practice effects into the analysis of results (Littleton, Register-Mihalik, & Guskiewicz, 2015).

A Validity

Validity refers to how well the model measures what it is intended to measure. McGonigle and Ruggeri (2014) in a review of CNS animal models, identify the challenges for validating animal models. The best known validated

CNS animal (and human) experimental models are those that are responsive to specific mechanisms for which there are approved drugs, rather than a model based on its pathophysiology. Biomedical experimental model (both human and animal) validity can be described using three subcategories:

B Construct Validity

Construct validity category refers to whether the model is based on similar biology (McGonigle & Ruggeri, 2014). The formal definition of construct validity is a test that is commonly defined as the accuracy with which the test measures what it is intended to measure (Van der Staay, Arndt, & Nordquist, 2009). In animal models, attempting to mimic human psychiatric disease and behavior, the construct validity become less clear. Belzung and Lemoine (2011) and Wilner (1990) summarize the debate of construct validation of CNS (including psychiatric) animal models. First, one should identify that there is similarity between biological dysfunctions in the clinical population and in the animal model, and, second, there should be homology between modeled processes. As many responses in behavioral studies are subjective, animal models have difficulty achieving construct validity (McGonigle & Ruggeri, 2014; Viaud-Delmon, Venault, & Chapouthier, 2010). Human EM models have an advantage over animal models for construct validity, because the EM model can primarily focus on the pathophysiology and disease etiology, unlike animal models where there is an additional need to show species homology plus similarity to human disease. With objective PD measurements, such as pharmacogenomics, electrophysiology, or brain imaging, construct validity could be achieved with reasonable correlations between animal to human experimental models. Additionally, one important purpose of human experimental models is to derisk a drug candidate prior to entering Phase 2 efficacy studies. Animal model value can be judged by comparing them with a human EM model analog, an exercise of using construct validity. Finally, regarding disease etiology, EM models are mainly used for evaluating PD responses to novel drug targets; therefore the question of disease etiology and construct validity of the model becomes less relevant.

C Face Validity

The next validity to consider is face validity, defined as the extent to which the measured variable appears to be an adequate measure of the conceptual variable (Van Selst, 2011). Researchers need to judge whether the model measurement "appears" to measure the same thing as the actual construct. In psychiatric/behavioral disease models, face validity refers how well the behavioral responses are similar to that of patients with disease. The use of animal models for guiding drug development relies heavily on the liberal interpretation of face validity. For example, the face validity of tail suspension test in rodents for evaluating antidepressant drug has been criticized because humans do not have tails [sic] (Bouwknecht, 2014). The actual response biomarker in this model is time to immobility, but this marker has poor face validity for major depression. Additionally the delay in patient response to antidepressant medication does not have good face validity and has not been well modeled (or mimicked) in animals (Wilner & Mitchell, 2006). Turning now to human experimental models, one example of an approach that has potential for meeting face validity criteria for major depression is the negative bias of emotional processing (Harmer et al., 2010; Pringle & Harmer, 2015), which can be evaluated in healthy volunteers. This model uses the cognitive neuropsychological model of antidepressant activity, resulting in a bias of processing affective information. In this model, Pringle and Harmer (2015) have shown that emotional processing in healthy volunteers may be more relevant to mood disorders than animal models, and their work has shown good face validity with various approved antidepressants, with multiple mechanisms of action.

D Predictive Validity

The third type of validity is predictive or the ability of the measurement to predict a future event. Predictive validity identifies that a similar observation will be seen in the target disease patient population as the human experimental model. Predictive validity is the most valuable for CNS drug development and the most challenging to obtain. Predictive validity in biomedical models is considered the "gold standard," since the perfect model should be insensitive to the compounds' mechanism of action (McGonigle & Ruggeri, 2014). As part of the validation process, a previously proved target should have multiple compounds tested by the model, which is then used to define a criterion, predicting the effect in patients with the disease. A "validated target" is defined as one where a safe and selective compound modulates a drug effector system resulting in therapeutic benefit to the target disease population (Mullane, Winquist, & Williams, 2014). Most often, human EM models are developed to compare compounds within a class of pharmacology against reference and lead drug candidates. Therefore researchers should qualify whether the predictive validity of the EM model is based on within a class or extends across several classes of drugs and effector systems.

Both animal models and human experimental models, unfortunately, have only a few examples of predictive validity within neuroscience. The dearth of validated

TABLE 5.2 Human Experimental Models That Have Predictive Validity

Model	Disease	Comment	Publication
PET RO of serotonin uptake transporter	Major depression	SSRIs/SNRIs need to occupy $\geq 80\%$ to have antidepressant efficacy	Meyer et al. (2004)
Sleep EEG	Major depression	SNRI/NRI increases in the onset latency of REM; decrease in total REM sleep duration	Chalon et al. (2005)
PET RO of dopamine D2 receptor in striatum	Schizophrenia	Dopamine D2 to occupy 60%–80% to have antipsychotic efficacy	Kapur (1998)
Scopolamine challenge	Alzheimer's disease	Cholinesterase inhibitors and other cholinergic agonists need to reverse scopolamine-induced cognitive/memory impairment for symptomatic memory gain	Klinkenberg and Blokland (2010)
Laser-evoked potential	Neuropathic pain	Laser-evoked potentials, measured by scalp electrodes predict analgesia efficacy in neuropathic and nociceptive pain	Chizh, Priestley, Rowbotham, and Schaffler (2009)
Saccadic eye movements	Sedation/ antianxiety	Most prominent with benzodiazepines	Visser et al. (2003)
CO2 inhalation	Antianxiety	Sensitive across various drug targets	Bailey, Dawson, Dourish, and Nutt (2011)
PSG	Sleep disorders	Drugs that improve PSG also improve sleep quality. However, orexin antagonists improve sleep quality without marked effects on PSG	Jobert et al. (2013), Ma et al. (2014)

Abbreviations: *PET*, positron emission tomography; *RO*, receptor occupancy; *EEG*, electroencephalography; *SSRI*, selective serotonin reuptake inhibitor; *SNRI*, serotonin, norepinephrine reuptake inhibitor; *CO2*, carbon dioxide; *PSG*, polysomnography.

EM models is probably related to the lack of construct validity and understanding of the pathophysiology of CNS disease. Predictive validation is difficult to prove as the literature may not contain negative human experimental model information. Biomedical publications are biased against negative results, and the translation information is incomplete for terminated drug projects (Fanelli, 2012). Also, it is important to point out that after treatments for psychiatric disorders, the response rate is approximately 50%, making it difficult to assess predictive validity for experimental models (Berton, Hahn, & Thase, 2012). In the succeeding text are selected examples of human experimental models that have positive predictive validation, chosen because they have demonstrated pharmacological validation either across various drug targets or within a pharmacological target, using numerous compounds. All of the examples predict clinical efficacy, as the targets have proved clinical validity (Table 5.2).

IV QUALIFICATION AND VALIDATION OF BIOMARKERS

EM models, by definition, utilize biomarker responses; therefore a short discussion of biomarker qualification is appropriate. The validation of a biomarker refers to the data demonstrating the biomarkers' assay performance, that is, the biomarker measurement methods, calibration standards, accuracy, precision, reproducibility, linearity, quality specifications, etc. Qualification refers to how the biomarker is interpreted and what actions will be taken with a given outcome. The FDA in their recent guidance (FDA, 2014) on drug development tools describes a "context of use" (COU) as a statement that defines how the biomarker should be used in the context of clinical management or drug development and includes the biomarker's purpose, its boundaries, the conditions of qualified use, and its interpretation. A biomarker's utility is fully realized when enough clinical data are collected for full characterization, including its positive and negative predictive value in specific patient populations.

As previously discussed, one of the primary uses of EM studies in CNS drug development is to quantify pharmacological responses for novel drug targets or to compare same pharmacological class with known reference drugs. As such, some biomarkers may not have full assay validation and certainly would not have qualification as their utility in patient management may not be known for years. Therefore biomarker validation goes through stages, starting as an exploratory research tool and ending as fully validated and qualified clinical endpoint. The validation for the EM study should have an appropriate stringency of experimental proof so that the measurements can be credible. This approach called "fit-for-purpose" biomarker validation and can be applied to novel biomarker measurements (Cummings, Raynaud, Jones, Sugar, & Dive, 2010).

In summary, there are both hope and resistance for the use of human EM models as valuable translation tools to aid in CNS drug development: hope, because with the EM approach, new biological technologies and pharmacological markers can be used to test the potential for novel compounds to modulate their intended targets in humans as a first step for derisking a project, and resistance, because there is a risk of false outcomes and misleading drug development decision-making because few biomarkers and EM models are clinically validated. Also, EM studies can be viewed as not necessary, adding time and budget prior to a POC study. EM approach is a powerful tool that should be used to better understand endophenotypes (patient subtypes), enhancing precision medicine opportunities, and facilitating development of novel CNS targeted drugs.

V EXAMPLES OF EM STUDIES IN CNS DRUG DEVELOPMENT

In the following examples, EM studies are conducted as a prelude to a Phase 2 proof-of-concept study. Importantly an EM study is not a substitute for a fully sized efficacy study. The EM study will use pharmacological endpoints to demonstrate that the research compound can bind to the target (e.g., receptor occupancy in imaging); postreceptor transduction occurs, resulting in a pharmacological or pharmacodynamic response. Although there is a tendency or an attraction to treat these studies as POC, especially if they are positive, this should not be the case.

VI EXAMPLE #1: ALPHA-7 NICOTINIC AGONISTS FOR COGNITIVE IMPAIRMENT ASSOCIATED WITH SCHIZOPHRENIA (CIAS)

Schizophrenia, a severe and chronic neuropsychiatric disorder, results in poor social and occupational functioning. The clinical features cluster into three symptom groups: (1) positive symptoms (e.g., hallucinations and delusions), (2) negative symptoms (e.g., impaired affective behavior and social withdrawal), and (3) poor cognition. The cognition impairment is centered around selective attention, working memory, executive function, episodic memory, language comprehension, and social-emotional processing (Healey, Bartholomeusz, & Penn, 2016; Keefe et al., 2006; Matza et al., 2006).

Positive symptoms are managed through the use of antipsychotic medications that have primarily a mechanism of action as dopamine 2 antagonists. However, there is currently no approved drug therapy for use in cognition impairment associated with schizophrenia (CIAS), and it remains a high unmet medical need.

One potential target for a CIAS drug is the alpha-7 nicotinic receptor (7nAChR); the receptor, which is located presynaptically, modulates the release of several neurotransmitters (Freedman, 2003; Hashimoto, 2014; Olincy & Stevens, 2007). The 7nAChRs are found in brain structures that are relevant to schizophrenia, including the hippocampus, lateral and medial geniculate nuclei, and reticular nucleus of the thalamus (Gündisch, 2005; Romanelli & Gualtieri, 2003). Animal studies demonstrate that blockade of 7nAChR causes failure of memory processes including sensory gating (Adams & Stevens, 2007; Singer, Feldon, & Yee, 2009; Stevens, Freedman, Collins, Hall, & Leonard, 1996).

A known characteristic of schizophrenia is a malfunctioning sensory gating mechanism, observed by an abnormally evoked potential response, including the P50 auditory evoked potential (Javitt & Freedman, 2015; Freedman et al., 2005; Light & Braff, 2000; McGhie & Chapman, 1961). Sensory gating deficits manifest as the inability of the patient to ignore sensory stimuli. Deficits in prepulse inhibition, continuous performance testing, and smooth pursuit eye tracking are all related to abnormal sensory gating and abnormalities in sustained attention (Bove, 2008; Geyer, Krebs-Thomson, & Braff, 2001). Attention deficit, especially sustained attention, is one of the cognitive characteristics of schizophrenia that may be related to sensory gating impairments (Adler et al., 1998; Judd, McAdams, Budnick, & Braff, 1992). Abnormalities in prepulse inhibition, including startle, continuous performance test, and various smooth pursuit eye tracking abnormalities, have all been interpreted as failures in sensory gating function in schizophrenia. They are overwhelmed by stimuli from their environment that normal subjects ignore.

The P50 evoked potential is a positive electroencephalographic waveform occurring 50 m after the auditory stimulus that appears to be particularly responsive to 7nAChR modulation. When pairs of auditory stimuli are presented, normal subjects show a significant diminution of the second response. However, schizophrenia patients fail to inhibit the P50 response to the second stimuli (Adler et al., 1998; Hashimoto, 2014; Olincy & Stevens, 2007).

Sensory gating mechanisms in animal models can be used as a translational tool for predicting drug responses in patients with schizophrenia. In rodents, hippocampal P20-N40 auditory evoked potentials are thought to be homologous to the human P50 response (Bickford-Wimer et al., 1990; Luntz-Leybman, Bickford, & Freedman, 1992).

Based on genome linkage studies, the P50 response is associated with the CHRNA7 gene. A single nucleotide polymorphism in the 5' core promoter region of the CHRNA7 gene is significantly associated with P50 suppression deficits, first identified by Freedman et al. (1997). Later studies found a positive genetic linkage of schizophrenia to the 15q13eq14 region; however, not all studies have shown consistent associations

(Curtis et al., 1999). Furthermore, it has been suggested that auditory P50 deficits could be a schizophrenia endophenotype (Earls, Curran, & Mittal, 2016).

Additionally, auditory "oddball" evoked potential models have been extensively studied on patients with schizophrenia and their relatives, also showing deficits in sustained attention, target detection, and attentional switching. The P300 ERP and mismatch negativity (MMN), like the P50, are also associated with these deficits, and these EP parameters are not consistently reversed by antipsychotic medications (Takahashi et al., 2013).

Therefore a human EM model, employing the P50 response, has translational value as a PD response marker of compounds acting on the 7nAChR pathway. A typical P50 EM study has, as its primary objective, to provide proof of mechanism for P50-mediated sensory gating. A secondary objective could be to evaluate dose-response. Nicotine is sometimes included as a positive control. There may also be additional electrophysiological measures including mismatch negativity (MMN) and P300. It is also common to include specific cognition measures as secondary pharmacological responses including attention; reaction time; visual/working memory; verbal memory; and attention switching, social/emotional, and executive function that are the deficits associated with schizophrenia.

The study subjects are patients who are diagnosed with schizophrenia and judged to be clinically stable and medicated as outpatients for at least 3 months. Smokers and nonsmokers can be eligible. Nonsmokers are preferred by some investigators as tobacco/nicotine will interfere with the 7nAChR mediated P50 responses (Olincy et al., 2006; Preskorn et al., 2014). However, enrolling nonsmoking patients can be difficult. If nicotine is chosen as a positive control, there may be confounding influence with smokers. Thus the study protocol needs to pay attention to the use of tobacco and smoking status. Background antipsychotic medications are maintained throughout the study period. The test drug can be given as a single- or repeat-dose administration, either as a crossover or a parallel treatment group trial design.

During the P50 procedure the patient's eyes are kept closed, and auditory stimulus tones are presented via headphones in a conditioning testing paradigm. A total of 100 pairs of identical brief tones are presented as double clicks: S1 and S2. The stimuli are short tone bursts of a single sine wave of 1500-Hz frequency and a duration of about 6–7 ms. The intrapair stimulus interval between S1 and S2 is fixed at 500 ms with a pseudorandomized interpair interval of 5–9 s. P50 response amplitude is calculated as the voltage difference between the P50 peak and the preceding negative trough. The P50 gating ratio is calculated as the amplitude of the P50 response to the second (test) stimulus divided by the amplitude of the P50 response to the first (conditioning) stimulus.

A gating ratio <0.5 is considered normal (Adler et al., 1998; Adler, Waldo, & Freedman, 1985; Preskorn et al., 2014; Shiina et al., 2010; Winterer et al., 2012).

There are a number of reports of employing P50 tests with 7nAChR modulation research compounds including DMXB-A (AKA: GTS-21) (Olincy et al., 2006), tropisetron (Shiina et al., 2010), EVP-6124 (encenicline) (Preskorn et al., 2014), and JNJ-39393406 (Winterer et al., 2012). These EM studies were used to demonstrate the translation of the P50 response from animal to human and to provide a guide for identifying pharmacologically active doses. Although alpha-7 nicotinic modulation has resulted in trends for improvements in cognition in the target schizophrenic patient population (Marder, 2016), no compound has a positive phase 2 study that would support regulatory registration for the improvement of cognition. The P50 EM model, now with the combined experience of several drug candidates, questions the unproved hypothesis (i.e., drug target) that normalizing gating abnormalities by alpha-7 nicotinic receptor agonism results in relevant improvements in cognition in patients with schizophrenia. Although the P50 EP model was positive with alpha-7 nicotinic agonism and there were improvements in certain cognition tests (i.e., attention and response speed), the model does not yet have predictive therapeutic validity because the drug target remains an unproved therapeutic. This highlights the gaps in our understanding of the relationship of gating abnormalities, cognitive pathophysiology, and alpha-7 nicotinic receptors.

VII EXAMPLE #2: KETAMINE CHALLENGE MODEL OF SCHIZOPHRENIA

A Schizophrenia Model in Healthy Volunteers

In this example, specific behavior characteristics of schizophrenia are induced in healthy volunteers by the infusion of low doses of the dissociative anesthetic, ketamine. The purpose of this experimental model for use in CNS drug development is (1) to evaluate the modulation of glutamatergic pathways by the research drug and (2) to demonstrate specific responses that were observed in the animal model compared with the human model of schizophrenia. Although there is evidence that the ketamine model of schizophrenia has face and construct validity, at present, it lacks definitive evidence of predictive validity. Therefore it is questionable if this model can be reliably used to predict efficacy (Goff, 2015) in patients with schizophrenia. Ketamine and related NMDA pharmacological probes are employed by neuroscientists to investigate disease pathophysiology and the abnormal neurocircuitry of schizophrenia. Additionally the ketamine model is used by some drug developers as a prelude to a definitive POC clinical trial for the treatment of patients with schizophrenia. Recently, ketamine has been used as a template NMDA drug for understanding its potent activity in rapid onset antidepressant actions in treatment-resistant depression (Abdallah, Sanacora, Duman, & Krystal, 2015; Duncan & Zarate, 2013; Zarate, Mathews, & Furey, 2013).

Dopaminergic treatments for schizophrenia and psychosis are widely used therapeutically. Dopamine D2 receptor antagonists alleviate positive symptoms and have led to the dopaminergic hypothesis of schizophrenia, developed in the 1960s with chlorpromazine (Howes & Kapur, 2009). However, the dopamine hypothesis does not easily explain negative symptoms and cognition abnormalities in schizophrenia. The glutamate hypothesis is a newer hypothesis that implicates glutamatergic dysfunction underlying positive, negative, and cognition impairment in schizophrenia. An important piece of the glutamate hypothesis is based on the observation that NMDA receptor antagonists (NMDAR) such as phencyclidine (PCP) and ketamine transiently induce positive symptoms and cognitive abnormalities similar to schizophrenia (Morris, Cochran, & Pratt, 2005; Newcomer et al., 1999; Olney, Newcomer, & Farber, 1999; Stone et al., 2008; Thornberg & Saklad, 1996). Since many subjective and behavioral effects of ketamine can be blocked by agents that decrease glutamate release, it is possible that psychotic symptoms are related to a hyperglutamatergic state (Deakin et al., 2008).

Ketamine (2-chlorophenyl-2-methylamino-cyclohexanone) is a noncompetitive NMDAR antagonist of the NMDA receptor, derived from PCP. Because of PCP's neurotoxicity and obvious reactions, such as hallucinations, confusion, and delirium, ketamine was intended as a safer alternative to PCP (Perry et al., 2007). NMDAR antagonists, including ketamine, have a net positive effect on excitatory transmission by inducing excessive release of glutamate (Kim et al., 2011; Stone et al., 2012) and acetylcholine (Giovannini, Camilli, Mundula, & Pepeu, 1994; Hasegawa, Kinoshita, Amano, et al., 1993).

Ketamine has the highest affinity against the NMDAR; however, it also has affinities for other receptors that may have relevance in schizophrenia. The most notable is its partial agonism for the D2 receptor. According to Kapur and Seeman (2002), the ketamine affinities to NMDA and D2 receptors are similar; the Ki is approximately 0.5–1 µM. Also, of interest is the affinity for the 5-HT2A receptor. Thus an argument can be made that ketamine may not be the most selective probe for evaluating glutamate pathways; rather, it is a mixed, semiselective agent that mimics the complex abnormalities of the dopaminergic, glutamatergic, and serotonergic pathways of schizophrenia. However, according to the glutamate hypothesis of schizophrenia, the dopamine and serotonin effects are secondary to a primary hypoglutamatergic

effect of ketamine. Additional receptors for which ketamine has potentially significant and clinically meaningful affinity include sigma-1, delta-, mu-, and kappa-opioid; muscarinic; and 5-HT2 (see review Frohlich & Horn, 2014).

Xu et al. (2015) performed a literature search on symptom dimension analyses of the positive and negative syndrome score (PANSS) for schizophrenia, comparing acute users and chronic abusers of ketamine-induced psychosis to early diagnosis and chronic states of schizophrenia. The purpose was to understand the quality (or face validity) of the ketamine's ability to mimic schizophrenia. The authors concluded that the chronic ketamine abuser symptoms bore a close resemblance to schizophrenia; however, the acute administration of ketamine showed less similarity, specifically in the domain that contained sedative and dissociative effects of acute ketamine. Thus one can make the case for ketamine (chronic) administration has face and construct validity as a model for schizophrenia.

The rate and dose of ketamine administered to healthy volunteers vary in EM protocols and can influence behavioral characteristics. There is no consensus on the optimal infusion parameters to induce the desired psychomimetic effects. Also, some investigators infuse the racemic mixture of ketamine, and others infuse the S (+) isomer. In healthy volunteers the most obvious effect of subanesthetic doses of both enantiomers is altered sensory perception. (S)-ketamine is four times as potent as (R)-ketamine in reducing pain perception and in causing auditory and visual disturbances (Øye, Paulsen, & Maurset, 1992). Kleinloog et al. (2015), using a concentration-controlled infusion approach, methodically varied the pseudo steady-state plasma concentration of S(+)-ketamine between 120 and 360 ng/mL. The optimum blood concentration, based on the PANSS scale, subject tolerability, various visual analogue scales, and prepulse inhibition, was between 100 and 200 ng/mL for up to 2 h. Deakin et al. (2008) administered a racemic ketamine as an initial 1-min infusion of 0.25 mg/kg followed by a sustained infusion of 0.25 mg/kg/h. Krystal et al. (1994) and (Perry et al., 2007), reviewing the collective experience of the Yale Ketamine Study Group, describe either an initial intravenous bolus followed by an infusion or a continuous infusion alone. Initial bolus doses ranged from 0.081 mg/kg over 10 min to 0.26 mg/kg over 1 min, and continuous infusion doses ranged from 0.04 mg/kg to 0.75 mg/kg over 60 to 120 min.

Ketamine infusions in healthy volunteer model are the homolog to PCP infusions in rodent models of schizophrenia. As a translational tool, one could identify specific behavioral attributes in the animal model and subsequently evaluate for those attributes in humans (Lahti, Holcomb, Gao, & Tamminga, 1999).

Examples of NMDA-dependent compounds that have a proved clinical utility in schizophrenia (i.e., data in both human ketamine and rodent PCP models), include olanzapine, lamotrigine, lorazepam, clozapine, and haloperidol. However, some of the outcomes are inconsistent, possibly due to the nonstandardized ketamine infusion regimen (Krystal et al., 1999, 1998). A glutamate metabotropic group II agonist, LY354740, showed activity in the ketamine model in humans (Krystal et al., 2005), but the metabotropic compounds failed as a monotherapy antipsychotic agent in Phase 2 studies (Wierońska, Zorn, Doller, & Pilc, 2015). Other than clozapine, the ketamine challenge has questionable predictive validity for antipsychotic efficacy (Goff, 2015) but has utility to evaluate glutamatergic processes.

In summary the ketamine model in healthy volunteers can be used to mimic cognitive and positive symptoms of schizophrenia. There is still debate whether ketamine is appropriate as a specific probe in understanding the role of NMDA in schizophrenia or whether the model can be used to test for initial activity of new drug compounds, prior to a Phase 2 POC. Assuming that it is to be used for derisking a drug prior to Phase 2, there still remain questions on how to use the human EM model results: choosing doses for Phase 2, proof of pharmacology or mechanism, or go/no-go decision-making.

VIII EXAMPLE #3: USE OF OCULOMOTOR AND SCHIZOTYPY MODELS

Eye movement can provide an important translational bridge between behavioral pharmacology research in animal models and clinical investigations. Eye movements are sensitive biomarkers of drug effects on specific sensorimotor and cognitive processes, perhaps more sensitive and reliable than behavioral responses. They are quite reproducible and have a greater accuracy than behavioral ratings or neuropsychological testing with pharmacological manipulations. As such, eye movements are used to construct precise PK/PD relationships for novel test compounds and to evaluate the sensitivity of drug response across species as a translational exercise. Reilly et al. (2008) comprehensively reviewed the attributes for use of oculomotor responses as a translational tool in drug characterization. First the neurophysiological and neurochemical basis involved in oculomotor control in nonhuman primate brain is well characterized from single unit recordings. The homology of involved brain regions between human and nonhuman primate species is supported by objective functional imaging studies. Behavioral pharmacology studies have well characterized the effects of numerous drugs that modulate specific brain regions involved with oculomotor movements. Their ease of the measurements lends themselves to perform as repeat measures without causing too much burden on the trial subjects or raters. With repeat measures, dose-response and time course relationships are constructed within each subject. Finally, recent studies, focusing on pharmacogenetics of oculomotor movements, have added another dimension for translatability.

Oculomotor measurements typically refer to two types of eye movements, smooth pursuit (SPEM) and saccadic, which are both observed in human and nonhuman primates. Smooth pursuit is where the image is kept within the fovea area of the retina during movement of an object. The process of initiating and maintaining smooth pursuit also involves attentional control. The primary response measure of pursuit accuracy is the velocity gain, corresponding to the ratio of smooth pursuit velocity over target velocity (Kumari et al., 2017). Saccades refer to the fast eye movements made to the sudden appearance of a visual target. Prosaccades require the subject to make a saccade to a single target stimulus soon after it appears. The antisaccade refers to the participant inhibiting the reflex-like saccade by initiating an eye movement opposite to the direction of the target. Response is assessed as the error rate, spatial accuracy, and latency of saccadic (or antisaccadic) movement (Hutton & Ettinger, 2006). Studies in healthy volunteers and schizophrenia patients demonstrate that SPEM is stable over time ranging from 1 week to 2 years. Practice effects, which can be an issue with cognition and behavioral testing, were also seen only with the antisaccade task. Test-retest reliabilities for SPEM and saccades are excellent with consistent within-session consistencies and between-session consistency (Ettinger et al., 2003). A high test-retest reliability, both within and between sessions, provides for the ability to construct predictive exposure response, PK/PD models for use in future clinical trials.

Oculomotor measurements, because of their high reliability and ease of performing, have been used across a variety of drug classes, in healthy volunteers, and a variety of disease conditions. The primary objective is to identify a dose range as an adjunct to other neurophysiological measurements. The oculomotor abnormalities of patients with schizophrenia have been well studied (Holahan & O'Driscoll, 2005) and in proband/family genetic studies of schizophrenia (Crawford et al., 1998).

Most prominent and consistent are the effect of benzodiazepines on eye movements and the pattern of abnormalities seen in schizophrenia and schizotypal patients. Benzodiazepines have been extensively tested on saccadic peak velocity and alertness in healthy volunteers. Visser et al. (2003) identified 56 different studies where saccadic movements and neuropsychological tests were used with benzodiazepines. Saccadic eye movements (i.e., peak velocity) and VAS for alertness showed statistical significance 100% and 79%, respectively.

Furthermore, dose-response relationships across various benzodiazepine compounds were constructed, and dose equivalencies were derived that correlated well to known potencies of the compounds (Visser et al., 2003). The decrease in peak velocity associated with benzodiazepines is generally thought to be related to the sedative properties; thus the oculomotor responses have predictive validity for sedative properties of drugs. Changes in oculomotor responses after benzodiazepines are reversed by specific antagonists of the benzodiazepine receptor (Ball, Glue, Wilson, & Nutt, 1991). Finally the patterns of effects on oculomotor movements are drug-specific; sedatives with differing mechanisms of action can have a specific pattern (i.e., drug signature) on eye movements (Busettini & Frölich, 2014).

Antipsychotic medications for use in schizophrenia have been extensively studied for oculomotor responses. Reilly et al. (2008) reviewed first-generation antipsychotic medications on several saccadic eye movements in schizophrenic and bipolar disorder patients. Patients were either not on medication for at least 6 months or have been on medications chronically. Both patient groups had a reduced gain on prosaccade compared with medication-free patients. Straube, Riedel, Eggert, and Müller (1999) and Muller, Riedel, Eggert, et al. (1999) methodically evaluated eye movements in first-episode schizophrenia patients who were medication-naive or medication-free for at least 1 month. Antipsychotic medications including first- and second-generation agents decreased peak saccade velocity for prosaccades and antisaccades.

There is potential for oculomotor movements as a translational model utilizing schizotypal patients as surrogates for schizophrenia patients. Schizotypal personality disorder is a heterogeneous group with a mix of social/interpersonal and perceptual symptoms empirically derived on the basis of symptoms exhibited by nonpsychotic relatives of schizophrenia patients (Cadenhead, Light, Geyer, McDowell, & Braff, 2002). Empirical studies investigate putative overlap with schizophrenia based on genetics, cognition, perception, motor control, and psychopharmacology. There is some evidence that schizotypal personality disorder represents an endophenotype, based on genetic linkage to schizophrenia. Therefore studies of schizotypal patients are a way to assess endophenotypic traits of the schizophrenia spectrum. As an experimental model, using schizotypal populations may show an enhanced drug sensitivity and have a higher predictive validation than healthy volunteers.

Specifically focusing on perceptual deficits, auditory discrimination (Cadenhead et al., 2002; Bates, 2005), smooth pursuit, and antisaccade eye movements are associated with schizotypy. The smooth pursuit and antisaccadic movement abnormalities, well described in schizophrenia, are also prominent in schizotypy. These include a reduced ability to match eye velocity to target velocity in the smooth pursuit task and the failure to inhibit automatic saccades to a peripheral target in the antisaccade task. On the pursuit task, both positive- and negative-symptom schizotypes have abnormalities. Antisaccade deficits may have a higher sensitivity for identifying high-risk subjects for positive symptoms, whereas pursuit deficits have equally sensitivity for positive- and negative-symptom schizotypes (Holahan & O'Driscoll, 2005).

Pharmacological studies with schizotypal patients show similarities to schizophrenia patients and have been used for providing arguments for the neural similarities. A key finding is that the administration of amphetamine produces an increase in striatal dopamine release in schizotypal patients, consistent with the overlap with schizophrenia (Ettinger, Meyhöfer, Steffens, Wagner, & Koutsouleris, 2014; Laruelle, Abi-Dargham, Gil, Kegeles, & Innis, 1999). Imaging studies show a reduction in dopamine D_2 and D_3 binding potential in schizotypal personality disorders, similar to schizophrenic patients (Breier et al., 1997). In a multicenter pharmacological study of 122 in each group, risperidone, amisulpride, nicotine, and placebo were administered to schizotypal patients with high or average on personality measures. Cognition tests also included the N-back and verbal fluency. The D2/D3 receptor antagonist, amisulpride, improved cognition responses with high schizotypy but worsened medium schizotypy controls (Koychev, McMullen, Lees, & Dadhiwala, 2012). The saccadic eye movement responses, in general, followed the pattern reported in schizophrenia patients, with some noted inconsistencies (Schmechtig et al., 2013). In an unrelated study, administration of risperidone led to an improvement in saccadic eye movements (Burke & Reveley, 2002).

To summarize, oculomotor assessments are an easy, reliable measurement that is thought to be related to the integrity of sensorimotor and processes of cognition. Oculomotor evaluation allows researchers to understand functional disturbances of brain systems in clinical disorders and to evaluate potential treatments. Oculomotor abnormalities have also been well characterized in a variety of disorders including schizophrenia, Parkinson's disease, multiple sclerosis, autism spectrum disorders, and Alzheimer's disease. In addition to benzodiazepines, whose oculomotor effects are well characterized, the literature has much information on the effects of numerous CNS-active compounds across many neurotransmitter systems.

IX EXAMPLE # 4 SCOPOLAMINE CHALLENGE FOR EVALUATION OF COGNITION

Scopolamine is a medication used in the treatment of motion sickness and postoperative nausea and vomiting.

Belladona or deadly nightshade plant contains the topane alkaloid, scopolamine, which acts as a competitive antagonist at the muscarinic cholinergic receptors. Although it is commonly thought of as a nonspecific antagonist, it has specificity for the M1 and possibly M5 muscarinic subtypes (Klinkenberg & Blokland, 2010). Additionally, there is possible involvement of NMDA receptor mechanisms, especially in the dorsal hippocampus (Khakpai et al., 2012). Administration of scopolamine produces deficits on tests of visual recognition memory, visuospatial praxis, verbal recall, visuospatial recall, psychomotor speed, and visuoperceptual function (see reviews: More, Kumar et al., 2016; Klinkenberg & Blokland, 2010).

Scopolamine has been well recognized for its ability to induce memory impairment that has certain face validity for (i.e., mimic) normal aging. It is a standard reference compound for inducing age- and dementia-related deficits in healthy young and healthy elderly humans and in animal models of abnormal cognition and aging. The amnesia, induced my scopolamine, mimics the memory loss of dementia (Flood & Cherkin, 1986). The hypothesis of age-related decline in cognition and the involvement of cholinergic integrity are based on the many studies involving scopolamine as a pharmacological tool. Scopolamine is extensively used to evaluate the role of the cholinergic integrity in animal models of dementia and is a standard model for the evaluation of novel drug treatments for Alzheimer's disease where cholinergic mechanisms are thought to be involved.

The role of acetylcholine in learning and memory has been summarized by Hasselmo (2006). Acetylcholine is key in hippocampal role of shifting between encoding and retrieval of memory. High acetylcholine would promote acquisition of new information, and low acetylcholine allows for retrieval of previously stored memories. The hypothesis predicts that low acetylcholine impairs encoding and consolidation and retrieval is impaired by high acetylcholine. Scopolamine, a cholinergic antagonist, predominately impairs acquisition and can even facilitate consolidation and is predicted to impair encoding (Winters, Bartko, Saksida, & Bussey, 2007). The main effect of scopolamine appears to be on verbal memory and possibly attentional processes, although there are varying reports on attention. Rusted and Warburton (1988) reported in a well-cited paper that the scopolamine-induced deficits in performance are restricted to components of working memory, problem solving, and tasks involving visuospatial processing. Administration of scopolamine to healthy elderly also shows deficits in executive problem solving and working memory (Laczó et al., 2016).

The value of the scopolamine challenge in humans as a translational tool for cognition in drug development depends greatly on the PD response measurements.

There are a number of neuropsychological test batteries available, focusing on specific cognitive abnormalities, and these each have their level of validation and ease of use in an EM clinical study setting. Computerized testing, as opposed to "pencil and paper," has been largely adopted by early clinical phase units. Computerized testing allows study personnel to efficiently conduct small focused clinical studies with repeat, postdose measurements. Furthermore the suppliers of these tests have developed disease-specific panels that are claimed to be sensitive to disease-specific abnormalities. The level of validation varies, and the clinical study designer needs to be aware of their individual strengths and weaknesses.

The Cambridge Neuropsychological Test Battery (CANTAB) has developed a system for measuring cognition that is built on the translation between animal models (rodent and mouse) and human. The stimulation object appears on a touchscreen monitor allowing the animal or human to select. The paired-associated learning task (PAL) is specifically geared toward Alzheimer's disease where there is a deficit in visual memory and new learning originating in the hippocampus (Bartko et al., 2011). The PAL test is reported to allow clinicians to identify and predict conversion of mild cognitive impairment to Alzheimer's disease (Blackwell et al., 2004; Sahakian et al., 1988; Swainson et al., 2001). PAL has been used with scopolamine in healthy adult humans, with differences showing in the quality of the impairment compared with diazepam and dementia (Robbins et al., 1997).

CogState Clinical Trials has also developed a computerized test battery that is customized for disease indication. For Alzheimer's disease the neuropsychological battery is composed of the evaluation of verbal memory international shopping list test (Lim et al., 2009), visual memory working memory, pattern separation (Maruff et al., 2013), and visual paired associate learning (Rentz et al., 2011).

Finally the cognitive drug research system (CDR) has a computerized testing system that is geared toward abnormalities in attention including tests of simple reaction time, choice reaction time, (and digit vigilance), working memory (numeric and spatial), episodic memory (immediate and delayed word recall), and picture recognition (Wesnes, Ward, McGinty, & Petrini, 2000).

The scopolamine challenge model in humans has been used in healthy young adults and healthy elderly who have normal cognition to gain basic knowledge about brain structures and cholinergic involvement in memory and aging. The model has been used to evaluate novel memory enhancing drugs. As a part of the evaluation, it is typical to reverse the scopolamine-induced amnesia with an acetylcholinesterase inhibitor, such as donepezil, to demonstrate the role of the cholinergic blockade (Lenz et al., 2012; Snyder, Bednar, Cromer, & Maruff, 2005).

In addition to neuropsychological responses, brain imaging studies, employing fMRI, are conducted with scopolamine to better understand the regions of brain activation involved with attention, learning, and memory (Antonova, Brammer, Williams, & Morris, 2010; Sperling et al., 2002; Thiel, Henson, Morris, Friston, & Dolan, 2001; Thienel et al., 2009; Voss et al., 2012). Also, these tests can be used translationally by comparing with animal studies as a means to understand the role of cholinergic drive in aging, memory, and behavior.

The memory tasks and endpoints vary between laboratories as does the administration of scopolamine. In one of the original published works, 1.0- and 0.5-mg scopolamine were administered intravenously to healthy young (Sitaram, Weingartner, & Gillin, 1978) and healthy elderly adults (Sunderland et al., 1986), respectively. Rusted and Warburton (1988) administered 0.6 mg subcutaneously according to their well-cited paper. Generally, memory testing is repeated for up to 2 h post dose, although the impairment may last for up to 7 h in this model. Robbins et al. (1997) and Fredrickson et al. (2008) constructed a dose-response with scopolamine ranging from 0.2 to 0.6 mg administered subcutaneously on paired associates and maze learning, respectively. Snyder et al. (2014) administered a low dose of 0.2 mg subcutaneously to healthy elderly adults who had risk factors for developing Alzheimer's disease and measured abnormal responses of executive functioning using the Groton Maze. This dose induced an impairment in executive functioning and working memory for only 5 h, compared with higher doses that cause memory impairment for up to 9 h. Also, they were also able to reverse the impairment with 5-mg donepezil.

In summary the scopolamine challenge model in humans serves as a valuable tool for translational scientists who are focused on the study of memory, learning, and involvement of the central cholinergic system. Behavioral tests provide homology in animal and human models of cognition supporting construct validity of the animal models. Cholinergic agonists and antagonists can be reliably tested in humans to evaluate PK/PD differences in various species allowing for comparative evaluations of procognitive drugs. Doses for future Phase 2 studies can be derived from comparisons using known cholinergic pharmacological probes and drug comparators. Alzheimer's disease is characterized by a loss of cholinergic tone that affects cognitive status (Whitehouse, 1986). Therefore the scopolamine challenge model in humans may be used as a model for preclinical dementia, those elderly patients who are at risk for developing dementia. The scopolamine challenge model, similar to other pharmacological challenges, has value for its ability to characterize the pharmacodynamics of new compounds with diverse mechanisms of action, if the pathways involve cholinergic tone. There is a robust publication history of about 30 years; however, there are still some differences between laboratories primarily in the administration of scopolamine and the choice of behavioral neuropsychological testing. Perhaps the use of neuroimaging can aide in validation by providing objective measurements of brain activation. Finally the scopolamine model should not be used as a predictor of efficacy, unless the compound mechanism of action primarily relies on the modulation of cholinergic tone (e.g., cholinergic agonist).

X END OF CHAPTER SUMMARY

In summary, EM studies provide an important translational approach for CNS drug developers. EM's greatest attraction is its low cost and speed, compared with a Phase 2 POC. The advantage of EM is that the information is obtained in humans, providing a strong construct validity. However, conducting EM studies requires good clinical planning so that the full value EM information is available in time to support a project decision to enter Phase 2. The examples in this chapter highlight the attraction for EM studies because they derisk programs by providing demonstrable target engagement, supporting animal model data, and providing a PK/PD rationale for dose selection to be used in subsequent Phase 2 studies. In this chapter, several examples are identified to highlight how EM studies, with varying levels of predictive validity, are used in drug development. However, like any modeled system, caution is required in overinterpreting the experimental system that may not have full predictive validity, regrettably leading to a false claim of patient efficacy. When a drug pipeline has multiple research compounds and drug targets, EM can be used to rank-order the drug candidates based on PK/PD attributes in comparison with a validated positive control. Also, EM data are used in a reverse translation with the potential of optimizing the next cycle of lead drug candidates. Finally, EM can use for furthering basic translational knowledge the relationship between drug action, pathophysiology, and therapeutics.

References

Abdallah, C., Sanacora, G., Duman, R., & Krystal, J. (2015). Ketamine and rapid-acting antidepressants: a window into a new neurobiology for mood disorder therapeutics. *Annual Review of Medicine*, 66(1), 1–15.

Adams, C. E., & Stevens, K. E. (2007). Evidence for a role of nicotinic acetylcholine receptors in schizophrenia. *Frontiers in Bioscience*, 12, 4755–4772.

Adler, L. E., Olincy, A., Waldo, M., Harris, J. G., Griffith, J., Stevens, K., et al. (1998). Schizophrenia, sensory gating, and nicotinic receptors. *Schizophrenia Bulletin*, 24(2), 189–202.

Adler, L. E., Waldo, M. C., & Freedman, R. (1985). Neurophysiologic studies of sensory gating in schizophrenia: comparison of auditory and visual responses. *Biological Psychiatry, 20*, 1284–1296.

Antonova, P., Brammer, S., Williams, D., & Morris. (2010). Scopolamine disrupts hippocampal activity during allocentric spatial memory in humans: an fMRI study using a virtual reality analogue of the Morris Water Maze. *Journal of Psychopharmacology, 25*(9), 1256–1265.

Bailey, J. E., Dawson, G. R., Dourish, C. T., & Nutt, D. J. (2011). Validating the inhalation of 7.5% CO(2) in healthy volunteers as a human experimental medicine: a model of generalized anxiety disorder (GAD). *Journal of Psychopharmacology, 25*(9), 1192–1198.

Ball, D. M., Glue, P., Wilson, S., & Nutt, D. J. (1991). Pharmacology of saccadic eye movements in man. Effects of the benzodiazepine receptor ligands midazolam and flumazenil. *Psychopharmacology, 105*(3), 361–367.

Bartko, S., Vendrell, I., Saksida, L., & Bussey, T. (2011). A computer-automated touchscreen paired-associates learning (PAL) task for mice: impairments following administration of scopolamine or dicyclomine and improvements following donepezil. *Psychopharmacology, 214*(2), 537–548.

Bates, T. C. (2005). The panmodal sensory imprecision hypothesis of schizophrenia: reduced auditory precision in schizotypy. *Personality and Individual Differences, 38*, 437–449.

Belzung, C., & Lemoine, M. (2011). Criteria of validity for animal models of psychiatric disorders: focus on anxiety disorders and depression. *Biology of Mood & Anxiety Disorders, 1*(1), 1–14.

Berton, O., Hahn, C. -G. G., & Thase, M. E. (2012). Are we getting closer to valid translational models for major depression? *Science (New York, N.Y.), 338*(6103), 75–79.

Bickford-Wimer, P. C., Nagamoto, H., Johnson, R., Adler, L. E., Egan, M., Rose, G. M., et al. (1990). Auditory sensory gating in hippocampal neurons: a model system in the rat. *Biological Psychiatry, 27*(2), 183–192.

Blackwell, A. D., Sahakian, B. J., Vesey, R., Semple, J. M., Robbins, T. W., & Hodges, J. R. (2004). Detecting dementia: novel neuropsychological markers of preclinical Alzheimer's disease. *Dementia and Geriatric Cognitive Disorders, 17*(1-2), 42–48.

Bouwknecht, J. (2014). Behavioral studies on anxiety and depression in a drug discovery environment: keys to a successful future. *European Journal of Pharmacology, 753*, 158–176.

Bove, E. A. (2008). Cognitive performance and basic symptoms in first-degree relatives of schizophrenic patients. *Comprehensive Psychiatry, 49*(4), 321–329.

Breier, A., Su, T. P., Saunders, R., Carson, R. E., Kolachana, B. S., de Bartolomeis, A., et al. (1997). Schizophrenia is associated with elevated amphetamine-induced synaptic dopamine concentrations: evidence from a novel positron emission tomography method. *Proceedings of the National Academy of Sciences of the United States of America, 94*, 2569–2574.

Burke, J. G., & Reveley, M. A. (2002). Improved antisaccade performance with risperidone in schizophrenia. *Journal of Neurology, Neurosurgery, and Psychiatry, 72*, 449–454.

Busettini, C., & Frölich, M. A. (2014). Effects of mild to moderate sedation on saccadic eye movements. *Behavioural Brain Research, 272*, 286–302.

Cadenhead, K. S., Light, G. A., Geyer, M. A., McDowell, J. E., & Braff, D. L. (2002). Neurobiological measures of schizotypal personality disorder: defining an inhibitory endophenotype? *The American Journal of Psychiatry, 159*(5), 869–\.

Chalon, S., Pereira, A., Lainey, E., Vandenhende, F., Watkin, J. G., Staner, L., et al. (2005). Comparative effects of duloxetine and desipramine on sleep EEG in healthy subjects. *Psychopharmacology, 177*, 357–365.

Chizh, B., Priestley, T., Rowbotham, M., & Schaffler, K. (2009). Predicting therapeutic efficacy—Experimental pain in human subjects. *Brain Research Reviews, 60*(1), 243–254.

Cook, D., Brown, D., Alexander, R., March, R., Morgan, P., Satterthwaite, G., et al. (2014). Lessons learned from the fate of AstraZeneca's drug pipeline: a five-dimensional framework. *Nature Reviews. Drug discovery, 13*(6), 419–431.

Crawford, T., Sharma, T., Puri, B., Murray, R., Berridge, D., & Lewis, S. (1998). Saccadic eye movements in families multiply affected with schizophrenia: the maudsley family study. *American Journal of Psychiatry, 155*(12), 1703–1710.

Cummings, J., Raynaud, F., Jones, L., Sugar, R., & Dive, C. (2010). Fit-for-purpose biomarker method validation for application in clinical trials of anticancer drugs. *British Journal of Cancer, 103*(9), 1313–1317.

Curtis, L., Blouin, J. -L., Radhakrishna, U., Gehrig, C., Lasseter, V. K., Wolyniec, P., et al. (1999). No evidence for linkage between schizophrenia and markers at chromosome 15q13-14. *American Journal of Medical Genetics, 88*, 109–112.

Deakin, J., Lees, J., McKie, S., Hallak, J., Williams, S., & Dursun, S. (2008). Glutamate and the neural basis of the subjective effects of ketamine: a pharmaco–magnetic resonance imaging study. *Archives of General Psychiatry, 65*(2), 154–164.

Duncan, W., & Zarate, C. (2013). Ketamine, sleep, and depression: current status and new questions. *Current Psychiatry Reports, 15*(9), 394.

Earls, H. A., Curran, T., & Mittal, V. (2016). A meta-analytic review of auditory event-related potential components as endophenotypes for schizophrenia: perspectives from first-degree relatives. *Schizophrenia Bulletin, 42*(6), 1504–1516.

Ettinger, U., Kumari, V., Crawford, T., Davis, R., Sharma, T., & Corr, P. (2003). Reliability of smooth pursuit, fixation, and saccadic eye movements. *Psychophysiology, 40*(4), 620–628.

Ettinger, U., Meyhöfer, I., Steffens, M., Wagner, M., & Koutsouleris, N. (2014). Genetics, cognition, and neurobiology of schizotypal personality: a review of the overlap with schizophrenia. *Frontiers in Psychiatry, 5*, 18.

Fanelli, D. (2012). Negative results are disappearing from most disciplines and countries. *Scientometric, 90*(3), 891–904.

FDA. (2014). In U.S. Department of Health and Human Services, F.D.A., & Center for Drug Evaluation and Research (CDER) (Eds.), *Guidance for Industry and FDA Staff- Qualification Process for Drug Development Tools.* MD: Silver Spring.

Flood, J. F., & Cherkin, A. (1986). Scopolamine effects on memory retention in mice: a model of dementia? *Behavioral and Neural Biology, 45*, 169–184.

Fredrickson, A., Snyder, P. J., Cromer, J., Thomas, E., Lewis, M., & Maruff, P. (2008). The use of effect sizes to characterize the nature of cognitive change in psychopharmacological studies: an example with scopolamine. *Human Psychopharmacology: Clinical and Experimental, 23*(5), 425–436.

Freedman, R. (2003). Schizophrenia. *The New England Journal of Medicine, 349*, 1738–1749.

Freedman, R., Coon, H., Myles-Worsley, M., Orr-Urtreger, A., Olincy, A., Davis, A., et al. (1997). Linkage of a neurophysiological deficit in schizophrenia to a chromosome 15 locus. *Proceedings of the National Academy of Sciences of the United States of America, 94*(2), 587–589.

Freedman, R., Ross, R., Leonard, S., Myles-Worsley, M., Adams, C. E., Waldo, M., et al. (2005). Early biomarkers of psychosis. *Dialogues in Clinical Neuroscience, 7*(1), 17–29.

Frohlich, J., & Horn, J. (2014). Reviewing the ketamine model for schizophrenia. *Journal of Psychopharmacology, 28*(4), 287–302.

Geyer, M. A., Krebs-Thomson, K., & Braff, D. L. (2001). Pharmacological studies of prepulse inhibition models of sensorimotor gating deficits in schizophrenia: a decade in review. *Psychopharmacology, 156*, 117–154.

Giovannini, M. G., Camilli, F., Mundula, A., & Pepeu, G. (1994). Glutamatergic regulation of acetylcholine output in different brain regions: a microdialysis study in the rat. *Neurochemistry International, 25*(1), 23–26.

Goff, D. (2015). Drug development in schizophrenia: are glutamatergic targets still worth aiming at? *Current Opinion in Psychiatry, 28*(3), 207.

Gündisch, D. (2005). Nicotinic acetylcholine receptor ligands as potential therapeutics. *Expert Opinion on Therapeutic Patents, 15*(9), 1221–1239.

Guttman, L. (1945). A basis for analyzing test-retest reliability. *Psychometrika, 10*(4), 255–282.

Harmer, C., Bodinat, C., Dawson, G., Dourish, C., Waldenmaier, L., Adams, S., et al. (2010). Agomelatine facilitates positive versus negative affective processing in healthy volunteer models. *Journal of Psychopharmacology, 25*(9), 1159–1167.

Hasegawa, M., Kinoshita, H., Amano, M., et al. (1993). MK-801 increases endogenous acetylcholine release in the rat parietal cortex: a study using brain microdialysis. *Neuroscience Letters, 150*(1), 53–56.

Hashimoto, K. (2014). Targeting of NMDA receptors in new treatments for schizophrenia. *Expert Opinion on Therapeutic Targets, 18*(9), 1049–1063.

Hasselmo, M. E. (2006). The role of acetylcholine in learning and memory. *Current Opinion in Neurobiology, 16*(6), 710–715.

Healey, K., Bartholomeusz, C., & Penn, D. (2016). Deficits in social cognition in first episode psychosis: a review of the literature. *Clinical Psychology Review, 50*, 108–137.

Holahan, A. -L. V., & O'Driscoll, G. A. (2005). Antisaccade and smooth pursuit performance in positive-and negative-symptom schizotypy. *Schizophrenia Research, 76*(1), 43–54.

Howes, O., & Kapur, S. (2009). The dopamine hypothesis of schizophrenia: version III—the final common pathway. *Schizophrenia Bulletin, 35*(3), 549–562.

Hutton, S., & Ettinger, U. (2006). The antisaccade task as a research tool in psychopathology: a critical review. *Psychophysiology, 43*(3), 302–313.

Javitt, D. C., & Freedman, R. (2015). Sensory processing dysfunction in the personal experience and neuronal machinery of schizophrenia. *The American Journal of Psychiatry, 172*(1), 17–31.

Jobert, M., Wilson, F. J., Roth, T., Ruigt, G. S., Anderer, P., Drinkenburg, W. H., et al. (2013). Guidelines for the recording and evaluation of pharmaco-sleep studies in man: the international pharmaco-EEG society (IPEG). *Neuropsychobiology, 67*(3), 127–167.

Judd, L., McAdams, L., Budnick, B., & Braff, D. L. (1992). Sensory gating deficits in schizophrenia: new results. *The American Journal of Psychiatry, 149*, 488–493.

Kaitin, K. I. (Ed.), (2014). CNS drugs take longer to develop, have lower success rates, than other drugs. *Tufts Center for the Study of Drug Development Impact Report, 16*(6).

Kapur, S. (1998). A new framework for investigating antipsychotic action in humans: lessons from PET imaging. *Molecular Psychiatry, 3*(2), 135–140.

Kapur, S., & Seeman, P. (2002). NMDA receptor antagonists ketamine and PCP have direct effects on the dopamine D(2) and serotonin 5-HT(2)receptors-implications for models of schizophrenia. *Molecular Psychiatry, 7*(8), 837–844.

Keefe, R. S., Bilder, R. M., Harvey, P. D., Davis, S. M., Palmer, B. W., Gold, J. M., et al. (2006). Baseline neurocognitive deficits in the CATIE schizophrenia trial. *Neuropsychopharmacology, 31*(9), 2033–2046.

Khakpai, F., Nasehi, M., Haeri-Rohani, A., Eidi, A., & Zarrindast, M. (2012). Scopolamine induced memory impairment; possible involvement of NMDA receptor mechanisms of dorsal hippocampus and/or septum. *Behavioural Brain Research, 231*(1), 1–10.

Kim, S. Y., Lee, H., Kim, H. J., Bang, E., Lee, S. H., Lee, D. W., et al. (2011). In vivo and ex vivo evidence for ketamine-induced hyperglutamatergic activity in the cerebral cortex of the rat: Potential relevance to schizophrenia. *NMR in Biomedicine, 24*(10), 1235–1242.

Kleinloog, D., Uit den Boogaard, A., Dahan, A., Mooren, R., Klaassen, E., Stevens, J., et al. (2015). Optimizing the glutamatergic challenge model for psychosis, using S+ -ketamine to induce psychomimetic

symptoms in healthy volunteers. *Journal of Psychopharmacology (Oxford, England), 29*(4), 401–413.

Klinkenberg, I., & Blokland, A. (2010). The validity of scopolamine as a pharmacological model for cognitive impairment: a review of animal behavioral studies. *Neuroscience & Biobehavioral Reviews, 34*(8), 1307–1350.

Koychev, I., McMullen, K., Lees, J., & Dadhiwala, R. (2012). A validation of cognitive biomarkers for the early identification of cognitive enhancing agents in schizotypy: a three-center double-blind placebo-controlled study. *European Journal of Neuropharmacology, 22*, 469–481.

Krystal, J., Karper, L., Bennett, A., D'Souza, D., Abi-Dargham, A., Morrissey, K., et al. (1998). Interactive effects of subanesthetic ketamine and subhypnotic lorazepam in humans. *Psychopharmacology, 135*(3), 213–229.

Krystal, J. H., Abi-Saab, W., Perry, E., D'Souza, D. C., Liu, N., Gueorguieva, R., et al. (2005). Preliminary evidence of attenuation of the disruptive effects of the NMDA glutamate receptor antagonist, ketamine, on working memory by pretreatment with the group II metabotropic glutamate receptor agonist, LY354740, in healthy human subjects. *Psychopharmacology, 179*(1), 303–309.

Krystal, J. H., D'Souza, D. C., Karper, L. P., Bennett, A., Abi-Dargham, A., Abi-Saab, D., et al. (1999). Interactive effects of subanesthetic ketamine and haloperidol in healthy humans. *Psychopharmacology, 145*(2), 193–204.

Krystal, J. H., Karper, L., Seibyl, J., Freeman, G., Delaney, R., Bremner, J., et al. (1994). Subanesthetic effects of the noncompetitive NMDA antagonist, ketamine, in humans—psychotomimetic, perceptual, cognitive, and neuroendocrine responses. *Archives of General Psychiatry, 51*(3), 199–214.

Kumari, V., Antonova, E., Wright, B., Hamid, A., Hernandez, E., Schmechtig, A., et al. (2017). The mindful eye: smooth pursuit and saccadic eye movements in meditators and non-meditators. *Consciousness and Cognition, 48*, 66–75.

Laczó, J., Markova, H., Lobellova, V., Gazova, I., Parizkova, M., Cerman, J., et al. (2016). Scopolamine disrupts place navigation in rats and humans: a translational validation of the hidden goal task in the morris water maze and a real maze for humans. *Psychopharmacology*, 1–13.

Lahti, A., Holcomb, H., Gao, X. -M., & Tamminga, C. (1999). NMDA-sensitive glutamate antagonism: a human model for psychosis. *Neuropsychopharmacology, 21*(6), S158–S169.

Laruelle, M., Abi-Dargham, A., Gil, R., Kegeles, L., & Innis, R. (1999). Increased dopamine transmission in schizophrenia: relationship to illness phases. *Biological Psychiatry, 46*, 56–72.

Lenz, R., Baker, J., Locke, C., Rueter, L., Mohler, E., Wesnes, K., et al. (2012). The scopolamine model as a pharmacodynamic marker in early drug development. *Psychopharmacology, 220*(1), 97–107.

Light, G., & Braff, D. (2000). Do self-reports of perceptual anomalies reflect gating deficits in schizophrenia patients? *Biological Psychiatry, 47*(5), 463–467.

Lim, Y. Y., Prang, K. H., Cysique, L., Pietrzak, R. H., Snyder, P. J., & Maruff, P. (2009). A method for cross-cultural adaptation of a verbal memory assessment. *Behavior Research Methods, 41*, 1190–1200.

Littleton, A. C., Register-Mihalik, J. K., & Guskiewicz, K. M. (2015). Test-retest reliability of a computerized concussion test: CNS vital signs. *Sports Health: A Multidisciplinary Approach, 7*(5), 443–447.

Littman, B. H., & Williams, S. A. (2005). The ultimate model organism: progress in experimental medicine. *Nature Reviews. Drug discovery, 4*(8), 631–638.

Luntz-Leybman, V., Bickford, P. C., & Freedman, R. (1992). Cholinergic gating of response to auditory stimuli in rat hippocampus. *Brain Research, 587*(1), 130–136.

Ma, J., Svetnik, V., Snyder, E., Lines, C., Roth, T., & Herring, W. (2014). Electroencephalographic power spectral density profile of the orexin receptor antagonist suvorexant in patients with primary insomnia and healthy subjects. *Sleep, 37*(10), 1609–1619.

Marder, S. (2016). Alpha-7 nicotinic agonist improves cognition in schizophrenia. *Evidence Based Mental Health, 19*(2), 60.

Maruff, P., Lim, Y. Y., Darby, D., Ellis, K. A., Pietrzak, R. H., Snyder, P. J., et al. (2013). Clinical utility of the cogstate brief battery in identifying cognitive impairment in mild cognitive impairment and Alzheimer's disease. *BMC Pharmacology and Toxicology, 1*(1), 1–11.

Matza, L. S., Buchanan, R., Purdon, S., Brewster-Jordan, J., Zhao, Y., & Revicki, D. A. (2006). Measuring changes in functional status among patients with schizophrenia: the link with cognitive impairment. *Schizophrenia Bulletin, 32*(4), 666–678.

McDowell, I. (2006). *Measuring health: A guide to rating scales and questionnaires.* USA: Oxford University Press.

McGhie, A., & Chapman, J. (1961). Disorders of attention and perception in early schizophrenia. *The British Journal of Medical Psychology, 34,* 103–116.

McGonigle, P., & Ruggeri, B. (2014). Animal models of human disease: challenges in enabling translation. *Biochemical Pharmacology, 87*(1), 162–171.

Meyer, J., Wilson, A., Sagrati, S., Hussey, D., Carella, A., Potter, W., et al. (2004). Serotonin transporter occupancy of five selective serotonin reuptake inhibitors at different doses: an [11C] DASB positron emission tomography study. *The American Journal of Psychiatry, 161*(5), 826–835.

More, S. V., Kumar, H., Cho, D. Y., & Yun, Y. S. (2016). Toxin-induced experimental models of learning and memory impairment. *International Journal of Molecular Sciences, 17*(9), E1147.

Morgan, P., Graaf, P., Arrowsmith, J., Feltner, D., Drummond, K., Wegner, C., et al. (2011). Can the flow of medicines be improved? Fundamental pharmacokinetic and pharmacological principles toward improving Phase II survival. *Drug Discovery Today, 17*(9-10), 419–424.

Morris, B. J., Cochran, S. M., & Pratt, J. A. (2005). PCP: from pharmacology to modelling schizophrenia. *Current Opinion in Pharmacology, 5*(1), 101–106.

Mullane, K., Winquist, R. J., & Williams, M. (2014). Translational paradigms in pharmacology and drug discovery. *Biochemical Pharmacology, 87*(1), 189–210.

Muller, N., Riedel, M., Eggert, T., et al. (1999). Internally and externally guided voluntary saccades in unmedicated and medicated schizophrenic patients. Part II. Saccadic latency, gain, and fixation suppression errors. *European Archives of Psychiatry and Clinical Neuroscience, 249*(1), 7–14.

Newcomer, J. W., Farber, N. B., Jevtovic-Todorovic, V., Selke, G., Melson, A. K., Hershey, T., et al. (1999). Ketamine-induced NMDA receptor hypofunction as a model of memory impairment and psychosis. *Neuropsychopharmacology, 20*(2), 106–118.

Olincy, A., Harris, J., Johnson, L., Pender, V., Kongs, S., Allensworth, D., et al. (2006). Proof-of-concept trial of an alpha7 nicotinic agonist in schizophrenia. *Archives of General Psychiatry, 63*(6), 630–638.

Olincy, A., & Stevens, K. E. (2007). Treating schizophrenia symptoms with an alpha7 nicotinic agonist, from mice to men. *Biochemical Pharmacology, 74*(8), 1192–1201.

Olney, J. W., Newcomer, J. W., & Farber, N. B. (1999). NMDA receptor hypofunction model of schizophrenia. *Journal of Psychiatric Research, 33*(6), 523–533.

Øye, I., Paulsen, O., & Maurset, A. (1992). Effects of ketamine on sensory perception: evidence for a role of N-methy-D-aspartate receptors. *The Journal Pharmacology and Experimental Therapeutics, 260,* 1209–1213.

Pangalos, M., Schechter, L., & Hurko, O. (2007). Drug development for CNS disorders: strategies for balancing risk and reducing attrition. *Nature Reviews. Drug Discovery, 6*(7), 521–532.

Perry, E., Cramer, J., Cho, H. -S., Petrakis, I., Karper, L., Genovese, A., et al. (2007). Psychiatric safety of ketamine in psychopharmacology research. *Psychopharmacology, 192*(2), 253–260.

Potter, W. (2012). New era for novel CNS drug development. *Neuropsychopharmacology, 37*(1), 278–280.

Preskorn, S., Gawryl, M., Dgetluck, N., Palfreyman, M., Uer, L., & Hilt, D. (2014). Normalizing effects of EVP-6124, an alpha-7 nicotinic partial agonist, on event-related potentials and cognition: a proof of concept, randomized trial in patients with schizophrenia. *Journal of Psychiatric Practice, 20*(1), 12.

Pringle, A., & Harmer, C. J. (2015). The effects of drugs on human models of emotional processing: an account of antidepressant drug treatment. *Dialogues in Clinical Neuroscience, 17*(4), 477–487.

Reilly, J. L., Lencer, R., Bishop, J. R., Keedy, S., & Sweeney, J. A. (2008). Pharmacological treatment effects on eye movement control. *Brain and Cognition, 68*(3), 415–435.

Rentz, D. M., Amariglio, R. E., Becker, J. A., Frey, M., Olson, L. E., Frishe, K., et al. (2011). Face-name associative memory performance is related to amyloid burden in normal elderly. *Neuropsychologia, 49,* 2776–2783.

Robbins, T. W., Semple, J., Kumar, R., Truman, M. I., Shorter, J., Ferraro, A., et al. (1997). Effects of scopolamine on delayed-matching-to-sample and paired associates tests of visual memory and learning in human subjects: comparison with diazepam and implications for dementia. *Psychopharmacology, 134,* 95–106.

Romanelli, M. N., & Gualtieri, F. (2003). Cholinergic nicotinic receptors: competitive ligands, allosteric modulators, and their potential applications. *Medical Research Reviews, 23,* 393–426.

Rusted, J. M., & Warburton, D. M. (1988). The effects of scopolamine on working memory in healthy young volunteers. *Psychopharmacology, 96*(2).

Sahakian, B., Morris, R., Evenden, J., Heald, A., Levy, R. I., & M, & Robbins, Tw. (1988). A comparative-study of visuospatial memory and learning in Alzheimer-type dementia and Parkinsons-disease. *Brain, 111*(3), 695–718.

Schmechtig, A., Lees, J., Grayson, L., Craig, K., Dadhiwala, R., Dawson, G., et al. (2013). Effects of risperidone, amisulpride and nicotine on eye movement control and their modulation by schizotypy. *Psychopharmacology, 227*(2), 331–345.

Shiina, A., Shirayama, Y., Niitsu, T., Hashimoto, T., Yoshida, T., Hasegawa, T., et al. (2010). A randomised, double-blind, placebo-controlled trial of tropisetron in patients with schizophrenia. *Annals of General Psychiatry, 9*(1), 1–10.

Shreiner, L. (1997). Learning versus confirming in clinical drug development. *Clinical Pharmacology and Therapeutics, 61*(3), 275–291.

Singer, P., Feldon, J., & Yee, B. K. (2009). Are DBA/2 mice associated with schizophrenia-like endophenotypes? A behavioral contrast with C57BL/6 mice. *Psychopharmacology, 206*(4), 677–698.

Sitaram, N., Weingartner, H., & Gillin, J. C. (1978). Human serial learning: enhancement with arecholine and choline and impairment with scopolamine. *Science, 201,* 274–276.

Snyder, P., Bednar, M., Cromer, J., & Maruff, P. (2005). Reversal of scopolamine-induced deficits with a single dose of donepezil, an acetylcholinesterase inhibitor. *Alzheimer's & Dementia: The Journal of the Alzheimer's Association, 1*(2), 126–135.

Snyder, P., Lim, Y., Schindler, R., Ott, B., Salloway, S., Daiello, L., et al. (2014). Microdosing of scopolamine as a "cognitive stress test": rationale and test of a very low dose in an at-risk cohort of older adults. *Alzheimer's & Dementia, 10*(2), 262–267.

Sperling, R., Greve, D., Dale, A., Killiany, R., Holmes, J., Rosas, H., et al. (2002). Functional MRI detection of pharmacologically induced memory impairment. *Proceedings of the National Academy of Sciences, 99*(1), 455–460.

Stevens, K. E., Freedman, R., Collins, A. C., Hall, M., & Leonard, S. (1996). Genetic correlation of inhibitory gating of hippocampal auditory evoked response and bungarotoxin-binding nicotinic cholinergic receptors in inbred mouse strains. *Neuropsychopharmacology, 15,* 152–162.

Stone, J. M., Dietrich, C., Edden, R., Mehta, M. A., De Simoni, S., Reed, L. J., et al. (2012). Ketamine effects on brain GABA and glutamate levels with 1H-MRS: relationship to ketamine-induced psychopathology. *Molecular Psychiatry, 17*(7), 664–665.

Stone, J. M., Erlandsson, K., Arstad, E., Squassante, L., Teneggi, V., Bressan, R. A., et al. (2008). Relationship between ketamine-induced psychotic symptoms and NMDA receptor occupancy. *Psychopharmacology, 197*(3), 401–408.

Straube, A., Riedel, M., Eggert, T., & Müller, N. (1999). Internally and externally guided voluntary saccades in unmedicated and medicated schizophrenic patients. Part I. Saccadic velocity. *European Archives of Psychiatry and Clinical Neuroscience, 249*(1), 1–6.

Sunderland, T., Tarlot, P. M., Weingartner, H., Murphy, D. L., Newhouse, P. A., Mueller, E. A., et al. (1986). Pharmacologic modelling of Alzheimer's disease. *Progress in Neuropsychopharmacology and Biological Psychiatry, 10*(3), 599–610.

Swainson, R., Hodges, J. R., Galton, C. J., Semple, J., Michael, A., Dunn, B. D., et al. (2001). Early detection and differential diagnosis of Alzheimer's disease and depression with neuropsychological tasks. *Dementia and Geriatric Cognitive Disorders, 12*(4), 265–280.

Takahashi, H., Rissling, A., Pascual-Marqui, R., Kirihara, K., Pela, M., Sprock, J., et al. (2013). Neural substrates of normal and impaired preattentive sensory discrimination in large cohorts of nonpsychiatric subjects and schizophrenia patients as indexed by MMN and P3a change detection responses. *NeuroImage, 66*, 594–603.

Thiel, C. M., Henson, R. N., Morris, J. S., Friston, K. J., & Dolan, R. J. (2001). Pharmacological modulation of behavioral and neuronal correlates of repetition priming. *Journal of Neuroscience, 21*, 6846–6852.

Thienel, R., Kellermann, T., Schall, U., Voss, B., Reske, M., Halfter, S., et al. (2009). Muscarinic antagonist effects on executive function of attention. *The International Journal of Neuropsychopharmacology, 12*(10), 1307–1317.

Thornberg, S. A., & Saklad, S. R. (1996). A review of NMDA receptors and the phencyclidine model of schizophrenia. *Pharmacotherapy, 16*(1), 82–93.

Vallance, A. (2007). A systematic review comparing the functional neuroanatomy of patients with depression who respond to placebo to those who recover spontaneously: is there a biological basis for the placebo effect in depression? *Journal of Affective Disorders, 98*(1-2), 177–185.

van der Staay, F. J., Arndt, S. S., & Nordquist, R. E. (2009). Evaluation of animal models of neurobehavioral disorders. *Behavioral and Brain Functions, 5*, 11. https://doi.org/10.1186/1744-9081-5-11.

Van Selst, M. (2011). *Chapter 5: Measurement concepts. In P. Cozby & S. Bates (Eds.), Methods in behavioral research.* (11th ed.). McGraw-Hill Higher Education. http://www.sjsu.edu/people/mark.vanselst/courses/psyc120/s2/cozby5measurement.pdf.

Viaud-Delmon, I., Venault, P., & Chapouthier, G. (2010). Behavioral models for anxiety and multisensory integration in animals and humans. *Progress in Neuro-Psychopharmacology & Biological Psychiatry, 35*(6), 1391–1399.

Visser, D. S., Post, V. J., Waal, D. P., Cornet, F., Cohen, A. F., & Gerven, J. (2003). Biomarkers for the effects of benzodiazepines in healthy volunteers. *British Journal of Clinical Pharmacology, 55*(1), 39–50.

Voss, B., Thienel, R., Reske, M., Kellermann, T., Sheldrick, A., Halfter, S., et al. (2012). Cholinergic blockade under working memory demands encountered by increased rehearsal strategies: evidence from fMRI in healthy subjects. *European Archives of Psychiatry and Clinical Neuroscience, 262*(4), 329–339.

Wehling, M. (2009). Assessing the translatability of drug projects: what needs to be scored to predict success? *Nature Reviews Drug Discovery, 8*(7), 541–546.

Wesnes, K. A., Ward, T., McGinty, A., & Petrini, O. (2000). The memory enhancing effects of a Ginkgo biloba/Panax ginseng combination in healthy middle-aged volunteers. *Psychopharmacology, 152*, 353–361.

Whitehouse, P. J. (1986). Cholinergic receptors in aging and Alzheimer's disease. *Progress in Neuropsychopharmacology and Biological Psychiatry, 10*, 665–676.

Wierońska, J., Zorn, S., Doller, D., & Pilc, A. (2015). Metabotropic glutamate receptors as targets for new antipsychotic drugs: historical perspective and critical comparative assessment. *Pharmacology & Therapeutics, 157*, 10–27.

Wilner, P. (1990). Animal models of depression: an overview. *Pharmacology & Therapeutics, 45*(3), 425–455.

Wilner, P., & Mitchell, P. J. (2006). Animal models of depression. In M. Koch (Ed.), *Animal models of neuropsychiatric diseases* (pp. 223–292). London, UK: Imperial College Press. Chapter 6.

Winterer, G., Gallinat, J., Brinkmeyer, J., Musso, F., Kornhuber, J., Thuerauf, N., et al. (2012). Allosteric alpha-7 nicotinic receptor modulation and P50 sensory gating in schizophrenia: a proof-of-mechanism study. *Neuropharmacology, 64*, 197–204.

Winters, B. D., Bartko, S. J., Saksida, L. M., & Bussey, T. J. (2007). Scopolamine infused into perirhinal cortex improves object recognition memory by blocking the acquisition of interfering object information. *Learning & Memory, 14*(9), 590–596.

Xu, K., Krystal, J., Ning, Y., Chen, D., He, H., Wang, D., et al. (2015). Preliminary analysis of positive and negative syndrome scale in ketamine-associated psychosis in comparison with schizophrenia. *Journal of Psychiatric Research, 61*, 64–72.

Zarate, C., Mathews, D., & Furey, M. (2013). Human biomarkers of rapid antidepressant effects. *Biological Psychiatry, 73*(12), 1142–1155.

New Approaches in Translational Medicine for Phase I Clinical Trials of CNS Drugs[☆]

Matthew Macaluso*, *Michael Krams*[†], *Jonathan Savitz*[‡], *Wayne C. Drevets*[†], *Sheldon H. Preskorn**

*Department of Psychiatry and Behavioral Sciences, University of Kansas School of Medicine-Wichita and KU-Wichita Clinical Trials Unit, Wichita, KS, United States [†]Janssen Research & Development, LLC, of Johnson & Johnson, Titusville, NJ, United States [‡]Laureate Institute for Brain Research, Tulsa, OK, United States

I INTRODUCTION

Phase I studies are traditionally focused on determining the safety, tolerability, and pharmacokinetics of a new molecular entity in young healthy volunteers—first in a single ascending dose (SAD) study and then in a multiple ascending dose (MAD) study (Spilker, 1991). The goal of this chapter is to help clinicians and researchers better understand the strengths and weaknesses of this classic approach. In this chapter, we first review the rationale for and design of Phase I studies. We then discuss how Phase I studies are expanding to examine the effects of drugs in specific target populations, exploring both traditional endpoints and newer biomarker endpoints selected specifically to test the potential utility of the drug on the target illness (Wong, Tauscher, & Gründer, 2009). We will focus especially on the expansion of Phase I studies to gather data in a time- and cost-effective fashion that will allow sponsors to determine whether or not to proceed ("go/no-go") with a new investigational molecule (Gallo, Chuang-Stein, Dragalin, et al., 2006). The second half of this chapter will discuss how the development of novel molecular entities (NMEs) designed specifically to affect the central nervous system (CNS) creates challenges that cannot be addressed solely by the traditional approach, which will need to be augmented by new methodological strategies (Paul, Mytelka, Dunwiddie, et al., 2010).

II TRADITIONAL EARLY PHASE DESIGN CONSIDERATIONS

The authors subscribe to Sheiner's proposal to distinguish between the "learn" and "confirm" phases of drug development (Sheiner, 1997). In the "learn" phase (Phases I and II, exploratory clinical development), the goal is first to establish whether the NME is safe and well tolerated and has an appropriate pharmacokinetic (PK) profile and then to move on to the initial proof-of-concept efficacy trials. At its simplest level the traditional Phase I development of a NME consists of one initial SAD study followed by one MAD study in normal volunteers.

Traditionally the first step in Phase I development is a single-dose administration of the NME. These studies typically involve cohorts of six to nine subjects who receive the NME and two to three additional subjects who receive placebo, with assignments to the groups made in a double-blind, randomized manner. This approach permits the investigators to document differences in serious or nuisance adverse effects that occur

---

[☆]Material in this chapter is adapted with permission from the chapter: "Phase I Trials: From Traditional to Newer Approaches," by Matthew Macaluso, DO; Michael Krams, MD, PhD; and Sheldon H. Preskorn, MD, in Kalali A, ed. *Essential CNS Drug Development*. Cambridge University Press, 2012. This material was also published with permission in the *Journal of Psychiatric Practice*:

1. Macaluso, M., Krams, M., Preskorn, S. H. (2012). Phase I Trials. In: *Essential CNS Drug Development*. Cambridge General Academic Press. May. ISBN: 0521766060.
2. Macaluso, M., Krams, M., Preskorn, S. H. (2011). Phase I trials: from traditional to newer approaches. Part II. *Journal of Psychiatric Practice*, 17(4), 277–284. PMID: 21775829.

with high frequency (i.e., in more than one in four patients). After each dose level is administered, safety and tolerability are assessed before determining whether to administer a higher dose and if so, how much higher. The goal of these studies is to establish the maximum tolerated single dose (MTD) of the NME (McConnell, 1989). The design typically includes clearly laid out decision rules on whether or not to escalate to the next dose level, given the observations at previous dose levels (Dragalin, 2006).

Key questions involved in the design of a traditional SAD study include the following:

- What is the starting dose? The initial dose chosen for the SAD study is typically below the no observable effect level (NOEL) in animals (Crane & Newman, 2000).
- What is the escalation scheme (i.e., what multiple of the last dose is used for the next dose)?
- What constitutes a dose-limiting safety or tolerability problem, given the indication that is ultimately being sought?

The goal is to learn about the following: (Arbuck, 1996)

- Are the pharmacokinetics (PK) of the compound suitable for use as a clinical agent, including time to peak concentration (Tmax), peak concentration (Cmax), half-life (T1/2), trough concentration (Cmin), and dose-drug concentration linearity over the dose range studied?
- Are the PK of the molecule related to its pharmacodynamics (PD), including onset of effect (Tmax), duration of effect (T1/2), and intensity of effect (Cmax and Cmin)?
- What is the nature of the drug's PD in terms of both desired and undesired effects?
- What dose should be chosen based on the SAD results and knowledge from preclinical models to take into the MAD study?

At the end of an SAD study, the following should ideally have been determined:

- The MTD
- PK/PD relationship for safety and tolerability

However, a number of important learning points may not be achieved by the simplest traditional SAD study, such as the following:

- Does the NME—assuming it is being developed for a CNS indication—penetrate into the brain?
- Does the NME achieve a concentration at its intended site of action that is likely to produce a therapeutic effect?
- Does the NME have an effect on a biomarker that should be predictive of efficacy in the therapeutic indication for which it is being developed?

All of these points can be studied in normal volunteers, but they may also need to be studied in patients with the target illness, either to assess for the desired effect or because the dose-response curve may shift in the population with the target illness compared with normal volunteers. In such cases, it may be worth planning the study so that one or more cohorts of subjects who approximate the target populations can be studied after the initial studies are done in normal volunteers (i.e., a seamless transition from normal to symptomatic volunteers). An example of this would be the development of an NME for dementia of the Alzheimer's type. This could involve a multistep process, going from healthy young volunteers to healthy elderly volunteers to elderly volunteers with a mild form of the target illness. Since Alzheimer's dementia occurs predominantly in elderly patients, it is important to determine if the safety, tolerability, and even the PK of the NME differ between young healthy volunteers and elderly healthy volunteers.

After completing an SAD study, an MAD study is conducted to better understand the PK and PD of multiple doses of the NME (Meinert & Tonascia, 1986). The designs and goals of the SAD and MAD studies are generally the same, except that in the MAD study, the drug is administered to the volunteers for 1 or more weeks on a daily basis, or even multiple times per day depending on the drug's half-life as determined in the SAD study. The initial doses for this study are based on the SAD study and are generally higher than the starting doses used in the SAD but almost always lower than the MTD determined in the SAD study. Otherwise the design is comparable in terms of number of subjects receiving the NME versus placebo, and the dose escalation is again based on the results of the previous dosing group. Again the MAD study may be designed to transition into the age range for the target population and/or into a mildly affected group having the target illness.

The distinction between late Phase I and early Phase II studies may not always be clearly demarcated, and a study may straddle the boundary between traditional Phase I and early Phase II. Thus the participants in the early groups in an SAD and/or MAD study are traditionally young healthy volunteers, whereas later volunteers may have the target illness. The goal of including these latter cohorts in these studies may be to provide an early test of mechanism and/or concept to determine whether the NME has the properties required to have a reasonable chance of success in Phase III studies (see Fig. 6.1). In the "confirm" phase (i.e., Phase III), the goal is to produce substantial evidence that a treatment regimen applied to a particular patient population achieves a specific endpoint that is accepted by regulatory agencies as being efficacious and will yield a sufficient effect size relative to any safety or tolerability concerns to warrant approval (Kraemer, Morgan, Leech, et al., 2003).

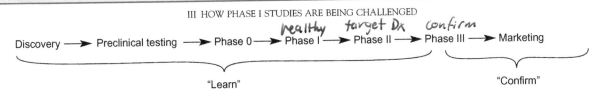

FIG. 6.1   "Learn and confirm" phases of drug development.

"Confirm" is about hypothesis testing, and the main customers are regulatory agencies (Sheiner, 1997). However, throughout the process, principally before entering Phase III testing, the goal is to efficiently identify viable and nonviable compounds and focus resources on the former and discontinue further development of the latter.

CNS drug development is facing challenges that will drastically affect the way both parts of the "learn" and "confirm" dichotomy will be conducted going forward (Paul et al., 2010). Essentially, psychiatric drug development has mainly been "living off" chlorpromazine and its derivatives for 50 years and now has entered the era of the human genome project, which is producing more targets than large pharma can afford to pursue efficiently. Hence, CNS drug development must "evolve," to be more efficient in making "go/no-go" decisions.

III HOW PHASE I STUDIES ARE BEING CHALLENGED

To understand the future, one must be cognizant of the past. A review of the last 30 years of CNS drug development reveals that CNS clinical trial designs have been largely static (Kaitin, 2010). Over this period, this approach has made sense because newer drugs were by and large a molecular refinement of previous molecules, with minor although sometimes clinically important changes in either PD or PK. However, the process of refining chlorpromazine and its derivatives appears to have been virtually exhausted, with the NMEs over the last decade having been predominantly "me-too" drugs with little, if any, clinically meaningful differences compared with already marketed drugs in the same therapeutic class. Over the last two to three decades, drugs were developed to treat the same set of symptoms treated by earlier drugs and were based on the same mechanisms of action as their predecessors. Nevertheless, advances were made by "cleaning up" the pharmacology of the older drugs. For example, selective serotonin reuptake inhibitors (SSRIs) were without a doubt "cleaner" versions of their predecessors, the tricyclic antidepressants (TCAs) (Preskorn, 1996a). While their mechanism of antidepressant action was not novel, they did not have the mechanisms responsible for the multiple adverse effects—ranging from nuisance to lethal—of the TCAs. It is fair to say that no psychiatric molecular entities with truly novel mechanisms of action have been marketed for the treatment of schizophrenia, major depression, or

anxiety disorders since the early 1960s (Preskorn, 2010). In this regard the "low-hanging fruit" appears to have been thoroughly picked.

Since the newer drugs worked via the same therapeutic mechanism of action as the older drugs, the designs of clinical trials over this period were essentially the same as those used for the original drugs. The advantage of this approach was that it followed a proved pathway to marketing approval by the US Food and Drug Administration. The disadvantage of drug development in recent decades is that the process did not yield novel compounds to address unmet needs in a new way. As most researchers involved with industry and academia believe, this "era" of refining pharmacology is coming to a close, as NMEs targeting specific illnesses are being developed. In fact, CNS drug development may see its first "disease-altering" compounds reach the market over the next 5–10 years with treatments for Alzheimer's dementia currently appearing to be earliest in line for likely approval (Panza, Frisardi, Imbimbo, et al., 2011).

This shift to developing truly novel compounds is likely to require new trial designs for a variety of reasons. First and foremost the same "rehashed" study designs that have been used to study older compounds, which had derivative mechanisms of action and acted symptomatically, are not likely to yield meaningful data in the study of truly novel compounds with different mechanisms of action that may be capable of altering the course of the illness. Therefore it will likely be necessary to revise Phase I studies to achieve a quicker, seamless transition from normal healthy volunteers to participants with the target illness, when the preliminary results warrant such a transition.

Such studies will aim to answer the following questions:

- Is the PK profile of the drug the same in normal healthy volunteers as in the participants with the target illness? There are many examples where the PK profile of a drug differs to a clinically meaningful extent between young healthy and elderly healthy patients.
- Is the dose-response curve in patients with the target illness the same as in young healthy volunteers? For example, a consensus exists that normal healthy volunteers do not tolerate dopamine-2 receptor antagonists (e.g., haloperidol) as well as volunteers with schizophrenia.
- Does the drug cross the blood-brain barrier, and does it reach/modulate its desired target to a desired degree?

To address these questions, Phase I studies must be expanded to include data that will allow sponsors to determine whether or not to proceed ("go/no-go" decisions) in a time- and cost-effective fashion.

The same principles apply to early Phase II studies, which are classically concerned with early proof of mechanism (POM), proof of concept (POC), and/or efficacy in the target illness. When developing a study to evaluate efficacy, the measurements used to evaluate specific efficacy endpoints must be carefully examined. If one is examining an NME, scales that were used to evaluate the efficacy of older molecules may not provide meaningful data. For example, the use of the positive and negative syndrome scale (PANSS) (Kay, Fiszbein, & Opler, 1987) to assess efficacy was effective in the late Phase I and early Phase II studies evaluating the pharmacology of "atypical" antipsychotics, which were in essence "reinventions" of chlorpromazine, which meets all or virtually all of the criteria for "atypicality." Using the PANSS as the primary outcome measure to examine a novel and potentially disease-altering treatment for schizophrenia may not yield meaningful information since the drug may be truly "novel" (Lepping, Sambhi, Whittington, et al., 2011).

IV CHALLENGES WITH THE TRADITIONAL APPROACH

Although the traditional process of drug development has led to the successful approval of thousands of currently marketed drugs, it has also resulted in large financial investment in testing compounds that eventually failed to be approved (Paul et al., 2010). As Phase I studies begin to examine truly novel molecules/drugs, many of the "classic" Phase I approaches will no longer be adequate or efficient. Nevertheless the traditional Phase I single ascending dose (SAD) and multiple ascending dose (MAD) studies have many merits, as was discussed in Part I of this series (Macaluso, Krams, & Preskorn, 2011). The goal in adapting Phase I studies is not to "throw the baby out with the bath water" but instead to keep what works and augment or modify the methodology when it does not address the challenges and demands of central nervous system (CNS) drug development in the 21st century.

As discussed earlier in this chapter, the stages of drug development have traditionally been divided into "learn" and "confirm" phases (Fig. 6.1). Thus the drug development process leading up to Phase III studies is designed to efficiently identify viable and nonviable compounds, to focus resources on the former and discontinue further development of the latter. In contrast the goal of the "confirm" phase (i.e., Phase III) is to produce substantial evidence that a treatment regimen applied to a particular patient population achieves a specific endpoint that is accepted by regulatory agencies as being efficacious and will yield a sufficient effect size relative to any safety or tolerability concerns to warrant approval (Kraemer et al., 2003). "Confirm" is about hypothesis testing, and the main customers are regulatory agencies (Spilker, 1991).

Some preliminary questions that Phase I studies may be called on to answer include the following:

- Do the safety, tolerability, and pharmacokinetics (PK) of the drug in normal volunteers generalize to the target population?
- What is the best dose, and what is the best dosing schedule to take into Phase II efficacy trials?
- In what subject population (i.e., patients or normal volunteers) and for what indication should the agent be studied?
- What measures should be used to assess the drug's efficacy?

The notion of using Phase I studies solely to explore PK and safety/tolerability in normal volunteers may be both inappropriate and inefficient. Should the earliest exploration of the PK and safety/tolerability of the NME instead be conducted in a population of otherwise healthy individuals with the target psychiatric illness? If that is possible, then one can more meaningfully explore other endpoints, including biomarker surrogates, potentially leading to greater efficiency and cost-effectiveness in the drug development process. Thus it is important to analyze to what extent data from subjects chosen for early clinical trials can be used to make appropriate inferences about subjects who will eventually be studied in "confirm" trials.

In CNS drug development, sponsors often start with Phase I studies in normal young healthy male volunteers. The reason only male volunteers are used is that the relevant toxicity data necessary to justify testing in women of child-bearing potential may not exist (i.e., that research may not yet have been done) (Sheiner, 1997). The main advantages of using normal volunteers are the speed with which subjects can be enrolled and generally lower costs. However, young healthy male subjects may not be good predictors of the PK and/or safety/tolerability of the drug in either females or patients with a particular illness. For instance, the maximum tolerated dose (MTD) of an antipsychotic when tested in normal subjects may be considerably lower than in patients with schizophrenia (Gilles & Luthringer, 2007). Hence, the use of otherwise healthy individuals with schizophrenia in Phase I studies might yield more meaningful PK and safety data than studies in normal healthy volunteers, particularly in terms of MTDs. It could therefore be argued that, even in early clinical trials, one should incorporate a seamless transition to participants with the target illness to characterize the dose-response curve in that specific population.

Along these lines, in oncology, individuals diagnosed with cancer are considered a "special population" (Agrawal & Emanuel, 2003). Phase I oncology studies are often conducted in patients rather than normal healthy volunteers for several reasons. First, cancer drugs are often inherently toxic to human cells, so that it may be considered unethical to expose normal healthy volunteers to the potential toxicity of the NME as the first step in its development (Emanuel, 1995). In addition, there is also the potential for an early efficacy signal in individuals who have likely exhausted all available treatments.

In a similar manner, psychiatric patients are also a "special population." As discussed earlier, doses of drugs that are commonly used in individuals with psychiatric illnesses are often intolerable to normal controls. However, we are also aware of a number of instances in which drugs with a CNS mechanism of action have dose–response curves that are skewed to the left in participants with the target illness compared with the healthy young volunteer population (e.g., dopamine D2 receptor antagonism in patients with Parkinson's disease or muscarinic receptor antagonism in patients with Alzheimer's disease) (Preskorn, 2011). These findings mean that the MTD determined in traditional Phase I studies may not be relevant to the dose needed in Phase II efficacy trials. Fig. 6.2 illustrates how dose-response curves can shift either to the right or left in different populations because of pharmacokinetic or pharmacodynamic variables (Preskorn, 1996b). In addition, the study of CNS drugs in Phase I involves a unique set of challenges that do not apply to other areas of drug investigation, including passage of drugs into the brain; targeting of a specific brain receptor area, transporter, or peptide; and a set of target illnesses that are described syndromically by a set of core symptoms, rather than by a biologically based marker of disease. Finally, certain biomarker endpoints, which are used as a proxy or surrogate for the illness in question, may only be manifested in individuals with the illness (e.g., abnormalities of P50 and P300 evoked potentials in individuals with schizophrenia) (de Visser, van der Post, Pieters, et al., 2001). Along the same lines a potential antipsychotic medication may target certain biological markers such as dopamine D2 or N-methyl-D-aspartate (NMDA) receptors. If that is the case, critical information to establish early on, even in Phase I, would be whether the drug enters the brain and whether it targets a physiologically relevant biomarker endpoint and to what degree (Conn & Roth, 2008). If these conditions are believed to be necessary for the drug to demonstrate efficacy in individuals with the target illness, why not establish such facts as early as possible in the development process? Further, if a drug is intended to target a mechanism that exists only in the diseased population (e.g., a receptor variant), then studying this question early makes good sense to determine the

Shift in the dose-response curve

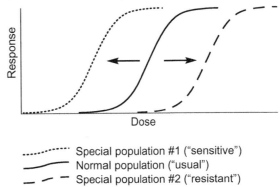

.... Special population #1 ("sensitive")
———— Normal population ("usual")
— — — Special population #2 ("resistant")

Pharmacokinetic explanation

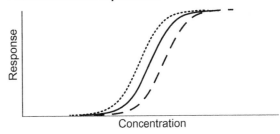

Pharmacodynamic explanation

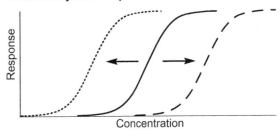

Top: population treated with different drug doses to establish dose-response curve. Data from normal patients, the largest subpopulation, represent the "usual" dose-response curve. The dose-response curve for the "more sensitive" group is shifted to the left and that of the "less sensitive" group is shifted to the right.
Middle: The "more sensitive" group is shifted to the left because they cleared the drug more slowly than normal (i.e., prescribed dose is functionally higher) and the "less sensitive" group is shifted to the right because they cleared the drug more rapidly than normal (i.e., prescribed dose is functionally lower).
Bottom: groups differ in "sensitivity" to mechanism of action of the drug. Pharmacodynamically "sensitive" patients are shifted to the left even after adjusting for the concentration of the drug at the regulatory protein of interest while the "resistant" patients are shifted to the right (i.e., they require a greater concentration because they are less sensitive to the drug's mechanism of action).

FIG. 6.2 Shifts in dose-response curves in special populations.

ultimate viability of the compound. Thus the population studied in Phase I may need to be critically assessed (Conn & Roth, 2008). The goal is to "learn" as much as possible in Phase I and II studies to support the best "go/no-go" decisions and to inform the design of the "confirm" studies in terms of population, dose(s), and dosing schedule to be studied and endpoints to be measured.

Thus we propose that the goal of early drug development should be to identify "necessary conditions" that, if not met, would warrant early termination of the drug's development. Even if these conditions are met, the drug may still face future challenges since all "sufficient conditions" must ultimately be met to achieve success. However, the emphasis on identifying "necessary conditions" is helpful in enabling early stopping decisions (Gallo et al., 2006). Establishing safety and tolerability is always necessary, but not generally sufficient if the goal is to develop an effective marketable drug. For example, a drug could be safe and well tolerated in populations of normal volunteers, demonstrating ideal PK, but it may be altogether inadequate as a CNS drug because it does not cross the blood-brain barrier or does not do so sufficiently to establish therapeutic concentrations (i.e., minimal predicted efficacious concentrations) in the brain region of interest. While crossing the blood-brain barrier can be addressed in preclinical pharmacology, those findings may not predict what happens in humans, nor do they indicate effective target modulation.[1]

If these types of questions can be addressed early, "go/no-go" decisions can be made safely in a more time- and cost-effective manner, adding a "litmus test" to the usual exploratory endpoints prior to entering the "confirm" phase. If such knowledge can be obtained in Phase I studies, meaningful amounts of time and money can be saved that can potentially be devoted to other compounds or areas of investigation.

A key point is that the approach to early clinical development should be tailored to the needs of the individual program. Studies investigating a first-in-class compound about which little preexisting knowledge is available will require a different approach than a "me-too" compound whose pharmacokinetic and pharmacodynamic profile is well understood.

Thus the drug developer must decide on a number of important issues in designing the early studies. For viable compounds the goal is to establish the correct treatment regimen, patient population, and endpoints to be taken into the "confirm" trials. This information also provides essential feedback on how to optimize follow-up compounds. In light of what has been learned about the NME, can effective target modulation be achieved? Can direct or indirect evidence of efficacy be demonstrated? How can the therapeutic index be optimized

relative to the exposure profile (i.e., does a high or low peak/trough ratio exist)?

In terms of dosing a much more accurate method of determining the MTD is the continuous reassessment method (CRM) (Goodman, Zahurak, & Piantadosi, 1995). The objective of the CRM is to use information gathered in real time to decide on how many more subjects should be allocated to the current dose and to decide what the next dose level should be. In some companies the CRM has superseded more traditional approaches. The CRM and other adaptive designs will be discussed in greater detail later in this chapter.

In gathering information about dosing, it is valuable to construct a "therapeutic index" (TI), which immediately focuses attention on both efficacy and safety. In early clinical trials, observations about efficacy may be limited to biomarker endpoints, but keeping efficacy in mind will strongly influence the design of these early trials. For instance, if an acceptable safety/tolerability profile cannot be established at exposures that are predicted to be in the minimal efficacious range, then there is no rationale for moving forward. The application of modeling techniques can efficiently establish which treatment regimen can achieve the best TI. Thus at least three key ingredients must be established as early as possible in the learn phase of the drug development process (Fig. 6.1):

1. Safety and tolerability
2. Pharmacodynamic measures of efficacy (biomarker-based)
3. Pharmacokinetic information relative to items 1 and 2

To gather information on how to administer the drug, the study should aim to identify the PK profile that will yield the optimal TI.

The questions posed earlier are as provocative as they are practical. If the early stage of the drug development process is not revised so that such questions can be answered efficiently in terms of both time and cost, the future of CNS drug development could be in jeopardy, as excessive dollars will be spent on compounds that are not viable, discouraging companies from developing NMEs, and biological agents that may in fact work. To better address these questions, we will now explore how adaptive trial designs can streamline this process.

V ADAPTIVE TRIALS DESIGN

What makes a trial design adaptive? The main tenet of adaptive trial design is the ability to examine data in "real time" and to "react" or adjust to the new information as the study is ongoing (Gallo et al., 2006). This approach allows a study to be terminated early for futility (i.e., the study is unable to meaningfully test efficacy) or for success earlier than would be possible in traditional

---

[1] Most drugs produce their beneficial and adverse effects by altering the functional status of a regulatory protein. Target modulation could, for example, involve a drug acting as an agonist, antagonist, or inverse agonist at a specific receptor. Effective target modulation means having the effect at the target (e.g., agonism or antagonism) necessary to produce the desired clinical effect with minimal adverse effects.

nonadaptive trials. The goal is to improve the quality and the speed of decision-making (e.g., the ability to determine which dose and dosing regimen are optimal). Adaptive designs also serve to minimize the number of individuals unnecessarily exposed to compounds that are not viable, either because they are ineffective or unsafe (Krams, Burman, Dragalin, et al., 2007). On the other hand, there is the potential to stop early for success, ultimately bringing beneficial therapies to patients more quickly. Hence, adaptive trial designs have the potential to produce valid decision-making more quickly while reducing costs and conserving valuable research resources.

The goal of adaptive designs is to learn from the accumulating trial data and to apply this newly gained knowledge as quickly as possible to optimize the ongoing execution of the study (Krams, Sharma, Dragalin, et al., 2009; Shen, Preskorn, Dragalin, et al., 2011). The study design allows for a number of potential modifications, including stopping the trial or increasing the likelihood of patient assignment to the optimal doses in terms of the balance of safety/tolerability and efficacy.

The adaptive approach must start with early study planning and should involve a multidisciplinary group ranging from statisticians to preclinical and clinical scientists and clinical operation personnel. This multidisciplinary team must develop an understanding of the specific patient population and appropriate study endpoints. The goal is to integrate real-time learning, including data acquisition, complex randomization schemes, and enhanced drug supply management requirements (Chow & Chang, 2008).

Adaptive designs involve looking at interim data during the course of the trial without having an impact on the trial's validity and integrity. One method of ensuring maintenance of the trial's validity and integrity is to establish the operating characteristics of the design in advance, either analytically or through simulations, and to demonstrate that there is no inflation of type 1 error. Strategic planning teams must preplan the setup/function of data monitoring committees and predetermine the decision tree for making adaptive "changes" to the study design. In addition, a clear process of how early data will be kept confidential must be defined and conceptualized in advance of study operationalization.

Such "real-time" adaptation can involve the following rules: (Chow & Chang, 2008)

- Allocation rule (how subjects will be allocated to which treatment arms)
- Sampling rule (how many subjects will be sampled at the next stage)
- Stopping rule (when to stop the trial, for efficacy, safety, and/or futility)

- Decision rule (decisions pertaining to design changes not covered by the previous three rules).

At any stage the data may be analyzed, and next stages redesigned taking into account all available data.

VI TYPES OF ADAPTIVE DESIGNS

First-in-human studies—ascending dose escalation designs. Adaptive model-based dose escalation designs, such as the CRM, can be applied to first-in-human and follow-up multiple ascending dose studies (O'Quigley, Pepe, & Fisher, 1990). In contrast to traditional dose escalation designs, adaptive dose escalation designs typically use modeling of the dose-response relationship to help identify the dose range closest to the MTD. These designs may also establish proof of target modulation, typically utilizing biomarker endpoints.

Determining optimal dose. In cases where dose-response is thought to be monotonic, design efficiency can be achieved by starting with the highest tolerated dose and placebo. Predetermined futility conditions can be included and should force early termination of the study if not met. If futility conditions are not met, the dose-response relationship can be assessed with the goal of determining "optimal dose."

Response adaptive dose-ranging study designs. Dose-response can be evaluated efficiently by using response-adaptive allocation designs in which the trial begins with a greater number of dose levels than are typically included in conventional clinical trials (U.S. Department of Health and Human Services, Food and Drug Administration (FDA), Center for Drug Evaluation and Research (CDER), Center for Biologics Evaluation and Research (CBER), 2008). Instead of using classic pairwise comparisons, these response-adaptive designs include real-time modeling of dose-response data to identify dose levels that fulfill predetermined set points. Data are applied immediately and affect the drug dose that future subjects in the study will receive. When futility is met, this may result in discontinuation of nonviable treatment arms or of the entire study.

Now that we have examined the applicability of adaptive trial designs, we will explore the use of biomarkers in clinical trials.

VII BIOMARKER ENDPOINTS

In an era in which 90% of investigational new drugs fail to make it to market, the use of biomarkers can efficiently and cost effectively identify NMEs that are unlikely to be viable, with the goal of not wasting any more resources on such NMEs than necessary to make a valid "no-go" decision. If the NME fails to reach its target or to appropriately modulate that target as assessed by a clinically meaningful biomarker, one can make a "no-go" decision.

The FDA defines a biomarker as a "measurable characteristic that is an indicator of normal biological processes, pathogenic processes, and/or response to therapeutic or other interventions" (U.S. Department of Health and Human Services, Food and Drug Administration (FDA), Center for Drug Evaluation and Research (CDER), Center for Biologics Evaluation and Research (CBER), 2008). Commonly used biomarkers in medicine include blood pressure as a predictor of strokes, congestive heart failure and serum cholesterol and triglyceride levels for atherosclerosis, serum glucose for complications of diabetes mellitus, various serum enzyme levels for disease and/or inflammation of various organs (e.g., heart, liver, and pancreas), and QTc interval for the risk of causing torsades de pointes (Chapel, Hutmacher, Haig, et al., 2009). The use of biomarkers in psychiatry has lagged behind their use in other areas of medicine because of the lack of information about the underlying pathophysiology and pathoetiology of psychiatric illnesses. Nevertheless a number of examples can be cited, including serum prolactin levels as a marker for the degree of dopamine D2 receptor antagonism achieved by a given dose of a given antipsychotic and changes in P50, N100, and P300 and mismatched negativity evoked potentials as markers for cognitive impairment associated with schizophrenia (Preskorn, D'Empaire, Baker, et al., 2009).

Most biomarkers are developed via pathway analysis, proteomics, and expression profiles. Researchers must use validated biomarkers that have been compared with already established clinical endpoints (i.e., antipsychotic efficacy). A review of all possible and/or validated biomarkers in psychiatry is beyond the scope of this chapter, and readers are referred to several recently released FDA documents on acceptable biomarkers in clinical trials for a more thorough review (U.S. Department of Health and Human Services, Food and Drug Administration (FDA), Center for Drug Evaluation and Research (CDER), Center for Biologics Evaluation and Research (CBER), 2008; U.S. Department of Health and Human Services, FDA, CDER, CBER, 2011a, 2011b).

One important approach to identifying validated biomarkers that is commonly applied in Phase 1 drug development involves the use of positron emission tomography (PET) to examine CNS drug binding using radioligands to target receptors or interaction with neurotransmitter systems (Paul et al., 2010). A now classic example of this approach is dopamine D2 receptor occupancy in the brain, which ultimately proved crucial for guiding the clinical dosing of antipsychotic drugs. In state-of-the-art neuroscience drug discovery, PET radioligands are routinely developed as biomarker companions for novel

therapeutic compounds, especially for compounds targeting mechanisms that are pharmaceutically unprecedented or may not have well characterized pharmacodynamic readouts in healthy humans. The application of such imaging tools is now commonly part of early human testing of an NME to establish central target engagement and to characterize dose-site occupancy relationships.

Nevertheless, limitations exist, as PET radioligand development for many molecular targets proves nonfeasible. In addition, while the development of PET radioligands that use F-18 as the radioisotope is preferred because of its 2-h half-life, for some ligands radiolabeling proves more feasible with C-11 as the radioisotope, which has a shorter half-life (20 min). This shorter half-life requires the radiosynthesis to occur on site, necessitating a cyclotron and radiochemistry group and substantially increasing the cost. Moreover, unless optimal radioligands are used, the brain exposure/concentration may be underestimated when compared with the drug's actual PK (Paul et al., 2010). Nevertheless, for many compound development programs, the use of suitable PET technology has proved invaluable for determining the drug doses needed to produce appropriate drug concentrations at the desired target in the desired brain regions to effectively modulate the target mechanism for optimal therapeutic efficacy.

Notably, in some neuropsychiatric conditions, PET radiotracers are also being used to establish the diagnosis for clinical trials participants. For example, some ongoing clinical trials for potential disease modifying treatments in Alzheimer's disease employ PET radioligands that bind to amyloid beta (e.g., [F-18]florbetapir) to provide biomarker evidence of Alzheimer's disease (Clark et al., 2011). Ensuring that drugs aimed at addressing pathological amyloid beta deposition are tested in patients who have abnormal amyloid levels in their brains is critical to the sensitivity of such trials.

Although diagnostic biomarkers have not been established for major psychiatric illnesses such as schizophrenia, bipolar disorder, and major depressive disorder, considerable work is underway to develop specific biomarkers that can be helpful in identifying a drug's potential therapeutic efficacy. For example, the use of saccadic eye movements and prolactin response in the development of antipsychotic drugs has been correlated with MTD and development of adverse events (Conn & Roth, 2008; Preskorn, 1996b). In fact, more than 80% of all studies utilizing such biomarkers have proved useful in moving the field and NMEs forward (Preskorn, 1996b).

In the area of antidepressant drugs, the biomarkers for the rapid identification of treatment effectiveness in major depression (BRITE-MD) study showed that prefrontal lobe activity after 1 week of treatment with escitalopram was correlated with efficacy 7 weeks later (Leuchter, Cook, Marangell, et al., 2009). Additional biomarkers in the development of antidepressant drugs include the effect of the drug on rapid eye movement (REM) sleep and EEG biomarkers (Steiger & Kimura, 2010). Finally a paradigm that appears particularly intriguing because of its ability to predict antidepressant efficacy across a variety of chemical classes involves the effect on behavioral and/or functional neuroimaging measures of the emotional processing bias in in attention, perception, and memory. Harmer, Cowen, and Goodwin (2011) have shown that the capacity of a drug to shift this bias toward the positive direction predicts its antidepressant efficacy in MDD and have extended this finding to include a variety of novel mechanisms extending beyond monoaminergic pathways (Harmer et al., 2011). If such predictive biomarkers prove reproducible across clinical trial sites and provide high predictive value for drug response prediction in the target clinical population, their use in early drug development could be extremely helpful in guiding antidepressant drug development.

The biomarkers discussed above for primary psychiatric disorders generally were discovered by assessing the effects of drugs on specific symptom domains. In contrast, CNS drug development for neurological disorders has progressed beyond symptoms to instead target specific disease processes. These advances have led to some treatments that can modify and intercept disease (e.g., multiple sclerosis). Similarly, psychiatric drug development will prove most effective if novel therapeutics can directly target key pathological constructs identified in psychiatric disorders (e.g., elevated limbic-thalamocortical activity in mood disorders) (Price & Drevets, 2012). The development of biomarkers that sensitively and specifically detect such constructs may ultimately prove most useful for guiding the drug development process for psychopathology. Moreover, since such constructs will more likely reflect pathophysiological processes that extend beyond the boundaries of the clinical syndrome-based DSM disorders, it is conceivable that such a drug discovery process also may guide advances in psychiatric nosology and diagnosis.

VIII BIOMARKERS FOR TARGET ENGAGEMENT AND ADHERENCE

The development of NMEs also may be accelerated by incorporating measurements of target engagement into Phase I or early Phase II studies. A marker of target engagement (also termed a pharmacodynamic biomarker) provides a readout of whether the NME is successfully targeting the intended disease pathway but is not itself necessarily predictive of a clinical outcome. Urine concentrations of 11-dehydrothromboxane B2 (11-D-TXB_2) serve as an example of this type of biomarker for the drug, aspirin. TXB_2 is a metabolite of thromboxane

A2 (TXA_2), which is in turn a metabolite of prostaglandin H2. Aspirin irreversibly inhibits platelet cyclooxygenase 1 (COX-1) preventing the formation of prostaglandin H2 and therefore by extension, TXA_2 and TXB_2. Thus decreases in the level of TXB_2 after treatment with aspirin provides evidence that aspirin is engaging COX-1. In this case TXB_2 also doubles as a useful biomarker of treatment adherence since daily use of aspirin has been shown to significantly decrease urinary concentrations of TXB_2 (Bruno et al., 2002).

IX BIOMARKERS FOR CLINICAL OUTCOME

Given the pathophysiological heterogeneity of many diseases, optimized drug therapy requires the identification of disease biotypes that show a preferential response to the NME. This need is particularly acute in the case of psychiatric disorders where the underlying pathophysiological mechanisms remain incompletely understood and specific treatments are usually selected on the basis of clinical intuition or trial and error. An example of a predictive biomarker can be found in the "immunopsychiatry" literature. It is clear that a subgroup of individuals with mood disorders shows evidence of immune dysregulation, and thus these patients may be candidates for treatment with immune-modulating medications. This hypothesis was borne out in a recent clinical trial of the TNF inhibitor, infliximab, for treatment-resistant depression (Miller & Raison, 2015; Raison et al., 2013). No overall difference in the change in depression ratings was detected between infliximab and placebo groups; however, infliximab-treated patients with a baseline high sensitivity C-reactive protein (hs-CRP) concentration greater than 5 mg/L had a greater decrease in depression ratings compared with the placebo group, whereas participants in the placebo group with a baseline CRP concentration of <5 mg/L showed a greater decrease in depression ratings over time than those individuals who received infliximab (Raison et al., 2013). Thus early clinical trial designs that adopt a stratified randomization approach based on a hypothesized biomarker of clinical response may be more likely to detect a therapeutic signal and avoid false negative results.

X CONCLUSION

CNS drug development is rapidly evolving to meet the unique and changing demands of the fields of psychiatry and neurology. To adapt to these unique challenges, Phase I clinical trials must incorporate new principles and methodology to more efficiently identify those drugs that are not viable before they enter late phase

TABLE 6.1   Key Points Concerning Early Phase-Drug Development Studies

---

Phase I Studies

Traditional Phase I studies involve

Normal voluteers

Single and multiple ascending doses used to determine whether the drug/biologic has appropriate pharmacokinetics and safety/tolerability

Expanded Phase I studies involve

Volunteers mildly symptomatic with the target illness

Single and multiple ascending doses used to determine whether the drug/biologic has similarly appropriate pharmacokinetics and safety/tolerability and whether the dose-response curves remain the same or shift to the right or left in the target population (Fig. 6.2)

Biomarker may be used to obtain early indication about whether the new drug is likely to be efficacious in individuals more symptomatic with the target illness

Studies in mildly symptomatic volunteers can be done (1) in the same protocol, with a seamless transition from normal volunteers to an approximation of the target population or (2) in separate protocols

Clinical Trial Designs

Traditional designs: predetermined and inflexible protocols that specify what conditions will be studied (e.g., placebo, active comparator, and three fixed doses of the investigational medication) with a predetermined number of subjects to be enrolled

Adaptive trial designs: design allows for modification of the protocol using prespecified algorithms and blinded analysis of data in real time. Common modifications that may be made during adaptive trials involve the doses studied, the number of individuals assigned to a given dose, and whether the study is terminated early for either success or failure (Kay et al., 1987)

(i.e., Phase III) testing. In this way the traditional approach to Phase I clinical trials is not "wrong" but instead is a starting point that must be modified and amplified to create a "customized" and efficient approach to the development of specific NMEs with truly novel mechanisms of action and the potential to modify disease rather than simply providing symptomatic relief. Key points covered in this chapter are summarized in Table 6.1. We have proposed that the potential target population be examined in Phase I trials after studies with normal volunteers have been completed to increase the efficiency and precision of decisions made based on Phase I data. The use of an adaptive design to examine data and modify the study in real time may also promote time- and cost-efficient identification of NMEs that have the characteristics necessary to be successful in "confirm" stages of drug development research. Finally, we discussed how use of biomarkers as surrogate endpoints can be incorporated into early clinical trials to determine if a drug modulates the appropriate targets, which, in turn, enhances the likelihood of demonstrating efficacy in the "confirm" phase of clinical testing.

References

Agrawal, M., & Emanuel, E. J. (2003). Ethics of phase 1 oncology studies: reexamining the arguments and data. *JAMA, 290*, 1075–1082.

Arbuck, S. G. (1996). Workshop on phase I study design—ninth NCI/EORTC new drug development symposium, Amsterdam, March 12, 1996. *Annals of Oncology, 7*, 567–573.

Bruno, A., McConnell, J. P., Mansbach, H. H., 3rd, Cohen, S. N., Tietjen, G. E., & Bang, N. U. (2002). Aspirin and urinary 11-dehydrothromboxane B(2) in African American stroke patients. *Stroke, 33*(1), 57–60.

Chapel, S., Hutmacher, M. M., Haig, G., et al. (2009). Exposure-response analysis in patients with schizophrenia to assess the effect of asenapine on QTc prolongation. *Journal of Clinical Pharmacology, 49*, 1297–1308.

Chow, S. C., & Chang, M. (2008). Adaptive design methods in clinical trials—a review. *Orphanet Journal of Rare Diseases, 3*, 11.

Clark, C. M., et al. (2011). Use of florbetapir-PET for imaging beta-amyloid pathology. *JAMA: The Journal of the American Medical Association, 305*, 275–283.

Conn, P. J., & Roth, B. L. (2008). Opportunities and challenges of psychiatric drug discovery: roles for scientists in academic, industry, and government settings. *Neuropsychopharmacology, 33*, 2048–2060.

Crane, M., & Newman, M. C. (2000). What level of effect is a no observed effect? *Environmental Toxicology and Chemistry, 19*, 516–519.

de Visser, S. J., van der Post, J., Pieters, M. S., et al. (2001). Biomarkers for the effects of antipsychotic drugs in healthy volunteers. *British Journal of Clinical Pharmacology, 512*, 119–132.

Dragalin, V. (2006). Terminology and classification of adaptive designs. *Drug Information Journal, 40*, 425–435.

Emanuel, E. J. (1995). A phase I trial on the ethics of phase I trials. *Journal of Clinical Oncology, 13*, 1049–1051.

Gallo, P., Chuang-Stein, C., Dragalin, V., et al. (2006). Executive summary of the PhRMA working group on adaptive designs in clinical drug development. *Journal of Biopharmaceutical Statistics, 16*, 275–283.

Gilles, C., & Luthringer, R. (2007). Pharmacological models in healthy volunteers: their use in the clinical development of psychotropic drugs. *Journal of Psychopharmacology, 21*, 272–282.

Goodman, S. N., Zahurak, M. L., & Piantadosi, S. (1995). Some practical improvements in the continual reassessment method for phase I studies. *Statistics in Medicine, 14*, 1149–1161.

Harmer, C. J., Cowen, P. J., & Goodwin, G. M. (2011). Efficacy markers in depression. *Journal of Psychopharmacology, 25*, 1148–1158.

Kaitin, K. I. (2010). Deconstructing the drug development process: the new face of innovation. *Clinical Pharmacology and Therapeutics, 87*, 356–361.

Kay, S. R., Fiszbein, A., & Opler, L. A. (1987). The positive and negative syndrome scale (PANSS) for schizophrenia. *Schizophrenia Bulletin, 13*, 261–276.

Kraemer, H. C., Morgan, G. A., Leech, N. L., et al. (2003). Measures of clinical significance. *Journal of the American Academy of Child and Adolescent Psychiatry, 42*, 1524–1529.

Krams, M., Burman, C. F., Dragalin, V., et al. (2007). Adaptive designs in clinical drug development: opportunities, challenges, and scope reflections following PhRMA's November 2006 workshop. *Journal of Biopharmaceutical Statistics, 17*, 957–964.

Krams, M., Sharma, S., Dragalin, V., et al. (2009). Adaptive approaches in clinical drug development: opportunities and challenges in design and implementation. *Pharmaceutical Medicine, 23*, 139–148.

Lepping, P., Sambhi, R. S., Whittington, R., et al. (2011). Clinical relevance of findings in trials of antipsychotics: systematic review. *The British Journal of Psychiatry, 198*, 341–345.

Leuchter, A. F., Cook, I. A., Marangell, L. B., et al. (2009). Comparative effectiveness of biomarkers and clinical indicators for predicting outcomes of SSRI treatment in major depressive disorder: results of the BRITE-MD study. *Psychiatry Research, 169*, 124–131.

Macaluso, M., Krams, M., & Preskorn, S. H. (2011). Phase I trials: from traditional to newer approaches. Part I. *Journal of Psychiatric Practice, 17*, 200–203.

McConnell, E. E. (1989). The maximum tolerated dose: the debate. *Journal of the American College of Toxicology, 8*, 1115–1120.

Meinert, C. L., & Tonascia, S. (1986). *Clinical trials: Design, conduct, and analysis* (p. 3). New York: Oxford University Press.

Miller, A. H., & Raison, C. L. (2015). The role of inflammation in depression: from evolutionary imperative to modern treatment target. *Nature Reviews. Immunology, 16*(1), 22–34. Mechawar, N., Savitz, J. (2016). Neuropathology of mood disorders: do we see the stigmata of inflammation? *Transcultural Psychiatry, 6*(11), e946.

O'Quigley, J., Pepe, M., & Fisher, L. (1990). Continual reassessment method: a practical design for phase 1 clinical trials in cancer. *Biometrics, 46*, 33–48.

Panza, F., Frisardi, V., Imbimbo, B. P., et al. (2011). Monoclonal antibodies against b-amyloid (Ab) for the treatment of Alzheimer's disease: the Ab target at a crossroads. *Expert Opinion on Biological Therapy, 11*, 679–686.

Paul, S. M., Mytelka, D. S., Dunwiddie, C. T., et al. (2010). How to improve R&D productivity: the pharmaceutical industry's grand challenge. *Nature Reviews. Drug Discovery, 9*, 203–214.

Preskorn, Personal communication, 2011.

Preskorn, S., D'Empaire, I., Baker, B., et al. (2009). EVP-6124, an alpha-7 nicotinic agonist produces normalizing effects on evoked response biomarkers and cognition in patients with chronic schizophrenia on stable antipsychotic therapy. In *Poster presented at the 2009 American College of Neuropsychopharmacology meeting, Hollywood, FL, December*.

Preskorn, S. H. (1996a). *Clinical pharmacology of SSRIs. Caddo, OK: Professional Communications*.

Preskorn, S. H. (1996b). If lack of concentration didn't cause the fall, what did? *Journal of Practical Psychiatry and Behavioral Health, 2*, 364–367.

Preskorn, S. H. (2010). CNS drug development: part I: the early period of CNS drugs. *Journal of Psychiatric Practice, 16*, 334–339.

Price, J. L., & Drevets, W. C. (2012). Neural circuits underlying the pathophysiology of mood disorders. *Trends in Cognitive Neuroscience, 16*(1), 61–71.

Raison, C. L., Rutherford, R. E., Woolwine, B. J., Shuo, C., Schettler, P., Drake, D. F., et al. (2013). A randomized controlled trial of the tumor necrosis factor antagonist infliximab for treatment-resistant depression: the role of baseline inflammatory biomarkers. *JAMA Psychiatry, 70*(1), 31–41.

Sheiner, L. B. (1997). Learning versus confirming in clinical drug development. *Clinical Pharmacology and Therapeutics, 61*, 275–291.

Shen, J., Preskorn, S., Dragalin, V., et al. (2011). How adaptive trial designs can increase efficiency in psychiatric drug development: a case study. *Innovations in Clinical Neuroscience, 8*, 26–34.

Spilker, B. (1991). *Guide to clinical trials*. Lippincott Williams & Wilkins.

Steiger, A., & Kimura, M. (2010). Wake and sleep EEG provide biomarkers in depression. *Journal of Psychiatric Research, 44*, 242–252.

U.S. Department of Health and Human Services, FDA, CDER, CBER. (2011a). *E16 genomic biomarkers related to drug response: Context, structure, and format of qualification submissions (draft guidance)*. Retrieved from www.fda.gov/downloads/Drugs/Guidance ComplianceRegulatoryInformation/Guidances/UCM174433.pdf. Accessed 7 July 2011.

U.S. Department of Health and Human Services, FDA, CDER, CBER. (2011b). *Guidance for industry: Pharmacogenomic data submissions*. Retrieved from www.fda.gov/downloads/Drugs/ GuidanceComplianceRegulatoryInformation/Guidances/ucm079849. pdf. Accessed 7 July 2011.

U.S. Department of Health and Human Services, Food and Drug Administration (FDA), Center for Drug Evaluation and Research (CDER), Center for Biologics Evaluation and Research (CBER). (2008). *E15 definitions for genomic biomarkers, pharmacogenetics, genomic data, and sample coding categories*. Retrieved from www.fda. gov/downloads/RegulatoryInformation/Guidances/ucm129296.pdf. Accessed 7 July 2011.

Wong, D. F., Tauscher, J., & Gründer, G. (2009). The role of imaging in proof of concept for CNS drug discovery and development. *Neuropsychopharmacology, 34*, 187–203.

7

Translational Approaches for Antidepressant Drug Development

Gerard J. Marek

Astellas Pharma Global Development, Inc., Development Medical Science, CNS and Pain, Northbrook, IL, United States

Despite over 50 years of progress in developing antidepressant drugs, major depressive disorder (MDD) continues to exact significant morbidity and mortality at both a patient and societal level. While the widespread introduction of selective serotonin reuptake inhibitors (SSRIs) and serotonin-norepinephrine reuptake inhibitors (SNRIs) appears to have resulted in decreased mortality due to suicide since a greater percentage of depressed patients are treated over an adequate duration of time with these later-generation medications compared with the tricyclic antidepressants (TCAs) and monoamine oxidase inhibitors, significant unmet medical needs remain. These unmet medical needs include greater antidepressant efficacy, dual antidepressant and anxiolytic efficacy, improved sleep, improvement of residual cognitive dysfunction, decreased suicidality, and effective therapies for treatment-resistant depression. The key feature in developing new chemical entity (NCE) is to provide differentiation from presently approved antidepressant drugs on one or more of these specific needs required by patients, prescribers, and especially payers. This chapter will discuss translational strategies that at the most basic level will provide increased confidence that the NCE will be an effective antidepressant. This chapter will also discuss translational strategies increasing confidence that important clinical differentiation for the NCE will be met compared with existing clinical standard of care. Some of these strategies are agnostic to the indication being studied, while others are specifically tailored for the NCE or the MDD indication. These strategies encompass approaches to minimize attrition in moving from Phase 1 to the Phase 2 "learn and confirm" period regarding therapeutic efficacy (Morgan et al., 2012).

I RECEPTOR OCCUPANCY AND TRANSLATIONAL APPROACHES FOR ONE DIMENSION OF MULTIMODAL DRUGS

One way to increase confidence that a novel antidepressant will possess demonstrable antidepressant activity is to provide specific translational evidence that previously has been found to be associated with therapeutic effects in patients with MDD. One of the most certain translational approaches is to first demonstrate in preclinical species that a target like the serotonin transporter (SERT) is occupied by 80% or more at doses/exposures that appear safe and tolerated with respect to toxicology studies (Grimwood & Hartig, 2008; Meyer et al., 2004). Then, SERT occupancy is determined in healthy volunteers during a Phase 1 Go/No Go decision-making study prior to moving to a Phase 2 proof-of-concept (PoC) randomized clinical trial (RCT) in patients with MDD. Two examples where this approach was utilized were the Bristol-Myers Squib (BMS) triple reuptake inhibitor BMS-820836 and the Lundbeck/Takeda SSRI plus medication vortioxetine. The weakness of this approach by itself is that simply demonstrating that a medication will occupy >80% of brain SERT does not guarantee that this NCE will exert clinically meaningful differentiation from SSRIs. However, this translational paradigm will significantly decrease the risk that the NCE will not possess clinical meaningful antidepressant activity. For example, the BMS triple reuptake inhibitor did possess therapeutic activity, though it did not appear to be more effective than a SSRI or SNRI (Bhagwagar et al., 2015; Zheng et al., 2015). This allowed the sponsor to make a strategic decision to terminate this program for a MDD indication.

Another recent example of using neuroimaging occupancy to inform drug development decision-making is the Lundbeck/Takeda vortioxetine program. Vortioxetine is a multimodal antidepressant drug acting on multiple serotonergic receptors including inhibiting SERT (Sanchez, Asin, & Artigas, 2015). Neuroimaging occupancy studies provided the sponsors with critical information that daily vortioxetine doses in the range of 20 mg result in at least 70% SERT occupancy, while daily 30-mg doses of vortioxetine did not occupy $5-HT_{1A}$ receptors (Stenkrona, Halldin, & Lundberg, 2013). This type of data provides confidence that antidepressant activity can be demonstrated with daily 20–30 mg doses. In fact, multiple meta-analyses of the registration data suggest that with adequate patient sample size, near-maximal efficacy is observed with daily 10- and 20-mg vortioxetine doses for patients with MDD with or without high levels of anxiety (Baldwin, Florea, Jacobsen, Zhong, & Nomikos, 2016; Thase, Mahableshwarkar, Dragheim, Loft, & Vieta, 2016). Preclinical studies raised the hypothesis that vortioxetine may enhance cognition in depressed patients compared with current standard of care (Sanchez et al., 2015). A portion of the pivotal studies also directly examined improved cognitive function, and a meta-analysis of these results found that 10- and 20-mg doses improved attention/speed of processing and executive function (Harrison, Lophaven, & Olsen, 2016). Understanding that a given compound will minimally possess antidepressant activity is an important achievement allowing the development team to focus on evidence that other pharmacological features that may provide clinically important differentiation compared with the existing standard of care.

II RECEPTOR OCCUPANCY FOR SINGLE TARGET DIRECTED PUTATIVE ANTIDEPRESSANT DRUGS

An additional example of using brain receptor occupancy is exemplified by the Merck MDD program for the neurokinin$_1$ (NK$_1$) receptor antagonist aprepitant. Having receptor occupancy in which to evaluate an inhibitor/antagonist is especially valuable given the uncertainty with how novel gerbil models of putative antidepressant-like pharmacodynamic action translate into human studies. These gerbil models were developed given pharmacological differences between rat and human NK$_1$ receptors in contrast with pharmacological similarities between gerbil and human NK$_1$

receptors. Exposures associated with greater than 90% aprepitant brain NK$_1$ receptor occupancy were established (Bergstrom et al., 2004; Keller et al., 2005). Five Phase 3 double-blind, placebo-controlled trials examined aprepitant for antidepressant efficacy in patients with MDD using doses/exposures resulting in greater than 90% NK$_1$ receptor occupancy. Despite positive Phase 2 trials with two different compounds, aprepitant did not result in antidepressant effects compared with placebo in any of these five trials (Keller et al., 2005). Three of these trials included the comparator SSRI paroxetine, which did statistically separate from placebo. Thus the brain NK$_1$ receptor occupancy work coupled with the negative Phase 3 trials confidence provided abundant confidence for the sponsor to walk away from aprepitant as a potential novel antidepressant medication. Similar understanding of brain NK$_1$ receptor antagonist occupancy allowed other sponsors to test an additional anxiety indication in Phase 2 with the understanding that the clinical trial was a test of the clinical hypothesis that the medication would work in this additional indication (Fig. 7.1).

In summary, development of antidepressant drugs are consistent with other CNS therapeutic indications where approximately 50%–90% receptor/transporter/enzyme occupancy is required for drugs that inhibit or block a single molecular target (Grimwood & Hartig, 2008). However, there is considerably less confidence how much CNS target occupancy is required for agonists or positive allosteric modulators (PAMs) to exert therapeutic effects. For example, efficacy of GABA$_A$ receptor PAMs, commonly known at benzodiazepines, in sleep and anxiety disorders is observed at with doses/exposures resulting in point estimates of less than 30% brain receptor occupancy (Grimwood & Hartig, 2008). Receptor/transporter/enzyme occupancy with agonists or PAMs can still be useful for dose selection in a program, although a single maximal dose with complete occupancy may not be the ideal single dose to test in a PoC trial. Desensitization or downregulation of the molecular target or differential recruitment of distinct neural circuitry countering an initial antidepressant-like response are examples requiring functional biomarkers assuming a single molecular target is engaged over the dose range testing complete target occupancy. Furthermore a NCE engaging a second molecular target invoking intracellular pathways or neural circuitry countering the first molecular target is an additional reason for developing functional biomarkers. Thus additional functional biomarkers are especially useful for developing agonists or PAMs as novel antidepressant medications for a number of reasons. These additional functional biomarkers also benefit development programs for antagonists/inhibitors where a radiotracer is not feasible either for technical reasons or development timelines for the NCE.

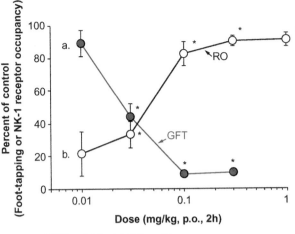

Comparison of effects of LY686017 on foot-tapping (GFT) and NK-1 receptor occupancy (RO) in Gerbils

ED50 (GFT) = 0.03 +/-0.004 mg/Kg
ED50 (RO) = 0.037 +/-0.008 mg/Kg

*= $P < .05$ vs. Vehicle

TABLE 7.1 Replicability of Commonly Tested SSRIs on REM Latency/Duration in Rats and Man

SSRI	REM latency increased	REM duration decreased	Species/condition
Fluoxetine	1 of 1[a]	2 of 2	Rats, cats
Fluoxetine	2 of 2	3 of 4	Humans/healthy
Fluoxetine	2 of 3	0 of 4	Humans/depressed
Citalopram	Not reported	2 of 2	Rats, mice
Citalopram	1 of 1	1 of 1	Humans/healthy
Paroxetine	1 of 1	3 of 3	Rats
Paroxetine	7 of 7	7 of 7	Humans/healthy
Paroxetine	2 of 2	4 of 4	Humans/depressed

[a] *Each cell indicates number of literature reports for change in REM latency or duration compared with total number of literature reports for that drug and species.*

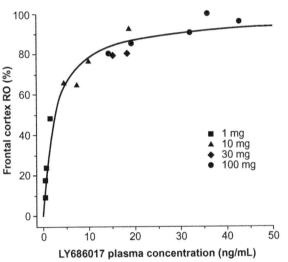

FIG. 7.1 The *top figure* displays the correlation between a pharmacodynamic model in gerbils (NK_1 receptor antagonist reversal of foot tapping) and NK_1 receptor occupancy in gerbils. The *bottom figure* shows the NK_1 receptor occupancy as a function of systemic exposure following doses of 1–100 mg. These data sets were the foundation for selecting a single dose of the NK_1 receptor antagonist LY686017 to test in a Phase 2 PoC social anxiety disorder trial after disclosures of the negative Merck antidepressant program for aprepitant (Tauscher, J., Kielbasa, W., Iyengar, S., Vandenhende, F., Peng, X., Mozley, D., et al. (2010). Development of the 2nd generation neurokinin-1 receptor antagonist LY686017 for social anxiety disorder. *European Neuropsychopharmacology, 20*, 80–87). The dose of LY686017 chosen for the study (50 mg) nearly completely occupied brain NK_1 receptors.

III INDIRECT BIOMARKERS FOR CONFIRMING CNS PENETRATION AND DOSE SELECTION: POLYSOMNOGRAPHY

Beyond understanding transporter or enzyme occupancy for SSRIs/SNRIs or monoamine oxidase inhibitors,

the next translational biomarker with the highest predictive probability of antidepressant activity are polysomnography (PSG) effects from sleep electroencephalogram (EEG) studies. Demonstrating shared actions with antidepressant drugs (prolonged REM onset and reduced REM sleep duration) in both rodents and humans is another strategy for developing novel antidepressant drugs. Wake-sleep cycles and the biology underlying these physiological and behavioral states are relatively conserved across all mammalian species including rodents and humans, though there may be some species divergent findings in relatively minor aspects of sleep biology (Brown, Basheer, McKenna, Strecker, & McCarley, 2012). Sleep PSG studies traditionally have taken on an important role as a translational approach given the prominent disruption of objective PSG and subjective sleep most patients with MDD (Winokur et al., 2001). Most antidepressant drugs have been found to decrease rapid eye movement (REM) sleep or prolong the latency toward REM sleep (Wilson & Argyropoulos, 2005; Winokur et al., 2001). This is true of SSRIs (Table 7.1), serotonin-norepinephrine reuptake inhibitors, tricyclic antidepressants, and monoamine oxidase inhibitors. During the 1960s, Organon had a tetracyclic compound, mianserin, that did not possess preclinical characteristics similar to most known antidepressants available during this period. Mianserin did not potently block monoamine reuptake or monoamine oxidase and was not active in most of the preclinical antidepressant tests in use during this period. This compound, a $5\text{-HT}_{2A/2C/3}$ and α_2-adrenergic receptor antagonist, did improve sleep suggesting that it might have antidepressant properties. Subsequent RCTs in patients with MDD confirmed this hypothesis. This led to mianserin being approved in a number of countries around the world as a novel antidepressant medication (Brogden, Heel, Speight, & Avery, 1978; Pinder & Fink, 1982). Mianserin was subsequently followed up with a structural and pharmacological analogue, mirtazapine, that was approved as an antidepressant by the FDA and throughout much of the world.

A more recent example of using sleep EEG studies as a translational biomarker in humans is the development of JNJ-18038683, a $5\text{-}HT_7$ receptor antagonist. JNJ-18038683 increased the latency to REM sleep while also decreasing the duration of REM sleep in rodents (Bonaventure et al., 2012, 2007). This $5\text{-}HT_7$ receptor antagonist prolonged REM sleep latency and also reduced REM sleep duration in healthy volunteers consistent with the hypothesis that these sleep-modulating dose/exposures resulted in adequate CNS penetration to proceed to a Phase 2a PoC trial (Bonaventure et al., 2012). A subsequent Phase 2 PoC trial failed to show separation from placebo for either JNJ-18038683 or the SSRI escitalopram (Bonaventure et al., 2012). However, a post hoc analysis eliminating sites that appeared to fail in separating from placebo did show a trend toward a statistical separation from placebo ($P < .057$). Regardless of whether the results with JNJ-18038683 represent a false positive result, this example doses demonstrate how PSG results can be used with Go/No Go decision-making criteria in advancing from Phase 1 to the Phase 2a PoC trial.

Similarly, Minerva Biosciences and Janssen conducted a Phase 1b single-dose PSG study (ClinicalTrials.gov; NCT02067299) for their sleep promoting $orexin_2$ receptor antagonist seltorexant (MIN-202 or JNJ42827922). An increase in sleep efficiency, decreased latency to persistent sleep, increased wake after sleep onset (WASO), and total sleep time all suggested CNS penetration for this compound and replicated preclinical findings for this class of drugs (Minerva Neurosciences, 2016, 2017). This provided confidence to conduct and select doses for a Phase 1b trial in patients with MDD. Based on data consistent with antidepressant efficacy of seltorexant from this Phase 1b trial, multiple Phase 2 trials for both MDD and insomnia have been planned. Essentially, in a modest 47 patient randomized, double-blind, placebo-controlled, diphenhydramine-controlled study of seltorexant in patients with depression (ClinicalTrials.gov; NCT02476058), an approximately two-point change in HAMD17 scores were observed both with and without the three sleep items of this HAMD scale (Minerva Neurosciences, 2016).

IV COGNITIVE BIAS AS EXPERIMENTAL MEDICINE PARADIGM FOR NOVEL ANTIDEPRESSANT DRUGS

Another distinct approach being explored at a human translational level during Phase 1 testing is to demonstrate effects of putative antidepressant drugs on positive

and negative emotional response bias to different stimuli as studied by Catherine Harmer and colleagues (Harmer, Duman, & Cowen, 2017; Warren, Pringle, & Harmer, 2015). The basic underlying hypothesis for this work is that affective information processing may itself be a target for improving MDD given long-standing theory advanced by Beck and associated practical treatment approaches such as cognitive behavioral therapy (CBT). SSRIs, NRIs, SNRIs, mirtazapine, and agomelatine modulate emotional processing of face recognition, word categorization, emotional word recall, and other additional paradigms (Warren et al., 2015). These effects generally are observed in healthy volunteers, depressed patients, and volunteers categorized as highly neurotic following either daily short-term (7 days) dosing or a single dose. Anxiolytic benzodiazepines have effects on threat processing but otherwise differ on emotional processing from the effects of antidepressants drugs (Murphy, Downham, Cowen, & Harmer, 2008). Rimonabant appears to have effects opposite to those of antidepressants, concordant with clinical data suggesting a drug-induced increase in depressive symptoms in obese patients (Horder, Cowen, Di Simplicio, Browning, & Harmer, 2009). The NMDA antagonist memantine had minimal effects on emotional processing that seem consistent with the lack of antidepressant efficacy for this approved Alzheimer's disease therapeutic agent (Pringle et al., 2012). Finally the emotional processing effects of NK_1 receptor antagonists are generally modest and restricted in range compared with approved antidepressants (Pringle et al., 2011). Thus drugs that are not antidepressants generally do not produce similar changes in emotional information processing as known antidepressant drugs. The primary weakness of this approach is that the positive and negative predictive validity of this approach can be challenged, especially with different endpoints for different mechanisms of action. However, the primary strength of this approach is increasing confidence that a central mechanism of action on brain circuitry involved in affective processes is present for the doses/exposures demonstrated during Phase 1 proof-of-pharmacology (PoP) studies.

V COGNITION: DEMONSTRATING PHARMACOLOGICAL POP OR A DIFFERENTIATING THERAPEUTIC EFFECT

In addition to investigating changes in cognitive biasing for affective information processing, frank cognitive impairment exists for many patients with MDD and is a potential translational medicine and therapeutic endpoint. Meta-analyses has suggested that while cognitive impairments involving attention, visual learning/memory, and executive function for many patients with MDD, these impairments appear accentuated in recurrent MDD (Lee, Hermens, Porter, & Redoblado-Hodge, 2012; McIntyre et al., 2013; Trivedi & Greer, 2014). Several different (CANTAB and CogState) cognitive batteries are available to assess whether NCEs modulate distinct cognitive domains in healthy volunteers or depressed patients (Davis et al., 2017; Rock, Roiser, Reidel, & Blackwell, 2013). Difficulties in conducting Phase 1 translational studies based on preclinical findings for distinct cognitive domains include likely ceiling effects for healthy volunteers and potential patient enrichment strategies involving recruiting patients with recurrent MDD exhibiting baseline cognitive deficits. The cut points chosen for demonstrating baseline cognitive deficits will obviously alter the study recruitment rates. Confidence that an NCE may result in clinical differentiation versus standard of care antidepressant drugs requires implementation of cognitive batteries in Phase 2 directed against a range of domains or focusing on single domains altered both by the NCE and MDD itself. Further information dealing with these issues and extending into regulatory interactions with the FDA has been detailed on 3 Feb. 2016 Psychopharmacologic Drug Advisory Committee discussing vortioxetine (https://www.fda.gov/AdvisoryCommittees/CommitteesMeeting Materials/Drugs/PsychopharmacologicDrugsAdvisory Committee/ucm.475314.htm).

VI SUICIDE: INTERSECTION OF COGNITION, MOOD AND BEHAVIOR

Decreasing suicidality, particularly decreasing actual suicides, is a key unmet medical need for patients with major neuropsychiatric illness including MDD, anxiety disorders, schizophrenia, and substance use disorders. Suicidality with respect to completed or near-fatal suicides is not a simple clinical endpoint that confidently can be predicted from preclinical models (Bortolato et al., 2013; Dumais et al., 2005; Mann, 2003; Pandey, 2013). Instead a range of preclinical paradigms addressing impulsivity, impaired decision-making, aggression, hypothalamic-pituitary adrenal (HPA) axis states, inflammation, low extracellular serotonin, and changes in serotonergic receptors may be used to explore NCEs (Gould et al., 2017; Malkesman et al., 2009). Several aspects of impulsivity will be discussed below.

Different aspects of impulsivity such as motor impulsivity or waiting impulsivity can be investigated in rodents using tasks like the 5-choice serial reaction time task (5-CSRTT) and the differential-reinforcement-of-low rate 72-s (DRL 72-s) schedule of reinforcement (Dalley, Everitt, & Robbins, 2011; Marek, Day, & Hudzik, 2016). While the 5-CSRTT for rodents was originally derived

from the continuous performance task used in humans, a human analogue of the rodent 5-CSRTT that is more sensitive to measuring premature responses from the discrete trials of a sustained attention task has recently been developed. Both the rodent and human paradigms address both sustained attention and executive function as the task requires subjects to correctly respond to the brief presentation of a stimulus in a predefined location of the subject's visual field for any given trial. Premature responses are the key endpoint measuring motoric impulsivity. This 4-choice serial reaction time (4-CSRT) task for human subjects is sensitive to lowering of brain 5-HT by dietary tryptophan depletion using approximately 20 subjects in a between-subject design (Worbe, Savulich, Voon, Fernandez-Egea, & Robbins, 2014). This result is analogous to increased impulsivity measured with the 5-CSRTT in rodents following lesions of serotonergic neurons (Winstanley, Dalley, Theobald, & Robbins, 2004). Similarly, methylphenidate will increase impulsivity in both rodents and humans measured with these "waiting impulsivity" tasks (D'Amour-Horvat & Leyton, 2014; Voon et al., 2015). Validation with treatments that clearly reduce motoric impulsivity in both rodents and humans is still required demonstrating this to be a useful Phase 1 experimental medicine paradigm. However, the potential utility of this approach is enhanced since motoric impulsivity relates to important clinical management issues across a range of indications including bipolar disorder, schizophrenia, MDD, and neurodegenerative disorders.

In addition to motor or "waiting" impulsivity, impulsive choice can be measured in both rodents and humans using delayed discounting of reward or temporal discounting paradigms (Dalley et al., 2011; Hamilton et al., 2015; Vanderveldt, Oliveira, & Green, 2016). In this paradigm an impulsive subject chooses an immediately available small reward rather than experiencing a delay for a larger magnitude of reward. Recently, with the use of a delay discounting task in comparing episodic MDD patients with remitted MDD patients and healthy control subjects ($n = 24$–29), it is found that impaired temporal discounting appears to be state-dependent and may be related to hopelessness (Puclu et al., 2014). Like for translational research with motor impulsivity, there is an urgent need to perform critical studies evaluating the feasibility of studying impulsive choice during Phase 1 studies ($n = 15$–20/group) by using medications that normalize or are expected to normalize impulsive choice. The utility of these types of studies, like for studying motoric impulsivity, is for suggesting doses (exposures) of an NCE that acts in the CNS on reasonably well-defined neurocircuitry. Given the multifaceted nature of suicide, it would appear challenging for either of these impulsivity measures or studies on aggression or other suicide-related features to provide robust predictions for surrogate-like efficacy.

VII CIRCULATING PLASMA MICRO RNAS: FUTURE PROMISE?

Peripheral indexes of CNS function to serve as biomarkers for windows into CNS function have long been sought though with few successful examples. One growing example showing promise in this area is the study of micro (mi)-ribonucleic acids (RNAs; miRs) to serve as disease and/or treatment-related biomarkers. These miRs are a class of small noncoding RNA molecules that regulate stability of mRNA through RNA degradation or translational repression. Independent replication from two new cohorts of patients treated with SSRIs, SNRIs, or TCAs have suggested that a number of miRs (miR-146a-5p; miR-146b-5p; miR-425-3p, and miR-24-3p) are expressed in circulating plasma of depressed patients on a differential basis with respect to treatment response (Lopez et al., 2017). These same authors reported some of these same miRs to be differentially expressed on an antidepressant treatment response basis in the rodent chronic social defeat stress model. Additional convergent data from Lopez et al. were the finding that a number of these miRs decreased in antidepressant drug treatment responders were found to be increased in the ventrolateral prefrontal cortex (vPFC) of depressed patients dying from suicide compared with control subjects free of psychiatric disease. The importance of these findings is further enhanced by miRNA target prediction and pathway analysis suggesting these treatment response biomarkers alter the *MAPK*, *Wnt*, calcium, endocytosis, and adherens junction signaling pathways (Lopez et al., 2017). These antidepressant treatment response-related miRs were previously suggested in a pilot study (Enatescu et al., 2016).

Investigators in this field must answer a number of questions before miRs confidently can be used for drug discovery. How many subjects are required to demonstrate significant effects between treatment responders and nonresponders? Are these changes observed for both humans and animal models for antidepressant treatments that do not block monoamine uptake like mirtazapine or electroconvulsive shock? Are these changes observed for nonresponsive patients treated with commonly used add-on strategies like second-generation antipsychotics? What is the nature of changes for these miRs when comparing known nonantidepressants using animal models?

VIII ADDITIONAL ELECTROPHYSIOLOGICAL BIOMARKERS

Other electrophysiological biomarkers have been examined as potential indicators of treatment response to therapeutic antidepressant drugs (Schmidt et al., 2017). This includes shifting oscillatory synchrony and recently has been reviewed elsewhere (Leuchter, Hunter, Krantz, & Cook, 2014). A suppression of EEG-vigilance may also be a biomarker for antidepressant treatment response (Schmidt et al., 2017).

IX INDIRECT EVIDENCE SUGGESTING CNS PENETRATION: PHYSIOLOGICALLY BASED PHARMACOKINETIC MODEL

Measuring cerebrospinal fluid (CSF) pharmacokinetics in humans with modeling from peripheral pharmacokinetic data, similar animal data, and key preclinical exposure data associated with receptor occupancy/preclinical assays measuring putative antidepressant-like activity is a translational approach of last resort and also can be combined with translational paradigms described earlier to move programs from Phase 1 to the Phase 2a PoC trial. This translational paradigm not only is independent of any specific planned indication but also comes with a number of limitations (de Lange, 2013). First, lumbar CSF drug concentrations imperfectly predict ventricular or cisterna magna CSF drug concentrations. Dissociations may also exist between lumbar CSF drug concentrations and the brain extracellular fluid compartment or the brain intracellular fluid compartment. Second, there may be other species-specific factors at both the blood-brain barrier (BBB) and the blood-cerebrospinal fluid barrier (BCSFB) that change distribution of drugs from the periphery into the brain including a range of active drug transporters. Third, occasionally preclinical receptor occupancy may reveal reasons that drug candidates fail during appropriate in vivo testing methods when preclinical brain levels suggest adequate target exposure. The power of using preclinical pharmacokinetic modeling informed by preclinical CSF drug levels or extracellular brain drug concentrations is that while relatively low throughput, this is a method of widespread utility as measuring drug concentration in a range of peripheral and central compartments is the basic core competency required (Ball, Bouzom, Scherrmann, Walther, & Decleves, 2014; Kielbasa & Stratford, 2012). A recent publication has highlighted the potential value of a rigorous physiologically based pharmacokinetic model using nine drugs with extensive preclinical and clinical data to critically evaluate the performance of such a model (Yamamoto et al., 2017). Thus, when other translational models are not available, this generic type of modeling activity can increase or decrease confidence that adequate doses/exposures can be tested during Phase 2a PoC trials.

X SUMMARY AND CONCLUSIONS

A number of translational paradigms exist to increase or decrease confidence that a given Phase 2 PoC trial in depressed patients will adequately test a novel NCE. The confidence and design of a PoC trial may depend on whether an antagonist/inhibitor is being tested compared with an orthosteric agonist/PAM. Use of brain target occupancy in both preclinical and clinical studies will maximize confidence that a given mechanism is being tested given that virtually all antagonists/inhibitors tested in neuropsychiatric indications possess a monotonically increasing dose(exposure)–response relationship for both preclinical models and patients with depression (where adequately low doses/exposures have been tested). In contrast, such confidence for a given mechanism is lower when using physiologically based pharmacokinetic mechanisms with CSF drug levels in humans. This range of testing refers simply to confidence that a given mechanism will be adequately tested from a single PoC trial.

In contrast, other paradigms may be increase confidence that an NCE is being tested at a CNS-penetrant dose/(exposures) and may also have useful activity in a domain of depression that remains an unmet medical need. Sleep EEG or PSG studies in humans can replicate preclinical findings. More importantly, improvements in sleep onset, reductions in REM sleep, or increases in restorative (slow-wave) sleep using both objective and subjective findings can be incorporated in Phase 1 Go/No Go decision-making criteria governing progress into Phase 2 PoC trials.

Other paradigms such as testing NCEs for effects in healthy or depressed humans of changes in emotional response bias suggest for positive drugs that a given dose/exposure may result in adequate CNS exposures. However, such activity may not provide evidence for clinical differentiation compared with previously approved antidepressant drugs. Beyond exploring emotional response bias, frank cognitive testing (memory or executive function or decreases in impulsivity) may suggest differentiation in the cognitive domain or for impulsivity that may translate into clinical reductions of suicide risk.

Finally, other paradigms such as changes in plasma miRs in depressed patients may be utilized later in the clinical program to complement clinical trials directed at examining efficacy of NCEs in treatment refractory depression. At this time, there is a lack of critical information to assess whether this approach can be applied prior to Phase 2 clinical trials. Other functional paradigms also exist, such as electrophysiological examination of EEG oscillatory synchrony that can be used on a case-by-case basis to follow up the most likely translational plans to provide meaningful data to stop or continue NCEs from Phase 1 studies to a Phase 2 PoC depression trial.

A key principle related to the evolution of biomarkers for developing antidepressant drugs is using the continued sharpened understanding of the pathophysiology underlying MDD itself. As sophistication improves at the clinical level, innovations in technology in the preclinical sphere will continue to increase the number and quality of translational paradigms that reasonably can be used both in rodents and humans. The discussions of cognition and impulsivity represent current examples where this laboratory bench to bedside translation appears to be developing new tools that may result in novel medications being developed for current unmet medical needs.

References

Baldwin, D. S., Florea, I., Jacobsen, P. L., Zhong, W., & Nomikos, G. G. (2016). A meta-analysis of the efficacy of vortioxetine in patients with major depressive disorder (MDD) and high levels of anxiety symptoms. *Journal of Affective Disorders, 206*, 140–150.

Ball, K., Bouzom, F., Scherrmann, J. -M., Walther, B., & Decleves, X. (2014). Comparing translational population-PBPK modelling of brain microdialysis with bottom-up prediction of brain-to-plasma distribution in rat and human. *Biopharmaceutics & Drug Disposition, 35*, 485–499.

Bergstrom, M., Hargreaves, R. J., Burnes, H. D., Goldberg, M. R., Sciberras, D., Reines, S. A., et al. (2004). Human positron emission tomograph studies of brain neurokinin 1 receptor occupancy by aprepitant. *Biological Psychiatry, 55*, 1007–1012.

Bhagwagar, Z., Torbeyns, A., Hennicken, D., Zheng, M., Dunlop, B. W., Mathew, S. J., et al. (2015). Assessment of the efficacy and safety of BMS-820836 in patients with treatment-resistant major depression. *Journal of Clinical Psychopharmacology, 35*, 454–459.

Bonaventure, P., Dugovic, C., Kramer, M., De Boer, P., Singh, J., Wilson, S., et al. (2012). Translational evaluation of JNJ-18038683, a 5-hydroxytryptamine type 7 receptor antagonist, on rapid eye movement sleep and in major depressive disorder. *Journal of Pharmacology and Experimental Therapeutics, 342*, 429–440.

Bonaventure, P., Kelly, L., Aluisio, L., Shelton, J., Lord, B., Galici, R., et al. (2007). Selective blockade of 5-hydroxytryptamine (5-HT)7 receptors enhances 5-HT transmission, antidepressant-like behavior, and rapid eye movement sleep suppression induced by citalopram in rodents. *The Journal of Pharmacology and Experimental Therapeutics, 321*(2), 690–698.

Bortolato, M., Pivac, N., Seler, D. M., Perkovic, M. N., Pessia, M., & Di Giovanni, G. (2013). The role of the serotonergic system at the interface of aggression and suicide. *Neuroscience, 236*, 160–185.

Brogden, R. N., Heel, R. C., Speight, T. M., & Avery, G. S. (1978). Mianserin: a review of its pharmacological properties and therapeutic efficacy in depressive illness. *Drugs, 16*, 273–301.

Brown, R. E., Basheer, R., McKenna, J. T., Strecker, R. E., & McCarley, R. W. (2012). Control of sleep and wakefulness. *Physiological Reviews, 92*, 1087–1187.

Dalley, J. W., Everitt, B. J., & Robbins, T. W. (2011). Impulsivity, compulsivity and top-down cognitive control. *Neuron, 69*, 680–694.

D'Amour-Horvat, V., & Leyton, M. (2014). Impulsive actions and choices in laboratory animals and humans: effects of high vs. low dopamine states produced by systemic treatments given to neurologically intact subjects. *Frontiers in Behavioral Neuroscience, 8*, 432.

Davis, M. T., DellaGiola, N., Matuskey, D., Harel, B., Maruff, P., Pietrzak, R. H., et al. (2017). Preliminary evidence concerning the pattern and magnitude of cognitive dysfunction in major depressive disorder using cogstate measures. *Journal of Affective Disorders, 218,* 82–85.

de Lange, E. C. (2013). Utility of CSF in translational neuroscience. *Journal of Pharmacokinetics and Pharmacodynamics, 40,* 315–326.

Dumais, A., Lesage, A. D., Alda, M., Rouleau, G., Dumont, G., Chawky, N., et al. (2005). Risk factors for suicide completion in major depression: a case-control study of impulsive and aggressive behaviors in men. *The American Journal of Psychiatry, 162,* 2116–2124.

Enatescu, V. R., Papava, I., Enatescu, I., Antonescu, M., Anghel, A., Seclaman, E., et al. (2016). Circulating plasma micro RNAs in patients with major depressive disorder treated with antidepressants: a pilot study. *Psychiatry Investigation, 13,* 549–557.

Gould, T. D., Georgiou, P., Brenner, L. A., Brundin, L., Can, A., Courtet, P., et al. (2017). Animal models to improve our understanding and treatment of suicidal behavior. *Translational Psychiatry, 7,* e1092.

Grimwood, S., & Hartig, P. R. (2008). Target site occupancy: emerging generalizations from clinical and preclinical studies. *Pharmacology & Therapeutics, 122,* 281–301.

Hamilton, K. R., Mitchell, M. R., Wing, V. C., Balodis, I. M., Bickel, W. T., Fillmore, M., et al. (2015). Choice impulsivity: definitions, measurement issues, and clinical implications. *Personal Disorders, 6,* 182–198.

Harmer, C. J., Duman, R. S., & Cowen, P. J. (2017). How do antidepressants work? New perspectives for refining future treatment approaches. *Lancet Psychiatry, 4,* 409–418.

Harrison, J. E., Lophaven, S., & Olsen, C. K. (2016). Which cognitive domains are improved by treatment with vortioxetine. *The International Journal of Neuropsychopharmacology, 19,* 1–6.

Horder, J., Cowen, P. J., Di Simplicio, M., Browning, M., & Harmer, C. J. (2009). Acute administration of the cannabinoid CB1 antagonist rimonabant impairs positive affective memory in healthy volunteers. *Psychopharmacology, 205,* 85–91.

Keller, M. B., Montgomery, S., Ball, W., Morrison, M., Snavely, D., Liu, G., et al. (2005). Lack of efficacy of the substance P (neurokinin$_1$ receptor). *Biological Psychiatry, 59,* 216–223.

Kielbasa, W., & Stratford, R. E. J. (2012). Exploratory translational modeling approach in drug development to predict human brain pharmacokinetics and pharmacologically relevant clinical doses. *Drug Metabolism and Disposition, 40,* 877–883.

Lee, R. S. C., Hermens, D. F., Porter, M. A., & Redoblado-Hodge, M. A. (2012). A meta-analysis of cognitive deficits in first-episode major depressive disorder. *Journal of Affective Disorders, 140,* 113–124.

Leuchter, A. F., Hunter, A. M., Krantz, D. E., & Cook, I. A. (2014). Intermediate phenotypes and biomarkers of treatment outcome in major depressive disorder. *Dialogues in Clinical Neuroscience, 16,* 525–537.

Lopez, J. P., Fiori, L. M., Cruceanu, C., Lin, R., Labonte, B., Cates, H. M., et al. (2017). Micro RNAs 146a/b-5 and 425-3p and 24-3p are markers of antidepressant response and regulate MAPK/Wnt-system genes. *Nature Communications, 8,* 15497.

Malkesman, O., Pine, D. S., Tragon, T., Austin, D. R., Henter, I. D., Chen, G., et al. (2009). Animal models of suicide-trait-related behaviors. *Trends in Pharmacological Sciences, 30,* 165–173.

Mann, J. J. (2003). Neurobiology of suicidal behavior. *Nature Reviews Neuroscience, 4,* 819–828.

Marek, G. J., Day, M., & Hudzik, T. J. (2016). The utility of impulsive bias and altered decision making as predictors of drug efficacy and target selection: rethinking behavioral screening for antidepressant drugs. *The Journal of Pharmacology and Experimental Therapeutics, 356,* 534–548.

McIntyre, R. S., Cha, D. S., Soczynska, J. K., Woldeyohannes, H. O., Gallaugher, L. A., Kudlow, P., et al. (2013). Cognitive deficits and functional outcomes in major depressive disorder: determinants, substrates, and treatment interventions. *Depression and Anxiety, 30,* 515–527.

Meyer, J. H., Wilson, A. A., Sagrati, S., Hussey, D., Carella, A., Potter, W. Z., et al. (2004). Serotonin transporter occupancy of five selective serotonin reuptake inhibitors at different doses: an [11C]DASB positron emission tomography study. *American Journal of Psychiatry, 161,* 826–835.

Minerva Neurosciences, I. (2016). *United states securities and exchange commission form 8-K.* .

Minerva Neurosciences, I. (2017). *MIN-202 for the treatment of insomnia and mood disorders.* .

Morgan, P., van der Graaf, P. H., Arrowsmith, J., Feltner, D. E., Drummond, K. S., Wegner, C. D., et al. (2012). Can the flow of medicines be improved? Fundamental pharmacokinetic and pharmacological principles toward improving Phase II survival. *Drug Discovery Today, 17,* 419–424.

Murphy, S. E., Downham, C., Cowen, P. J., & Harmer, C. J. (2008). Direct effects of diazepam on emotional processing in healthy volunteers. *Psychopharmacology, 199,* 503–513.

Pandey, G. N. (2013). Biological basis of suicide and suicidal behavior. *Bipolar Disorders, 15,* 524–541.

Pinder, R. M., & Fink, M. (1982). Mianserin. *Modern Problems of Pharmacopsychiatry, 18,* 70–101.

Pringle, A., McTavish, S. F., Williams, C., Smith, R., Cowen, P. J., & Harmer, C. J. (2011). Short-term NK1 receptor antagonism and emotional processing in healthy volunteers. *Psychopharmacology, 215,* 239–246.

Pringle, A., Parsons, E., Cowen, L. G., McTavish, S. F., Cowen, P. J., & Harmer, C. J. (2012). Using an experimental medicine model to understand the antidepressant potential of the N-methyl-D-aspartic acid (NMDA) receptor antagonist memantine. *Journal of Psychopharmacology, 26,* 1417–1423.

Puclu, E., Trotter, P. D., Thomas, D. R., McFarquhar, M., Juhasz, G., Sahakian, B. J., et al. (2014). Temporal discounting in major depressive disorder. *Psychological Medicine, 44,* 1825–1834.

Rock, P., Roiser, J., Reidel, W., & Blackwell, A. (2013). Cognitive impairment in depression: a systematic review and meta-analysis. *Psychological Medicine, 29,* 1–12.

Sanchez, c., Asin, K. E., & Artigas, F. (2015). Vortioxetine, a novel antidepressant with multimodal activity: review of preclinical and clinical data. *Pharmacology & Therapeutics, 145,* 43–57.

Schmidt, F. M., Sander, C., Dietz, M. -E., Nowak, C., Schroder, T., Mergl, R., et al. (2017). Brain arousal regulation as response predictor for antidepressant therapy in major depression. *Scientific Reports, 7,* 45187.

Stenkrona, P., Halldin, C., & Lundberg, J. (2013). 5-HTT and 5-HT1A receptor occupancy of the novel substance vortioxetine (Lu AA21004). A PET study in control subjects. *European Neuropsychopharmacology, 23,* 1190–1198.

Thase, M. E., Mahableshwarkar, A. R., Dragheim, M., Loft, H., & Vieta, E. (2016). A meta-analysis of randomized, placebo-controlled trials of vortioxetine for the treatment of major depressive disorder in adults. *European Neuropsychopharmacology, 26,* 979–993.

Trivedi, M. H., & Greer, T. L. (2014). Cognitive dysfunction in unipolar depression: implications for treatment. *Journal of Affective Disorders, 152-154,* 19–27.

Vanderveldt, A., Oliveira, L., & Green, L. (2016). Delay discounting: pigeon, rat, human—does it matter? *Journal of Experimental Psychology. Animal Learning and Cognition, 42,* 141–162.

Voon, V., Chang-Webb, Y. C., Morris, L. S., Cooper, E., Sethi, A., Baek, K., et al. (2015). Waiting impulsivity: the influence of acute methylphenidate and feedback. *The International Journal of Neuropsychopharmacology, 19,* 1–10.

Warren, M. B., Pringle, A., & Harmer, C. J. (2015). A neurocognitive model for understanding treatment action in depression. *Philosophical Transactions of the Royal Society B, 370,* 20140213.

Wilson, S., & Argyropoulos, S. (2005). Antidepressants and sleep: a qualitative review of the literature. *Drugs, 65,* 927–947.

Winokur, A., Gary, K. A., Rodner, S., Rae-Red, C., Fernando, A. T., & Szuba, M. P. (2001). Depression, sleep physiology, and antidepressant drugs. *Depression and Anxiety, 14*, 19–28.

Winstanley, C. A., Dalley, J. W., Theobald, D. E. H., & Robbins, T. W. (2004). Fractionating impulsivity: contrasting effects of central 5-HT depletion on different measures of impulsive behavior. *Neuropsychopharmacology, 29*, 1331–1343.

Worbe, Y., Savulich, G., Voon, V., Fernandez-Egea, E., & Robbins, T. W. (2014). Serotonin depletion induces "waiting impulsivity" on the human four-choice reaction time task: cross-species translational significance. *Neuropsychopharmacology, 39*, 1519–1526.

Yamamoto, Y., Valitalo, P. A., van den Berg, D. -J., Hartman, R., van den Brink, W., Wong, Y. C., et al. (2017). A generic multi-compartmental CNS distribution model structure for 9 drugs allows prediction of human brain target site concentrations. *Pharmaceutical Research, 34*, 333–351.

Zheng, M., Appel, L., Luo, F., Lane, R., Burt, D., Risinger, R., et al. (2015). Safety, pharmacokinetic, and positron emission tomography evaluation of serotonin and dopamine transporter occupancy following multiple-dose administration of the triple reuptake inhibitor BMS-820836. *Psychopharmacology, 232*, 529–540.

8

Biomarker Opportunities to Enrich Clinical Trial Populations for Drug Development in Schizophrenia and Depression

Bruce J. Kinon[a]

Lundbeck North America, Deerfield, IL, United States

I INTRODUCTION

Patients suffering from complex psychiatric disorders such as schizophrenia and major depressive disorder present with much heterogeneity in the many factors associated with their illness. This heterogeneity is multi-determined and in part due to diagnostic imprecision, genetic variance, epigenetic influence, and comorbid illness. Such heterogeneity challenges the development of disease-specific treatments since a myriad of targets presumably mediate the pathophysiology and pheno-type of the many individuals making up a disease class and lead, in part, to interpatient variance in treatment outcome and disease course. Our current diagnostic nosology for psychiatric disorders relies on the presence, duration, and functional impact of specific signs and symptoms. There is no identified brain pathology or lab-oratory tests to confirm diagnosis and guide treatment selection as there are in other medical illnesses. Unfortu-nately, these signs and symptoms are nonspecific and do overlap diagnostic boundaries (Kennedy et al., 2012).

The "one size fits all" approach to developing new drugs for patients suffering from such broad indications as schizophrenia or major depression may not be the most effective strategy for finding new treatments for patients with significant interindividual variability in drug response and in whom few reliable predictors of this variability exist (Center for Health Policy at Brookings, 2015). The challenge of this variability is highlighted when one considers the estimation that most drugs used in the clinic may actually be effective in only 25%–60% of patients (Spear, Heath-Chiozzi, & Huff, 2001).

The success rate in developing new CNS drugs has not increased for many years. It has become increasingly more challenging to demonstrate efficacy in clinical trials designed to test newly discovered compounds in part due to disease heterogeneity within diagnostic groups. This has led to the exit of many pharmaceutical compa-nies from developing new psychiatric and neurological treatments (Suhara et al., 2016). Developing drugs that offer robust benefits in more homogeneous pharmacolog-ically relevant subsets of patients may be a more effective pathway for bringing drugs to market (Pacanowski, Leptak, & Zineh, 2014).

The aspiration of "personalized" or "precision" medi-cine is to match individual patients to a specific treatment whose "pharmacology" or mechanism of action is rele-vant to an individual's particular characteristics. The greater homogeneity achieved between patient and dis-ease characterization would possibly benefit not only more efficient drug development but also the efficacious delivery of those treatments with optimal likelihood of response in the individual patient (Dunlop & Mayberg, 2014). Homogeneous patient subgroups would share biological targets that mediate similar aspects of disease expression. Biomarkers for targeted therapies would have the potential to enable the selection of the most beneficial treatment that may mediate response in an indi-vidual patient's disease (Graig, Phillips, & Moses, 2016).

Clinical biomarkers can be classified into at least two of several categories: prognostic biomarkers and predictive biomarkers. Prognostic biomarkers approximate disease

---

[a] The opinions expressed are solely those of the author and not necessarily those of Lundbeck.

progression in untreated patients; predictive biomarkers identify subpopulations of patients that are more likely to respond to treatment (Lassere et al., 2007). These biomarkers may provide an approach to trial enrichment that entails the prospective use of patient characteristics to select a study population in which a drug effect is more likely to be observed than it would be in the general population (Center for Health Policy at Brookings, 2015). The term enrichment is defined as the prospective use of any patient characteristic to select a study population in which detection of a drug effect is more likely than it would be in an unselected population (US Food and Drug Administration, 2012).

Enrichment strategies fall into three broad categories:

1. General strategies employed in clinical trials to select an appropriate subset of the overall population through inclusion and exclusion criteria in which to detect a treatment effect if one exists and/or minimizes risk (Pacanowski et al., 2014).
2. Prognostic enrichment strategies in which a prognostic biomarker is employed to assist in choosing patients with a greater likelihood of having a disease-related endpoint event or a substantial worsening in condition. This prognostic approach to enrichment utilizes prognostic biomarkers to screen for and enroll high-risk patients with more severe forms of the disease of interest increasing the chances of detecting a treatment response or a larger net change in an endpoint of interest. This in turn allows for an increased study power to facilitate the early demonstration of proof of concept (US Food and Drug Administration, 2012).
3. Predictive enrichment strategies choose patients more likely to respond to the drug treatment than other patients with the condition being treated. Such selection can lead to a larger effect size and permit use of a smaller study population. Selection of patients

could be based on a specific aspect of a patient's physiology or a disease characteristic that is related in some manner to the study of drug's mechanism, or it could be empirical (e.g., the patient has previously appeared to respond to a drug in the same class) (US Food and Drug Administration, 2012). Predictive enrichment relies on biomarkers to identify and enroll patients most likely to respond to treatment allowing study designs in which one may be looking for a difference in response between biomarker-positive and biomarker-negative arms of the study. Predictive enrichment can be especially important in cases where biomarker-positive patients make up a small fraction of the overall population whose treatment response may fail to be detected through signal dilution due to testing in a general population (Center for Health Policy at Brookings, 2015).

Some examples of drugs/tests of FDA-approved enrichment strategies are provided in Table 8.1.

Trial enrichment through the use of appropriate biomarkers holds the potential to reduce the time and cost of drug development by improving the efficiency and success rate of the clinical trials required to demonstrate the safety and efficacy of new therapeutics by selecting subsets of patients with shared biological characteristics optimal for testing candidate compounds rather than based upon symptomatic diagnosis, thus reducing heterogeneity in subject cohorts (Engelberg Center for Health Care, 2014; Suhara et al., 2016).

II ENRICHMENT OPPORTUNITIES IN SCHIZOPHRENIA

It is challenging to determine which enrichment strategies may be useful in schizophrenia clinical trials owing to the complex nature of the pathophysiology of this disease

TABLE 8.1　Examples of Biomarker Enrichment Strategies

Drug/diagnostic test	Indication	Enrichment strategy	Biomarker
Atorvastatin	Prevention of cardiovascular disease	Prognostic	Multiple risk factors for coronary heart disease
Memantine	Moderate to severe Alzheimer's disease (AD)	Prognostic	Minimental status examination
Tamoxifen	Metastatic breast cancer	Prognostic	Metastatic breast cancer; women at high breast cancer risk; estrogen receptor positive tumors
MammaPrint	Prognostic test for certain breast cancer patients	Prognostic	Seventy gene array identifies risk of breast cancer recurrence/metastasis
Clozapine	Treatment-resistant schizophrenia	Predictive	Previous APD treatment failures
Ivacaftor	Cystic fibrosis	Predictive	Carriers of CFTR G551D mutation
Imatinib	Chronic myelogenous leukemia (CML)	Predictive	Philadelphia chromosome-positive CML
Abacavir	Prevention and treatment HIV/AIDS	Predictive	Contraindicated in carriers of HLA-B*5701 allele
Trastuzumab	Metastatic breast cancer	Predictive	HER2 receptor positive tumors

and our limited knowledge of relevant biomarkers. We will consider several areas where such strategies have been utilized or may be further developed.

Individuals who present a clinically high risk (CHR) for conversion to schizophrenia are in need of treatment interventions to mitigate this risk. Development of an appropriate preventative treatment is dependent upon the identification of a prognostic biomarker(s) that may be linked to a pathophysiological mechanism that is associated with the onset or progression of schizophrenia to select the appropriate at-risk population and also the identification of a predictive biomarker through which an intervention strategy is believed to work to forestall or inhibit the pathophysiological process leading to conversion to psychosis in at-risk individuals (Millan et al., 2016). Therefore an enrichment strategy based upon prognostic and/or predictive biomarkers may be useful to improve the effectiveness of treatment development in this area of need. In addition, a surrogate endpoint biomarker that reflects a reduction in risk of a conversion that might only become apparent years later may also be necessary. Research strategies to identify any of the noted biomarkers not only would greatly contribute to the development of a treatment intervention for CHR individuals but also might significantly advance our understanding of the pathophysiology and improved treatment of schizophrenia.

Prognostic biomarkers may include those clinical features associated with a high risk of potentially transitioning to psychosis in CHR individuals such as unusual thought content, cognitive impairment, suspiciousness and paranoid ideation, substance abuse, migrant status, isolation, poor quality of life, and neurological soft signs (Millan et al., 2016). Numerous neurobiological features may be associated with the progression to psychosis including alterations in dopaminergic, glutamatergic, and GABAergic transmission; loss of cortical gray matter; disruption of network connectivity; disruption in event-related potentials; and immune markers (Millan et al., 2016). A possible prognostic biomarker to consider in future trials may involve the selection of at-risk subjects who are found to have increased striatal glutamate on proton-MRS, as these patients have been reported to have a greater risk of psychosis conversion compared with subjects who did not convert over a 2-year follow-up period (de la Fuente-Sandoval et al., 2013). In an effort to better predict an individual's risk for conversion to schizophrenia, a risk calculator has been developed through the North American Prodrome Longitudinal Study (NAPLS-2) (Cannon et al., 2016) that utilizes key demographic (age and family history of psychosis), clinical (unusual thought content and suspiciousness), neurocognitive (verbal learning and memory and speed of processing), and psychosocial (traumas, stressful life events, and decline in social functioning) predictor variables to generate a number representing the probability of transition to psychosis given a particular profile of input variables. This multivariable risk predictor may be used to enrich investigations into the CHR state. Recent research from the NAPLS consortium has found a steeper rate of brain gray matter loss in at-risk individuals who convert compared with those who do not and that this loss may be associated with a process of neuroinflammation (Cannon et al., 2015). Several risk calculators have recently been developed in an attempt to identify individuals at imminent risk of conversion to schizophrenia (Fusar-Poli et al., 2017; Schmidt et al., 2017).

Although antipsychotic treatment has demonstrated some reduction in preventing conversion, the results have either been nonsignificant as compared with placebo control or if significant initially, this advantage is then lost upon long-term follow-up (Millan et al., 2016). Based upon the observation that omega-3 polyunsaturated fatty acids (omega-3 PUFAs) may be reduced in schizophrenia and their potential neuroprotective role (Horrobin, Glen, & Vaddadi, 1994), omega-3 PUFAs have also been studied as preventive treatment for psychosis conversion in ultrahigh-risk (UHR) individuals (Amminger et al., 2010). An initial controlled trial in 81 individuals demonstrated a positive effect that was maintained for up to 12 months. Unfortunately a subsequent larger trial ($n = 304$) failed to demonstrate any benefit for omega-3 PUFAs in preventing conversion when studied out to 6 months (McGorry et al., 2017). In their commentary of these results, Kane and Correll (Kane & Correll, 2017) offer the opinion that the failure to replicate the initial omega-3 PUFAs trial findings may in part be due to a lack of predictive enrichment in the latter trial for a subgroup of patients who might be selected based upon a biomarker relevant to the proposed target of omega-3 PUFAs' action.

The urgency to provide the optimal treatment as early as possible, as discussed with the CHR state, is also a need close to the time of psychosis onset where appropriate early intervention has been found to be associated with improved symptoms and functioning compared with traditional care (Álvarez-Jiménez, Parker, Hetrick, PD, & Gleeson, 2011; Bird et al., 2010). The Recovery After an Initial Schizophrenia Episode (RAISE) initiative was designed to compare the effectiveness of a comprehensive, team-based treatment (NAVIGATE) compared with usual care in US community treatment centers in first-episode psychosis subjects (Kane et al., 2015). The 2-year controlled trial determined that the comprehensive intervention led to greater improvement in functional and clinical outcomes. Therefore a predictive enrichment strategy utilizing first-episode patients hypothesized to benefit from a targeted therapy in fact did demonstrate the predicted favorable response. Furthermore, patients with shorter duration of untreated psychosis derived substantially more benefit from NAVIGATE. Prolonged duration of untreated psychosis (DUP) after the development of psychosis is a risk factor for poor disease outcome, and research efforts to minimize this period are encouraged (Bertolote & McGorry, 2005). These findings from RAISE suggest that clinical research to further understand and improve treatment for first-episode psychosis may benefit from a predictive enrichment strategy and possibly also from a prognostic enrichment strategy in which enrollment is enriched, and stratified, with patients having a shorter (<1 year) or longer (>1 year) history of DUP with the presumption that DUP may be a prognostic risk for disease outcome in a first-episode population.

One of the most challenging issues in treating psychotic exacerbations in schizophrenia is deciding how long an initial trial of an antipsychotic should last and what response metrics should be used to inform that decision. A large body of recent research has shown that the onset of response can occur rapidly within the first week or two (Agid, Kapur, Arenovich, & Zipursky, 2003; Kinon et al., 2010; Leucht, Busch, Hamann, Kissling, & Kane, 2005; Samara et al., 2015). To date, all of the studies on early response/nonresponse have shown that early nonresponse is a robust predictor of subsequent nonresponse with continued treatment of the same medication (Ascher-Svanum et al., 2008; Correll et al., 2003; Kinon et al., 2008; Leucht, Busch, Kissling, & Kane, 2007; Leucht, Shamsi, Busch, Kissling, & Kane, 2008; Kinon et al., 2010; Samara et al., 2015). These studies also show that a majority of patients (approximately 70%) do not achieve this "early response" criterion (generally defined as a 20% or greater improvement in PANSS total score at 2 weeks compared with baseline) with either a typical or atypical antipsychotic drug (Ascher-Svanum et al., 2008; Kinon et al., 2008, 2010). The clinical dilemma in these patients who do not show early response is whether one should "switch" or "stay" in regard to their current treatment. The conclusion reached in a metaanalysis of 34 clinical trials was that patients not even minimally improved by week 2 of antipsychotic treatment are unlikely to respond later and may benefit from a treatment change (Samara et al., 2015).

Unfortunately, there are no unequivocal predictors for those patients who will be early responders versus early nonresponders. Patients are therefore obliged to initiate an antipsychotic treatment trial to subsequently determine their early response status. Once established, early response/nonresponse is a robust prognostic biomarker of treatment outcome that has relevance in clinical practice and potentially in clinical research where phase 2 acute efficacy proof-of-concept trials may be expeditiously designed to be run shorter (e.g., 2–4 weeks) rather than longer (e.g., 6–8 weeks). As concluded by Samara et al. (2015) in their metaanalysis, early response/ nonresponse to an antipsychotic drug trial is the strongest predictor of treatment improvement. And among many

potential predictors identified in the literature, such as early subjective response, severity of illness, homovanillic acid level, structural brain imaging, and APD receptor-related genotyping, it is well-replicated and of direct value in clinical practice.

Early response to antipsychotic treatment may therefore be considered a prognostic biomarker that can be utilized to enrich a clinical trial with either patients likely to respond (i.e., early responders) to antipsychotic drug treatment or, conversely, unlikely to respond (i.e., early nonresponders) to treatment. Such enrichment strategies may identify a suitable cohort to investigate issues related to treatment response, including new drugs, or issues related to treatment resistance, including potential innovative therapies.

There exists a great unmet need to find new treatments for schizophrenia that go beyond dopamine receptor antagonism. Accumulating evidence implicates a dysregulation of CNS glutamatergic activity, perhaps early in the disease, contributing to the pathophysiology of schizophrenia (Goff & Coyle, 2001; Jentsch & Roth, 1999; Krystal et al., 1999; Patil et al., 2007; Paz, Tardito, Atzori, & Tseng, 2008). Pomaglumetad methionil (pomaglumetad), a metabotropic glutamate 2/3 (mGlu2/3) receptor agonist, which may normalize this heightened activity of cortical pyramidal neurons, was developed as a drug with a novel, nondopamine D2 receptor antagonist-dependent mechanism to treat schizophrenia (Patil et al., 2007). Unfortunately, extensive clinical testing of pomaglumetad through phase 3 in acutely exacerbated patients suffering from chronic schizophrenia failed to demonstrate antipsychotic efficacy ultimately leading to the discontinuation of its schizophrenia development program.

Several reasons were posited as possible explanations for these negative results. One questioned whether the targeted population (neuroleptic-responsive patients) may have in fact been suffering from a hyperglutamatergic disease state responsible for their acute symptoms, as opposed to a hyperdopaminergic disease state. Another questioned whether the patients' prior atypical antipsychotic treatment history may have led to an epigenetics-driven downregulation of the mGlu2 receptor (a target of LY404039, the active parent of pomaglumetad) (Kurita et al., 2012; Kurita, Holloway, & González-Maeso, 2013), thus diminishing the effectiveness of pomaglumetad. Therefore a series of post hoc analyses were conducted to specifically test these hypotheses in patient subgroups from the pomaglumetad trials who were identified by critical characteristics related to duration of illness (i.e., early in disease, a presumed hyperglutamatergic schizophrenia disease state vs late in disease, a presumed normoglutamatergic schizophrenia disease state) or to previous medication use that may downregulate the mGlu2 receptor (i.e., prior exposure vs nonexposure to CNS drugs with prominent 5-HT2A receptor antagonist activity (Kurita et al., 2012).

The first targeted post hoc analysis did demonstrate symptom improvement in the early-in-disease patients consistent with research suggesting that early-in-disease patients may be experiencing pathological hyperglutamatergic activity that may be normalized by pomaglumetad. These results were interesting though speculative as there was no direct data from the trial to prove that these apparently responsive patients were in fact in a hyperglutamatergic illness state. Nonetheless, this finding did raise the speculation that evidence of central hyperglutamatergic activity (e.g., increased glutamate MRS signal in anterior cingulate cortex) may be a predictive marker for pomaglumetad response in schizophrenia. In addition, the second post hoc analysis demonstrated that patients chronically treated with neuroleptics with nearly exclusive dominant D2-antagonism such as haloperidol, but not patients with chronic exposure to drugs with potent 5-HT2A receptor antagonism properties, such as most atypical antipsychotics, exhibited a therapeutic response to pomaglumetad, which suggested that preservation of the mGlu2 receptor target, through avoidance of downregulation, was necessary to effectuate an antipsychotic response to pomaglumetad. In this case, increased suspicion of mGlu2 receptor subsensitivity, inferred by treatment history, would be a predictive marker on which to exclude a potentially nonresponsive subgroup from further pomaglumetad trials.

These post hoc results, though tentative, are suggestive, at least in regard to pomaglumetad, that the demonstration of antipsychotic efficacy of a potential glutamate-based pharmacotherapy for schizophrenia, with no dopamine antagonist properties, may require clinical trial testing in an appropriate patient subgroup whose treatment responsiveness may be fundamentally related to a dysregulation of CNS glutamatergic tone. This predictive enrichment strategy is consistent with the growing awareness in CNS drug development that new potential therapies targeted to a specific neurobiological target should be clinically investigated in a patient subgroup presumed to have a disease state modulated by that target.

The benefit afforded by predicting antipsychotic drug response includes the ability to expedite the often lengthy trial and error process that delays patients from receiving effective treatment and also to better understand the neurobiology of an individual's antipsychotic drug response. Sarpal et al. (2015) have recently reported that improvement of psychotic symptoms was associated with changes in striatal functional connectivity over the course of treatment in first-episode patients. Their results indicated that the psychotic state was associated with corticostriatal functional dysconnectivity that diminished, along with improvement in symptoms, with APD treatment. They subsequently developed a predictive biomarker based upon a baseline resting-state fMRI assessment of functional connectivity of the striatum that was validity tested in a cohort of patients with chronic schizophrenia experiencing an acute psychotic episode (Sarpal et al., 2016). Their striatal connectivity index showed a significant separation at baseline between subsequent responders and nonresponders in this chronic patient cohort. In addition, connectivity improvement on the index was associated with a shorter length of psychiatric hospital stay.

The researchers concluded that although the striatal connectivity index, as a predictive biomarker, may have too limited sensitivity and specificity, at this time in its early development, to precisely target antipsychotic drug treatment to specific patients, it may provide an important index of "target engagement" for the development of novel antipsychotic therapeutics and provide further insight into possible mechanisms that underlie psychotic symptoms.

The genetics of schizophrenia are complex. A few large copy number variants (CNVs) may impart a significant effect on the illness as indicated by odd ratios compared with controls of up to 70-fold for the 22q11.21 deletion, but these CNVs are rare. The Psychiatric Genomics Consortium (PGC) has recently (CNV and Schizophrenia Working Groups of the Psychiatric Genomics Consortium, 2017) identified that the eight schizophrenia CNV risk loci that surpass genome-wide-significance are carried by a small fraction (1.4%) of schizophrenia cases in the PGC sample and are estimated to account for 0.85% of the variance in disease liability. In comparison, the 108 genome-wide significant loci identified in the PGC genome-wide analysis study were calculated to contribute 3.4% of the variance in schizophrenia liability. Combined, the CNV and SNP loci that have been identified to date explain a small proportion (<5%) of heritability (CNV and Schizophrenia Working Groups of the Psychiatric Genomics Consortium, 2017). Therefore the influence of genetics on precision medicine in schizophrenia remains quite limited.

A pharmacogenetic prognostic biomarker that is associated with an increased risk for clozapine-induced agranulocytosis, and whose effect has been replicated in an independent sample, is the C allele of the 6672G > C single-nucleotide polymorphism (SNP) in the HLA-DQB1 region (Athanasiou et al., 2011). The CC individuals have a greatly increased risk of agranulocytosis (O.R. 16.8). Through their systematic review of biomarkers in schizophrenia, Prata, Mechelli, and Kapur (2014) have singularly identified this as the best replicated, potentially clinically significant biomarker.

Despite its high OR and specificity of 99.7%, this biomarker suffers from a limited sensitivity of 21% and therefore probably will not be able to change the current clinical monitoring for agranulocytosis.

Although genetic markers for the dopamine D2 receptor are prominent in schizophrenia genetic surveys, a predictive relationship between variation in the dopamine D2 receptor gene (DRD2) and APD response has not been clearly established (Malhotra, Murphy Jr, & Kennedy, 2004) perhaps due to the effects of chronic disease and treatment in the cohorts studied. Therefore Lencz et al. (2006) have studied this relationship in first-episode schizophrenia patients. Based upon previous work, two DRD2 promoter region polymorphisms (A-241G and -141 Ins/Del) were examined in a cohort of 61 first-episode schizophrenia patients, with limited previous APD exposure, who received treatment for 16 weeks with either risperidone or olanzapine. Time to sustained APD response was found to be sooner in the A-241G carriers and in the -141Ins/Ins homozygotes. This was the first report of a significant relationship between a DRD2 genetic variation, a predictive biomarker, and treatment response in first-episode schizophrenia patients. Lencz et al. (2006) advise that replication is required to confirm the potential clinical utility of this finding. This study further illustrates that the prognostic enrichment strategy of utilizing first-episode patients as a potentially more homogeneous population, as compared with chronic schizophrenia, may have contributed to the positive pharmacogenetic findings.

The identified predictive genetic biomarkers may provide the basis to identify future patient subgroups to further advance research into the mechanism of APD action and the basis for treatment responsiveness. Genetic association studies from the pomaglumetad development program provide an additional example of the identification and replication of a potential useful genetic marker for schizophrenia research with the potential for a predictive enrichment strategy to enhance drug development efforts.

Although pomaglumetad did not demonstrate efficacy in the phase 3 general patient population with schizophrenia, it did in the original proof-of-concept study (Patil et al., 2007). Pharmacogenetic results from that study identified a significant association between 16 tightly linked, single-nucleotide polymorphisms (SNPs) in the serotonin 2A receptor gene, HTR2A, the most significant of which was rs7330461, and response to treatment with pomaglumetad in Caucasian patients with schizophrenia (Liu et al., 2012). Caucasian patients carrying the T/T genotype at rs7330461 showed a more favorable response to pomaglumetad compared with patients with either the A/T or A/A genotype at rs7330461 (Liu et al., 2012). The validity of this observation

was supported by previous studies suggesting potential epigenetic interactions between 5-HT2A and mGlu2/3 receptors (Wischhof & Koch, 2015). Based on these findings a planned strategy was made to examine the association of rs7330461 with efficacy outcome in each subsequent pomaglumetad clinical trial in an attempt to replicate this initial finding.

Genetic association data from three individual clinical studies and integrated analyses of genetic association data from these three studies plus a fourth additional smaller study provided a sample of 1115 Caucasian patients for whom genotyping information for rs7330461 was available, consisting of 513 A/A homozygous, 466 A/T heterozygous, and 136 T/T homozygous patients (Nisenbaum et al., 2016). Caucasian T/T homozygous patients showed significantly ($P \leq .05$) greater improvement in positive and negative syndrome scale (PANSS) total scores during treatment with pomaglumetad 40 mg twice daily compared with A/A homozygous patients in each individual study and in the integrated analysis. Additionally, T/T homozygous patients receiving pomaglumetad had significantly ($P \leq .05$) greater improvements in PANSS total scores compared with placebo and similar improvements as T/T homozygous patients receiving standard-of-care (SOC) treatment. These findings are in contrast to the general patient population in which a significant treatment response had not been observed in patients treated with pomaglumetad compared with placebo (Downing et al., 2014). This appears to be the first reported pharmacogenetic study to identify and replicate in multiple studies a genetic predictive biomarker of efficacy for the treatment of schizophrenia that does not rely on drug metabolism genes.

The pharmacogenetic results from the pomaglumetad drug development program not only highlight the possibility of realizing a personalized psychiatric medicine based upon a predictive genetic marker but also emphasize the importance of including, if possible, a plausible biomarker strategy in at least proof-of-concept trials that may allow for targeted therapies to be sought even after a negative outcome of the originally planned study. A logical extension of the pomaglumetad findings may be a prospective trial that purposefully pursues a predictive enrichment strategy to include the T/T homozygous patients (see Fig. 8.1).

The examples reviewed in the Schizophrenia section were selected to illustrate how biomarkers may be utilized to potentially reduce heterogeneity in clinical cohorts, to begin to better understand the neurobiology of schizophrenia and its response to treatment, and to explore strategies to expeditiously predict nonresponding patients to avoid the time-consuming and disappointing trial and error approach to finding a satisfactory treatment.

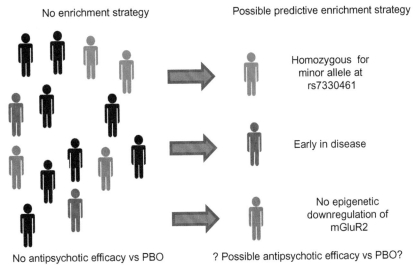

No enrichment strategy   Possible predictive enrichment strategy

Homozygous for minor allele at rs7330461

Early in disease

No epigenetic downregulation of mGluR2

No antipsychotic efficacy vs PBO   ? Possible antipsychotic efficacy vs PBO?

FIG. 8.1   Can predictive enrichment reveal antipsychotic efficacy for pomaglumentad?

III ENRICHMENT OPPORTUNITIES IN DEPRESSION

The identification of relevant prognostic and predictive biomarkers may also enhance signal detection in clinical trial research for new depression treatments. In addition, such markers could help to foster a personalized medicine approach to the use of such treatments in a target patient population with a higher likelihood to respond than a general disease population prescribed treatment on a trial and error basis (Leuchter et al., 2010). It would also reduce the patient suffering and despair associated with successive disappointing treatment failures.

Research continues to identify a reliable predictive biomarker, the treatment-selection biomarker (TSB), to guide selection of the initial treatment for depression (Dunlop & Mayberg, 2014). First-line treatments for major depressive disorder (MDD) consist of either antidepressant medication or an evidence-based psychotherapy (American Psychiatric Association, 2010), and these two types of treatments have fundamentally different mechanisms of action. Dunlop and Mayberg (2014) propose that if independent brain states differentially mediate the response to each of these distinct treatment modalities, then a TSB could be developed that indicates whether psychotherapy or medication represents the best treatment for particular patients, based on their brain state at the time of treatment initiation.

Determination of metabolism in the anterior insula, which functions in processing risk, reward, consciousness, and performance monitoring (Craig, 2009; Zhang, Chang, Guo, Zhang, & Wang, 2013), may be utilized as a TSB (Dunlop & Mayberg, 2014). The potential value of insula metabolism as a predictive TSB is supported by findings from several research groups

(McGrath, Kelley, Holtzheimer, et al., 2013; Sliz & Hawley, 2012), which found that the FDG-PET neuroimaging signal was able to predict differential outcomes in MDD patients randomly assigned to 12 weeks of treatment with either a structured psychotherapy (cognitive behavior therapy (CBT)) or antidepressant medication (escitalopram). A brain region was considered a potential TSB if it differentiated both the remitter-nonresponder differences (by treatment) and the escitalopram-CBT differences (by outcome). In addition, a low insula resting-state activity was found to be an indicator that the patient may be a good candidate for psychotherapy (Dunlop & Mayberg, 2014). Conversely, reduced volume of the right hippocampus and reduced gray matter volume in the left DLPFC have been identified as structural imaging predictors of nonresponse to antidepressant medication (Fu, Steiner, & Costafreda, 2012). These results provide support for the concept that pretreatment brain states, as identified by neuroimaging, may be used in the prediction of treatment outcomes for MDD (Dunlop & Mayberg, 2014). Other imaging studies have indicated that amygdala activation in response to emotional facial expressions in MDD patients also predicts symptom resolution (Canli, Cooney, Goldin, et al., 2005; DeRubeis, Siegle, & Hollon, 2008; Fu, Williams, Cleare, et al., 2008). In a recent prospective study (Dunlop et al., 2017) utilizing fMRI to identify resting-state functional connectivity differential predictors of outcomes among adults with treatment-naive major depression randomly assigned to receive 12 weeks of treatment with either CBT or an antidepressant medication, negative connectivity scores at baseline were associated with remission to medication and treatment failure with CBT, whereas positive connectivity scores were associated with remission to CBT and treatment failure with medication. These robust findings indicate that neuroimaging may have an important role in the application of precision medicine for depression by identifying neural signatures of brain states that are differentially responsive to treatments with differing mechanisms of action. The present results, in conjunction with a prior CBT compared with antidepressant medication study using FDG- PET (Dunlop, Kelley, McGrath, et al., 2015; McGrath et al., 2013), argue strongly that brain state subtypes of heterogeneous major depressive disorder patients may serve as predictive markers reflecting the biological capacity to benefit differentially from treatments with differing mechanisms of action (Dunlop et al., 2017). These findings support consideration of a predictive enrichment strategy based on these resting-state connectivity markers to stratify patients in proof-of-concept treatment trials investigating new potential therapies for depression.

Inflammatory disease markers may also provide for a predictive enrichment strategy for the investigation of potential new depression treatments that may work through an antiinflammation mechanism (Kappelmann, Lewis, Dantzer, Jones, & Khandaker, 2016). Inflammation has been proposed as a causal or contributing mechanism for the development of depressive symptoms. Proinflammatory cytokines, like the interleukins IL-1β and IL-6 and tumor necrosis factor alpha (TNF-α), are known to induce "sickness behavior," a set of behavioral changes overlapping with depressive symptomatology (Maes et al., 2012). Cytokine-mediated communication between the immune system and the brain has been implicated in the pathogenesis of depression (Dantzer, O'Connor, Freund, Johnson, & Kelley, 2008; Maes, 1995; Raison, Capuron, & Miller, 2006). Major depression commonly develops after interferon treatment, a potent inducer of cytokines, in patients affected by hepatitis C virus. Experimental immune activation in healthy volunteers leads to depressive symptoms and poor cognitive performance (Harrison et al., 2009; Reichenberg et al., 2001). Kappelmann et al. (2016) carried out a systematic review and metaanalysis of antidepressant activity of anticytokine treatment using clinical trials of chronic inflammatory conditions where depressive symptoms were measured as a secondary outcome. This review of 20 studies including metaanalyses of 16 studies totaling 5063 participants indicated that anticytokine treatment improved depressive symptoms and that the antidepressant effect was associated with severity of depressive symptoms at baseline, but not with improvement in physical illness.

The relevance of inflammatory markers as predictive biomarkers in depression treatment trials is further supported by the finding that depression patients who demonstrated an antidepressant response to acute ketamine infusion were found, at baseline, to have significantly higher serum levels of IL-6 than those of the control and nonresponder groups (Yang et al., 2015).

A genetic contribution to mood disorders is well established. Family studies show an increased risk of depression in individuals with shared genes (Shih, Belmonte, & Zandi, 2004). Twin studies suggest that genetic influences in depression are moderate, perhaps explaining up to 40% of the variance in adult MDD (Shih et al., 2004; Sullivan, Neale, & Kendler, 2000) and approximately 70% in bipolar disorder (McGuffin et al., 2003). Although numerous candidate gene and single-nucleotide polymorphism (SNP) associations have also been identified in MDD, generally replication is poor that may be related to heterogeneity inherent in MDD (Bosker, Hartman, Nolte, et al., 2011).

Genetic predictors of treatment outcome in 1,953 patients with major depressive disorder who were treated with the antidepressant citalopram in the sequenced treatment alternatives for depression (STAR*D) study were prospectively assessed (McMahon et al., 2006). A significant and reproducible association between treatment outcome and a marker in serotonin HTR2A was found. The clinical impact of HTR2A on treatment outcome was modest. Participants who were homozygous for the A allele had an 18% reduction in absolute risk of having no response to treatment, compared with those homozygous for the other allele. This genetic finding was replicated with escitalopram response in the GENDEP study (Kennedy et al., 2012). Additional alleles predictive of treatment outcome will need to be discovered before clinically more relevant effect sizes are obtainable.

Other areas where a biomarker strategy for patient and outcome enrichment may be useful but in which no definitive studies are necessarily available include the following:

- Stress response as a prognostic marker for depression risk and a predictive marker of treatment response
- Glutamatergic activity as a prognostic marker for depression risk and a predictive marker of treatment response
- Irritability as a prognostic and/or predictive marker in MDD
- Cognitive impairment as a prognostic or predictive marker in MDD
- Treatment-resistant depression (TRD) as a prognostic or predictive marker in depression

The identification of reliable prognostic and predictive biomarkers in MDD remains a challenge due to the clinical heterogeneity of this disorder (Lener et al., 2016). Advances in our understanding of the underlying neural systems that are unique to subgroups of patients that mediate their illness and treatment response will hopefully inform the enrichment strategies that may illuminate research into new treatments for depression.

IV CONCLUSION

The identification of prognostic and predictive biomarkers in schizophrenia and depression is an active area of investigation that will potentially improve the probability of successfully developing new disease treatments. Biomarkers that can identify the appropriate patient for the appropriate treatment intervention will provide the basis for rational tailored therapy in schizophrenia and depression that will hopefully result in greater expedience in treatment service delivery and greater effectiveness in treatment outcome.

In addition to improved alleviation of the symptoms of schizophrenia and depression, it may eventually become possible to target the underlying pathophysiology and to meaningfully delay, prevent, or moderate its progress.

References

Agid, O., Kapur, S., Arenovich, T., & Zipursky, R. B. (2003). Delayed-onset hypothesis of antipsychotic action. *Archives of General Psychiatry, 60*, 1228–1235.

Álvarez-Jiménez, M., Parker, A. G., Hetrick, S. E., PD, M. G., & Gleeson, J. F. (2011). Preventing the second episode: a systematic review and meta-analysis of psychosocial and pharmacological trials in first-episode psychosis. *Schizophrenia Bulletin, 37*(3), 619–630.

American Psychiatric Association (2010). *Practice guideline for the treatment of patients with major depressive disorder* (3rd ed.). Arlington, VA: American Psychiatric Association.

Amminger, G. P., Schäfer, M. R., Papageorgiou, K., Klier, C. M., Cotton, S. M., Harrigan, S. M., et al. (2010). Long-chain w-3 fatty acids for indicated prevention of psychotic disorders. *Archives of General Psychiatry, 67*(2), 146–154.

Ascher-Svanum, H., Nyhuis, A. W., Faries, D. E., Kinon, B. J., Baker, R. W., & Shekhar, A. (2008). Clinical, functional, and economic ramifications of early nonresponse to antipsychotics in the naturalistic treatment of schizophrenia. *Schizophrenia Bulletin, 34*, 1163–1171.

Athanasiou, M. C., Dettling, M., Cascorbi, I., Mosyagin, I., Salisbury, B. A., Pierz, K. A., et al. (2011). Candidate gene analysis identifies a polymorphism in HLA- DQB1 associated with clozapine-induced agranulocytosis. *The Journal of Clinical Psychiatry, 72*, 458–463.

Bertolote, J., & McGorry, P. (2005). Early intervention and recovery for young people with early psychosis: consensus statement. *The British Journal of Psychiatry. Supplement, 48*, s116–s119.

Bird, V., Premkumar, P., Kendall, T., Whittington, C., Mitchell, J., & Kuipers, E. (2010). Early intervention services, cognitive–behavioural therapy and family intervention in early psychosis: systematic review. *The British Journal of Psychiatry, 197*, 350–356.

Bosker, F. J., Hartman, C. A., Nolte, I. M., et al. (2011). Poor replication of candidate genes for major depressive disorder using genome-wide association data. *Molecular Psychiatry, 16*(5), 516–532.

Canli, T., Cooney, R. E., Goldin, P., et al. (2005). Amygdala reactivity to emotional faces predicts improvement in major depression. *Neuroreport, 16*(12), 1267–1270.

Cannon, T. D., Chung, Y., He, G., Sun, D., Jacobson, A., van Erp, T. G. M., et al. (2015). Progressive reduction in cortical thickness as psychosis develops: a multisite longitudinal neuroimaging study of youth at elevated clinical risk. *Biological Psychiatry, 77*, 147–157.

Cannon, T. D., Yu, C., Addington, J., Bearden, C. E., Cadenhead, K. S., Cornblatt, B. A., et al. (2016). An individualized risk calculator for research in prodromal psychosis. *The American Journal of Psychiatry*. https://doi.org/10.1176/appi.ajp.2016.15070890.

Center for Health Policy at Brookings. (2015). *Improving productivity in pharmaceutical research and development: The role of clinical pharmacology and experimental medicine*. Discussion Guide Washington, DC: Center for Health Policy at Brookings. July 28, 2015.

CNV and Schizophrenia Working Groups of the Psychiatric Genomics Consortium. (2017). Contribution of copy number variants to schizophrenia from a genome-wide study of 41,321 subjects. *Nature Genetics, 49*, 27–35.

Correll, C. U., Malhotra, A. K., Kaushik, S., McMeniman, M., & Kane, J. M. (2003). Early prediction of antipsychotic response in schizophrenia. *The American Journal of Psychiatry, 160*, 2063–2065.

Craig, A. D. (2009). How do you feel—now? The anterior insula and human awareness. *Nature Reviews. Neuroscience, 10*, 59–70.

Dantzer, R., O'Connor, J. C., Freund, G. G., Johnson, R. W., & Kelley, K. W. (2008). From inflammation to sickness and depression: when the immune system subjugates the brain. *Nature Reviews Neuroscience, 9*, 46–56.

de la Fuente-Sandoval, C., León-Ortiz, P., Azcárraga, M., Favila, R., Stephano, S., & Graff-Guerrero, A. (2013). Striatal glutamate and the conversion to psychosis: a prospective 1H-MRS imaging study. *International Journal of Neuropsychopharmacology, 16*, 471–475.

DeRubeis, R. J., Siegle, G. J., & Hollon, S. D. (2008). Cognitive therapy versus medication for depression: treatment outcomes and neural mechanisms. *Nature Reviews. Neuroscience, 9*(10), 788–796.

Downing, A. M., Kinon, B. J., Millen, B. A., Zhang, L., Liu, L., Morozova, M. A., et al. (2014). A double-blind, placebo-controlled comparator study of LY2140023 monohydrate in patients with schizophrenia. *BMC Psychiatry, 14*, 351. https://doi.org/10.1186/s12888-014-0351-3.

Dunlop, B. W., Kelley, M. E., McGrath, C. L., et al. (2015). Preliminary findings supporting insula metabolic activity as a predictor of outcome to psychotherapy and medication treatments for depression. *The Journal of Neuropsychiatry and Clinical Neurosciences, 27*, 237–239.

Dunlop, B. W., & Mayberg, H. S. (2014). Neuroimaging-based biomarkers for treatment selection in major depressive disorder. *Dialogues in Clinical Neuroscience, 16*(4), 479–490.

Dunlop, B. W., Rajendra, J. K., Craighead, W. E., Kelley, M. E., McGrath, C. L., Choi, K. S., et al. (2017). Functional connectivity of the subcallosal cingulate cortex and differential outcomes to treatment with cognitive-behavioral therapy or antidepressant medication for major depressive disorder. *The American Journal of Psychiatry*. https://doi.org/10.1176/appi.ajp.2016.16050518.

Engelberg Center for Health Care Reform. (2014). *Advancing the use of biomarkers and pharmacogenomics in drug development*. September 5, Discussion Guide.

Fu, C. H., Steiner, H., & Costafreda, S. G. (2012). Predictive neural biomarkers of clinical response in depression: a meta-analysis of functional and structural neuroimaging studies of pharmacological and psychological therapies. *Neurobiology of Disease, 52*, 75–83.

Fu, C. H., Williams, S. C., Cleare, A. J., et al. (2008). Neural responses to sad facial expressions in major depression following cognitive behavioral therapy. *Biological Psychiatry, 64*(6), 505–512.

Fusar-Poli, P., Rutigliano, G., Stahl, D., Davies, C., Bonoldi, I., Reilly, T., et al. (2017). Development and validation of a clinically based risk calculator for the transdiagnostic prediction of psychosis. *JAMA Psychiatry*. https://doi.org/10.1001/jamapsychiatry.2017.0284.

Goff, D. C., & Coyle, J. T. (2001). The emerging role of glutamate in the pathophysiology and treatment of schizophrenia. *The American Journal of Psychiatry, 158*, 1367–1377.

Graig, L. A., Phillips, J. K., & Moses, H. L. (Eds.), (2016). *Biomarker tests for molecularly targeted therapies*. Washington DC: The National Academies Press.

Harrison, N. A., Brydon, L., Walker, C., Gray, M. A., Steptoe, A., & Critchley, H. D. (2009). Inflammation causes mood changes through alterations in subgenual cingulate activity and mesolimbic connectivity. *Biological Psychiatry, 66*, 407–414.

Horrobin, D. F., Glen, A. I., & Vaddadi, K. (1994). The membrane hypothesis of schizophrenia. *Schizophrenia Research, 13*(3), 195–207.

Jentsch, J. D., & Roth, R. H. (1999). The neuropsychopharmacology of phencyclidine: from NMDA receptor hypofunction to the dopamine hypothesis of schizophrenia. *Neuropsychopharmacology, 20,* 201–225.

Kane, J. M., & Correll, C. U. (2017). ω-3 Polyunsaturated fatty acids to prevent psychosis. The importance of replication studies. *JAMA Psychiatry, 74*(1), 11–12.

Kane, J. M., Robinson, D. G., Schooler, N. R., Mueser, K. T., Penn, D. L., Rosenheck, R. A., et al. (2015). Comprehensive versus usual community care for first-episode psychosis: 2-year outcomes from the NIMH RAISE early treatment program. *The American Journal of Psychiatry.* https://doi.org/10.1176/appi.ajp.2015.15050632.

Kappelmann, N., Lewis, G., Dantzer, R., Jones, P. B., & Khandaker, G. M. (2016). Antidepressant activity of anti-cytokine treatment: a systematic review and meta-analysis of clinical trials of chronic inflammatory conditions. *Molecular Psychiatry.* https://doi.org/10.1038/mp.2016.167.

Kennedy, S. H., Downar, J., Evans, K. R., Feilotter, H., Lam, R. W., MacQueen, G. M., et al. (2012). The Canadian biomarker integration network in depression (CAN-BIND): advances in response prediction. *Current Pharmaceutical Design, 18,* 5976–5989.

Kinon, B. J., Chen, L., Ascher-Svanum, H., Stauffer, V. L., Kollack-Walker, S., Lambert, J. L., et al. (2008). Predicting response to atypical antipsychotics based on early response in the treatment of schizophrenia. *Schizophrenia Research, 102,* 230–240.

Kinon, B. J., Chen, L., Ascher-Svanum, H., Stauffer, V. L., Kollack-Walker, S., Zhou, W., et al. (2010). Early response to antipsychotic drug therapy as a clinical marker of subsequent response in the treatment of schizophrenia. *Neuropsychopharmacology, 35,* 581–590.

Krystal, J. H., D'Souza, D. C., Petrakis, I. L., Belger, A., Berman, R. M., Charney, D. S., et al. (1999). NMDA agonists and antagonists as probes of glutamatergic dysfunction and pharmacotherapies in neuropsychiatric disorders. *Harvard Review of Psychiatry, 7,* 125–143.

Kurita, M., Holloway, T., García-Bea, A., Kozlenkov, A., Friedman, A. K., Moreno, J. L., et al. (2012). HDAC2 regulates atypical antipsychotic responses through the modulation of mGlu2 promoter activity. *Nature Neuroscience, 15,* 1245–1254.

Kurita, M., Holloway, T., & González-Maeso, J. (2013). HDAC2 as a new target to improve schizophrenia treatment. *Expert Review of Neurotherapeutics, 13,* 1–3.

Lassere, M. N., et al. (2007). Definitions and validation criteria for biomarkers and surrogate endpoints: development and testing of a quantitative hierarchical levels of evidence schema. *The Journal of Rheumatology, 34*(3), 607–615.

Lencz, T., Robinson, D. G., Xu, K., Ekholm, J., Sevy, S., Gunduz-Bruce, H., et al. (2006). DRD2 promoter region variation as a predictor of sustained response to antipsychotic medication in first-episode schizophrenia patients. *The American Journal of Psychiatry, 163,* 529–531.

Lener, M. S., Niciu, M. J., Ballard, E. D., Park, M., Park, L. T., Nugent, A., et al. (2016). Glutamate and GABA systems in the pathophysiology of major depression and antidepressant response to ketamine glutamate and GABA in depression. *Biological Psychiatry.* https://doi.org/10.1016/j.biopsych.2016.05.005.

Leucht, S., Busch, R., Hamann, J., Kissling, W., & Kane, J. M. (2005). Early-onset hypothesis of antipsychotic drug action: a hypothesis tested, confirmed and extended. *Biological Psychiatry, 57,* 1543–1549.

Leucht, S., Busch, R., Kissling, W., & Kane, J. M. (2007). Early prediction of antipsychotic nonresponse among patients with schizophrenia. *The Journal of Clinical Psychiatry, 68,* 352–360.

Leucht, S., Shamsi, S. A., Busch, R., Kissling, W., & Kane, J. M. (2008). Predicting antipsychotic drug response. Replication and extension to six weeks in an international olanzapine study. *Schizophrenia Research, 101,* 312–319.

Leuchter, A. F., Cook, I. A., Hamilton, S. P., Narr, K. L., Toga, A., Hunter, A. M., et al. (2010). Biomarkers to predict antidepressant response. *Current Psychiatry Reports, 12,* 553–562.

Liu, W., Downing, A. M., Munsie, L. M., Chen, P., Reed, M. R., Ruble, C. L., et al. (2012). Pharmacogenetic analysis of the mGlu2/3 agonist LY2140023 monohydrate in the treatment of schizophrenia. *The Pharmacogenomics Journal, 12,* 246–254.

Maes, M. (1995). Evidence for an immune response in major depression: a review and hypothesis. *Progress in Neuro-Psychopharmacology and Biological Psychiatry, 19,* 11–38.

Maes, M., Berk, M., Goehler, L., Song, C., Anderson, G., Gałecki, P., et al. (2012). Depression and sickness behavior are Janus-faced responses to shared inflammatory pathways. *BMC Medicine, 10,* 66. https://doi.org/10.1186/1741-7015-10-66.

Malhotra, A. K., Murphy, G. M., Jr., & Kennedy, J. L. (2004). Pharmacogenetics of psychotropic drug response. *The American Journal of Psychiatry, 161,* 780–796.

McGorry, P. D., Nelson, B., Markulev, C., Yuen, H. P., Schäfer, M. R., Mossaheb, N., et al. (2017). Effect of ω-3 polyunsaturated fatty acids in young people at ultrahigh risk for psychotic disorders: the NEURAPRO randomized clinical trial. *JAMA Psychiatry, 74*(1), 19–27.

McGrath, C. L., Kelley, M. E., Holtzheimer, P. E., et al. (2013). Toward a neuroimaging treatment selection biomarker for major depressive disorder. *JAMA Psychiatry, 70,* 821–829.

McGuffin, P., Rijsdijk, F., Andrew, M., Sham, P., Katz, R., & Cardno, A. (2003). The heritability of bipolar affective disorder and the genetic relationship to unipolar depression. *Archives of General Psychiatry, 60*(5), 497–502.

McMahon, F. J., Buervenich, S., Charney, D., Lipsky, R., Rush, A. J., Wilson, A. F., et al. (2006). Variation in the gene encoding the serotonin 2A receptor is associated with outcome of antidepressant treatment. *American Journal of Human Genetics, 78,* 804–814.

Millan, M. J., Andrieux, A., Bartzokis, G., Cadenhead, K., Dazzan, P., Fusar-Poli, P., et al. (2016). Altering the course of schizophrenia: progress and perspectives. *Nature Reviews Drug Discovery, 15,* 485–515. Advance Online Publication.

Nisenbaum, L. K., Downing, A. C. M., Zhao, F., Millen, B. A., Munsie, L., Kinon, B. J., et al. (2016). Serotonin 2A receptor SNP rs7330461 association with treatment response to pomaglumetad methionil in patients with schizophrenia. *Journal of Personalized Medicine, 6,* 9. https://doi.org/10.3390/jpm6010009.

Pacanowski, M. A., Leptak, C., & Zineh, I. (2014). Next-generation medicines: past regulatory experience and considerations for the future. *Clinical Pharmacology & Therapeutics, 95*(3), 247–249.

Patil, S. T., Zhang, L., Martenyi, F., Lowe, S. L., Jackson, K. A., Andreev, B. V., et al. (2007). Activation of mGlu2/3 receptors as a new approach to treat schizophrenia: a randomized phase 2 clinical trial. *Nature Medicine, 13,* 1102–1107.

Paz, R. D., Tardito, S., Atzori, M., & Tseng, K. Y. (2008). Glutamatergic dysfunction in schizophrenia: from basic neuroscience to clinical psychopharmacology. *European Neuropsychopharmacology, 18,* 773–786.

Prata, D., Mechelli, A., & Kapur, S. (2014). Clinically meaningful biomarkers for psychosis: a systematic and quantitative review. *Neuroscience and Biobehavioral Reviews, 45,* 134–141.

Raison, C. L., Capuron, L., & Miller, A. H. (2006). Cytokines sing the blues: inflammation and the pathogenesis of depression. *Trends in Immunology, 27,* 24–31.

Reichenberg, A., Yirmiya, R., Schuld, A., Kraus, T., Haack, M., Morag, A., et al. (2001). Cytokine- associated emotional and cognitive disturbances in humans. *Archives of General Psychiatry, 58,* 445–452.

Samara, M. T., Leucht, C., Leeflang, M. M., Anghelescu, I. G., Chung, Y. C., Crespo-Facorro, B., et al. (2015). Early improvement as a predictor of later response to antipsychotics in schizophrenia: a diagnostic test review. *The American Journal of Psychiatry, 172*(7), 617–629.

Sarpal, D. K., Argyelan, M., Robinson, D. G., Szeszko, P. R., Karlsgodt, K. H., John, M., et al. (2016). Baseline striatal functional connectivity as a predictor of response to antipsychotic drug treatment. *The American Journal of Psychiatry, 173*, 69–77.

Sarpal, D. K., Robinson, D. G., Lencz, T., Argyelan, M., Ikuta, T., Karlsgodt, K., et al. (2015). Antipsychotic treatment and functional connectivity of the striatum in first-episode schizophrenia. *JAMA Psychiatry, 72*(1), 5–13.

Schmidt, A., Cappucciati, M., Radua, J., Rutigliano, G., Rocchetti, M., Dell'Osso, L., et al. (2017). Improving prognostic accuracy in subjects at clinical high risk for psychosis: systematic review of predictive models and meta-analytical sequential testing simulation. *Schizophrenia Bulletin, 43*(2), 375–388.

Shih, R. A., Belmonte, P. L., & Zandi, P. P. (2004). A review of the evidence from family, twin and adoption studies for a genetic contribution to adult psychiatric disorders. *International Review of Psychiatry, 16*(4), 260–283.

Sliz, D., & Hawley, S. (2012). Major depressive disorder and alternations in insular cortical activity: a review of current functional magnetic imaging research. *Frontiers in Human Neuroscience, 6*, 323.

Spear, B. B., Heath-Chiozzi, M., & Huff, J. (2001). Clinical application of pharmacogenetics. *Trends in Molecular Medicine, 7*(5), 201–204.

Suhara, T., Chaki, S., Kimura, H., Furusawa, M., Matsumoto, M., Ogura, H., et al. (2016). Indications of success: strategies for utilizing neuroimaging biomarkers in CNS drug discovery and development—CINP/JSNP working group report. *The International Journal of Neuropsychopharmacology, 20*(4), 285–294.

Sullivan, P. F., Neale, M. C., & Kendler, K. S. (2000). Genetic epidemiology of major depression: review and meta-analysis. *The American Journal of Psychiatry, 157*(10), 1552–1562.

U.S. Food and Drug Administration. (2012). *Guidance for industry: Enrichment strategies for clinical trials to support approval of human drugs and biological products.*

Wischhof, L., & Koch, M. (2015). 5-HT2A and mGlu2/3 receptor interactions: on their relevance to cognitive function and psychosis. *Behavioural Pharmacology, 27*, 1–11.

Yang, J. J., Wang, N., Yang, C., Shi, J. Y., Yu, H. Y., & Hashimoto, K. (2015). Serum interleukin-6 is a predictive biomarker for ketamine's antidepressant effect in treatment-resistant patients with major depression. *Biological Psychiatry, 77*, e19–e20.

Zhang, W. -N., Chang, S. -H., Guo, L. -Y., Zhang, K. -L., & Wang, J. (2013). The neural correlates of reward-related processing in major depressive disorder: a meta- analysis of functional magnetic resonance imaging studies. *Journal of Affective Disorders, 151*, 531–539.

9

Applications of Neuroimaging Biomarkers in CNS Drug Development

Patricia E. Cole*, Steven G. Einstein†

*Cole Imaging and Biomarker Consulting, LLC, Glenview, IL, United States †Janssen Research and Development, Titusville, NJ, United States

I INTRODUCTION

Imaging biomarkers are increasingly important for decision-making in CNS drug development. This chapter will focus on the major imaging techniques and technologies used to interrogate the brain in translational CNS imaging in neurodegenerative diseases and psychiatry, namely, positron-emission tomography (PET)/single-photon emission computed tomography (SPECT) and magnetic resonance imaging (MRI), including structural, functional, perfusion, and diffusion MRI and MR spectroscopy.

Key questions to be answered during early drug development include the following: Does the drug reach its intended target? Is the drug safe? What is the optimal dose and schedule? What are the downstream pharmacodynamic and biological consequences of target engagement, and do those confirm the mechanism of action of the drug or the expected effect on disease pathology? Although Phase 3 trials are designed to confirm efficacy, early evidence of clinically relevant change is highly desirable to derisk late development. Patient enrichment/selection biomarkers are also critical in order to better define the clinical trial population and to maximize the opportunity to detect drug signals in the clinical trial population of interest. Review of Phase 2 studies indicates a success rate of 28% in 2006–07 and an even poorer rate in 2008–09; Pharmaceutical Benchmarking Forum data from 2005 to 2010 indicates that the highest attrition rate across pharma is in Phase 2 and is 71 % (Arrowsmith, 2011; Morgan et al., 2012). Pfizer reviewed 44 of their programs and have described three pillars of survival for a candidate to succeed in Phase 2 and improve likelihood of progression to Phase 3: (1) exposure at the target site of action over desired period of time, (2) binding to the pharmacological target as expected for its mode of action, and (3) expression of pharmacological activity commensurate with the demonstrated target exposure and binding (Morgan et al., 2012). Suhara et al., 2017 have added two additional "tiers" onto Pfizer's initial three in their report on strategies of utilizing biomarkers in CNS drug discovery and development, namely, Tier 4/patient stratification and Tier 5/clinical efficacy prediction/disease-related changes to estimate clinical proof of concept (PoC) (Suhara et al., 2017). Astra Zeneca, in reviewing their Phase 2 pipeline, provided the five most important technical determinants of project success and pipeline quality, the five R's: right target, right patient, right tissue, right safety, and right commercial potential (Cook et al., 2014). Given the recognized attrition rate in early drug development, the appropriate use of imaging biomarkers in early trials allows an improved ability to assess key questions and achieve features of success. These biomarkers may also provide a better understanding of the reasons for failure, inadequate drug versus incorrect concept, and so guide decision-making on the fate of a drug development program. This chapter will provide examples of how various CNS imaging biomarkers can help with improved decision-making in CNS drug development.

II PET/SPECT

PET and SPECT are radionuclide-based imaging techniques, both of which provide insight into functional and molecular processes in clinical and animal tissues and both of which can be used to track small-molecule drugs and biologics. PET is currently used more than SPECT in CNS clinical trials.

A PET Radionuclides and Tracers

PET tracers (not only mostly small molecules but also peptides, proteins, and nanoparticles) are labeled with positron-emitting radionuclides. These radionuclides (corresponding half-life in minutes) are most commonly ^{11}C-carbon (20.4 m), ^{18}F-fluorine (110 m), ^{15}O-oxygen (2 m), and ^{13}N-nitrogen (10 m); they are isotopes of naturally occurring nuclei (^{12}C, ^{19}F, ^{16}O, and ^{14}N), thus allowing the replacement of the nonradioactive isotope with a PET radionuclide. Most small-molecule PET tracers for CNS use are labeled with either ^{11}C or ^{18}F. ^{15}O ligands have been used in brain PET for assessment of blood flow (^{15}O-water) and oxygen consumption; ^{13}N-ammonia PET has been used to assess ammonia metabolism in hepatic encephalopathy. ^{124}I-iodine, ^{64}Cu-copper, ^{68}Ga-gallium, and ^{89}Zr-zirconium with half-lives of 4.2 d, 12.7 h, 68 min, and 3.3 days, respectively, are also positron-emitting radioisotopes but are mostly used outside of the brain when a longer half-life is required and/or for labeling peptides, proteins, antibodies, or nanoparticles. The PET isotopes used for neuroimaging require a cyclotron for production. For clinical use, these can be generated as a precursor at an academic or commercial production facility under good manufacturing practice (GMP) conditions and provided to the PET center where further chemical synthesis incorporating the radionuclide into a small molecule under GMP conditions may be required. Commercial synthesis and distribution is available for some ^{18}F-containing tracers (e.g., ^{18}F-FDG and ^{18}F-amyloid ligands), and distribution can be made within a few hour radius of the facility. Production of ^{11}C ligands or ^{11}C precursors for small-molecule labeling is usually done using an on-site or very nearby cyclotron production facility. Academic and commercial production cyclotron facilities also supply PET precursors and some ligands manufactured under good laboratory practice (GLP) to preclinical imaging laboratories.

B SPECT Radionuclides and Tracers

These ligands contain a heavy metal radionuclide rather than a naturally occurring isotope, and their radioactive decay is by a single gamma photon emission rather than the dual photon emission in positron decay. The most common radionuclides used in SPECT in the brain are ^{99m}Tc (technetium) and ^{123}I (iodine) with half-lives of 6 and 13.1 h, respectively. The ^{99m}Tc radionuclide is typically synthesized in a commercial generator in the site nuclear pharmacy as $^{99m}TcO_4^-$ and combined with compound to be labeled using a commercial kit (e.g., ^{99m}HM-PAO/exametazime). ^{123}I-based radiotracers are typically distributed by the manufacturer from regional production centers (e.g., DaTscan).

C PET Scanners and Image Reconstruction

Once a positron is generated from the decay of a positron-emitting radionuclide, the positron travels a very short distance in tissue (about 2 mm), and then the positron itself annihilates/decays into two gamma photons of equal energy that are emitted in opposite directions from that decay event. This unique decay allows tracking of the PET tracer emission in the body by using coincident detection of the gamma photons, with the set of emissions forming a line of response. The lines of response represent spatial distribution of the data. These data are transformed into a frequency distribution called a sinogram that is further corrected before being converted back to a spatial image. The emission scan is created in either 2-D or 3-D depending on the scanner. Using coincident detection allows more decays to be captured by the scanner detectors so that PET has a much higher sensitivity than SPECT (at least two orders of magnitude). This detection sensitivity leads to improved image quality and improved temporal resolution with less time needed to acquire adequate decay events/counts. Since adequate images can be collected in a shorter time, PET also has a better ability to study dynamic processes compared with SPECT. Because the annihilation photons are attenuated by absorption and scattering as they pass through tissue and bone before reaching the scanner detectors, attenuation correction, which takes into account the amount and type of tissue traversed, has to be performed on the acquired data (the sinograms) before images are reconstructed. To perform the attenuation correction, a transmission scan is performed. The type of transmission scan depends on the scanner type. There are three general types of PET scanners. So-called dedicated PET scanners produce an attenuation-corrected reconstructed PET image using a transmission scan produced by a positron-emitter rod (e.g., $^{68}Germanium$) that rotates around the bore of the scanner. PET/CT hybrid scanners produce both PET and CT images; they use a low-dose CT scan as the transmission scan for attenuation correction. PET/MR hybrid scanners generate both PET and MRI images and utilize the MR image data to perform attenuation correction of the emission data. Other corrections such as scatter corrections, random corrections, and dead-time corrections are applied to the PET data, and the data are reconstructed to create an image with intensity and spatial information. The reconstruction from frequency to spatial dimensions can use multiple algorithms from filtered back projections to iterative reconstruction; the particular algorithm varies with scanner and scanner manufacturer. See PET and SPECT physics publications and texts for further technical details (Dahlbom, 2017; Rahmim & Zaidi, 2008; Saha, 2016). Typically, a 3-D TI-weighted MRI scan is used for coregistration for brain PET scans; this MRI is acquired separately on an MR scanner if dedicated PET or PET/CT scanners are used for the PET or during the same scanning session if a PET/MR scanner is used. PET and PET/CT scanners for animal imaging (micro- and nano-PET/CT) are available and thus allow for translational imaging. The resolution of these small nano-PET/CT animal scanners is <0.7 mm. Resolution of current clinical PET scanners is about 4 mm.

D SPECT Scanners and Image Reconstruction

The SPECT technique captures a 3-D distribution of the gamma (γ) rays emitted from the decay of the single-photon emitting radionuclide within the tracer molecule. The γ-rays are detected using a so-called gamma camera that consists of scintillation crystals optically coupled to a photomultiplier tube array that converts the capture of the γ-ray into an electrical current. These detectors rotate around the subject, and multiple 2-D images (called projections) at multiple angles are acquired. These projections are topographically reconstructed by computer to provide a 3-D data set that can be manipulated to provide as a set of 2-D images along any chosen axis. Because the emission of the γ-rays is isotropic in single-photon emission (in contrast to positron decay), geometric collimators are used so that only photons traveling in the direction specified by the apertures in the collimators are detected/counted. The detection efficiency of SPECT and SPECT/CT scanners is therefore much less than that of PET scanners so sensitivity is lower. Quantification of tracer uptake in brain SPECT images requires attenuation correction. The most widely used method is that of Chang that requires delineation of the outer contour of the skull for which automated methods have recently been developed (Lange et al., 2015). The resolution of clinical SPECT scanners is about 8 mm, less than that of PET. Small-animal SPECT and SPECT/CT scanners exist, and because of advances with pinhole collimators, the resolution of nano-SPECT/CT scanner is about 0.8 mm, which is better than micro-PET instruments and comparable with the new nano-PET animal scanner.

E Biodistribution

The isotopic substitution of an ^{11}C or ^{18}F PET radioisotope into a CNS drug candidate creates a molecule that can be imaged by PET without changing the properties of the drug itself. A dynamic brain PET performed following injection of the radiolabeled drug provides information on drug delivery into the brain. The brain signal allows the determination of the equilibrium partition coefficient for free and nonspecifically bound drug between brain tissue and plasma. Typically, isotopically labeled drug candidates are not optimal tracers for assessing target engagement so efforts are usually focused on generating a target engagement probe since the information to be gained is much more informative for the drug development process (Gunn & Rabiner, 2017).

F Target Engagement and Tissue Kinetics

A critical aim of both preclinical and clinical drug development is to demonstrate that a drug candidate interacts with its intended target. Novel molecules are designed and characterized with both in vitro assays and animal models to optimize the engagement with a human brain target. A powerful imaging approach for demonstrating direct target engagement in human brain tissue is a target occupancy study in which a radioactive PET (or SPECT) tracer that binds to the target is displaced by the small-molecule drug candidate that targets the same binding site. The target is often a receptor but can also be an enzyme, ion channel protein, or neurotransmitter transporter protein. PET tracers are more commonly used than SPECT tracers because of the improved spatial and temporal resolution of PET scans and the increased radiolabeling possibilities with PET radionuclides. Brain occupancy studies allow for a proof of target engagement and proof that the drug candidate crosses the blood-brain barrier (BBB) and reaches brain tissue. Target occupancy data in conjunction with PK data allow for modeling to optimize dose and schedule of dosing for subsequent clinical trials.

At a basic level, percent occupancy is typically determined in healthy volunteers at several single doses of the drug candidate and at two or more times after single-dose administration, for example, at the time of peak plasma drug exposures and drug plasma trough levels (known from single-ascending-dose PK data). Occupancy describes the drug candidate-induced change in the concentration of available target. The quantification of available concentration requires fitting to a mathematical model that defines the relationship of tracer input (using arterial plasma concentration of the radioligand or a reference region in the brain if an area exists that is free of target) to brain tissue response to the radiotracer. Fig. 9.1 illustrates a family of commonly used PET compartment models. The choice of model is supported by statistical techniques such as Akaike or Bayesian information criteria; further details of tracer kinetic modeling are described in Gunn, Guo, Salinas, Tziortzi, & Searle, 2011; Lammertsma & Hume, 1996; Cunningham, Rabiner, Slifstein, Laruelle, & Gunn, 2010; Laruelle, Slifstein, & Huang, 2002). Binding potential parameters derived from the modeling can be used to calculate occupancy. For example, BP_{ND} at baseline and postdrug, where BP_{ND} is the ratio at equilibrium of specifically bound radionuclide to that of nondisplaceable

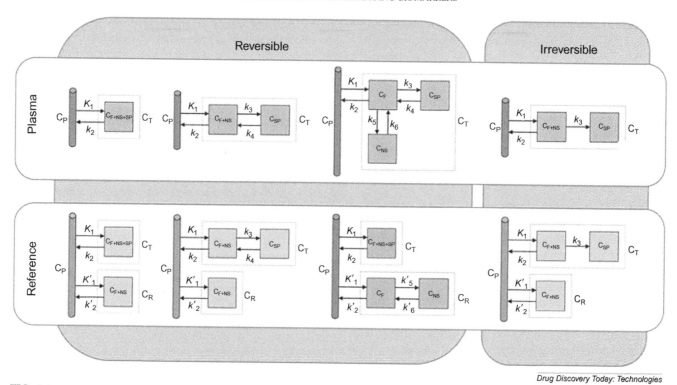

Drug Discovery Today: Technologies

FIG. 9.1   Family of commonly used PET compartment models. Models are described by (A) whether the input function is derived from plasma sampling (requiring arterial sampling) or from a reference tissue (if there is a brain region identified as being devoid of the target of interest) and by (B) whether tracer kinetics are reversible or irreversible (i.e., at least one exchange is in one direction only). *With permission from Gunn, R. N., Guo, Q., Salinas, C. A., Tziortzi, A. C., & Searle, G. E. (2011). Advances in biomathematical modeling for PET neuroreceptor imaging. Drug Discovery Today: Technologies, 8, 45–51.*

radioligand (unbound and nonspecifically bound tracer), is related to occupancy as shown in Eq. (9.1):

$$\%\text{occupancy} = \left[\text{BP}_{\text{ND}}^{\text{Baseline}} - \text{BP}_{\text{ND}}^{\text{Postdrug}} \right] / \text{BP}_{\text{ND}}^{\text{Baseline}} \quad (9.1)$$

BP_{ND} is typically used when reference tissue models can be applied (i.e., when there are areas that are target-rich and areas that are target-free) (Gunn et al., 2011; Innis et al., 2007).

A good example of the impact of receptor occupancy studies in CNS drug development is provided by the development of neurokinin 1 receptor (NK-1R) antagonists and PET tracers for NK-1R. NK-1R and its ligand substance P are present in brain regions involved in the regulation of affective behavior and stress response and in brain regions responsible for the regulation of the vomiting reflex. A PET ligand for NK-1R (^{18}F-SPA-RQ) was developed and used to demonstrate that the NK-1R antagonist aprecipant crossed the BBB and blocked the receptor in human brain. The receptor occupancy (RO) curve is shown in Fig. 9.2 as a function of trough plasma drug levels. (Bergström et al., 2004). The RO data were successfully used to select the lowest efficacious dose in Phase 3 for full receptor blockade for the indication of chemotherapy-induced vomiting, where the lowest effective biological dose is key to minimize the likelihood

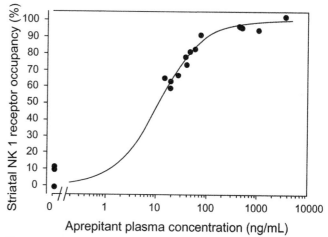

FIG. 9.2   Estimated relationship between % occupancy of neurokinin 1 (NK-1) receptors in the striatum and the plasma trough concentration of NK-1 inhibitor aprepitant. *With permission from Bergström, M., Hargreaves, R. J., Burns, H. D., Goldberg, M. R., Sciberras, D., Reines, S. A., et al. (2004). Human positron emission tomography studies of brain neurokinin 1 receptor occupancy by aprepitant. Biological Psychiatry, 55, 1007–1014.*

of drug-drug interactions in the oncology setting where multiple drugs are used (Warr et al., 2005). The same NK-1R PET tracer was more recently used in an occupancy study to compare the original single oral dose of aprepitant with a single IV dose of fosaprepitant, the prodrug of aprecipant. Illustrating another use of PET tracers, this occupancy study showed central bioequivalence equivalent for both drugs and provided significant support for registration of fosaprepitant (Van Laere et al., 2012). In contrast to the success of aprepitant in chemotherapy-induced emesis, Phase 3 trials of aprepitant in depression with dosing guided by PET occupancy failed (Keller et al., 2006). In addition, a proof-of-concept trial in social anxiety disorder with another NK-1 antagonist (LY686017) also failed despite dose guidance by PET occupancy data using an NK-1R tracer (Tauscher et al., 2010). Because the PET occupancy data ensured adequate receptor blockade, proper dose, and pharmacokinetics, the failures of aprepitant and LY686017 suggest that the concept of antagonizing dysregulated substance P blockade at NK-1Rs as a therapeutic approach in depression and anxiety was faulty.

Although the majority of PET tracers have been developed for cell surface receptors or transporters, tracers have been developed for several brain enzymes including acetylcholinesterase (AChE) and phosphodiesterases (PDEs) (including 10a, 4, and 2a). Loss of cholinergic neurons contributes to cognitive deficits in Alzheimer's disease (AD). Some of the symptom-controlling drugs in AD are targeted at inhibiting brain AChE with resultant increase in acetylcholine and improvement in cholinergic synaptic transmission. Brain AChE inhibition by donepezil and rivastigmine was demonstrated in proof-of-mechanism studies using PET tracers analogues of acetyl choline (including ^{11}C-N-methyl-4-piperadyl acetate [^{11}C-MP4A], ^{11}C-N-methyl-piperidin-4-yl propionate [^{11}C-PMP], and ^{11}C-donepezil) (Shinotoh, Fukushi, Nagatsuka, & Irie, 2004). These tracers were also used to explore brain AChE enzyme occupancy as a function of plasma exposure of AChE inhibitor in primates and in AD subjects (Ota et al., 2010; Shiraishi et al., 2005).

Another set of brain-relevant enzymes are the phosphodiesterases that are classified into 11 families (PDE1-PDE11). These enzymes regulate the extracellular signal from G-protein-coupled receptors (GPCRs) to intracellular signaling cascades by inactivating the second messengers cyclic adenosine monophosphate (CAMP) and cyclic guanosine monophosphate (cGMP) (via hydrolyzing their phosphodiester bonds). A number of PDEs (including PDE4, PDE2A, and PDE10A) are found in the brain, and PDE inhibitors are being developed as novel therapeutics for both psychiatric and neurodegenerative disorders. PDE10A is highly expressed in medium spiny striatal neurons, and PDE10A inhibitors are potential therapeutics in diseases with a reduced activity of these neurons (e.g., schizophrenia, Huntington's disease, and Parkinson's disease). Brain areas of prominent expression of PDE2A protein include limbic and basal ganglia areas, and PDE2 inhibitors increase cyclic nucleotides in these areas;

PDE2A inhibitors result in improved long-term potentiation in hippocampal slices and improved learning and memory in animal models (Boess et al., 2004). These data suggest PDE2A inhibitors may improve cognitive processes in schizophrenia. Several PDE10A PET tracers have been developed and validated in for use in humans including ^{18}F-JNJ42259152, ^{18}F-MNI659, ^{11}C-IMA107, ^{11}C-T773, and ^{11}C-Lu AE92686, the latter possibly being useful for target assessment in the substantia nigra and the striatum (Barret et al., 2014; Plisson et al., 2014; Takano et al., 2016; Van Laere et al., 2013; Yang et al., 2017). A PDE2A tracer is also validated in humans, ^{18}F-PF05270430 (Naganawa et al., 2016). For PDE10A and PDE2A inhibitors in development, these tracers will allow confirmation that a PDE inhibitor drug candidate crosses the BBB in humans and will allow occupancy data to be generated to guide dosing and schedule in subsequent patient studies.

An important feature of occupancy studies is that they can provide information on drug pharmacokinetics and duration of action at the level of the target tissue; this allows the dosing and schedule/interval to be based on target tissue engagement rather than plasma kinetics. Studies assessing the time course of PET occupancy versus plasma kinetics have shown a dissociation between receptor kinetics and plasma kinetics for a number of drugs (e.g., olanzapine and risperidone) at their receptors as illustrated in Fig. 9.3 (Tauscher, Jones, Remington, Zipursky, & Kapur, 2002). When occupancy data are available, clinical dosing schedules based on time of onset and duration of brain target occupancy should be used.

With regard to the design of PET occupancy studies, studies should use a sequential adaptive design in healthy volunteers that optimizes the ability to assess the relationship between plasma exposure, occupancy, and tissue kinetics and minimizes the number of subjects needed (Zamuner et al., 2010). Studies typically use 3–5 single doses of drug candidate, 3–4 time points per dose level (baseline and 2–3 postdose time points), and 2–4 subjects per dose level. The initial dose and timing of the postdose PET scans is based on human PK data and preclinical data on target occupancy kinetics. Subsequent doses and times of postdose scans are then data-driven using analyses from initial/preceding PET scans.

Another design consideration is whether single-dose occupancy studies are sufficient or if occupancy needs to be measured after repeat/multidosing for each dose level tested to ensure that steady state level of the drug candidate is achieved. If the free concentration of the drug in the brain can be assumed to be in equilibrium with the free plasma concentration at all times (so-called direct drug kinetics where plasma PK matches brain PK), a classic Emax model of occupancy as a direct function of

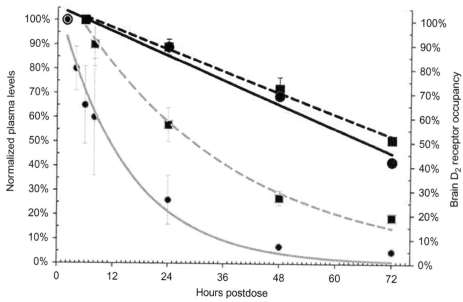

FIG. 9.3   Time course of striatal D2 receptor occupancy as compared with time course of plasma levels of olanzapine and risperidone plus 9-OH-risperidone following discontinuation of drug in patients treated for psychosis. Results are normalized to 100% of peak value, and error bars represent 1 standard deviation (sd). For D2 striatal occupancy: olanzapine, black dashed line; risperidone, black solid line. For concentration levels in plasma: olanzapine, gray dashed line; risperidone and 9-OH-risperidone, gray solid line. *With permission from Tauscher, J., Jones, C., Remington, G., Zipursky, R. B., & Kapur, S. (2002). Significant dissociation of brain and plasma kinetics with antipsychotics.* Molecular Psychiatry, 7, 317–321.

free plasma concentration can be used (Zhang, Beal, & Sheiner, 2003). The single-dose occupancy model can then be combined with repeat-dose PK data to predict occupancy achieved with repeat dosing. If there is a dynamic lag/dissociation between the plasma and target-bound concentrations of the drug candidate, the "direct model" does not apply because the drug concentrations in the different tissue compartments are not in equilibrium. The drug kinetics are so-called indirect, and to determine occupancy at repeat dosing, occupancy needs to be (a) measured in a multidose steady-state study or (b) determined from a single-dose occupancy study with multiple postdose time points per dose level and use of an advanced "indirect" model (Abanades et al., 2011; Gunn et al., 2011; Gunn & Rabiner, 2017; Lim et al., 2007; Salinas et al., 2013). An important reason for using an adaptive design for a single-dose PET occupancy study is that the data allow an assessment of whether an indirect relationship is present and thus ensures accurate characterization of the PK-occupancy relationship.

G Development of Novel Brain PET Ligands

If a PET ligand for a target of interest is not available for use and a new PET ligand needs to be developed, it is important to incorporate the timelines for development into the timelines for the drug candidate development plan. Even when tracer development is

undertaken by a pharmaceutical company with access to compound libraries, discovery timeline for a lead drug candidate including an accompanying PET ligand is about 18 months, and tracer development timelines should parallel drug candidate development. Although the toxicology required for tracer development is less since only microdoses are used, the process is time-consuming and expensive and has had a high attrition rate. In recent years, biomathematical approaches have been developed to better predict the in vivo performance of a potential ligand from in silico and in vitro data (Guo, Brady, & Gunn, 2009). Additionally, mass spectrometry enables testing of potential tracers that do not contain a radiolabel, avoiding the need of complex labeling chemistry efforts before in vivo testing. Analysis of rodent brain tissue harvested at different times postdose by liquid chromatography coupled with tandem mass spectrometry (LC-MS/MS) allows a rapid determination of whether a compound distributes to target-rich regions versus tissue with little or no target (Barth & Need, 2014). This LC-MS/MS approach is more cost-effective and time-efficient than the classic approach (Fig. 9.4). Successful tracer candidates from this process can then move through the radiolabeling process followed by the preclinical (rodent or nonhuman primate) PET and human PET tracer studies (dynamic brain uptake with specific, blockable binding; test-retest; and dosimetry) needed to validate a tracer for use in human occupancy studies.

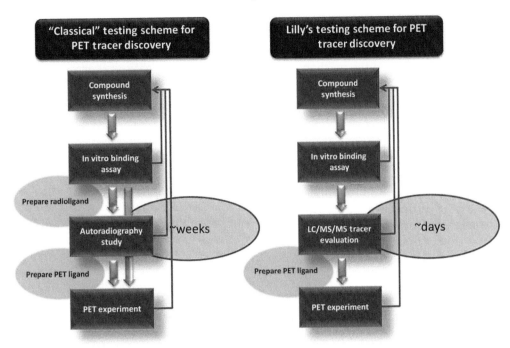

FIG. 9.4 Comparison of flow schemes for the identification of tracers using traditional versus LC-MS/MS approaches. *Reprinted with permission from Barth and Need, Identifying novel radiotracers for PET imaging of the brain: application of LC-MS/MS to tracer identification. Copyright 2014 American Chemical Society (2014, Fig. 2, p. 1150).*

H Pharmacodynamic Effects of CNS Target Engagement Assessed With PET Imaging

Although proofs that a drug candidate reaches the tissue and engages the target with known occupancy-exposure relationship are major milestones achievable with target engagement studies, the next key challenge in the clinic is establishing the downstream pharmacodynamic (PD) and biological consequences of engagement and determining if those PD effects confirm the expected mechanism of action of the drug or anticipated effect on disease pathology that hopefully was demonstrated preclinically. Measurements of PD effects resultant from target engagement can be particularly useful when a PET tracer for the target is not available: functional information about CNS effects infer brain penetration and target engagement. Such PD studies are particularly useful if they can guide dosing regimen and schedule, provide a proof of mechanism, or allow a decision to stop development if a PD signal is not detected at any of the doses tested. Two examples of the use of PET tracers to assess PD effects that will be discussed in this section are the measurement of neurotransmitter flux and use of glucose metabolism to probe drug effects on synaptic transmission. A third example is provided by the use of amyloid PET probes to show removal of brain amyloid by antiamyloid therapeutics (Section II.I.2).

1 PET Measurements of Neurotransmitter Flux

PET imaging can be used to examine a drug-induced synaptic release of neurotransmitter. The concept is the following: at baseline a PET tracer is injected that binds to a synaptic neurotransmitter receptor; when a drug is then given that increases (or decreases) the endogenous synaptic levels of the neurotransmitter, there is a change in the binding of the radiolabeled tracer at the receptor because of changes in concentration of the endogenous ligand that competes with the bound PET ligand at the receptor. The change in binding of the PET tracer in the brain region of interest thus reflects the change in synaptic endogenous neurotransmitter. Neurotransmitter fluxes have been studied in the dopaminergic system with antagonist D2 tracers like ^{11}C-raclopride and ^{123}I-IBZM in both nonhuman primates and humans. Using amphetamine to increase endogenous dopamine (DA), studies have shown a dose dependence between amphetamine dose and changes in tracer binding and a reproducible relationship between synaptic dopamine change and change in radioligand binding (Laruelle et al., 2002). The problem is that large changes of intrasynaptic DA result in relative small changes in tracer binding. Recently the use of a newer tracer ^{11}C-(+)-PHNO, which is an agonist binding preferentially to D2/D3 (high-affinity) receptors, has shown a gain in sensitivity to changes in synaptic DA induced by both nicotine and amphetamine in the stratum compared with ^{11}C-raclopride (Gallezot et al., 2014; Shotbolt et al., 2012). As an example, ^{11}C-PHNO PET has been used to provide new information on the mechanism of action of varenicline, a nicotinic partial agonist used for tobacco use disorder. ^{11}C-PHNO PET before and after varenicline demonstrates that the drug is able to increase DA levels in the dorsal caudate region of the brain (Di Ciano et al., 2016).

For drugs that are shown preclinically by microdialysis to significantly modulate amphetamine-induced DA, ^{11}C-PHNO PET in conjunction with amphetamine allows the translational possibility to demonstrate such an effect of the drug in human brain. The ability to estimate neurotransmitter release for other systems such as serotonin, glutamate, norepinephrine, and GABA is an area of active research with progress recently reviewed by Finnema et al. (2015).

2 Pharmacodynamic Effects on Brain Activity Using ^{18}F-FDG PET

Brain tissue metabolic activity can be measured with ^{18}F-FDG, which is a radiolabeled analogue of glucose. Cerebral glucose metabolic rate is considered a measure of synaptic activity given that the main metabolic activity of the brain relates to neurotransmission (Fowler & Volkow, 2001). FDG PET thus provides a downstream pharmacodynamic measure for evaluating drug-induced alterations in neurotransmitter flux by assessing the impact of synaptic transmission changes on cerebral metabolism. FDG PET was used to study the effect of AChE inhibitor rivastigmine in patients with AD using a picture-naming task. Patients showing a clinical response to the drug showed increased FDG uptake in several hypometabolic areas of the brain during the task, whereas patients without a clinical response did not show such an increase in brain metabolism (Potkin et al., 2001). In a study of donepezil, FDG uptake in frontal, parietal, and temporal cortical regions decreased more slowly in AD subjects on donepezil compared with controls (Tune et al., 2003). These findings are consistent with convergent preclinical and clinical evidence linking increased cholinergic transmission to improved/maintained brain glucose metabolism and reduced memory deficit.

FDG PET can also be used to assess the direct effects of a drug on resting-state metabolic activity. For example, the corticotropin-releasing factor receptor 1 antagonist R317573 was shown using FDG PET to acutely and dose-dependently modulate regional cerebral glucose uptake in healthy volunteers (Schmidt et al., 2010). The lack of impact on resting-state FDG uptake has also been used to infer that a drug candidate does NOT cross the BBB. Arterial spin labeling (ASL), an advanced MRI perfusion technique, can also be used to assess the effects of drugs on brain perfusion and to determine whether drugs cross the BBB (Section III.E).

I PET/SPECT Imaging of Neuropathology

The availability of molecular imaging agents targeted to particular features of neuropathology not only excitingly allows in vivo detection of the pathology/pathophysiology in both humans and animal models but also provides the possibility to monitor the disease process in its full spatial extent longitudinally from presymptomatic stages to full-blown clinical disease. Importantly for drug development, these imaging agents along with soluble biomarkers of neuropathology allow the effects of therapeutic interventions on disease progression to be assessed. In dementias, neuropathology studies indicate that multiple pathologies coexist and, in the State of Florida Alzheimer's Disease Initiative Brain Bank, evaluation of 1242 brains from memory disorder clinics revealed that fewer than 50% of subjects with a primary diagnosis of Alzheimer's disease (AD) had AD pathology only; vascular dementia (VaD) and Lewy body dementia (LBD) were the most common copathologies (Rabinovici et al., 2017). Imaging and soluble biomarkers will be key in understanding how copathologies may influence therapeutic approaches, novel targets, trial designs, and impact on endpoints along with regulatory approvals of new therapeutics. Another point of major interest is temporal dimension of disease pathology. Fig. 9.5 is a model for AD that integrates the detectability of disease pathologies with progression of the disease. Given the major trend in neurodegenerative disease drug development to direct disease-modification interventions to the predementia and presymptomatic stages of the disease process, imaging and soluble biomarkers will play a critical role in patient selection and in disease progression monitoring in clinical drug trials. This section will review imaging markers of amyloid, tau, inflammation, metabolic function/neurodegeneration, synaptic density, dopaminergic deficit in Parkinson's disease (PD), and some additional imaging markers for Huntington's disease pathology. Efforts are ongoing to generate PET tracers for α-synuclein and transactive response DNA-binding protein 43 (TDP-43) but not yet successful. Some evidence suggests that several proteins that misfold, oligomerize, fibrillize, and aggregate (α-synuclein in LBD, PD, and Parkinson's disease dementia [PDD]; TDP-43 in amyotrophic lateral sclerosis [ALS] and certain types of frontotemporal lobe dementia [FTLD]; and tau in AD and other tauopathies) spread in a prion-like fashion in the brain so that PET imaging agents that can assess spatiotemporal progression and inhibition of that by drugs will likely be very useful in drug development.

1 FDG PET for Assessment of Neurodegeneration

In neurodegenerative diseases, FDG PET is widely used in the differential diagnosis of dementia since the regional distribution of hypometabolism is often characteristic of the type of dementia (Jagust, Reed, Mungas, Ellis, & Decarli, 2007). The AD hypometabolic pattern

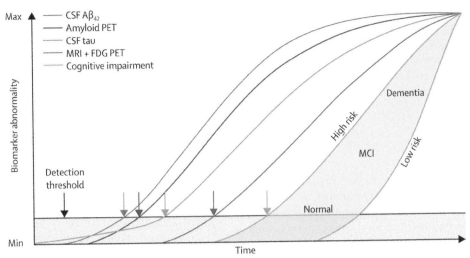

FIG. 9.5 Model integrating detection of AD pathology/biomarker abnormalities with the temporal course of AD disease progression. The black horizontal line represents the threshold for biomarker detection of pathophysiological changes, and the gray area beneath the black line is the zone in which abnormal pathophysiological changes are below the biomarker detection threshold. Tau pathology precedes Aβ but only early and at a sub-threshold biomarker detection level. Aβ deposition then occurs and rises above the biomarker detection level (*purple arrow* for CSF Aβ42 and red arrow for amyloid PET). The Aβ deposition induces acceleration of tauopathy, and CSF tau rises above the detection level (*blue arrow*). Later, FDG PET and MRI *(blue arrow)*, neurodegeneration markers, rise above the detection level. Finally, mild cognitive impairment (MCI) is detectable *(green arrow)*. The light green—filled area represents a range of cognitive responses that depend on an individual's risk profile. *With permission from Jack, C. R., Jr., Knopman, D. S., Jagust, W. J., Petersen, R. C., Weiner, M. W., Aisen, P. S., et al. (2013). Tracking pathophysiological processes in Alzheimer's disease: an updated hypothetical model of dynamic biomarkers.* Lancet Neurology, 12, 207–216.

commonly involves the parietotemporal, posterior cingulate, and medial temporal lobes with late involvement of the frontal association cortices. In contrast, FTD shows predominantly frontal hypometabolism, whereas parieto-occipital hypometabolism is seen in LBD.

The basis for the cerebral glucose hypometabolism is considered to be neuronal degeneration/synaptic loss. The extent of hypometabolism worsens with the severity of the disease process and is highly correlated with the severity of cognitive impairment in AD. FDG PET along with atrophy by MRI (Section III.B.1) are widely accepted as imaging biomarkers for tau-related neurodegeneration and occur in a similar timeframe in the temporal modeling of biomarker changes in AD (Fig. 9.5) (Jack et al., 2013; Sperling et al., 2011). Jack et al. (2016) has proposed a descriptive classification scheme for AD where seven major biomarkers are divided into three binary (+ or −) categories, A for a β-amyloid biomarkers, T for a tau biomarker, and N for a biomarker of neurodegeneration when N can be FDG PET, structural MRI, or CSF total tau.

FDG PET metrics have been shown to predict conversion from MCI to AD, and several studies have compared the number of subjects N required to achieve a 25% reduction in disease progression ($\alpha = 0.5$) over a 2-year period (Chen et al., 2010; Landau et al., 2011). Use of a statistical region of interest (ROI) with a single multiregion volume of interest was the analysis approach that yielded the smallest N, and this N was similar to that needed if structural MRI atrophy was used (Chen et al., 2010).

2 Amyloid PET Imaging

Amyloid plaques are, along with tau neurofibrillary tangles, the pathological hallmark of Alzheimer's disease. One of the earliest tracers developed that became a gold standard for amyloid PET was a thioflavin T derivative developed at the University of Pittsburgh that was radio-labeled with ^{11}C and named ^{11}C-PiB (as in Pittsburgh compound B). It was studied in subjects with AD versus healthy elderly and young controls and showed significantly higher retention in frontal, temporal, and parietal cortex compared with controls (Klunk et al., 2004); it binds to neuritic plaque but not diffuse amyloid. Several test-retest studies have been performed, and the test-retest variability ranges from 3% to 10% depending on analysis methodology (Tolboom et al., 2009). This tracer has had a huge impact on the understanding of the pathophysiology of AD and the time course of amyloid deposition that essentially plateaus by the time dementia sets in and is present a decade or more before symptoms are evident (Fig. 9.5). Because an ^{18}F ligand is commercially more viable given its longer half-life, the race was on to generate ^{18}F-labeled amyloid tracers. There are now three commercially available ^{18}F-amyloid tracers and a fourth in late development:

1. ^{18}F-AV-45/florbetapir/Amyvid developed by Avid, now part of Eli Lilly.
2. ^{18}F-Bayer94-9172 (AV1)/florbetaben/Neuraceq developed by Bayer AG and licenced by Piramal Healthcare.

3. ^{18}F-PiB/flutemetamol/Vizamyl developed by GE Healthcare.
4. ^{18}F-AZD4694/NAV4694/flutafuranol developed by AstraZeneca then licenced to Navidea who has sublicenced to Cerveau Technologies to complete commercial development.

In order to be able to standardize quantitative amyloid plaque estimation by PET, a Centiloid (CL) Project was initiated with the aim of scaling the outcome of an analysis method or tracer to a 0–100 scale with 0 CL units tagged to an average of young controls ≤45 years and 100 CL tagged to an average of mild AD patients. Scale units for the Centiloid scale are in CLs. A standard method for analyzing ^{11}C-PiB data was described, and a method was also described for adapting any "nonstandard" method of PiB analysis or any other tracer analysis to the Centiloid scale (Klunk et al., 2015). This approach has the advantage that quantitative values from amyloid PET images can be expressed in the universal CL units that will allow the integration of multiple tracers and analysis methods. The comparison with the PiB standard method has been completed for florbetapir, florbetaben, and ^{18}F-NAV4694 tracers, and quantification of their amyloid scans can be expressed in CL units (Jagust et al., 2015; Rowe et al., 2017, 2016).

The first way in which amyloid PET has been used in DD clinical trials is for patient selection. According to the NIA-AA research guidelines for mild cognitive impairment (MCI) due to AD and the guidelines for defining the preclinical/presymptomatic stages of AD, either amyloid imaging or CSF levels of Aβ42 are recommended for defining an amyloid-positive status, while recommended biomarkers of neuronal injury include the presence of tau (originally by tau and phospho-tau from CSF (but now tau PET as well) or atrophy by MRI or hypometabolism by FDG PET (Albert et al., 2011; Sperling, Aisen, et al., 2011). Clinical trials have used, within a given trial, a combination of CSF Aβ42 and a mixture of amyloid tracers with cutoffs for positivity to allow flexibility in site capabilities. Clearly the presence of amyloid pathology is a critical subject selection criteria for trials of antiamyloid agents. An interesting example of the importance of patient selection in reducing noise came from the Biogen Phase 1b trial of their antiamyloid antibody where, of 278 patients with an evaluable PET scan, 170 (61%) and 185 (67%) were amyloid-positive by visual reading and quantitative analysis, respectively; 39% were excluded from the treatment phase of the study due to an amyloid-negative scan based on visual readings (there was 92% concordance on visual compared with quantitative reads) (Sevigny et al., 2016). The use of amyloid PET for patient selection has typically been binary (amyloid-positive or amyloid-negative) either based on a visual read or, if quantitative,

based on a cut point established for the particular tracer. Recently a 4-year longitudinal study of 174 subjects aged 40–89 who were cognitively normal at baseline and who had a florbetapir PET at baseline plus cognitive testing at baseline and 4 years was completed. Analysis suggested that the magnitude of amyloid burden at baseline is associated with the rate of cognitive decline over 4 years, information not available from a dichotomous use of the amyloid PET data (Farrell et al., 2017). This potential information about future decline may be very helpful in patient characterization in presymptomatic treatment trials.

The second way that amyloid PET has been used is to assess the effects of antiamyloid agents in removing brain amyloid. To date, proof-of-mechanism (PoM) studies have been performed with amyloid PET for three monoclonal antibodies targeting Aβ: bapineuzumab (PiB), gantenerumab, (PiB) and aducanumab (florbetapir) (Ostrowitzki et al., 2012; Rinne et al., 2010; Sevigny et al., 2016). Amyloid burden was reduced by all three antibodies. With bapineuzumab, in mild/moderate AD, the pooled treatment group (three dose levels) showed decreases in PiB uptake, whereas uptake in the placebo group increased (Fig. 9.6A and B). The higher-dose group for gantenerumab showed a reduction in PiB uptake, while the placebo group had an increase in PiB over 7 months in mild/moderate AD patients. Aducanumab treatment over 1 year, in prodromal and mild AD, resulted in the reduction of amyloid burden by PET in a dose- and time-dependent manner; clinical decline slowed as assessed by the Mini-Mental State Examination (MMSE) and Clinical Dementia Rating-Sum of Boxes (CDR-SB) disease progression score. These studies illustrate that amyloid PET can be used for PoM for amyloid removal agents; dose guidance; and, in the case of aducanumab, early proof of efficacy.

A third way to use amyloid PET is to assess perfusion by using the early frames of amyloid PET. Neuronal damage is reflected as cerebral hypoperfusion by PET where measures of a PET ligand's influx rate are related to regional cerebral blood flow (rCBF) (Albert et al., 2011; Forsberg, Engler, Blomquist, Långström, & Nordberg, 2012). The standard amyloid scan for amyloid pathology is performed at quasi-equilibrium, about 50–70 min postinjection for ^{18}F-florbetapir and 90–110 min for ^{18}F-florbetaben. By adding a short scan period right after injection for 6–15 min and collecting early frame images, model-based influx rates can be determined. This dual-phase imaging has been done cross-sectionally for ^{11}C-PiB, ^{18}F-florbetapir, and ^{18}F-florbetaben and allows dual biomarkers to be assessed, namely, perfusion estimates as a biomarker of neurodegeneration and the presence/extent of amyloid burden as a biomarker of disease pathology (Daerr et al., 2016; Devous et al., 2014; Forsberg et al., 2012; Matthews et al., 2016;

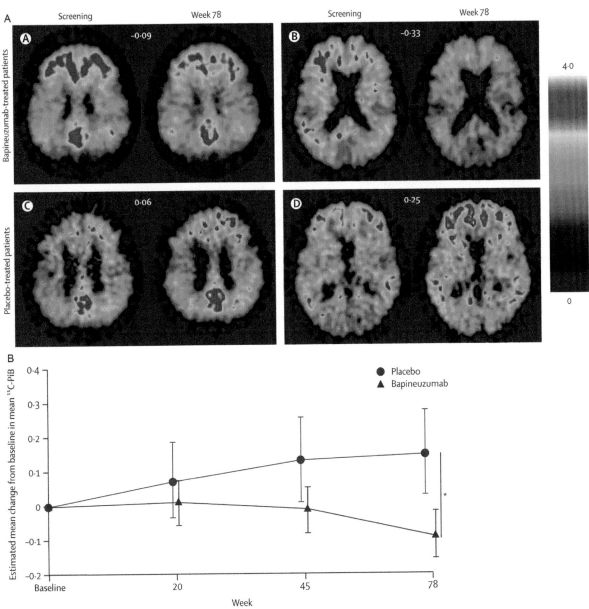

FIG. 9.6   Effect of bapineuzumab on brain amyloid assessed by ^{11}C-PiB PET. (A) ^{11}C-PiB images at screening and 72 weeks from subjects with AD treated with bapineuzumab (labeled A and B) or subjects with AD treated with placebo (labeled C and D). The scale bar shows the PiB uptake ratios relative to the cerebellum by color. (B) Graph of estimated mean change from baseline in mean ^{11}C-PiB uptake as a function of study treatment time in weeks for placebo group and bapineuzumab group. The data points are least squares means, and bars are 95% confidence intervals. The difference in uptake between placebo and bapineuzumab groups at week 78 is −0.24 (P = 0.003). *With permission from Rinne, J. O., Brooks, D. J., Rossor, M. N., Fox, N. C., Bullock, R., Klunk, W. E., et al. (2010). 11C-PiB PET assessment of change in fibrillar amyloid-beta load in patients with Alzheimer's disease treated with bapineuzumab: a phase 2, double-blind, placebo-controlled, ascending-dose study.* Lancet Neurology, 9, 363-372.

Shcherbinin, Eads, Schwarz, & Sims, 2016). These studies have also shown that rCBF from amyloid scans correlates very strongly with voxel-wise and regional FDG metrics of hypometabolism so that the early phase of an amyloid scan can substitute for a separate ^{18}F-FDG PET scan with less burden to the patient and improved cost-effectiveness for the information gained (Daerr et al., 2016; Devous et al., 2014). What remains an active area of investigation is the acquisition and analysis of follow-up early frame amyloid scans so that the possibility of using longitudinal change in early frame amyloid perfusion metrics as a potential outcome measure of change in brain function can be assessed.

An important point to consider when using quantitative amyloid PET for longitudinal multicenter trials is the factors that influence the variability in the data and how to minimize that variability. Reliable quantification requires protocol-specific imaging site training, standardization of the acquisition and tracer administration and reconstruction methods, image quality control, and central standardized analysis. Schmidt et al. (2013) have identified ways in which technical factors can contribute to variability and provide recommendations for mitigating sources of noise.

3 Tau PET Imaging

Tubulin-associated unit (tau) is an intracellular protein that is key in stabilizing microtubules in axons and regulating axonal transport. Abnormal aggregates of tau are a feature of a number of neurodegenerative diseases called tauopathies that include AD, progressive supranuclear palsy (PSP), corticobasilar degeneration (CBD), Pick's disease (PiD), frontotemporal lobar degeneration with tau inclusions (FTLD-tau), argyrophilic grain disease (AGD), and chronic traumatic encephalopathy (CTE). The spatial distribution of tau deposition as well as the isoform composition and morphology of tau aggregates and the ultrastructural characteristics of the tau filaments (paired helical filaments in AD vs. straight filaments in PSP, CBD, and PiD) differs across tauopathies (Dickson, Kouri, Murray, & Josephs, 2011; Wang & Mandelkow, 2016) In AD, tau aggregation and spreading appear to be facilitated by beta amyloid aggregation. The neuropathological topographic progression of tau deposition as described by Braak stages I–VI appears to start in the transentorhinal area (I–II) and then progresses to hippocampus, to paralimbic and adjacent mediobasal temporal cortex (III–IV), then to cortical association areas, and finally to primary sensory-motor and visual areas (V–VI) (Braak, Alafuzoff, Arzberger, Kretzschmar, & Del Tredici, 2006; Braak & Braak, 1991). Clinical autopsy studies show a much tighter correlation between tau aggregation/neurofibrillary tangles (NFTs) and cognitive impairment than between amyloid and cognitive impairment (extent and distribution of NFTs are closely correlated with cognitive decline (Nelson et al., 2012)). For purposes of a biomarker for disease-modifying therapeutics in AD (and potentially other tauopathies), quantification of the spatiotemporal distribution of tau aggregation in vivo would thus provide a way to assess early disease progression and intervention-induced slowing of progression likely to be related to clinical outcomes. Following the concept of what was done with amyloid PET tracers, a PET tracer targeting tau protein aggregates in principle can provide this type of information and be a powerful tool to better understand the longitudinal progression of the disease noninvasively.

As background to understanding the potential properties and clinical uses of current tau tracers (which will be summarized later), their development compared with that of amyloid tracers was challenging because of the complexity of tau proteins and tau pathology. Tau is encoded by the microtubule-associated protein tau (MAPT) gene that has 16 exons. The human brain has six main isoforms that arise form alternate splicing of certain exons. One way of describing the isoforms is by whether they have three or four carboxy-terminal repeat domains of the microtubule binding domain, the so-called 3R or 4R. Tauopathies can be classified by the tau isoforms found in the tau aggregates: AD has a mixture of 3R + 4R as does CTE; PSP, CBD, and AGD contain 4R; Pick's disease aggregates contain 3R. Tau deposits are both intracellular and extracellular. Morphologically, tangles, neuropil threads, and dystrophic neurites are seen in AD; neuronal inclusions called Pick bodies in PiD; oligodendrial coiled bodies, globose tangles, and tufted astrocytes in PSP; and astrocytic plaques, coiled bodies, and neuropil threads in CBD. The ultrastructural characteristics of tau filaments are paired helical filaments (PHF) in AD versus straight filaments in PSP, CBD, and PiD (Wang & Mandelkow, 2016). Of the current tau PET tracers, selectivity for isoform and tau filament ultrastructure varies, and more characterization needs to be done to understand their pathological specificity (or lack of).

As of early 2017, five tau PET imaging tracers are being assessed in humans, and some are already being used in clinical trials. None are FDA-approved for use at this time. They all have a high selectivity over amyloid plaques. One additional advanced tracer was dropped from commercial development in early 2017 because of off-target binding. The five tracers being used currently in human studies are:

1. Lilly/Avid ^{18}F-AV-145/T807/flortaucipir: most widely studied tau tracer to date; binds to PHF and the NFTs (both intra and extraneuronal) of AD and some tau mutations in FTD; does not bind to the 3R and 4R isoforms of PiD, PSP, or CBD; does not bind to α-synuclein. AV-1451 images cross-sectionally across the AD spectrum were able to be classified into patterns similar to Braak staging (Schwarz et al., 2016). In a study of 217 subjects (healthy young and old, MCI and AD) who underwent both Amyvid and AV-1451 PET, results suggested that deposition of tau beyond the mesial temporal lobe is associated with amyloid deposition, and the results suggested cortical tau is associated with cognitive impairment (Pontecorvo et al., 2017). Off-target binding to melanin-containing cells has been demonstrated along with off-target binding to MAO-A, for which early reports said affinity was weak (μM); it appears that the MAO-A affinity is too low to interfere with measuring tau (Hostetler et al., 2016; Marquié et al., 2015). Flortaucipir is being used in the A4 secondary prevention trial (Sperling et al., 2014). It is unavailable for use for most pharma groups as of early 2017.

2. ^{18}F-GTP1: Genentech in-house tau tracer binding selectively to NFTs in early development. Studies to date presented at 2017 conferences indicate expected

distribution of tracer by Braak stage in a spectrum of AD subjects; the tracer also distinguished controls from prodromal from AD. The tracer is being used currently in a longitudinal 18-month natural history study (GN30009) (Bohorquez et al., 2017). Genentech will share its tracer with academic investigators but does not plan commercial distribution on a larger scale.

3. ^{18}F-RO6958948: Roche in-house tau tracer binding to AD tau aggregates; does not bind to PiD, PSP, or CBD (Gobbi et al., 2017). Phase 1 study in AD is ongoing; is a longitudinal, follow-up study in participants with Alzheimer's disease (AD) who previously participated in study BP29409 (NCT02187627); and is designed to assess the longitudinal change of tau pathology in the brain of participants with AD using the PET ligand ^{18}F-RO6958948. Roche plans to use this tracer in its investigational AD therapeutics and plans to make it available to academic investigators and consortia such as the European Prevention of Alzheimer's Dementia Consortium (EPAD).

4. ^{18}F-MK-6240: Merck-developed tracer binding to PHF and NFTs and brain uptake consistent with tau Braak staging; no significant specific binding in non-AD brain tissues; highly selective for tau over amyloid; fivefold higher affinity to NFTs compared with AV-1451; no significant off-target binding (assessed for MAO-A and B) (Hostetler et al., 2016). Phase 1 study in controls and AD subjects is completed. Clinical development of the tracer has been licenced and sale of ^{18}F-MK-6240 made to Cerveau Technologies so that tracer can be widely available for use in therapeutics trials.

5. ^{18}F-PI-2620: Piramal tracer developed in conjunction with AC Immune. Shows high-affinity binding to PHP and excellent selectivity with no off-target binding to amyloid or MAO-A or -B. Detects both 3R and 4R isoforms in brain homogenate. Initial human images showed expected uptake in AD subjects versus controls and no uptake in AD or controls in the choroid plexus, amygdala, or striatum as seen with first-generation tau tracers. In a subject with PCP, images showed subcortical uptake in the substantia nigra and globus pallidus, expected for PCP tau pathology (Mueller et al., 2017). Piramal plans to widely distribute ^{18}F-PI-2620 as a tau imaging agent.

Development has stopped for the most advanced of the Tohoku University tau tracers, ^{18}F-THK5351, which was licenced to GE Healthcare for commercialization and had been used quite extensively as an investigation agent in human studies and trials. Ng et al. (2017) tested the ability of MAO-B inhibitor selegiline to block uptake of ^{18}F-THK5351 in eight subjects (five MCI, two AD, and

one PSP). Regional SUVs were blocked by 36%–51%. Interpretation of scans for tau binding is thus confounded by the high off-target binding and clearly compromised the use of the tracer as a tau-binding ligand.

An area of active research in tau PET imaging is how best to quantify the tau PET signal so that tau PET imaging can be used as a biomarker for tracking disease progression. This is an area in which consensus has not yet been reached. Jagust and coworkers evaluated 10 different tau PET measures in two different cohorts of cognitively normal, MCI, and AD subjects who underwent flortaucipir PET. Whole-brain tau PET measures may be adequate to detect AD tau pathology, but regional measures of AD-vulnerable regions seem to increase the sensitivity to early tau PET signal, atrophy, and memory decline (Maass et al., 2017).

4 PET Imaging of Neuroinflammation

Neuroinflammation is a complex interaction of the immune elements of the brain, the pathophysiology of the particular CNS disease or state (neurodegenerative, psychiatric, and central pain response), and the peripheral immune system and is increasingly been shown to be a component of many CNS diseases. From an imaging perspective, potential approaches to assessing inflammation and the effect of drug interventions on the process involve the following:

1. Imaging of the actors in the inflammation process such as microglia and astrocytes or inflammatory cytokines.
2. Imaging of peripheral immune cells that access the CNS by a broken-down BBB (along with imaging evidence of a compromised BBB).
3. Imaging of the downstream consequences of neuroinflammation such as neuronal loss or dysfunction and demyelination.

This section will give some examples of the current state of PET tracers in imaging the neuroinflammation from the perspective of approach (1). Section II.I.5 provides a PET imaging example of approach (3), and MRI techniques for approaches (2) and (3) are described in Section III.F.

PET IMAGING OF TRANSLOCATOR PROTEIN (TSPO)

PET imaging allows the visualization of inflammatory cells in vivo with molecularly targeted tracers. Conceptually the intent was to use these tracers to assess for the presence of inflammation and to determine if antiinflammatory drug candidates altered inflammation as a proof of mechanism and/or concept. A number of probes have been developed that target TSPO, which is a protein found in the outer mitochondrial membrane. TSPO was previously known as the peripheral benzodiazepine receptor (PBR). TSPO has been viewed as an important marker of inflammation since it is present at low levels in normal CNS, but the switch of microglia to an activated phenotype results in the upregulation of TSPO expression that is seen in a large number of CNS diseases including AD, multiple sclerosis (MS), PD, HD, stroke, CTE, schizophrenia, major depressive disorder, and obsessive compulsive disorder (Albrecht, Granziera, Hooker, & Loggia, 2016; Calsolaro & Edison, 2016; Dupont et al., 2017). The first-generation TSPO tracer was ^{11}C-PK11195 that has a very low signal to noise. Subsequently, a number of second- and third-generation tracers have been developed including ^{11}C-PBR28, ^{18}F-PBR111, ^{11}C-DPA-713, ^{18}F-DPA-714, ^{18}F-GE-180 (^{18}F-flutriciclamide), and ^{11}C-ER176. None of these tracers are currently commercially available so clinical trials have been done at centers experienced with GMP synthesis of the tracer of interest. Although these later-generation tracers have improved signal-to-noise and less nonspecific binding, TSPO tracers suffer from a number of limitations that limit their current use and value and require further research. First, analysis of tracer uptake is complicated by the existence of a genetic polymorphism in exon 4 of the TSPO gene (rs6971) in which there is an alanine-to-threonine substitution (A14T7). This polymorphism affects the binding affinity of the tracer, and sensitivity to the polymorphism varies amongst the tracers. Three binding states have been identified: high-affinity binders (HAB; A/A; ~50% of subjects of European ancestry), mixed-affinity binders (MAB; A/T; ~40%), and low-affinity binders (LAB; T/T; ~10%). ^{11}C-PK11195 binding is insensitive in the brain to the two different binding sites for the tracer, whereas ^{11}C-PBR28, for example, is very sensitive and does not allow the detection of LAB subjects. ^{11}C-ER176 is sensitive to the two binding sites but is able to detect TSPO in LAB subjects. Binding status of subjects can be determined by genetic analysis to ensure the proper selection of subjects (and the omission of LABs pending the tracer); binding status needs to be accounted for in the quantification of the tracer uptake data (Owen, Guo, Rabiner, & Gunn, 2015). A second issue with TSPO tracers is the complexity of analysis since TSPO is present throughout the brain in pathological states. Unless a reference region has been validated (as is the case with ^{11}C-PBR28 in AD) or a pseudoreference region has been validated (as with MS; Datta et al., 2017), dynamic scans with plasma input data from arterial lines need to be used, and in addition, modeling to account for vascular binding may be needed (Albrecht et al., 2016). A third issue is that TSPO presence reflects a broader inflammatory state since TSPO is also expressed to a lesser extent on astrocytes and on peripheral immune cells in the brain if the BBB is broken. Fourth, current TSPO tracers bind to both the M1 (proinflammatory/toxic) and M2 (antiinflammatory/reparative) phenotypes of activated microglia so it is not possible to use these imaging probes to understand the pro- versus antiinflammatory functions of activated

microglia in the pathophysiology of the various diseases and the potential effects of therapeutic interventions on these functions.

In multiple sclerosis (MS), MRI has classically been used to demonstrate focal inflammatory lesions in relapsing remitting MS, and these are shown as well with ^{11}C-PK11195 PET. A more unique application of microglial PET imaging is likely to be in the area of progressive MS (both secondary [SPMS] and primary [PPMS]). Neuropathology studies in SPMS and PPMS have shown that there is a diffuse inflammatory process with the activation of microglial cells within plaques and also outside of plaques in the normal-appearing white matter (NAWM) that is associated with axonal damage (Frischer et al., 2009). The diffuse inflammation is not assessed with conventional MRI (but may be with diffusion basis spectral imaging; see Section III.F) but has been detected with ^{11}C-PK11195, ^{11}C-PBR28, and ^{18}F-PRB11 PET. Correlation between TSPO uptake and disability has been shown in some but not all TSPO PET studies to date and needs further investigation (Airas, Rissanen, & Rinne, 2015; Datta et al., 2017). For ^{11}C-PBR28, test-retest was performed in MS patients (RRMS and SPMS) and healthy controls and showed a mean absolute test-retest variability of 7%–9% across gray-matter, NAWM, and MS lesions indicating the potential of a reliable assessment of signal (Park et al., 2015).

In Parkinson's disease patients, imaging with ^{11}C-PBR28 was used to assess the effects of an 8-week treatment with AZD3241, an irreversible inhibitor of myeloperoxidase that is a reactive oxygen-generating enzyme expressed by microglia. The proposed mechanism of action for AZD3241 was that the drug would reduce oxidative stress and result in a reduction of the neuroinflammation evident in Parkinson's disease patients. Results showed that AZD3241 reduced the total distribution volume of ^{11}C-PBR28 across all brain regions including nigrostriatal regions at both 4 and 8 weeks, whereas there was no change in the volume of distribution of the tracer in the placebo group. The study provided support for the hypothesis that the drug has an effect on microglia (Jucaite et al., 2015)

In AD and MCI patients, results with TSPO PET are sometimes contradictory, but most suggest that microglia activation detected by TSPO imaging occurs temporally close to conversion to dementia. ^{11}C-PK11195 PET has shown tracer uptake in the association cortex similar to that of amyloid plaque distribution with about 60% of AD subjects showing binding compared with 40% of amnestic MCI subjects in one study. ^{11}C-PBR28 PET showed uptake in subjects with AD but not in subjects with MCI. Subjects with early-onset AD (before 65years) have more tracer uptake than subjects with late-onset-AD (Dupont et al., 2017; Casolaro et al., 2016). Fan, Brooks, Okello, and Edison (2017) have recently evaluated the longitudinal changes over 14 ± 4 months with ^{11}C-PK11195 in subjects with MCI and AD versus controls; ^{11}C-PiB PET was also performed at baseline. Both MCI and AD subjects had increased microglial activation compared with controls (41% and 38%, respectively). There was a longitudinal decrease in microglial activation over the study period for MCI subjects, while AD subjects showed an increase. The authors speculate that there might be two peaks in microglial activation, with the early peak being neuroprotective phenotype and the second peak being a proinflammatory phenotype. Clearly the availability of TSPO or other tracers selective for microglial phenotypes M1 and M2 would be very helpful as well as additional longitudinal data with more sensitive TSPO tracers.

Sandiego et al. (2015) have utilized a systemic administration of lipopolysaccharide (LPS) in conjunction with ^{11}C-PBR28 PET in humans to demonstrate and measure microglial activation in healthy volunteers. LPS administration resulted in increased ^{11}C-PBR28 brain binding along with increases in peripheral inflammatory cytokine levels and self-reported sickness symptoms. This LPS-TSPO PET model may be useful in assessing antiinflammatory and neuroprotective drug candidates for their effect on microglial activation.

OTHER PET TRACERS FOR MICROGLIA

Although most efforts to image microglia have focused on TSPO PET, the limitations of the tracers for this target indicate the need for additional approaches to PET imaging assessment of inflammation (Tronel et al., 2017). An interesting candidate is the purinergic ion channel 7 receptor (P2X7R), a cationic, ATP-mediated ion channel that is expressed peripherally and in microglia, astrocytes, and Schwann cells in the CNS. It is a key player in the inflammasome activation and maturation and release of the proinflammatory cytokine IL-1β as well as microglial proliferation. It has been proposed as a potential marker of M1 microglia. Its mechanism of action is of interest in a broad range of neurodegenerative neuroinflammatory diseases including AD, PD MS, HD, ALS, and FTD (Burnstock, 2008). P2X7R inhibitors are being developed as drug candidates for neuroinflammation, and P2X7R PET probes are being developed as well. ^{11}C-JNJ-54173717 has undergone successful preclinical evaluation in rat and monkey brain and will undergo further evaluation in the clinic to assess P2X7R expression in neurodegenerative diseases (Ory et al., 2016). Another P2X7R tracer, ^{11}C-GSK1482160, has been evaluated in vitro, in vivo in both mice and EAE rat models and also appears to have the potential to monitor neuroinflammation (Han et al., 2017) Other targets of interest for PET imaging agents include cyclooxygenase (Cox) 1 and 2, cannabinoid receptor type 2 (CBR2), and nicotinic acetylcholine receptors α7 and α4β2; their development status is summarized in Tronel et al., 2017.

MAO-B PET IMAGING OF REACTIVE ASTROCYTES

Astrocytosis is a complex response of astrocytes to injury and disease, and reactive astrocytes are part of the neuroinflammatory and neurodegenerative process (Sofroniew, 2015). Monoamine oxidase B is upregulated in reactive astrocytes, and selective MAO-B PET probes have been developed to image reactive astrocytes, including ^{11}C-deprenyl-D2 (^{11}C-DED) and a new fluorinated derivative, ^{18}F-fluorodeprenyl-D2 (Nag et al., 2016). A recent longitudinal study with ^{11}C-DED has shown that, compared with controls, astrocytosis was elevated more in the presymptomatic autosomal dominant AD (ADAD) subjects from 17 years prior to expected symptom onset than in the amyloid-positive MCI subjects of the sporadic AD type. In the ADAD subjects, ^{11}C-DED uptake steadily declined, whereas amyloid deposition increased; in the sporadic MCI subjects, amyloid plaque deposition increased, and ^{11}C-DED uptake did not change significantly over the average 3-year period (Rodriguez-Vieitez et al., 2016). The mechanisms at play for the temporal differences between ADAD and sporadic MCI subjects in ^{11}C-DED change are unexplained. Aside from the limited data on the MAO-B astrocyte probes, it is important to note that MAO-B is not selective to astrocytes—it is also present in serotonergic and histaminergic neurons (Albrecht et al., 2016). Results with MAO-B tracers thus currently need to be interpreted with caution, and more exploratory research is needed to clarify their potential use in drug development.

INMiND

A much-needed precompetitive effort to progress the quality of imaging biomarkers for neuroinflammation comes in the form of the European Union Concerted Action "Imaging of Neuroinflammation in Neurodegeneration." This project was established and funded for 5 years starting in 2012 by the EU 7th Framework Programme. The aim of the project is to identify novel biomarkers for activated microglia for both diagnostic and therapeutic purposes. The focus is on cellular and in vivo studies to assess the dynamic pattern of microglial activation and its relation to M1/M2 phenotype/polarization status (toxicity and repair) at various stages of various CNS diseases (AD, PD, HD, MS, and ALS). Novel animal models and imaging biomarkers are being developed and validated to assess microglial function and activity in vivo to validate the outcome of known and new neuroprotective strategies in both preclinical models and in patients (Mohammadi, 2013).

5 PET Imaging of Synaptic Density

Disruption or the loss of synapses is associated with and part of the pathophysiology of many CNS diseases. A noninvasive method for quantifying synaptic density in patients in vivo would be a powerful tool. Recently,

Carson and coworkers have developed a synapse-specific PET radioligand for the synaptic vesicle glycoprotein 2A (SV2A), ^{11}C-UCB-J. SV2A is ubiquitously and homogenously found throughout the brain in the presynaptic membrane of vertebrates. In brain tissue, quantification of synaptic density is often performed using immunohistochemistry (IHC) with antibodies targeting key pre- or postsynaptic proteins such as synaptophysin (SYN). Finnema et al. (2016) validated that SV2A was an alternate marker of synaptic density by comparing uptake of ^{11}C-UCB-J in baboon brain with IHC analysis of the baboon brain tissue for SYN and demonstrating high correlation of cellular and regional distributions. Assessment of ^{11}C-UCB-J in healthy volunteers showed specific binding to SV2A with favorable tracer kinetics and a coefficient of variation for regional binding potential values (BP_{ND}) of 12% ± 2%. The tracer was also evaluated in patients with refractory temporal lobe epilepsy and showed unilateral loss of neurons in the epileptogenic zone. This early work suggests that SVA PET could be a powerful tool for assessing synaptic loss and for monitoring treatments directed at altering neuronal loss in CNS diseases such as AD. UCB-J has three fluorine atoms in its structure so that a fluorinated version of the tracer may be able to be developed to allow for distribution from central production facilities.

6 SPECT/PET Imaging of Dopaminergic Deficit in Parkinson's Disease

Dopaminergic deficit is a key feature of PD and related parkinsonian diseases (PDD, LBD, PSP, and MSA). Both SPECT and PET imaging agents are available for assessing this deficit that has a specific signal pattern, namely, earliest signal loss in the posterior putamen relative to the anterior putamen and caudate. The loss is initially asymmetric, opposite to the side of movement abnormalities, and then progresses. The various tracers evaluate different aspects of cellular function. ^{18}F-DOPA is a marker of dopamine synthesis. Dopamine transporter (DAT) agents bind to a transporter on the presynaptic membrane. These DAT tracers include SPECT tracers (^{123}I-ioflupane/^{123}I-FP-CIT/ DaTscan™, ^{123}I-β-CIT, ^{123}I-altropane, and ^{99m}Tc-Trodat) and PET tracers (^{18}F-PE21 and ^{18}F-FP-CIT). Dopaminergic loss can also be assessed with imaging of the VMAT2 vesicular transported located intracellularly on vesicles in the presynaptic nerve terminal. PET tracers for VMAT2 include ^{11}C-DTZB and ^{18}F-AV-133 (Brooks, 2016). The most widely used and commercially available tracer is DaTscan™. For clinical trials in early PD, confirmation of DA deficit with imaging is important in subject selection since up to 15% of subjects may present clinically with PD but in fact lack a DA deficit; these subjects have been called subjects without evidence of a DA deficit (SWEDDs) and do not progress as typical PD subjects (Brooks, 2016).

With respect to prodromal PD (early signs or symptoms of PD neurodegeneration without clinical diagnosis based on motor parkinsonism), MDS research criteria for prodromal PD includes dopaminergic imaging with PET or SPECT as part of the diagnostic information needed (Berg et al., 2015). As a means to identify subjects at risk for PD in the prodromal stage and with the knowledge that hyposmia is a consistent finding in 80%–90% of early PD patients, the Parkinson Associated Risk Study (PARS) utilized a combination smell testing followed by DAT scanning with ^{123}I-β-CIT at baseline and at 2 and 4 years. 67% of subjects with hyposmia and a DAT deficit (≤65% of age related norm) at baseline converted in a 4-year period to PD versus 9% of those with DAT in an indeterminate range and 2.8% of those with no DAT deficit. Subjects with a baseline DAT deficit progressed much more rapidly (20% decline in in striatal binding ratio by 4 years) than those with indeterminate or no DAT loss. This combination of olfactory dysfunction and DAT imaging provides a useful approach to selecting subjects for testing disease-modification therapies (Jennings et al., 2017).

7 PET Translational Imaging in Huntington's Disease

Several neurotransmitters and brain enzymes have been reported to be affected in HD in the premanifest stage, including the dopamine system, PDE10A, and the serotonin system. Changes in neurotransmitter systems and enzymes related to HD progression were assessed using micro-PET in a recently reported knock-in animal model for HD, zQ175, and wild-type mice at 6 and 9 months of age. The zQ175 mice had a marked reduction compared with wild-type mice in tracer binding to D1, D2, and 5-HT2a receptors using ^{11}C-NNC112, ^{11}C-raclopride, and ^{11}C-MDL-10097 respective as tracers for the receptors and a loss of PDE10A in the striatum using ^{18}F-MNI-659 as a tracer (Häggkvist et al., 2017). The findings parallel those in HD patients. The mouse model can be used to assess disease-modification therapies and provides a nice example of potential translation of assessment of novel therapeutics from preclinical to clinical assessment HD subjects since the PET tracer studies can be run clinically as well.

III MRI

A Overview of MR Techniques

MRI was initially developed and perceived as a nonionizing radiation analogue to x-ray-based CT scanning in the 1980s. Whereas the underlying mechanism of CT contrast is tissue density, MRI tissue contrast mechanisms rely primarily on the magnetization properties of atomic nuclei (such as the protons in the hydrogen atoms in water).

When an external magnetic field is applied (i.e., the main magnetic field in the bore of an MRI scanner), the randomly oriented protons of water in tissue are aligned. Their alignment is then perturbed/excited by a radio frequency (RF) pulse in the scanner, following which the protons return to their original alignment in the external magnetic field by various relaxation processes and emit RF energy that is captured as frequency information from each location and then Fourier-transformed into an intensity level that is then displayed on a grayscale matrix of pixels creating an image. Magnetic field gradients are used to map the location of the signal in the body. Depending on the sequence of pulses, images with different characteristics are generated. Tissue can be characterized by two different relaxation/decay times. T1 is the time constant (longitudinal) for excited protons to realignment with the external magnetic field. T2 is the time constant (transverse) for the protons spinning perpendicular to the main field to go out of phase with each other. T2* is the time constant for decay of the transverse magnetization and is a combination of T2 relaxation and magnetic field inhomogeneity: magnetic susceptibility occurs when specific body tissue or structure distorts the magnetic field that it resides in causing magnetic field inhomogeneity and degradation of the local signal level. This effect is captured in T2*. T2* is the main determinant of contrast in gradient-echo (GRE) pulse sequences.

It is the density of ^{1}H protons and the rate in which the signal from these protons, when excited, decays in water, fat, and other body tissues that account for the exquisite soft tissue contrast in MRI compared with CT. The classic MRI sequences are T1-weighted (T1-w), T2-weighted (T2-w), and proton density (PD). PD images reflect the density of protons. The contrast and brightness of a T1-w image are essentially determined by the T1 properties of the tissue being imaged, similarly for T2-w images. For example, CSF is dark on T1-w and bright on T2-w images. Three-dimensional T1 (3DT1) images provide excellent tissue delineation of gray and white matter and CSF for tissue volume measurements. The acquisition of a 3DT1 sequence following intravenous injection of an exogenous contrast agent such as a gadolinium-based chelate agent (GBCA) allows for the visualization of vascular structures and breakdown of the BBB with GBCA leakage into interstitial tissue space. The paramagnetic gadolinium decreases the T1 relaxation time of adjacent water molecules in the interstitial space and increases the signal intensity on T1-w images in the region of GBCA leakage. Because most tissue pathologies have a higher water content than normal tissue, T2-w images provide good visualization of disease processes, and the affected areas (such as edema, inflammation, or white-matter lesions) appear bright on T2-w images. A fluid-attenuated inversion recovery (FLAIR) sequence is a variation of a T2 weighting in which an additional type of RF pulse is used to null the signal from CSF so that subtle lesions and periventricular or perisulcal lesions are better detected. GRE sequences are sensitive to T2*, and areas of tissue with magnetic field in homogeneities (such as hemorrhage, microhemorrhage, and hemosiderosis) appear dark on these images. More advanced types of sequences and scanner field strengths have been developed over the years that are of interest in CNS drug development, and these include the following:

1. Functional MRI (fMRI): fMRI detects the blood oxygen level-dependent (BOLD) changes that occur when changes in neuronal activity are produced either by a task (task-based fMRI) or by the spontaneous fluctuations in brain activity that are present in the absence of a task (resting-state (rs)-fMRI). The physical origin of the BOLD signal is a magnetic susceptibility effect. Stimulation of a brain area (i.e., visual cortex stimulation by a flashing light) causes increased cerebral blood flow in the capillaries of the activated area that exceeds the rate of cerebral oxygen utilization. The BOLD signal, a proxy for neuronal activity, is a measure of neurovascular coupling reflecting a local vascular effect in response to neuronal activity and depends on three physiological parameters: regional cerebral blood (rCBF), cerebral blood volume (CBV), and the cerebral metabolic rate of oxygen consumption. Since oxyhemoglobin (oxyHb) is diamagnetic and deoxyhemoglobin (deoxyHb) is paramagnetic relative to the essentially diamagnetic surrounding tissue, deoxyHb reduces the T2* time constant for signal decay of an MR pulse because of the microscopic magnetic field inhomogeneities deoxyHb creates in the microvasculature. Because local blood is more oxygenated during a brain stimulus, there is a relative decrease in deoxyHb and an increase in oxyHb resulting in an increase in T2* during activation compared with pretask. The MR signal thus decays less rapidly causing a stronger/increased MR signal when the signal is recorded. The signal change due to the task is averaged over multiple repetitions of the task. Statistical processing converts signal intensity change into color maps that are overlaid on a structural MR image indicating the regions of the brain affected. A typical sequence used in acquiring fMRI data is a GRE echoplanar sequence, which is a fast T2*-weighted sequence. The temporal scale for BOLD imaging effects is limited by the time course of the hemodynamic response and is about 6 seconds.

2. Diffusion MRI: On a more microscopic and cellular level, diffusion MRI utilizes sequences that are sensitive to water motion. Diffusion is restricted in tissue with high cellular density (inflammation), in

fluids rich in protein material, and in cytotoxic cell edema. Restricted diffusion causes high signal on diffusion-weighted imaging (DWI) sequences. Diffusion can be quantified by calculating the apparent diffusion coefficient (ADC). Restricted diffusion gives a low ADC value and low signal on an ADC map of the brain. A common application of DWI is in stroke imaging where DWI enhancement occurs within 10 min of stroke onset. A mismatch between amount of tissue abnormal on diffusion (infarcted tissue) versus a larger tissue volume with abnormal hypoperfusion may indicate salvageable tissue if reperfusion is carried out (Vilela & Rowley, 2017).

An important parameter in diffusion MRI is the so-called b-value that is selected in setting up diffusion-weighted sequences and is determined by the strength and timing of the gradients being used to generate diffusion-weighted images. Typical b values on current scanners range from $b = 0$ to 4000 s/mm^2; the higher the b-value, the stronger the diffusion effects.

Diffusion is isotropic if random motion of water molecules is equal in all directions and anisotropic if water motion is directionally restricted as it is in axons. In diffusion tensor imaging (DTI), diffusion is measured in multiple directions (at least six noncolinear directions), and the diffusion coefficients of the tensors allow the calculation of the following quantities: axial diffusivity (describing water diffusion parallel to axons), radial diffusivity (describing water diffusion perpendicular to axons), fractional anisotropy (FA) (reflecting the directionality of water displacement by diffusion with a value of 1 indicating infinite anisotropic diffusion), and mean diffusivity (MD) (representing the average magnitude of water displacement by diffusion with larger MD values indicating a more isotropic medium). These parameters are useful for assessing for demyelination and axonal injury. Voxel by voxel maps of MD and FA can be created, and FA data can also be used to visualize white-matter fiber tracts and assess structural connectivity in the brain.

Extensive work has gone into optimizing the DTI acquisition and data processing in the Human Connectome Project (HCP) (Sotiropoulos et al., 2013) for mapping brain connections. Key features include use of 3T scanners, multiple b-value shells, and simultaneous multislice echoplanar imaging with multiband excitation. These technical advances result in substantial improvement in fiber orientation and tractography and in resolution of complex fiber geometries. DTI data being acquired in ADNI 3 as part of the advanced MRI imaging utilize the HCP acquisition approach, and the DTI data will be evaluated to assess the usefulness of DTI parameters as biomarkers of neuropathology across the spectrum of AD.

One of the limitations of DTI modeling is that it assumes Gaussian behavior of water movement and monoexponential decays of diffusion displacement, a model that does not fully reflect the restricted behavior of water in complex tissue microstructure, particularly when the tissues are diseased with areas of focal myelin breakdown and diffuse tissue structure alterations including edema and cellular infiltrates (inflammatory cells). As a consequence, more advanced models of tissue architecture and more non-Gaussian advanced DWI techniques have been developed including diffusion kurtosis imaging (DKI), q-space imaging (QSI), neurite orientation dispersion and density imaging (NODDI), and diffusion basis spectral imaging (DBSI). Although these techniques and their applications to neuropathology are still in development, the quantitative parameters obtained from them have significant potential value in being able to monitor the effects of disease and interventions on brain tissue microstructure. Further information on these advanced diffusion MRI techniques is provided in Schneider et al. (2017), Hori et al. (2012), and Wang et al. (2014).

Examples of MR diffusion imaging are discussed in the context of MS imaging in Section III.F.

3. Magnetization transfer ratio (MTR) imaging and myelin water imaging: MTR imaging is a semiquantitative MRI technique that exploits the exchange of magnetization between two pools of water, the free water proton pool and bound/restricted water in the macromolecular proton pool in tissue. The technique is sensitive to the relative degree of myelination/remyelination in brain tissue but is influenced by s combination of edema, inflammation, and axon content.

Typically two sets of images are obtained, one with and one without an off-resonance saturation pulse. Most commercial 3T MR scanners can perform these sequences within 15 min. The relative signal difference between the two images (with and without an off-resonance saturation pulse) is obtained by subtracting one image from the other and is quantified by the MTR (Amann et al., 2015; Brown et al., 2016; Brown, Narayanan, & Arnold, 2014). A lower MTR indicates a reduced exchange between free and bound protons and suggests neuroaxonal breakdown/myelin damage. As confirmed by animal experimental autoimmune encephalomyelitis (EAE) models, increased MTR is evidence for possible remyelination or resolution of edema. The presence of inflammation and edema (increased free water) confound the specificity of this measure for myelin since MTR is also sensitive to increased water content from inflammation (Vavasour, Laule, Li, Traboulsee, & MacKay, 2011). Implementation of MTR in single-center trials or substudies as an exploratory biomarker of myelin is feasible; use in multicenter clinical trials is also feasible but requires careful standardization of the pulse sequences, scanners, and analyses (Enzinger et al., 2015).

Advanced methods exist that improve on the specificity of the imaging readouts for myelin by assessing the fundamental pool properties of the free and bound water pools in tissue. In quantitative magnetization transfer (qMT) imaging, several images (rather than just 2) are acquired with different MT weighting, leading to an increased scanning time. The data from these images are used to model the MT effect, and the model allows more quantitative parameters including the restricted/bound pool proton fraction f to be calculated. The parameter f has been shown to correlate mainly with myelin content both preclinically and clinically and has been suggested as an emerging biomarker for demyelination (Mallik, Samson, Wheeler-Kingshott, & Miller, 2014; Ou, Sun, Liang, Song, & Gochberg, 2009). The qMT technique has not been practical in clinical trials because of the long scan time and the need for sequences that are not as widely available as the MTR sequences, but recent advances are improving the likelihood of using this approach (Enzinger et al., 2015).

Another advanced method for assessing myelin is myelin water imaging (MWI). The aim of this technique is to quantify the relative amount of water trapped in myelin bilayers, a metric with the potential of being a specific biomarker of myelin. The T2 water signal in tissue has three components/proton pools due to compartmentalization with different T2 properties on MRI, namely, (i) a long T2 component due to CSF, (ii) an intermediate T2 component from intracellular and extracellular water, and (iii) a short T2 component arising from water trapped in the myelin bilayers. An advanced multiecho T2 sequence acquisition allows these components to be analyzed, and the ratio of the fast T2 pool (myelin bilayer water) to the proton water pool is called the myelin water fraction (MWF) (Enzinger et al., 2015; Poloni, Minagar, Haacke, & Zivadinov, 2011). Imaging acquisitions for myelin water imaging have been very challenging with long acquisition times and limited brain coverage. Recent advances reduce the acquisition time and improve on brain coverage; spinal cord imaging is also possible. In preparation for use of MWF in an MS clinical trial, a multicenter test-retest study with healthy volunteers using the same model and vendor 3T scanner and standardized acquisition protocol at sic sites showed reproducible MWF measures between scans and across sites. The intrasite mean coefficient of variability (COV) for MWF was 3.99%, and the mean intersite COV was 4.68% for MWF supporting use of the MWF in carefully designed small multicenter trials as an exploratory marker of myelination (Meyers et al., 2013). Examples of MTR imaging can be found in Section III.F.

4. Dynamic susceptibility contrast-enhanced MR (DSC-MR) perfusion is an MR technique in which the first pass of a paramagnetic gadolinium-containing small-molecule contrast MR agent through the brain is monitored dynamically with a series of T2- or T2*-weighed images using spin-echo pulse sequences or GRE pulse sequences, respectively. The susceptibility effect from the first pass of paramagnetic contrast agent leads to a loss of signal intensity in the signal intensity versus time curve that is then converted to a contrast concentration versus time curve on a pixel basis. Parametric maps of CBV, CBF, and mean transit time (MTT) are then created, and regional values of these metrics are generated by region of interest analysis (Østergaard, 2005). Recently, Ostergaard and colleagues have developed a statistical methods using DSC-MRI data to determine capillary time heterogeneity, a metric that permits assessment of tissue with impaired oxygen delivery in disorders such as stroke and AD (Mouridsen, Hansen, Østergaard, & Jespersen, 2014) (see Section III.E.1).

5. Dynamic contrast enhancement MRI (DCE-MRI) is a technique based on the dynamic acquisition of T1-weighed images before, during, and after administration of gadolinium-based small-molecule MR contrast agent (GBCA). The resultant signal versus time intensity curve is a composite of tissue perfusion, vascular permeability, and the interstitial (extracellular-extravascular space). In contrast to conventional static contrast-enhanced T1 MRI that shows contrast enhancement at a single point in time (e.g., the type of imaging used in MS (Section III.F)), this technique shows the wash-in, plateau, and wash-out kinetics of the tissue. Pharmacokinetic modeling allows several parameters to be derived including k^{trans} that reflects a combination of blood flow and permeability, the fractional volume of the interstitial space (v_e), and the fractional volume of the plasma space (v_p) (Tofts et al., 1999). The predominant use of DCE-MRI in the CNS has been in the evaluation of the BBB in neurodegenerative diseases (Montagne et al., 2016) and in the assessment of antiangiogenic and antivascular therapeutics for brain tumors (Leach et al., 2005). Application of DCE-MRI in AD are described in Sections III.C and III.E.1.

6. ASL-MRI is an MRI technique for assessing cerebral blood flow. Blood flow is assessed by magnetically labeling water in the blood so that no exogenous contrast agent is needed. So-called "labeled" brain images are obtained with rapid imaging techniques after a time delay to allow the labeled blood to reach the brain tissue. Control brain images without labeling are also acquired. The signal difference between the labeled and control images can be quantified on a voxel basis and reflects the labeled blood delivered by perfusion. Assuming that all of the labeled blood has arrived in the imaging voxel by the time the image is acquired, the signal difference is proportional to CBF. CBF determined using ASL is an absolute measure of CBF expressed in physiological units (mL/100g/min). The recommended technique for the magnetic blood labeling is pseudocontinuous arterial spin labeling (PCASL) (Alsop et al., 2015). With current technologies, a whole-brain perfusion study can be acquired in about 6–7 min. Much effort and progress has already been made on issues and recommendations regarding the standardization of acquisition sequence, use of parallel imaging techniques and multichannel head coils, optimal field strength (at least 3T), quality assessments of generated images, data processing and quantification/analysis methods, control of physiological variables that contribute to signal variability, the effect of the time delay in the labeling (time delay/transit time being quantified by the postlabeling time (PLD), the time between the end of the PCASL labeling pulse train and image acquisition), the reproducibility in healthy volunteers and patients, and differences in perfusion maps across vendors (Alsop et al., 2015; Clement et al., 2017; Gevers et al., 2011; Harston et al., 2017; Mezue et al., 2014; Mutsaerts et al., 2015, 2017; Steketee et al., 2015; Wu, Lou, Wu, & Ma, 2014). Scanner vendors are in the process of implementing recommended sequence and hardware improvements. The quantitative biomarker working groups of European and US radiology societies (EIBALL and QIBA) are working together to further optimize the technical features of acquisition and analysis for quantitation of ASL in multicenter clinical trials and practice (Golay, 2017). The consensus document on clinical implementation guidelines for ASL generated by the International Society for Magnetic Resonance in Medicine (ISMRM) Perfusion Study Group and the European Consortium for ASL in Dementia is also viewed as a living document, to be periodically updated (Alsop et al., 2015). These activities are key for proper implementation of ASL in both single-center and multicenter trials. Applications are discussed in Section III.E.2.

7. MR spectroscopy (MRS): Proton (^{1}H) MR spectroscopy provides complementary information to that obtained with MRI and can be acquired on preclinical and clinical MRI scanners that have MRS capability. ^{1}H-MRS provides a spectrum of the interrogated tissue rather than an image and has peaks that occur at a series of frequencies (see Fig. 9.14 in Section III.G). The signal intensity of the peaks represents the protons attached to carbon atoms in various brain metabolites that are mobile and present at concentrations of at least micromoles/gram in the brain. The spectra can be used to assess levels of metabolites in normal and diseased brain tissue. Peak locations are typically expressed in parts per million, a scale that is independent of field strength. Spectra can be obtained from a selected single volume of the brain, so-called single-voxel spectroscopy (SVS) or from multiple brain regions, so-called MR spectroscopic imaging (MRSI). Commonly used SVS localization acquisition techniques are point-resolved spectroscopy (PRESS); stimulated echo acquisition mode (STEAM); and, more recently, an advanced MRS acquisition technique called semilocalization by adiabatic selective refocusing (sLASER) (Deelchand, Kantarci, & Öz, 2017). SVS is useful for probing a defined area of pathology/pathophysiology. MRSI is useful for assessing tissue heterogeneity or comparing different brain regions. The ability to quantify the metabolite concentrations depends on spectral resolution and noise. Spectral resolution increases with increasing field strength with 3T scanner becoming most commonly used clinically although 7T scanners are better for low-concentration metabolites such as gamma-aminobutyric acid (GABA; Emir, Tuite, & Öz, 2012). Tools for automatic basic quantification of metabolite ratios exist on current scanners, but widely used tools such as the linear combination of model spectra (LCModel) provide metabolite quantification with quantitative error estimates (Provencher, 2001). The COV of metabolite levels in human volunteers has been measured in test-retest studies at 1.5, 3, and 7T. Using SVS at 3T, COV is ≤6% for total n-acetylaspartate (tNAA), total creatinine (tCr, including creatine and phosphocreatine), total choline-containing compounds (tCho, including glycerophosphocholine, phosphocholine, and choline), myoinositol (mI or mIns), and glutamate (Glu). For MRSI at 3T, COV is <10% for tNAA, tCho, tCr, and mI (Öz et al., 2014). Knowledge of the COVs for acquisition and the quantification errors allows for high-quality measurements in clinical trial data.

   A 2014 consensus statement confirmed the readiness of ^{1}H-MRS on current MR scanners for clinical trial and clinical practice use and provides guidelines for calibration schemes, acquisition protocols for high-quality artifact-free spectra, analysis methodology, and recommended quality control of spectra to ensure technical adequacy (Öz et al., 2014). Recent work with a more advanced MRS acquisition protocol sLASER has shown improved intra- and intersite reproducibility at 3 and 7T (Deelchand et al., 2015; Van de Bank et al., 2015). Additionally, for the five major metabolites listed earlier, the test-retest COV at 3 and 7T improves to ≤5% for spectra averaged over 5-min acquisition (Terpstra et al., 2016).

Deelchand et al. (2017) has compared sLASER and PRESS in the same session and scanner with different MR technologists and showed that the translation into the clinical setting is feasible. Optimal implementation in multicenter trials will benefit from further automation and standardization across platforms. Examples of ^{1}H-MRS are provided in Section III.G.

MRS can be performed with nonradioactive nuclei other than ^{1}H including fluorine 19 (^{19}F), phosphorus 31(^{31}P), and carbon 13 (^{13}C) (Mason & Krystal, 2006). Additional scanner hardware, however, is required for multinuclear MRS acquisitions. ^{19}F is the naturally occurring isotope of fluorine in nature so MRS can be used to determine drug concentration and kinetics in the brain as has been done for fluoxetine, for example (Henry et al., 2000). ^{31}P MRS can measure high-energy phosphate levels (including phosphocreatine (PCr), inorganic phosphate, and nucleoside triphosphates) along with phospholipid-associated phosphomono- and diesters. These spectra contain information related to tissue bioenergetics, pH, and phospholipid metabolism. ^{13}C MRS is used to study metabolic fluxes in animal and human brain and has been extensively reviewed by Rothman, De Feyter, de Graaf, Mason, and Behar (2011). Glutamate, glutamine, and GABA are rapidly synthesized and can be monitored by incorporating glucose labeled with ^{13}C at C1 to assess the rates of glucose utilization in the tricarboxylic acid (TCA) cycle and glutamate-glutamine neurotransmitter cycling. Labeling glucose at C2 with ^{13}C allows estimates of the astrocyte TCA cycle flow compared with the rate of glutamine synthesis. Examples of ^{31}P and ^{13}C MRS in drug development are provided in Section III.G.

8. Scanner field strengths: The main magnetic field in an MR scanner is characterized by its field strength in Tesla (T). Clinical scanners range from 0.2T to 7T. The most common field strength for neuroapplications is currently 3T, and the more advanced techniques described earlier should be performed on a 3T (with the caveat that 7T scanners are becoming more common, and some high-resolution structural and MRS scans are better at 7T).

In summary, MR techniques provide 2-D multiplanar and 3-D structural information; exquisite endogenous tissue contrast; contrast enhancement with exogenous contrast agents; and many functional, physiological, and biochemical measures. The additional power of MRI/MRS techniques is that they are multiparametric so that imaging is cost-effective and less inconvenient for patients. Within a single scanning session (an hour or less), one can obtain multiple biologically unique measurements of brain pathology/pathophysiology to inform the effects of a potential therapeutic in drug development. In the sections to follow, examples of current uses and future applications of MRI/MRS in drug development are given.

B Structural MRI for Brain Atrophy

Brain atrophy is a feature of many neurodegenerative diseases and is analyzed with MRI using 3DT1-type sequences. Typical sequences are a magnetization-prepared rapid acquisition with gradient echo (MP-RAGE) or an inversion recovery-spoiled fast gradient echo (IR-SPGR). These sequences provide superior gray-/white-matter contrast and spatial resolution in the brain. The Alzheimer's Disease Neuroimaging Initiative (ADNI, 2017) has harmonized the 3DT1 imaging acquisition at 3T across vendors and scanner versions for use in multicenter clinical trials. Details of the acquisition are available on the ADNI website: (https://adni.loni.usc.edu/methods/documents/mri-protocols/). Once the images are acquired, they are corrected automatically on the scanner platform for artifacts such as image intensity nonuniformity and image distortions arising from nonlinearities in the gradient coils. The images are then put through an imaging processing pipeline with the purpose of segmenting the intracranial contents from the skull and segmenting intracranial content into white matter (WM), gray matter (GM), and cerebrospinal fluid (CSF). The images must also be registered, either to a common template, to multiple templates, or for longitudinal measurements to a baseline image of the same individual. Use of a reference space allows the quantification of differences between individuals or groups at a single time point or quantification of tissue/CSF volume change over time for a single individual. A wide variety of morphometric methods have been applied ranging from manual segmentation, automated region of interest (ROI) analyses, surface mapping defined by tissue boundaries (such as boundary shift integral (BSI) and SIENA techniques), voxel-based morphometry (VBM) techniques, and tensor-based morphometry (TBM) techniques (Ashburner & Friston, 2000; Cash et al., 2015; Cover et al., 2011; Fischl, 2012; Fox & Freeborough, 1997; Hua et al., 2013;

McEvoy et al., 2009; Wolz et al., 2010) In VBM, a voxel-level statistical analysis of the entire brain tissue is performed between groups of subjects or longitudinally without any assumptions about which structures should be assessed. TBM is an image analysis technique that measures brain structural differences from the gradients of deformation fields that align one image to another. Since volumes of intracranial structures are proportional to the total intracranial volume, whole-brain and substructure volumes are usually normalized to total intracranial volume (TIV) (Malone et al., 2015).

1 MRI of Brain Atrophy in AD

The extensive work with brain volumetrics in AD illustrates the use and potential of MRI volumetric approaches in clinical trials. Brain atrophy measures derived from MRI have been used in three general ways: as an aid to diagnosis and classification, as a prognostic tool for likelihood of progression and patient enrichment, and as a longitudinal marker of disease progression.

Brain atrophy by MRI is part of the recent National Institute of Aging-Alzheimer's Association (NIA-AA) diagnostic criteria for MCI/selection criteria for MCI trials and one of the biomarkers for neurodegeneration/neuronal damage in the recent proposal by Jack et al. (2016) to use amyloid/tau/neurodegeneration (A/T/N) as an unbiased classification scheme for AD (see Section II.I.1) (Albert et al., 2011; Jack et al., 2016; Sperling, Aisen, et al., 2011). The atrophy pattern in AD is characterized by early medial temporal lobe and temporoparietal association cortex atrophy; hippocampal volume (HCV) and entorhinal cortex thickness have been used as common measures of AD-like atrophy. In contrast, FTD and LBD can show differences in atrophy patterns that aid in differentiating those dementias from AD although the overlap in areas of atrophy is substantial. For example, patients with behavioral variant FTD have bilateral frontal atrophy, while those with FLD and motor neuron disease have more paracentral atrophy. The focal atrophy in DLB showed little cortical involvement with most of the GM atrophy occurring in the dorsal midbrain, substantia nigra, and hypothalamus and a relative sparing of hippocampal and temporoparietal cortices, whereas atrophy of the latter is prominent in AD (Whitwell et al., 2007; Whitwell, Jack, Senjem, & Josephs, 2006).

A second use of MRI volumetrics is as a prognostic biomarker for AD disease progression. For clinical trials in prodromal/predementia AD where there is mild cognitive impairment, a major issue is how to enrich for subjects likely to progress within the time frame of a typical interventional disease-modification trial since clinical criteria alone are not sufficient. Decreased HCV) is a well-documented biomarker of disease progression. Changes in HCV correlate with Braak NFT progression; occur in prodromal phases of the disease; are most rapid around

the time of onset of dementia; and are functionally related to one of the early symptoms of the disease, memory impairment (Jack et al., 2013; Jack, 2011; Jedynak et al., 2012). Importantly, the reproducibility of HCV measurements is very good (test-retest of $<3\%$) within and across measurements at 1.5 and 3T, assuming the same automated image analysis algorithm is used for all subjects (Cavedo et al., 2017; Wolz et al., 2014). This technical performance in HCV determination along with standardized 3DT1 acquisitions is critical for using low HCV as an enrichment biomarker.

In 2011, the Coalition Against Major Diseases (CAMD), a consortium of the Critical Path Initiative focusing on the development of methods and tools to expedite development of therapeutics in AD and PD, submitted a dossier to the Scientific Advice Working Party (SAWP) of the Committee for Medicinal Products for Human Use (CHMP) of the European Medicines Agency (EMA) requesting a qualification opinion on the use of low HCV as a biomarker for in predementia/prodromal AD trials. Following responses to questions and discussion points raised by SAWP and a public comment period, a positive response opinion was adapted by the CHMP on November 11, 2011 (European Medicines Agency, 2011; Hill et al., 2014). As a part of the submitted dossier, CAMD presented the results of a literature review they had performed on longitudinal studies of at least 18 months of duration in predementia AD/MCI that had incorporated HCV measurements by MRI at baseline. Twenty-five of the 27 studies meeting the search criteria showed that subjects who converted to dementia had significantly smaller HCV at baseline compared with those who did not convert. CAMD also analyzed the results of four different HCV analysis algorithms that were applied to the 24-month longitudinal MRI data for ADNI MCI subjects. This analysis showed, with receiver-operating characteristic (ROC) curves, that the HCV quantification method did not impact the utility of predictive performance of HCV as an enrichment biomarker. The general process to follow in using HCV as an enrichment biomarker is illustrated in Fig. 9.7. The CHMP also stated that all steps (imaging acquisition, analysis, etc.) be standardized and follow international guidelines.

Yu et al. (2014) has provided an example of operationalizing HCV as an enrichment biomarker and choosing HCV cut points using ADNI data. Both sample size and trial costs were significantly reduced (by 40%–60% and 30%–40%, respectively) across a wide range of endpoints. The EMA/CHMP opinion on low HCV as am enrichment biomarker also pointed out that the concomitant assessment of two enrichment biomarkers would also be of great value. Of note, the CHMP has also given a positive opinion for CSF $A\beta_{1-42}$ biomarkers and amyloid PET as enrichment biomarkers for prodromal AD trials. Wolz et al. (2016) has recently described the effects of

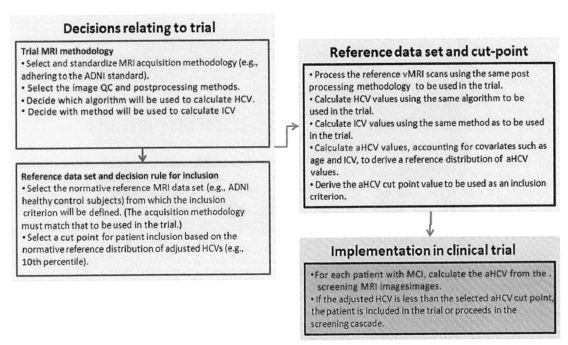

Decisions relating to trial

Trial MRI methodology
• Select and standardize MRI acquisition methodology (e.g., adhering to the ADNI standard).
• Select the image QC and postprocessing methods.
• Decide which algorithm will be used to calculate HCV.
• Decide with method will be used to calculate ICV

Reference data set and decision rule for inclusion
• Select the normative reference MRI data set (e.g., ADNI healthy control subjects) from which the inclusion criterion will be defined. (The acquisition methodology must match that to be used in the trial.)
• Select a cut point for patient inclusion based on the normative reference distribution of adjusted HCVs (e.g., 10th percentile).

Reference data set and cut-point
• Process the reference vMRI scans using the same post processing methodology to be used in the trial.
• Calculate HCV values using the same algorithm to be used in the trial.
• Calculate ICV values using the same method as to be used in the trial.
• Calculate aHCV values, accounting for covariates such as age and ICV, to derive a reference distribution of aHCV values.
• Derive the aHCV cut point value to be used as an inclusion criterion.

Implementation in clinical trial
• For each patient with MCI, calculate the aHCV from the screening MRI imagesimages.
• If the adjusted HCV is less than the selected aHCV cut point, the patient is included in the trial or proceeds in the screening cascade.

FIG. 9.7   Diagram illustrating an operational algorithm for use of hippocampal volume in a clinical trial. *With permission from Hill, D. L., Schwarz, A. J., Isaac, M., Pani, L., Vamvakas, S., Hemmings, R., Carrillo, M. C., et al. (2014). Coalition against major diseases/European Medicines Agency biomarker qualification of hippocampal volume for enrichment of clinical trials in predementia stages of Alzheimer's disease.* Alzheimer's & Dementia, 10, 421–429.

combining amyloid markers (CSF $A\beta_{1-42}$ above the designated cut point or amyloid PET positivity) and neurodegeneration markers (low HCV) using ADNI data sets. This operationalizes the A+/N+ criteria of the NIA-AA guidelines for predementia trials and substantially improves the power of predementia trials by identifying a more rapidly progressing population.

Volumetric MRI biomarkers may also be useful for subject selection in clinical trials enrolling presymptomatic AD subjects. The Dominantly Inherited Alzheimer Network (DIAN) is a multicenter observational study in which subjects are enrolled who are members of families known to carry mutations that cause autosomal dominant AD (ADAD). The mutations are nearly 100% penetrant, and approximately 50% of family members are carriers and at risk for ADAD. The age of onset of dementia is early compared with sporadic AD and within a family, the age of symptom onset is quite predictable. The serial MRIs performed in DIAN are providing data to determine when atrophy rates diverge from noncarriers, a so-called change point. (Kinnunen et al., 2017). Whole-brain, ventricular, and HCVs were determined at baseline, and volume changes in these structures were measured using BSI for noncarriers, presymptomatic carriers, and very mild and overtly symptomatic carriers. Nonlinear mixed-effect models (either assuming a single step change in atrophy rate to a new stable increased value or a gradually acceleration in atrophy rate after the change point) were used to estimate the timing of the change point relative to symptom onset and atrophy rates before and after the change

points. Both models showed that atrophy rates increased after the change point, and the "gradual acceleration" model showed that all evaluated regions underwent a change in atrophy rate before symptom onset. Such models of atrophy may be useful in predicting the time to clinical onset and in selecting subjects with autosomal dominant mutations for prevention trials.

If drug interventions are to be assessed in prevention trials in presymptomatic sporadic AD, the challenge will be how to identify subjects at risk for such trials, and newer volumetric MRI techniques may be helpful. In the presymptomatic stage, much less tissue damage has occurred so that arresting neurodegeneration at this stage is appealing for clinical outcome. The current methods for identifying subjects in the target population, however, may not be sensitive enough to detect more subtle disease. One approach is to use more advanced methods to assess the anatomical locations where histopathologic evidence suggests that the disease starts. Certain subfields in the hippocampus and adjacent medial temporal lobe su-regions such as entorhinal and perirhinal cortex first demonstrate the deposition of neurofibrillary tangles in the early stages of Alzheimer's disease (Braak stages 1–3) (Braak & Del Tredici, 2015). In order to examine structural changes in these regions, extremely high-spatial resolution scanning is required, perhaps up to five times the resolution of the existing 3D T1-weighted volumetric scans. In addition, the inherent T1 tissue contrast is inadequate to provide accurate volumetric/segmentation analyses so a T2-weighted pulse sequence is needed.

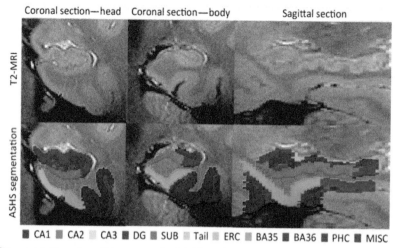

FIG. 9.8   High-resolution T2-weighted imaging of hippocampal subfields and medial temporal lobe subregions at 7T. Scans were acquired using a 2-D turbo spin-echo sequence at a resolution of 0.40 mm × 0.40 mm × 1.0 mm (slices). Some of these anatomical regions show the very earliest volume changes and are the first to form neurofibrillary tangles (NFTs) in Alzheimer's disease patients, including Brodmann's area 35 (BA35); Brodmann's area 36 (BA36); entorhinal cortex (ERC); and hippocampal subfields CA1, CA2, CA3, dentate gyrus (DG), and subiculum (SUB). *Images courtesy David Wolk, Penn Memory Center, University of Pennsylvania.*

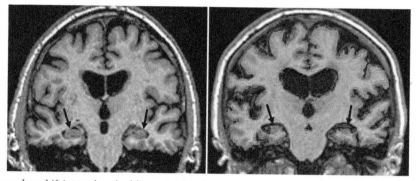

FIG. 9.9   Illustration of boundary shift integral method for measuring change in whole-brain volume. The images represent changes from a screening time point to a 1-year follow-up. The image on the left is the change in a normal control, and the one on the right is an AD subject. Pixels overlaid in red represent a transformation from brain to CSF (i.e., atrophy), while those labeled in green indicate a change in the opposite direction (and may be an indication of the variability of the method since new brain tissue is unlikely to occur). There is no change in the brain of the normal volunteer (*left*), while there is a significant loss of brain tissue in the AD subject (*right*). The arrows in the brain indicate the location of the hippocampus, and the amount of red pixel overlays in that region. *Images courtesy Nick Fox, Dementia Research Centre, London, UK.*

An example of a very high-resolution T2 scan is presented in Fig. 9.8. Yushkevich et al. (2015) has presented evidence at 3T that certain brain regions experience early, subtle atrophy in MCI compared with normal subjects (CA1 and Brodmann area 35 of the perirhinal cortex). Automated techniques are also being developed for segmentation of the entorhinal cortex and hippocampal subfields at 7T where spatial resolution is higher than at 3T (Wisse et al., 2016). Although atrophy in the entorhinal cortex and hippocampal subfields and the relationship between atrophy and the onset of symptoms/cognitive decline are active areas of research in both academic and consortia efforts, high-resolution substructure segmentation and associated atrophy metrics of volume change and thickness have the potential to play a role in subject selection for prevention trials in sporadic AD.

A third way that brain atrophy by MRI has been used in AD clinical trials is as longitudinal biomarker of disease progression, the concept being that an effective treatment would lead to a slowing of neuronal degeneration with less volume loss and a slowing of atrophy compared with a placebo control arm. Fig. 9.9 illustrates the automated BSI method introduced by Fox and Freeborough (1997) to assess changes in whole-brain volume and ventricular volume. This method defines the change in the boundary between brain tissue and CSF and then computes difference in the tissue and fluid compartments. As discussed at the beginning of Section III.B, numerous other methodologies have been used, from manual tracing to volumetric TBM approaches. Because of the precision of longitudinal MRI volume change measurements compared with clinical measures, the sample size for a clinical trial using a change in a volumetric MRI atrophy metric versus a clinical assessment as an outcome measure is markedly reduced; the actual value depends on the type of atrophy measure, the trial design, and statistical methods (Jack, 2011). For example, advanced volumetric TBM methods require only 56 (44, 64) subjects with MCI to detect a 25% slowing of atrophy with 80% power and 95% confidence (Gutman et al., 2015).

Although atrophy of the whole brain or hippocampus and ventricular enlargement metrics have been incorporated as outcome measures into trials with putative disease-modifying agents, no disease-modifying agents have been effective to date. While atrophy by MRI is a robust biomarker for the progression of the disease and correlates with cognitive functional decline in untreated patients, confirmation that brain atrophy metrics are good biomarkers for monitoring the efficacy of pharmacological interventions AD and MCI clinically awaits data on their use with successful disease-modifying agents. The brain atrophy results in the trial that assessed immunization with aggregated human $A\beta_{1-42}$ (AN1792) were puzzling. The trial was halted because of meningoencephalitis after 300 subjects had received at least one dose of ANI792. Not only was the atrophy hypothesis not confirmed, but also instead antibody responders had greater reductions in brain BSI and increases in ventricular BSI compared with placebo subjects. This paradox of accelerated atrophy in responders remains unexplained; reductions in amyloid burden and reduction of inflammation associated with the disease itself have been proposed as potential reasons for the paradoxical results (Fox et al., 2005).

C MRI as a Safety Biomarker in Antiamyloid AD Therapeutics

Aside from the use of standard MRI sequences during enrollment to ensure that subjects do not have exclusionary brain pathology such as tumor or stroke, MR imaging has added major value to the safety monitoring of novel therapeutics in AD. In trials of amyloid-modifying agents, both immunologic and small molecule-based, vasogenic edema and/or the development of new microhemorrhages were observed as early as a few weeks after dosing. These effects were described in detail by Sperling et al. (2011) and designated with a specific name, ARIA, an acronym for *a*myloid-*r*elated *i*maging *a*bnormalities, differentiating this observation from other vasogenic edema and microhemorrhage etiologies. Two forms of the phenomena are defined and seem to share some common pathophysiology mechanisms related to increased permeability of brain and leptomeningeal vessels with leakage of either serum proteins (ARIA-E (for edema)) or red blood cells (called ARIA-H for the hemosiderin residual of microhemorrhage). The appearance of ARIA-E is usually transient and seems to occur early in the treatment regimen. While there are often no clinical symptoms and even on the MRI images the ARIA may resolve fairly quickly, the impact that ARIA imposes on clinical trial design can be very significant. To reduce the risk of initiating an ARIA event, dose may be reduced or the dosing regimen modified. This dose reduction may impact the efficacy of the compound during the trial. ARIA findings have led to increased safety monitoring with MRI in trials of amyloid-modifying agents. A T2 FLAIR sequence is used to demonstrate the ARIA-E parenchymal vasogenic edema/sulcal effusions as high signal intensity (Fig. 9.10), while a T2* GRE sequence shows the hemosiderin deposits from microhemorrhage and superficial siderosis as signal voids (Fig. 9.11). The Alzheimer's Association Research Roundtable Working Group provided recommendations endorsed by the FDA for minimum standards for MRI acquisitions and for the frequency of scans that is to be based on the dosing frequency, pharmacokinetics, and exposure duration of the drug. Given the possibility that the number of

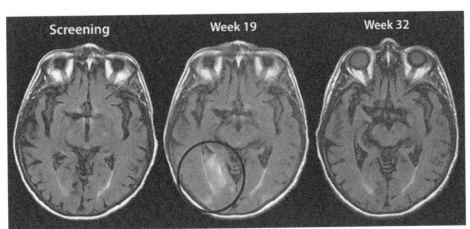

FIG. 9.10 ARIA-E. FLAIR MRI images obtained at screening and after 19 and 32 weeks of bapineuzumab treatment show the development of ARIA-E on the week 19 scan *(red circle)* and resolution of ARIA-E on the 32-week scan. *Images courtesy Janssen R&D, adapted from "Guidelines for MRI Interpretation in Bapineuzumab Clinical Trials," p. 12.*

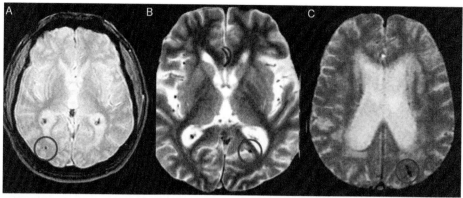

FIG. 9.11   ARIA-H. Gradient-echo (GRE)/T2* MRI images illustrating examples of microhemorrhages or small hemosiderin deposits (<10 mm in diameter) in left and middle images *(red circles)*. Larger linear hemorrhages (≥10 mm in diameter) are depicted in the right hand image *(red circle)*. The T2* images are very sensitive to the distortion of the local magnetic fields that are created by the iron in the deposited hemosiderin and the distortion creates voids in the image. *Images courtesy Janssen R&D, adapted from "Guidelines for MRI Interpretation in Bapineuzumab Clinical Trials," pp. 9–11.*

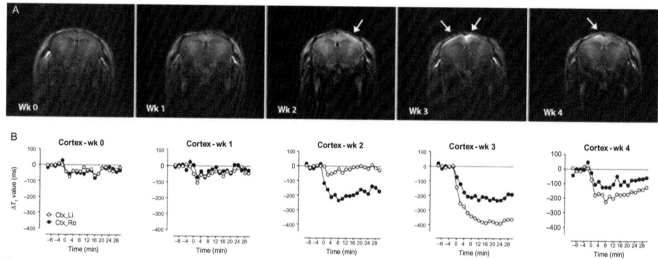

FIG. 9.12   Dynamic T1-w brain MRIs in PDAPP mice showing gadolinium leakage following antiamyloid treatment. (A) T1-weighted images following gadolinium-DOTA administration for pretreatment and posttreatment images and (B) time course curves of calculated T1 values versus time for pretreatment (week 0) and weeks 1–4, post antiamyloid treatment in transgenic PDAPP mice. PDAPP mice are engineered to deposit large amounts of amyloid plaque. T1 values were derived from T1 maps of cortical ROIs (○: left and ●: right). BBB leakage events occurred at weeks 2, 3, and 4 and can be seen on T1-weighted images as a signal enhancement in superficial cortical areas *(white arrows)*. After calculation of the T1 values, leakages into the brain *(white arrows in (A))* are seen as a drop in T1 relaxation values (ms) in (B). The time courses of the individual cortical ROIs (weeks 0–4) illustrate the asymmetry and support the effects seen on the images. The nonoverlapping enhancement patterns in weeks 2, 3, and 4 imply independent events, suggesting that the initiation, evolution, and resolution of a BBB event occurred ≤1 week. *Images and graphs reprinted from Journal of Alzheimer's Disease, vol. 54, no. 2, Ines Blockx, et al., "Monitoring blood-brain barrier integrity following amyloid-immunotherapy using gadolinium-enhanced MRI in a PDAPP mouse model," pp. 723–35, 2016, with permission from IOS Press (2017, Fig. 6AB, p. 730).*

baseline microhemorrhages may reflect the severity of underlying vessel amyloid angiopathy and increase the risk of ARIA-E, the Working Group also recommended that subjects with more than four microhemorrhages be excluded from trials with amyloid-modifying agents (Sperling, Jack, et al., 2011).

Numerous attempts to understand the underlying causes of ARIA were made using transgenic animal models that expressed amyloid pathology. Zago et al. (2013) were able to analyze vascular wall structure from excised sections of PDAPP mouse brains following treatment with an antiamyloid compound. These findings

demonstrated physical holes generated in the vessel lumen immediately after treatment that were quickly repaired. Essentially the blood-brain barrier was compromised by the removal of imbedded vascular amyloid, but these wall defects were quickly repaired, suggesting an explanation for the early development and transient nature observed in ARIA. As a follow-up to this work, Blockx et al. (2016) applied dynamic contrast-enhanced (gadolinium) T1-weighted scans to the same PDAPP model using the same antiamyloid treatment. Numerous, transient gadolinium leakage events were observed using this method (Fig. 9.12). Given that the leaks seem to occur randomly and resolve in a matter of days, clinical translation of this technique would be difficult since MRI scans are typically performed about 12 weeks apart. However, this model may be very useful in the development of future clinical protocols by helping to define treatment dosing strategies that reduce the risk of developing ARIA. It may also be possible to define an individual's ARIA risk using DCE-MRI methods.

D fMRI Tools for Assessing the Pharmacodynamic Effects of Potential Therapeutics

While PET occupancy studies are a powerful method for directly demonstrating target engagement and for providing dose guidance, the demonstration of downstream effects of target engagement with imaging can be valuable in several ways. For example, if a PET ligand is unavailable, PD measurements can provide proof of brain penetration and can provide dose guidance when performed as a function of dose. PD imaging effects can also provide evidence in both animal models and humans for the hypothesized mechanism and confirmation of its translatability from mouse to man. Additionally, if the rationale underlying the development of a drug candidate is based on a clear hypothesis relating the action of the drug to a specific biological effect, measurement of alterations in brain function with well-characterized imaging may provide support for the central biological action. Sections II.H and II.I provided some examples of how PET imaging measurements and assessments provided PD evidence on biological effects that link the mechanism of action/biological effect to preclinical data and thus provide a translational bridge between preclinical data and a clinical response. Within the MRI tool box, fMRI techniques with BOLD imaging have the potential to provide some useful PD assessments of how drug candidates may alter brain biological activity and potentially can also inform patient selection criteria. This section will describe some examples of these fMRI applications. These applications involve three types of approaches that in principle allow the effect of a drug on brain circuitry to be assessed in normal and/or diseased brains.

(a) Drug-induced change in the BOLD resting-state pattern of the brain without a task or stimulus, so-called pharmacological MRI (phMRI).
(b) Task-based fMRI: the effects of a stimulus such as sensory (heat-induced pain, auditory or visual), emotional (pictures of sad or happy faces), or cognitive tasks (working memory and executive function tasks) on the BOLD signal are measured and also compared with the effect of a drug on task performance.
(c) Resting-state or task-free functional connectivity (fc): analysis of the rs-fMRI BOLD signals that reflect low level spontaneous fluctuations in brain activity allows the identification of brain pathways that are temporally connected together and synchronously fire together, so-called resting-state networks (RSNs). The recent development of multiband accelerated echoplanar imaging in which multiple sections are acquired simultaneously allows brain coverage in 0.5–1 s (versus 2–3 s for single-slice acquisitions) and isotropic resolutions of 2 mm at 3T (Smith et al., 2013). A 10-min acquisition provides satisfactory data quality. The rs-fMRI data have commonly been analyzed by seed-based approaches in which the signal in a seed region is correlated with that in all other brain voxels and by independent component analysis (ICA) in which all voxels are considered at once, and a mathematical algorithm is used to separate the data set into networks that are temporally correlated in their BOLD signal but maximally independent in the spatial domain. More recently, graph theory analyses are being applied to rs-fMRI data, an approach in which the brain is treated as a single interrelated network (Bullmore & Sporns, 2009; Smith et al., 2011). At least 10 functional connectivity networks/resting-state networks have been consistently and commonly detected in humans (Pievani, Filippini, van den Heuvel, Cappa, & Frisoni, 2014). They include the default mode network (DMN) (precuneus, posterior cingulate, bilateral inferior-lateral-parietal, and ventromedial frontal cortices) that is deactivated during specific task behaviors and is involved in memory and other cognitive functions; the salience network (dorsal anterior cingulate, bilateral orbitofrontal, and insular cortices) integrates limbic and autonomic information and relates it to social rules; the executive network (bilateral dorsolateral and medial prefrontal cortices) that is involved in executive control, working memory, action inhibition, and emotion. The test-retest reproducibility of rs-fMRI data is variable within and across the RSNs but is good for several (including DMN, frontoparietal, salience, and visual RSNs) if careful attention to standardization of acquisition, analysis, and testing conditions (Barkhof, Haller, & Rombouts, 2014). Functional connectivity networks /circuits can be altered in disease states, and the effects of drugs on these networks can be assessed. The advantage of rs-fMRI over task-based MRI is that it is simpler and faster to implement and thus applicable to patients who cannot perform a task, can be used to look at multiple cortical systems and their interactions instead of the single system activated by a given task, and avoids the confounds of using tasks (task performance variability and practice effects).

An example of the use of phMRI using BOLD imaging is provided in the area of analgesics by a study in which two drugs that had been developed as potential analgesics were each assessed in 12 healthy volunteers: one drug (buprenorphine, a μ-opioid partial agonist and kappa-opioid antagonist) was efficacious for pain, and the other (aprepitant, an NK-1 antagonist) had failed as an analgesic in clinical trials. phMRI showed direct effects of each drug on the pretreatment resting BOLD signal consistent with the known target receptor distribution of each drug. However, only buprenorphine showed significant modulation of the functional connectivity CNS circuitry relevant to pain processing, and the changes in these regions were also dependent on the plasma concentration of buprenorphine. The effect of each acute drug treatment on a task-based fMRI BOLD signal was also assessed using a noxious heat stimulus. Buprenorphine but not aprepitant attenuated the evoked BOLD response to acute pain in CNS structures where the sensation to pain is mediated (Upadhyay et al., 2011).

A variation on the phMRI approach is to use a known compound to provoke a strong CNS response and then assess how that response is modulated by pretreatment with a test compound. N-methyl-D-aspartate (NMDA) receptor antagonists such as ketamine at a subanesthetic dose level induce psychomimetic symptoms similar to schizophrenia (SCZ) and have been used to investigate the role of glutaminergic dysfunction in humans; both ketamine and phencyclidine, another NMDA receptor antagonist, have been used in rodent models of schizophrenia (Chin et al., 2011; Deakin et al., 2008; Gozzi et al., 2008; Långsjö et al., 2003; Littlewood et al., 2006). Subanesthetic doses of ketamine induce robust BOLD fMRI changes with good reliability in human brain regions that are consistent with changes seen in the rat (De Simoni et al., 2013; Deakin et al., 2008). Using pretreatment with two mechanistically different agents, lamotrigine and risperidone, both of which are hypothesized to reduce ketamine-induced glutamate release, the attenuation of the ketamine phMRI BOLD response was quantified and shown to be reduced by each drug compared with placebo in human volunteers (Doyle et al., 2013). This study also suggested that serotonergic mechanisms play a role in ketamine-induced subgenual (sg) anterior cingulate cortex (ACC) changes, a finding that is potentially relevant for the antidepressant effects of ketamine. Joules et al. (2015) also evaluated the functional connectivity effects of ketamine and the modulation of those effects by lamotrigine and risperidone. Ketamine altered the functional connectivity pattern of the brain, an effect that was significantly modulated by risperidone but much less so by lamotrigine. Using this acute ketamine phMRI imaging assay to assess the modulation by novel drugs of both the magnitude of the BOLD signal and the connectivity pattern both preclinically and clinically provides powerful and robust translational imaging biomarkers to assess the mechanistic actions of novel compounds in psychiatric disorders and may also be used to provide dose guidance.

Task-based BOLD fMRI has been used to identify abnormalities in brain circuit function in disease states as elicited by relevant tasks and to probe the effects of drugs on those circuits and disease symptoms. In a metaanalysis of 44 task-based fMRI studies, healthy adults and SCZ patients were shown to activate a similar cortical control network using executive cognition tasks such as N-back, sequence recall and Stroop tasks. SCZ subjects, however, showed reduced activity in the dorsolateral prefrontal cortex (DLPFC), the ACC, and the thalamic mediodorsal nucleus with increases in activity in other PFC areas (Minzenberg, Laird, Thelen, Carter, & Glahn, 2009). In depression, task-based fMRI, most commonly with emotional faces, has shown the activation of the ACC (particularly the rostral and sgACC) and the amygdala. Task-based fMRI biomarkers have been used to predict drug and behavioral intervention response in mood and anxiety disorders. In some cases, the deactivation of the sgACC to negative stimuli has predicted worse response to antidepressants, whereas, with behavioral modification, the deactivation of the sgACC to negative stimuli predicted better response. Increased activation of the amygdala during a negative information processing task has predicted response to drugs and behavioral therapy. In these studies, robust activation is seen across groups and distinguishes responder and nonresponder groups (Frodl et al., 2011; Keedwell et al., 2009; Siegle, Thompson, Carter, Steinhauer, & Thase, 2007). As pointed out by Fu, Steiner, and Costafreda (2013), an optimal clinically relevant prognostic biomarker requires good precision across time at the individual level to be used for individual patient selection and outcome assessments. Nord, Gray, Charpentier, Robinson, and Roiser (2017) has looked in healthy volunteers at the test-retest performance of sgACC and amygdala activation (and the right fusiform face (FFA) area as a control region) by responses to emotional faces using three different tasks and two scanning session separated by 9–21 days. Each task was performed twice during a scanning session. Of concern, the study showed robust group activations of the amygdala and sgACC but within subject reliability was (between day and within day) was poor for both regions, while the right FFA control region showed excellent reliability. Further research (including assessment in patients, use of other tasks, and habituation effects) is needed to optimize choices of prognostic and response biomarkers with task-based fMRI.

Analyses of normal control versus disease group differences in rs-fMRI BOLD signals have demonstrated alterations in functional connectivity (fc) in RSNs in numerous neurologic and psychiatric diseases (Barkhof et al., 2014; Fox & Greicius, 2010; Greicius, 2008; Pievani et al., 2014). The consistency of network abnormalities depends on the particular disease state with AD studies (AD, MCI, and amyloid-positive subjects) and behavioral variant (bv) FTD studies showing the most consistency and correlation with disease severity, whereas schizophrenia studies have shown inconsistent results (Barkhof et al., 2014; Pievani et al., 2014). Although still a developing area of active research, rs-fc-fMRI measures have the potential to aid in the identification of early disease markers and to add to our understanding of the pathophysiology of brain diseases in terms of neuronal networks and their disintegration. An example in the AD area comes from the work of Jones and colleagues (Jones et al., 2016; Wiepert et al., 2017). Using clinical phenotyping, metadata and multimodality imaging (including rs-fc-fMRI) from 128 ADNI 2 subjects and a system-based conceptualization of AD, they investigated the pattern of DMN subsystem connectivity changes across the AD disease spectrum. They showed that the posterior DMN fails before amyloid plaques are measurable with PET, that this failure appears to initiate a connectivity cascade that continues across the entire disease spectrum, and that there is high connectivity between the posterior DMN and hubs of high connectivity (many of these being in the frontal lobe) that is associated with amyloid accumulation. They propose a cascading network failure model in which a connectivity overload precedes structural and functional decline and suggest that, rather than being a positive compensatory process, high connectivity is a load-shifting process that is transiently compensatory. They acknowledge that it is unknown whether system-level pathophysiology drives downstream molecular events related to synaptic function (Jones et al., 2016). In a subsequent paper (Wiepert et al., 2017), they proceed to optimize the network failure quotient (NFQ), a candidate biomarker of large-scale network failure in AD, using rs-fMRI data from ADNI on 43 clinically normal and 28 AD subjects. The optimized NFQ (oNFQ) is a summary metric of network failure and is ratio of increases in connectivity to decreases in connectivity that was then validated on a similar sample of AD and cognitively normal subjects from the Mayo Clinic. They then showed a correlation of the oNFQ to other biomarkers of AD severity (tau PET, amyloid PET, cortical thickness FDG PET, and auditory verbal learning test scores using 218 Mayo subjects). While not currently suitable for discriminating clinical diagnostic categories, a biomarker of the large-scale effects of AD pathophysiology such as the NFQ has the potential to provide more detailed descriptions of the AD disease process and attempted interventions. Since no other biomarker examined was associated with preclinical/presymptomatic network changes once age was controlled for, the authors suggest further development of NFQ as a biomarker tool to detect some of the earliest manifestations of AD pathophysiology in the preclinical disease phase of AD in prevention trials.

E MR Measures of Brain Perfusion and BBB Breakdown in Neurovascular Dysfunction

Brain perfusion can be assessed by two types of MR techniques, dynamic susceptibility contrast (DSC) MRI and ASL-MRI. Breakdown of the BBB can be evaluated using DCE-MRI. Applications of these techniques are discussed in this section.

1 DSC- and DCE-MRI

In addition to DSC-MR perfusion measures of CBV being used as common metrics for assessing perfusion and the effects of drugs on brain tumors, DSC (along with ASL, see Section III.E.2) has been used for assessing hypoperfusion/reduced CBF in AD, other neurodegenerative diseases, stroke, and MS (Inglese et al., 2007; Montagne et al., 2016). Recently, Ostergaard and colleagues have developed a statistical methods using DSC-MRI data to determine capillary time heterogeneity (CTH), a metric that permits assessment of tissue with impaired oxygen delivery in disorders such as stroke and AD (Mouridsen et al., 2014). In a single-center study, they have used the DSC-MRI with advanced perfusion analysis to show, in AD subjects, that poor cognitive performance and regional cortical thinning correlated with lower CBF and CBV, with higher MTT and CTH and with low oxygen tension across the cortex. Cognitive decline over time was associated with diminishing microvascular cortical perfusion and with disruptions of cortical microvascular flow patterns, as shown by increasing whole-brain relative transit time heterogeneity (RTH) where RTH = CTH/MTT. They suggest the deterioration of microvascular hemodynamics may precipitate hypoxia that in turn promotes amyloid deposition (Nielsen et al., 2017).

DCE-MRI has been used to show BBB breakdown during normal aging with worsening in the hippocampus and cortical and subcortical regions in subjects with early AD and vascular cognitive impairment (Montagne et al., 2016). The DCE-MRI changes occur prior to changes in HCV. Dynamic T1-weighed contrast-enhanced MRI has also provided insight into the effect of antiamyloid agents on the BBB in animal models (Section III.C; Blockx et al., 2016)

Given the active research into the mechanisms by which the brain neurovascular unit (vascular cells [endothelial cells, pericytes, and vascular smooth muscle cells], glia [astrocytes, microglia, and oligodendrocytes], and neurons) contributes to CBF control and to neurovascular dysfunction in neurodegenerative disease, the advanced perfusion metrics obtained from DSC-MRI (and ASL metrics; see Section III.E.2) and BBB metrics from DCE-MRI may emerge as detection biomarkers of early AD and as a means to assess the effects of interventions on the neurovascular aspects of neurodegenerative diseases (Kisler, Nelson, Montagne, & Zlokovic, 2017; Montagne et al., 2016). More data from longitudinal studies are needed to further understand the temporal relationship of these imaging biomarkers of neurovascular dysfunction to structural and functional connectivity, the molecular pathology of neurodegeneration, and cognitive function.

2 ASL-MRI

ASL-MRI has improved signal to noise on the now widely available 3T MRI scanners as a result of the higher field strength, improved pulse sequences, and multichannel head coils. As a consequence, ASL is increasingly being used to assess CBF in neurologic and psychiatric indications (Haller et al., 2016; Wolk & Detre, 2012). A few examples will be discussed here.

It has been suggested that ASL could replace FDG PET as a marker of neuronal damage given the tight coupling of perfusion and neuronal metabolism. Significant overlap has been shown between the areas of hypometabolism on FDG PET and the areas of hypoperfusion using ASL-generated CBF in AD and MCI patients. Some studies show discordance in the medial temporal lobe in MCI and early AD with hyperperfusion/increased CBD in the MTL structures compared with hypoperfusion reported in FDG PET (Wolk & Detre, 2012). Ongoing longitudinal data from studies such as ADNI where FDG PET, ASL, BOLD fMRI, and Tau PET are being performed across the AD spectrum may provide insight into the differences seen to date.

A recent single-center study used CBF from ASL, apoprotein E genotype, and amyloid PET to compare 27 cognitively normal with 16 amnestic MCI subjects to evaluate the effects of ApoE on brain perfusion in amyloid-positive MCI subjects. Global CBF was lower in E4+ carriers than in E4− carriers across all subjects and within the cognitively normal. Regional analysis showed more widespread reduced CBF in the E4+ versus E4− group compared with the amyloid + versus amyloid − group (Michels et al., 2016). The data suggest apoE genotype may exert some of its influence on CBF independent of Abeta deposition, a possibility supported by the emerging data on the effects of ApoE on the cerebrovascular system (Zlokovic, 2013). Certainly, clinical studies utilizing brain perfusion as a disease biomarker in the AD spectrum should control for ApoE status.

A randomized controlled double-blind crossover single-center Phase 2 trial designed to assess the effect of single-dose tadalafil on increasing CBF in white matter in small-vessel disease (PASTIS) is utilizing ASL (Pauls et al., 2017).

In assessing the effects of psychiatric drugs, ASL measures the actual perfusion effects in an unambiguous way compared with BOLD fMRI (see Section III.A). In three-way crossover study in healthy volunteers assessing two drugs used for attention-deficit/hyperactivity disorder versus placebo, multiclass recognition analysis of whole-brain ASL data accurately discriminated all three drug conditions from each other and showed differential effects of the two drugs in striatocerebellar circuits, the thalamus, and the midbrain/substantia nigra (Marquand et al., 2012). This approach, applied early in drug development, can provide a relative comparison of drug-induced brain activities, show the effects of potential drug combinations, and provide pharmacodynamic information on an individual basis about the brain networks that drug candidates may be modulating. Compared with BOLD imaging, ASL perfusion has the advantage that it can be better used to assess drug effects across multiple time points given the complex nature of the task-based BOLD signal and the better repeatability of the ASL quantification (Marquand et al., 2012).

In acute ischemic stroke (AIS), imaging plays a key role in the selection and monitoring of outcomes for drug and/or device-based endovascular interventions/revascularizations in both clinical trials and clinical practice. Both CT and MR techniques are used, with CT perfusion assessment of the infarct core and penumbra being most commonly used in AIS trials (Vilela & Rowley, 2017). ADC maps derived from DWI-MRI (Section III.A) show reduced ADC values for infarcted tissue and are the most sensitive imaging method for assessing the ischemic core within the first 6 h post AIS. DSC-MRI can be used for assessing hypoperfusion volume and the possible presence of a penumbra, a rim of viable tissue surrounding the infarcted core with decreased CBF and a target of revascularization. Assessment of intracranial collateral flow is another important component for patient selection and outcomes and can be performed with CT angiography (CTA), MR angiography (MRA), ASL, or digital angiography. The improved ASL whole-brain imaging techniques have allowed ASL to be validated for measuring perfusion in AIS, and ASL metrics are concordant with DSC-MRI metrics (Haller et al., 2016; Harston et al., 2017). ASL has the advantage of being noncontrast technique easily combined with DWI and T2-weighted FLAIR sequences and MRA for patient selection and easily repeated longitudinally for outcome measures (Harston et al., 2017). ASL has the unique noninvasive capability of vessel-selective ASL that is more sensitive for assessing perfusion changes with collateral flow (Haller et al., 2017). Further details on the use of ASL, DWI, DSC, MRA, and CT techniques in AIS are described in Haller et al. (2016), Harston et al. (2017), and Vilela and Rowley (2017).

F Role of MRI Techniques in Multiple Sclerosis Therapeutics

MRI assessments have played a major role in the diagnosis of MS, in the understanding of MS pathophysiology, and in the development of MS therapeutics. In the latter, MR readouts have frequently served as key endpoints. Both the conventional MR assessments used in these trials and some of the nonconventional MR imaging approaches being used as exploratory endpoints will be summarized in this section. They illustrate the multiparametric information content available with MRI techniques. PET probes of neuroinflammation in MS have been discussed earlier (section "PET Imaging of Translocator Protein (TSPO)"). MRS applications are discussed in Section III.G.

Standard brain imaging for MS includes fast spin-echo T2-w, PD, T2-w FLAIR, and pre- and postcontrast 3DT1-w sequences, in accord with MRI consensus guidelines (Filippi et al., 2017; Traboulsee et al., 2016; Wattjes et al., 2015). The postcontrast T1 imaging with gadolinium (Gd) chelates demonstrates bright areas of enhancement that represent acute active inflammatory MS lesions/plaques with a leaky BBB that allows for the visualization of the plaques on this imaging sequence (Section III.A). These so-called T1-enhancing lesions are the hallmark of RRMS but can been seen to a lesser extent in SPMS and PPMS. The enhancement is transient and lasts only a few weeks. On unenhanced T1 images, the lesions are usually hypointense. On T2, PD, and FLAIR sequences, the lesions appear hyperintense, representing increased water but not specifically distinguishing if this water is due to an increased amount of edema, inflammation, demyelination, or axonal loss. Residual T2 enhancement persists after the postcontrast T1 enhancement resolves. In clinical trials, MRI metrics commonly used to assess efficacy in RRMS are based on the macroscopic T1 and T2 properties of white-matter lesions and include the cumulative number or volume of new or enlarging lesions on T2 or the number or volume of newly enhancing T1 lesions (Montalban et al., 2017; Sormani & Bruzzi, 2013). Unfortunately, there is only a weak correlation between the T2 lesion load, T1-enhancing lesions, and the disability outcomes and progression of cognitive impairment in MS. These types of biomarkers therefore do not serve as a true surrogate endpoint in RRMS clinical trials. Such measures, however, have played a key role as primary endpoints in Phase 2 studies and as secondary endpoints in Phase 3 MS therapeutics trials.

Efforts have turned to more advanced imaging approaches to find imaging biomarkers that better correlate with disability progression and that provide better insight into the pathophysiologic aspects of MS in MS lesions and in GM and normal-appearing WM (Enzinger et al., 2015; Poloni et al., 2011; Sormani & Pardini, 2017). Brain atrophy metrics analyzed from 3DT1 whole-brain MRI sequences have been shown to correlate with the expanded disability status scale (EDSS) across all subtypes of MS (Popescu et al., 2013; Radue et al., 2015) and are currently the most robust imaging biomarkers for disease progression and disability. In the Phase 3 clinical trial of ocrelizumab versus placebo in PPMS (ORATORIO), percent change in brain volume from week 24 to week 120 was used as a secondary endpoint along with the change in the total volume of T2 lesions from baseline to week 120 (Montalban et al., 2017). Ocrelizumab met its primary endpoint of a significant reduction in EDSS progression with ocrelizumab compared with placebo; the percentage of brain volume was 0.90% with ocrelizumab versus 1.09% ($P = 0.02$) with placebo; the total volume of T2 lesions was reduced by 3.4% with ocrelizumab and increased by 7.4% with placebo (Fig. 9.13). Ocrelizumab is the first MS therapeutic to be FDA-approved for PPMS. Regional atrophy measures such as cortical gray-matter volume or deep gray-matter volume changes also correlate with disability and are also potential biomarkers for PPMS/SPMS trials. Several Phase 2 trials in PPMS and SPMS are using total brain volume changes. Two ongoing Phase 2 trials in progressive MS assessing potential neuroprotection drugs, SPRINT-MS evaluating ibudilast in both PPMS and SPMS and MS-SMART evaluating amiloride, riluzole, and fluoxetine in SPMS, have completed enrollment and are using brain atrophy metrics as primary outcome measures (Chataway, 2016; Fox et al., 2016). In addition to convention metrics for inflammation (number of new T1-enhancing lesions and/or new and enlarging T2 lesions), both trials have multiple secondary and exploratory endpoints evaluating metrics from more advanced imaging including DTI, MTR, volumetric MRI for cortical atrophy, optical coherence tomography (OCT) for retinal nerve fiber layer (RNFL) thickness, and MRS for measurement of several brain metabolites (see Section III.G) as well as a number of new hypointense T1 lesions (black holes), a marker of axonal loss. Imaging assessments in these trials are being performed at baseline and then every 24 weeks for 96 weeks using standardized imaging acquisitions and centralized analyses. The imaging procedures used and analysis results should provide valuable guidance for future clinical trials.

Advanced MR techniques such as MTR, MWI, and DTI and more advanced MR diffusion techniques (Section III.A) have the potential for increasing the understanding of microstructural alterations underlying disease progression and for providing patient selection/stratification and trial outcome measures for therapeutics targeting tissue repair/remyelination. MTR studies of acute and chronic WM MS lesions show reductions in MTR particularly in chronic hypointense T1 lesions (black holes). The evolution of MS lesions has been monitored temporally with MTR and showed both

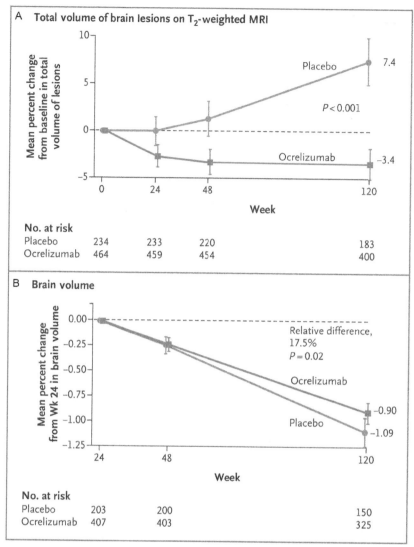

FIG. 9.13   MRI secondary endpoint results in ocrelizumab PPMS trial for intent-to-treat population. (A) The percent change in total volume of T2-w MRI lesions from baseline to week 120 for placebo versus ocrelizumab. (B) Mean percent change in brain volume using 3DT1 MRI scans from week 24 to week 120 for placebo versus ocrelizumab. The brain volume change at 24 weeks (instead of baseline) was used to account for a pseudoatrophy effect (due to reduction in inflammation/edema and resultant brain volume loss) that may occur after start of antiinflammatory therapy. *With permission from Montalban, X., Hauser, S. L., Kappos, L., Arnold, D. L., Bar-Or, A., Comi, G., et al. (2017). Ocrelizumab versus placebo in primary progressive multiple sclerosis.* The New England Journal of Medicine, 376, 209–220.

demyelination and remyelination changes (Chen et al., 2008). MTR is more sensitive than conventional MRI sequences in detecting microstructural abnormality in normal-appearing white matter (NAWM) and normal-appearing gray matter (NAGM). A progressive decrease in MTR in NAWM precedes the development of T1-enhancing lesions (Goodkin et al., 1998). MTR abnormalities in NAGM are correlated with disability and are greater in patients with PPMS and SPMS (Filippi et al., 2013; Filippi & Rocca, 2007).

MR diffusion techniques have also been used to show changes in NAWM and WM lesions. Histopathology shows strong correlations between reduced DTI-derived FA and MD and demyelination/axonal density in both WM lesions and NAWM (Scewann et al., 2009). Using

DTI, lower FA in the corpus callosum predicts a greater progression of disability at 5 years in PPMS patients (Bodini et al., 2013). Although still a research tool, DSBI, using multiple-tensor modeling of diffusion MR data that are acquired in about 15 min on a 3T scanner, appears to have analysis measures in humans that separately quantify and distinguish axon injury/loss and demyelination from inflammation (increased cellularity and edema) (Wang et al., 2015). These metrics hopefully should allow the temporal relationship of inflammation and myelin injury to be assessed once further validation work and test-retest assessments are completed (Cross & Song, 2017).

While MTR and DTI acquisitions have been standardized for use in multicenter trials, actual use in clinical trials is still exploratory with limited use of DTI and MTR metrics as outcome measures. A single-center MTR study comparing MTR in subjects treated with alemtuzumab with untreated control subjects showed a reduction in both NAWM and NAGM MTRs per year in the control group, whereas MTR was stabilized in the alemtuzumab-treated subjects, suggesting tissue protection by alemtuzumab (Button et al., 2013). High level results of the MS-SPRINT trial were presented at the seventh joint ECTRIMS-ACTRIMS meeting in October 2017. Treatment with ibudilast compared with placebo showed a 48% relative reduction in brain atrophy, a 77%–82% reduction in rate of decline in MTR in overall brain tissue and gray matter and no change in DTI metrics versus placebo (https://consultqd.clevelandclinic.org/2017/10/). The usefulness of MTR, WMF, and diffusion MR metrics as putative biomarkers of myelin damage and tissue repair awaits additional well-controlled imaging data and clinical outcomes from well-designed treatment trials for further qualification.

As the understanding of MS has evolved to recognize the importance of GM pathology, newer MR techniques such as 3-D FLAIR and 3-D double inversion recovery (DIR) sequences are increasingly being used to improve the detection of cortical lesions (CL) over 2-D FLAIR and T2-w sequences. Cortical lesion volume at baseline is an independent predictor of disability accumulation in PPMS (Calabrese et al., 2009). There are three types of cortical lesions. Type 1/cortical-juxtacortical lesions are at GM/WM junctions, type 2 lesions are intracortical, and type 3 lesions are cortical subpial lesions. 3-D FLAIR has higher spatial resolution and better signal to noise compared with 2-D FLAIR. 3-D DIR allows a better delineation of the anatomic border between GM and WM by the suppression of signal from both CSF and WM. Both sequences detect more lesions than T2-w or 2-D FLAIR sequences with DIR being the most sensitive (Poloni et al., 2011). A recent evaluation of the reliability of CL detection on DIR MRI longitudinally at multiple subsequent time points applying the MAGNIMs scoring criteria for CLs suggests further improvements of both technical conditions and scoring recommendations are needed for reliable use of cortical lesion detection (Faizy et al., 2017). In addition, there is still poor harmonization of the acquisition across centers and scanners. Currently, CL volumes are not a fully qualified biomarker for assessing outcomes of therapeutic interventions.

G CNS Applications of MR Spectroscopy

A number of brain metabolites measurable by ^{1}H MRS can be used to assess brain physiology and pathophysiology and can measure changes in cell density, cell type, and neurochemistry (Licata & Renshaw, 2010; Mason & Krystal, 2006; Öz, Tkáč, & Uğurbil, 2013). NAA in the mature brain is found almost exclusively in neurons,

axons, and dendrites (mainly glutaminergic neurons) and is a measure of neuronal integrity/neuronal dysfunction and may also reflect mitochondrial dysfunction. mI functions as an osmotic regulator of brain tissue volume and is found predominantly in glia; elevated mI is considered a marker for gliosis. tCho changes are seen in diseases with membrane turnover and cellular proliferation such as MS. Although tCr is a marker of energy systems, its components (PCr and Cr) are in equilibrium so the tCr (also called Cr) peak is fairly stable and often used as an internal reference standard. Elevated lactate (Lac) indicates anaerobic glycolysis and is considered a marker of hypoxia/ischemia. Glu and GABA are the primary excitatory and inhibitory neurotransmitters, respectively, and along with GABA, these compounds regulate neuronal energy metabolism. Glu and Gln resonances can overlap, and the combined resonance peak is called "Glx." At 3T and 7T, Glu and Gln can be separated with the appropriate acquisition technique.

MRS studies in spinocerebellar ataxia 1 (SCA1) provide an example of the use of MRS in neurodegenerative disease (Öz, Hutter, et al., 2010; Öz, Nelson, et al., 2010; Öz et al., 2011, 2015). SCA1 is a fatal movement disorder with cerebellar degeneration. It is caused by an autosomal dominant mutation of an expanded CAG repeat that causes a polyglutamine expansion in the ataxin-1 protein. Pathologically, SCA 1 is characterized by the loss of Purkinje cells in the cerebellum and neurodegeneration in the brainstem, pons, and spinocerebellar tracts. Clinical manifestations include ataxia, dysarthria, and bulbar dysfunction. Transgenic mouse models of SCA1 have shown that neuronal loss and motor dysfunction are reversible by suppressing expression of mutant ataxin-1. These data suggest that therapies directed at reducing the production of mutant ataxin-1 (such as gene silencing with RNA interference (RNAi)) may be promising approaches (Xia et al., 2004; Zu et al., 2004). MRS has been explored in patients and SCA-1 mouse models to determine if neurochemical biomarkers measured with MRS can be used to monitor cerebellar abnormalities and disease progression.

MRS at 4T was used to determine the neurochemical profile in the cerebellum, pons, and brainstem for 11 patients with early to moderate SCA1 compared with 15 healthy controls (Fig. 9.14). Total NAA and Glu levels were lower, while mI and tCr levels were higher in patients compared with controls. These findings are consistent with neuronal dysfunction, gliosis, and cycling alterations in Glu-Gln in SCA1 patients. Levels of tNAA, mI, and Glu in the cerebellum correlated with SCA1 patient ataxia scores using the Scale for the Assessment and Rating of Ataxia (SARA) (Öz, Hutter, et al., 2010).

MRS neurochemical spectra of a transgenic mouse model overexpressing mutant ataxin-1 (SCA1[82Q]) were obtained longitudinally and compared with control mice. Alterations in NAA, mI, and Glu in individual mice were

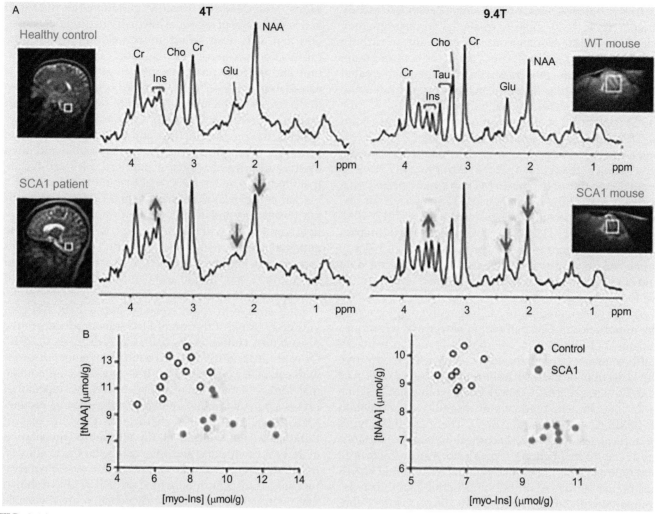

FIG. 9.14   MRS neurochemical profiles in a SCA1 patient and a SCA1 transgenic mouse. Single-voxel ¹H MR spectra from the cerebellum of a SCA1 patient and a transgenic mouse with the same mutation show parallel neurochemical alterations (decreased tNAA and Glu with elevated mI) compared with their respective healthy controls. *WT*, wild-type. *With permission from Öz, G., Tkáč, I., & Uğurbil, K. (2013). Animal models and high field imaging and spectroscopy.* Dialogues in Clinical Neuroscience, 15, 263–278.

evident at 6 weeks (an age at which there were no signs of ataxia), worsened over time, and were significantly correlated with disease progression by histology (molecular layer thickness and overall severity score) (Öz, Nelson, et al., 2010). These data show that the same metabolites that correlated with clinical status in patients (NAA, mI, and Glu) also correlate with progressive disease in the mouse (Fig. 9.14) and that these metabolites in the SCA1[82Q] mice were abnormal even when the mice were presymptomatic. A study using a conditional SCA1[82Q] mouse model, in which the cerebellar pathology and ataxia phenotype can be reversed by doxycycline, had an early doxycycline treatment group and a midstage doxycycline treatment group along with an untreated conditional group and wild-type (WT) control group. MRS was performed pre- and posttreatment as was histology. NAA and mI trended towards normalization to WT levels in both early and midstage groups and

correlated with histology in all groups indicating that MRS biomarkers detect rescue from pathology and reflect treatment efficacy (Öz et al., 2011). An additional study with conditional SCA1[82Q] mice assessed the relative sensitivities of MRS, ataxin-1 transgene expression with quantitative polymerase chain reaction (qPCR), histology, and rotarod motor behavior to disease reversal with doxycycline. Receiver operating characteristic (ROC) analyses were used to distinguish treated from untreated groups and showed high areas under the curve (AUC) for MRS, qPCR, and histology (0.97–0.98) with a lower AUC for Rotarod (AUC = 0.72) (Öz et al., 2015).

Taken together, the MRS data in SCA1 indicate that MRS-detected neurochemicals NAA and mI are measures that could translate from preclinical models to patients and serve clinically as outcome measures for disease progression and the effect of investigational therapies on disease progression. The efforts to standardize acquisition of single-voxel cerebellar ^{1}H MRS at multiple clinical sites should allow implementation of MRS in multicenter trials at those sites (Öz et al., 2014).

In MS clinical trials, MRS metrics have seen limited use, but ongoing clinical trials are assessing changes in CNS metabolites determined by MRS as exploratory markers of treatment outcome (see Section III.F). Recent work by Llufriu et al. (2014) has shown that the mI-NAA metabolite ratio in NAWM is able to predict brain volume loss as well as change in EDSS score over time and the 12-month sustained EDSS score progression. This metric may provide a useful baseline marker for patient selection and stratification in trials and merits further evaluation as an exploratory outcome measure in PoC trials evaluating therapies aimed at neuroprotection (Miller, 2014).

Multiple studies have shown that Glu and GABA play a major role in numerous psychiatric disorders, including SCZ, mood disorders, and addiction (Coyle & Konopaske, 2016; Licata & Renshaw, 2010; Mason & Krystal, 2006; Merritt, Egerton, Kempton, Taylor, & McGuire, 2016). From a metaanalysis of MRS data on Glu, Gln, and Glx in SCZ, Merritt et al. (2016) concludes that there is increased neurotransmitter Glu release in SCZ. The simple use of changes in steady-state Glu/Gln/Glx levels as biomarkers for disease progression and treatment effects is, however, problematic. As discussed by Coyle and Konopaske (2016), glutamate is involved in metabolic processes, not just as a component of the neurotransmitter pool. These other functions can therefore affect steady-state glutamate levels. Additionally, ketamine challenges in normal rodents and healthy humans may be different from challenges in SCZ where synaptic density is different. The complex genetics of SCZ also raises questions about treating SCZ patients as a single disorder. Mouse genetic models of SCZ could aid in dissecting the differences in the amino acid metabolite biology seen with MRS in SCZ patients versus healthy humans. ^{13}C MRS can be performed in rodents and healthy humans and patients with psychiatry disorders to better understand the turnover of synaptic Glu and GABA and facilitate the interpretation of steady-state changes in these metabolites in SCZ (Rothman et al., 2011).

Abdallah et al. (2014) used ^{1}HMRS to measure total Glu and GABA and ^{13}C MRS (with an infusion of [1-^{13}C] glucose; see Section III.A) to measure neuronal TCA cycling for mitochondrial energy production, GABA synthesis, and Glu/Gln cycling in MDD subjects relative to healthy controls. They found that there was reduced oxidative energy production within the glutaminergic neurons in MDD subjects, a finding that they propose that could be related to several pathophysiologic processes such as mitochondrial dysfunction, reduced levels of synaptic activity, and/or altered coupling of neuronal-astroglial metabolism. Rodent models should prove useful to further understand the pathophysiologic changes related to reduced oxidative energy production and to identify possible treatment targets.

As an example of the use of ^{31}P-MRS, that technique was recently used to assess the effect of creatine treatment in SSRI-resistant adolescents with major depressive disorder. The study demonstrated a dose-dependent increase compared with placebo in high-energy brain metabolites and that the increase in frontal lobe ^{31}PCr was inversely correlated with depression scores (Kondo et al., 2016).

In summary, MRS with ^{13}C and ^{31}P can provide important mechanistic information in early clinical POM-type trials when implemented at sites with the appropriate MRS expertise.

IV CHALLENGES AND PITFALLS IN USING IMAGING BIOMARKERS IN CNS DRUG DEVELOPMENT

To realize the full potential and value of imaging biomarkers in CNS drug development, careful attention to the characteristics of the biomarker and the details of implementation is required. In general, the chosen imaging biomarker needs to have an adequate dynamic range so that pharmacodynamic and pathophysiological effects of interventions can be detected. A standardized/harmonized acquisition needs to be implemented across all imaging sites and trial time points. This standardization should use consensus guidelines if they are available and harmonize differences in scanner properties for the scanners being used in the study (i.e., use equivalent acquisition protocols and sequences and use comparable scanner resolutions). The image analysis process should be prospectively specified in accord with the clinical trial protocol endpoints and standardized across all subjects and time points as well. The purpose of standardized acquisitions and centralized standardized analysis is to minimize the variability in the imaging outputs. The precision and variability of the biomarker metric/measurand resultant from the acquisition through analysis chain should be known before that metric is used in assaying a novel compound. Test-retest studies of repeatability and reproducibility allow both intrasubject and intersubject/intersite variability to be determined (see examples in Sections II.G, II.I.5, III.A, III.B.1, III.D, III.G and section "PET Imaging of Translocator Protein (TSPO)"). Comparison of this information with the anticipated pharmacological effect size allows an understanding of when a biological signal detection or change is real and above the noise level and permits the proper sample size for powering an imaging study/substudy to be determined.

Some studies, such as a PET receptor occupancy study, are typically performed at a single academic imaging center or specialized imaging contract research organization (CRO). Standardization of imaging acquisition and analysis is fairly straight forward in this setting where a single scanner is being used. Imaging-specific documents should prospectively document the planned acquisition details and analysis plans. Quality control (QC) of the imaging acquisitions should be done in real time in accord with a prespecified list of quality checks. Multicenter trials with imaging typically utilize a central core imaging laboratory (either commercial or academic) to assist with implementation and to perform a centralized analysis of the imaging. The imaging core lab evaluates the imaging center/department used by each clinical site to ensure that the center has the necessary imaging equipment. The core lab trains the imaging centers in the trial-specific imaging acquisition and typically has the imaging site submit anonymized sample scans or phantom scans to confirm both scan quality and the center's ability to implement the standardized scan. This process also confirms the functionality of the secure electronic transmission system that will be used for the submission of subject scans to the core lab. These activities are completed prior to the scanning of trial subjects. Study scans are submitted to the core lab within 1–2 days of acquisition and undergo rapid turn-around QC to ensure baseline and follow-up scans are of acceptable quality. An imaging acquisition manual is provided to the sites by the core lab, and an imaging review charter (or an equivalent document) is prepared by the imaging core lab and reviewed and approved by the sponsor. The charter must be consistent with the trial protocol and statistical analysis plan. It summarizes the role of imaging in the trial and the endpoints to which the imaging contributes. Other components are a summary of the acquisition or link to the imaging manual, description of the QC and image archiving processes, description of the image analysis deliverables, and the methodology of the analysis process. The charter should follow the FDA guidelines for charters addressed in Clinical Trial Imaging Endpoint Process Standards; Draft Guidance for Industry (https://www.regulations.gov/document?D=FDA-2011-D-0586-0026). Schwarz and coworkers describe in detail a procedural framework for good imaging practice in pharmacological fMRI studies applied to drug development (Schwarz et al., 2011a, 2011b). These guidelines in a general conceptual manner also apply to the implementation of many of the advanced imaging approaches used in drug development.

Precompetitive consortia and working groups are serving a critical role in guiding the implementation of imaging biomarkers. They are defining standards for acquisition and analysis, establishing best practices for imaging in clinical trials and qualifying biomarkers for a specified context of use. For example, ADNI has provided standardized imaging acquisition protocols (including structural MRI, 3D FLAIR, T2* GRE, DTI, ASL, and rs-BOLD fMRI sequences and standardized acquisitions for FDG, amyloid, and tau PET) for quantitative imaging in AD clinical trials. ADNI has also generated a landmark publically available longitudinal data set of imaging and nonimaging biomarkers across the disease spectrum of AD (Weiner et al., 2017). The ADNI data are widely used in the development and qualification of imaging biomarkers. The Quantitative Imaging Biomarker Alliance (QIBA) engages academic researchers and health-care professionals as well as representatives from pharma, imaging equipment manufacturers, software developers, and imaging CROs to identify needs, barriers, and solutions to be able to develop and test consistent, reliable, valid, and achievable quantification of imaging data across imaging vendor platforms, clinical sites, and time (Buckler et al., 2011; Sullivan et al., 2015). CNS-related biomarker committees within QIBA focus on DTI, ASL, fMRI in surgical resection planning, DSC-MRI, amyloid PET, and neuroSPECT (https://qibawiki.rsna.org). The Human Genome Project is delivering robust acquisitions for DTI and rs-BOLD fMRI (Glasser et al., 2016). INMIND (section "INMiND") is designed to progress imaging biomarkers for neuroinflammation (Mohammadi, 2013). CAMD has qualified low HCV as an enrichment biomarker for AD prodromal trials with EMA and is working on a similar qualification with FDA (Hill et al., 2014; Section III.B.1). CAMD has also qualified dopaminergic deficit detected with DAT SPECT scans as an enrichment biomarker in early Parkinson's disease trials (EMA et al., 2016). The successes of these various consortia speak to the value of continuing and additional cooperative efforts to test and further qualify promising imaging biomarker candidates for drug development.

References

Abanades, S., van der Aart, J., Barletta, J. A., Marzano, C., Searle, G. E., Salinas, C. A., et al. (2011). Prediction of repeat-dose occupancy from single-dose data: characterization of the relationship between plasma pharmacokinetics and brain target occupancy. *Journal of Cerebral Blood Flow and Metabolism, 31,* 944–952.

Abdallah, C. G., Jiang, L., De Feyter, H. M., Fasula, M., Krystal, J. H., Rothman, D. L., et al. (2014). Glutamate metabolism in major depressive disorder. *The American Journal of Psychiatry, 171,* 1320–1327.

Airas, L., Rissanen, E., & Rinne, J. O. (2015). Imaging neuroinflammation in multiple sclerosis using TSPO-PET. *Clinical and Translational Imaging, 3,* 461–473.

Albert, M. S., DeKosky, S. T., Dickson, D., Dubois, B., Feldman, H. H., Fox, N. C., et al. (2011). The diagnosis of mild cognitive impairment due to Alzheimer's disease: recommendations from the National Institute on Aging-Alzheimer's Association workgroups on diagnostic guidelines for Alzheimer's disease. *Alzheimer's & Dementia, 7,* 270–279.

Albrecht, D. S., Granziera, C., Hooker, J. M., & Loggia, M. L. (2016). In vivo imaging of human neuroinflammation. *ACS Chemical Neuroscience, 7,* 470–483.

Alsop, D. C., Detre, J. A., Golay, X., Günther, M., Hendrikse, J., Hernandez-Garcia, L., et al. (2015). Recommended implementation of arterial spin-labeled perfusion MRI for clinical applications: a consensus of the ISMRM perfusion study group and the European consortium for ASL in dementia. *Magnetic Resonance in Medicine*, 73, 102–116.

Alzheimer's Disease Neuroimaging Initiative, MRI scanner protocols. Available from: https://adni.loni.usc.edu/methods/documents/mri-protocols/. 2017.

Amann, M., Sprenger, T., Naegelin, Y., Reinhardt, J., Kuster, P., Hirsch, J. G., et al. (2015). Comparison between balanced steady-state free precession and standard spoiled gradient echo magnetization transfer ratio imaging in multiple sclerosis: methodical and clinical considerations. *NeuroImage*, 108, 87–94.

Arrowsmith, J. (2011). Phase II failures: 2008–2010. *Nature Reviews. Drug Discovery*, 10, 328–329.

Ashburner, J., & Friston, K. J. (2000). Voxel-based morphometry—the methods. *NeuroImage*, 11, 805–821.

Barkhof, F., Haller, S., & Rombouts, S. A. (2014). Resting-state functional MR imaging: a new window to the brain. *Radiology*, 272, 29–49.

Barret, O., Thomae, D., Tavares, A., Alagille, D., Papin, C., Waterhouse, R., et al. (2014). In vivo assessment and dosimetry of 2 novel PDE10A PET radiotracers in humans: 18F-MNI-659 and 18F-MNI-654. *Journal of Nuclear Medicine*, 55, 1297–1304.

Barth, V., & Need, A. (2014). Identifying novel radiotracers for PET imaging of the brain: application of LC-MS/MS to tracer identification. *ACS Chemical Neuroscience*, 5, 1148–1153.

Berg, D., Postuma, R. B., Adler, C. H., Bloem, B. R., Chan, P., Dubois, B., et al. (2015). MDS research criteria for prodromal Parkinson's disease. *Movement Disorders*, 30, 1600–1611.

Bergström, M., Hargreaves, R. J., Burns, H. D., Goldberg, M. R., Sciberras, D., Reines, S. A., et al. (2004). Human positron emission tomography studies of brain neurokinin 1 receptor occupancy by aprepitant. *Biological Psychiatry*, 55, 1007–1014.

Blockx, I., Einstein, S., Guns, P. J., Van Audekerke, J., Guglielmetti, C., Zago, W., et al. (2016). Monitoring blood-brain barrier integrity following amyloid-β immunotherapy using gadolinium-enhanced MRI in a PDAPP mouse model. *Journal of Alzheimer's Disease*, 54, 723–735.

Bodini, B., Cercignani, M., Khaleeli, Z., Miller, D. H., Ron, M., Penny, S., et al. (2013). Corpus callosum damage predicts disability progression and cognitive dysfunction in primary-progressive MS after five years. *Human Brain Mapping*, 34, 1163–1172.

Boess, F. G., Hendrix, M., van der Staay, F.-J., Erb, C., Schreiber, R., van Staveren, W., et al. (2004). Inhibition of phosphodiesterase 2 increases neuronal cGMP, synaptic plasticity and memory performance. *Neuropharmacology*, 47, 1081–1092.

Bohorquez, S. S., Barret, O., Tamagnan, G., Alagille, D., Seibyl, J., et al. (2017). Assessing optimal injected dose for tau PET imaging using [18F]GTP1 (Genentech Tau Probe 1). *Journal of Nuclear Medicine*, 58, 848.

Braak, H., & Braak, E. (1991). Neuropathological staging of Alzheimer-related changes. *Acta Neuropathologica*, 82, 239–259.

Braak, H., Alafuzoff, I., Arzberger, T., Kretzschmar, H., & Del Tredici, K. (2006). Staging of Alzheimer disease-associated neurofibrillary pathology using paraffin sections and immunocytochemistry. *Acta Neuropathologica*, 112, 389–404.

Braak, H., & Del Tredici, K. (2015). The preclinical phase of the pathological process underlying sporadic Alzheimer's disease. *Brain*, 138, 2814–2833.

Brooks, D. J. (2016). Molecular imaging of dopamine transporters. *Ageing Research Reviews*, 30, 114–121.

Brown, R. A., Narayanan, S., & Arnold, D. L. (2014). Imaging of repeated episodes of demyelination and remyelination in multiple sclerosis. *NeuroImage Clinical*, 6, 20–25.

Brown, R. A., Narayanan, S., Stikov, N., Cook, S., Cadavid, D., Wolansky, L., et al. (2016). MTR recovery in brain lesions in the BECOME study of glatiramer acetate vs interferon β-1b. *Neurology*, 87, 905–911.

Buckler, A. J., Bresolin, L., Dunnick, N. R., Sullivan, D. C., & Group. (2011). A collaborative enterprise for multi-stakeholder participation in the advancement of quantitative imaging. *Radiology*, 258, 906–914.

Bullmore, E., & Sporns, O. (2009). Complex brain networks: graph theoretical analysis of structural and functional systems. *Nature Reviews. Neuroscience*, 10, 186–198.

Burnstock, G. (2008). Purinergic signaling and disorders of the central nervous system. *Nature Reviews. Drug Discovery*, 7, 575–590.

Button, T., Altmann, D., Tozer, D., Dalton, C., Hunter, K., Compston, A., et al. (2013). Magnetization transfer imaging in multiple sclerosis treated with alemtuzumab. *Multiple Sclerosis*, 19, 241–244.

Calabrese, M., Rocca, M. A., Atzori, M., Mattisi, I., Bernardi, V., Favaretto, A., et al. (2009). Cortical lesions in primary progressive multiple sclerosis: a 2-year longitudinal MR study. *Neurology*, 72, 1330–1336.

Calsolaro, V., & Edison, P. (2016). Neuroinflammation in Alzheimer's disease: current evidence and future directions. *Alzheimer's & Dementia*, 12, 719–732.

Cash, D. M., Frost, C., Iheme, L. O., Ünay, D., Kandemir, M., Fripp, J., et al. (2015). Assessing atrophy measurement techniques in dementia: results from the MIRIAD atrophy challenge. *NeuroImage*, 123, 149–164.

Cavedo, E., Suppa, P., Lange, C., Opfer, R., Lista, S., Galluzzi, S., et al. (2017). Fully automatic MRI-based hippocampus volumetry using FSL-FIRST: intra-scanner test-retest stability, inter-field strength variability, and performance as enrichment biomarker for clinical trials using prodromal target populations at risk for Alzheimer's disease. *Journal of Alzheimer's Disease*. https://doi.org/10.3233/JAD-161108 [Epub ahead of print].

Chataway, J. (2016). *MS-SMART: Multiple sclerosis-secondary progressive multi-arm randomization trial*. https://clinicaltrials.gov/ct2/show/study/NCT01910259?term=ms-smart&rank=1. Accessed 12 October 2017.

Chen, J. T., Collins, D. L., Atkins, H. L., Freedman, M. S., Arnold, D. L., & Canadian MS/BMT Study Group. (2008). Magnetization transfer ratio evolution with demyelination and remyelination in multiple sclerosis lesions. *Annals of Neurology*, 63, 254–262.

Chen, K., Langbaum, J. B., Fleisher, A. S., Ayutyanont, N., Reschke, C., Lee, W., et al. (2010). Twelve-month metabolic declines in probable Alzheimer's disease and amnestic mild cognitive impairment assessed using an empirically pre-defined statistical region-of-interest: findings from the Alzheimer's Disease Neuroimaging Initiative. *NeuroImage*, 51, 654–664.

Chin, C. L., Upadhyay, J., Marek, G. J., Baker, S. J., Zhang, M., Mezler, M., et al. (2011). Awake rat pharmacological magnetic resonance imaging as a translational pharmacodynamic biomarker: metabotropic glutamate 2/3 agonist modulation of ketamine-induced blood oxygenation level dependence signals. *The Journal of Pharmacology and Experimental Therapeutics*, 336, 709–715.

Clement, P., Mutsaerts, H. J., Václavů, L., Ghariq, E., Pizzini, F. B., Smits, M., et al. (2017). Variability of physiological brain perfusion in healthy subjects—a systematic review of modifiers. Considerations for multi-center ASL studies. *Journal of Cerebral Blood Flow and Metabolism*. https://doi.org/10.1177/0271678X17702156 2017 Jan 1:271678X17702156. (Epub ahead of print).

Cook, D., Brown, D., Alexander, R., March, R., Morgan, P., Satterthwaite, G., et al. (2014). Lessons learned from the fate of AstraZeneca's drug pipeline: a five-dimensional framework. *Nature Reviews. Drug Discovery*, 13, 419–431.

Cover, K. S., van Schijndel, R. A., van Dijk, B. W., Redolfi, A., Knol, D. L., Frisoni, G. B., et al. (2011). Assessing the reproducibility of the SienaX and Siena brain atrophy measures using the ADNI back-to-back MP-RAGE MRI scans. *Psychiatry Research*, 193, 182–190.

Coyle, J. T., & Konopaske, G. (2016). Glutamatergic dysfunction in schizophrenia evaluated with magnetic resonance spectroscopy. *JAMA Psychiatry*, 73, 549–650.

Cross, A. H., & Song, S. K. (2017). A new imaging modality to non-invasively assess multiple sclerosis pathology. *Journal of Neuroimmunology, 304,* 81–85.

Cunningham, V. J., Rabiner, E. A., Slifstein, M., Laruelle, M., & Gunn, R. N. (2010). Measuring drug occupancy in the absence of a reference region: the Lassen plot re-visited. *Journal of Cerebral Blood Flow and Metabolism, 30,* 46–50.

Daerr, S., Brendel, M., Zach, C., Mille, E., Schilling, D., Zacherl, M. J., et al. (2016). Evaluation of early-phase [18F]-florbetaben PET acquisition in clinical routine cases. *NeuroImage Clinical, 14,* 77–86.

Dahlbom, M. (Ed.), (2017). *Physics of PET and SPECT imaging.* Boca Raton, FL: CRC Press.

Datta, G., Colasanti, A., Kalk, N., Owen, D. R., Scott, G., Rabiner, E. I., et al. (2017). [11C]PBR28 or [18F]PBR111 detect white matter inflammatory heterogeneity in multiple sclerosis. *Journal of Nuclear Medicine.* https://doi.org/10.2967/jnumed.116.187161 [Epub ahead of print].

Deakin, J. F., Lees, J., McKie, S., Hallak, J. E., Williams, S. R., & Dursun, S. M. (2008). Glutamate and the neural basis of the subjective effects of ketamine: a pharmaco-magnetic resonance imaging study. *Archives of General Psychiatry, 65,* 154–164.

Deelchand, D. K., Adanyeguh, I. M., Emir, U. E., Nguyen, T. -M., Valabregue, R., Henry, P. -G., et al. (2015). Two-site reproducibility of cerebellar and brainstem neurochemical profiles with short-echo, single-voxel MRS at 3T. *Magnetic Resonance in Medicine, 73,* 1718–1725.

Deelchand, D. K., Kantarci, K., & Öz, G. (2017). Improved localization, spectral quality, and repeatability with advanced MRS methodology in the clinical setting. *Magnetic Resonance in Medicine.* https://doi.org/10.1002/mrm.26788.

De Simoni, S., Schwarz, A. J., O'Daly, O. G., Marquand, A. F., Brittain, C., Gonzales, C., et al. (2013). Test-retest reliability of the BOLD pharmacological MRI response to ketamine in healthy volunteers. *NeuroImage, 64,* 75–90.

Devous, M. D., Joshi, A. D., Kennedy, I., Navitsky, M., Pontecorvo, M. J., Skovronsky, D. J., et al. (2014). Employing early uptake data from 18F-florbetaben scans as an early estimate of regional cerebral blood flow: comparison to 18F-FDG. *Alzheimer's & Dementia, 10,* 102.

Di Ciano, P., Guranda, M., Lagzdins, D., Tyndale, R. F., Gamaleddin, I., & Selby, P. (2016). Varenicline-induced elevation of dopamine in smokers: a preliminary [(11)C]-(+)-PHNO PET study. *Neuropsychopharmacology, 41,* 1513–1520.

Dickson, D. W., Kouri, N., Murray, M. E., & Josephs, K. A. (2011). Neuropathology of frontotemporal lobar degeneration-tau (FTLD-tau). *Journal of Molecular Neuroscience, 45,* 384–389.

Doyle, O. M., De Simoni, S., Schwarz, A. J., Brittain, C., O'Daly, O. G., Williams, S. C., et al. (2013). Quantifying the attenuation of the ketamine pharmacological magnetic resonance imaging response in humans: a validation using antipsychotic and glutamatergic agents. *The Journal of Pharmacology and Experimental Therapeutics, 345,* 151–160.

Dupont, A. C., Largeau, B., Santiago-Ribeiro, M. J., Guilloteau, D., Tronel, C., & Arlicot, N. (2017). Translocator protein-18 kDa (TSPO) positron emission tomography (PET) imaging and its clinical impact in neurodegenerative diseases. *International Journal of Molecular Sciences, 18,* 785–822.

Emir, U. E., Tuite, P. J., & Öz, G. (2012). Elevated pontine and putamenal GABA levels in mild-moderate Parkinson disease detected by 7 Tesla proton MRS. *PLoS One, 7*(1), e30918. https://doi.org/10.1371/journal.pone.0030918.

Enzinger, C., Barkhof, F., Ciccarelli, O., Filippi, M., Kappos, L., Rocca, M. A., et al. (2015). Nonconventional MRI and microstructural cerebral changes in multiple sclerosis. *Nature Reviews. Neurology, 11,* 676–686.

European Medicines Agency. (2011). *Qualification opinion of low hippocampal volume (atrophy) by MRI for use in regulatory clinical trials: In pre-dementia stage of Alzheimer's disease.* Available from: http://www.ema.europa.

eu/docs/en_GB/document_library/Regulatory_and_procedural_guideline/2011/10/WC500116264.pdf.2011. Accessed 25 August 2017.

European Medicines Agency. (2016). *Letter of support for molecular imaging of the dopamine transporter biomarker as an enrichment biomarker for clinical trials for early Parkinson's disease.* Available from: www.ema.europa.eu/docs/en_GB/document_library/Other/.../WC500213914.pdf. Accessed 31 October 2017.

Faizy, T. D., Thaler, C., Ceyrowski, T., Broocks, G., Treffler, N., Sedlacik, J., et al. (2017). Reliability of cortical lesion detection on double inversion recovery MRI applying the MAGNIMS-criteria in multiple sclerosis patients within a 16-months period. *PLoS One, 12*(2), e0172923. https://doi.org/10.1371/journal.pone.0172923.

Fan, Z., Brooks, D. J., Okello, A., & Edison, P. (2017). An early and late peak in microglial activation in Alzheimer's disease trajectory. *Brain, 140,* 792–803.

Farrell, M. E., Kennedy, K. M., Rodrigue, K. M., Wig, G., Bischof, G. N., Rieck, J. R., et al. (2017). Association of longitudinal cognitive decline with amyloid burden in middle-aged and older adults: evidence for a dose-response relationship. *JAMA Neurology, 74,* 830–838.

Filippi, M., & Rocca, M. A. (2007). Magnetization transfer magnetic resonance imaging of the brain, spinal cord, and optic nerve. *Neurotherapeutics, 4,* 401–413.

Filippi, M., Preziosa, P., Copetti, M., Riccitelli, G., Horsfield, M. A., Martinelli, V., et al. (2013). Gray matter damage predicts the accumulation of disability 13 years later in MS. *Neurology, 81,* 1759–1767.

Filippi, M., Rocca, M. A., Ciccarelli, O., De Stefano, N., Evangelou, N., Kappos, L., et al. (2017). MRI criteria for the diagnosis of multiple sclerosis: MAGNIMS consensus guidelines. *Lancet Neurology, 15,* 292–303.

Finnema, S. J., Scheinin, M., Shahid, M., Lehto, J., Borroni, E., Bang-Andersen, B., et al. (2015). Application of cross-species PET imaging to assess neurotransmitter release in brain. *Psychopharmacology, 232,* 4129–4157.

Finnema, S. J., Nabulsi, N. B., Eid, T., Detyniecki, K., Lin, S. F., Chen, M. K., et al. (2016). Imaging synaptic density in the living human brain. *Science Translational Medicine, 8,* 348ra96https://doi.org/10.1126/scitranslmed.aaf6667.

Fischl, B. (2012). FreeSurfer. *NeuroImage, 62,* 774–781.

Forsberg, A., Engler, H., Blomquist, G., Långström, B., & Nordberg, A. (2012). The use of PIB-PET as a dual pathological and functional biomarker in AD. *Biochimica et Biophysica Acta, 1822,* 380–385.

Fowler, J. S., & Volkow, N. D. (2001). 18FDG for the study of central nervous system drugs. *Journal of Clinical Pharmacology, 41*(Suppl), 9S–10S.

Fox, M. D., & Greicius, M. (2010). Clinical applications of resting state functional connectivity. *Frontiers in Systems Neuroscience, 4,* 19. https://doi.org/10.3389/fnsys.2010.00019.

Fox, N. C., & Freeborough, P. A. (1997). Brain atrophy progression measured from registered serial MRI: validation and application to Alzheimer's disease. *Journal of Magnetic Resonance Imaging, 7,* 1069–1075.

Fox, N. C., Black, R. S., Gilman, S., Rossor, M. N., Griffith, S. G., Jenkins, L., et al. (2005). Effects of Abeta immunization (AN1792) on MRI measures of cerebral volume in Alzheimer disease. *Neurology, 64,* 1563–1572.

Fox, R. J., Coffey, C. S., Cudkowicz, M. E., Gleason, T., Goodman, A., Klawiter, E. C., et al. (2016). Design, rationale, and baseline characteristics of the randomized double-blind phase II clinical trial of ibudilast in progressive multiple sclerosis. *Contemporary Clinical Trials, 50,* 166–177.

Frischer, J. M., Bramow, S., Dal-Bianco, A., Lucchinetti, C. F., Rauschka, H., Schmidbauer, M., et al. (2009). The relation between inflammation and neurodegeneration in multiple sclerosis brains. *Brain, 132,* 1175–1189.

Frodl, T., Scheuerecker, J., Schoepf, V., Linn, J., Koutsouleris, N., Bokde, A. L., et al. (2011). Different effects of mirtazapine and venlafaxine on brain activation: an open randomized controlled fMRI study. *The Journal of Clinical Psychiatry, 72*, 448–457.

Fu, C. H., Steiner, H., & Costafreda, S. G. (2013). Predictive neural biomarkers of clinical response in depression: a meta-analysis of functional and structural neuroimaging studies of pharmacological and psychological therapies. *Neurobiology of Disease, 52*, 75–83.

Gallezot, J. D., Kloczynski, T., Weinzimmer, D., Labaree, D., Zheng, M. Q., Lim, K., et al. (2014). Imaging nicotine- and amphetamine-induced dopamine release in rhesus monkeys with [(11)C]PHNO vs [(11)C]raclopride PET. *Neuropsychopharmacology, 39*, 866–874.

Gevers, S., van Osch, M. J., Bokkers, R. P., Kies, D. A., Teeuwisse, W. M., Majoie, C. B., et al. (2011). Intra- and multicenter reproducibility of pulsed, continuous and pseudo-continuous arterial spin labeling methods for measuring cerebral perfusion. *Journal of Cerebral Blood Flow and Metabolism, 31*, 1706–1715.

Glasser, M. F., Smith, S. M., Marcus, D. S., Andersson, J. L., Auerbach, E. J., Behrens, T. E., et al. (2016). The Human Connectome Project's neuroimaging approach. *Nature Neuroscience, 19*, 1175–1187.

Gobbi, L. C., Knust, H., Körner, M., Honer, M., Czech, C., Belli, S., et al. (2017). Identification of three novel radiotracers for imaging aggregated tau in Alzheimer's disease with positron emission tomography. *Journal of Medicinal Chemistry.* https://doi.org/10.1021/acs.jmedchem.7b00632 [Epub ahead of print].

Golay, X. (2017). The long and winding road to translation for imaging biomarker development: the case for arterial spin labelling (ASL). *European Radiology Experimental, 1*, 3–6.

Goodkin, D. E., Rooney, W. D., Sloan, R., Bacchetti, P., Gee, L., Vermathen, M., et al. (1998). A serial study of new MS lesions and the white matter from which they arise. *Neurology, 51*, 1689–1697.

Gozzi, A., Large, C. H., Schwarz, A., Bertani, S., Crestan, V., & Bifone, A. (2008). Differential effects of antipsychotic and glutamatergic agents on the phMRI response to phencyclidine. *Neuropsychopharmacology, 33*, 1690–1703.

Greicius, M. (2008). Resting-state functional connectivity in neuropsychiatric disorders. *Current Opinion in Neurology, 21*, 424–430.

Gunn, R. N., Guo, Q., Salinas, C. A., Tziortzi, A. C., & Searle, G. E. (2011). Advances in biomathematical modeling for PET neuroreceptor imaging. *Drug Discovery Today: Technologies, 8*, 45–51.

Gunn, R. N., & Rabiner, E. A. (2017). Imaging in central nervous system drug discovery. *Seminars in Nuclear Medicine, 47*, 89–98.

Guo, Q., Brady, M., & Gunn, R. N. (2009). A biomathematical modeling approach to central nervous system radioligand discovery and development. *Journal of Nuclear Medicine, 50*, 1715–1723.

Gutman, B. A., Wang, Y., Yanovsky, I., Hua, X., Toga, A. W., Jack, C. R., Jr., et al. (2015). Empowering imaging biomarkers of Alzheimer's disease. *Neurobiology of Aging, 36*(Suppl. 1), S69–S80.

Häggkvist, J., Tóth, M., Tari, L., Varnäs, K., Svedberg, M., Forsberg, A., et al. (2017). Longitudinal small-animal PET imaging of the zQ175 mouse model of Huntington disease shows in vivo changes of molecular targets in the striatum and cerebral cortex. *Journal of Nuclear Medicine, 58*, 617–622.

Haller, S., Zaharchuk, G., Thomas, D. L., Lovblad, K. O., Barkhof, F., & Golay, X. (2016). Arterial spin labeling perfusion of the brain: emerging clinical applications. *Radiology, 281*, 337–356.

Han, J., Liu, H., Liu, C., Jin, H., Perlmutter, J. S., Egan, T. M., et al. (2017). Pharmacologic characterizations of a P2X7 receptor-specific radioligand, [11C]GSK1482160, for neuroinflammatory response. *Nuclear Medicine Communications, 38*, 372–382.

Harston, G. W., Okell, T. W., Sheerin, F., Schulz, U., Mathieson, P., Reckless, I., et al. (2017). Quantification of serial cerebral blood flow in acute stroke using arterial spin labeling. *Stroke, 48*, 123–130.

Henry, M. E., Moore, C. M., Kaufman, M. J., Michelson, D., Schmidt, M. E., Stoddard, E., et al. (2000). Brain kinetics of paroxetine and fluoxetine on the third day of placebo substitution: a fluorine MRS study. *The American Journal of Psychiatry, 157*, 1506–1508.

Hill, D. L., Schwarz, A. J., Isaac, M., Pani, L., Vamvakas, S., Hemmings, R., et al. (2014). Coalition against major diseases/European Medicines Agency biomarker qualification of hippocampal volume for enrichment of clinical trials in predementia stages of Alzheimer's disease. *Alzheimer's & Dementia, 10*, 421–429.

Hori, M., Fukunaga, I., Masutani, Y., Taoka, T., Kamagata, K., Suzuki, Y., et al. (2012). Visualizing non-Gaussian diffusion: clinical application of q-space imaging and diffusional kurtosis imaging of the brain and spine. *Magnetic Resonance in Medical Sciences, 11*, 221–233.

Hostetler, E. D., Walji, A. M., Zeng, Z., Miller, P., Bennacef, I., Salinas, C., et al. (2016). Preclinical characterization of ^{18}F-MK-6240, a promising PET tracer for in vivo quantification of human neurofibrillary tangles. *Journal of Nuclear Medicine, 57*, 1599–1606.

Hua, X., Hibar, D. P., Ching, C. R., Boyle, C. P., Rajagopalan, P., Gutman, B. A., et al. (2013). Unbiased tensor-based morphometry: improved robustness and sample size estimates for Alzheimer's disease clinical trials. *NeuroImage, 66*, 648–661.

Inglese, M., Park, S. J., Johnson, G., Babb, J. S., Miles, L., Jaggi, H., et al. (2007). Deep gray matter perfusion in multiple sclerosis: dynamic susceptibility contrast perfusion magnetic resonance imaging at 3 T. *Archives of Neurology, 64*, 196–202.

Innis, R. B., Cunningham, V. J., Delforge, J., Fujita, M., Gjedde, A., Gunn, R. N., et al. (2007). Consensus nomenclature for in vivo imaging of reversibly binding ligands. *Journal of Cerebral Blood Flow and Metabolism, 27*, 1533–1539.

Jack, C. R., Jr. (2011). Alliance for aging research AD Biomarkers Work Group: structural MRI. *Neurobiology of Aging, 32*, S48–S57.

Jack, C. R., Jr., Knopman, D. S., Jagust, W. J., Petersen, R. C., Weiner, M. W., Aisen, P. S., et al. (2013). Tracking pathophysiological processes in Alzheimer's disease: an updated hypothetical model of dynamic biomarkers. *Lancet Neurology, 12*, 207–216.

Jack, C. R., Jr., Bennett, D. A., Blennow, K., Carrillo, M. C., Feldman, H. H., Frisoni, G. B., et al. (2016). A/T/N: an unbiased descriptive classification scheme for Alzheimer disease biomarkers. *Neurology, 87*, 539–547.

Jagust, W., Reed, B., Mungas, D., Ellis, W., & Decarli, C. (2007). What does fluorodeoxyglucose PET imaging add to a clinical diagnosis of dementia? *Neurology, 69*, 871–877.

Jagust, W. J., Landau, S. M., Koeppe, R. A., Reiman, E. M., Chen, K., Mathis, C. A., et al. (2015). The Alzheimer's disease neuroimaging initiative 2 PET core: 2015. *Alzheimer's & Dementia, 11*, 757–771.

Jedynak, B. M., Lang, A., Liu, B., Katz, E., Zhang, Y., Wyman, B. T., et al. (2012). A computational neurodegenerative disease progression score: method and results with the Alzheimer's disease neuroimaging initiative cohort. *NeuroImage, 63*, 1478–1486.

Jennings, D., Siderowf, A., Stern, M., Seibyl, J., Eberly, S., Oakes, D., et al. (2017). Conversion to Parkinson disease in the PARS hyposmic and dopamine transporter-deficit prodromal cohort. *JAMA Neurology.* https://doi.org/10.1001/jamaneurol.2017.0985.

Jones, D. T., Knopman, D. S., Gunter, J. L., Graff-Radford, J., Vemuri, P., Boeve, B. F., et al. (2016). Cascading network failure across the Alzheimer's disease spectrum. *Brain, 139*, 547–562.

Joules, R., Doyle, O. M., Schwarz, A. J., O'Daly, O. G., Brammer, M., & Williams, S. C. (2015). Ketamine induces a robust whole-brain connectivity pattern that can be differentially modulated by drugs of different mechanism and clinical profile. *Psychopharmacology, 232*, 4205–4218.

Jucaite, A., Svenningsson, P., Rinne, J. O., Cselényi, Z., Varnäs, K., Johnström, P., et al. (2015). Effect of the myeloperoxidase inhibitor AZD3241 on microglia: a PET study in Parkinson's disease. *Brain, 138*, 2687–2700.

Keedwell, P., Drapier, D., Surguladze, S., Giampietro, V., Brammer, M., & Phillips, M. (2009). Neural markers of symptomatic improvement during antidepressant therapy in severe depression: subgenual cingulate and visual cortical responses to sad, but not happy, facial stimuli are correlated with changes in symptom score. *Journal of Psychopharmacology, 23*, 775–788.

Keller, M., Montgomery, S., Ball, W., Morrison, M., Snavely, D., & Liu, G. (2006). Lack of efficacy of the substance p (neurokinin1 receptor) antagonist aprepitant in the treatment of major depressive disorder. *Biological Psychiatry, 59*, 216–223.

Kinnunen, K. M., Cash, D. M., Poole, T., Frost, C., Benzinger, T. L. S., Ahsan, R. L., et al. (2017). Presymptomatic atrophy in autosomal dominant Alzheimer's disease: a serial MRI study. *Alzheimer's & Dementia*. https://doi.org/10.1016/j.jalz.2017.06.2268. pii: S1552-5260(17)32521-9. (Epub ahead of print).

Kisler, K., Nelson, A. R., Montagne, A., & Zlokovic, B. V. (2017). Cerebral blood flow regulation and neurovascular dysfunction in Alzheimer disease. *Nature Reviews. Neuroscience, 18*, 419–434.

Klunk, W. E., Engler, H., Nordberg, A., Wang, Y., Blomqvist, G., Holt, D. P., et al. (2004). Imaging brain amyloid in Alzheimer's disease with Pittsburgh Compound-B. *Annals of Neurology, 55*, 306–319.

Klunk, W. E., Koeppe, R. A., Price, J. C., Benzinger, T. L., Devous, M. D., Sr., Jagust, W. J., et al. (2015). The Centiloid Project: standardizing quantitative amyloid plaque estimation by PET. *Alzheimer's & Dementia, 11*, 1–15.

Kondo, D. G., Forrest, L. N., Shi, X., Sung, Y. H., Hellem, T. L., Huber, R. S., et al. (2016). Creatine target engagement with brain bioenergetics: a dose-ranging phosphorus-31 magnetic resonance spectroscopy study of adolescent females with SSRI-resistant depression. *Amino Acids, 48*, 1941–1954.

Lammertsma, A. A., & Hume, S. P. (1996). Simplified reference tissue model for PET receptor studies. *NeuroImage, 4*, 153–158.

Landau, S. M., Harvey, D., Madison, C. M., Koeppe, R. A., Reiman, E. M., Foster, N. L., et al. (2011). Associations between cognitive, functional, and FDG-PET measures of decline in AD and MCI. *Neurobiology of Aging, 32*, 1207–1218.

Lange, C., Kurth, J., Sesse, A., Schwarzenböck, S., Steinhoff, K., Umland-Seidler, B., et al. (2015). Robust, fully automatic delineation of the head contour by stereotactic normalization for attenuation correction according to Chang in dopamine transporter scintigraphy. *European Radiology, 25*, 2709–2717.

Långsjö, J. W., Kaisti, K. K., Aalto, S., Hinkka, S., Aantaa, R., Oikonen, V., et al. (2003). Effects of subanesthetic doses of ketamine on regional cerebral blood flow, oxygen consumption, and blood volume in humans. *Anesthesiology, 99*, 614–623.

Laruelle, M., Slifstein, M., & Huang, Y. (2002). Positron emission tomography: imaging and quantification of neurotransporter availability. *Methods, 27*, 287–299.

Leach, M. O., Brindle, K. M., Evelhoch, J. L., Griffiths, J. R., Horsman, M. R., Jackson, A., et al. (2005). The assessment of antiangiogenic and antivascular therapies in early-stage clinical trials using magnetic resonance imaging: issues and recommendations. *British Journal of Cancer, 92*, 1599–1610.

Licata, S. C., & Renshaw, P. F. (2010). Neurochemistry of drug action: insights from proton magnetic resonance spectroscopic imaging and their relevance to addiction. *Annals of the New York Academy of Sciences, 1187*, 148–171.

Lim, K. S., Kwon, J. S., Jang, I. J., Jeong, J. M., Lee, J. S., Kim, H. W., et al. (2007). Modeling of brain D2 receptor occupancy-plasma concentration relationships with a novel antipsychotic, YKP1358, using serial PET scans in healthy volunteers. *Clinical Pharmacology and Therapeutics, 81*, 252–258.

Littlewood, C. L., Jones, N., O'Neill, M. J., Mitchell, S. N., Tricklebank, M., & Williams, S. C. (2006). Mapping the central effects of ketamine in the rat using pharmacological MRI. *Psychopharmacology, 186*, 64–81.

Llufriu, S., Kornak, J., Ratiney, H., Oh, J., Brenneman, D., Cree, B. A., et al. (2014). Magnetic resonance spectroscopy markers of disease progression in multiple sclerosis. *JAMA Neurology, 71*, 840–847.

Maass, A., Landau, S., Baker, S. L., Horng, A., Lockhart, S. N., La Joie, R., et al. (2017). Comparison of multiple tau-PET measures as biomarkers in aging and Alzheimer's disease. *NeuroImage, 157*, 448–463.

Mallik, S., Samson, R. S., Wheeler-Kingshott, C. A., & Miller, D. H. (2014). Imaging outcomes for trials of remyelination in multiple sclerosis. *Journal of Neurology, Neurosurgery, and Psychiatry, 85*, 1396–1404.

Malone, I. B., Leung, K. K., Clegg, S., Barnes, J., Whitwell, J. L., Ashburner, J., et al. (2015). Accurate automatic estimation of total intracranial volume: a nuisance variable with less nuisance. *NeuroImage, 104*, 366–372.

Marquié, M., Normandin, M. D., Vanderburg, C. R., Costantino, I. M., Bien, E. A., Rycyna, L. G., et al. (2015). Validating novel tau positron emission tomography tracer [F-18]-AV-1451 (T807) on postmortem brain tissue. *Annals of Neurology, 78*, 787–800.

Marquand, A. F., O'Daly, O. G., De Simoni, S., Alsop, D. C., Maguire, R. P., Williams, S. C., et al. (2012). Dissociable effects of methylphenidate, atomoxetine and placebo on regional cerebral blood flow in healthy volunteers at rest: a multi-class pattern recognition approach. *NeuroImage, 60*, 1015–1024.

Mason, G. F., & Krystal, J. H. (2006). MR spectroscopy: its potential role for drug development for the treatment of psychiatric disease. *NMR in Biomedicine, 19*, 690–701.

Matthews, D. C., Lukic, A. S., Andrews, R. D., Wernick, M. N., Strother, S. C., Schmidt, M. E., et al. (2016). Combining neurodegenerative characterization with amyloid measurement using an early frame amyloid PET multivariate classifier. *Alzheimer's & Dementia, 12*, 1082–1083.

McEvoy, L. K., Fennema-Notestine, C., Roddey, J. C., Hagler, D. J., Jr., Holland, D., Karow, D. S., et al. (2009). Alzheimer disease: quantitative structural neuroimaging for detection and prediction of clinical and structural changes in mild cognitive impairment. *Radiology, 251*, 195–205.

Merritt, K., Egerton, A., Kempton, M. J., Taylor, M. J., & McGuire, P. K. (2016). Nature of glutamate alterations in schizophrenia: a meta-analysis of proton magnetic resonance spectroscopy studies. *JAMA Psychiatry, 73*, 665–674.

Meyers, S. M., Vavasour, I. M., Mädler, B., Harris, T., Fu, E., Li, D. K., et al. (2013). Multicenter measurements of myelin water fraction and geometric mean T2: intra- and intersite reproducibility. *Journal of Magnetic Resonance Imaging, 38*, 1445–1453.

Mezue, M., Segerdahl, A. R., Okell, T. W., Chappell, M. A., Kelly, M. E., & Tracey, I. (2014). Optimization and reliability of multiple postlabeling delay pseudo-continuous arterial spin labeling during rest and stimulus-induced functional task activation. *Journal of Cerebral Blood Flow and Metabolism, 34*, 1919–1927.

Michels, L., Warnock, G., Buck, A., Macauda, G., Leh, S. E., Kaelin, A. M., et al. (2016). Arterial spin labeling imaging reveals widespread and Aβ-independent reductions in cerebral blood flow in elderly apolipoprotein epsilon-4 carriers. *Journal of Cerebral Blood Flow and Metabolism, 36*, 581–595.

Miller, D. H. (2014). Magnetic resonance spectroscopy: a possible in vivo marker of disease progression for multiple sclerosis? *JAMA Neurology, 71*, 828–830.

Minzenberg, M. J., Laird, A. R., Thelen, S., Carter, C. S., & Glahn, D. C. (2009). Meta-analysis of 41 functional neuroimaging studies of executive function in schizophrenia. *Archives of General Psychiatry, 66*, 811–822.

Mohammadi, D. (2013). INMiND: getting to the bottom of neuroinflammation. *Lancet Neurology, 12*, 1135–1136.

Montagne, A., Nation, D. A., Pa, J., Sweeney, M. D., Toga, A. W., & Zlokovic, B. V. (2016). Brain imaging of neurovascular dysfunction in Alzheimer's disease. *Acta Neuropathologica, 131*, 687–707.

Montalban, X., Hauser, S. L., Kappos, L., Arnold, D. L., Bar-Or, A., Comi, G., et al. (2017). Ocrelizumab versus placebo in primary progressive multiple sclerosis. *The New England Journal of Medicine, 376*, 209–220.

Morgan, P., Van Der Graaf, P. H., Arrowsmith, J., Feltner, D. E., Drummond, K. S., Wegner, C. D., et al. (2012). Can the flow of medicines be improved? Fundamental pharmacokinetic and pharmacological principles towards improving phase II survival. *Drug Discovery Today, 17*, 419–424.

Mouridsen, K., Hansen, M. B., Østergaard, L., & Jespersen, S. N. (2014). Reliable estimation of capillary transit time distributions using DSC-MRI. *Journal of Cerebral Blood Flow and Metabolism, 34*, 1511–1521.

Mueller, A., Kroth, H., Berndt, M., Capotosti, F., Molette, J., Schieferstein, H., et al. (2017). Characterization of the novel PET tracer PI-2620 for the assessment of tau pathology in Alzheimer's disease and other tauopathies. *Journal of Nuclear Medicine, 58*(Suppl. 1), 847.

Mutsaerts, H. J., Steketee, R. M., Heijtel, D. F., Kuijer, J. P., van Osch, M. J., Majoie, C. B., et al. (2015). Reproducibility of pharmacological ASL using sequences from different vendors: implications for multicenter drug studies. *Magma, 28*, 427–436.

Mutsaerts, H. J. M. M., Petr, J., Thomas, D. L., De Vita, E., Cash, D. M., van Osch, M. J. P., et al. (2017). Comparison of arterial spin labeling registration strategies in the multi-center GENetic frontotemporal dementia Initiative (GENFI). *Journal of Magnetic Resonance Imaging.* https://doi.org/10.1002/jmri.25751 [Epub ahead of print].

Nag, S., Fazio, P., Lehmann, L., Kettschau, G., Heinrich, T., Thiele, A., et al. (2016). In vivo and in vitro characterization of a novel MAO-B inhibitor radioligand, ^{18}F-labeled deuterated fluorodeprenyl. *Journal of Nuclear Medicine, 57*, 315–320.

Naganawa, M., Waterhouse, R. N., Nabulsi, N., Lin, S. F., Labaree, D., Ropchan, J., et al. (2016). First-in-human assessment of the novel PDE2A PET radiotracer 18F-PF-05270430. *Journal of Nuclear Medicine, 57*, 1388–1395.

Nelson, P. T., Alafuzoff, I., Bigio, E. H., Bouras, C., Braak, H., Cairns, N. J., et al. (2012). Correlation of Alzheimer disease neuropathologic changes with cognitive status: a review of the literature. *Journal of Neuropathology and Experimental Neurology, 71*, 362–381.

Nielsen, R. B., Egefjord, L., Angleys, H., Mouridsen, K., Gejl, M., Møller, A., et al. (2017). Capillary dysfunction is associated with symptom severity and neurodegeneration in Alzheimer's disease. *Alzheimer's & Dementia, 13*, 1143–1153.

Ng, K. P., Pascoal, T. A., Mathotaarachchi, S., Therriault, J., Kang, M. S., Shin, M., et al. (2017). Monoamine oxidase B inhibitor, selegiline, reduces ^{18}F-THK5351 uptake in the human brain. *Alzheimer's Research & Therapy, 9*, 25–34.

Nord, C. L., Gray, A., Charpentier, C. J., Robinson, O. J., & Roiser, J. P. (2017). Unreliability of putative fMRI biomarkers during emotional face processing. *NeuroImage, 156*, 119–127.

Ory, D., Celen, S., Gijsbers, R., Van Den Haute, C., Postnov, A., Koole, M., et al. (2016). Preclinical evaluation of a P2X7 receptor-selective radiotracer: PET studies in a rat model with local overexpression of the human P2X7 receptor and in nonhuman primates. *Journal of Nuclear Medicine, 57*, 1436–1441.

Østergaard, L. (2005). Principles of cerebral perfusion imaging by bolus tracking. *Journal of Magnetic Resonance Imaging, 22*, 710–717.

Ostrowitzki, S., Deptula, D., Thurfjell, L., Barkhof, F., Bohrmann, B., Brooks, D. J., et al. (2012). Mechanism of amyloid removal in patients with Alzheimer disease treated with gantenerumab. *Archives of Neurology, 69*, 198–207.

Ota, T., Shinotoh, H., Fukushi, K., Kikuchi, T., Sato, K., Tanaka, N., et al. (2010). Estimation of plasma IC50 of donepezil for cerebral acetylcholinesterase inhibition in patients with Alzheimer disease using positron emission tomography. *Clinical Neuropharmacology, 33*, 74–78.

Ou, X., Sun, S. W., Liang, H. F., Song, S. K., & Gochberg, D. F. (2009). Quantitative magnetization transfer measured pool-size ratio reflects optic nerve myelin content in ex vivo mice. *Magnetic Resonance in Medicine, 61*, 364–371.

Owen, D. R., Guo, Q., Rabiner, E. A., & Gunn, R. N. (2015). The impact of the rs6971 polymorphism in TSPO for quantification and study design. *Clinical and Translational Imaging, 3*, 417–422.

Öz, G., Hutter, D., Tkáč, I., Clark, H. B., Gross, M. D., Jiang, H., et al. (2010). Neurochemical alterations in spinocerebellar ataxia type 1 and their correlations with clinical status. *Movement Disorders, 25*, 1253–1261.

Öz, G., Nelson, C. D., Koski, D. M., Henry, P. G., Marjanska, M., Deelchand, D. K., et al. (2010). Noninvasive detection of presymptomatic and progressive neurodegeneration in a mouse model of spinocerebellar ataxia type 1. *The Journal of Neuroscience, 30*, 3831–3838.

Öz, G., Vollmers, M. L., Nelson, C. D., Shanley, R., Eberly, L. E., Orr, H. T., et al. (2011). In vivo monitoring of recovery from neurodegeneration in conditional transgenic SCA1 mice. *Experimental Neurology, 232*, 290–298.

Öz, G., Tkáč, I., & Uğurbil, K. (2013). Animal models and high field imaging and spectroscopy. *Dialogues in Clinical Neuroscience, 15*, 263–278.

Öz, G., Alger, J. R., Barker, P. B., Bartha, R., Bizzi, A., Boesch, C., et al. (2014). Clinical proton MR spectroscopy in central nervous system disorders. *Radiology, 270*, 658–679.

Öz, G., Kittelson, E., Demirgöz, D., Rainwater, O., Eberly, L. E., Orr, H. T., et al. (2015). Assessing recovery from neurodegeneration in spinocerebellar ataxia 1: comparison of in vivo magnetic resonance spectroscopy with motor testing, gene expression and histology. *Neurobiology of Disease, 74*, 158–166.

Park, E., Gallezot, J. D., Delgadillo, A., Liu, S., Planeta, B., Lin, S. F., et al. (2015). (11)C-PBR28 imaging in multiple sclerosis patients and healthy controls: test-retest reproducibility and focal visualization of active white matter areas. *European Journal of Nuclear Medicine and Molecular Imaging, 42*, 1081–1092.

Pauls, M. M. H., Clarke, N., Trippier, S., Betteridge, S., Howe, F. A., Khan, U., et al. (2017). Perfusion by arterial spin labelling following single dose tadalafil in small vessel disease (PASTIS): study protocol for a randomised controlled trial. *Trials, 18*, 229. https://doi.org/10.1186/s13063-017-1973-9.

Pievani, M., Filippini, N., van den Heuvel, M. P., Cappa, S. F., & Frisoni, G. B. (2014). Brain connectivity in neurodegenerative diseases—from phenotype to proteinopathy. *Nature Reviews. Neurology, 10*, 620–633.

Plisson, C., Weinzimmer, D., Jakobsen, S., Natesan, S., Salinas, C., Lin, S. F., et al. (2014). Phosphodiesterase 10A PET radioligand development program: from pig to human. *Journal of Nuclear Medicine, 55*, 595–601.

Poloni, G., Minagar, A., Haacke, E. M., & Zivadinov, R. (2011). Recent developments in imaging of multiple sclerosis. *The Neurologist, 17*, 185–204.

Pontecorvo, M. J., Devous, M. D., Sr., Navitsky, M., Lu, M., Salloway, S., Schaerf, F. W., et al. (2017). Relationships between flortaucipir PET tau binding and amyloid burden, clinical diagnosis, age and cognition. *Brain, 140*, 748–763.

Popescu, V., Agosta, F., Hulst, H. E., Sluimer, I. C., Knol, D. L., Sormani, M. P., et al. (2013). Brain atrophy and lesion load predict long term disability in multiple sclerosis. *Journal of Neurology, Neurosurgery, and Psychiatry, 84*, 1082–1091.

Potkin, S. G., Anand, R., Fleming, K., Alva, G., Keator, D., Carreon, D., et al. (2001). Brain metabolic and clinical effects of rivastigmine in Alzheimer's disease. *The International Journal of Neuropsychopharmacology, 4*, 223–230.

Provencher, S. W. (2001). Automatic quantitation of localized in vivo 1H spectra with LCModel. *NMR in Biomedicine, 14*, 260–264.

Rabinovici, G. D., Carrillo, M. C., Formanc, M., DeSanti, S., Miller, D. S., Kozauer, N., et al. (2017). Multiple comorbid neuropathologies in the setting of Alzheimer's disease neuropathology and implications for drug development. *Alzheimers & Dementia: Translational Research Clinical Interventions, 3*, 83–91.

Radue, E. W., Barkhof, F., Kappos, L., Sprenger, T., Häring, D. A., de Vera, A., et al. (2015). Correlation between brain volume loss and clinical and MRI outcomes in multiple sclerosis. *Neurology, 84*, 784–793.

Rahmim, A., & Zaidi, H. (2008). PET versus SPECT: strengths, limitations and challenges. *Nuclear Medicine Communications, 29*, 193–207.

Rinne, J. O., Brooks, D. J., Rossor, M. N., Fox, N. C., Bullock, R., Klunk, W. E., et al. (2010). ^{11}C-PiB PET assessment of change in fibrillar amyloid-beta load in patients with Alzheimer's disease treated with bapineuzumab: a phase 2, double-blind, placebo-controlled, ascending-dose study. *Lancet Neurology, 9*, 363–372.

Rodriguez-Vieitez, E., Saint-Aubert, L., Carter, S. F., Almkvist, O., Farid, K., Schöll, M., et al. (2016). Diverging longitudinal changes in astrocytosis and amyloid PET in autosomal dominant Alzheimer's disease. *Brain, 139*, 922–936.

Rothman, D. L., De Feyter, H. M., de Graaf, R. A., Mason, G. F., & Behar, K. L. (2011). ^{13}C MRS studies of neuroenergetics and neurotransmitter cycling in humans. *NMR in Biomedicine, 24*, 943–957.

Rowe, C. C., Jones, G., Doré, V., Pejoska, S., Margison, L., Mulligan, R. S., et al. (2016). Standardized expression of ^{18}F-NAV4694 and ^{11}C-PiB β-amyloid PET results with the Centiloid scale. *Journal of Nuclear Medicine, 57*, 1233–1237.

Rowe, C. C., Doré, V., Jones, G., Baxendale, D., Mulligan, R. S., Bullich, S., et al. (2017). ^{18}F-Florbetaben PET beta-amyloid binding expressed in Centiloids. *European Journal of Nuclear Medicine and Molecular Imaging.* https://doi.org/10.1007/s00259-017-3749-6 [Epub ahead of print].

Saha, G. B. (2016). *Basics of PET imaging* (3rd ed.). Switzerland: Springer International Publishing AG.

Salinas, C., Weinzimmer, D., Searle, G., Labaree, D., Ropchan, J., Huang, Y., et al. (2013). Kinetic analysis of drug-target interactions with PET for characterization of pharmacological hysteresis. *Journal of Cerebral Blood Flow and Metabolism, 33*, 700–707.

Sandiego, C. M., Gallezot, J. D., Pittman, B., Nabulsi, N., Lim, K., Lin, S. F., et al. (2015). Imaging robust microglial activation after lipopolysaccharide administration in humans with PET. *Proceedings of the National Academy of Sciences of the United States of America, 112*, 12468–12473.

Schmidt, M. E., Andrews, R. D., van der Ark, P., Brown, T., Mannaert, E., Steckler, T., et al. (2010). Dose-dependent effects of the CRF (1) receptor antagonist R317573 on regional brain activity in healthy male subjects. *Psychopharmacology, 208*, 109–119.

Schmidt, M. E., Chiao, P., Klein, G., Matthews, D., Thurfjell, L., Cole, P. E., et al. (2013). The influence of biological and technical factors on quantitative analysis of amyloid PET: points to consider and recommendations for controlling variability in longitudinal data. *Alzheimer's & Dementia, 11*, 1050–1068.

Schneider, T., Brownlee, W., Zhang, H., Ciccarelli, O., Miller, D. H., & Wheeler-Kingshott, C. G. (2017). Sensitivity of multi-shell NODDI to multiple sclerosis white matter changes: a pilot study. *Functional Neurology, 32*, 97–101.

Schwarz, A. J., Becerra, L., Upadhyay, J., Anderson, J., Baumgartner, R., Coimbra, A., et al. (2011a). A procedural framework for good imaging practice in pharmacological fMRI studies applied to drug development #1: processes and requirements. *Drug Discovery Today, 16*, 583–593.

Schwarz, A. J., Becerra, L., Upadhyay, J., Anderson, J., Baumgartner, R., Coimbra, A., et al. (2011b). A procedural framework for good imaging practice in pharmacological fMRI studies applied to drug development #2: protocol optimization and best practices. *Drug Discovery Today, 16*, 671–682.

Schwarz, A. J., Yu, P., Miller, B. B., Shcherbinin, S., Dickson, J., Navitsky, M., et al. (2016). Regional profiles of the candidate tau PET ligand 18F-AV-1451 recapitulate key features of Braak histopathological stages. *Brain, 139*, 1539–1550.

Seewann, A., Vrenken, H., van der Valk, P., Blezer, E. L., Knol, D. L., Castelijns, J. A., et al. (2009). Diffusely abnormal white matter in chronic multiple sclerosis: imaging and histopathologic analysis. *Archives of Neurology, 66*, 601–609.

Sevigny, J., Suhy, J., Chiao, P., Chen, T., Klein, G., Purcell, D., et al. (2016). Amyloid PET screening for enrichment of early-stage Alzheimer disease clinical trials: experience in a phase 1b clinical trial. *Alzheimer Disease and Associated Disorders, 30*, 1–7.

Sevigny, J., Chiao, P., Bussière, T., Weinreb, P. H., Williams, L., Maier, M., et al. (2016). The antibody aducanumab reduces Aβ plaques in Alzheimer's disease. *Nature, 537*, 50–56.

Shcherbinin, S., Eads, J. A., Schwarz, A. J., & Sims, J. R. (2016). Florbetapir F 18 PET: from dual-phase to dual-biomarker imaging. In: *Clinical trials on Alzheimer's disease conference, OC16 for ADNI*.

Shinotoh, H., Fukushi, K., Nagatsuka, S., & Irie, T. (2004). Acetylcholinesterase imaging: its use in therapy evaluation and drug design. *Current Pharmaceutical Design, 10*, 1505–1517.

Shiraishi, T., Kikuchi, T., Fukushi, K., Shinotoh, H., Nagatsuka, S., Tanaka, N., et al. (2005). Estimation of plasma IC50 of donepezil hydrochloride for brain acetylcholinesterase inhibition in monkey using N-[11C]methylpiperidin-4-yl acetate ([11C]MP4A) and PET. *Neuropsychopharmacology, 30*, 2154–2161.

Shotbolt, P., Tziortzi, A. C., Searle, G. E., Colasanti, A., van der Aart, J., Abanades, S., et al. (2012). Within-subject comparison of [(11)C]-(+)-PHNO and [(11)C]raclopride sensitivity to acute amphetamine challenge in healthy humans. *Journal of Cerebral Blood Flow and Metabolism, 32*, 127–136.

Siegle, G. J., Thompson, W., Carter, C. S., Steinhauer, S. R., & Thase, M. E. (2007). Increased amygdala and decreased dorsolateral prefrontal BOLD responses in unipolar depression: related and independent features. *Biological Psychiatry, 61*, 198–209.

Smith, S. M., Miller, K. L., Salimi-Khorshidi, G., Webster, M., Beckmann, C. F., Nichols, T. E., et al. (2011). Network modelling methods for FMRI. *NeuroImage, 54*, 875–891.

Smith, S. M., Beckmann, C. F., Andersson, J., Auerbach, E. J., Bijsterbosch, J., Douaud, G., et al. (2013). Resting-state fMRI in the Human Connectome Project. *NeuroImage, 80*, 144–168.

Sofroniew, M. W. (2015). Astrocyte barriers to neurotoxic inflammation. *Nature Reviews. Neuroscience, 16*, 249–263.

Sormani, M. P., & Bruzzi, P. (2013). MRI lesions as a surrogate for relapses in multiple sclerosis: a meta-analysis of randomised trials. *Lancet Neurology, 12*, 669–676.

Sormani, M. P., & Pardini, M. (2017). Assessing repair in multiple sclerosis: outcomes for phase II clinical trials. *Neurotherapeutics.* https://doi.org/10.1007/s13311-017-0558-3 [Epub ahead of print].

Sotiropoulos, S. N., Jbabdi, S., Xu, J., Andersson, J. L., Moeller, S., Auerbach, E. J., et al. (2013). Advances in diffusion MRI acquisition and processing in the Human Connectome Project. *NeuroImage, 80*, 125–143.

Sperling, R. A., Aisen, P. S., Beckett, L. A., Bennett, D. A., Craft, S., Fagan, A. M., et al. (2011). Toward defining the preclinical stages of Alzheimer's disease: recommendations from the National Institute on Aging-Alzheimer's Association workgroups on diagnostic guidelines for Alzheimer's disease. *Alzheimer's & Dementia, 7*, 280–292.

Sperling, R. A., Jack, C. R., Jr., Black, S. E., Frosch, M. P., Greenberg, S. M., Hyman, B. T., et al. (2011). Amyloid-related imaging abnormalities in amyloid-modifying therapeutic trials: recommendations from the Alzheimer's Association Research Roundtable Workgroup. *Alzheimer's & Dementia, 7*, 367–385.

Sperling, R. A., Rentz, D. M., Johnson, K. A., Karlawish, J., Donohue, M., Salmon, D. P., et al. (2014). The A4 study: stopping AD before symptoms begin? *Science Translational Medicine, 6*, 228–234.

Steketee, R. M., Mutsaerts, H. J., Bron, E. E., van Osch, M. J., Majoie, C. B., & van der Lugt, A. (2015). Quantitative functional arterial spin labeling (fASL) MRI—sensitivity and reproducibility of regional CBF changes using pseudo-continuous ASL product sequences. *PLoS One. 10*(7), e0132929. https://doi.org/10.1371/journal.pone.0132929.

Suhara, T., Chaki, S., Kimura, H., Furusawa, M., Matsumoto, M., Ogura, H., et al. (2017). Strategies for utilizing neuroimaging biomarkers in CNS drug discovery and development: CINP/JSNP working group report. *The International Journal of Neuropsychopharmacology, 20*, 285–294.

Sullivan, D. C., Obuchowski, N. A., Kessler, L. G., Raunig, D. L., Gatsonis, C., Huang, E. P., et al. (2015). Metrology standards for quantitative imaging biomarkers. *Radiology, 277*, 813–825.

Takano, A., Stenkrona, P., Stepanov, V., Amini, N., Martinsson, S., Tsai, M., et al. (2016). A human [(11)C]T-773 PET study of PDE10A binding after oral administration of TAK-063, a PDE10A inhibitor. *NeuroImage, 141*, 10–17.

Tauscher, J., Jones, C., Remington, G., Zipursky, R. B., & Kapur, S. (2002). Significant dissociation of brain and plasma kinetics with antipsychotics. *Molecular Psychiatry, 7*, 317–321.

Tauscher, J., Kielbasa, W., Iyengar, S., Vandenhende, F., Peng, X., & Mozley, D. (2010). Development of the 2nd generation neurokinin-1 receptor antagonist LY686017 for social anxiety disorder. *European Neuropsychopharmacology, 20*, 80–87.

Terpstra, M., Cheong, I., Lyu, T., Deelchand, D. K., Emir, U. E., Bednarık, P., et al. (2016). Test-retest reproducibility of neurochemical profiles with short-echo, single-voxel MR spectroscopy at 3T and 7T. *Magnetic Resonance in Medicine, 76*, 1083–1091.

Tofts, P. S., Brix, G., Buckley, D. L., Evelhoch, J. L., Henderson, E., Knopp, M. V., et al. (1999). Estimating kinetic parameters from dynamic contrast-enhanced T(1)-weighted MRI of a diffusable tracer: standardized quantities and symbols. *Journal of Magnetic Resonance Imaging, 10*, 223–232.

Tolboom, N., Yaqub, M., Boellaard, R., Luurtsema, G., Windhorst, A. D., Scheltens, P., et al. (2009). Test-retest variability of quantitative [11C] PIB studies in Alzheimer's disease. *European Journal of Nuclear Medicine and Molecular Imaging, 36*, 1629–1638.

Traboulsee, A., Simon, J. H., Stone, L., Fisher, E., Jones, D. E., Malhotra, A., et al. (2016). Revised recommendations of the Consortium of MS Centers Task Force for a standardized MRI protocol and clinical guidelines for the diagnosis and follow-up of multiple sclerosis. *American Journal of Neuroradiology, 37*, 394–401.

Tronel, C., Largeau, B., Santiago-Ribeiro, M. J., Guilloteau, D., Dupont, A. C., & Arlicot, N. (2017). Molecular targets for PET imaging of activated microglia: the current situation and future expectations. *International Journal of Molecular Sciences, 18*, 802–824.

Tune, L., Tiseo, P. J., Ieni, J., Perdomo, C., Pratt, R. D., Votaw, J. R., et al. (2003). Donepezil HCl (E2020) maintains functional brain activity in patients with Alzheimer disease: results of a 24-week, double-blind, placebo-controlled study. *The American Journal of Geriatric Psychiatry, 11*, 169–177.

Upadhyay, J., Anderson, J., Schwarz, A. J., Coimbra, A., Baumgartner, R., Pendse, G., et al. (2011). Imaging drugs with and without clinical analgesic efficacy. *Neuropsychopharmacology, 36*, 2659–2673.

Van de Bank, B. L., Emir, U. E., Boer, V. O., van Asten, J. J. A., Maas, M. C., Wijnen, J. P., et al. (2015). Multi-center reproducibility of neurochemical profiles in the human brain at 7 Tesla. *NMR in Biomedicine, 28*, 306–316.

Van Laere, K., De Hoon, J., Bormans, G., Koole, M., Derdelinckx, I., De Lepeleire, I., et al. (2012). Equivalent dynamic human brain NK1-receptor occupancy following single-dose i.v. fosaprepitant vs. oral aprepitant as assessed by PET imaging. *Clinical Pharmacology and Therapeutics, 92*, 243–250.

Van Laere, K., Ahmad, R. U., Hudyana, H., Dubois, K., Schmidt, M. E., Celen, S., et al. (2013). Quantification of ^{18}F-JNJ-42259152, a novel phosphodiesterase 10A PET tracer: kinetic modeling and test-retest study in human brain. *Journal of Nuclear Medicine, 54*, 1285–1293.

Vavasour, I. M., Laule, C., Li, D. K., Traboulsee, A. L., & MacKay, A. L. (2011). Is the magnetization transfer ratio a marker for myelin in multiple sclerosis? *Journal of Magnetic Resonance Imaging, 33*, 713–718.

Vilela, P., & Rowley, H. A. (2017). Brain ischemia: CT and MRI techniques in acute ischemic stroke. *European Journal of Radiology, 96*, 162–172.

Wang, X., Cusick, M. F., Wang, Y., Sun, P., Libbey, J. E., Trinkaus, K., et al. (2014). Diffusion basis spectrum imaging detects and distinguishes coexisting subclinical inflammation, demyelination and axonal injury in experimental autoimmune encephalomyelitis mice. *NMR in Biomedicine, 27*, 843–852.

Wang, Y., Sun, P., Wang, Q., Trinkaus, K., Schmidt, R. E., et al. (2015). Differentiation and quantification of inflammation, demyelination and axon injury or loss in multiple sclerosis. *Brain, 138*, 1223–1238.

Wang, Y., & Mandelkow, E. (2016). Tau in physiology and pathology. *Nature Reviews. Neuroscience, 17*, 5–21.

Warr, D. G., Hesketh, P. J., Gralla, R. J., Muss, H. B., Herrstedt, J., Eisenberg, P. D., et al. (2005). Efficacy and tolerability of aprepitant for the prevention of chemotherapy-induced nausea and vomiting in patients with breast cancer after moderately emetogenic chemotherapy. *Journal of Clinical Oncology, 23*, 2822–2830.

Wattjes, M. P., Ro, A., Miller, D., Yousry, T. A., Sormani, M. P., de Stefano, M. P., et al. (2015). Evidence-based guidelines: MAGNIMS consensus guidelines on the use of MRI in multiple sclerosis—establishing disease prognosis and monitoring patients. *Nature Reviews. Neurology, 11*, 597–606.

Weiner, M. W., Veitch, D. P., Aisen, P. S., Beckett, L. A., Cairns, N. J., Green, R. C., et al. (2017). Recent publications from the Alzheimer's Disease Neuroimaging Initiative: reviewing progress toward improved AD clinical trials. *Alzheimer's & Dementia, 13*(4), e1–e85. https://doi.org/10.1016/j.jalz.2016.11.007.

Whitwell, J. L., Jack, C. R., Jr., Senjem, M. L., & Josephs, K. A. (2006). Patterns of atrophy in pathologically confirmed FTLD with and without motor neuron degeneration. *Neurology, 66*, 102–104.

Whitwell, J. L., Weigand, S. D., Shiung, M. M., Boeve, B. F., Ferman, T. J., Smith, G. E., et al. (2007). Focal atrophy in dementia with Lewy bodies on MRI: a distinct pattern from Alzheimer's disease. *Brain, 130*, 708–719.

Wiepert, D. A., Lowe, V. J., Knopman, D. S., Boeve, B. F., Graff-Radford, J., Petersen, R. C., et al. (2017). A robust biomarker of large-scale network failure in Alzheimer's disease. *Alzheimer's & Dementia: Diagnosis, Assessment & Disease Monitoring, 6*, 152–161.

Wisse, L. E., Kuijf, H. J., Honingh, A. M., Wang, H., Pluta, J. B., Das, S. R., et al. (2016). Automated hippocampal subfield segmentation at 7T MRI. *American Journal of Neuroradiology, 37*, 1050–1057.

Wolk, D. A., & Detre, J. A. (2012). Arterial spin labeling MRI: an emerging biomarker for Alzheimer's disease and other neurodegenerative conditions. *Current Opinion in Neurology, 25*, 421–428.

Wolz, R., Aljabar, P., Hajnal, J. V., Hammers, A., Rueckert, D., & Alzheimer's Disease Neuroimaging Initiative. (2010). LEAP: learning embeddings for atlas propagation. *NeuroImage, 49,* 1316–1325.

Wolz, R., Schwarz, A. J., Yu, P., Cole, P. E., Rueckert, D., Jack, C. R., Jr., et al. (2014). Robustness of automated hippocampal volumetry across magnetic resonance field strengths and repeat images. *Alzheimer's & Dementia, 10,* 430–438.

Wolz, R., Schwarz, A. J., Gray, K. R., Yu, P., Hill, D. L., & Alzheimer's Disease Neuroimaging Initiative. (2016). Enrichment of clinical trials in MCI due to AD using markers of amyloid and neurodegeneration. *Neurology, 87,* 1235–1241.

Wu, B., Lou, X., Wu, X., & Ma, L. (2014). Intra- and inter-scanner reliability and reproducibility of 3D whole-brain pseudo-continuous arterial spin-labeling MR perfusion at 3T. *Journal of Magnetic Resonance Imaging, 39,* 402–409.

Xia, H., Mao, Q., Eliason, S. L., Harper, S. Q., Martins, I. H., Orr, H. T., et al. (2004). RNAi suppresses polyglutamine-induced neurodegeneration in a model of spinocerebellar ataxia. *Nature Medicine, 10,* 816–820.

Yang, K. -C., Stepanov, V., Amini, N., Martinsson, S., Takano, A., Nielsen, J., et al. (2017). Characterization of [11C]Lu AE92686 as a PET radioligand for phosphodiesterase 10A in the nonhuman primate brain. *European Journal of Nuclear Medicine and Molecular Imaging, 44,* 308–320.

Yu, P., Sun, J., Wolz, R., Stephenson, D., Brewer, J., Fox, N. C., et al. (2014). Operationalizing hippocampal volume as an enrichment biomarker for amnestic mild cognitive impairment trials: effect of algorithm, test-retest variability, and cut point on trial cost, duration and sample size. *Neurobiology of Aging, 35,* 808–818.

Yushkevich, P. A., Pluta, J. B., Wang, H., Xie, L., Ding, S. L., Gertje, E. C., et al. (2015). Automated volumetry and regional thickness analysis of hippocampal subfields and medial temporal cortical structures in mild cognitive impairment. *Hum. Brain Mapping, 36,* 258–287.

Zago, W., Schroeter, S., Guido, T., Khan, K., Seubert, P., Yednock, T., et al. (2013). Vascular alterations in PDAPP mice after anti-Abeta immunotherapy: implications for amyloid-related imaging abnormalities. *Alzheimer's & Dementia, 9*(5 Suppl), S105–S115.

Zamuner, S., Di Iorio, V. L., Nyberg, J., Gunn, R. N., Cunningham, V. J., Gomeni, R., et al. (2010). Adaptive-optimal design in PET occupancy studies. *Clinical Pharmacology and Therapeutics, 87,* 563–571.

Zhang, L., Beal, S. L., & Sheiner, L. B. (2003). Simultaneous vs. sequential analysis for population PK/PD data I: best-case performance. *Journal of Pharmacokinetics and Pharmacodynamics, 30,* 387–404.

Zlokovic, B. V. (2013). Cerebrovascular effects of apolipoprotein E: implications for Alzheimer disease. *JAMA Neurology, 70,* 440–444.

Zu, T., Duvick, L. A., Kaytor, M. D., Berlinger, M. S., Zoghbi, H. Y., Clark, H. B., et al. (2004). Recovery from polyglutamine-induced neurodegeneration in conditional SCA1 transgenic mice. *The Journal of Neuroscience, 24,* 8853–8861.

10

PET Occupancy and Competition in Translational Medicine and CNS Drug Development

Paul Cumming*,†, Gerhard Gründer‡

*Department of Nuclear Medicine, Inselspital, University of Bern, Bern, Switzerland †School of Psychology and Counselling and IHBI, Queensland University of Technology, and QIMR Berghofer Medical Research Institute, Brisbane, QLD, Australia ‡Department of Molecular Neuroimaging, Central Institute of Mental Health (CIMH), Mannheim, Germany

I INTRODUCTION

In a famous story related by Henry Moseley, George de Hevesy was inspired while drinking a cup of tea in 1913 to speculate about the fate of the individual water molecules as they were absorbed into his body. Moseley did not survive the Great War, but de Hevesy went on to interrogate his tea problem by conducting the first isotope dilution experiment, using heavy water to calculate the aqueous volume of the human body. Stemming from this early application of deuterium, a stable isotope, radionuclides have become standard tools for experimental biochemistry since the 1950s. The more complete realization of de Hevesy's reverie emerged in the 1970s with the advent of single-photon emission tomography (SPECT) and positron emission tomography (PET) imaging of cerebral perfusion, using a radionuclide such as xenon-133 that mixes perfectly with water or using water itself labeled with oxygen-15. In subsequent years, radiotracer imaging has extended far beyond its humble beginnings in perfusion studies and is now providing transformational insights into the molecular basis of normal brain function, pathophysiology, and drug action.

Molecular imaging is an autoradiographic technique applied for the living brain; whereas the abundance of a neuroreceptor or other binding site might formerly have been measured in tissue samples obtained postmortem or as biopsy specimens, molecular imaging with SPECT or PET radiopharmaceuticals makes it possible to visualize and even quantify particular classes of binding sites in the living brain. Radiopharmaceutical imaging often uses derivatives of well-characterized drugs modified by incorporation into their structure of appropriate radionuclides. SPECT images of tracer uptake are obtained by mapping single gamma emissions derived from radionuclides such as iodine-123, whereas PET relies upon the detection of pairs of gamma rays originating at the point of annihilation of a positron derived from the decay of short-lived radionuclides such as carbon-11 or fluorine-18. Although the particular instrumentation and image reconstruction procedures differ, molecular imaging experiments are always a matter of following the tracer from the point of injection into venous circulation to its deposition into the brain or other target tissue. So, we can now follow the fate not only of water from the tea but also perhaps of the caffeine, as it is absorbed in the gut and passes through the liver and into the arterial circulation and thence across the blood-brain barrier to interact with its binding sites, the adenosine receptors.

II COMPETITION FROM ENDOGENOUS AGONISTS

Molecular brain imaging began with studies of dopamine synthesis using the tracer [^{18}F]fluorodopa and investigations of dopamine receptors. In the earliest phase of dopamine receptor imaging in the living brain, PET examinations were made using either the benzamide D2/3 antagonist [^{11}C]raclopride or the butyrophenone D2/3/4 antagonist N-[^{11}C]methylspiperone ([^{11}C]NMSP). Seminal studies showed increased striatal [^{11}C]NMSP binding in untreated patients with schizophrenia (Wong et al., 1986), whereas [^{11}C]raclopride PET showed no such increase (Farde et al., 1990). This conundrum was partially resolved by Philip Seeman, who noted that whereas raclopride binding was vulnerable to competition from endogenous dopamine, this competition was less evident for butyrophenone ligands (Seeman, 1988; Seeman, Guan, & Niznik, 1989). As such the butyrophenone imaging study may give a clearer depiction of the abundance of dopamine D2-like receptors, uncomplicated by effects of competition from brain dopamine. In addition to this issue of vulnerability to competition, the 10-fold higher affinity of NMSP was initially presented as a factor contributing to the discrepant PET findings in schizophrenia (Hall, Wedel, Halldin, & Farde, 1990). However, according to current understanding, competitive displacement is not really a matter of the radiopharmaceutical affinity per se, so long as it is administered at an essentially massless tracer dose. Rather, it is the prevailing occupancy of binding sites by the invisible partner (in this case endogenous dopamine) that determines the fraction of the total number of dopamine D2/3 receptors detectable by [^{11}C]raclopride in the living brain. Still other factors muddy the waters of a simple competition model, notably the favored internalization of the [^{3}H]spiperone receptor complex into an intracellular compartment presumably invulnerable to competition from endogenous dopamine (Chugani, Ackermann, & Phelps, 1988). Indeed, aspects of the competition model are firmly linked to the dependence of a receptor-ligand on the cellular environment experienced by plasma membrane receptors versus internalized receptors, now residing in an endosomal compartment and thus invulnerable to endogenous dopamine (Quelch, Withey, Nutt, Tyacke, & Parker, 2014). Other lines of research suggest that some 80% of dopamine D2/3 receptors in the striatum normally reside in an intracellular compartment (Cumming, 2011).

The prototypic PET tracers [^{11}C]NMSP and [^{11}C]raclopride are pharmacologically defined as antagonists, since their binding to dopamine receptors blocks activation of dopamine or other agonists. As with other seven transmembrane receptors, binding of dopamine to D2-like receptors normally activates intracellular second messenger systems by causing an intracellular component of the receptor complex to exchange GDP for GTP; this event provokes the receptor to shift to a low affinity state with respect to its agonists. In contrast, antagonist ligands are indifferent to the GTP shift in agonist affinity state of the receptor. This distinction is of fundamental importance to medicinal chemistry, as some 40% of all modern drugs act at G protein-coupled receptors. In the context of molecular imaging, the distinction can influence the nature of the binding competition occurring in the living brain, as

agonist ligands may only recognize a subset of their receptors occurring in a high-affinity state at the time of scanning.

Despite these caveats and reservations, the competition model between endogenous dopamine and radiopharmaceuticals binding to dopamine D2/3 receptors has been of immense heuristic value in clinical psychiatry research, as reviewed by Laruelle (2000). Amphetamine is a psychostimulant that releases large amounts of intracellular dopamine by a facilitated exchange diffusion process. Whereas amphetamine has little or no inherent affinity for dopamine D2/3 receptors, the surge of dopamine release provoked by amphetamine decreases the availability of binding sites for [^{11}]raclopride and other benzamide ligands by 10–30% in humans and nonhuman primates, in measure of the increased D2/3 receptor occupancy by endogenous dopamine. Remarkably a low dose of amphetamine displacing only 10% of [^{123}I]IBZM binding in healthy humans caused a twofold higher displacement in patients with schizophrenia (Laruelle et al., 1996; Laruelle, Abi-Dargham, Gil, Kegeles, & Innis, 1999). The converse experiment entails transient dopamine depletion with the tyrosine hydroxylase inhibitor AMPT; this pharmacological treatment increased the [^{123}I]IBZM binding by 9% in healthy humans relative to their own baseline SPECT scan versus a 19% increase in patients with schizophrenia (Abi-Dargham et al., 2000). Together, these robust findings indicate a greater dynamic range for dopamine receptor occupancy in schizophrenia in conjunction with supernormal occupancy at baseline, although it is not entirely clear if the difference reflects only tonic dopamine levels and the amount of amphetamine-releasable dopamine at hand or is also related to the affinity state of the dopamine receptors.

Notably the scatter plots of individual SPECT results clearly show evidence for heterogeneity; nearly one-half of patients with schizophrenia had amphetamine-induced displacement close to the normal mean of 10%, whereas the remainder had a nearly 30% decrease in [^{123}I]IBZM binding. This heterogeneous profile was also seen in the AMPT depletion studies and for that matter also in [^{18}F]fluorodopa PET and other tracers of dopamine synthesis in schizophrenia (Fusar-Poli & Meyer-Lindenberg, 2013). While increased dopamine synthesis capacity is a consistent finding in many PET studies, there is never complete separation of the findings in groups of patients and healthy controls. This heterogeneity has implications that have yet to be played out; dopamine antagonists such as haloperidol might be expected to have greater efficacy in those schizophrenia patients with a hyperactive dopamine system, but what rationale is there for blocking dopamine transmission in patients for whom molecular markers are normal? If nothing is amiss with dopamine in half of schizophrenia patients, what therapeutic channel might be more justified?

In the present context, population heterogeneity will place demands on the minimum group size required to draw firm conclusions from occupancy studies.

The competition between endogenous dopamine and benzamide antagonists for dopamine D2/3 binding sites in the living brain is well established. However, there seems to be a ceiling effect on the fraction of sites that is vulnerable to amphetamine-evoked dopamine release. In mice, only some 40% occupancy could be obtained after amphetamine treatment, based on the maximal displacement of the antagonist [11C]raclopride. However, in the same mice, approximately 70% of the D2/3 sites binding the agonist ligand N-[3H]propylnorapomorphine were displaced under the amphetamine condition (Cumming et al., 2002). Similar results have been reported in human PET studies comparing the displaceabilities of agonist and antagonist ligands by amphetamine-evoked dopamine release (Narendran et al., 2010; Shotbolt et al., 2012). Amphetamine is among the most potent releasers of dopamine, and most other pharmacological challenges provoke relatively mild effects on interstitial dopamine levels. Thus a high dose of nicotine caused only a modest (5%–10%) displacement of [11C]raclopride binding in the striatum of living pigs (Cumming et al., 2003), no doubt indicative of its indirect pharmacological action at D2/3 sites. Other studies in monkeys showed that binging of the D3-prefering agonist [11C]PHNO was fitter than the antagonist [11C]raclopride for detecting the dopamine release evoked by nicotine challenge (Gallezot et al., 2014). These consistent findings across species support a model in which some 50% of dopamine D2/3 binding sites are normally occurring in a high-affinity state for binding dopamine and exogenous agonists, whereas the remaining receptors are refractory to agonist competition. This fraction presumably determines the ceiling effect on displaceability of radiopharmaceuticals by dopamine after an amphetamine challenge.

Efforts to generalize the endogenous agonist occupancy model to other neurotransmitter systems have been less successful than the case of dopamine. For example, a PET study with the serotonin 5-HT2A antagonist [18F]altanserin showed only 10%–20% displacement of the specific binding in the human brain after treatment with the potent serotonin releaser dexfenfluramine (Quednow et al., 2012). Based on arguments presented earlier, one might suppose that the relatively modest displacement by serotonin could indicate a lower agonist binding fraction of the case of serotonin 5-HT2A receptors. This explanation was invoked to explain the invulnerability of the serotonin 5-HT1A antagonist [18F]MPPF binding in monkey brain to challenge with the serotonin releaser fenfluramine (Udo de Haes, Harada, Elsinga, Maguire, & Tsukada, 2006). However, others have reported a dose-dependent displacement of hippocampal [18F]MPPF binding in rats

treated with fenfluramine (Zimmer et al., 2002). Significant changes in competition from endogenous serotonin might be attainable under the rather extreme pharmacological challenge conditions that are possible in animal studies but might well be unsafe in humans.

Given the experience with dopamine D2/3 receptors, one might predict that a serotonin agonist ligand would be fitter for detecting fluctuations in the serotonin receptor occupancy by endogenous serotonin. However, the serotonin 5-HT1A agonist ligand [11C]CUMI-101 proved to be invulnerable to challenge with intravenous citalopram, a serotonin selective reuptake inhibitor (SSRI; Pinborg et al., 2012), although others found 13%–30% [11C]CUMI-101 binding reduction in the brain of a monkey after a challenge with a high intravenous dose of citalopram (Milak et al., 2011). These discrepancies may be in part a matter of the pharmacodynamics of drugs acting to increase synaptic serotonin levels by blockade of plasma membrane transporters. Thus calculations based on serotonin receptor affinity and expected brain concentrations after a single oral dose of citalopram suggest that the increase in synaptic serotonin may simply be too small to be discernible in the PET competition paradigm (Tyacke & Nutt, 2015).

PET studies with [11C]yohimbine reveal the distribution of α_2-adrenergic receptors in the brain of living pigs (Jakobsen et al., 2006). This binding proved to be partially sensitive to displacement following amphetamine challenge (Landau, Doudet, & Jakobsen, 2012) and was furthermore responsive to electric stimulation of the vagus nerve (Landau et al., 2015). These findings in pig may present an avenue for investigating noradrenaline transmission in the human brain, notably in relation to the antidepressant and antiseizure actions attributed to vagal nerve stimulation, but there have not yet been any translational clinical PET studies exploiting this paradigm in humans.

Other than the biogenic monoamines, the best case for the endogenous agonist competition model is presented by opioid receptors. Here again the agonist binding fraction of receptors may be a factor influencing the sensitivity of the method. Thus pharmacological challenge with an opiate full agonist had little effect on binding of the weak partial opiate agonist [11C]diprenorphine in a rat brain (Hume et al., 2007). However, others report that smoking nicotine-containing cigarettes resulted in displacement of the opioid agonist ligand [11C]carfentanil to an extent that was sensitive to a genetic polymorphism of the human opioid receptor (Domino, Hirasawa-Fujita, Ni, Guthrie, & Zubieta, 2015). While the pharmacological mechanism of the nicotinic/opioid interaction was not established, this kind of result may eventually illuminate the pharmacogenetic basis of individual vulnerability to substance abuse. However, as in the case of pramipexole competition against [18F]fallypride binding discussed in the succeeding text, opioid competitors and PET ligands may be recognizing different, albeit overlapping, populations of receptors in the living brain.

Furthermore, we do not yet know what factors are intervening in the competition model for the case of opioid peptide receptors, nor has there been a formal demonstration that endogenous opioid peptides can actually displace any PET ligand, agonist or antagonist. This would need experimental confirmation by combining small-animal PET with cerebral microdialysis to link binding changes with altered concentration of specific opioid peptides in the interstitial medium, a standard of proof hitherto met only for the cases of dopamine and serotonin. The competition model with opioid receptors is even more problematic for the case of nonpharmacological challenges. While acute administration of a painful stimulus causes some displacement of the opioid agonist [11C]carfentanil in the brain of awake humans (Scott, Stohler, Koeppe, & Zubieta, 2007), it seems rather a matter of conjecture that this should be attributed to increased opioid receptor release. To post a rhetorical question, if pain causes release of opioid peptides in the brain to an extent causing significant occupancy at opioid receptors, why do we feel pain at all? Similarly the changes in [11C]diprenorphine binding in the human brain developing after successful surgical treatment for neuralgic pain may be adaptive changes rather than indication of altered occupancy by opioid peptides at their binding sites (Jones et al., 1999).

III COMPETITIVE BINDING FROM PHARMACEUTICALS

PET studies of receptor occupancy by endogenous agonists such as dopamine and serotonin potentially provide information about neurotransmission that is otherwise unobtainable without invasive procedures. While this prospect has attracted considerable interest in the research community, the more common form of the PET competition paradigm entails measuring the occupancy by a therapeutic medication at binding sites in the brain. Here the pharmaceutical displaces the binding of the radiotracer at its sites in the living brain. From the observed binding change relative to a nonmedicated baseline condition, one calculates the occupancy as a percentage. In the pioneering instance of this approach, [11C]raclopride PET was used to determine therapeutic occupancy at dopamine D2/3 receptors of antipsychotic medications used to treat schizophrenia (Farde, Hall, Ehrin, & Sedvall, 1986). In that study the occupancy at striatal D2/3 receptors was as high as 90%, which was initially believed to mark the threshold for effective treatment. However, there is now general acceptance that striatal D2/3 occupancy of 65%–80% is associated with effective treatment with most antipsychotics, albeit

with a few exceptions (Farde et al., 1992; Gründer, Hippius, & Carlsson, 2009). The risk for extrapyramidal motor side effects increases greatly above the 80% threshold; excessive blockade of dopamine receptors provokes a distressing, uncomfortable, and even disabling syndrome of iatrogenic parkinsonism, so it is most important to establish an appropriate dosage regimen, optimizing benefits while minimizing side effects. A metaanalysis of 51 occupancy studies with different antipsychotic medications shows that therapeutic occupancy at striatal receptors extends over a considerable range, from only 49% for quetiapine and 62% for clozapine to 92% for haloperidol and risperidone and even 96% for olanzapine (Lako, van den Heuvel, Knegtering, Bruggeman, & Taxis, 2013). This is with the caveat that this study reported neither the doses nor the mean plasma drug concentrations at the time of scanning. However, the overall results illustrate that striatal D2/3 receptor occupancy does not in itself represent a biomarker distinguishing first-generation (FGAs or "conventional" and "classical") from second-generation (SGAs or "atypical") antipsychotics, although some of the SGAs, specifically clozapine, quetiapine, and aripiprazole, are characterized by occupancies at the extreme ends of the spectrum (Gründer et al., 2009).

Whereas [11C]raclopride PET is suited for quantitation of dopamine D2/3 sites where they are most abundant, higher affinity ligands such as [18F]fallypride can also detect dopamine receptors in the cortex and other brain regions of lower receptor density (see Fig. 10.1). Like other benzamide PET ligands, [18F]fallypride reveals the composite of D2 and D3 sites, but quantitation requires prolonged PET recordings lasting several hours to attain a condition of equilibrium binding in regions of high receptor density. Nonetheless, [18F]fallypride has proven useful for clinical research, as therapeutic and side effect profiles of antipsychotic drugs may relate to actions at cortical receptors or other extrastriatal sites that cannot be detected with [11C]raclopride PET or [123I]IBZM SPECT. Indeed, Pilowsky et al. using SPECT and the high-affinity tracer [123I]epidepride reported markedly higher binding of the prototypical SGA clozapine in extrastriatal brain regions than in the striatum, suggesting that clozapine exerts "limbic selectivity" (Pilowsky et al., 1997). This finding could later be confirmed in a larger patient sample with [18F]fallypride PET. Mean D2/3 receptor occupancy by clozapine was significantly higher in cortical (inferior temporal cortex 55%) than in striatal regions (putamen 36% and caudate 43%; Gründer et al., 2006). Like clozapine the SGA quetiapine had the highest D2/3 occupancy in the cerebral cortex (44%) versus the putamen (only 26%) of patients with schizophrenia (Vernaleken et al., 2010), where these occupancies were calculated relative to receptor availability in a healthy population (which may result in bias, if receptor abundances are affected by disease or treatment history). The "preferential" extrastriatal binding was less pronounced

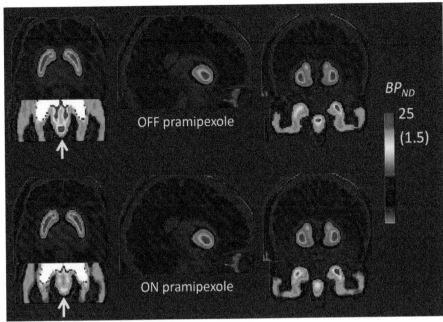

FIG. 10.1   Mean parametric maps of [^{18}F]fallypride binding potential (BP$_{ND}$) in a group of patients with Parkinson's disease scanned after cessation of pramipexole monotherapy (upper row) and in the condition of monotherapy (lower row). The parametric maps are projected upon an MR template, with the scale set at 25 units of BP$_{ND}$ for visualizing the striatum, and rescaled to 1.5 for the inserts so as to show binding changes in regions of low receptor abundance, that is, the midbrain (arrow) and temporal cortex. Evidently the method was not sensitive to occupancy by pramipexole (3 × 0.7 mg p.d.) in the striatum, where D2 binding predominates, but did reveal a 25% displacement in the midbrain; this may reflect preferential occupancy of the D3-prefering agonist in the substantia nigra, where all or most of the [^{18}F]fallypride represents D3 receptors. Modified from Deutschländer, A., la Fougère, C., Boetzel, K., Albert, N. L., Gildehaus, F. J., Bartenstein, P., et al. (2016). Occupancy of pramipexole (Sifrol) at cerebral dopamine D2/3 receptors in Parkinson's disease patients. NeuroImage Clinical, 12, 41–46.

with the SGA ziprasidone (Vernaleken et al., 2008), which has a somewhat higher affinity for D2/3 receptors than clozapine and quetiapine. Indeed the latter two SGAs have among the lowest affinity for D2-like dopamine receptors of any antipsychotic in clinical use. Antipsychotic efficacy of clozapine and quetiapine is usually associated with D2/3 receptor occupancies markedly below the "therapeutic occupancy window" of 65%–80% in the striatum that characterizes most other antipsychotics (Gründer et al., 2006, 2009). The observation that these SGAs block extrastriatal D2 receptors to a considerably higher extent implicates the cortical receptors as a crucial site of antipsychotic drug action.

In the ziprasidone occupancy study, there was clear evidence for time dependence of the occupancy at striatal receptors, with higher occupancy at briefer intervals after dosing (Vernaleken et al., 2008). Furthermore, occupancy of amisulpride at striatal dopamine D2/3 receptors of patients with schizophrenia correlated with the plasma concentration at the time of tracer injection, rather than the dosage regimen per se (Vernaleken et al., 2004). A more systematic investigation of this temporal dependence showed that occupancy at striatal [^{11}C]raclopride binding sites in patients with schizophrenia declined from 68% to 11% in 24 h after cessation of treatment with ziprasidone, indicating an occupancy half-life of 8 h

(Suzuki et al., 2013). As in most patient studies, occupancy was estimated relative to findings in healthy age-matched controls, which (as noted earlier) assumes that baseline D2/3 receptor availability is negligibly different between patients and healthy controls.

While Gründer et al. (2008) could not detect differential striatal and extrastriatal binding of the partial agonist aripiprazole in their PET study with [^{18}F]fallypride (Gründer et al., 2008), another [^{18}F]fallypride PET study of similar design showed occupancy by aripiprazole was slightly higher at cortical binding sites than in the striatum (Kegeles et al., 2008). Conversely a dual-tracer study with [^{11}C]raclopride and also [^{11}C]FLB457, an alternate high-affinity ligand suitable for cortical receptor quantitation, did not show any regional differences in occupancy by clozapine (Talvik et al., 2001), albeit in a rather small group of patients. Here, some caution may be required, as striatal D2/3 receptor occupancy was determined with [^{11}C]raclopride, while extrastriatal occupancy was quantified with [^{11}C]FLB457, which makes assessments difficult to compare. In addition, the PET occupancy method is prone to bias and low sensitivity in quantifying [^{18}F]fallypride or [^{11}C]FLB457 in regions of low receptor abundance such as the cortex, especially perhaps in a pharmacological condition of substantial blocking.

The concept of typicality (FGAs vs SGAs) is muddied by pharmacological distinctions, which are relevant to the therapeutic occupancy in relation to the risk for extrapyramidal side effects, as noted earlier. However, the atypical antipsychotic aripiprazole showed dose-dependent occupancy at dopamine D2/3 sites in an [^{11}C]raclopride PET study, attaining occupancy higher than 90% without provoking any extrapyramidal symptoms at all (Yokoi et al., 2002). This favorable property of aripiprazole is certainly due to its pharmacological nature as a partial agonist, as distinct from a full receptor blocker. Even with nearly complete occupancy in the striatum, there is evidently sufficient activation of the D2/3 receptors to avoid motor symptoms. Evidently the intrinsic partial agonist property of the drug "clamps" dopaminergic signaling at a level that disfavors positive symptoms of schizophrenia. This issue is of great importance to patient compliance; switching from risperidone or olanzapine to aripiprazole resulted in a significant increase in subject well-being, despite the very high occupancy (Mizrahi et al., 2009). Another off-target aspect of antipsychotic medication is the risk of hyperprolactinemia due to dopamine D2/3 antagonism at lactotrophs of the pituitary gland, which can result in loss of libido, amenorrhea, and galactorrhea. A [^{11}C]raclopride PET study established a threshold occupancy by olanzapine or risperidone of 66% for provoking iatrogenic hyperprolactinemia in older patients with schizophrenia (Iwata et al., 2016); this threshold is less than that provoking extrapyramidal motor symptoms. Indeed, physicochemical properties of an antipsychotic drug seem to play a large role in its propensity to induce hyperprolactinemia, since the two drugs with the most pronounced effect on pituitary function, amisulpride and risperidone, do so at quite low doses and plasma concentrations.

Other aspects of aripiprazole therapy have been studied in a PET study using the high-affinity dopamine D2/3 antagonist [^{18}F]fallypride, with scanning at various times after cessation of treatment (Gründer et al., 2008). The occupancy was equally high in the striatum over a wide range of plasma drug levels, and high occupancy persisted long after the last dose of medication, even at 1 week after the last medication, depending on the attained plasma concentration. This likely reflects aripiprazole's very long plasma elimination half-life of about 3 days and the very high affinity for D2/3 receptors. Of course, aspects of aripiprazole's pharmacological profile may well be mediated by binding at sites other than dopamine D2/3 receptors. Since there are so many neuroreceptor types in the brain, PET studies aiming to ascertain binding at multiple sites must be informed by prior knowledge of drug specificity.

"Atypicality" of aripiprazole and other antipsychotic medications has been the topic of a number of PET studies employing ligands for dopamine and serotonin receptors. In an early dual-tracer study, Nordström et al. (1995) showed that occupancy of the prototypic atypical antipsychotic clozapine at 5-HT2 receptors exceeded that at D2/3 receptors. In a subsequent study, aripiprazole showed 90% occupancy at striatal dopamine D2/3 receptors labeled with [^{11}C]raclopride, 54%–60% occupancy at serotonin 5-HT2 receptors labeled with [^{18}F]setoperone, and only 16% at 5-HT1A receptors labeled with [^{11}C]WAY100635 (Mamo et al., 2007). In another example the occupancy of lurasidone at cortical serotonin 5-HT2A sites in the monkey brain was 40% at a dose blocking nearly 90% of striatal dopamine D2/3 sites, whereas in the same study, olanzapine was equipotent at the two classes of receptors (Nakazawa et al., 2013). These results highlight the importance of considering assumptions about the sites of action of a given drug; while the D2/3 receptor occupancy may be decisive for antipsychotic efficacy, occupancy at serotonin 5-HT2A or other receptor types may have important consequences for side effects or indeed may contribute to the therapeutic action.

As reviewed earlier the occupancy model has drawn considerable attention in relation to occupancy by antipsychotic medications at dopamine D2/3 receptors due to the key issue of obtaining sufficient blockade to relieve symptoms of schizophrenia while not provoking extrapyramidal side effects due to excessive dopamine antagonism. However, the converse situation, that is to say the therapeutic occupancy by a dopamine agonist, has been investigated in only a very few PET studies. Directly acting dopamine D2/3 receptor agonists are widely used for the symptomatic treatment of Parkinson's disease, either as monotherapy or in conjunction with the indirect agonist L-DOPA. Here the danger of excessive occupancy lies in the risk of provoking nausea, dizziness, and agitation or more disturbing side effects such as hallucinations and impulse control disorders. From the 80% threshold of dopamine receptor blockade seen in antipsychotic studies, one might suppose that agonists normally activate only a small fraction of dopamine receptors. Indeed, in one of the few such studies in Parkinson's disease patients, acute treatment with the dopamine agonist apomorphine (0.03 mg/kg) decreased [^{11}C]raclopride binding by only 9% in the more intact putamen and by 15% on the side contralateral to the main symptoms. This asymmetrical effect suggested "sensitization," that is to say an increase in affinity state of dopamine receptors for agonists on the side with the greatest loss of dopamine (de la Fuente-Fernández et al., 2001). In a [^{18}F]fallypride PET study, treatment with a clinically effective dose of the D3-prefering dopamine agonist pramipexole failed to elicit a measurable displacement in the striatum (Fig. 10.1; Deutschländer et al., 2016). This negative result is hard to interpret, although it may be related to the paucity of dopamine D3 receptors in the dorsal striatum; perhaps, one-third of the [^{18}F]fallypride binding in the ventral striatum corresponds to D3 sites, and the D3 fraction is even higher in some extrastriatal brain regions. Indeed the study revealed occupancies by pramipexole in the range of 10%–20% in the thalamus and other extrastriatal regions where D3 receptors may constitute all or most of the total binding revealed by [^{18}F]fallypride, which does not inherently distinguish between D2 and D3 subtypes. Nonetheless the finding of some antiparkinsonian effects obtained without discernible agonist occupancy in the putamen needs close scrutiny; putamen is, after all, the site of main nigrostriatal dopamine degeneration in Parkinson's disease.

The pramipexole study highlights the need to consider bias arising from displacement of a less specific PET ligand by a more pharmacologically specific competitor; more specific investigations require more specific probes, such as the D3-preferring agonist [^{11}C]-(+)-PHNO. While the D3/D2 selectivity of this tracer is also imperfect, its binding can be attributed mainly to D3 sites in some brain regions such as the thalamus, whereas the binding in the putamen is mostly attributable to D2 sites, thus potentially affording two channels of occupancy information in a single PET scan. One such study investigated the brain occupancy by buspirone, an anxiolytic medication presumably acting via partial agonism at 5-HT1A serotonin receptors, but with some affinity as a dopamine antagonist. In humans treated with a high dose of buspirone, there was only 25% occupancy at D2 and likewise D3 receptors, thus excluding important effects via dopamine blockade (Le Foll et al., 2016). In another such study, occupancy of the atypical antipsychotic compound cariprazine was measured in patients with schizophrenia; at the highest dosage, there was nearly complete occupancy at D2 and D3 sites (consistent with its partial agonist profile), but at lower doses, some preference for D3 binding was observed (Girgis et al., 2016). This is in line with preclinical studies characterizing cariprazine as a partial agonist, with six- to eightfold higher affinity for human dopamine D3 over D2 receptors (Veselinović, Paulzen, & Gründer, 2013). This study in patients complemented an earlier multitracer study in nonhuman primate, which showed rather scant occupancy at serotonin 5-HT1A sites labeled with [^{11}C]WAY-100635 at a cariprazine dose evoking substantial occupancy at dopamine receptors (Seneca et al., 2011).

In the study by Seneca et al. (2011), the drug occupancy at 5-HT1A receptors was not uniform throughout the brain, but attained its highest value (30%) in the dorsal raphé. Here the majority of 5-HTA1 binding sites are somatodendritic autoreceptors of serotonin neurons. Thus, even for a supposedly homogeneous class of neuroreceptors, drug occupancy may differ between regions. This cannot be due to regional differences in drug concentration, but must reflect other factors in the living brain. In the case of serotonin autoreceptors, there is reason to expect higher affinity for serotonin agonists, since autoreceptors are positioned to provide feedback modulation of serotonin release. Thus the partial agonist property of cariprazine may explain its preferential binding to 5-HT1A autoreceptors in the dorsal raphé, with less occupancy at postsynaptic 5-HT1A receptors in the hippocampus. The case is clearer in the [^{18}F]fallypride study discussed earlier (Deutschländer et al., 2016), where the agonist pramipexole had highest occupancy in the substantia nigra, a region where the D3 binding sites are autoreceptors on dopamine neurons, and thus apt to have higher affinity for dopamine and other agonists so as best to exert a regulatory function.

If schizophrenia research has been the main driver for dopamine occupancy studies, depression research has placed more emphasis on serotonin. As distinct from neuroreceptors, the serotonin transporter (SERT) in the plasma membrane of serotonin neurons has drawn particular attention due to its importance as the site of action of many clinically used antidepressant compounds, notably the SSRIs. Meyer and colleagues have done extensive and very elaborate work to characterize the binding of the SSRIs (and SSNRIs) to the SERT (Meyer, 2007; Meyer et al., 2004). They determined the SERT occupancy over a wide range of SSRI doses and plasma concentrations, respectively, in large numbers of healthy controls and patients suffering from depression. Concentrations of citalopram, paroxetine, fluoxetine, and sertraline in blood all correlated well with SERT occupancy; at least 80% occupancy was required for optimal clinical outcome (Meyer, 2007; Meyer et al., 2004). Serial PET scanning with [^{11}C]DASB has been used to determine the time course of SSRI occupancy of SERT in the living brain (Arakawa et al., 2016); a few hours after drug administration, the occupancy was 75% for escitalopram (10 mg) and sertraline (50 mg) and remained at 60% some 48 h later, despite the fourfold declines in plasma drug levels. In contrast, paroxetine (20 mg) initially evoked only 40% occupancy, which declined rapidly in the follow-up scans, apparently in relation to the more rapid plasma kinetics. On the other hand, in a β-[^{123}I]CIT SPECT study, occupancy at SERT was only 42% in depressed patients treated with citalopram for 6 weeks (Rominger et al., 2015). The lower occupancy may reflect quantitation biases inherent in the less sensitive SPECT method. The occupancy by escitalopram at SERT in the brain of healthy volunteers has been measured in relation to plasma drug concentrations during [^{11}C]DASB PET recordings; the occupancy was significantly higher in the dorsal raphé than in other brain regions (Kim et al., 2016). As in the case of serotonin and dopamine autoreceptors, some additional factors may influence the nature of competition at transporters lying near the cell bodies. There are exceptions, as in the case of the novel serotonin 5-HT1 antagonist GSK588045, now under development as an anxiolytic and antidepressant, based on its joint SERT blockade and serotonin autoreceptor antagonism. In a [^{11}C]WAY100635 PET study in nonhuman primates, GSK588045 blocked 5-HT1A sites with similar potency at somatodendritic autoreceptors and postsynaptic heteroreceptors (Comley et al., 2015).

In the study of Kim et al. (2016), pharmacokinetic/pharmacodynamic modeling predicted that a daily citalopram dose of 20 mg/kg would be necessary to obtain occupancy exceeding 80%. As usual in such studies, it is not clear what occupancy at SERT is actually required to enhance significantly serotonin transmission. This is also true for the case of tramadol, an opiate analgesic with additional actions as a blocker of SERT and noradrenaline uptake sites (NET). In a [^{11}C]DASB PET study of healthy volunteers, 100 mg tramadol produced 50% occupancy at SERT, suggesting (but not proving) that serotonin may be a factor in its mechanism of action (Ogawa et al., 2014). Clomipramine has been shown to occupy 80% of the SERT at doses as low as 10 mg, with a calculated median effective dose (ED50) of less than 3 mg and an EC50 (plasma concentration estimated to provide half-maximal occupancy) of 1.42 ng/mL (Suhara et al., 2003); doses of 25 mg daily had very high SERT occupancy. These observations are in sharp contrast to the fact that the clinically used clomipramine doses are 50–150 mg/day for depression and sometime much higher doses for patients suffering from obsessive-compulsive disorder (Foa et al., 2005). Therapeutic plasma concentrations for clomipramine are traditionally in the range of 175–450 ng/mL (Hiemke et al., 2011), predicting 99% occupancy at SERT in the brain, based on the findings of Suhara et al. (2003). These profound discrepancies call into question the validity of the clinical studies upon which therapeutic doses and plasma concentrations of the TCAs are based. Another theoretical option to explain the discrepant findings is that the pharmacological principle of TCA (and probably SSRI) action may not be exclusively attributable to the inhibition of serotonin reuptake. In this scenario, binding to the SERT could represent an epiphenomenon of the treatment with this class of drugs (Gründer, Hiemke, Paulzen, Veselinovic, & Vernaleken, 2011). Alternately the optimal therapeutic effects might be obtained by synergism between SERT blockade and some other pharmacological action yet to be identified.

Rat studies with DA-8031, an SSRI being developed for treatment of premature ejaculation, showed a relationship between increased serotonin levels measured by microdialysis and occupancy at SERT sites measured by [^{11}C]DASB PET (Park et al., 2014). As noted earlier, there have not been enough such studies to draw clear conclusions about the threshold of occupancy required to enhance serotonin neurotransmission. Ultimately, it is the functional and behavioral responses that are decisive, rather than the occupancy per se. For example, serotonergic signaling in the amygdala may mediate some aspects of the antidepressant response to SSRI treatment; clinical improvement has been linked to attenuated BOLD signal activation in the amygdala upon provocation of the patient with images of hostile faces. In a group of patients treated for depression, the occupancy of β-[^{123}I]CIT sites in the amygdala by paroxetine medication correlated (weakly) with the individual attenuation of the BOLD signal effect upon prolonged treatment (Ruhé et al., 2014). The authors linked these findings to the changes in attentional bias associated with response to antidepressant treatments via SERT blockade.

Drug occupancy at NET is less well documented than for the other biogenic amine transporters, perhaps due to the inherently low abundance of this transporter in the brain. Using (S,S)-[^{18}F]FMeNER-D(2), perhaps the best of the available NET tracers, the occupancy by the dual-uptake blocker (NET/SERT) milnacipran was 25% at 25 mg and 50% at 200 mg; similar dose-response was seen for occupancy at SERT in the depressed patients (Nogami et al., 2013). Using the same tracer the dose-occupancy relationship was established for the tricyclic antidepressant nortriptyline in a group of depressed patients (Takano et al., 2014). In that study, occupancy in the patients was calculated relative to data from healthy controls. As noted earlier for the case of dopamine receptor occupancy, this approach may be problematic if baseline NET levels are perturbed by the illness. While NET availability is still undocumented in depression, a metaanalysis of SERT studies showed a global 10% reduction in availability of patients with depression, compared with healthy age-matched controls (Gryglewski, Lanzenberger, Kranz, & Cumming, 2014). As such, calculating SERT occupancy in patients relative to control values would result in some overestimation. Taken together the data from PET occupancy studies at biogenic amine transporters are inconclusive in that some studies, for example, those characterizing bupropion, raise the question as to whether clinical effects are obtained through mechanisms other than monoamine reuptake inhibition (Gründer et al., 2011).

BMS-820836 is another novel antidepressant, described as a triple reuptake blocker with affinity for plasma membrane dopamine transporters (DAT) and SERT and NET. In a group of individuals treated under steady-state conditions, BMS-820836 had 80% occupancy at SERT that persisted at least 24 h after the last dose, whereas the occupancy at DAT was more like 25% and tended to decline rapidly after the last dose (Zheng et al., 2015). In that study, DAT availability was measured with [^{11}C]PE2I, a tropane derivative with high affinity and specificity for DAT. This agent has also been used for assessing the therapeutic occupancy of the antinarcolepsy medication modafinil in healthy volunteers relative to a baseline condition; 50% occupancy was obtained with a single dose of 200 mg modafinil, and a relationship between plasma levels and occupancy was established (Kim et al., 2014). This occupancy at clinically relevant doses was comparable with that reported earlier for methylphenidate (Volkow et al., 1998) and cocaine itself (Volkow et al., 1999) at [11C] cocaine binding sites in the human striatum, which mainly reflect binding to DAT, with negligible contribution from SERT and NET. These results may call into question claims that modafinil has low abuse potential despite its cocaine-like action (modafinil is used in some circles as a cognitive performance enhancing agent). However, it is not only the peak occupancy at DAT that defines abuse potential but also the time delay to peak occupancy after drug taking. Indeed, this is part of the rationale for slow-release formulations of methylphenidate for the treatment of attention deficit hyperactivity disorder (ADHD). However, still other factors intervene in the living organism; microdialysis studies and concurrent [18F]FECNT PET scanning in spider monkeys showed a partial dissociation between the time course of occupancy of striatal DAT by tropane compounds and the actual increase in dopamine overflow (Kimmel et al., 2012). Such discrepancies likely relate to additional factors such as autoreceptor-mediated adaptive responses, which would tend to abbreviate the increased dopamine signaling during an interval of blocked reuptake. Also at issue is the pharmacological selectivity of a psychostimulant; while methylphenidate acts at NET and DAT, the occupancy at NET sites labeled with (S,S)-[11C]methylreboxetine was higher at clinically relevant doses, such that the presumably therapeutic potentiation of noradrenaline transmission may be obtained with less effect on dopamine (Hannestad et al., 2010).

In addition to the classic biogenic amines dopamine, serotonin, and noradrenaline, histamine also has a neurotransmitter function. Histamine neurons of the hypothalamus provide a widespread innervation of the neuraxis, acting at several pharmacologically distinct receptors types, which are also expressed in peripheral organs. The histamine H1 receptors in peripheral vasculature are an important target for symptomatic relief of allergies and inflammation. However, the classical H1 antihistamines taken as over-the-counter medications can evoke notable sedation through unintended effects in the central nervous system. This problem led to the development of new structures, which retain their effects on peripheral H1 sites, but with reduced passage across the blood-brain barrier due to active extrusion mediated by the P-glycoprotein. In the first such PET occupancy study with the H1 antagonist ligand [11C]doxepin, oral treatment of the classic antihistamine chlorpheniramine (2 mg) evoked 77% occupancy in the frontal cortex versus only 17% for the second-generation, nonsedating antihistamine terfenadine (60 mg); there was no relationship between occupancy and individual plasma terfenadine concentration (Yanai et al., 1995). Using the same [11C] doxepin PET method, the relatively nonsedating antihistamine cetirizine provoked 30% occupancy of H1 sites in the brain of healthy volunteers, whereas fexofenadine at a therapeutic dose had no detectable occupancy (Tashiro et al., 2004); consistent with the PET results, reaction time and accuracy in cognitive tests were impaired in the cetirizine group but unaffected with fexofenadine treatment. Similarly, occupancies of oral diphenhydramine (56%) and bepotastine (15%) at H1 sites in the brain of healthy young adults matched with subjective reports of sleepiness at the time of scanning (Tashiro et al., 2008). However, escalating dosages of cetirizine showed a nonlinear relationship with occupancy at [11C]doxepin binding sites in the brain of volunteers; the occupancy was 13% for 10 mg and 25% for 20 mg but increased sharply to 68% for 30 mg, which evoked sedation (Tashiro et al., 2009). Presumably the P-glycoprotein extrusion could be overwhelmed by the highest cetirizine dosage, although peripheral pharmacokinetic effects were not excluded in that study. Histamine H1 occupancy measured the morning after an oral administration of diphenhydramine (45%) or bepotastine (17%) provided an explanation for the hangover effect of first-generation antihistamines (Zhang et al., 2010).

Occupancy at H1 sites in the brain may contribute to the side effects or indeed therapeutic actions of psychiatric drugs. Thus the SSRI fluvoxamine (25 mg) did not elicit any displacement of [11C]doxepin binding, whereas mirtazapine (15 mg) had occupancy as high as 90% (Sato et al., 2013). The mirtazapine occupancy correlated with the individual plasma concentration (AUC) during the 3 h after dosing, consistent with the well-known sedative property of that antidepressant. However, the antidepressant effects of mirtazapine have been attributed to substantial antagonism at α_2-adrenergic receptors, as attested by PET studies in pig (Smith et al., 2006), and it may be that part of the sedative profile is also related to adrenergic blockade in synergism with H1 antagonism. Single doses of the atypical antipsychotic medications olanzapine and quetiapine produced 60%–80% H1 occupancy in the brain of healthy volunteers (Sato et al., 2015). This result not only seems relevant to sedation but also has bearing on the weight gain often experienced with chronic administration of these drugs, which has been linked specifically to histaminergic regulation of appetite.

Nicotinic receptors in the brain are relevant to the reinforcing properties of nicotine and tobacco abuse and are emerging targets for therapeutics. The first successful molecular imaging tracer for nicotinic receptors was the SPECT ligand 5-[123I]iodo-85380, which binds to the $\beta2$ subunit of the common $\alpha4\beta2$ receptor subtype. Despite rather slow kinetics requiring long or interrupted SPECT recordings, it serves to establish occupancy after smoking to satiety. This occupancy proved to be nearly 70% (Esterlis et al., 2010); such remarkably high occupancy persisted for hours after one session of smoking, despite the rapid metabolism of nicotine. Indeed a SPECT study with the same ligand revealed 19% occupancy in nonsmokers exposed to secondhand smoke for 1 h in the confines of an automobile and showed that this moderate occupancy provoked craving in smokers, suggesting a priming effect (Brody et al., 2011). Even one or two puffs of a cigarette evoked 50% occupancy measured after 3 h, and smoking an entire cigarette produced almost complete occupancy at 5-[^{123}I]iodo-85380 binding sites (Brody et al., 2006). In a PET study using a structurally related α4β2 receptor subtype ligand, a single dose of varenicline (0.5 mg) produced near saturation in the brain of withdrawn smokers (Lotfipour, Mandelkern, Alvarez-Estrada, & Brody, 2012). Varenicline (Champix), which was developed as an aid to smoking cessation, is a full agonist at the homopentameric α7 receptor and a partial agonist at α4β2 and other heteropentameric nicotinic acetylcholine receptors. The repertoire of molecular imaging agents for nicotinic receptors has not yet matched their known pharmacological diversity. However, in recent years, a number of α7-selective PET tracers have been developed and applied for occupancy studies in experimental animals, that is, [^{3}H]AZ11637326 in rats (Maier et al., 2011) and [^{18}F]ASEM in mice (Wong et al., 2014). At the time of writing, the only such study in humans used [^{11}C]CHIBA-1001 to measure occupancy at α7 receptors by tropisetron, which is better known as a serotonin 5-HT3 receptor antagonist with antinausea properties (Ishikawa et al., 2011); a single oral tropisetron dose evoked occupancy as high as 20% in the human brain. This approach stands to find further applications, given the growing interest in α7 nicotinic receptors as a target for procognitive and neuroprotective therapies.

IV CONCLUSIONS

In this review of the use of molecular imaging for measuring receptor and transport occupancies, we have emphasized a limited number of applications, notably the special case of occupancy of dopamine and other biogenic amine receptors by their endogenous neurotransmitter. We also emphasize the more widely used methods for assessing occupancy and biogenic amine receptors and transporters during treatment with medications or exposure to drugs of abuse. This survey does not do justice to the broad range of molecular targets now assessed in PET occupancy studies. In a sense the conventional neurotransmitter systems served as a learning platform for establishing methods that are becoming widely accepted as endpoints in pharmaceutical development.

This chapter began with a reference to George de Hevesy's whimsical reverie about following the tea as it passed into his body. Tea is more than just hot water but is a vehicle for delivering the xanthine alkaloids caffeine and theophylline to the body, provoking mild stimulant effects on the heart and central nervous system via blockade of the inhibitory purinergic receptors. In a full realization of that reverie, the interaction of caffeine at its receptors in the nervous system is now evident through the PET occupancy method. In one such study, changes in the availability of adenosine A1 binding sites labeled with [^{18}F]-CPFPX were monitored after intravenous infusions of caffeine; an extrapolation of measured data predicted 50% occupancy after drinking the equivalent of five cups of coffee (Elmenhorst, Meyer, Matusch, Winz, & Bauer, 2012). Caffeine not only is the most widely used stimulant in the world but also has considerable potential to guide pharmaceutical development. Thus, for example, adenosine 2A antagonists such as tozadenant are underdevelopment as treatments for Parkinson's disease, based on their capacity to disinhibit the activity of residual dopamine neurons. PET studies with the adenosine 2A ligand [^{18}F]-MNI-444 in nonhuman primates in conjunction with pharmacokinetic modeling established the expected relationship between tozadenant plasma levels and occupancy at brain receptors, which should be directly translatable to designing trials in humans (Barret et al., 2014). In a rare example relating receptor occupancy to pharmacodynamic response, [^{11}C] SCH442416 PET was used to measure adenosine 2A occupancy in the striatum of awake nonhuman primates. During the PET recordings the experimental adenosine antagonist ASP5854 was administered while also measuring the catalepsy induced by the antipsychotic compound haloperidol. This experiment established an 85% threshold of adenosine 2A occupancy to be sufficient for blocking catalepsy.

PET occupancy studies at caffeine binding sites are not merely a matter of academic interest, but illuminate important clinical questions about pharmacokinetics of medications. Indeed, this approach has emerged in the past decade as an invaluable tool for investigating the relationship between drug dosage and receptor occupancy. There can be some ambiguity about occupancy results in relation to assumptions about the steady-state condition. In particular, PET endpoints such as binding potential (BP$_{ND}$) are calculated assuming that binding site affinity and availability are constant for the duration of the PET recording. Indeed, changes in availability are best interpreted relative to a steady-state brain concentration of the competing drug, which cannot easily be predicted from the plasma concentration prevailing during the PET recording. Some caution is also required in relation to affinity states of G protein-coupled receptors and also with respect to the possibility of off-target binding.

Studies in patient populations can be logistically difficult due to the need to obtain data in an untreated baseline condition, and drug occupancy calculations relative to data from healthy age-matched controls can result in bias if the disease has independent effects on receptor availability. Finally, there is a certain risk of circularity in occupancy studies, as there may be no prior information about the amount of occupancy that is relevant, either to pharmaceutical action or undesired side effects. For example, it is not entirely clear what degree of blockade of SERT is necessary or sufficient for antidepressant action or if partial SERT blockade can contribute to side effects of a medication targeting some other site. Occupancy is a matter of pharmacokinetics, but therapeutic effects are more a matter of pharmacodynamics.

References

Abi-Dargham, A., Rodenhiser, J., Printz, D., Zea-Ponce, Y., Gil, R., Kegeles, L. S., et al. (2000). Increased baseline occupancy of D2 receptors by dopamine in schizophrenia. *Proceedings of the National Academy of Sciences of the United States of America, 97*(14), 8104–8109.

Arakawa, R., Tateno, A., Kim, W., Sakayori, T., Ogawa, K., & Okubo, Y. (2016). Time-course of serotonin transporter occupancy by single dose of three SSRIs in human brain: a positron emission tomography study with [(11)C]DASB. *Psychiatry Research, 251*, 1–6.

Barret, O., Hannestad, J., Alagille, D., Vala, C., Tavares, A., Papin, C., et al. (2014). Adenosine 2A receptor occupancy by tozadenant and preladenant in rhesus monkeys. *Journal of Nuclear Medicine, 55*(10), 1712–1718.

Brody, A. L., Mandelkern, M. A., London, E. D., Khan, A., Kozman, D., Costello, M. R., et al. (2011). Effect of secondhand smoke on occupancy of nicotinic acetylcholine receptors in brain. *Archives of General Psychiatry, 68*(9), 953–960.

Brody, A. L., Mandelkern, M. A., London, E. D., Olmstead, R. E., Farahi, J., Scheibal, D., et al. (2006). Cigarette smoking saturates brain alpha 4 beta 2 nicotinic acetylcholine receptors. *Archives of General Psychiatry, 63*(8), 907–915.

Chugani, D. C., Ackermann, R. F., & Phelps, M. E. (1988). In vivo [3H] spiperone binding: evidence for accumulation in corpus striatum by agonist mediated receptor internalization. *Journal of Cerebral Blood Flow and Metabolism, 8*, 291–303.

Comley, R. A., van der Aart, J., Gulyás, B., Garnier, M., Iavarone, L., Halldin, C., et al. (2015). In vivo occupancy of the 5-HT1A receptor by a novel pan 5-HT1(A/B/D) receptor antagonist, GSK588045, using positron emission tomography. *Neuropharmacology, 92*, 44–48.

Cumming, P. (2011). Absolute abundances and affinity states of dopamine receptors in mammalian brain: a review. *Synapse, 65*(9), 892–909.

Cumming, P., Rosa-Neto, P., Watanabe, H., Smith, D., Bender, D., Clarke, P. B., et al. (2003). Effects of acute nicotine on hemodynamics and binding of [11C]raclopride to dopamine D2,3 receptors in pig brain. *NeuroImage, 19*(3), 1127–1136.

Cumming, P., Wong, D. F., Gillings, N., Hilton, J., Scheffel, U., & Gjedde, A. (2002). Specific binding of [(11)C]raclopride and N-[(3)H]propyl-norapomorphine to dopamine receptors in living mouse striatum: occupancy by endogenous dopamine and guanosine triphosphate-free G protein. *Journal of Cerebral Blood Flow and Metabolism, 22*(5), 596–604.

de la Fuente-Fernández, R., Lim, A. S., Sossi, V., Holden, J. E., Calne, D. B., Ruth, T. J., et al. (2001). Apomorphine-induced changes in synaptic dopamine levels: positron emission tomography evidence

for presynaptic inhibition. *Journal of Cerebral Blood Flow and Metabolism, 21*(10), 1151–1159.

Deutschländer, A., la Fougère, C., Boetzel, K., Albert, N. L., Gildehaus, F. J., Bartenstein, P., et al. (2016). Occupancy of pramipexole (Sifrol) at cerebral dopamine D2/3 receptors in Parkinson's disease patients. *NeuroImage Clinical, 12*, 41–46.

Domino, E. F., Hirasawa-Fujita, M., Ni, L., Guthrie, S. K., & Zubieta, J. K. (2015). Regional brain [(11)C]carfentanil binding following tobacco smoking. *Progress in Neuro-Psychopharmacology & Biological Psychiatry, 59*, 100–104.

Elmenhorst, D., Meyer, P. T., Matusch, A., Winz, O. H., & Bauer, A. (2012). Caffeine occupancy of human cerebral A1 adenosine receptors: in vivo quantification with 18F-CPFPX and PET. *Journal of Nuclear Medicine, 53*(11), 1723–1729.

Esterlis, I., Cosgrove, K. P., Batis, J. C., Bois, F., Stiklus, S. M., Perkins, E., et al. (2010). Quantification of smoking-induced occupancy of beta2-nicotinic acetylcholine receptors: estimation of nondisplaceable binding. *Journal of Nuclear Medicine, 51*(8), 1226–1233.

Farde, L., Hall, H., Ehrin, E., & Sedvall, G. (1986). Quantitative analysis of D2 dopamine receptor binding in the living human brain by PET. *Science, 231*(4735), 258–261.

Farde, L., Nordström, A. L., Wiesel, F. A., Pauli, S., Halldin, C., & Sedvall, G. (1992). Positron emission tomographic analysis of central D1 and D2 dopamine receptor occupancy in patients treated with classical neuroleptics and clozapine. Relation to extrapyramidal side effects. *Archives of General Psychiatry, 49*(7), 538–544.

Farde, L., Wiesel, F., Stone-Elander, S., Halldin, C., Nordsrom, A. L., Hall, H., et al. (1990). D2 dopamine receptors in neuroleptic-naïve schizophrenic patients: a positron emission tomography study with [11C]raclopride. *Archives of General Psychiatry, 47*, 213–219.

Foa, E. B., Liebowitz, M. R., Kozak, M. J., Davies, S., Campeas, R., Franklin, M. E., et al. (2005). Randomized, placebo-controlled trial of exposure and ritual prevention, clomipramine, and their combination in the treatment of obsessive-compulsive disorder. *The American Journal of Psychiatry, 162*(1), 151–161.

Fusar-Poli, P., & Meyer-Lindenberg, A. (2013). Striatal presynaptic dopamine in schizophrenia, part II: meta-analysis of [(18)F/(11)C]-DOPA PET studies. *Schizophrenia Bulletin, 39*(1), 33–42.

Gallezot, J. D., Kloczynski, T., Weinzimmer, D., Labaree, D., Zheng, M. Q., Lim, K., et al. (2014). Imaging nicotine- and amphetamine-induced dopamine release in rhesus monkeys with [(11)C]PHNO vs [(11)C]raclopride PET. *Neuropsychopharmacology, 39*(4), 866–874.

Girgis, R. R., Slifstein, M., D'Souza, D., Lee, Y., Periclou, A., Ghahramani, P., et al. (2016). Preferential binding to dopamine D3 over D2 receptors by cariprazine in patients with schizophrenia using PET with the D3/D2 receptor ligand [(11)C]-(+)-PHNO. *Psychopharmacology, 233*(19–20), 3503–3512.

Gründer, G., Fellows, C., Janouschek, H., Veselinovic, T., Boy, C., Bröcheler, A., et al. (2008). Brain and plasma pharmacokinetics of aripiprazole in patients with schizophrenia: an [18F]fallypride PET study. *The American Journal of Psychiatry, 165*(8), 988–995.

Gründer, G., Hiemke, C., Paulzen, M., Veselinovic, T., & Vernaleken, I. (2011). Therapeutic plasma concentrations of antidepressants and antipsychotics: lessons from PET imaging. *Pharmacopsychiatry, 44*(6), 236–248.

Gründer, G., Hippius, H., & Carlsson, A. (2009). The 'atypicality' of antipsychotics: a concept re-examined and re-defined. *Nature Reviews. Drug Discovery, 8*(3), 197–202.

Gründer, G., Landvogt, C., Vernaleken, I., Buchholz, H. G., Ondracek, J., Siessmeier, T., et al. (2006). The striatal and extrastriatal D2/D3 receptor-binding profile of clozapine in patients with schizophrenia. *Neuropsychopharmacology, 31*(5), 1027–1035.

Gryglewski, G., Lanzenberger, R., Kranz, G. S., & Cumming, P. (2014). Meta-analysis of molecular imaging of serotonin transporters in major depression. *Journal of Cerebral Blood Flow and Metabolism, 34*(7), 1096–1103.

Hall, H., Wedel, I., Halldin, C. J. K., & Farde, L. (1990). Comparison of the in vitro binding properties of *N*-[3H]methylspiperone and [3H] raclopride to rat and human brain membranes. *Journal of Neurochemistry, 55*, 2048–2057.

Hannestad, J., Gallezot, J. D., Planeta-Wilson, B., Lin, S. F., Williams, W. A., van Dyck, C. H., et al. (2010). Clinically relevant doses of methylphenidate significantly occupy norepinephrine transporters in humans in vivo. *Biological Psychiatry, 68*(9), 854–860.

Hiemke, C., Baumann, P., Bergemann, N., Conca, A., Dietmaier, O., Egberts, K., et al. (2011). AGNP consensus guidelines for therapeutic drug monitoring in psychiatry: update 2011. *Pharmacopsychiatry, 44*(6), 195–235.

Hume, S. P., Lingford-Hughes, A. R., Nataf, V., Hirani, E., Ahmad, R., Davies, A. N., et al. (2007). Low sensitivity of the positron emission tomography ligand [11C]diprenorphine to agonist opiates. *The Journal of Pharmacology and Experimental Therapeutics, 322*(2), 661–667.

Ishikawa, M., Sakata, M., Toyohara, J., Oda, K., Ishii, K., Wu, J., et al. (2011). Occupancy of α7 nicotinic acetylcholine receptors in the brain by tropisetron: a positron emission tomography study using [(11)C]CHIBA-1001 in healthy human subjects. *Clinical Psychopharmacology and Neuroscience, 9*(3), 111–116.

Iwata, Y., Nakajima, S., Caravaggio, F., Suzuki, T., Uchida, H., Plitman, E., et al. (2016). Threshold of dopamine D2/3 receptor occupancy for hyperprolactinemia in older patients with schizophrenia. *The Journal of Clinical Psychiatry, 77*(12), e1557–e1563.

Jakobsen, S., Pedersen, K., Smith, D. F., Jensen, S. B., Munk, O. L., & Cumming, P. (2006). Detection of alpha2-adrenergic receptors in brain of living pig with 11C-yohimbine. *Journal of Nuclear Medicine, 47*(12), 2008–2015.

Jones, A. K., Kitchen, N. D., Watabe, H., Cunningham, V. J., Jones, T., Luthra, S. K., et al. (1999). Measurement of changes in opioid receptor binding in vivo during trigeminal neuralgic pain using [11C]diprenorphine and positron emission tomography. *Journal of Cerebral Blood Flow and Metabolism, 19*(7), 803–808.

Kegeles, L. S., Slifstein, M., Frankle, W. G., Xu, X., Hackett, E., Bae, S. A., et al. (2008). Dose-occupancy study of striatal and extrastriatal dopamine D2 receptors by aripiprazole in schizophrenia with PET and [18F]fallypride. *Neuropsychopharmacology, 33*(13), 3111–3125.

Kim, E., Howes, O. D., Kim, B. H., Chon, M. W., Seo, S., Turkheimer, F. E., et al. (2016). Regional differences in serotonin transporter occupancy by escitalopram: an [11C]DASB PK-PD study. *Clinical Pharmacokinetics, 56*, 371–381 [Epub ahead of print].

Kim, W., Tateno, A., Arakawa, R., Sakayori, T., Ikeda, Y., Suzuki, H., et al. (2014). In vivo activity of modafinil on dopamine transporter measured with positron emission tomography and [18F]FE-PE2I. *The International Journal of Neuropsychopharmacology, 17*(5), 697–703.

Kimmel, H. L., Nye, J. A., Voll, R., Mun, J., Stehouwer, J., Goodman, M. M., et al. (2012). Simultaneous measurement of extracellular dopamine and dopamine transporter occupancy by cocaine analogs in squirrel monkeys. *Synapse, 66*(6), 501–508.

Lako, I. M., van den Heuvel, E. R., Knegtering, H., Bruggeman, R., & Taxis, K. (2013). Estimating dopamine D2 receptor occupancy for doses of 8 antipsychotics: a meta-analysis. *Journal of Clinical Psychopharmacology, 33*(5), 675–681.

Landau, A. M., Doudet, D. J., & Jakobsen, S. (2012). Amphetamine challenge decreases yohimbine binding to α2 adrenoceptors in Landrace pig brain. *Psychopharmacology, 222*(1), 155–163.

Landau, A. M., Dyve, S., Jakobsen, S., Alstrup, A. K., Gjedde, A., & Doudet, D. J. (2015). Acute vagal nerve stimulation lowers α2 adrenoceptor availability: possible mechanism of therapeutic action. *Brain Stimulation, 8*(4), 702–707.

Laruelle, M. (2000). Imaging synaptic neurotransmission with in vivo binding competition techniques: a critical review. *Journal of Cerebral Blood Flow and Metabolism, 20*(3), 423–451.

Laruelle, M., Abi-Dargham, A., Gil, R., Kegeles, L., & Innis, R. (1999). Increased dopamine transmission in schizophrenia: relationship to illness phases. *Biological Psychiatry, 46*(1), 56–72.

Laruelle, M., Abi-Dargham, A., van Dyck, C. H., Gil, R., D'Souza, C. D., Erdos, J., et al. (1996). Single photon emission computerized tomography imaging of amphetamine-induced dopamine release in drug-free schizophrenic subjects. *Proceedings of the National Academy of Sciences of the United States of America, 93*(17), 9235–9240.

Le Foll, B., Payer, D., Di Ciano, P., Guranda, M., Nakajima, S., Tong, J., et al. (2016). Occupancy of dopamine D3 and D2 receptors by buspirone: a [11C]-(+)-PHNO PET study in humans. *Neuropsychopharmacology, 41*(2), 529–537.

Lotfipour, S., Mandelkern, M., Alvarez-Estrada, M., & Brody, A. L. (2012). A single administration of low-dose varenicline saturates α4β2* nicotinic acetylcholine receptors in the human brain. *Neuropsychopharmacology, 37*(7), 1738–1748.

Maier, D. L., Hill, G., Ding, M., Tuke, D., Einstein, E., Gurley, D., et al. (2011). Pre-clinical validation of a novel alpha-7 nicotinic receptor radiotracer, [(3)H]AZ11637326: target localization, biodistribution and ligand occupancy in the rat brain. *Neuropharmacology, 61*(1-2), 161–171.

Mamo, D., Graff, A., Mizrahi, R., Shammi, C. M., Romeyer, F., & Kapur, S. (2007). Differential effects of aripiprazole on D(2), 5-HT(2), and 5-HT(1A) receptor occupancy in patients with schizophrenia: a triple tracer PET study. *The American Journal of Psychiatry, 164*(9), 1411–1417.

Meyer, J. H. (2007). Imaging the serotonin transporter during major depressive disorder and antidepressant treatment. *Journal of Psychiatry & Neuroscience, 32*, 86–102.

Meyer, J. H., Wilson, A. A., Sagrati, S., Hussey, D., Carella, A., Potter, W. Z., et al. (2004). Serotonin transporter occupancy of five selective serotonin reuptake inhibitors at different doses: an [11C]DASB positron emission tomography study. *The American Journal of Psychiatry, 161*, 826–835.

Milak, M. S., Severance, A. J., Prabhakaran, J., Kumar, J. S., Majo, V. J., Ogden, R. T., et al. (2011). In vivo serotonin-sensitive binding of [11C]CUMI-101: a serotonin 1A receptor agonist positron emission tomography radiotracer. *Journal of Cerebral Blood Flow and Metabolism, 31*(1), 243–249.

Mizrahi, R., Mamo, D., Rusjan, P., Graff, A., Houle, S., & Kapur, S. (2009). The relationship between subjective well-being and dopamine D2 receptors in patients treated with a dopamine partial agonist and full antagonist antipsychotics. *The International Journal of Neuropsychopharmacology, 12*(5), 715–721.

Nakazawa, S., Yokoyama, C., Nishimura, N., Horisawa, T., Kawasaki, A., Mizuma, H., et al. (2013). Evaluation of dopamine D2/D3 and serotonin 5-HT2A receptor occupancy for a novel antipsychotic, lurasidone, in conscious common marmosets using small-animal positron emission tomography. *Psychopharmacology, 225*(2), 329–339.

Narendran, R., Mason, N. S., Laymon, C. M., Lopresti, B. J., Velasquez, N. D., May, M. A., et al. (2010). A comparative evaluation of the dopamine D(2/3) agonist radiotracer [11C](−)-N-propyl-norapomorphine and antagonist [11C]raclopride to measure amphetamine-induced dopamine release in the human striatum. *The Journal of Pharmacology and Experimental Therapeutics, 333*(2), 533–539.

Nogami, T., Takano, H., Arakawa, R., Ichimiya, T., Fujiwara, H., Kimura, Y., et al. (2013). Occupancy of serotonin and norepinephrine transporter by milnacipran in patients with major depressive disorder: a positron emission tomography study with [(11)C]DASB and (S,S)-[(18)F]FMeNER-D(2). *The International Journal of Neuropsychopharmacology, 16*(5), 937–943.

Nordström, A. L., Farde, L., Nyberg, S., Karlsson, P., Halldin, C., & Sedvall, G. (1995). D1, D2, and 5-HT2 receptor occupancy in relation to clozapine serum concentration: a PET study of schizophrenic patients. *The American Journal of Psychiatry, 152*, 1444–1449.

Ogawa, K., Tateno, A., Arakawa, R., Sakayori, T., Ikeda, Y., Suzuki, H., et al. (2014). Occupancy of serotonin transporter by tramadol: a positron emission tomography study with [11C]DASB. *The International Journal of Neuropsychopharmacology*, 17(6), 845–850.

Park, H. S., Jung, I. S., Lim, N. H., Sung, J. H., Lee, S., Moon, B. S., et al. (2014). Proof of mechanism study of a novel serotonin transporter blocker, DA-8031, using [11C]DASB positron emission tomography and in vivo microdialysis. *Urology*, 84(1), 245 e1–7.

Pilowsky, L. S., Mulligan, R. S., Acton, P. D., Ell, P. J., Costa, D. C., & Kerwin, R. W. (1997). Limbic selectivity of clozapine. *Lancet*, 350(9076), 490–491.

Pinborg, L. H., Feng, L., Haahr, M. E., Gillings, N., Dyssegaard, A., Madsen, J., et al. (2012). No change in [^{11}C]CUMI-101 binding to 5-HT(1A) receptors after intravenous citalopram in human. *Synapse*, 66(10), 880–884.

Quednow, B. B., Treyer, V., Hasler, F., Dörig, N., Wyss, M. T., Burger, C., et al. (2012). Assessment of serotonin release capacity in the human brain using dexfenfluramine challenge and [18F]altanserin positron emission tomography. *NeuroImage*, 59(4), 3922–3932.

Quelch, D. R., Withey, S. L., Nutt, D. J., Tyacke, R. J., & Parker, C. A. (2014). The influence of different cellular environments on PET radioligand binding: an application to D2/3-dopamine receptor imaging. *Neuropharmacology*, 85, 305–313.

Rominger, A., Cumming, P., Brendel, M., Xiong, G., Zach, C., Karch, S., et al. (2015). Altered serotonin and dopamine transporter availabilities in brain of depressed patients upon treatment with escitalopram: a [123I]β-CIT SPECT study. *European Neuropsychopharmacology*, 25(6), 873–881.

Ruhé, H. G., Koster, M., Booij, J., van Herk, M., Veltman, D. J., & Schene, A. H. (2014). Occupancy of serotonin transporters in the amygdala by paroxetine in association with attenuation of left amygdala activation by negative faces in major depressive disorder. *Psychiatry Research*, 221(2), 155–161.

Sato, H., Ito, C., Tashiro, M., Hiraoka, K., Shibuya, K., Funaki, Y., et al. (2013). Histamine H₁ receptor occupancy by the new-generation antidepressants fluvoxamine and mirtazapine: a positron emission tomography study in healthy volunteers. *Psychopharmacology*, 230(2), 227–234.

Scott, D. J., Stohler, C. S., Koeppe, R. A., & Zubieta, J. K. (2007). Time-course of change in [11C]carfentanil and [11C]raclopride binding potential after a nonpharmacological challenge. *Synapse*, 61(9), 707–714.

Seeman, P. (1988). Brain dopamine receptors in schizophrenia: PET problems. *Archives of General Psychiatry*, 45, 560–598.

Seeman, P., Guan, H. -C., & Niznik, H. B. (1989). Endogenous dopamine lowers the dopamine D2 receptor density as measured by [3H] raclopride: implications for positron emission tomography of the human brain. *Synapse*, 3, 96–97.

Seneca, N., Finnema, S. J., Laszlovszky, I., Kiss, B., Horváth, A., Pásztor, G., et al. (2011). Occupancy of dopamine D₂ and D₃ and serotonin 5-HT₁ A receptors by the novel antipsychotic drug candidate, cariprazine (RGH-188), in monkey brain measured using positron emission tomography. *Psychopharmacology (Berlin)*, 218(3), 579–587.

Shotbolt, P., Tziortzi, A. C., Searle, G. E., Colasanti, A., van der Aart, J., Abanades, S., et al. (2012). Within-subject comparison of [(11)C]-(+)-PHNO and [(11)C]raclopride sensitivity to acute amphetamine challenge in healthy humans. *Journal of Cerebral Blood Flow and Metabolism*, 32(1), 127–136.

Smith, D. F., Dyve, S., Minuzzi, L., Jakobsen, S., Munk, O. L., & Marthi, K., & Cumming, P. (2006). Inhibition of [11C]mirtazapine binding by alpha2-adrenoceptor antagonists studied by positron emission tomography in living porcine brain. *Synapse*, 59(8), 463–471.

Suhara, T., Takano, A., Sudo, Y., Ichimiya, T., Inoue, M., Yasuno, F., et al. (2003). High levels of serotonin transporter occupancy with low-dose clomipramine in comparative occupancy study with fluvoxamine using positron emission tomography. *Archives of General Psychiatry*, 60(4), 386–391.

Suzuki, T., Graff-Guerrero, A., Uchida, H., Remington, G., Caravaggio, F., Borlido, C., et al. (2013). Dopamine D2/3 occupancy of ziprasidone across a day: a within-subject PET study. *Psychopharmacology*, 228(1), 43–51.

Takano, H., Arakawa, R., Nogami, T., Suzuki, M., Nagashima, T., Fujiwara, H., et al. (2014). Norepinephrine transporter occupancy by nortriptyline in patients with depression: a positron emission tomography study with (S,S)-[^{18}F]FMeNER-D₂. *The International Journal of Neuropsychopharmacology*, 17(4), 553–560.

Talvik, M., Nordström, A. L., Nyberg, S., Olsson, H., Halldin, C., & Farde, L. (2001). No support for regional selectivity in clozapine-treated patients: a PET study with [(11)C]raclopride and [(11)C] FLB 457. *The American Journal of Psychiatry*, 158(6), 926–930.

Tashiro, M., Duan, X., Kato, M., Miyake, M., Watanuki, S., Ishikawa, Y., et al. (2008). Brain histamine H1 receptor occupancy of orally administered antihistamines, bepotastine and diphenhydramine, measured by PET with 11C-doxepin. *British Journal of Clinical Pharmacology*, 65(6), 811–821.

Tashiro, M., Kato, M., Miyake, M., Watanuki, S., Funaki, Y., Ishikawa, Y., et al. (2009). Dose dependency of brain histamine H (1) receptor occupancy following oral administration of cetirizine hydrochloride measured using PET with [11C]doxepin. *Human Psychopharmacology*, 24(7), 540–548.

Tashiro, M., Sakurada, Y., Iwabuchi, K., Mochizuki, H., Kato, M., Aoki, M., et al. (2004). Central effects of fexofenadine and cetirizine: measurement of psychomotor performance, subjective sleepiness, and brain histamine H1-receptor occupancy using 11C-doxepin positron emission tomography. *Journal of Clinical Pharmacology*, 44(8), 890–900.

Tyacke, R. J., & Nutt, D. J. (2015). Optimising PET approaches to measuring 5-HT release in human brain. *Synapse*, 69(10), 505–511.

Udo de Haes, J. I., Harada, N., Elsinga, P. H., Maguire, R. P., & Tsukada, H. (2006). Effect of fenfluramine-induced increases in serotonin release on [18F]MPPF binding: a continuous infusion PET study in conscious monkeys. *Synapse*, 59(1), 18–26.

Vernaleken, I., Fellows, C., Janouschek, H., Bröcheler, A., Veselinovic, T., Landvogt, C., et al. (2008). Striatal and extrastriatal D2/D3-receptor-binding properties of ziprasidone: a positron emission tomography study with [18F]fallypride and [11C] raclopride (D2/D3-receptor occupancy of ziprasidone). *Journal of Clinical Psychopharmacology*, 28(6), 608–617.

Vernaleken, I., Janouschek, H., Raptis, M., Hellmann, S., Veselinovic, T., Bröcheler, A., et al. (2010). Dopamine D2/3 receptor occupancy by quetiapine in striatal and extrastriatal areas. *The International Journal of Neuropsychopharmacology*, 13(7), 951–960.

Vernaleken, I., Siessmeier, T., Buchholz, H. G., Härtter, S., Hiemke, C., Stoeter, P., et al. (2004). High striatal occupancy of D2-like dopamine receptors by amisulpride in the brain of patients with schizophrenia. *The International Journal of Neuropsychopharmacology*, 7(4), 421–430.

Veselinović, T., Paulzen, M., & Gründer, G. (2013). Cariprazine, a new, orally active dopamine D2/3 receptor partial agonist for the treatment of schizophrenia, bipolar mania and depression. *Expert Review of Neurotherapeutics*, 13(11), 1141–1159.

Volkow, N. D., Wang, G. J., Fowler, J. S., Fischman, M., Foltin, R., Abumrad, N. N., et al. (1999). Methylphenidate and cocaine have a similar in vivo potency to block dopamine transporters in the human brain. *Life Sciences*, 65(1), PL7–12.

Volkow, N. D., Wang, G. J., Fowler, J. S., Gatley, S. J., Logan, J., Ding, Y. S., et al. (1998). Dopamine transporter occupancies in the human brain induced by therapeutic doses of oral methylphenidate. *The American Journal of Psychiatry*, 155, 1325–1331.

Wong, D. F., Kuwabara, H., Pomper, M., Holt, D. P., Brasic, J. R., George, N., et al. (2014). Human brain imaging of α7 nAChR with [(18)F]ASEM: a new PET radiotracer for neuropsychiatry and determination of drug occupancy. *Molecular Imaging and Biology*, 16(5), 730–738.

Wong, O. F., Wagner, H. N., Tune, L. E., Dannals, R. F., Pearlson, G. O., Links, J. M., et al. (1986). Positron emission tomography reveals elevated D2 dopamine receptors in drug-naive schizophrenics. *Science, 234*, 1558–1563.

Yanai, K., Ryu, J. H., Watanabe, T., Iwata, R., Ido, T., Sawai, Y., et al. (1995). Histamine H1 receptor occupancy in human brains after single oral doses of histamine H1 antagonists measured by positron emission tomography. *British Journal of Pharmacology, 116*(1), 1649–1655.

Yokoi, F., Gründer, G., Biziere, K., Stephane, M., Dogan, A. S., Dannals, R. F., et al. (2002). Dopamine D2 and D3 receptor occupancy in normal humans treated with the antipsychotic drug aripiprazole (OPC 14597): a study using positron emission tomography and [11C] raclopride. *Neuropsychopharmacology, 27*(2), 248–259.

Zhang, D., Tashiro, M., Shibuya, K., Okamura, N., Funaki, Y., Yoshikawa, T., et al. (2010). Next-day residual sedative effect after nighttime administration of an over-the-counter antihistamine sleep aid, diphenhydramine, measured by positron emission tomography. *Journal of Clinical Psychopharmacology, 30*(6), 694–701.

Zheng, M., Appel, L., Luo, F., Lane, R., Burt, D., Risinger, R., et al. (2015). Safety, pharmacokinetic, and positron emission tomography evaluation of serotonin and dopamine transporter occupancy following multiple-dose administration of the triple monoamine reuptake inhibitor BMS-820836. *Psychopharmacology, 232*(3), 529–540.

Zimmer, L., Mauger, G., Le Bars, D., Bonmarchand, G., Luxen, A., & Pujol, J. F. (2002). Effect of endogenous serotonin on the binding of the 5-HT1A PET ligand 18F-MPPF in the rat hippocampus: kinetic beta measurements combined with microdialysis. *Journal of Neurochemistry, 80*(2), 278–286.

11

Stable Isotope Labeling Kinetics in CNS Translational Medicine: Introduction to SILK Technology

Randall J. Bateman*, Tim West[†], Kevin Yarasheski[†],
Bruce W. Patterson[‡], Brendan Lucey*, John R. Cirrito*,
Sylvain Lehmann[§], Christophe Hirtz[§], Audrey Gabelle[¶],
Timothy Miller*, Nicolas Barthelemy*, Chihiro Sato*, James G. Bollinger*,
Paul Kotzbauer*, Katrina Paumier*

*Department of Neurology, Washington University School of Medicine, St. Louis, MO, United States [†]C2N
Diagnostics, LLC, St. Louis, MO, United States [‡]Department of Internal Medicine, Washington University School of
Medicine, St. Louis, MO, United States [§]Laboratoire de Biochimie Protéomique Clinique, CHU de Montpellier,
Université de Montpellier, INSERM U1183, Montpellier, France [¶]Memory Research and Resources Center,
Gui de Chauliac Hospital, Université de Montpellier, Montpellier, France

I OVERVIEW

Alzheimer's disease (AD) is the most common cause of dementia and is an increasingly important public health problem. It is currently estimated to affect 5 million people in the United States, with an expected increase in this number to 13 million by the year 2050 (Brookmeyer et al., 2011). AD leads to impairments in memory, cognitive functions, and, ultimately, the lack of independence in activities of daily function causing a heavy personal toll on the patient and the family.

Highly effective therapies and accurate diagnostic tests are not currently available. Several targets have been identified as contributors to AD pathophysiology (i.e., amyloid-beta (Aβ), tau, and inflammation) with most current therapeutic approaches targeting Aβ (Blennow, de Leon, & Zetterberg, 2006). However, Aβ pathophysiology is not fully understood at present, and treatment of AD during the mild-to-moderate stage of dementia in therapeutic trials may be too late as 50% of hippocampal neurons are dead at this point (Gomez-Isla et al., 1996; Price et al., 2001). Furthermore, clinical diagnostic accuracy is insufficient without the aid of relevant, precise AD biomarkers. The ideal diagnostic test would be simple, reflect the underlying pathophysiology, and accurately detect AD pathology even at the earliest stages. Thus, better understanding of the pathophysiology of Aβ and diagnostic tests based on this are needed to offer anti-Aβ therapeutic strategies their best chance of success.

The amyloid hypothesis proposes that Aβ overproduction or underclearance—that is, dysfunctional Aβ kinetics—leads to a common pathophysiology resulting in a cascade of events culminating in neuronal cell death, manifesting as progressive clinical dementia of the Alzheimer's disease type (Hardy & Selkoe, 2002). In order to understand Aβ kinetics in the pathophysiology of AD, we developed a novel method to metabolically label and quantify CNS proteins. We used the technique of stable isotope labeling kinetics (SILK) to successfully measure the production and clearance of Aβ in the human CNS (Bateman et al., 2006). We detected very rapid CSF Aβ kinetics in healthy, younger patients, with an Aβ half-life of approximately 4 hours. We then demonstrated decreased Aβ production, using a gamma-secretase inhibitor designed to block Aβ generation (Bateman et al., 2009). We found that Aβ clearance slows with aging from the third to seventh decade of life, resulting in a remarkable 250% slower turnover in those in their seventies compared with younger people in their twenties. Aβ42 is specifically altered with an increase in rates of aggregation 50% higher in those with amyloidosis. Our lab has now advanced the SILK technology application from measuring proteins in the CSF to measuring Aβ kinetics directly in the blood. Ongoing studies will determine the relationship among the brain, CSF, and blood Aβ kinetics.

Although the current method of CNS Aβ-SILK with intravenous labeling and CSF collection has proven robust, it takes up disproportionate time and resources and requires a physician to place lumbar catheters. A more convenient method is needed to enable other researchers to more easily apply the approach and to enable the possibility of a simplified clinical test for treatment trials or diagnostic tests.

The study of blood Aβ kinetics will contribute to a better understanding of its production, transport, and breakdown within and between the brain, CSF, and blood supply. These fundamental measurements of Aβ kinetics will help to determine the effects of peripheral Aβ metabolism on pathophysiologic changes in AD. This information alone will provide key insights into whole-body Aβ metabolism and be useful for defining the causes of AD. Furthermore, these results may lead to a specific blood-derived biomarker for AD.

II UNDERLYING PRINCIPLES FOR SILK TECHNOLOGY

Kevin Yarasheski, Tim West

Stable isotopes are naturally occurring, nondecaying versions of atoms that differ in the number of neutrons in their nucleus. This leads to small differences in their molecular weight compared with other isotopes. For example, carbon exists in two stable isotope configurations (^{12}C and ^{13}C), with the lighter isotope being the most abundant (99%) in nature. Other than the small difference in molecular weight, the ^{12}C and ^{13}C have the same physiochemical properties and are biologically indistinguishable. By enriching the low-abundance versions of stable isotopes and incorporating these into biomolecules, it is possible to produce biomolecules that contain a heavy isotope content that is stable, yet unlike the natural isotope abundance present in biological systems. For example, the amino acid leucine contains six carbon atoms, and the chance of all six naturally occurring as the heavy, low-abundance isotope (i.e., $[^{13}C_6]$ leucine) is 0.01^6. Stable isotope labeling kinetics (SILK) uses a method

of administering $[^{13}C_6]$ leucine to an organism, sampling the protein products and then using high-resolution mass spectrometry to quantify in vivo heavy leucine incorporation into protein synthesis pathways.

At trace amounts in biological systems, $[^{13}C_6]$ leucine (or "heavy leucine") is metabolically indistinguishable from the more abundant $[^{12}C_6]$ leucine. Heavy leucine is recognized by its corresponding tRNA at the same rate as naturally abundant ^{12}C-leucine. Both enter canonical cellular protein translation processes, where the ratio of heavy-to-light leucine incorporated into a protein over a known time period reflects the synthesis and clearance rate (or kinetics) for recently formed protein biomolecules. The $[^{13}C_6]$ leucine-labeled protein will follow the same biological pathways and is thus used to quantify the in vivo flux (or turnover) of proteins through a cell or system. The newly synthesized protein exists in two forms; a ($^{13}C_6$) heavy and ($^{12}C_6$) light form, differing only in their respective molecular weights, so the production and clearance rates for newly generated proteins are quantified after taking samples from the organism, isolating the desired protein(s), enzymatically digesting the protein(s) into predictable peptides, and using mass spectrometry (MS) to quantify heavy-to-light ratios in the leucine-containing peptides derived from the protein(s) of interest. The multidimensionality of MS analysis allows investigators to quantify heavy-to-light ratios in several metabolic intermediates and proteins simultaneously. Therefore it is possible to quantify the in vivo production, transport, and degradation rates—that is, kinetics of many biomolecular species after administering a heavy labeled precursor.

One advantage of SILK is that it assesses the effect that metabolic modulators such as production inhibitors or clearance activators have on a target biomolecule. The quantification of turnover for newly synthesized biomolecules has several advantages over static measurements of concentration. The therapeutic efficacy or target engagement for a novel compound (e.g., a protein production inhibitor or clearance activator) is not adequately assessed by simply quantifying the targeted protein concentration before and after intervention. This may lead to the premature dismissal of new therapies that can be proven efficacious through more appropriate analysis. However, SILK quantifies the metabolism of target proteins and reflects which side of the production-clearance equation has been affected by the intervention. This information may reveal that a therapeutic intervention effectively engages its target, even though the overall protein concentration is unchanged. In other words, SILK analysis can result in a more accurate and convincing demonstration of whether rates of protein production or clearance have been favorably altered or not.

III SPECIFIC PROTEIN APPLICATIONS OF SILK

A Stable Isotope Labeling of Aβ

This chapter is dedicated to the application of SILK to the quantification of Aβ metabolic turnover and other proteins generated in the central nervous system. Aβ is produced by cleavage of the amyloid precursor protein (APP) by two endoproteinases. Aβ is then released into the extracellular space. CNS Aβ samples can be obtained using brain biopsies and brain parenchymal or intracerebroventricular catheters or sampling the cerebrospinal fluid (CSF). These methods are listed in the order of proximity to the actual site of Aβ production that is the neuron. Lumbar catheters to collect CSF samples have been regularly used to take repeat brain Aβ samples in human participants. This method has been used to collect continual CSF samples for up to 48 hours. The quantification of human CSF Aβ kinetics involves intravenous [^{13}C$_6$] leucine administration, followed by sample extraction at various intervals. The Aβ is purified from the CSF, proteolytically digested into leucine-containing peptides, and mass spectrometry (MS) is used to determine the heavy-to-light ratio in Aβ peptides (Fig. 11.1).

Aβ metabolism was first described by Bateman et al. in 2006. For this initial study, participants were infused with [^{13}C$_6$] leucine over a 9-hour time period, and CSF samples were collected hourly for 36 hours, beginning right before the start of the [^{13}C$_6$] leucine infusion. This first study demonstrated that (i) ^{13}C$_6$-labeled Aβ appeared in the CSF around 4–6 hours after the start of the infusion and peaked around 16–20 hours, (ii) both Aβ production and clearance could be measured using this study protocol, and (iii) the variability in Aβ turnover between study participants was much lower than the variability in absolute Aβ concentrations. This made the SILK-Aβ assay ideal for investigating the role of the metabolism of this protein in the underling pathophysiology of Alzheimer's disease. In addition, it could directly evaluate the effects of investigational therapeutics with Aβ metabolism targeting mechanisms of action on CSF Aβ kinetics production and clearance rates.

B Amyloid-Beta Kinetics in Aging and Amyloid Deposition in Alzheimer's Disease

Bruce Patterson, Randall Bateman

The soluble Aβ peptides in CSF are in constant flux, being generated through a series of events (i.e., synthesis, intracellular transport, export, and fluid flow) and disappearing through myriad processes (i.e., interstitial/CSF fluid transport and resorption, reuptake by brain tissue, in situ proteolysis to smaller peptides and/or amino acids

and deposition onto amyloid plaques). SILK tracer kinetics is undertaken to quantify this process of turnover. When a tracer is introduced to the system (such as by intravenous infusion of [^{13}C$_6$] leucine), the isotopic enrichment of the Aβ peptides increases over time as unlabeled peptides are removed and replaced by newly synthesized, isotopically labeled peptides. As the tracer administration is discontinued and the plasma leucine enrichment returns to natural abundance, the concentration of labeled Aβ peptides decreases accordingly as the labeled peptides are removed and replaced by newly synthesized, unlabeled peptides.

Mathematical models provide frameworks to quantify the rates of Aβ kinetics from such enrichment time-course data and kinetic parameters that can be compared with groups of subjects. We formerly used a simplified approach to estimate the fractional synthesis rate (FSR) and fractional clearance rate (FCR) of CSF Aβ peptides from the upslope and monoexponential downslope, respectively, of these curves (Bateman et al., 2006; Mawuenyega et al., 2010). The FSR and FCR are fractional rates of synthesis and clearance that represent the fraction of the soluble Aβ pool synthesized or cleared per unit of time (typically reported as pools/hour). Using this approach, we reported that the FSR of Aβ40 and Aβ42 did not differ between AD and controls, whereas the FCR for both Aβ40 and Aβ42 was significantly slower in AD than controls (Mawuenyega et al., 2010). In addition, the FCR was significantly slower than the FSR in AD subjects, whereas the FSR and FCR were not significantly different in controls (Mawuenyega et al., 2010). This supported the concept that amyloid plaques develop due to an impaired clearance mechanism.

Such simplified FSR and FCR parameters can provide accurate fractional turnover rates (FTRs) of the system if certain assumptions about the system are valid. By definition, the fraction of the pool synthesized and cleared per unit time must be the same for a system at steady state. However, the FSR and FCR values may be significantly different (Mawuenyega et al., 2010), which suggests a violation of the assumptions. In this case, FSR and FCR parameters may not provide accurate values for the true FTRs. Given the limitations of this approach, we developed a compartmental model that comprehensively describes the entire SILK tracer time-course curves for Aβ peptide turnover as a continuous process rather than artificially dividing the SILK time course into "synthesis" and "clearance" phases (Patterson et al., 2015; Potter et al., 2013). Our model (Fig. 11.2) provides an excellent fit to the entire Aβ38, Aβ40, and Aβ42 SILK time course in late-onset AD, autosomal dominant AD, and control subjects of all ages (Patterson et al., 2015; Potter et al., 2013).

The compartmental model provides a mathematical framework to transform the shape and magnitude of

FIG. 11.1   Sample collection and MS determination of heavy-to-light ratios Aß peptides.

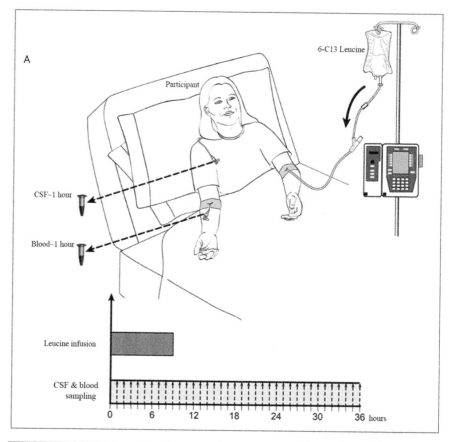

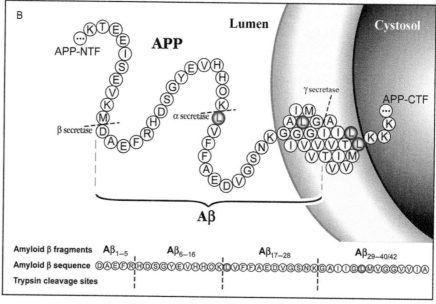

the plasma leucine tracer time course (using a 9-hour primed constant infusion of $[^{13}C_6]$ leucine) into the shapes and magnitudes of the enrichment time courses of the three Aβ isoforms (as above) over the full 36-hour SILK protocol. It does this by varying a set of adjustable parameters to optimize the curve fit to each individual subject's SILK data. In the early stages of the model's development, we found that a minimally complex model to

describe the SILK time course for each isoform comprised a single compartment linked to a delay chain that consisted of five identical subcompartments. It was assumed that the single compartment represented unique turnover kinetics for each isoform and that the delay process represented all processes common to all Aβ isoforms. Therefore two of the five delay process subcompartments were assumed to represent precursors to the soluble Aβ

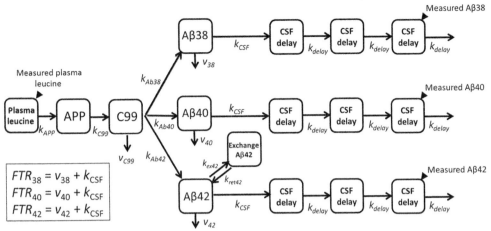

FIG. 11.2 Compartmental model of CSF Aβ38, Aβ40, and Aβ42 SILK kinetics. *From Patterson, B. W., Elbert, D. L., Mawuenyega, K. G., Kasten, T., Ovod, V., Ma, S., et al. (2015). Age and amyloid effects on human central nervous system amyloid-beta kinetics.* Annals of Neurology, 78(3), 439–453. *doi:10.1002/ana.24454.*

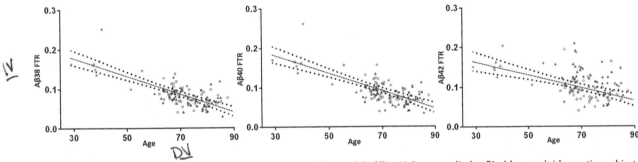

FIG. 11.3 Aβ isoform fractional turnover rates decrease with age. The model of Fig. 11.2 was applied to 51 older amyloid-negative subjects (*blue circles*), 49 older amyloid-positive subjects (*red triangles*), and 12 younger amyloid-negative subjects (*green asterisks*), to determine the fractional turnover rate (FTR, pools/hour) for Aβ38 (left), Aβ40 (middle), and Aβ42 (right). The *solid line* represents a linear fit with 95% confidence interval. *From Patterson, B. W., Elbert, D. L., Mawuenyega, K. G., Kasten, T., Ovod, V., Ma, S., et al. (2015). Age and amyloid effects on human central nervous system amyloid-beta kinetics.* Annals of Neurology, 78 (3), 439–453, https://doi.org/10.1002/ana.24454.

peptides in the brain (i.e., APP and C99) and that the remaining three represented fluid transport processes that should transport all three isoforms through the CSF in an identical manner. In Fig. 11.2, the FTR of each peptide is the sum of the rate constant for soluble peptide transfer into the CSF delay chain (k_{CSF}, equal for all peptides) plus a rate constant (independently adjustable v_{38}, v_{40}, and v_{42}) that represents irreversible losses to all other pathways.

Given the assumption that all isoforms shared a common delay process, another feature was needed to optimally describe the Aβ42 SILK time course in subjects with amyloid plaques: an exchange process whereby isotopically labeled Aβ42 was temporarily removed from and then reentered into the throughput stream, causing a relative flattening out of the terminal tail of the Aβ42 SILK curve relative to Aβ38 and Aβ40. This compartment was only necessary to fit the Aβ42 SILK curves in subjects with amyloid plaques and may represent temporary reversible binding of isotopically labeled Aβ42 to the surface of plaques. Given the set of rate constants that optimally describe the SILK tracer curve shape for a given subject and define a differential amount of irreversible losses from the system for individual peptides, another

set of three adjustable parameters defines the absolute rate of production into CSF for each isoform to account for the concentration of each isoform as measured by MS per subject.

We have noted three key outcomes in the application of this model to a wide range subjects: carriers and noncarriers of PSEN mutations (Potter et al., 2013), older participants with or without late-onset AD (Patterson et al., 2015), and younger normal controls (Patterson et al., 2015).

First, carriers of PSEN mutations had a 24% higher production rate of Aβ42 relative to Aβ40 compared with noncarriers of PSEN, providing in vivo confirmation of in vitro studies showing that PSEN mutations alter the amount of Aβ42 relative to Aβ40 produced from the differential cleavage of C99 (Potter et al., 2013). These studies support the utility of directly measuring kinetics and production rates. Further, the group of mutation carriers with amyloidosis as measured by PIB PET had similar findings to late-onset sporadic AD with regard to increased Aβ42 irreversible loss rates from the system (see in the succeeding text).

Second, aging has a profound influence on all Aβ isoform turnover rates and half-lives. *The FTR of Aβ peptides slows down considerably with age* (Fig. 11.3). This rate decreased 60% between the ages 30 (0.184 pools/hour) and 80 (0.074 pools/hour), which corresponded to a 2.5-fold longer half-life (from 3.8 to 9.4 hour) over five decades (Patterson et al., 2015). This slowing of the soluble Aβ peptide turnover rate with age may be due to a slowing down of clearance processes such as brain interstitial and CSF fluid flow and could allow more time for protein aggregation to occur such that a greater amount of the irreversible loss of soluble peptides could be directed toward deposition into plaques with increasing age (Patterson et al., 2015).

Third, *amyloidosis is tightly associated with altered Aβ42 kinetics*. Aβ42 turnover relative to Aβ40 is faster in subjects with amyloid plaques (Patterson et al., 2015; Potter et al., 2013) (see Fig. 11.4). The SILK time courses for Aβ40 and

Aβ42 are virtually completely superimposable in "super normal" subjects without amyloid plaques who have high Aβ42/Aβ40 concentration ratios (left panel). On the opposite end of the spectrum, Aβ42 peaks about 3 hours earlier than Aβ40 indicating faster turnover in subjects with amyloid plaques and low Aβ42/Aβ40 concentration ratios (right panel). Subjects with intermediate concentration ratios (between 0.10 and 0.16) show a less pronounced Aβ42 turnover anomaly (middle panel). Aβ38 and Aβ40 SILK curves are virtually superimposable regardless of amyloid status in all subjects.

The association between the Aβ42/Aβ40 FTR ratio and the CSF Aβ42/Aβ40 concentration ratio reveals that a number of subjects who were classified as amyloid negative on the basis of PET PIB scans exhibited clear

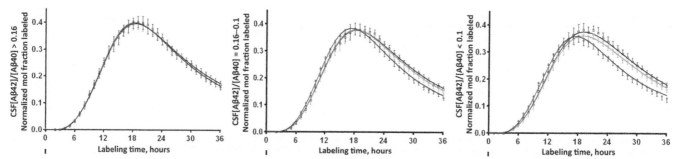

FIG. 11.4   Effect of amyloid status on Aβ peptide kinetics. SILK time-course enrichment curves (normalized to plasma leucine plateau enrichment) for Aβ38 (*blue*), Aβ40 (*green*), and Aβ42 (*red*) for subjects stratified by CSF Aβ42/Aβ40 concentration ratios. From left to right, the panels represent (A) normal subjects without amyloid plaques and elevated Aβ42/40 concentration ratios, >0.16; (B) a mixture of subjects with and without amyloid plaques with Aβ42/40 concentration ratios between 0.10 and 0.16; and (C) subjects with amyloid plaques and reduced Aβ42/40 concentration ratios, <0.10. Data points represent group means with 95% confidence intervals. The *solid lines* represent the mean of model fits to all subjects within each group. *From Patterson, B. W., Elbert, D. L., Mawuenyega, K. G., Kasten, T., Ovod, V., Ma, S., et al. (2015). Age and amyloid effects on human central nervous system amyloid-beta kinetics. Annals of Neurology, 78 (3), 439–453, https://doi.org/10.1002/ana.24454.*

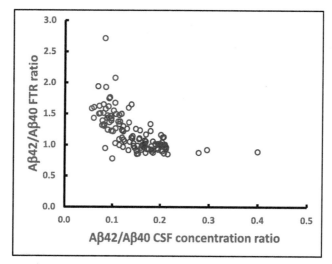

FIG. 11.5   Association between the Aβ42/Aβ40 FTR ratio and the CSF Aβ42/Aβ40 concentration ratio. The compartmental model of Fig. 11.2 was applied to 73 subjects without amyloid plaques (age 24–87) and 48 subjects with amyloid plaques (age 60–85). All subjects were noncarriers of PSEN mutations. Amyloid status was defined by PET PIB score when available (amyloid positive if PET PIB MCBP score >0.18) and by CSF Aβ42/Aβ40 concentration ratio if PET PIB was not available (amyloid positive if Aβ42/Aβ40 <0.12). It is particularly noteworthy that a number of subjects who were classified as amyloid negative based on PET PIB scans show evidence of elevated Aβ42 turnover, that is, Aβ42/Aβ40 FTR ratio clearly elevated, >1.2.

evidence of the faster Aβ42 kinetic anomaly with elevated Aβ42/Aβ40 FTR ratios (Fig. 11.5). This figure shows an inverse relationship between the FTR and concentration ratios such that faster turnover of Aβ42 relative to Aβ40 is associated with a decrease in Aβ42 concentrations relative to Aβ40 in CSF. The model assumed that all kinetic processes in the model are identical for all three isoforms except for the rate of irreversible loss. The model therefore provides a mechanistic link between these observations: the decreased concentration of Aβ42 relative to Aβ40 in CSF is associated with faster irreversible loss of Aβ42 from the system, that is, v_{42} increases in relation to v_{40} (Fig. 11.2) as plaques develop, causing an increase in the Aβ42/Aβ40 FTR ratio and a decreased Aβ42/Aβ40 concentration ratio. The kinetic anomaly of faster Aβ42 turnover thus may be a more sensitive marker that may precede PET PIB scans for the presence of amyloid plaques and may be a useful early marker for plaque development.

The comprehensive model applied to the full SILK time course to a larger cohort of subjects shows that soluble Aβ42 FTR is *faster* in subjects with plaques. This faster Aβ42 turnover is associated with and possibly causative of a decreased CSF Aβ42 concentration relative to Aβ40 through a mechanism of increased soluble Aβ42 irreversible loss that correlates with rate of amyloid growth as measured by PIB PET. This increased soluble Aβ42 turnover rate represents a measure that may indicate the aggregation rate from the soluble to insoluble form of Aβ42. Ultimately, this increased aggregation rate decreases the clearance from the CNS as it aggregates in the brain in cerebral Aβ amyloidosis.

SILK time-course enrichment curves (normalized to plasma leucine plateau enrichment) for Aβ38 (blue), Aβ40 (green), and Aβ42 (red) for subjects stratified by CSF Aβ42/Aβ40 concentration ratios. From left to right, the panels represent (A) normal subjects without amyloid plaques and elevated Aβ42/40 concentration ratios, >0.16; (B) a mixture of subjects with and without amyloid plaques with Aβ42/40 concentration ratios between 0.10 and 0.16; and (C) subjects with amyloid plaques and reduced Aβ42/40 concentration ratios, <0.10. Data points represent group means with 95% confidence intervals. The *solid lines* represent the mean of model fits to all subjects within each group. From Patterson et al. (2015).

C Sleep-Wake Patterns, Aβ and AD

Brendan Lucey

The assessment of sleep-wake patterns is an additional example where SILK can be used to further elucidate the effects of natural sleep process on Aβ kinetics in humans. More specifically, as SILK technology advancements allow Aβ kinetics to be measured directly in the blood, future research including sleep studies can be conducted by using

the SILK technique to better characterize sleep impact on Aβ production and clearance in humans.

Recent studies in rodents and humans have found that the concentration of Aβ oscillates with the sleep-wake cycle, that is, as a diurnal pattern (Huang et al., 2012; Kang et al., 2009). Two processes are hypothesized to underlie the changes in Aβ metabolism that produce this oscillation. First, neuronal activity releases Aβ into the brain's interstitial fluid (Cirrito et al., 2005). Soluble APP, which is cleaved into Aβ, also oscillates with the sleep-wake cycle (Dobrowolska et al., 2014). Therefore increased neuronal activity during waking has been proposed to produce an increase in Aβ concentration, while decreased neuronal activity during sleep results in a relative decrease in Aβ production. Second, Aβ may be cleared through sleep-associated changes in interstitial fluid flows (i.e., via the "glymphatic" system) (Xie et al., 2013). Although both of these proposed mechanisms have been reported to drive changes in Aβ concentrations in rodents, the evidence in humans is less clear. Sleep deprivation has been suggested to increase Aβ concentrations (Ooms et al., 2014). Further, increased Aβ concentrations have been correlated with recovery in patients with neurological injury (Brody et al., 2008). Both of these studies suggest processes that increased wakefulness (i.e., behavioral sleep deprivation, and recovery from brain injury) increase Aβ production.

There are no studies demonstrating Aβ clearance through the glymphatic system in humans. However, there is indirect evidence regarding its importance in the protein's diurnal oscillation. Individuals with cerebral amyloid deposition show a loss of amplitude of the diurnal Aβ pattern, regardless of their sleep-wake activity (Huang et al., 2012; Lucey et al., 2017) (Fig 11.6). A similar pattern has been observed in individuals with autosomal dominant Alzheimer's disease mutations (Roh et al., 2012). This amplitude attenuation is hypothesized to be due to cerebral plaque deposition as a "sinkhole" of Aβ clearance, leading to reduced fluctuation.

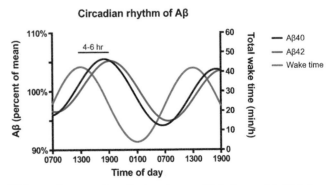

FIG. 11.6 Diurnal oscillations in two Aβ isoforms. *Modified from Huang, Y., Potter, R., Sigurdson, W., Santacruz, A., Shih, S., Ju, Y.-E., et al., (2012). Effects of age and amyloid deposition on Aβ dynamics in the human central nervous system.* Archives of Neurology, 69(1), 51–58.

Since Aβ deposition in the brain is concentration-dependent, these studies linking sleep-wake activity to Aβ metabolism suggest a potential novel therapy to prevent or delay Alzheimer's disease through sleep (Lucey & Bateman, 2014). Improvements in sleep quality and other parameters may be able to decrease Aβ levels and prevent or delay amyloid deposition in the brain, a key first step in the pathogenesis of Alzheimer's disease. Experiments using Aβ SILK in individuals under different sleep conditions, such as sleep deprivation versus normal sleep controls, or comparing individuals with poor sleep quality treated with a sleep hypnotic drug or placebo, may reveal sleep-driven changes in Aβ kinetics. Further study is needed to establish the feasibility of this exciting potential anti-Alzheimer's disease therapy.

D SSRI Antidepressants Slow Aβ Production

John Cirrito

Synaptic transmission regulates Aβ production via two independent mechanisms. Dr. Roberto Malinow's group was the first to directly demonstrate that increasing or decreasing synaptic activity within a mouse brain slice had direct and proportional effects on extracellular Aβ concentration (Kamenetz et al., 2003). Later studies utilized an in vivo microdialysis method to measure extracellular Aβ within the brain interstitial fluid (ISF) over time in awake and freely moving mice (Cirrito et al., 2005). Electrically stimulating the brain to generate seizures caused ISF Aβ levels to increase by 30%–40% within 30 minutes. Conversely, blocking synaptic activity substantially reduced Aβ levels. These pathways cause Aβ to be produced at or near the presynaptic terminal (Cirrito et al., 2008). In the human brain, regions that have high baseline levels of neuronal activity, that is, the "default mode network," are also the most vulnerable to developing Aβ plaques (Buckner et al., 2009; Sheline & Raichle, 2013).

In an independent synaptic mechanism, activation of some neurotransmitter receptors lead to altered APP processing to regulate Aβ generation. Muscarinic acetylcholine receptor agonists increase α-secretase APP cleavage, thus reducing Aβ production (Caccamo et al., 2006; Hock et al., 2003). Whereas β-secretase and γ-secretase cleave APP to generate Aβ, α-secretase cleaves APP within the Aβ sequence to prevent the peptide's formation. The activation of NMDA receptors and serotonin receptors induce second messenger signaling pathways, including the extracellular regulated kinase (ERK), to also increase α-secretase activity and reduce Aβ levels (Cirrito et al., 2011; Verges, Restivo, Goebel, Holtzman, & Cirrito, 2011). While the presynaptic mechanism and signaling mechanism can occur even within the same neuron, they appear to regulate Aβ levels independently.

There are 15 subtypes of serotonin receptors; however, the activation of 5HT$_4$-R, 5HT$_6$-R, or 5HT$_7$-R subtypes only reduces Aβ generation in mouse models of AD (Fisher, Wallace, Tripoli, Sheline, & Cirrito, 2016). The broad activation of all receptor subtypes using selective serotonin reuptake inhibitors (SSRIs) causes a significant 25% reduction in ISF Aβ for 24 hours (Cirrito et al., 2011). Increasing serotonin signaling, such as with SSRIs, has been one of the primary treatments for depression since the 1950s. In contrast, other types of antidepressants such as selective norepinephrine reuptake inhibitors (SNRIs) do not alter Aβ levels. Chronic administration of the SSRI citalopram to an AD mouse model starting at young ages dramatically reduced Aβ plaque burden, as well as CSF Aβ levels (Cirrito et al., 2011). Citalopram treatment in older mice that had already developed amyloid plaques completely arrested further plaque growth but failed to reverse the pathology (Sheline et al., 2014). Many aspects of the cell signaling pathway that links serotonin receptor activation to changes in Aβ levels and, ultimately, AD-like pathophysiology have been determined in animal models (Cirrito et al., 2011; Fisher et al., 2016).

SILK analysis of CSF Aβ provided a means to directly test the role of serotonin in Aβ kinetics in humans. Using a proven protocol to label Aβ with [^{13}C$_6$] leucine (Bateman et al., 2006), cognitively normal research participants who were naïve to treatment with any antidepressants were admitted to the hospital and given a bolus injection of citalopram (60 mg) or placebo, and a lumbar catheter was placed to take hourly samples of CSF starting 8 hours after drug administration. Interestingly, citalopram caused a 30% decrease in CSF Aβ$_{40}$ and Aβ$_{42}$ even at the initial 8-hour time point that persisted for at least 36 hours compared with placebo-treated participants (Sheline et al., 2014). The generation rate of newly synthesized Aβ was significantly slower in the citalopram group, while the clearance rate was not altered. This was entirely consistent with the prior animal model studies that found a similar magnitude decrease in brain Aβ levels following SSRIs administration that was due to reduced production.

The factor of depression, independent of treatment, appears to increase the risk of AD development (Geerlings, den Heijer, Koudstaal, Hofman, & Breteler, 2008; Green et al., 2003; Ownby, Crocco, Acevedo, John, & Loewenstein, 2006). Several retrospective studies have found that SSRIs reduce the risk of AD symptoms in depressed individuals (Geda, 2010; Norum, Hart, & Levy, 2003; St George-Hyslop & Morris, 2008). Depressed individuals taking SSRIs are at lower risk of AD compared with untreated depressed individuals, though still at higher risk compared with nondepressed controls. The ultimate question is this: Are individuals with a history of SSRI use, but not depression, protected from AD or Aβ pathologies? While initial studies in mice and humans are promising that SSRIs could be used to lower Aβ and possibly reduce

AD pathology, more studies are necessary to determine if chronic SSRI use can cause a persistent reduction in brain Aβ levels to ultimately reduce plaques and have a positive effect on AD cognitive symptoms.

E Tau in AD and Kinetics

Nicolas Barthelemy, Chihiro Sato, Randall Bateman

Tau is predominantly an intracellular soluble protein that associates with and stabilizes microtubules in axons. Tau is known to undergo alternative splicing producing 3R (repeat) and 4R isoforms that are present in human adults and to be phosphorylated at many sites. Tauopathies are defined as those neurodegenerative diseases with tau pathology, the most common pathological manifestation in neurodegenerative diseases, and include Alzheimer's disease (AD), corticobasal degeneration (CBD), progressive supranuclear palsy (PSP), and frontotemporal dementia (FTD). Tauopathies can be classified by the specific tau isoforms (3R vs 4R) that are aggregated and their locations in the brain (Buee, Bussiere, Buee-Scherrer, Delacourte, & Hof, 2000). AD is the most common tauopathy, with abnormalities including the presence of tau tangles, tau neurites, and altered tau phosphorylation, as well as increased levels of cerebrospinal fluid (CSF) total tau and phosphorylated tau (p-tau). Previously, it was assumed that tau is passively released into the extracellular space after cell death or injury, but recent studies indicate that tau is also actively secreted into the extracellular space in a regulated manner under physiological and pathological conditions (Kanmert et al., 2015). Growing evidence suggests that tau spreads in anatomically connected areas in disease and propagates from cell to cell or trans-synaptically in a prion-like manner (Holmes et al., 2014). However, the underlying mechanism of tau release, aggregation, and spreading including basic tau kinetics remains unknown.

The understanding of normal and pathophysiological tau processing in the various tauopathies is central to understanding how tau contributes to disease and for the development of tau-targeted therapeutics. Tau-targeted therapy has recently gained considerable interest in the AD field, including a phase 3 trial of LMT-X, proposed to prevent tau aggregation in AD or FTD patients, and a phase 1 trial of an active tau vaccine AADvac-1. For example, questions such as "Why is soluble CSF tau increased in AD? Why does tau aggregate in tauopathies? What is the order of tau concentration and aggregation changes? and Does Aβ aggregation induce tau production?" are unanswered. Given these remaining questions about tau metabolism, tau appears as a highly relevant target for SILK monitoring. Alteration of tau production and clearance in AD measured by SILK could address questions about active or passive release of tau and potentially detect tau kinetic changes linked to intraneuronal tau aggregate accumulation in AD and other tauopathies. In order to evaluate the efficacy of tau-targeted therapeutics, methods to evaluate tau kinetics in living humans are needed. Recent developments to monitor tau deposition in vivo by PET allow us to envision promising combinations with tau SILK analysis and tau PET of both tau deposition and kinetics.

Measuring tau kinetics in humans with the SILK method in combination with monitoring tau aggregation with recently developed tau positron emission tomography (PET) imaging will be beneficial in addressing these questions. Previous mouse studies suggest that tau protein has a half-life of a few weeks, much longer than that of amyloid-beta (9 hours in the human CNS). We have recently developed strategies to label humans at a higher dose and sample longer times (Crisp et al., 2015) (Fig. 11.7A-C). Our preliminary results in 10 healthy controls suggest that the half-life of tau in the human CNS is approximately 23 days (Sato et al., 2018) (Fig. 11.7D-E). Using tau SILK and tau PET imaging, we are currently testing the hypothesis that the tau kinetics is altered in disease using larger cohorts of AD and other tauopathy patients.

F Apolipoprotein E, its Kinetics and Their Role in AD

James Bollinger, Randall Bateman

Apolipoprotein E (ApoE) is a ∼34 kD polymorphic lipoprotein composed of 299 amino acids. The human ApoE gene is located on chromosome 19 and encodes for three major allelic isoforms: ApoE2, ApoE3, and ApoE4. The differences between the expressed isoforms are limited to amino acid residues 112 and 158. The ApoE2 allele (7% prevalence in European Americans) encodes for ApoE2 isoform (Cys112 and Cys158), while ApoE3 allele (78% prevalence in European Americans) encodes the wild-type ApoE3 isoform (Cys112 and Arg158) and the ApoE4 allele (15% prevalence in European Americans) encodes for the ApoE4 isoform (Arg112 and Arg158). These subtle differences in amino acid sequence impact the total charge and structure of ApoE that affects its association with cognate cellular receptors and lipoprotein particles. Although expressed in a variety of tissues, ApoE expression is highest in the liver followed by the brain. ApoE in the brain is derived exclusively from within the blood-brain barrier and is present in the CSF at concentrations of ∼5 mg/L. The CNS pool of ApoE is produced primarily by astrocytes although microglia and neurons can also contribute expression. Among other roles, CNS-derived ApoE is implicated in the uptake of lipids after neuronal degradation and the redistribution of these lipids to other cells for the purpose of proliferation, membrane repair, and remyelination of new axons.

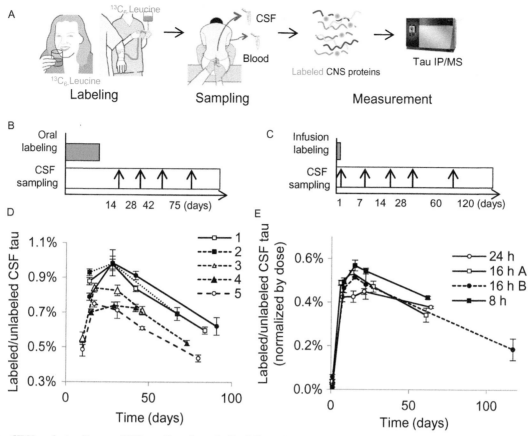

FIG. 11.7   SILK analysis of human CNS tau. *From* Sato, C., Barthélemy, N. R., Mawuenyega, K. G., et al. (2018). Tau kinetics in neurons and the human central nervous system [published correction appears in Neuron. 2018 May 16;98 (4):861-864]. Neuron, 97 (6), 1284–1298.e7, doi:10.1016/j. neuron.2018.02.015.

As the single largest genetic risk factor for LOAD, ApoE has been the focal point of numerous studies investigating its potential role in AD's clinical progression and underlying molecular neuropathology (Bales et al., 1999; Bellosta et al., 1995; Holtzman, 2001; Huang et al., 2001; Nathan et al., 1994; Shibata et al., 2000; Strittmatter et al., 1993, 1994). Population studies have demonstrated that ApoE4 significantly enhances risk for developing AD (Farrer et al., 1997), while ApoE2 decreases this risk (Corder et al., 1994). Furthermore, ApoE isoforms have been demonstrated to significantly impact amyloid-β (Aβ) deposition in animal models and in the human brain. Importantly, animal studies suggest that not only the ApoE isoform but also the amount of ApoE influence aspects of AD pathology such as the amount of Aβ deposition (Bales et al., 1999; DeMattos et al., 2004; Holtzman, Bales, et al., 2000; Holtzman, Fagan, et al., 2000). Collectively, these studies indicate that ApoE4 is a major contributing factor for AD via impaired Aβ clearance possibly caused by impaired ApoE4 clearance or decreased amounts of ApoE. However, due to conflicting results in the literature, it is not clear if or how the ApoE4 isoform differs in amount and/or turnover compared with the ApoE3 isoform in human samples.

When applied in conjunction with a recently developed assay that is independent of isoform-specific antibodies, the SILK approach offers significant appeal to address some of the discrepancies in the ELISA-based literature on relative amounts of the ApoE isoforms and provide further insight into differences in their metabolism. Thus far the technique has been applied in two key studies. The first looked to quantify the FTRs of ApoE isoforms in 18 cognitively normal adults and in ApoE3 and ApoE4 targeted-replacement mice (Wildsmith et al., 2012). The study observed no isoform-specific differences in CNS ApoE3 and ApoE4 turnover rates in human CSF (Fig. 11.8). However, a significant difference in turnover rates was observed between CNS and peripheral ApoE, which adds considerable weight to the hypothesis that these two pools of ApoE are subject to different metabolism. Interestingly, this study also noted a slower turnover rate for CSF ApoE than for Abeta. A second study applied the SILK approach to measure the effect of an RXR agonist on CSF ApoE and Aβ levels. The study observed a slight elevation of CSF ApoE levels but no difference in Aβ metabolism and attributed these observations to low CSF penetrance of the RXR agonist (Ghosal et al., 2016).

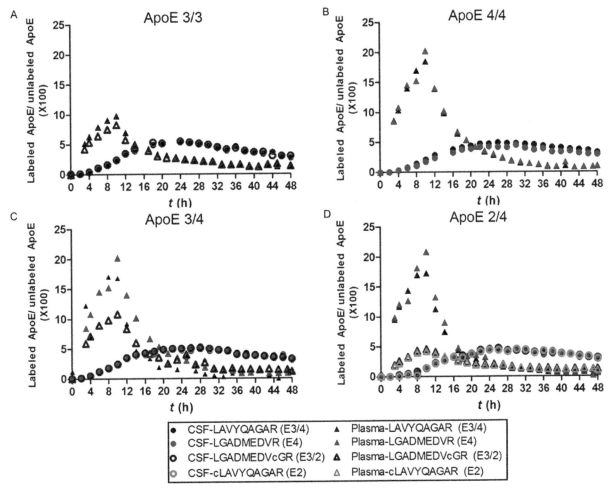

FIG. 11.8 The turnover rates of different ApoE isoforms in human plasma and CSF.

Collectively, these studies highlight the potential for SILK analysis to help elucidate the role of ApoE isoforms in AD pathology and to measure the pharmacodynamic effects of proposed disease-modifying therapies.

G SOD1 SILK

Tim Miller

This section will introduce the implications of using SILK for diseases other than Alzheimer's disease and will begin with a discussion of the manner in which superoxide dismutase 1 (SOD1) contributes to the onset of amyotrophic lateral sclerosis (ALS). SOD1 is a well-characterized enzyme that catalyzes a dismutation reaction using copper and zinc to detoxify reactive species by simultaneous reducing and oxidizing superoxide. Decades after this important enzymatic function was described, mutations in SOD1 were recognized as a causative factor in 1% of ALS through a dominant gain of function mechanism, unrelated to its enzymatic function. Based on the recognition of a dominant gain of function

mechanism, therapeutic strategies for this genetic subset have focused on methods to lower levels of SOD1 mRNA and protein (Smith et al., 2006). Some of these efforts have moved toward clinical development (Miller et al., 2013). As a result of mRNA-targeting therapeutic strategies (such as antisense oligonucleotides or siRNA), the protein concentration will decline based on its half-life. Since SOD1 protein reduction will be a key outcome in clinical studies, SOD1 half-life is an important variable for understanding the magnitude and timing of SOD1 lowering in clinical trial participants.

Developing a method for the measurement of SOD1 half-life by SILK required novel development methods. These included the identification of antibodies to reliably immunoprecipitate the protein and the identification of proteolytic fragments that may be used to quantify both unlabeled and $[^{13}C_6]$ leucine-labeled proteins. Following this method development, SOD1 SILK was applied to existing human CSF samples from participants who received $[^{13}C_6]$ 6C-13 leucine intravenously over 9 hours. As shown in this initial experiment, $[^{13}C_6]$ leucine remained barely detectable after 36 hours, despite abundant counts of the nonlabeled SOD1 molecules. The best explanation for this was that SOD1 had a long half-life, thus requiring sampling periods over longer periods of time.

Efforts in human SOD1 expressing rats subsequently demonstrated that an oral [^{13}C$_6$] leucine-labeling paradigm for 7 days yielded adequate C-13 leucine labeling of SOD1 (Crisp et al., 2015). With this newly established protocol, Miller's team was able to define SOD1 half-life for the first time in vivo. There were several interesting conclusions from these initial animal studies that are likely of general interest. First, tissue culture half-life of proteins such as SOD1 (about 24 hours) may grossly underestimate in vivo half-life (15 days). Second, methods can be established for long-lived proteins. Third, though described previously, half-life of proteins in different tissues may vary greatly. SOD1 in the central nervous system (brain and spinal cord) half-life was about 15 days, while SOD1 half-life in the liver was 3 days.

The labeling method used in the animal models was directly translated to humans with an oral labeling paradigm of 330 mg of 6C-13 leucine three times per day for 10 days. Participants were also asked to partake in a controlled leucine diet, which met all leucine requirements, but minimized the amount of extra leucine intake. Whether or not this controlled leucine, in fact, influenced total label has not been tested. CSF was taken at four different time points spaced out over 2.5 months. In four participants, CSF SOD1 half-life was determined to be 23 days. The number of participants is too low to draw many conclusions from the relatively wide range, but there were no obvious relationship to age, gender, or CSF total protein half-life. As SILK measurements expand to other CSF proteins in the CSF, one interesting point will be to determine whether there are relationships between half-lives of different proteins in different people.

While as yet untested in animal models or people, SILK should theoretically be an important pharmacodynamic marker for any mRNA-lowering therapeutic, since the synthesis rate of SOD1 protein should be affected first and most dramatically if its mRNA is targeted. We anticipate that this type of pharmacokinetics approach may be incorporated into future clinical studies.

H The Rationale for Pursuing Alpha-Syn SILK Studies in Parkinson's Disease

Paul Kotzbauer, Katrina Paumier

Parkinson's disease (PD) is defined by the accumulation of alpha-synuclein (α-syn) fibrils in neuronal cytoplasmic and neuritic inclusions known as Lewy bodies and Lewy neurites (Forno, 1996). PD typically presents as a movement disorder characterized by tremor, bradykinesia, rigidity, and impaired postural reflexes. Up to

80% of patients develop dementia over time. Postmortem tissue studies indicate that disease progression correlates with an ascending pattern of Lewy body deposition in the brain, where early stage cases are defined by involvement of nuclei in the medulla and pons and late stage cases are defined by widespread involvement of neocortex. The development of dementia in PD is associated with neocortical deposition of α-syn (Compta et al., 2011; Hely, Reid, Adena, Halliday, & Morris, 2008; Hurtig et al., 2000).

The role of α-syn in pathogenesis is supported by the identification of dominant mutations in the SNCA gene, encoding α-syn, in rare familial versions of PD (Appel-Cresswell et al., 2013; Golbe, Di Iorio, Bonavita, Miller, & Duvoisin, 1990; Kruger et al., 1998; Lesage et al., 2013; Polymeropoulos et al., 1997; Proukakis et al., 2013; Singleton et al., 2003; Zarranz et al., 2004). Mutations associated with familial PD include duplication and triplication of the SNCA gene locus, which are likely to increase protein synthesis rates. In addition, multiple single-nucleotide mutations producing missense amino acid substitutions have also been identified in familial PD cases. Some missense mutations increase the rate of fibril formation in vitro. However, other mutations slow the rate of fibril formation and may lead to pathological alpha-syn accumulation by other mechanisms such as changes in turnover.

SILK could be utilized to determine whether changes in synthesis and turnover rates of α-syn are associated with accumulation of fibrillar α-syn in both sporadic and familial PD. Furthermore, multiple therapeutic approaches targeting α-syn accumulation are being pursued. These include approaches to inhibit synthesis of α-syn protein, approaches to prevent or reverse α-syn aggregation, and approaches to increase clearance of either monomeric or aggregated α-syn. Alpha-syn SILK could be a valuable approach to assess target engagement and pharmacodynamics at both the preclinical and clinical stages of development for these α-syn-targeted therapies.

IV GLOBAL PROTEOME ANALYSIS AND COMPARTMENT TRANSFER FOLLOWING LABELING

Sylvain Lehmann, Christophe Hirtz, Audrey Gabelle

For the last decade, the development of quantitative mass spectrometry (MS) approaches (Ong & Mann, 2005) has demonstrated its major interest in many fields of physiology and pathophysiology of the CNS (Becker & Hoofnagle, 2012; Pan, Zhang, Zhang, & Li, 2014). MS allows for high specificity and absolute quantification and access to numerous proteins and their posttranslational modifications using its multiplex format (Domon & Aebersold, 2006). In general, quantitative data on the global proteome are generated at steady state by studying the area of peptides generated after enzymatic digestion of protein extracts (bottom-up approach). Longitudinal studies can also be conducted including series of measurements at different time points to evaluate the relevance of a biomarker over time. Importantly, another dimension of proteomics was reached with the measurements for specific targets (i.e., Aβ, tau, or ApoE) in terms of protein turnover, synthesis, and breakdown in vivo after ^{13}C-leucine intravenous infusion (Bateman et al., 2006; Hinkson & Elias, 2011; Lehmann et al., 2015).

It is possible to combine the SILK approach, which focuses on specific targets, with large-scale bottom-up MS. This may allow for the determination of synthesis and clearance rates of up to hundreds of proteins derived from human biological fluids. We implemented this new approach, called "stable isotope labeling by amino acids in vivo" (SILAV), which is an expansion of the SILK technology, toward a large-scale proteome analysis. In brief, clinical batches of $[^{13}C_6]$ leucine prepared in agreement with the European Pharmacopeia (Tall et al., 2015) were used. We followed a classical intravenous infusion protocol in patients, which included a 10-minute initial bolus at 2 mg/kg and a 530-minute infusion regimen at 2mg/kg/h. The fluids collected were not limited to CSF (lumbar or ventricular) but included blood, urine, or saliva. After processing in the clinical laboratory, samples were aliquoted and stored at –80°C until analysis.

For bottom-up MS, sample proteins were denatured, reduced, and alkylated prior to LysC/trypsin digestion. The digestion products were desalted using C18 tips and then fractionated using strong cation-exchange chromatography (SCX) to have access to the "deep proteome." For CSF samples, five different SCX fractions were generated and then analyzed using nano-RSLC coupled to Q-TOF Impact II (Bruker Daltonics) in data-dependent acquisition (DDA) mode. This allowed us to generate a library associated with the MS identification of the peptides. From 40uL of the CSF analyzed using this protocol, we could identify a total of 6398 peptides corresponding to 1226 proteins. 4528 of these peptides, corresponding to 1064 proteins, contained leucine and were therefore susceptible to labeling. The appearance and clearance rates of proteins were estimated using the Skyline software. In the first series of analyses, approximately 10% (424) of the leucine-containing peptides, corresponding to 209 proteins, showed significant $[^{13}C_6]$ leucine incorporation (Fig. 11.9).

This approach has the potential to provide information related to metabolic and pathophysiological pathways by analyzing many proteins in parallel. It can evaluate the metabolism of proteins and precursors belonging to the same families, such as apolipoproteins (APOA1, A2, A4, B, D, E, H, or J) or complement (C1QA, C1QB, C1QC, C1R, C1RL, C1S, CFAB, CFAH, CO1A1, CO2,

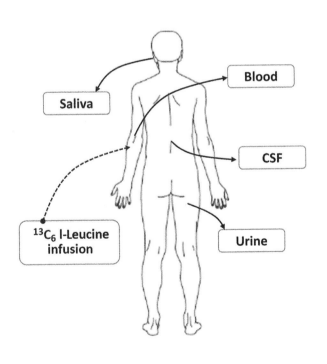
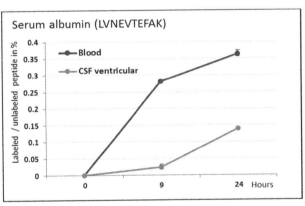
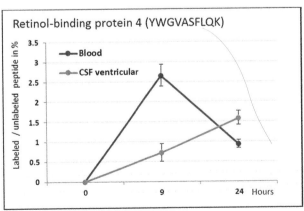

FIG. 11.9 Leucine-containing peptides corresponding to proteins that showed significant leucine incorporation.

CO3, CO3A1, CO4A, CO4B, CO5, CO6A1, CO7, or CO8A). This could be useful to assess the impact of a therapeutic intervention on multiple targets. In addition, the method can follow multiple peptides belonging to a single protein, giving access to the differential kinetic behavior of different protein domains or feature (i.e., N- or C-terminus) or different isoforms (i.e., ApoE2, ApoE3, or ApoE4).

The parallel SILAV analysis of multiple biological fluids also represents an unprecedented method to study human physiology and pathology. For example, the comparison of blood and CSF protein turnover can lead to the exploration of choroid plexus and blood-brain barrier (BBB) functions and disorders (Torbett, Baird, & Eliceiri, 2015). This may be achieved by analyzing the labeled leucine incorporation rate of the same peptides from two fluids in parallel. The interpretation of these results requires multiple information types, such as the origin of protein synthesis, protein migration between tissues, or the catabolic rates in different organs. In the example highlighted in Fig. 11.10, albumin (ALBU) and retinal-binding protein 4 (RET4) presented very different patterns. Interestingly, these two proteins differ slightly in their synthesis pattern, with ALBU originating mainly from the liver and RET4 from both liver and adipose tissue.

However, additional explanations and models of compartment transfer rate are necessary to explain the very different pattern of these two proteins. We are currently generating this type of data for many peptides. Some had a rapid increase of the percentage of label uptake in 9 hours and then exhibited a subsequent decrease. Others showed a continuous increase for up to 24 or 36 hours. The pattern of uptake was sometimes, but not always, similar for two fluids. Analyzing these patterns represents an important challenge in the generation of compartment and metabolic models. It could help elucidate the mechanisms behind the development of BBB dysfunctions and will aid our understanding of protein leakages from blood to the CNS.

In conclusion, the SILK/SILAV paradigm gives access to unique information on human protein metabolism. The comparison of the rates of synthesis, degradation, and posttranslational modifications between proteins and between the same proteins in different tissues is of great importance. In particular, previous research has elucidated many aspects of the cell signaling pathway that links serotonin activity to changes in Aβ levels (Cirrito et al., 2011; Fisher et al., 2016), and SILK analysis of CSF Aβ has demonstrated its efficacy as a means of directly testing the role of serotonin in Aβ kinetics in humans (Sheline et al., 2014). Measuring tau kinetics in humans with SILK method in combination with monitoring tau aggregation with recently developed tau positron emission tomography (PET) imaging is another beneficial aspect of this technique that may demonstrate that tau kinetic changes are

linked to intraneuronal aggregates that accumulate in Alzheimer's disease (AD) and other tauopathies. Novel research also highlights the potential for SILK analysis to help elucidate the role of apolipoprotein E (ApoE) isoforms in AD pathology and the ability to use this technique to measure the pharmacodynamic effects of proposed disease-modifying therapies (Ghosal et al., 2016).

Research has even demonstrated that the use of SILK can be expanded for the evaluation of diseases other than AD. For instance, SILK technology was combined with mass spectrometric detection to assess the in vivo kinetics of superoxide dismutase 1 (SOD1) mutation that causes amyotrophic lateral sclerosis (ALS) (Crisp et al., 2015). The results of this particular study provide support for the application of SILK toward the design and implementation of clinical trials involving ALS patients. Finally, the aforementioned applications of SILK indicate that this technology may also be utilized to determine whether changes in synthesis and turnover rates of alpha-synuclein are associated with the accumulation of fibrillar alpha-synuclein in both sporadic and familial Parkinson's disease (PD). Therefore, SILK may be an especially beneficial approach to assessing target engagement and pharmacodynamics at the preclinical and clinical stages of development for various medical therapies.

Finally, when applied to CNS disorders, this approach could give informative data on the role and involvement of the blood-brain and blood-CSF barrier. Active or passive transfer of subclasses of proteins in different pathologic conditions (e.g. Alzheimer's or Parkinson's disease) could also provide mechanistic information and help identify important factors and biomarkers. By simultaneously studying the metabolism of many CSF proteins, including isoforms of protein as illustrated for ApoE, it will be also possible to build more complex and likely more relevant models of the pathophysiological events occurring in the CNS. Such network analysis will be also important to evaluate the global effect of therapeutic approaches that may have off-target impacts.

V SUMMARY

In summary, SILK has provided clear insights into Alzheimer's disease pathophysiology with altered synthesis and clearance of amyloid-beta, drug effects on amyloid-beta synthesis, and profound changes in amyloid-beta clearance with aging. The initial work in AD has been expanded to specific proteins involved in the pathogenesis of other CNS disorders including ALS (e.g., SOD) and Parkinson's disease (e.g., alpha-synuclein). The SILK work with apolipoprotein E and SILAV has expanded the use to addressing questions in the peripheral compartment outside of the CNS. Understanding the kinetics adds the

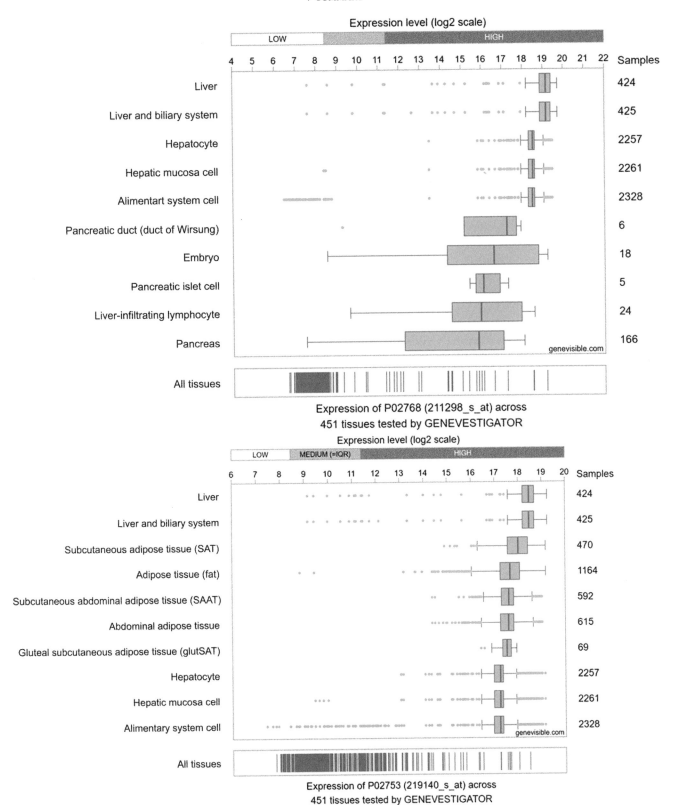

FIG. 11.10   The metabolic patterns of human albumin (ALBU, upper) and retinal-binding protein 4 (RET4, lower).

dimension of time and flux to studies of proteins and is an integral part of understanding physiology and the pathophysiology of neurological diseases. This increased understanding has assisted in diagnostic and therapeutic developments for CNS proteinopathies.

References

Appel-Cresswell, S., Vilarino-Guell, C., Encarnacion, M., Sherman, H., Yu, I., Shah, B., et al. (2013). Alpha-synuclein p.H50Q, a novel pathogenic mutation for Parkinson's disease. *Movement Disorders*, *28*(6), 811–813. https://doi.org/10.1002/mds.25421.

Bales, K. R., Verina, T., Cummins, D. J., Du, Y., Dodel, R. C., Saura, J., et al. (1999). Apolipoprotein E is essential for amyloid deposition in the APP(V717F) transgenic mouse model of Alzheimer's disease. *Proceedings of the National Academy of Sciences of the United States of America*, *96*(26), 15233–15238.

Bateman, R. J., Munsell, L. Y., Morris, J. C., Swarm, R., Yarasheski, K. E., & Holtzman, D. M. (2006). Human amyloid-beta synthesis and clearance rates as measured in cerebrospinal fluid in vivo. *Nature Medicine*, *12*(7), 856–861.

Bateman, R. J., Siemers, E. R., Mawuenyega, K. G., Wen, G., Browning, K. R., Sigurdson, W. C., et al. (2009). A gamma-secretase inhibitor decreases amyloid-beta production in the central nervous system. *Annals of Neurology*, *66*(1), 48–54. https://doi.org/10.1002/ana.21623.

Becker, J. O., & Hoofnagle, A. N. (2012). Replacing immunoassays with tryptic digestion-peptide immunoaffinity enrichment and LC-MS/MS. *Bioanalysis*, *4*(3), 281–290. https://doi.org/10.4155/bio.11.319.

Bellosta, S., Mahley, R. W., Sanan, D. A., Murata, J., Newland, D. L., Taylor, J. M., et al. (1995). Macrophage-specific expression of human apolipoprotein E reduces atherosclerosis in hypercholesterolemic apolipoprotein E-null mice. *The Journal of Clinical Investigation*, *96*(5), 2170–2179. https://doi.org/10.1172/JCI118271.

Blennow, K., de Leon, M. J., & Zetterberg, H. (2006). Alzheimer's disease. *Lancet*, *368*(9533), 387–403. https://doi.org/10.1016/S0140-6736(06)69113-7.

Brody, D. L., Magnoni, S., Schwetye, K. E., Spinner, M. L., Esparza, T. J., Stocchetti, N., et al. (2008). Amyloid-β dynamics correlate with neurological status in the injured human brain. *Science*, *321*(5893), 1221–1224.

Brookmeyer, R., Evans, D. A., Hebert, L., Langa, K. M., Heeringa, S. G., Plassman, B. L., et al. (2011). National estimates of the prevalence of Alzheimer's disease in the United States. *Alzheimers Dement*, *7*(1), 61–73. https://doi.org/10.1016/j.jalz.2010.11.007.

Buckner, R. L., Sepulcre, J., Talukdar, T., Krienen, F. M., Liu, H., Hedden, T., et al. (2009). Cortical hubs revealed by intrinsic functional connectivity: mapping, assessment of stability, and relation to Alzheimer's disease. *The Journal of Neuroscience*, *29*(6), 1860–1873. https://doi.org/10.1523/JNEUROSCI.5062-08.2009.

Buee, L., Bussiere, T., Buee-Scherrer, V., Delacourte, A., & Hof, P. R. (2000). Tau protein isoforms, phosphorylation and role in neurodegenerative disorders. *Brain Research. Brain Research Reviews*, *33*(1), 95–130.

Caccamo, A., Oddo, S., Billings, L. M., Green, K. N., Martinez-Coria, H., Fisher, A., et al. (2006). M1 receptors play a central role in modulating AD-like pathology in transgenic mice. *Neuron*, *49*(5), 671–682.

Cirrito, J. R., Disabato, B. M., Restivo, J. L., Verges, D. K., Goebel, W. D., Sathyan, A., et al. (2011). Serotonin signaling is associated with lower amyloid-{beta} levels and plaques in transgenic mice and humans.

Proceedings of the National Academy of Sciences of the United States of America, *108*(36), 14968–14973. https://doi.org/10.1073/pnas.1107411108.

Cirrito, J. R., Kang, J. E., Lee, J., Stewart, F. R., Verges, D. K., Silverio, L. M., et al. (2008). Endocytosis is required for synaptic activity-dependent release of amyloid-beta in vivo. *Neuron*, *58*(1), 42–51. doi:S0896-6273(08)00124-4 [pii]. https://doi.org/10.1016/j.neuron.2008.02.003.

Cirrito, J. R., Yamada, K. A., Finn, M. B., Sloviter, R. S., Bales, K. R., May, P. C., et al. (2005). Synaptic activity regulates interstitial fluid amyloid-beta levels in vivo. *Neuron*, *48*(6), 913–922.

Compta, Y., Parkkinen, L., O'Sullivan, S. S., Vandrovcova, J., Holton, J. L., Collins, C., et al. (2011). Lewy- and Alzheimer-type pathologies in Parkinson's disease dementia: which is more important? *Brain*, *134*(Pt 5), 1493–1505.

Corder, E. H., Saunders, A. M., Risch, N. J., Strittmatter, W. J., Schmechel, D. E., Gaskell, P. C., Jr., et al. (1994). Protective effect of apolipoprotein E type 2 allele for late onset Alzheimer disease. *Nature Genetics*, *7*(2), 180–184.

Crisp, M. J., Mawuenyega, K. G., Patterson, B. W., Reddy, N. C., Chott, R., Self, W. K., et al. (2015). In vivo kinetic approach reveals slow SOD1 turnover in the CNS. *The Journal of Clinical Investigation*, *125*(7), 2772–2780. https://doi.org/10.1172/jci80705.

DeMattos, R. B., Cirrito, J. R., Parsadanian, M., May, P. C., O'Dell, M. A., Taylor, J. W., et al. (2004). ApoE and clusterin cooperatively suppress abeta levels and deposition. Evidence that ApoE regulates extracellular Abeta metabolism in vivo. *Neuron*, *41*(2), 193–202.

Dobrowolska, J. A., Kasten, T., Huang, Y., Benzinger, T. L., Sigurdson, W., Ovod, V., et al. (2014). Diurnal patterns of soluble amyloid precursor protein metabolites in the human central nervous system. *PLoS ONE*, *9*(3), e89998.

Domon, B., & Aebersold, R. (2006). Mass spectrometry and protein analysis. *Science*, *312*(5771), 212–217. https://doi.org/10.1126/science.1124619.

Farrer, L. A., Cupples, L. A., Haines, J. L., Hyman, B., Kukull, W. A., Mayeux, R., et al. (1997). Effects of age, sex, and ethnicity on the association between apolipoprotein E genotype and Alzheimer disease. A meta-analysis. APOE and Alzheimer disease meta analysis consortium. *JAMA*, *278*(16), 1349–1356.

Fisher, J. R., Wallace, C. E., Tripoli, D. L., Sheline, Y. I., & Cirrito, J. R. (2016). Redundant Gs-coupled serotonin receptors regulate amyloid-beta metabolism in vivo. *Molecular Neurodegeneration*, *11*(1), 45. https://doi.org/10.1186/s13024-016-0112-5.

Forno, L. S. (1996). Neuropathology of Parkinson's disease. *Journal of Neuropathology and Experimental Neurology*, *55*(3), 259–272.

Geda, Y. E. (2010). Blowing hot and cold over depression and cognitive impairment. *Neurology*, *75*(1), 12–14. https://doi.org/10.1212/WNL.0b013e3181e8cc2f.

Geerlings, M. I., den Heijer, T., Koudstaal, P. J., Hofman, A., & Breteler, M. M. (2008). History of depression, depressive symptoms, and medial temporal lobe atrophy and the risk of Alzheimer disease. *Neurology*, *70*(15), 1258–1264. https://doi.org/10.1212/01.wnl.0000308937.30473.d1.

Ghosal, K., Haag, M., Verghese, P. B., West, T., Veenstra, T., Braunstein, J. B., et al. (2016). A randomized controlled study to evaluate the effect of bexarotene on amyloid-B and apolipoprotein E metabolism in healthy subjects. *Alzheimer's & Dementia*, *2*(2), 110–120. https://doi.org/10.1016/j.trci.2016.06.001.

Golbe, L. I., Di Iorio, G., Bonavita, V., Miller, D. C., & Duvoisin, R. C. (1990). A large kindred with autosomal dominant Parkinson's disease. *Annals of Neurology*, *27*(3), 276–282.

Gomez-Isla, T., Price, J. L., McKeel, D. W., Jr., Morris, J. C., Growdon, J. H., & Hyman, B. T. (1996). Profound loss of layer II entorhinal cortex neurons occurs in very mild Alzheimer's disease. *The Journal of Neuroscience*, *16*(14), 4491–4500.

Green, R. C., Cupples, L. A., Kurz, A., Auerbach, S., Go, R., Sadovnick, D., et al. (2003). Depression as a risk factor for Alzheimer disease: the MIRAGE Study. *Archives of Neurology, 60*(5), 753–759. https://doi.org/10.1001/archneur.60.5.75360/5/753.

Hardy, J., & Selkoe, D. J. (2002). The amyloid hypothesis of Alzheimer's disease: progress and problems on the road to therapeutics. *Science, 297*(5580), 353–356.

Hely, M. A., Reid, W. G., Adena, M. A., Halliday, G. M., & Morris, J. G. (2008). The Sydney multicenter study of Parkinson's disease: the inevitability of dementia at 20 years. *Movement Disorders, 23*(6), 837–844.

Hinkson, I. V., & Elias, J. E. (2011). The dynamic state of protein turnover: it's about time. *Trends in Cell Biology, 21*(5), 293–303. https://doi.org/10.1016/j.tcb.2011.02.002.

Hock, C., Maddalena, A., Raschig, A., Muller-Spahn, F., Eschweiler, G., Hager, K., et al. (2003). Treatment with the selective muscarinic m1 agonist talsaclidine decreases cerebrospinal fluid levels of A beta 42 in patients with Alzheimer's disease. *Amyloid, 10*(1), 1–6.

Holmes, B. B., Furman, J. L., Mahan, T. E., Yamasaki, T. R., Mirbaha, H., Eades, W. C., et al. (2014). Proteopathic tau seeding predicts tauopathy in vivo. *Proceedings of the National Academy of Sciences of the United States of America, 111*(41), E4376–E4385. https://doi.org/10.1073/pnas.1411649111.

Holtzman, D. M. (2001). Role of apoe/Abeta interactions in the pathogenesis of Alzheimer's disease and cerebral amyloid angiopathy. *Journal of Molecular Neuroscience, 17*(2), 147–155.

Holtzman, D. M., Bales, K. R., Tenkova, T., Fagan, A. M., Parsadanian, M., Sartorius, L. J., et al. (2000). Apolipoprotein E isoform-dependent amyloid deposition and neuritic degeneration in a mouse model of Alzheimer's disease. *Proceedings of the National Academy of Sciences of the United States of America, 97*(6), 2892–2897.

Holtzman, D. M., Fagan, A. M., Mackey, B., Tenkova, T., Sartorius, L., Paul, S. M., et al. (2000). Apolipoprotein E facilitates neuritic and cerebrovascular plaque formation in an Alzheimer's disease model. *Annals of Neurology, 47*(6), 739–747.

Huang, Y., Liu, X. Q., Wyss-Coray, T., Brecht, W. J., Sanan, D. A., & Mahley, R. W. (2001). Apolipoprotein E fragments present in Alzheimer's disease brains induce neurofibrillary tangle-like intracellular inclusions in neurons. *Proceedings of the National Academy of Sciences of the United States of America, 98*(15), 8838–8843. https://doi.org/10.1073/pnas.151254698.

Huang, Y., Potter, R., Sigurdson, W., Santacruz, A., Shih, S., Ju, Y. -E., et al. (2012). Effects of age and amyloid deposition on Aβ dynamics in the human central nervous system. *Archives of Neurology, 69*(1), 51–58.

Hurtig, H. I., Trojanowski, J. Q., Galvin, J., Ewbank, D., Schmidt, M. L., Lee, V. M., et al. (2000). Alpha-synuclein cortical Lewy bodies correlate with dementia in Parkinson's disease. *Neurology, 54*(10), 1916–1921.

Kamenetz, F., Tomita, T., Hsieh, H., Seabrook, G., Borchelt, D., Iwatsubo, T., et al. (2003). APP processing and synaptic function. *Neuron, 37*(6), 925–937.

Kang, J. -E., Lim, M. M., Bateman, R. J., Lee, J. J., Smyth, L. P., Cirrito, J. R., et al. (2009). Amyloid-β dynamics are regulated by orexin and the sleep-wake cycle. *Science, 326*(5955), 1005–1007.

Kanmert, D., Cantlon, A., Muratore, C. R., Jin, M., O'Malley, T. T., Lee, G., et al. (2015). C-terminally truncated forms of tau, but not full-length tau or its C-terminal fragments, are released from neurons independently of cell death. *The Journal of Neuroscience, 35*(30), 10851–10865. https://doi.org/10.1523/jneurosci.0387-15.2015.

Kruger, R., Kuhn, W., Muller, T., Woitalla, D., Graeber, M., Kosel, S., et al. (1998). Ala30Pro mutation in the gene encoding alpha-synuclein in Parkinson's disease. *Nature Genetics, 18*(2), 106–108.

Lehmann, S., Vialaret, J., Combe, G. G., Bauchet, L., Hanon, O., Girard, M., et al. (2015). Stable isotope labeling by amino acid in Vivo (SILAV): a new method to explore protein metabolism. *Rapid Communications in Mass Spectrometry, 29*(20), 1917–1925. https://doi.org/10.1002/rcm.7289.

Lesage, S., Anheim, M., Letournel, F., Bousset, L., Honore, A., Rozas, N., et al. (2013). G51D alpha-synuclein mutation causes a novel parkinsonian-pyramidal syndrome. *Annals of Neurology, 73*(4), 459–471. https://doi.org/10.1002/ana.23894.

Lucey, B. P., & Bateman, R. J. (2014). Amyloid-β diurnal pattern: possible role of sleep in Alzheimer's disease pathogenesis. *Neurobiology of Aging, 35*, S29–S34.

Lucey, B. P., Mawuenyega, K. G., Patterson, B. W., Elbert, D. L., Ovod, V., Kasten, T., et al. (2017). Associations between β-amyloid kinetics and the β-amyloid diurnal pattern in the central nervous system. *JAMA Neurology, 74*(2), 207–215. https://doi.org/10.1001/jamaneurol.2016.4202.

Mawuenyega, K. G., Sigurdson, W., Ovod, V., Munsell, L., Kasten, T., Morris, J. C., et al. (2010). Decreased clearance of CNS beta-amyloid in Alzheimer's disease. *Science, 330*(6012), 1774. https://doi.org/10.1126/science.1197623.

Miller, T. M., Pestronk, A., David, W., Rothstein, J., Simpson, E., Appel, S. H., et al. (2013). An antisense oligonucleotide against SOD1 delivered intrathecally for patients with SOD1 familial amyotrophic lateral sclerosis: a phase 1, randomised, first-in-man study. *Lancet Neurology, 12*(5), 435–442. https://doi.org/10.1016/s1474-4422(13)70061-9.

Nathan, B. P., Bellosta, S., Sanan, D. A., Weisgraber, K. H., Mahley, R. W., & Pitas, R. E. (1994). Differential effects of apolipoproteins E3 and E4 on neuronal growth in vitro. *Science, 264*(5160), 850–852.

Norum, J. H., Hart, K., & Levy, F. O. (2003). Ras-dependent ERK activation by the human G(s)-coupled serotonin receptors 5-HT4 (b) and 5-HT7(a). *The Journal of Biological Chemistry, 278*(5), 3098–3104. https://doi.org/10.1074/jbc.M206237200.

Ong, S. E., & Mann, M. (2005). Mass spectrometry-based proteomics turns quantitative. *Nature Chemical Biology, 1*(5), 252–262. https://doi.org/10.1038/nchembio736.

Ooms, S., Overeem, S., Besse, K., Rikkert, M. O., Verbeek, M., & Claassen, J. A. (2014). Effect of 1 night of total sleep deprivation on cerebrospinal fluid β-amyloid 42 in healthy middle-aged men: a randomized clinical trial. *JAMA Neurology, 71*(8), 971–977.

Ownby, R. L., Crocco, E., Acevedo, A., John, V., & Loewenstein, D. (2006). Depression and risk for Alzheimer disease: systematic review, meta-analysis, and metaregression analysis. *Archives of General Psychiatry, 63*(5), 530–538. https://doi.org/10.1001/archpsyc.63.5.530.

Pan, J., Zhang, C., Zhang, Z., & Li, G. (2014). Review of online coupling of sample preparation techniques with liquid chromatography. *Analytica Chimica Acta, 815*, 1–15. https://doi.org/10.1016/j.aca.2014.01.017.

Patterson, B. W., Elbert, D. L., Mawuenyega, K. G., Kasten, T., Ovod, V., Ma, S., et al. (2015). Age and amyloid effects on human central nervous system amyloid-beta kinetics. *Annals of Neurology, 78*(3), 439–453. https://doi.org/10.1002/ana.24454.

Polymeropoulos, M. H., Lavedan, C., Leroy, E., Ide, S. E., Dehejia, A., Dutra, A., et al. (1997). Mutation in the alpha-synuclein gene identified in families with Parkinson's disease. *Science, 276*(5321), 2045–2047.

Potter, R., Patterson, B. W., Elbert, D. L., Ovod, V., Kasten, T., Sigurdson, W., et al. (2013). Increased in vivo amyloid-beta42 production, exchange, and loss in presenilin mutation carriers. *Science Translational Medicine, 5*(189), 189ra177. https://doi.org/10.1126/scitranslmed.3005615.

Price, J. L., Ko, A. I., Wade, M. J., Tsou, S. K., McKeel, D. W., & Morris, J. C. (2001). Neuron number in the entorhinal cortex and

CA1 in preclinical Alzheimer disease. *Archives of Neurology, 58*(9), 1395–1402.

Proukakis, C., Dudzik, C. G., Brier, T., MacKay, D. S., Cooper, J. M., Millhauser, G. L., et al. (2013). A novel alpha-synuclein missense mutation in Parkinson disease. *Neurology, 80*(11), 1062–1064. https://doi.org/10.1212/WNL.0b013e31828727ba.

Roh, J. H., Huang, Y., Bero, A. W., Kasten, T., Stewart, F. R., Bateman, R. J., et al. (2012). Disruption of the sleep-wake cycle and diurnal fluctuation of amyloid-β in mice with Alzheimer's disease pathology. *Science Translational Medicine, 4*(150), 150ra122.

Sato, C., Barthélemy, N. R., Mawuenyega, K. G., Patterson, B. W., Gordon, B. A., Jockel-Balsarotti, J., et al. (2018). Tau kinetics in neurons and the human central nervous system. *Neuron, 97*(6), 1284–1298.e7. https://doi.org/10.1016/j.neuron.2018.02.015.

Sheline, Y. I., & Raichle, M. E. (2013). Resting state functional connectivity in preclinical Alzheimer's disease. *Biological Psychiatry.* https://doi.org/10.1016/j.biopsych.2012.11.028.

Sheline, Y. I., West, T., Yarasheski, K., Swarm, R., Jasielec, M. S., Fisher, J. R., et al. (2014). An antidepressant decreases CSF Abeta production in healthy individuals and in transgenic AD mice. *Science Translational Medicine, 6*(236), 236re234. https://doi.org/10.1126/scitranslmed.3008169.

Shibata, M., Yamada, S., Kumar, S. R., Calero, M., Bading, J., Frangione, B., et al. (2000). Clearance of Alzheimer's amyloid-ss (1-40) peptide from brain by LDL receptor-related protein-1 at the blood-brain barrier. *The Journal of Clinical Investigation, 106*(12), 1489–1499.

Singleton, A. B., Farrer, M., Johnson, J., Singleton, A., Hague, S., Kachergus, J., et al. (2003). Alpha-Synuclein locus triplication causes Parkinson's disease. *Science, 302*(5646), 841.

Smith, R. A., Miller, T. M., Yamanaka, K., Monia, B. P., Condon, T. P., Hung, G., et al. (2006). Antisense oligonucleotide therapy for neurodegenerative disease. *The Journal of Clinical Investigation, 116*(8), 2290–2296. https://doi.org/10.1172/jci25424.

St George-Hyslop, P. H., & Morris, J. C. (2008). Will anti-amyloid therapies work for Alzheimer's disease? *Lancet, 372*(9634), 180–182. https://doi.org/10.1016/S0140-6736(08)61047-8.

Strittmatter, W. J., Saunders, A. M., Goedert, M., Weisgraber, K. H., Dong, L. M., Jakes, R., et al. (1994). Isoform-specific interactions of apolipoprotein E with microtubule-associated protein tau: implications for Alzheimer disease. *Proceedings of the National Academy of Sciences of the United States of America, 91*(23), 11183–11186.

Strittmatter, W. J., Weisgraber, K. H., Huang, D. Y., Dong, L. M., Salvesen, G. S., Pericak-Vance, M., et al. (1993). Binding of human apolipoprotein E to synthetic amyloid beta peptide: isoform-specific effects and implications for late-onset Alzheimer disease. *Proceedings of the National Academy of Sciences of the United States of America, 90*(17), 8098–8102.

Tall, M. L., Lehmann, S., Diouf, E., Gerard, C., Filali, S., Gabelle, A., et al. (2015). Injectable preparation of labeled leucine with the carbon 13 for a clinical research program on the Alzheimer disease: pharmaceutical control of raw materials and the finished product and stability study. *Annales Pharmaceutiques Françaises, 73*(1), 43–59. https://doi.org/10.1016/j.pharma.2014.06.002.

Torbett, B. E., Baird, A., & Eliceiri, B. P. (2015). Understanding the rules of the road: proteomic approaches to interrogate the blood brain barrier. *Frontiers in Neuroscience, 9*, 70. https://doi.org/10.3389/fnins.2015.00070.

Verges, D. K., Restivo, J. L., Goebel, W. D., Holtzman, D. M., & Cirrito, J. R. (2011). Opposing synaptic regulation of amyloid-beta metabolism by NMDA receptors in vivo. *The Journal of Neuroscience, 31*(31), 11328–11337. https://doi.org/10.1523/JNEUROSCI.0607-11.2011.

Wildsmith, K. R., Basak, J. M., Patterson, B. W., Pyatkivskyy, Y., Kim, J., Yarasheski, K. E., et al. (2012). In vivo human apolipoprotein E isoform fractional turnover rates in the CNS. *PLoS One, 7*(6), e38013.

Xie, L., Kang, H., Xu, Q., Chen, M. J., Liao, Y., Thiyagarajan, M., et al. (2013). Sleep drives metabolite clearance from the adult brain. *Science, 342*, 373–377.

Zarranz, J. J., Alegre, J., Gomez-Esteban, J. C., Lezcano, E., Ros, R., Ampuero, I., et al. (2004). The new mutation, E46K, of alpha-synuclein causes Parkinson and Lewy body dementia. *Annals of Neurology, 55*(2), 164–173.

12

Applications of Neurophysiological Biomarkers in CNS Drug Development: Focus on Psychoses

Yash B. Joshi, Melissa A. Tarasenko, Gregory A. Light

Department of Psychiatry, University of California, San Diego, CA, United States

I INTRODUCTION

Profound advancements in the understanding of central nervous system (CNS) disease have occurred over the last century. In stark contrast the number of pharmacological agents targeting CNS disease successfully completing all phases of clinical trials and advancing to market has trickled, with diminishing interest by drug companies due to the immense cost, lack of diagnostic clarity in recruiting participants, and ambiguous endpoints that often are not able to approximate clinical effectiveness. Although these observations apply to multiple CNS diseases, this is particularly true for psychiatric conditions. Psychiatric illnesses are phenomenologically complex, with appropriate classification requiring interview by skilled clinicians using criterion-based checklists such as those found in the *Diagnostic and Statistical Manual of Mental Disorders* (DSM), which results in patient populations that have tremendous heterogeneity, often without any links to the underlying etiological basis (Nemeroff et al., 2013). Despite these challenges a combination of genetic, imaging, and neurophysiological studies has revealed important clues to the pathophysiology of psychiatric diseases. A new wave of biomarker-informed strategies for demonstrating the engagement of cognitively relevant brain networks (i.e., target engagement) following even initial drug exposure and providing early readouts of efficacy signals will weigh heavily into future CNS drug development. These biomarkers hold considerable promise for contributing to more reliable and "go-no go" trial decisions and ultimately for predicting and/or monitoring individual response to treatments and perhaps ultimately be used to guide the personalized assignments to treatments. In this chapter, we aim to highlight how neurophysiological biomarkers captured by electroencephalogram (EEG) can be used in CNS drug development, using schizophrenia as a model disease and mismatch negativity (MMN) as a leading candidate biomarker exemplar.

II SCHIZOPHRENIA AND NEUROCOGNITIVE IMPAIRMENT: A CHALLENGE FOR DRUG DEVELOPMENT

The diagnosis of schizophrenia requires the presence of positive symptoms (e.g., hallucinations and delusions), negative symptoms (e.g., diminished emotional expressivity and avolition), and social/occupational dysfunction over a period of time, as defined by the DSM (American Psychiatric Association, 2013). Dozens of studies have also identified neurocognitive impairment as a core component of the disorder, characterized by deficits in global cognition, problem solving, and learning and memory. Despite decades of clinical trials leading to the development of dozens of antipsychotics that can be effective in treating positive and negative symptoms, there is little evidence that these agents can modulate neurocognitive impairment in an enduring, meaningful way. This limitation presents a significant obstacle for treatment as neurocognitive impairment in schizophrenia is linked to functional disability and poor outcomes. Fortunately, many groups have identified and validated neurophysiological biomarkers, discussed in the succeeding text, which show great promise in aiding development of "procognitive" agents aimed at targeting neurocognitive impairment in schizophrenia.

III WHAT MAKES A GOOD NEUROPHYSIOLOGICAL BIOMARKER?

The utility of neurophysiological measures to guide CNS drug development is critically dependent on their ability to serve as biomarkers (see Table 12.1). Useful biomarkers must be accessible and measurable in preclinical models of disease (from rodents to nonhuman primates), sufficiently well studied such that those biomarkers are linked to relevant underlying neural circuits and known mechanisms of cognitive dysfunction in CNS disease, and reliably assessed in both healthy subjects and affected individuals. For utility in human trials, biomarkers must also be suitable for use as a repeated measure (i.e., insensitivity to order or practice effects) and responsive to interventions. If these candidate biomarkers ultimately prove useful for predicting or monitoring clinical effectiveness, biomarker acquisition should also be low-cost, scalable, and suitable for use in multicenter studies that can be conducted in real-world clinical settings.

In schizophrenia research, all of the above criteria have been identified for biomarker development by a variety of expert consensus panels. The Measurement and Treatment Research to Improve Cognition in Schizophrenia (MATRICS) initiative was called for by the National Institute of Mental Health and brought together academics, the pharmaceutical industry, and the US Food and Drug Administration (FDA) to address the disabling neurocognitive impairment associated with schizophrenia (Green et al., 2004). The panel agreed that there was a lack of consensus on a well-accepted instrument for measuring neurocognition in clinical trials, the best molecular targets for drug development, the optimal trial design for studies of those targets, and how regulatory agencies should approve and label novel agents. This initiative identified the following criteria as desirable in an FDA-approved

TABLE 12.1   Desirable Qualities in Neurophysiological Biomarkers

Valid	Reliable
	Insensitive to order or practice effects
	Related to daily psychosocial functioning
	Responsive to pharmacological agents or cognitive training interventions
Translatable	Available animal models
	Links to cognitive mechanisms
	Links to neural circuits
Scalable	Low cost
	Can be administered by nonspecialists
	Does not need specialized testing environment
	Suitable for use in multicenter studies

battery for use in clinical outcome measures: (1) high test-retest reliability, (2) utility as a repeated measure, (3) relationship to functional outcome, (4) tolerability and practicality, and (5) responsivity to therapeutic agents.

The Cognitive Neuroscience Treatment Research to Improve Cognition in Schizophrenia (CNTRICS) initiative, launched after MATRICS, further expanded on the MATRICS criteria by adding that these measures should have (6) construct validity, (7) a clear link to neural circuits and cognitive mechanisms, and (8) available animal models (Carter et al., 2008).

At the time of CNTRICS, mismatch negativity (MMN), (among other biomarkers, such as prepulse inhibition), was already considered a "mature" neurophysiological biomarker based on the criteria outlined earlier and ready for widespread implementation in clinical trial studies (Butler et al., 2012; Green et al., 2009). However, having biomarkers that meet these criteria in selected academic research settings would be of little value without the ability to implement them in a real-world, clinical setting. To that end the Consortium on the Genetics of Schizophrenia (COGS) study (Light et al., 2015) recently demonstrated that MMN could be easily recorded, without the use of highly trained specialists, in this multicenter study.

IV MISMATCH NEGATIVITY

Mismatch negativity (MMN) is an event-related potential (ERP) that is evoked when a train of "standard" stimuli is interrupted by an oddball or "deviant" stimulus that differs from the standards (Näätänen & Alho, 1997; Näätänen, Tederm, Alho, & Lavikainen, 1992). MMN is preattentive; primarily reflects an automatic response to sensory stimuli; and is able to be elicited without any effort, behavioral response, or even conscious awareness on the part of the subject (Light et al., 2012; Näätänen, 1995; Näätänen et al., 1992; Nagai et al., 2013; Rissling et al., 2012).

MMN may be evoked both by auditory and visual stimuli. For visual stimuli the deviant stimuli can differ from standards in qualities such as size, color, contrast, and duration. With visual stimuli, MMN is associated with negative potentials in occipital and temporal regions (for a review, see Czigler, 2014; Kremláček et al., 2016). MMN evoked by auditory stimuli, the paradigm that has the most experimental data and the focus of this chapter, is evoked by deviant qualities such as tone, volume, timbre, and duration. After auditory deviant stimuli presentation, MMN onset occurs approximately 50 ms, peaking after an additional 100–150 ms (see Fig. 12.1; Näätänen, Paavilainen, Rinne, & Alho, 2007; Todd, Michie, Schall, Ward, & Catts, 2012). Localization studies have consistently revealed cortical sources

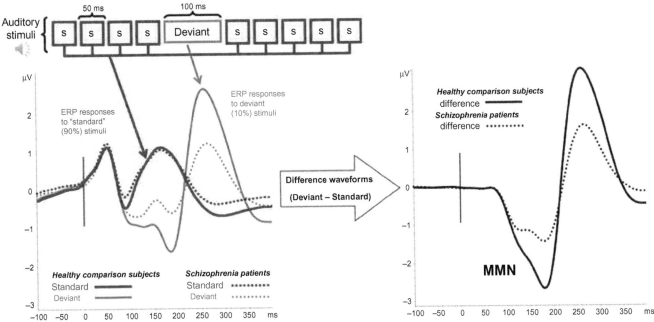

FIG. 12.1   Mismatch negativity. The earlier waveform represents the difference between event-related potentials in response to repeated standard stimuli (S) and a rare, deviant stimulus. The negative inflection denotes mismatch negativity (MMN), with prototypical responses from healthy volunteers *(solid lines)* and with a reduction in MMN amplitude in patients with schizophrenia *(dotted lines)*. From Light, G. A., & Swerdlow, N. R. (2015). *Future clinical uses of neurophysiological biomarkers to predict and monitor treatment response to schizophrenia.* Annals of the New York Academy of Sciences, 1344, 105–119.

located in broadly distributed temporal, frontal, and even parietal brain regions (Alho, 1995; Rissling et al., 2014; Takahashi et al., 2012).

Many studies have shown that MMN is a sensitive index of N-methyl-D-aspartate receptor (NMDA) functioning (Catts, Lai, Weickert, Weickert, & Catts, 2016; Michie, Malmierca, Harms, & Todd, 2016). This is particularly relevant to drug development as NMDA receptor dysfunction is thought by many to be a critical underlying pathophysiological *sine qua non* in psychotic disorders. NMDA receptor antagonists diminish MMN in nonhuman primates, and ketamine, an NMDA antagonist, reduces MMN in healthy control human subjects (Gil-da-Costa, Stoner, Fung, & Albright, 2013; Javitt, Zukin, Heresco-Levy, & Umbricht, 2012; Kreitschmann-Andermahr et al., 2001; Rosburg & Kreitschmann-Andermahr, 2016; Umbricht, Koller, Vollenweider, & Schmid, 2002). Lower baseline MMN is also associated with psychotic-like behavioral effects experienced by healthy subjects when exposed to ketamine (Umbricht et al., 2002).

Despite experimental evidence strongly linking the NMDA receptor system to MMN, the theoretical constructs underlying auditory MMN remain contested (Garrido, Kilner, Stephan, & Friston, 2009). One line of evidence suggests that MMN is evoked when there is a break in regularity, with a temporoprefrontal network generating a response to an auditory input based on its similarity to the memory trace of previous stimuli. This view—called the *model-adjustment hypothesis*—implies

that MMN is an automatic index of auditory change detection that triggers a switch in the focus of attention. A competing view posits that auditory MMN reflects neuronal adaptation in the auditory cortex. This *adaptation hypothesis* suggests that MMN results from the electrophysiological differences in responses to novel and nonnovel stimuli, where responses to nonnovel sounds are delayed and suppressed. More recently, predictive coding models have been proposed, formulated in terms of empirical Bayesian models of learning and inference. These models bridge both accommodation and adaptation hypotheses by positing that MMN can be mechanistically understood as a function of both plasticity in connections across multiple hierarchical levels of auditory information processing and local adaptation within primary auditory cortices. While future work will clarify this question, applications of MMN to CNS drug development are fortunately not limited by the absence of a complete mechanistic understanding.

V MMN AND RELEVANCE TO DRUG DEVELOPMENT IN SCHIZOPHRENIA: OBSERVATIONS FROM HUMAN STUDIES

Reduction of MMN amplitude in patients with schizophrenia was reported over two decades ago and has been replicated numerous times (Shelley et al., 1991). A multitude of human studies have shown that MMN deficits are found in patients with chronic psychosis

(Brockhaus-Dumke et al., 2005; Catts et al., 1995; Javitt, Shelley, Silipo, & Lieberman, 2000; Javitt, Steinschneider, Schroeder, Vaughan Jr., & Arezzo, 1994; Michie, 2001; Oades et al., 2006; Oknina et al., 2005; Salisbury, Kuroki, Kasai, Shenton, & McCarley, 2007; Salisbury, Shenton, Griggs, Bonner-Jackson, & McCarley, 2002; Shelley et al., 1991; Umbricht et al., 2003; Umbricht & Krljes, 2005). MMN deficits are seen in unmedicated schizophrenia patients (Bodatsch et al., 2011; Brockhaus-Dumke et al., 2005; Catts et al., 1995; Kirino & Inoue, 1999; Light et al., 2015; Rissling et al., 2012) and are shown to be resistant to traditional psychotropic medications indicated for psychosis, that is, antipsychotics (Düring, Glenthøj, & Oranje, 2016; Fawzy, Gado, Abdalla, & Ibrahim, 2015; Schall, Catts, Karayanidis, & Ward, 1999; Umbricht et al., 1998, 1999). With this in mind, it is readily apparent that neurophysiological biomarkers like MMN could better clarify subject recruitment and assist in screening participants for clinical trials testing novel agents. Given that clinical trials in psychiatry are experiencing a rise in the magnitude of the placebo effect (Alphs, Benedetti, Fleischhacker, & Kane, 2012), the use of objective, reliable, and direct assays of brain function, including those that are outside of conscious effortful control, is a critical next step for future CNS development programs.

In addition to refining the selection of participants for clinical trials, MMN is a useful biomarker to study genetics and for therapies aimed at individuals who are genetically loaded to be at high risk for developing disease. MMN has been shown to be highly heritable with amplitude reductions present in asymptomatic first-degree relatives of those with schizophrenia. MMN deficits are also found in patients with chromosome 22q deletion, which result in congenital syndromes that are associated with schizophrenia-like psychoses (Zarchi et al., 2013). With this in mind, biomarker-informed studies could aid in preventative strategies in families with known genetic risk, including rare gene mutations and loci conferring elevated risk of schizophrenia development (an "endophenotype" approach, for a review see Turetsky et al., 2017), and could be another avenue for novel therapeutic testing.

Apart from genetic studies of risk, EEG biomarkers also hold significant promise for use in studies of individuals at high clinical risk for psychosis. Abnormal MMN is also found in recent-onset psychosis and prodromal illness. Of those who are considered at high clinical risk to develop schizophrenia, only about a third go on to develop a psychotic illness after several years of follow-up (Nagai et al., 2013; Perez et al., 2014). In those studies, baseline MMN amplitude appears to be smaller in those who convert to psychosis in follow-up, and MMN in those who do not convert appear to be similar to age-matched controls (Bodatsch et al., 2011). Strikingly, work has also shown that MMN amplitude appears to

anticipate time-to-convert to psychosis—more severe MMN deficits relate to shorter time for psychosis conversion (Atkinson, Michie, & Schall, 2012; Higuchi et al., 2013; Jahshan et al., 2012; Shaikh et al., 2012). Use of neurophysiological biomarkers could therefore be useful for the identification of those at imminent risk and in need of intensive intervention (Belger, Yucel, & Donkers, 2012).

Arguably the most important metric of treatment in CNS illness is the ability of interventions to improve functional outcome. In schizophrenia and related illnesses, biomarkers that are associated with functional outcomes can be particularly powerful outcome metrics in CNS drug development. In people with schizophrenia, MMN deficits account for substantial portions of variance in cognitive and psychosocial functioning and independent living (Kawakubo & Kasai, 2006; Light & Braff, 2005; Rasser et al., 2011; Rissling et al., 2014; Wynn, Sugar, Horan, Kern, & Green, 2010). Therefore MMN would be useful for tracking outcome in addition to standard assessments and instruments that measure clinical severity. In this way, EEG biomarkers can help to better capture objective measures of functional success.

VI TARGET IDENTIFICATION AND APPLICATION IN PRECLINICAL SETTINGS: MMN IN ANIMAL MODELS OF SCHIZOPHRENIA

EEG biomarkers used in humans would have limited use in drug development if they were not able to be evoked and manipulated in lower model systems. This application is essential for drug discovery and development. Fortunately, event-related potentials analogous to those found in humans can be evoked from rodents to nonhuman primates.

As with human subjects, MMN is modulated by ketamine, an NMDA antagonist, in wild-type mice. Aside from pharmacological inhibition in wild-type rodents, MMN is also found to be aberrant in genetic models of the disease (Ehrlichman et al., 2009; Ehrlichman, Maxwell, Majumdar, & Siegel, 2008; Harms et al., 2018). The pioneering work of Siegel and others has shown that mice lacking one copy of neuregulin-1, a susceptibility gene for schizophrenia, show impaired MMN, and replicate an important subset of the electrophysiological phenomenology found in patients (Amann et al., 2010; Ehrlichman et al., 2009; Hahn et al., 2006). MMN deficits are also seen in mice with reduced levels of the NMDA receptor subunit NR1, a putative model for increased vulnerability to schizophrenia (Featherstone et al., 2015; Halene et al., 2009). In addition to rodents, MMN has been observed in cat and nonhuman primate models as well, and MMN may be modulated by ketamine in these systems as well (Gil-da-Costa et al., 2013;

Javitt, Steinschneider, Schroeder, & Arrezzo, 1996; Pincze, Lakatos, Rajkai, Ulbert, & Karmos, 2001).

A platform for translating novel molecular agents for procognitive drug development in schizophrenia would therefore involve screening molecular targets rapidly with animal models to elicit an MMN response before proceeding to early phase clinical trials where corresponding electrophysiological changes in human MMN could be assayed. In a similar manner, targets for other disorders with response to electrophysiological biomarkers could be easily and effectively screened.

VII BIOMARKERS IN ACTION: USE OF MISMATCH NEGATIVITY IN CLINICAL TRIALS FOR PROCOGNITIVE AGENTS

Use of MMN in clinical trials assessing putative procognitive agents in schizophrenia has led to observations with important implications for drug development. Our group's work on memantine, approved for used in Alzheimer's disease, serves as an illustrative example (Bhakta et al., 2016; Swerdlow, Bhakta, et al., 2016; Swerdlow, Tarasenko, et al., 2016; Light & Swerdlow, 2015).

Memantine is a moderate affinity, noncompetitive, voltage-dependent NMDA antagonist that preferentially blocks excessive NMDA activation without altering normal activity (Johnson & Kotermanski, 2005; Kishi & Iwata, 2013). As indicated earlier, while memantine would be predicted to be clinically effective in schizophrenia, where abnormal NMDA functioning is key to known disease pathophysiology, clinical trials using memantine as a procognitive agent have yielded variable outcomes (de Lucena et al., 2009; Krivoy et al., 2007; Lee et al., 2012; Lieberman et al., 2009; Zdanys & Tampi, 2008). However, in the majority of studies, neurocognitive performance (and the neural substrates that underlie them) was not specifically interrogated, with outcome measures primarily consisting of assessments of positive and negative symptoms. With these data alone, judging the procognitive benefits of memantine for use in schizophrenia is difficult. However, in a related series of experiments, our group has found that memantine enhances MMN in a dose-dependent way in patients with schizophrenia using single-dose exposure designs (Bhakta et al., 2016; Swerdlow, Bhakta, et al., 2016; Swerdlow, Tarasenko, et al., 2016). While future studies will clarify if neurocognitive improvement is observed with longer-term exposure to memantine, our work has demonstrated target engagement in the appropriate clinical population. Beyond assurance of target engagement, MMN could be utilized to *specifically select those subjects that have shown MMN improvement after single doses of memantine*, enriching the subject pool to improve signal in subsequent trials.

This type of experimental medicine approach is similarly being utilized with other agents, including N-acetyl-cysteine (Lavoie et al., 2008), ketamine analogues (Schmidt et al., 2012), cholinergic agents (Preskorn et al., 2014), and even stimulants such as amphetamines (Swerdlow, Bhakta, et al., 2016; Swerdlow, Tarasenko, et al., 2016), promising to factor heavily in future trial design.

VIII MISMATCH NEGATIVITY: BEYOND SCHIZOPHRENIA

Beyond the observations and uses listed earlier for schizophrenia, MMN has useful applications in other CNS contexts (Näätänen & Escera, 2000). MMN elicited by deviant syllable stimuli is reduced in children with dyslexia compared with that elicited by deviant tone stimuli (Schulte-Körne, Deimel, Bartling, & Remschmidt, 1998), indicating that MMN reflects disruption of phonological processing, a potentially useful platform for therapeutic selection since MMN deficits can also be seen in 2-month-old infants at high risk for dyslexia (Van Leeuwen et al., 2006). Aside from neurodevelopmental disorders, MMN deficits are seen in normal aging when evoked by longer intervals between stimuli and are more dramatically pronounced in those with dementing illnesses like Alzheimer's disease (Pekkonen, 2000). MMN-informed biomarkers could be a boon in recruitment of subjects for clinical trials and tracking outcomes in Alzheimer disease, complementing brain imaging and cerebrospinal biomarkers of disease severity and treatment response. Use of these biomarkers would be especially salient since recent trials of disease-modifying agents have failed and have brought into question commonly held pathophysiological ground of Alzheimer's disease itself (Gauthier et al., 2016; Karran & Hardy, 2014). In studies of stroke and poststroke aphasia (primarily cortical, left, and right hemispheric lesions), MMN is attenuated acutely following stroke and normalization of MMN amplitude correlates with functional improvement over days and is associated with speech comprehension improvement on cognitive screens of aphasia up to several months later (Ilvonen et al., 2003). Numerous other exploratory studies have shown disease-specific associations with multiple sclerosis, Parkinson's disease, Huntington's disease, coma, and sleep disorders, among others (Näätänen et al., 2011). While the relevance of these studies is not yet as clear as with schizophrenia, these are likely fertile ground for investigation as it relates to biomarker application. Beyond disease contexts, MMN has been used to index various aspects of music and language learning (Näätänen, 2003), and the ways in which MMN can be used to direct procognitive agent discovery for nondiseased populations are an open and enticing question.

IX CONCLUSION

Drug development for central nervous system applications has been hampered by a lack of etiopathological clarity, signal-to-noise thresholds of unclear importance for target engagement, and costly failures when drugs fail to achieve clinically relevant outcomes in large-scale trials. Neurophysiological biomarkers have the possibility of altering the trajectory of central nervous system drug development by improving screening for new compounds in preclinical systems, helping to clarify which populations may benefit from specific targets engaged by novel therapeutics, and correlating target engagement to functional outcome. In schizophrenia, this goal is crystallizing with EEG biomarkers, particularly for mismatch negativity. With further work appropriately incorporating EEG biomarkers in early stage trial designs, there is significant promise and hope that drug development will be accelerated for central nervous system indications, in both health and disease.

References

Alho, K. (1995). Cerebral generators of mismatch negativity (MMN) and its magnetic counterpart (MMNm) elicited by sound changes. *Ear and Hearing, 16*, 38–51.

Alphs, L., Benedetti, F., Fleischhacker, W. W., & Kane, J. M. (2012). Placebo-related effects in clinical trials in schizophrenia: what is driving this phenomenon and what can be done to minimize it? *International Journal of Neuropsychopharmacology, 15*, 1003–1014.

Amann, L. C., Gandal, M. J., Halene, T. B., Ehrlichman, R. S., White, S. L., McCarren, H. S., et al. (2010). Mouse behavioral endophenotypes for schizophrenia. *Brain Research Bulletin, 83*, 147–161.

American Psychiatric Association. (2013). *Diagnostic and statistical manual of mental disorders* (5th ed.). Arlington, VA: American Psychiatric Publishing.

Atkinson, R. J., Michie, P. T., & Schall, U. (2012). Duration mismatch negativity and P3a in first-episode psychosis and individuals at ultra-high risk of psychosis. *Biological Psychiatry, 71*, 98–104.

Belger, A., Yucel, G. H., & Donkers, F. C. (2012). In search of psychosis biomarkers in high-risk populations: is the mismatch negativity the one we've been waiting for? *Biological Psychiatry, 71*, 94–95.

Bhakta, S. G., Chou, H. H., Rana, B., Talledo, J. A., Balvanenda, B., Gaddis, L., et al. (2016). Effects of acute memantine administration on MATRICS Consensus Cognitive Battery performance in psychosis: testing an experimental medicine strategy. *Psychopharmacology (Berlin), 233*, 2339–2410.

Bodatsch, M., Ruhrmann, S., Wagner, M., Müller, R., Schultze-Lutter, F., Frommann, I., et al. (2011). Prediction of psychosis by mismatch negativity. *Biological Psychiatry, 69*, 959–966.

Brockhaus-Dumke, A., Tendolkar, I., Pukrop, R., Schultze-Lutter, F., Klosterkötter, J., & Ruhrmann, S. (2005). Impaired mismatch negativity generation in prodromal subjects and patients with schizophrenia. *Schizophrenia Research, 73*, 297–310.

Butler, P. D., Chen, Y., Ford, J. M., Geyer, M. A., Silverstein, S. M., & Green, M. F. (2012). Perceptual measurement in schizophrenia: promising electrophysiology and neuroimaging paradigms from CNTRICS. *Schizophrenia Bulletin, 38*, 81–91.

Carter, C. S., Barch, D. M., Buchanan, R. W., Bullmore, E., Krystal, J. H., Cohen, J., et al. (2008). Identifying cognitive mechanisms targeted for treatment development in schizophrenia: an overview of the first meeting of the Cognitive Neuroscience Treatment Research to Improve Cognition in Schizophrenia Initiative. *Biological Psychiatry, 64*, 4–10.

Catts, S. V., Shelley, A. M., Ward, P. B., Liebert, B., McConaghy, N., Andrews, S., et al. (1995). Brain potential evidence for an auditory sensory memory deficit in schizophrenia. *The American Journal of Psychiatry, 152*, 213–219.

Catts, V. S., Lai, Y. L., Weickert, C. S., Weickert, T. W., & Catts, S. V. (2016). A quantitative review of the postmortem evidence for decreased cortical N-methyl-d-aspartate receptor expression levels in schizophrenia: how can we link molecular abnormalities to mismatch negativity deficits? *Biological Psychology, 116*, 57–67.

Czigler, I. (2014). Visual mismatch negativity and categorization. *Brain Topography, 27*, 590–598.

de Lucena, D., Fernandes, B. S., Berk, M., Dodd, S., Medeiros, D. W., Pedrini, M., et al. (2009). Improvement of negative and positive symptoms in treatment refractory schizophrenia: a double-blind, randomized, placebo-controlled trial with memantine as add-on therapy to clozapine. *Journal of Clinical Psychiatry, 70*, 1416–1423.

Düring, S., Glenthøj, B. Y., & Oranje, B. (2016). Effects of blocking D2/D3 receptors on mismatch negativity and P3a amplitude of initially antipsychotic naïve, first episode schizophrenia patients. *International Journal of Neuropsychopharmacology, 19*(3), pyv109.

Ehrlichman, R. S., Gandal, M. J., Maxwell, C. R., Lazarewicz, M. T., Finkel, L. H., Contreras, D., et al. (2009). N-methyl-d-aspartic acid receptor antagonist-induced frequency oscillations in mice recreate pattern of electrophysiological deficits in schizophrenia. *Neuroscience, 158*, 705–712.

Ehrlichman, R. S., Liminais, S. N., White, S. L., Rudnick, N. D., Ma, N., Dow, H. C., et al. (2009). Neuregulin 1 transgenic mice display reduced mismatch negativity, contextual fear conditioning and social interactions. *Brain Research, 1294*, 116–127.

Ehrlichman, R. S., Maxwell, C. R., Majumdar, S., & Siegel, S. J. (2008). Deviance-elicited changes in event-related potentials are attenuated by ketamine in mice. *Journal of Cognitive Neuroscience, 20*, 1403–1414.

Fawzy, N., Gado, O., Abdalla, A. M., & Ibrahim, W. M. (2015). Auditory mismatch negativity, P300, and disability among first-episode schizophrenia patients without auditory hallucinations. *The Egyptian Journal of Psychiatry, 36*, 112.

Featherstone, R. E., Shin, R., Kogan, J. H., Liang, Y., Matsumoto, M., & Siegel, S. J. (2015). Mice with subtle reduction of NMDA NR1 receptor subunit expression have a selective decrease in mismatch negativity: implications for schizophrenia prodromal population. *Neurobiology of Disease, 73*, 289–295.

Garrido, M. I., Kilner, J. M., Stephan, K. E., & Friston, K. J. (2009). The mismatch negativity: a review of underlying mechanisms. *Clinical Neurophysiology, 120*, 453–463.

Gauthier, S., Albert, M., Fox, N., Goedert, M., Kivipelto, M., Mestre-Ferrandiz, J., et al. (2016). Why has therapy for development for dementia failed in the last two decades? *Alzheimers Dement, 12*, 60–64.

Gil-da-Costa, R., Stoner, G. R., Fung, R., & Albright, T. D. (2013). Nonhuman primate model of schizophrenia using a noninvasive EEG method. *Proceedings of the National Academy of Sciences of the United States of America, 110*, 15425–15430.

Green, M. F., Butler, P. D., Chen, Y., Geyer, M. A., Silverstein, S., Wynn, J. K., et al. (2009). Perception measurement in clinical trials of schizophrenia: promising paradigms from CNTRICS. *Schizophrenia Bulletin, 35*, 163–181.

Green, M. F., Nuechterlein, K. H., Gold, J. M., Barch, D. M., Cohen, J., Essock, S., et al. (2004). Approaching a consensus cognitive battery for clinical trials in schizophrenia: the NIMH-MATRICS conference to select cognitive domains and test criteria. *Biological Psychiatry, 56*, 301–307.

Hahn, C. G., Wang, H. Y., Cho, D. S., Talbot, K., Gur, R. E., Berrettini, W. H., et al. (2006). Altered neuregulin 1-erbB4 signaling contributes to NMDA receptor hypofunction in schizophrenia. *Nature Medicine*, 12, 824–828.

Halene, T. B., Ehrlichman, R. S., Yliang, Y., Christian, E. P., Jonak, G. J., Gur, T. L., et al. (2009). Assessment of NMDA receptor NR1 subunit hypofunction in mice as a model for schizophrenia. *Genes, Brain, and Behavior*, 8, 661–675.

Harms, L., Fulham, W. R., Todd, J., Meehan, C., Schall, U., Hodgson, D. M., et al. (2018). Late deviance detection in rats is reduced, while early deviance detection is augmented by the NMDA receptor antagonist MK-801. *Schizophrenia Research*, 191, 43–50.

Higuchi, Y., Sumiyoshi, T., Seo, T., Miyanishi, T., Kawasaki, Y., & Suzuki, M. (2013). Mismatch negativity and cognitive performance for the prediction of psychosis in subjects with at-risk mental state. *PLoS ONE*, 8, e54080.

Ilvonen, T. M., Kujala, T., Kiesiläinen, A., Salonen, O., Kozou, H., Pekkonen, E., et al. (2003). Auditory discrimination after left-hemisphere stroke a mismatch negativity follow-up study. *Stroke*, 34, 1746–1751.

Jahshan, C., Cadenhead, K. S., Rissling, A. J., Kirihara, K., Braff, D. L., & Light, G. A. (2012). Automatic Sensory Information Processing Abnormalities across the Illness Course of Schizophrenia. *Psychological Medicine*, 42, 85–97.

Javitt, D. C., Shelley, A. M., Silipo, G., & Lieberman, J. A. (2000). Deficits in auditory and visual context-dependent processing in schizophrenia: defining the pattern. *Archives of General Psychiatry*, 57, 1131–1137.

Javitt, D. C., Steinschneider, M., Schroeder, C. E., & Arrezzo, J. C. (1996). Role of cortical N-methyl-D-aspartate receptors in auditory sensory memory and mismatch negativity generation: implications for schizophrenia. *Proceedings of the National Academy of Sciences of the United States of America*, 93, 11962–11967.

Javitt, D. C., Steinschneider, M., Schroeder, C. E., Vaughan, H. G., Jr., & Arezzo, J. C. (1994). Detection of stimulus deviance within primate primary auditory cortex: intracortical mechanisms of mismatch negativity (MMN) generation. *Brain Research*, 667, 192–200.

Javitt, D. C., Zukin, S. R., Heresco-Levy, U., & Umbricht, D. (2012). Has an angel shown the way? Etiological and therapeutic implications of the PCP/NMDA model of schizophrenia. *Schizophrenia Bulletin*, 38, 958–966.

Johnson, J. W., & Kotermanski, S. E. (2005). Mechanism of action of memantine. *Current Opinion in Pharmacology*, 6, 61–67.

Karran, E., & Hardy, J. (2014). A critique of the drug discovery and phase 3 clinical programs targeting the amyloid hypothesis for Alzheimer disease. *Annals of Neurology*, 76, 185–205.

Kawakubo, Y., & Kasai, K. (2006). Support for an association between mismatch negativity and social functioning in schizophrenia. *Progress in Neuro-Psychopharmacology and Biological Psychiatry*, 30, 1267–1268.

Kirino, E., & Inoue, R. (1999). The relationship of mismatch negativity to quantitative EEG and morphological findings in schizophrenia. *Journal of Psychiatric Research*, 33, 445–456.

Kishi, T., & Iwata, N. (2013). NMDA receptor antagonist interventions in schizophrenia: meta-analysis of randomized, placebo-controlled trials. *Journal of Psychiatric Research*, 47, 1143–1149.

Kreitschmann-Andermahr, I., Rosburg, T., Demme, U., Gaser, E., Nowak, H., & Sauer, H. (2001). Effect of ketamine on the neuromagnetic mismatch field in healthy humans. *Cognitive Brain Research*, 12, 109–116.

Kremláček, J., Kreegipuu, K., Tales, A., Astikainen, P., Põldver, N., Näätänen, R., et al. (2016). Visual mismatch negativity (vMMN): a review and meta-analysis of studies in psychiatric and neurological disorders. *Cortex*, 80, 76–112.

Krivoy, A., Weizman, A., Laor, L., Hellinger, N., Zemishlany, Z., & Fischel, T. (2007). Addition of memantine to antipsychotic treatment in schizophrenia inpatients with residual symptoms: a preliminary study. *European Neuropsychopharmacology*, 18, 117–121.

Lavoie, S., Murray, M. M., Deppen, P., Knyazeva, M. G., Berk, M., Boulat, O., et al. (2008). Glutatione precursor, N-acetyl-cysteine, improves mismatch negativity in schizophrenia patients. *Neuropsychopharmacology*, 33, 2187–2199.

Lee, J. G., Lee, S. W., Lee, B. J., Park, S. W., Kim, G. M., & Kim, Y. H. (2012). Adjunctive memantine therapy for cognitive impairment in chronic schizophrenia: a placebo-controlled pilot study. *Psychiatry Investigation*, 9, 166–173.

Lieberman, J. A., Papadakis, K., Csernansky, J., Litman, R., Volavka, J., Jia, X. D., et al. (2009). A randomized, placebo-controlled study of memantine as adjunctive treatment in patients with schizophrenia. *Neuropsychopharmacology*, 34, 1322–1329.

Light, G. A., & Braff, D. L. (2005). Stability of mismatch negativity deficits and their relationship to functional impairments in chronic schizophrenia. *The American Journal of Psychiatry*, 162, 1741–1743.

Light, G. A., & Swerdlow, N. R. (2015). Future clinical uses of neurophysiological biomarkers to predict and monitor treatment response to schizophrenia. *Annals of the New York Academy of Sciences*, 1344, 105–119.

Light, G. A., Swerdlow, N. R., Rissling, A. J., Radant, A., Sugar, C. A., Sprock, J., et al. (2012). Characterization of neurophysiologic and neurocognitive biomarkers for use in genomic and clinical outcome studies of schizophrenia. *PLoS One*, 7, e39434.

Light, G. A., Swerdlow, N. R., Thomas, M. L., Calkins, M. E., Green, M. F., Greenwood, T. A., et al. (2015). Validation of mismatch negativity and P3a for use in Multi-site studies of schizophrenia: characterization of demographic, clinical, cognitive, and functional correlates in COGS-2. *Schizophrenia Research*, 163, 63–72.

Michie, P. T. (2001). What has MMN revealed about the auditory system in schizophrenia? *International Journal of Psychophysiology*, 42, 177–194.

Michie, P. T., Malmierca, M. S., Harms, L., & Todd, J. (2016). The neurobiology of MMN and implications for schizophrenia. *Biological Psychology*, 116, 90–97.

Näätänen, R. (1995). The mismatch negativity: a powerful tool for cognitive neuroscience. *Ear and Hearing*, 16, 6–18.

Näätänen, R. (2003). Mismatch negativity: clinical research and possible applications. *International Journal of Psychophysiology*, 48, 179–188.

Näätänen, R., & Alho, K. (1997). Mismatch negativity – the measure for central sound representation accuracy. *Audiology & Neuro-Otology*, 2, 341–353.

Näätänen, R., & Escera, C. (2000). Mismatch negativity: clinical and other applications. *Audiology & Neuro-Otology*, 5, 105–110.

Näätänen, R., Kujala, T., Kreegipuu, K., Carlson, S., Escera, C., Baldeweg, T., et al. (2011). The mismatch negativity: an index of cognitive decline in neuropsychiatric and neurological diseases and in ageing. *Brain*, 134, 3435–3453.

Näätänen, R., Paavilainen, P., Rinne, T., & Alho, K. (2007). The mismatch negativity (MMN) in basic research of central auditory processing: a review. *Clinical Neurophysiology*, 118(12), 2544–2590.

Näätänen, R., Tederm, W., Alho, K., & Lavikainen, J. (1992). Auditory attention and selective input modulation: a topographical ERP study. *Neuroreport*, 3, 493–496.

Nagai, T., Tada, M., Kirihara, K., Araki, T., Jinde, S., & Kasai, K. (2013). Mismatch negativity as a "Translatable" brain marker toward early intervention for psychosis: a review. *Frontiers in Psychiatry*, 4, 115.

Nemeroff, C. B., Weinberger, D., Rutter, M., MacMillan, H. L., Bryant, R. A., Wessely, S., et al. (2013). DSM-5: a collection of psychiatrist views on the changes, controversies, and future directions. *BMC Medicine*, 11, 202.

Oades, R. D., Wild-Wall, N., Juran, S. A., Sachsse, J., Oknina, L. B., & Röpcke, B. (2006). Auditory change detection in schizophrenia: sources of activity, related neuropsychological function and symptoms in patients with a first episode in adolescence, and patients 14 years after an adolescent illness-onset. *BMC Psychiatry*, *6*, 7.

Oknina, L. B., Wild-Wall, N., Oades, R. D., Juran, S. A., Ropcke, B., Pfueller, U., et al. (2005). Frontal and temporal sources of mismatch negativity in healthy controls, patients at onset of schizophrenia in adolescence and others at 15 years after onset. *Schizophrenia Research*, *76*, 25–41.

Pekkonen, E. (2000). Mismatch negativity in aging and in Alzheimer's and Parkinson's diseases. *Audiology & Neuro-Otology*, *5*, 216–224.

Perez, V. B., Woods, S. W., Roach, B. J., Ford, J. M., McGlashan, T. H., Srihari, V. H., et al. (2014). Automatic auditory processing deficits in schizophrenia and clinical high-risk patients: forecasting psychosis risk with mismatch negativity. *Biological Psychiatry*, *75*, 459–469.

Pincze, Z., Lakatos, P., Rajkai, C., Ulbert, I., & Karmos, G. (2001). Separation of mismatch negativity and the N1 wave in the auditory cortex of the cat: a topographic study. *Clinical Neurophysiology*, *112*, 778–784.

Preskorn, S. H., Gawryl, M., Dgetluck, N., Palfreyman, M., Bauer, L. O., & Hilt, D. C. (2014). Normalizing effects of EVP-6124, an alpha-7 nicotinic partial agonist, on event-related potentials and cognition: a proof of concept, randomized trial in patients with schizophrenia. *Journal of Psychiatric Practice*, *20*, 12–24.

Rasser, P. E., Schall, U., Todd, J., Michie, P. T., Ward, P. B., Johnston, P., et al. (2011). Gray matter deficits, mismatch negativity, and outcomes in schizophrenia. *Schizophrenia Bulletin*, *37*, 131–140.

Rissling, A. J., Braff, D. L., Swerdlow, N. R., Hellemann, G., Rassovsky, Y., Sprock, J., et al. (2012). Disentangling early sensory information processing deficits in schizophrenia. *Clinical Neurophysiology: Official Journal of the International Federation of Clinical Neurophysiology*, *123*, 1942–1949.

Rissling, A. J., Miyakoshi, M., Sugar, C. A., Braff, D. L., Makeig, S., & Light, G. A. (2014). Cortical substrates and functional correlates of auditory deviance processing deficits in schizophrenia. *NeuroImage*, *6*, 424–437.

Rosburg, T., & Kreitschmann-Andermahr, I. (2016). The effects of ketamine on the mismatch negativity (MMN) in humans–A meta-analysis. *Clinical Neurophysiology*, *127*, 1387–1394.

Salisbury, D. F., Kuroki, N., Kasai, K., Shenton, M. E., & McCarley, R. W. (2007). Progressive and interrelated functional and structural evidence of post-onset brain reduction in schizophrenia. *Archives of General Psychiatry*, *64*, 521–529.

Salisbury, D. F., Shenton, M. E., Griggs, C. B., Bonner-Jackson, A., & McCarley, R. W. (2002). Mismatch negativity in chronic schizophrenia and first-episode schizophrenia. *Archives of General Psychiatry*, *59*, 686–694.

Schall, U., Catts, S. V., Karayanidis, F., & Ward, P. B. (1999). Auditory event-related potential indices of fronto-temporal information processing in schizophrenia syndromes: valid outcome prediction of clozapine therapy in a three-year follow-up. *The International Journal of Neuropsychopharmacology*, *2*, 83–93.

Schmidt, A., Bachmann, R., Kometer, M., Csomor, P. A., Stephan, K. E., Seifritz, E., et al. (2012). Mismatch negativity encoding of prediction errors predicts S-ketamine-induced cognitive impairments. *Neuropsychopharmacology*, *37*, 865–875.

Schulte-Körne, G., Deimel, W., Bartling, J., & Remschmidt, H. (1998). Auditory processing and dyslexia: evidence for a specific speech processing deficit. *Neuroreport*, *9*, 337–340.

Shaikh, M., Valmaggia, L., Broome, M. R., Dutt, A., Lappin, J., Day, F., et al. (2012). Reduced mismatch negativity predates the onset of psychosis. *Schizophrenia Research*, *134*, 42–48.

Shelley, A. M., Ward, P. B., Catts, S. V., Michie, P. T., Andrews, S., & McConaghy, N. (1991). Mismatch negativity: an index of a preattentive processing deficit in schizophrenia. *Biological Psychiatry*, *30*, 1059–1062.

Swerdlow, N. R., Bhakta, S., Chou, H. H., Talledo, J. A., Balvaneda, B., & Light, G. A. (2016). Memantine effects on sensorimotor gating and mismatch negativity in patients with chronic psychosis. *Neuropsychopharmacology*, *41*, 419–430.

Swerdlow, N. R., Tarasenko, M., Bhakta, S. G., Talledo, J., Alverez, A. I., Hughes, E. L., et al. (2016). Amphetamine enhances gains in auditory discrimination training in adult schizophrenia patients. *Schizophrenia Bulletin*, *43*, 872–880.

Takahashi, H., Rissling, A. J., Pascual-Marqui, R., Kirihara, K., Pela, M., Sprock, J., et al. (2012). Neural substrates of normal and impaired preattentive sensory discrimination in large cohorts of nonpsychiatric subjects and schizophrenia patients. *NeuroImage*, *66C*, 594–603.

Todd, J., Michie, P. T., Schall, U., Ward, P. B., & Catts, S. V. (2012). Mismatch negativity (MMN) reduction in schizophrenia—impaired prediction-error generation, estimation or salience? *International Journal of Psychophysiology*, *83*, 222–231.

Turetsky, B. I., Calkins, M. E., Light, G. A., Olincy, A., Radant, A. D., & Swerdlow, N. R. (2017). Neurophysiological endophenotypes of schizophrenia: the viability of selected candidate measures. *Schizophrenia Bulletin*, *33*, 69–94.

Umbricht, D., Javitt, D., Novak, G., Bates, J., Pollack, S., Lieberman, J., et al. (1998). Effects of clozapine on auditory event-related potentials in schizophrenia. *Biological Psychiatry*, *44*, 716–725.

Umbricht, D., Javitt, D., Novak, G., Bates, J., Pollack, S., Lieberman, J., et al. (1999). Effects of risperidone on auditory event-related potentials in schizophrenia. *International Journal of Neuropsychopharmacology*, *2*, 299–304.

Umbricht, D., Koller, R., Schmid, L., Skrabo, A., Grubel, C., Huber, T., et al. (2003). How specific are deficits in mismatch negativity generation to schizophrenia? *Biological Psychiatry*, *53*, 1120–1131.

Umbricht, D., Koller, R., Vollenweider, F. X., & Schmid, L. (2002). Mismatch negativity predicts psychotic experiences induced by NMDA receptor antagonist in healthy volunteers. *Biological Psychiatry*, *51*, 400–406.

Umbricht, D., & Krljes, S. (2005). Mismatch negativity in schizophrenia: a meta-analysis. *Schizophrenia Research*, *76*, 1–23.

Van Leeuwen, T., Been, P., Kuijpers, C., Zwarts, F., Maassen, B., & van der Leij, A. (2006). Mismatch response is absent in 2-month-old infants at risk for dyslexia. *Neuroreport*, *17*, 351–355.

Wynn, J. K., Sugar, C., Horan, W. P., Kern, R., & Green, M. F. (2010). Mismatch negativity, social cognition, and functioning in schizophrenia patients. *Biological Psychiatry*, *67*, 940–947.

Zarchi, O., Carmel, M., Avni, C., Attias, J., Frisch, A., Michaelovsky, E., et al. (2013). Schizophrenia-like neurophysiological abnormalities in 22q11. 2 deletion syndrome and their association to COMT and PRODH genotypes. *Journal of Psychiatric Research*, *47*, 1623–1629.

Zdanys, K., & Tampi, R. (2008). A systematic review of off-label uses of memantine for psychiatric disorders. *Progress in Neuropsychopharmacology and Biological Psychiatry*, *32*, 1362–1374.

Heart Rate Variability as a Translational Biomarker for Emotional and Cognitive Deficits

Bart A. Ellenbroek, Meyrick Kidwell, Joyce Colussi-Mas, Jiun Youn

Victoria University of Wellington, School of Psychology, Behavioural Neurogenetics Group, Wellington, New Zealand

I INTRODUCTION

About 30%–35% of individuals in the developed world will, in their lifetime, suffer from a brain disorder (Andlin-Sobocki, Jonsson, Wittchen, & Olesen, 2005). Given that many of these disorders typically develop early in life (especially psychiatric disorders, which generally develop before the age of 25), and have proven therapeutically very challenging, central nervous system (CNS) disorders are the costliest class of disorders in terms of years lived with disability (Vos et al., 2012). This does not only put a heavy personal burden on the patient and the family, but also on society as a whole. Indeed, a recent study from the European Union estimated the overall cost of brain disorders at almost 800 billion Euro, more than the costs for lung, cardiovascular, endocrine and cancer disorders combined (Gustavsson et al., 2011). This cost is a combination of direct medical costs (including costs for hospitalization and medication), direct nonmedical costs (such as adjustments to housing, social services) and, the largest factors, indirect costs (this includes, among others, early retirement, loss of productivity).

Given the substantial costs for patients, their families and society, and the high unmet medical need, one would expect an enormous investment from the pharmaceutical industry to improve the treatment of CNS conditions. However, in recent years many pharmaceutical companies, including but not limited to AstraZeneca, GSK and Novartis, and most recently Pfizer, have either substantially reduced investment or actively withdrawn from the CNS market (Chandler, 2013; Miller, 2010).

While there is still discussion about the reasons for this massive withdrawal, one important contributor is the lack of confidence in the prediction of the therapeutic properties of new compounds. In line with this, a detailed market analysis showed that the attrition rate of CNS drugs (together with cancer drugs) is the highest among all therapeutic areas, with only 8% of all drugs entering phase I successfully reaching registration (Kola & Landis, 2004). Further analysis showed that the failure rate was particularly high in phase III and at registration, when the clinical efficacy is assessed. Moreover, between 1990 and 2004, the success rates have declined most sharply in phase II and III (Pammolli, Magazzini, & Riccaboni, 2011), a trend that has only increased in more recent years, with 55% of all failures in 2012 due to lack of efficacy (Arrowsmith & Miller, 2013).

There are undoubtedly several reasons why drugs do not show the expected clinical improvements in patients and it has been suggested that the decline in productivity coincided with (and by extension is causally related to) the shift from a more phenotype to a more target based approach (Sams-Dodd, 2005, 2006). While this is hard to prove, a study analysing all the FDA approved new molecular entities between 1999 and 2008 confirmed that, especially with first-in-class type drugs, more drugs were approved based on a phenotype (37%) than a target based (23%) approach (Swinney & Anthony, 2011). Perhaps a more important reason why drugs fail is, however, the limited predictive validity of animal models for psychiatric illnesses (Ellenbroek & Youn, 2016; Kaffman & Krystal, 2012). This is particularly well illustrated in a recent analysis of the failure of AZD8529, a mGluR2 modulator for the treatment of schizophrenia (Cook et al., 2014). In spite of the fact that the drug was effective in 7 preclinical models for antipsychotic activity (as well as two models for anxiety), the drug failed in a 4-week placebo controlled phase II clinical trial with over 150 patients.

It is well accepted that animal models have only limited predictive validity and there are again multiple reasons for this, including the heterogeneity of patient

ISSN: 1569-7339
https://doi.org/10.1016/B978-0-12-803161-2.00013-8

populations with likely multiple different causes, and different symptoms. Moreover, most animal models are based on either a single genetic or environmental factor, while the overwhelming clinical evidence shows that all psychiatric disorders are due to a combination of multiple genetic and environmental factors (Ellenbroek & Youn, 2016). For the present paper, we would like to focus on another limitation of animal models, namely the lack of translationally valid symptoms. The vast majority of animal models have focussed on changes in behaviour to assess aspects of psychiatric disorders, and while not totally without merit, this significantly limits the possibility to study complex psychological and emotional symptoms in animals. Even within the field of cognition, it has been realized that a more fundamental translational approach was needed. This led to the CNTRICS initiative (Cognitive Neuroscience Treatment Research to Improve Cognition in Schizophrenia) which, among others, developed a framework for translationally valid animal models for different domains of cognition, such as perceptual processing, attentional control, working memory, executive control, etc. (Dudchenko, Talpos, Young, & Baxter, 2013; Gilmour et al., 2013; Lustig, Kozak, Sarter, Young, & Robbins, 2013; Siegel, Talpos, & Geyer, 2013). Unfortunately, less progress has been made with respect to other aspects of psychiatric disorders, and especially in relation to emotional (dys)regulation (ED), in spite of the fact that aspects of ED are present in virtually all psychiatric disorders. In the present paper, we will make a case for using heart rate variability (HRV) as a translational marker for ED. As will be shown in later sections, HRV is also implicated in cognitive aspects, such as attention. Considering that there is a growing body of evidence that ED and cognitive deficit impact each other, HRV might offer the possibility of identifying the common pathology in both domains.

II HEART RATE VARIABILITY AS A PARAMETER FOR EMOTIONAL STATES

Heart Rate Variability (HRV) refers to the beat-to-beat variability in intervals between individual heart beats. The overall heart rate (HR), and by extension HRV, is determined by the balance between the sympathetic and parasympathetic branch of the autonomous nervous system (sANS and pANS). sANS, generally referred to as the flight-or-fight system, is typically activated in response to stressors leading to an increase in HR, while the pANS (sometimes referred to as the rest-and-digest system) is activated during periods of rest, allowing for recuperation, and is associated with a reduction in HR. In line with the opposite effects on heart rate, both branches of the ANS also have an opposite effect on

HRV, with both decreases in pANS and increases in sANS leading to a reduction in HRV.

Thus, HR and HRV represent an objective and quantifiable measure of the overall state of the ANS. Ever since the classical theory of William James and Carl Lange at the end of the 19th century, it has been accepted that emotions are associated with a physiological response in the body, mediated by the ANS. While the James-Lange theory that the emotional experience only occurs as a consequence of an ANS change is generally considered to be incorrect, there is no doubt that emotional responses have both a central and peripheral component. In a recent intriguing study, over 700 participants were exposed to different emotional stimuli and asked to colour body maps with warm colours when they felt a specific body part become warmer with a specific emotion, and with cool colours when they felt the body part became less active (Nummenmaa, Glerean, Hari, & Hietanen, 2014). This resulted in detailed colour maps that differed depending on the emotional stimulus. Interestingly, virtually without exception changes in the heart area were identified with every emotion. This clearly indicates that different emotions are associated with specific changes in the ANS and consequently that measurements of the ANS can give insight in the emotional state and emotional regulation of the subject.

One of the most influential theories regarding the relationship between the ANS and emotion is the polyvagal theory proposed by Stephen Porges (Porges, 2007, 2009). In its simplest form, the theory states that in mammals the myelinated fibres originating in the nucleus ambiguus function as an active vagal (pANS) braking system on the heart. As a result, rapid inhibition and disinhibition of vagal tone to the heart can quickly excite or calm an individual. Functionally, the vagal brake enables an individual to rapidly engage and disengage with the external world (including the social environment) to promote a state of calmness and adjustment. Under normal circumstances, the vagal influence onto the sino-arterial (SA) node in the heart is relatively strong, which explains why the basal HR is substantially lower than the intrinsic rate of the SA pacemaker cells. The vagal influence is reduced to support the metabolic requirements for mobilization (i.e., in case of the flight-or-fight response), leading to an increase in HR and a decrease in HRV. Conversely, vagal influence is increased to sustain social engagement (Porges, 2007), leading to a decrease in HR and an increase in HRV. Thus, the polyvagal theory provides a framework explaining the influence of the ANS on behavioural and psychological processes, and changes in HRV can be used to quantifiably assess these processes.

There is substantial evidence to show that changes in HRV are indeed associated with disturbances in social and emotional regulation. Emotional regulation has been defined as the ability of an individual to adjust the expression and experience of positive and negative emotions in accordance with the context (Gross, 2001). Several laboratories have reported that children with emotional dysregulation (ED) have lower baseline HRV, as well as a blunted HRV response during test sessions. For instance, children with a lower HRV across the preschool period were emotionally more negative, had more behavioural problems and less social skills (Calkins & Keane, 2004). This pattern is not only clear in children but extends into adulthood. A recent review convincingly showed that healthy individuals with a high HRV show more stable adaptation to emotionally salient stimuli than individuals with lower HRV (Park & Thayer, 2014). Other studies have extended these findings, showing that baseline HRV also affects attentional bias for affective stimuli, such that individuals with lower HRV showed a biased attention towards angry faces (Miskovic & Schmidt, 2010). Moreover, such individuals showed faster attentional engagement and slower disengagement to fearful stimuli, were less accurate in discriminating between fearful and neutral faces, and were more distracted by fearful distractors (Park, Van Bavel, Vasey, Egan, & Thayer, 2012; Park, Van Bavel, Vasey, & Thayer, 2013).

There is also evidence that baseline HRV is related to emotional reactivity. These studies have shown that individuals with high baseline HRV react less negatively to unpleasant stimuli, although others have argued that baseline HRV is more related to emotional flexibility (Balzarotti, Biassoni, Colombo, & Ciceri, 2017). In other words, individuals with high HRV are expected to have a high emotional reactivity, being able to more flexibly adapt to environmental stressors.

While the majority of studies investigating the relationship between ANS and emotional regulation have focused on resting patterns of HRV, there is evidence that the changes in HRV seen during test conditions are not necessarily correlated to resting HRV levels, suggesting they may represent an additional measure of emotional regulation (Balzarotti et al., 2017). According to the polyvagal theory, HRV is expected to decrease when an individual is exposed to stressful (negative) stimuli, and multiple lines of research have shown decreases in individuals preforming a stressful cognitive task, or when experiencing helplessness, worry or social stress (Balzarotti et al., 2017). Conversely, some studies have found increases in phasic HRV when experiencing positive emotional states (Martens et al., 2010; Matsunaga et al., 2009; Muroni, Crnjar, & Barbarossa, 2011). However, it should be kept in mind that, in contrast to the studies on baseline HRV, the results of phasic HRV changes are considerably less consistent (Balzarotti et al., 2017). One reason for this is that the phasic changes may actually be more related to the response of the individual to the stressors, rather than the stressor itself. Subsequently, differences in HRV

reactivity may be more related to coping styles than the emotional properties of stimuli per se. For instance, in one study, individuals were asked to exert a high or low degree of self-control using a food manipulation task. In situation of high control (eat carrots but resist cookies), phasic HRV was significantly larger than in situation of low control (eat cookies, resist carrots) (Segerstrom & Nes, 2007). Similarly, in a study with recovering alcoholics, an increase in HRV was seen upon exposure to an imaginary drink, but only in those individuals with relatively strong self-control (Ingjaldsson, Laberg, & Thayer, 2003).

Overall, these data clearly show that (especially resting state) HRV is a clear indicator of emotional stability and flexibility. While the results of the phasic changes in HRV are more controversial, there is little doubt that emotional laden stimuli alter HRV, perhaps dependent on the coping and/or personality style of the individual which are partly under cognitive control.

III HEART RATE VARIABILITY AS A PARAMETER FOR COGNITIVE FUNCTIONING

As discussed in the previous section there is strong evidence linking HRV to emotional disturbances, and in particular emotional flexibility. However, there is also persuasive evidence that HRV is related to cognitive performance, and especially cognitive flexibility. In a series of studies on military personnel, subjects were divided into a high and low HRV cohort, based on a median split and subsequently subjected to different cognitive tasks. These tasks included a simple reaction time and latency tasks, as well as more complex executive functioning and working memory tasks (Hansen, Johnsen, & Thayer, 2003). The results showed superior performance for the high HRV group. However, this was only seen in the executive function and working memory tests, but not in the nonexecutive function tests. In a subsequent follow-up study, these data were confirmed and it was shown that only low HRV subjects improved their performance under stress (threat of electric shock), while the performance of the high HRV group remained the same (Hansen, Johnsen, & Thayer, 2009). While this may sound counterintuitive, the authors suggested the results may be due to a ceiling effect, with the high HRV group already performing at a very high level, while the threat may have increased arousal in the low HRV group.

Other studies have also shown a link between resting HRV and cognitive performance, including a negative correlation between HRV and intra-individual reaction time variability in an attention task (Williams, Thayer, & Koenig, 2016). In a recent extensive study in elderly individuals including over 2500 individuals with a mean age of 75, individuals with low HRV (lowest 1/3 of the total group) had significantly worse performance in a reaction time task compared to high HRV subjects (highest 1/3 of the total group). Intriguingly, these results were already evident with only a 10 second HRV analysis (Mahinrad et al., 2016). In another study in elderly hypertensive subjects, different HRV parameters (see below) were predictors of frontal lobe cognitive functioning (Santos et al., 2015). Finally, a recent meta-analysis found a significant association between cognitive perseveration and low HRV (Ottaviani et al., 2016). Overall, these studies seem to confirm an association between low HRV and cognitive dysfunction, especially in relation to tasks relying on frontal cortical activation. However, it should be kept in mind that many variables can influence HRV and a recent structure equation modelling approach using data from the MIDUS II (Midlife in the United States study II) found that while HRV was significantly associated with executive functioning, this association disappeared after correcting for age (Mann, Selby, Bates, & Contrada, 2015). Thus, more research is certainly required to elucidate the exact relationship between HRV and cognitive functioning.

IV THE MEASUREMENT OF HEART RATE VARIABILITY

One potential source of conflicting data is related to the measurement of HRV. While HRV, as discussed above, is defined as the beat-to-beat variation in intervals, there are multiple sources of fluctuations in heart rate, including respiration, thermoregulation, renin activity and, most relevant for the present paper, central nervous activation (Berntson et al., 1997). As a result of this complex modulation, there are a multitude of ways to quantify HRV, and there is convincing evidence that different aspects of HRV are related to different neurobiological processes, most notably in the relative contribution of the sANS and the pANS. It would be beyond the scope of this paper to describe all the different HRV measures, and the reader is referred to several excellent sources on this (Acharya, Joseph, Kannathal, Lim, & Suri, 2006; Task Force of the European Society of Cardiology, 1996). Nonetheless, in order to fully understand the applicability of HRV as a translational biomarker, a brief description is necessary.

As HRV is based on fluctuations in heart beat (R-R) intervals, the first step in the analysis is to determine the R peaks. Since artefact (such as missed or premature beats) may occur, pre-processing of the electrocardiogram (ECG) signal is usually necessary to reduce/remove type A (premature QRS detection) and type B (failure to detect R waves) errors. This then leads to a table containing the consecutive so-called normal-to-normal (NN) intervals.

A Linear Measurements of HRV

In general, HRV can be measured using either linear or nonlinear techniques. The vast majority of studies that have investigated HRV have focussed largely or exclusively on linear measurements as they are more easily calculated and interpreted. Within the linear techniques, both time and frequency domain analyses are generally applied. Time domain analysis generally focus directly on the NN intervals or on the variability in NN intervals. The simplest time-domain parameters are the pNN_{50}: defined as the percentage of interval differences of successive NN intervals larger than 50 ms, the SDNN: the standard deviation of NN intervals, and the RMSSD: the root mean square of successive differences in NN Intervals. The latter (RMSSD) is one of the most often used HRV parameters as it has been suggested to primarily assess pANS activity (DeGiorgio et al., 2010).

In addition to time domain analyses, frequency domain analyses are very popular as well. Most of these frequency analyses are based on a fast Fourier transformation (FFT) of the NN intervals, although other techniques (such as continuous wavelet transform and short-time Fourier transform) have also been employed (Vandeput, 2010). Following an FFT analysis of HRV, several different spectral bands can be distinguished: High Frequency (HF) contains the power between 0.15 and 0.4 Hz, Low Frequency (LF) contains the power between 0.04 and 0.15 Hz and very low frequency (VLF) which contains the spectral information below 0.04 Hz. When recordings are extended for 24 h (normally 15 min are sufficient for assessing HRV), a further ultra-low frequency (ULF) band is often distinguished containing the spectral information below 0.003 Hz. In these long recordings, the VLF band is accordingly adjusted to include only the range between 0.003 and 0.04 Hz. In addition, total power (TP) is usually reported as the total area under the power density curve between 0 and 1 Hz.

When comparing linear HRV measures, there is more theoretical knowledge on the interpretation of frequency domain measures compared to time domain measures. For instance, it is generally accepted that the HF frequency band is virtually exclusively determined by the pANS, while both pANS and sANS contribute to the LF (Berntson et al., 1997). However, when assessed over a prolonged period, many time and frequency variables are highly correlated, either because of mathematical or physiological similarity. Thus, SDNN correlates strongly with total power, and RMSSD and pNN_{50} correlate with HF (Vandeput, 2010).

B Nonlinear Measurements of HRV

While linear measures of HRV are relatively simple to calculate (and interpret), they have one important drawback. The cardiac system is inherently dynamic, nonlinear and nonstationary, with performance continually fluctuating as extrinsic and intrinsic stimuli simultaneously influence the state of the system (Christini et al., 2001). Therefore, pure linear measures are unable to capture the true complexity of the cardiac system and most notably will be unable to account for the subtle interactions between the various central and peripheral control mechanisms regulating cardiac function (Malpas, 2002). As with linear methods, a multitude of different nonlinear methods for assessing HRV have been developed (Acharya et al., 2006; Vandeput, 2010; Voss, Schulz, Schroeder, Baumert, & Caminal, 2009). The most frequently used nonlinear techniques are the Poincare plot, the approximate entropy (ApEn) and the detrended fluctuation analysis (DFA).

The Poincare plot offers a two-dimensional graph plotting RR_n against RR_{n+1}, mostly leading to a cigar-shaped point-cloud centered around the $X = Y$ diagonal. Generally, three different indices are calculated: SD1, short-term RR-interval variability (minor axis of the cloud); SD2, long-term RR interval variability (major axis of the cloud) and the SD1/SD2 ratio. There is some evidence to suggest that SD1 at least partly correlates with RMSSD and HF, while SD2 correlates partly with LF.

ApEn represents a simple index of complexity/predictability of time series and basically quantifies the likelihood that series of patterns that are close remain similar for subsequent incremental comparisons (Voss et al., 2009). One important limitation of ApEn is that stationarity and noise-free data are required. According to Voss and colleagues, ApEn is also partly correlated with pANS measures (RMSSD, HF and pNN_{50}).

DFA is based on fractal analysis and typically involves the estimation of a short-term fractal scaling exponent ($\alpha 1$, over a range of 4–16 heartbeats) and a long-term scaling component ($\alpha 2$, over a range of 16–64 heartbeats). While a useful measure, it has been suggested that at least 8000 data points are needed to reliably calculate $\alpha 1$ and $\alpha 2$. However, in contrast to the Poincare and ApEn parameters, both $\alpha 1$ and $\alpha 2$ seems to be more correlated with LF and LF/HF ratio (or VLF/(HF + LF) ratio in the case of $\alpha 2$). In addition, $\alpha 1$ seems to be correlated with the SD1/SD2 ratio of the Poincare Plot.

V HEART RATE VARIABILITY AND PSYCHIATRIC ILLNESS

Given the above-mentioned relationship between HRV and emotional and cognitive functioning, it will not be surprising that many mental disorders are associated with changes in HRV (Fig. 13.1). Moreover, a recent meta-analysis (Vargas, Soros, Shoemaker, & Hachinski, 2016) identified several key structures involved in HRV, including the anterior cingulate and prefrontal cortex, the insula, the amygdala, and the striatum. These structures are also implicated in most major psychiatric disorders (Ellenbroek & Youn, 2016; Goodkind et al., 2015).

A Schizophrenia

Within the field of psychiatry, HRV abnormalities have probably been most studied in patients with schizophrenia (Montaquila, Trachik, & Bedwell, 2015). In one of the earliest studies, 20 paranoid unmedicated schizophrenic patients (half of which were first episode patients) were compared with 20 healthy controls (Rechlin, Claus, & Weis, 1994). The authors did not find any significant differences in either LF- or HF-HRV between the groups. However, subsequent research has clearly identified differences in ANS control, especially reduced pANS activity. A recent meta-analysis (Clamor, Lincoln, Thayer, & Koenig, 2016) combined the data from 34 studies, with 29 papers including data on HF-HRV (1353 patients and 1702 controls) and 24 papers including data on RMSSD (1016 patients and 1469 controls). The study found a significantly reduced HRV in patients, which was obvious in both HF-HRV ($Z = 3.35$; $P = .0008$) and in RMSSD ($Z = 6.18$; $P = .00001$). Further covariate analysis showed that factors such as age, duration of illness, medication and inpatient/outpatient status did not influence the effect. These findings are in line with other studies that found reduced HRV in patients with schizophrenia. For instance, Chang and colleagues found altered HRV in drug naïve first episode patients (Chang et al., 2013). Interestingly, in patients with schizophrenia, HRV abnormalities seem to be correlated especially with the negative symptoms (Chung et al., 2013), in line with the above mentioned relation between HRV and emotionality. Moreover, while less severe than in patients, reductions in HRV have also been reported in first degree relatives of patients (Bar et al., 2012; Jauregui et al., 2011). Together, these studies give a clear indication that low vagal activity may be an endophenotype for schizophrenia (Clamor et al., 2016).

B Major Depressive Disorders (MDD)

Changes in the cardiovascular system are prevalent in MDD, and there appears to be a bi-directional relationship between cardiovascular disorders and MDD. Thus, rates of depression in patients with cardiovascular disease range from 20% to 40% (Carney et al., 1987). Conversely, a recent meta-analysis including over 120,000 patients showed that depression increased the risk of cardiovascular incidents by 80%–90% (Nicholson, Kuper, & Hemingway, 2006). Since MDD also increased the risk of cardiovascular mortality and contributes

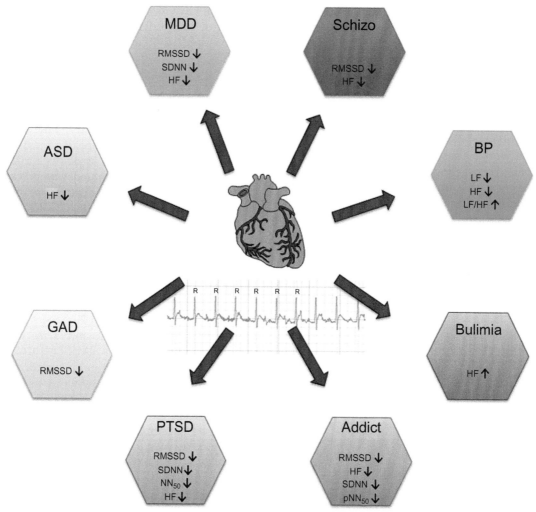

FIG. 13.1   HRV abnormalities in psychiatric disorders. HRV abnormalities have been described in many psychiatric disorders, including schizo-phrenia (schizo), major depressive disorders (MDD), bipolar disorders (BP), autism spectrum disorders (ASD), general anxiety disorders (GAD), post-traumatic stress disorder (PTSD), addiction (Addict) and bulimia. This figure summarized the most often reported changes.

to the progression and prognosis of cardiovascular disease, the relationship between MDD and cardiovascular disease has been described as a downward spiral in which both reinforce each other (Penninx, 2017). One factor that has been suggested to mediate the relationship between MDD and cardiovascular disease is reduced HRV (Musselman, Evans, & Nemeroff, 1998). However, alterations in HRV are also critically related to the MDD disease process itself (Kidwell & Ellenbroek, 2018). Thus, several meta-analyses have found a reduction in HRV, especially HF-HRV, in MDD. While in one of the first of these analyses the effects were considered "suggestive rather than conclusive" (Rottenberg, 2007), a subsequent analysis found significantly reduced HRV, both in the time and the frequency domain (HF) (Kemp et al., 2010). In addition, this study found an increase in the LF/HF ratio (likely a direct consequence of the reduction in HF) as well as some reductions in nonlinear measures. In subsequent studies, reductions

in RMSSD, SDNN and HF were confirmed (Kemp, Quintana, Felmingham, Matthews, & Jelinek, 2012; Wang et al., 2013).

While baseline reductions in HRV have been largely confirmed in MDD patients, there is still a discussion with respect to whether HRV changes correlate with symptoms. Rottenberg and colleagues could not find a significant relationship between resting HF-HRV or alterations in HF-HRV and overall clinical improvements in MDD patients treated with acupuncture for 167 weeks, although resting HF was correlated with sad mood and crying (Rottenberg, Chambers, Allen, & Manber, 2007). On the other hand, several studies found that the severity of symptoms was inversely correlated to RMSSD, SDNN and HF parameters (Kemp et al., 2010; Nahshoni et al., 2001; Udupa et al., 2007; Yeh, Chung, Hsu, & Hung, 2017). One possible explanation for these divergent findings may be related to the intrinsic effects specific antidepressant drugs have on the functioning of the heart. For instance, selective serotonin reuptake inhibitors generally do not affect cardiac functioning per se and overall seem to reverse some of the MDD specific reduction in HRV, while monoamine oxidase inhibitors and tricyclic antidepressants by themselves seem to reduce HRV (Kidwell & Ellenbroek, 2018). As a result, even though such tricyclics improve depressive symptoms, their cardiotoxic effects lead to a further reduction in HRV, rather than a normalization.

C Bipolar Disorders (BP)

BP are characterized by depressive periods interspersed with (hypo)manic episodes. Given the similarity between the depressive episodes in BP and in MDD, it is not surprising that reductions in HRV are also present in BP. This was confirmed by a recent meta-analysis which included 15 studies with a total of 572 BP patients and 1183 healthy controls. The sample also included 683 patients with MDD (Faurholt-Jepsen, Kessing, & Munkholm, 2017). The study revealed reduced HRV, especially in LF, although there was also some indication for a reduction in HF and an increase in LF/HF ratio. However, the meta-analysis also showed a very high heterogeneity between the different studies, which might be due to the lack of differentiation between the depressive and manic stage. A recent study from the same research team investigated the influence of the affective state of HRV in BP patients (Faurholt-Jepsen, Brage, Kessing, & Munkholm, 2017). The study used a measure of HRV that differed from those typically used, namely the difference between the second-shortest and the second-longest inter-beat-interval collected over a 30-second epoch. Using this measure, a group of 16 bipolar patients were followed longitudinally for a 12-week period. The results showed that HRV was significantly increased during manic periods (18% vs. depressed and 17% vs. euthymic states). There were no significantly differences (at least in this measure of HRV) between euthymic and depressed states. However, there was a significant negative correlation between HRV and the severity of depressive symptoms (measured using the Hamilton Depression Rating Scale-17), as well as a positive correlation between HRV and manic symptoms (assessed with the Young Mania Rating Scale).

While the previous study showed that BP patients show more HRV abnormalities during manic than depressive states, it should be realized that most of the measures used to assess HRV in this study are quite uncommon and therefore the results are difficult to compare to other studies. A recent study tried to shed light on this and aimed to investigate whether there were HRV differences between BP and MDD patients (Chang, Chang, Kuo, & Huang, 2015), comparing a variety of HRV data from 116 BP-II patients with 591 MDD patients

and 421 healthy volunteers. Compared to controls as well as to MDD patients, depressed BP-II patients showed reduced VLF, LF and HF and increased LF/HF ratio. These data were recently replicated in a study that also included BP-I and BP not otherwise specified patients, again reporting lower LF-HRV and slightly lower HF-HRV in BP patients (Hage et al., 2017). Additionally, although the groups were small, HF-HRV seemed highest in the BP group treated with antipsychotics and lowest in those treated with a combination of antipsychotics and mood stabilizers. These two studies therefore clearly suggest that HRV parameters can differentiate between BP and MDD, suggesting the promising aspect of HRV as a diagnostic tool.

D Autism Spectrum Disorders (ASD)

ASD is one of the earliest diagnosable psychiatric illnesses, typically identified in children before the age of 4 (Levy, Mandell, & Schultz, 2009). Patients with ASD suffer from deficits in social interaction and communication, as well as from an increase in perseverative behaviour and thought, often described as a preference for sameness. However, in addition to these "core" symptoms, many patients also show developmental, behavioural, gastrointestinal, psychiatric and neurological abnormalities (Levy et al., 2009).

While the exact psychopathology is far from elucidated, there is general consensus that patients with ASD suffer from emotional dysregulation (Mazefsky & White, 2014). One of the earliest theories around ASD suggest patients suffer from "hypo-arousal" (Tinbergen & Tinbergen, 1972). Resulting from an underactivity of the amygdala, patients experience a reduced perception of the rewarding properties of the eyes and face, and therefore avoid social interaction. Intriguingly, another school of thought actually suggests that ASD patients suffer from "hyper-arousal" (Senju & Johnson, 2009). In this model, eyes and face are considered strongly aversive signals for patients and hence are avoided as a means to prevent overstimulation. It is perhaps illustrative of the heterogeneity and complexity of ASD that both theories still exist side by side (Ellenbroek & Sengul, 2017).

Irrespective of this, there is mounting evidence of HRV abnormalities in ASD, again mainly with respect to a reduction in pANS. Thus, reduced baseline HF-HRV compared to typically developing children has been reported in most studies (Cohen, Masyn, Mastergeorge, & Hessl, 2015; Edmiston, Jones, & Corbett, 2016; Harder et al., 2016; Matsushima et al., 2016; Neuhaus, Bernier, & Beauchaine, 2014), although some studies failed to find this effect. Most notably, Pace and colleagues, assessing HRV during sleep, found an increase in HF-HRV, as well as increases in RMSSD, while failing to find a reduction in HF

(Pace, Dumortier, Favre-Juvin, Guinot, & Bricout, 2016). On the other hand, an increase in LF-HRV during sleep has also been reported (Harder et al., 2016).

While studies have shown that baseline HRV is related to social skills and internalizing (Edmiston et al., 2016; Neuhaus et al., 2014), some studies have also investigated how HRV changes in response to either sensory stimuli or treatment. Thus, some have reported a blunted HF-HRV response in ASD patients to (social) stressful stimuli (Smeekens, Didden, & Verhoeven, 2015). In a very interesting study, HF-HRV changes were analysed during a social interaction with both a familiar and an unfamiliar partner in 10-year-old boys with ASD and compared to typically developing boys (Neuhaus, Bernier, & Beauchaine, 2016). Especially in relation to an unfamiliar partner, HF-HRV rapidly increased in typically developing boys, in line with the above mentioned polyvagal theory that increases in pANS are important for social adaptation. In boys with ASD, on the other hand, this increase was not seen. While treatment of ASD has proven very difficult, one recent study used regional transcranial magnetic stimulation (rTMS) of the dorsolateral prefrontal cortex in 33 patients (mean age: 13 years). During the 12 weeks of treatment, the authors found a significant reduction in all aspects of stereotyped, perseverative behaviour and thought and, in parallel, found an increase in several measures of HRV, most notably HF, RMSSD, and pNN_{50} (Wang, Hensley, Casanova, & Sokhadze, 2015). Interestingly, a positive effect of rTMS over the same cortical region on HRV was also observed in patients with MDD (Udupa et al., 2007), in line with earlier suggestions that rTMS may normalize ANS imbalances (Yoshida et al., 2001). Taken together, the studies strongly indicate that HRV plays a key role in the pathology of ASD.

E Other Psychiatric Disorders

In addition to the above-mentioned disorders, HRV has also been studied in patients suffering from other psychiatric conditions, albeit to a lesser extent. Like MDD, patients suffering from generalized anxiety disorder (GAD) show significant signs of emotional dysregulation and, associated with this, a reduced HRV. In a recent study, reduced RMSSD was found in GAD patients (Makovac et al., 2016). Interestingly, the changes in HRV after exposure to a scenario that induced worrisome and ruminative thoughts were negatively correlated with baseline amygdala connectivity. This provides further evidence that reduced prefrontal control over the amygdala may be a link between emotional dysregulation and ANS imbalance.

Altered amygdala connectivity, and associated with this a reduced prefrontal control, has also been implicated in drug addiction, and several studies have recently investigated the sympathovagal balance in substance use disorder. Baseline reductions in virtually all HRV parameters were seen in nicotine dependent users as well as in nicotine and alcohol addicts (Yuksel, Yuksel, Sengezer, & Dane, 2016), with the effects generally being stronger in the combined group compared to the nicotine only group. One problem with studying HRV changes in addiction, is the fact that drugs of abuse may directly affect the cardiac system, and thus the HRV changes may be less related to the emotional disturbances seen in these patients. However, recent studies with individuals addicted to internet gaming also found reduced HRV, again across a broad range of parameters (Kim, Hughes, Park, Quinn, & Kong, 2016). Likewise, while there were no baseline HRV differences between chronic pain patients that use opiates in a controlled and uncontrolled manner, the misusers showed a blunted HRV response to positive and negative emotional stimuli (Garland, Bryan, Nakamura, Froeliger, & Howard, 2017). Finally, several studies have reported beneficial effects of HRV biofeedback on abstinence (Eddie, Kim, Lehrer, Deneke, & Bates, 2014; Penzlin et al., 2017). In one study, treatment as usual (TAU) was compared to TAU and HRV biofeedback which aids a voluntary control on maximizing HRV (Eddie et al., 2014). While both groups did not differ at baseline, low baseline HRV levels were associated with increased craving in the TAU group, whereas higher baseline levels were associated with greater reduction in craving. This effect was seen in several HRV parameters, most convincingly in SDNN and RMSSD, but also significant in pNN_{50} and HF.

Finally, lower HRV (especially HF-HRV) was reported in a meta-analysis in borderline personality disorder (Koenig, Kemp, Feeling, Thayer, & Kaess, 2016), and, in post-traumatic stress disorder, lower RMSSD, SDNN, NN_{50} and HF, but not LF have been reported (Meyer et al., 2016). Interestingly, in contrast to virtually all the disorders mentioned before, bulimia nervosa seems to be associated with an increase in HF-HRV, although there is quite a large variability in individual studies (Peschel et al., 2016), while no convincing changes in HRV have been reported in a meta-analysis of children with attention deficit hyperactivity disorder (Koenig et al., 2017).

VI HEART RATE VARIABILITY MEASUREMENTS IN RODENTS

So far, we have shown that many psychiatric illnesses are accompanied by changes in HRV, and that these changes may represent a biomarker, or perhaps even an endophenotype for emotional and cognitive dysregulation. However, HRV has the important additional advantage that it can also be (relatively easily) measured in rodents using virtually identical techniques.

Basically, there are two different ways of measuring heart rate, and by extension HRV in rodents: telemetric and nontelemetric techniques. The most often used technique involves the implantation of a small telemetry probe that measures biopotentials (Olivier et al., 2008; Stiedl, Jansen, Pieneman, Ogren, & Meyer, 2009). The probe is connected to two leads, is usually implanted in the peritoneal cavity and the leads are either connected to the muscles of the fore- and hindlimb (Olivier et al., 2008), or to the xiphoid process which is the lower cartilage portion of the sternum and the hindlimb (Sgoifo et al., 1996). The latter method requires a more complicated surgery, but the resulted heart rate signal is stronger and show less movement related artefacts. After the animals have recovered, experiments can start and the probe sends heart rate data (usually combined with temperature and movement) wirelessly to a receiver placed underneath the cage, thus allowing measurements in truly freely moving animals. Several different systems exist, with the DataScience (see Table 13.1), probably being the oldest and most commonly used. However, there are multiple alternative systems that all have slightly different characteristics, such as storing the data on the probe itself, which removes the necessity of having a receiver (and thus allowing for larger open fields to be used). Moreover, some of these techniques allow for the measurement of HRV in multiple animals in the same environment, thus making it possible to investigate ANS changes in a social context.

TABLE 13.1 Some of the Most Commonly Used Systems for Measuring HRV in Rodents

Telemetry	
DSI	www.datasci.com/
Indus	www.indusinstruments.com/products/small-animal-implantable-telemetry/
Millar	www.millar.com/research/products/telemetry-systems
Starr	www.starrlifesciences.com/products/e-mitter-mouse-telemetry
Taconic	www.taconic.com/prepare-your-model/preconditioning-solutions/telemetry/
Transonic	www.transonic.com/product/endogear3/
TSE/Stellar	www.tse-systems.com/products/implantable-telemetry/animal-telemetry-system.htm
Nontelemetry	
Emka	www.emka.fr/product/non-invasive-systems-for-rodents/
MouseSpecifics	www.mousespecifics.com/heart-monitoring/ecgenie/

While telemetry has been the most used technique for measuring heart rate in rodents, it has several disadvantages, most notably the requirement for surgery and the price of the probes. Moreover, the probes can only be used for a number of different animals after which they generally need to be refurbished, which adds to the running costs.

To circumvent these disadvantages, nontelemetric techniques have also been developed, generally based on a pulse rate monitors, based on recording ECG signals through the skin. Generally, the animals' ECG is assessed through disposable footpads. These systems are substantially cheaper than the telemetry systems, especially in running costs, and do not require surgery. However, for accurate signal detection the animals should remain more or less still, and thus ECG and HRV analyses are more or less restricted to baseline assessment (or to drug induced changes). Because no probe needs to be implanted, the system does offer the advantage of being able to record from very young animals. ECG have been successfully recorded from mice as young as postnatal day 2 (Heier, Hampton, Wang, & Didonato, 2010).

In contrast to the very large human literature on HRV and psychiatric illnesses, the animal literature is much more restricted. This may in part be due to the above-mentioned costs, especially when telemetric systems are used. Nonetheless some groups have investigated changes in HRV in rats and mice. The High and Low anxiety Lines (HAB, LAB) have been selectively bred for their anxiety response in the elevated plus maze (Landgraf & Wigger, 2002). In line with the above-mentioned study in humans, HAB rats showed reduced baseline RMSSD, HF and total power (Carnevali et al., 2014). Interestingly, while there were no major differences in response to restraint, the HAB rats showed a slowed recovery, both in RMSSD and HF. Intriguingly, a study in mice found opposite results. Compared to C57Bl/6 mice, Balb/c mice showed increased anxiety-like behaviour in the open field, coupled with higher HRV, with long-term HRV being negatively correlated with the time spent in the centre of the open field (Depino & Gross, 2007). In this study, a somewhat unusual HRV analysis was performed and no data were presented on RMSSD, LF or HF, which may have contributed to the contradictory results.

In another study, males rats had significantly lower HRV than female rats and both sexes showed a drop in HRV after exposure to predator scent (Koresh et al., 2016). More importantly, all animals were tested 6 days later, to assess the impact of the predator scent exposure using the elevated plus maze and the acoustic startle response. Animals that were only minimally affected by the scent exposure had significantly higher HRV and more rapid recovery than animals that were substantially affected. Thus, these data, again, seem to confirm the theory that emotional flexibility (or resilience) is associated with HRV flexibility.

Within the field of depression, rats exposed to 4 weeks of chronic mild stress showed a nonsignificant reduction in HRV (Grippo, Beltz, Weiss, & Johnson, 2006), which seemed unaffected by repeated fluoxetine treatment. Several other studies have investigated the physiological and behavioural consequences of social isolation in prairie voles. Since prairie voles are a monogamous species, they are particularly affected by social isolation and show signs of depression, such as an increase in anhedonia (Grippo et al., 2011). In addition, lower SDNN and HF-HRV were found in 2 and 4 weeks isolated female prairie voles (Grippo et al., 2011). In a follow-up of this study a much shorter period of social isolation (after a social bonding period in which males and females were housed together for 5 days) increased immobility in both isolated males and females. In addition, while HRV was not significantly altered, pharmacological experiments suggested an imbalance between the sANS and the pANS (McNeal et al., 2014), as atenolol (a β blocker selectively reducing sANS) affected only the isolated animals, and atropine (a muscarinic antagonist selectively blocking the pANS) was more effective in socially housed voles.

VII HEART RATE VARIABILITY AS A TRANSLATIONAL BIOMARKER: CONCLUSIONS AND OUTLOOK

In this review, we have shown that HRV is disrupted in many psychiatric disorders, including affective disorders, schizophrenia and ASD. While close similarities in HRV abnormalities between these disorders have been reported (see Fig. 13.1), more recent studies indicate that differences can be detected as well (Moon, Lee, Kim, & Hwang, 2013). Nonetheless, the full extent of the HRV alterations in most psychiatric disorders is still not entirely known. For instance, very few studies have so far used the more advanced nonlinear measures of HRV. In fact a recent study clearly showed that nonlinear methods were superior in assessing mood and cognition, at least in healthy volunteers (Young & Benton, 2015). Likewise, few studies have looked at the relationship between (baseline) HRV and the severity or specificity of symptoms. As discussed above, HRV changes have been associated with cognitive and emotional abnormalities as well as with personality traits and coping styles. Thus, it seems likely that the HRV alterations in psychiatric patients are also associated with specific symptoms or symptom clusters. So far, the vast majority of HRV studies has only investigated baseline HRV, while ignoring the adaptive nature of HRV. Given the study in ASD mentioned above (Neuhaus et al., 2016), this aspect also needs substantially more attention, as it may directly tap into the vulnerability or resilience of the individual to (stressful) change. Another aspect which has received

relatively little attention is the influence of psychotropic drugs. Thus, some drugs directly affect the cardiac system and often worsen HRV abnormalities, while others seem to improve both clinical symptoms and HRV changes. For example, a recent review analysed, among others, the effects of antipsychotics on HRV (Alvares, Quintana, Hickie, & Guastella, 2016) and found that while clozapine reduced HRV, olanzapine and amisulpride had little effect, while sertindole seemed to (slightly) worsen HRV. Table 13.2 gives a brief overview of the effects of several (classes of) psychotropic drugs in HRV. While most effects listed in this table have been replicated in sevberal studies, some of the results are based on a single or only a few studies. However, it would be beyond the scope of this paper to describe all studies in detail. Finally, from a more theoretical point of view, more information is needed with respect to potential changes in first-degree relatives. Some studies, reviewed above, in schizophrenia have found mild (but significant) HRV reductions in healthy family members, and a recent genome wide association study identified 17 single nucleotide polymorphisms in eight loci related to HRV (Nolte et al., 2017), most of which were associated with RMSSD.

Although many aspects of HRV still need to be further evaluated, one of the most exciting features of HRV is that it can be measured in humans and rodents with virtually identical techniques. Moreover, HRV (especially resting HRV) can be repeatedly measured in the same animal throughout life, and, with the above mentioned noninvasive methods from as young as postnatal day 2. Additionally, the basic regulation of heart rate and HRV, i.e., the balance between the sANS and the

TABLE 13.2 The Effects of Some Drugs and Classes of Drugs on HRV in Humans

Drug	Effect
TCA	Decrease HF, SDNN, RMSSD and increase LF
SSRI	No effect to a slight increase in HRV
SNRI	No effect on HRV
Amisulpride	No effect on HRV
Sertindole	No effect to slight increase in HRV
Olanzapine	No effect
Clozapine	Reduced HRV
Alcohol	Reduced RMSSD, pNN_{50}, LF and HF
Propofol	Reduced LF, HF and total power
Ketamine	Reduced LF, HF and total power
Caffeine	No effect on HRV
Diazepam	No effect on HRV
Nicotine	Reduced HF, increased LF, increased LF/HF ratio

TCA, tricyclic antidepressants; SSRI, selective serotonin reuptake inhibitors; SNRI, serotonin noradrenaline reuptake inhibitors.

pANS, and the regulatory systems within the brain stem and medulla are evolutionarily old and very similar between different species (Porges, 2007). Finally, the fact that HRV can be assessed in health volunteers as well, opens up the possibility that it can already be used in phase I studies to assess whether a novel drug engages successfully with the target.

As discussed above, there is a dearth of animal research in the field of HRV, particularly in relation to animal models for psychiatric disorders. While some evidence in the field of depression and anxiety has been published, no studies have, to date, investigated HRV abnormalities in models for schizophrenia, ASD or addiction. Moreover, as in clinical studies, more research is needed to investigate the flexibility of the HRV system and its response to stressful environmental stimuli. Finally, more fundamental research is required, both with respect to investigating the neurobiological and pharmacological mechanisms underlying both linear and especially nonlinear aspects of HRV.

While this section focussed predominantly on future research, we feel that HRV represents a unique and very useful biomarker for emotional and cognitive deficits commonly seen in psychiatric patients. Given the fact that it can be assessed in humans and animals with the same techniques and that HRV changes occur in multiple disorders, HRV holds the potential of not only being a translational but also a transdiagnostic biomarker.

References

Acharya, U. R., Joseph, K. P., Kannathal, N., Lim, C. M., & Suri, J. S. (2006). Heart rate variability: a review. *Medical & Biological Engineering & Computing, 44*, 1031–1051.

Alvares, G. A., Quintana, D. S., Hickie, I. B., & Guastella, A. J. (2016). Autonomic nervous system dysfunction in psychiatric disorders and the impact of psychotropic medications: a systematic review and meta-analysis. *Journal of Psychiatry & Neuroscience, 41*, 89–104.

Andlin-Sobocki, P., Jonsson, B., Wittchen, H. U., & Olesen, J. (2005). Cost of disorders of the brain in Europe. *European Journal of Neurology, 12*, 1–27.

Arrowsmith, J., & Miller, P. (2013). Trial watch: phase II and phase III attrition rates 2011-2012. *Nature Reviews. Drug Discovery, 12*, 569.

Balzarotti, S., Biassoni, F., Colombo, B., & Ciceri, M. R. (2017). Cardiac vagal control as a marker of emotion regulation in healthy adults: a review. *Biological Psychology, 130*, 54–66.

Bar, K. J., Rachow, T., Schulz, S., Bassarab, K., Haufe, S., Berger, S., et al. (2012). The phrenic component of acute schizophrenia—a name and its physiological reality. *PLoS One, 7*, e33459.

Berntson, G. G., Bigger, J. T., Eckberg, D. L., Grossman, P., Kaufmann, P. G., Malik, M., et al. (1997). Heart rate variability: origins, methods, and interpretive caveats. *Psychophysiology, 34*, 623–648.

Calkins, S. D., & Keane, S. P. (2004). Cardiac vagal regulation across the preschool period: stability, continuity, and implications for childhood adjustment. *Developmental Psychobiology, 45*, 101–112.

Carnevali, L., Trombini, M., Graiani, G., Madeddu, D., Quaini, F., Landgraf, R., et al. (2014). Low vagally-mediated heart rate variability and increased susceptibility to ventricular arrhythmias in rats bred for high anxiety. *Physiology & Behavior, 128*, 16–25.

Carney, R. M., Rich, M. W., Tevelde, A., Saini, J., Clark, K., & Jaffe, A. S. (1987). Major depressive disorder in coronary-artery disease. *American Journal of Cardiology, 60*, 1273–1275.

Chandler, D. J. (2013). Something's got to give: psychiatric disease on the rise and novel drug development on the decline. *Drug Discovery Today, 18*, 202–206.

Chang, H. A., Chang, C. C., Kuo, T. B., & Huang, S. Y. (2015). Distinguishing bipolar II depression from unipolar major depressive disorder: differences in heart rate variability. *The World Journal of Biological Psychiatry, 16*, 351–360.

Chang, H. A., Chang, C. C., Tzeng, N. S., Kuo, T. B., Lu, R. B., & Huang, S. Y. (2013). Cardiac autonomic dysregulation in acute schizophrenia. *Acta Neuropsychiatrica, 25*, 155–164.

Christini, D. J., Stein, K. M., Markowitz, S. M., Mittal, S., Slotwiner, D. J., Scheiner, M. A., et al. (2001). Nonlinear-dynamical arrhythmia control in humans. *Proceedings of the National Academy of Sciences of the United States of America, 98*, 5827–5832.

Chung, M. S., Yang, A. C., Lin, Y. C., Lin, C. N., Chang, F. R., Shen, S. H., et al. (2013). Association of altered cardiac autonomic function with psychopathology and metabolic profiles in schizophrenia. *Psychiatry Research, 210*, 710–715.

Clamor, A., Lincoln, T. M., Thayer, J. F., & Koenig, J. (2016). Resting vagal activity in schizophrenia: meta-analysis of heart rate variability as a potential endophenotype. *British Journal of Psychiatry, 208*, 9–16.

Cohen, S., Masyn, K., Mastergeorge, A., & Hessl, D. (2015). Psychophysiological responses to emotional stimuli in children and adolescents with autism and fragile X syndrome. *Journal of Clinical Child and Adolescent Psychology, 44*, 250–263.

Cook, D., Brown, D., Alexander, R., March, R., Morgan, P., Satterthwaite, G., et al. (2014). Lessons learned from the fate of AstraZeneca's drug pipeline: a five-dimensional framework. *Nature Reviews. Drug Discovery, 13*, 419–431.

DeGiorgio, C. M., Miller, P., Meymandi, S., Chin, A., Epps, J., Gordon, S., et al. (2010). RMSSD, a measure of vagus-mediated heart rate variability, is associated with risk factors for SUDEP: The SUDEP-7 Inventory. *Epilepsy & Behavior, 19*, 78–81.

Depino, A. M., & Gross, C. (2007). Simultaneous assessment of autonomic function and anxiety-related behavior in BALB/c and C57BL/6 mice. *Behavioural Brain Research, 177*, 254–260.

Dudchenko, P. A., Talpos, J., Young, J., & Baxter, M. G. (2013). Animal models of working memory: a review of tasks that might be used in screening drug treatments for the memory impairments found in schizophrenia. *Neuroscience and Biobehavioral Reviews, 37*, 2111–2124.

Eddie, D., Kim, C., Lehrer, P., Deneke, E., & Bates, M. E. (2014). A pilot study of brief heart rate variability biofeedback to reduce craving in young adult men receiving inpatient treatment for substance use disorders. *Applied Psychophysiology and Biofeedback, 39*, 181–192.

Edmiston, E. K., Jones, R. M., & Corbett, B. A. (2016). Physiological response to social evaluative threat in adolescents with autism spectrum disorder. *Journal of Autism and Developmental Disorders, 46*, 2992–3005.

Ellenbroek, B. A., & Sengul, H. K. (2017). Autism spectrum disorders: autonomic alterations with a special focus on the heart. *Heart and Mind, 1*, 78–83.

Ellenbroek, B. A., & Youn, J. (2016). Gene—environment interactions in psychiatry. In *Nature, nurture, neuroscience*. London, UK: Academic Press.

Faurholt-Jepsen, M., Brage, S., Kessing, L. V., & Munkholm, K. (2017). State-related differences in heart rate variability in bipolar disorder. *Journal of Psychiatric Research, 84*, 169–173.

Faurholt-Jepsen, M., Kessing, L. V., & Munkholm, K. (2017). Heart rate variability in bipolar disorder: a systematic review and meta-analysis. *Neuroscience and Biobehavioral Reviews, 73*, 68–80.

Garland, E. L., Bryan, C. J., Nakamura, Y., Froeliger, B., & Howard, M. O. (2017). Deficits in autonomic indices of emotion regulation and reward processing associated with prescription opioid use and misuse. *Psychopharmacology, 234*, 621–629.

Gilmour, G., Arguello, A., Bari, A., Brown, V. J., Carter, C., Floresco, S. B., et al. (2013). Measuring the construct of executive control in schizophrenia: defining and validating translational animal paradigms for discovery research. *Neuroscience and Biobehavioral Reviews, 37*, 2125–2140.

Goodkind, M., Eickhoff, S. B., Oathes, D. J., Jiang, Y., Chang, A., Jones-Hagata, L. B., et al. (2015). Identification of a common neurobiological substrate for mental illness. *JAMA Psychiatry, 72*, 305–315.

Grippo, A. J., Beltz, T. G., Weiss, R. M., & Johnson, A. K. (2006). The effects of chronic fluoxetine treatment on chronic mild stress-induced cardiovascular changes and anhedonia. *Biological Psychiatry, 59*, 309–316.

Grippo, A. J., Carter, C. S., McNeal, N., Chandler, D. L., Larocca, M. A., Bates, S. L., et al. (2011). 24-hour autonomic dysfunction and depressive behaviors in an animal model of social isolation: implications for the study of depression and cardiovascular disease. *Psychosomatic Medicine, 73*, 59–66.

Gross, J. J. (2001). Emotion regulation in adulthood: timing is everything. *Current Directions in Psychological Science, 10*, 214–219.

Gustavsson, A., Svensson, M., Jacobi, F., Allgulander, C., Alonso, J., Beghi, E., et al. (2011). Cost of disorders of the brain in Europe 2010. *European Neuropsychopharmacology, 21*, 718–779.

Hage, B., Britton, B., Daniels, D., Heilman, K., Porges, S. W., & Halaris, A. (2017). Low cardiac vagal tone index by heart rate variability differentiates bipolar from major depression. *The World Journal of Biological Psychiatry*, 1–9.

Hansen, A. L., Johnsen, B. H., & Thayer, J. F. (2003). Vagal influence on working memory and attention. *International Journal of Psychophysiology, 48*, 263–274.

Hansen, A. L., Johnsen, B. H., & Thayer, J. F. (2009). Relationship between heart rate variability and cognitive function during threat of shock. *Anxiety, Stress, and Coping, 22*, 77–89.

Harder, R., Malow, B. A., Goodpaster, R. L., Iqbal, F., Halbower, A., Goldman, S. E., et al. (2016). Heart rate variability during sleep in children with autism spectrum disorder. *Clinical Autonomic Research, 26*, 423–432.

Heier, C. R., Hampton, T. G., Wang, D., & Didonato, C. J. (2010). Development of electrocardiogram intervals during growth of FVB/N neonate mice. *BMC Physiology, 10*, 16.

Ingjaldsson, J. T., Laberg, J. C., & Thayer, J. F. (2003). Reduced heart rate variability in chronic alcohol abuse: relationship with negative mood, chronic thought suppression, and compulsive drinking. *Biological Psychiatry, 54*, 1427–1436.

Jauregui, O. I., Costanzo, E. Y., de Achaval, D., Villarreal, M. F., Chu, E., Mora, M. C., et al. (2011). Autonomic nervous system activation during social cognition tasks in patients with schizophrenia and their unaffected relatives. *Cognitive and Behavioral Neurology, 24*, 194–203.

Kaffman, A., & Krystal, J. H. (2012). New frontiers in animal research of psychiatric illness. *Methods in Molecular Biology, 829*, 3–30.

Kemp, A. H., Quintana, D. S., Felmingham, K. L., Matthews, S., & Jelinek, H. F. (2012). Depression, comorbid anxiety disorders, and heart rate variability in physically healthy, unmedicated patients: implications for cardiovascular risk. *PLoS One, 7*, e30777.

Kemp, A. H., Quintana, D. S., Gray, M. A., Felmingham, K. L., Brown, K., & Gatt, J. M. (2010). Impact of depression and antidepressant treatment on heart rate variability: a review and meta-analysis. *Biological Psychiatry, 67*, 1067–1074.

Kidwell, M., & Ellenbroek, B. A. (2018). Heart and soul: heart rate variability and major depression. *Behavioural Pharmacology, 29*, 152–164.

Kim, N., Hughes, T. L., Park, C. G., Quinn, L., & Kong, I. D. (2016). Altered autonomic functions and distressed personality traits in male adolescents with internet gaming addiction. *Cyberpsychology, Behavior and Social Networking, 19*, 667–673.

Koenig, J., Kemp, A. H., Feeling, N. R., Thayer, J. F., & Kaess, M. (2016). Resting state vagal tone in borderline personality disorder: a meta-analysis. *Progress in Neuro-Psychopharmacology & Biological Psychiatry, 64*, 18–26.

Koenig, J., Rash, J. A., Kemp, A. H., Buchhorn, R., Thayer, J. F., & Kaess, M. (2017). Resting state vagal tone in attention deficit (hyperactivity) disorder: a meta-analysis. *World Journal of Biological Psychiatry, 18*, 256–267.

Kola, I., & Landis, J. (2004). Can the pharmaceutical industry reduce attrition rates? *Nature Reviews. Drug Discovery, 3*, 711–715.

Koresh, O., Kaplan, Z., Zohar, J., Matar, M. A., Geva, A. B., & Cohen, H. (2016). Distinctive cardiac autonomic dysfunction following stress exposure in both sexes in an animal model of PTSD. *Behavioural Brain Research, 308*, 128–142.

Landgraf, R., & Wigger, A. (2002). High vs low anxiety-related behavior rats: an animal model of extremes in trait anxiety. *Behavior Genetics, 32*, 301–314.

Levy, S. E., Mandell, D. S., & Schultz, R. T. (2009). Autism. *Lancet, 374*, 1627–1638.

Lustig, C., Kozak, R., Sarter, M., Young, J. W., & Robbins, T. W. (2013). CNTRICS final animal model task selection: control of attention. *Neuroscience and Biobehavioral Reviews, 37*, 2099–2110.

Mahinrad, S., Jukema, J. W., van Heemst, D., Macfarlane, P. W., Clark, E. N., de Craen, A. J. M., et al. (2016). 10-Second heart rate variability and cognitive function in old age. *Neurology, 86*, 1120–1127.

Makovac, E., Meeten, F., Watson, D. R., Herman, A., Garfinkel, S. N., Critchley, H. D., et al. (2016). Alterations in amygdala-prefrontal functional connectivity account for excessive worry and autonomic dysregulation in generalized anxiety disorder. *Biological Psychiatry, 80*, 786–795.

Malpas, S. C. (2002). Neural influences on cardiovascular variability: possibilities and pitfalls. *American Journal of Physiology. Heart and Circulatory Physiology, 282*, H6–20.

Mann, S. L., Selby, E. A., Bates, M. E., & Contrada, R. J. (2015). Integrating affective and cognitive correlates of heart rate variability: a structural equation modeling approach. *International Journal of Psychophysiology, 98*, 76–86.

Martens, A., Greenberg, J., Allen, J. J. B., Hayes, J., Schirnel, J., & Johns, M. (2010). Self-esteem and autonomic physiology: self-esteem levels predict cardiac vagal tone. *Journal of Research in Personality, 44*, 573–584.

Matsunaga, M., Isowa, T., Kimura, K., Miyakoshi, M., Kanayama, N., Murakami, H., et al. (2009). Associations among positive mood, brain, and cardiovascular activities in an affectively positive situation. *Brain Research, 1263*, 93–103.

Matsushima, K., Matsubayashi, J., Toichi, M., Funabiki, Y., Kato, T., Awaya, T., et al. (2016). Unusual sensory features are related to resting-state cardiac vagus nerve activity in autism spectrum disorders. *Research in Autism Spectrum Disorders, 25*, 37–46.

Mazefsky, C. A., & White, S. W. (2014). Emotion regulation: concepts & practice in autism spectrum disorder. *Child and Adolescent Psychiatric Clinics Of North America, 23*, 15–24.

McNeal, N., Scotti, M. A., Wardwell, J., Chandler, D. L., Bates, S. L., Larocca, M., et al. (2014). Disruption of social bonds induces behavioral and physiological dysregulation in male and female prairie voles. *Autonomic Neuroscience, 180*, 9–16.

Meyer, P. W., Muller, L. E., Zastrow, A., Schmidinger, I., Bohus, M., Herpertz, S. C., et al. (2016). Heart rate variability in patients with post-traumatic stress disorder or borderline personality disorder: relationship to early life maltreatment. *Journal of Neural Transmission, 123*, 1107–1118.

Miller, G. (2010). Is pharma running out of brainy ideas? *Science, 329*, 502–504.

Miskovic, V., & Schmidt, L. A. (2010). Frontal brain electrical asymmetry and cardiac vagal tone predict biased attention to social threat. *International Journal of Psychophysiology, 75*, 332–338.

Montaquila, J. M., Trachik, B. J., & Bedwell, J. S. (2015). Heart rate variability and vagal tone in schizophrenia: a review. *Journal of Psychiatric Research, 69*, 57–66.

Moon, E., Lee, S. H., Kim, D. H., & Hwang, B. (2013). Comparative study of heart rate variability in patients with schizophrenia, bipolar disorder, post-traumatic stress disorder, or major depressive disorder. *Clinical Psychopharmacology & Neuroscience, 11*, 137–143.

Muroni, P., Crnjar, R., & Barbarossa, I. T. (2011). Emotional responses to pleasant and unpleasant oral flavour stimuli. *Chemosensory Perception, 4*, 65–71.

Musselman, D. L., Evans, D. L., & Nemeroff, C. B. (1998). The relationship of depression to cardiovascular disease—epidemiology, biology, and treatment. *Archives of General Psychiatry, 55*, 580–592.

Nahshoni, E., Aizenberg, D., Sigler, M., Zalsman, G., Strasberg, B., Imbar, S., et al. (2001). Heart rate variability in elderly patients before and after electroconvulsive therapy. *The American Journal of Geriatric Psychiatry, 9*, 255–260.

Neuhaus, E., Bernier, R., & Beauchaine, T. P. (2014). Brief report: social skills, internalizing and externalizing symptoms, and respiratory sinus arrhythmia in autism. *Journal of Autism and Developmental Disorders, 44*, 730–737.

Neuhaus, E., Bernier, R. A., & Beauchaine, T. P. (2016). Children with autism show altered autonomic adaptation to novel and familiar social partners. *Autism Research, 9*, 579–591.

Nicholson, A., Kuper, H., & Hemingway, H. (2006). Depression as an aetiologic and prognostic factor in coronary heart disease: a meta-analysis of 6362 events among 146 538 participants in 54 observational studies. *European Heart Journal, 27*, 2763–2774.

Nolte, I. M., Munoz, M. L., Tragante, V., Amare, A. T., Jansen, R., von der Vaez, A., et al. (2017). Genetic loci associated with heart rate variability and their effects on cardiac disease risk. *Nature Communications, 8*, 15805.

Nummenmaa, L., Glerean, E., Hari, R., & Hietanen, J. K. (2014). Bodily maps of emotions. *Proceedings of the National Academy of Sciences of the United States of America, 111*, 646–651.

Olivier, J. D. A., Cools, A. R., Olivier, B., Homberg, J. R., Cuppen, E., & Ellenbroek, B. A. (2008). Stress-induced hyperthermia and basal body temperature are mediated by different 5-HT1A receptor populations: a study in SERT knockout rats. *European Journal of Pharmacology, 590*, 190–197.

Ottaviani, C., Thayer, J. F., Verkuil, B., Lonigro, A., Medea, B., Couyoumdjian, A., et al. (2016). Physiological concomitants of perseverative cognition: a systematic review and meta-analysis. *Psychological Bulletin, 142*, 231–259.

Pace, M., Dumortier, L., Favre-Juvin, A., Guinot, M., & Bricout, V. A. (2016). Heart rate variability during sleep in children with autism spectrum disorders. *Physiology & Behavior, 167*, 309–312.

Pammolli, F., Magazzini, L., & Riccaboni, M. (2011). The productivity crisis in pharmaceutical R&D. *Nature Reviews. Drug Discovery, 10*, 428–438.

Park, G., & Thayer, J. F. (2014). From the heart to the mind: cardiac vagal tone modulates top-down and bottom-up visual perception and attention to emotional stimuli. *Frontiers in Psychology, 5*, 278.

Park, G., Van Bavel, J. J., Vasey, M. W., Egan, E. J. L., & Thayer, J. F. (2012). From the heart to the mind's eye: cardiac vagal tone is related to visual perception of fearful faces at high spatial frequency. *Biological Psychology, 90*, 171–178.

Park, G., Van Bavel, J. J., Vasey, M. W., & Thayer, J. F. (2013). Cardiac vagal tone predicts attentional engagement to and disengagement from fearful faces. *Emotion, 13*, 645–656.

Penninx, B. W. (2017). Depression and cardiovascular disease: epidemiological evidence on their linking mechanisms. *Neuroscience and Biobehavioral Reviews, 74*, 277–286.

Penzlin, A. I., Barlinn, K., Illigens, B., Weidner, K., Siepmann, M., & Siepmann, T. (2017). Effect of short-term heart rate variability biofeedback on long-term abstinence in alcohol dependent patients—a one-year follow-up. *BMC Psychiatry, 17*, 325.

Peschel, S. K. V., Feeling, N. R., Vogele, C., Kaess, M., Thayer, J. F., & Koenig, J. (2016). A systematic review on heart rate variability in Bulimia Nervosa. *Neuroscience and Biobehavioral Reviews, 63*, 78–97.

Porges, S. W. (2007). The polyvagal perspective. *Biological Psychology, 74*, 116–143.

Porges, S. W. (2009). The polyvagal theory: new insights into adaptive reactions of the autonomic nervous system. *Cleveland Clinic Journal of Medicine, 76*(Suppl 2), S86–S90.

Rechlin, T., Claus, D., & Weis, M. (1994). Heart-rate-variability in schizophrenic-patients and changes of autonomic heart-rate parameters during treatment with clozapine. *Biological Psychiatry, 35*, 888–892.

Rottenberg, J. (2007). Cardiac vagal control in depression: a critical analysis. *Biological Psychology, 74*, 200–211.

Rottenberg, J., Chambers, A. S., Allen, J. J. B., & Manber, R. (2007). Cardiac vagal control in the severity and course of depression: the importance of symptomatic heterogeneity. *Journal of Affective Disorders, 103*, 173–179.

Sams-Dodd, F. (2005). Target-based drug discovery: is something wrong? *Drug Discovery Today, 10*, 139–147.

Sams-Dodd, F. (2006). Strategies to optimize the validity of disease models in the drug discovery process. *Drug Discovery Today, 11*, 355–363.

Santos, W. B., Matoso, J. M., Maltez, M., Goncalves, T., Casanova, M., Moreira, I. F., et al. (2015). Spectral analyses of systolic blood pressure and heart rate variability and their association with cognitive performance in elderly hypertensive subjects. *Journal of Human Hypertension, 29*, 488–494.

Segerstrom, S. C., & Nes, L. S. (2007). Heart rate variability reflects self-regulatory strength, effort, and fatigue. *Psychological Science, 18*, 275–281.

Senju, A., & Johnson, M. H. (2009). Atypical eye contact in autism: models, mechanisms and development. *Neuroscience and Biobehavioral Reviews, 33*, 1204–1214.

Sgoifo, A., Stilli, D., Medici, D., Gallo, P., Aimi, B., & Musso, E. (1996). Electrode positioning for reliable telemetry ECG recordings during social stress in unrestrained rats. *Physiology & Behavior, 60*, 1397–1401.

Siegel, S. J., Talpos, J. C., & Geyer, M. A. (2013). Animal models and measures of perceptual processing in Schizophrenia. *Neuroscience and Biobehavioral Reviews, 37*, 2092–2098.

Smeekens, I., Didden, R., & Verhoeven, E. W. M. (2015). Exploring the relationship of autonomic and endocrine activity with social functioning in adults with autism spectrum disorders. *Journal of Autism and Developmental Disorders, 45*, 495–505.

Stiedl, O., Jansen, R. F., Pieneman, A. W., Ogren, S. O., & Meyer, M. (2009). Assessing aversive emotional states through the heart in mice: implications for cardiovascular dysregulation in affective disorders. *Neuroscience and Biobehavioral Reviews, 33*, 181–190.

Swinney, D. C., & Anthony, J. (2011). How were new medicines discovered? *Nature Reviews. Drug Discovery, 10*, 507–519.

Task Force of the European Society of Cardiology. (1996). Heart rate variability. Standards of measurement, physiological interpretation, and clinical use. Task Force of the European Society of Cardiology and the North American Society of Pacing and Electrophysiology. *European Heart Journal, 17*, 354–381.

Tinbergen, E. A., & Tinbergen, N. (1972). *Early childhood autism: An ethological approach*. Berlin: Parey.

Udupa, K., Sathyaprabha, T. N., Thirthalli, J., Kishore, K. R., Raju, T. R., & Gangadhar, B. N. (2007). Modulation of cardiac autonomic functions in patients with major depression treated with repetitive transcranial magnetic stimulation. *Journal of Affective Disorders, 104*, 231–236.

Vandeput, S. (2010). *Heart rate variability: linear and nonlinear analysis with applications in human physiology*. Leuven: Katholieke Universiteit Leuven.

Vargas, E. R., Soros, P., Shoemaker, J. K., & Hachinski, V. (2016). Human cerebral circuitry related to cardiac control: a neuroimaging meta-analysis. *Annals of Neurology, 79*, 709–716.

Vos, T., Flaxman, A. D., Naghavi, M., Lozano, R., Michaud, C., Ezzati, M., et al. (2012). Years lived with disability (YLDs) for 1160 sequelae of 289 diseases and injuries 1990-2010: a systematic analysis for the Global Burden of Disease Study 2010. *Lancet, 380,* 2163–2196.

Voss, A., Schulz, S., Schroeder, R., Baumert, M., & Caminal, P. (2009). Methods derived from nonlinear dynamics for analysing heart rate variability. *Philosophical Transactions. Series A, Mathematical, Physical, and Engineering Sciences, 367,* 277–296.

Wang, Y., Hensley, M. K., Casanova, M. F., & Sokhadze, E. (2015). Prefrontal rTMS treatment effects on autonomic activity in children with autism. *Applied Psychophysiology and Biofeedback, 40,* 122.

Wang, Y., Zhao, X., O'Neil, A., Turner, A., Liu, X., & Berk, M. (2013). Altered cardiac autonomic nervous function in depression. *BMC Psychiatry, 13,* 187.

Williams, D. P., Thayer, J. F., & Koenig, J. (2016). Resting cardiac vagal tone predicts intraindividual reaction time variability during an attention task in a sample of young and healthy adults. *Psychophysiology, 53,* 1843–1851.

Yeh, M. L., Chung, Y. C., Hsu, L. C., & Hung, S. H. (2017). Effect of transcutaneous acupoint electrical stimulation on post-hemorrhoidectomy-associated pain, anxiety, and heart rate variability. *Clinical Nursing Research, 27*(4), 450–466.

Yoshida, T., Yoshino, A., Kobayashi, Y., Inoue, M., Kamakura, K., & Nomura, S. (2001). Effects of slow repetitive transcranial magnetic stimulation on heart rate variability according to power spectrum analysis. *Journal of the Neurological Sciences, 184,* 77–80.

Young, H., & Benton, D. (2015). We should be using nonlinear indices when relating heart-rate dynamics to cognition and mood. *Scientific Reports, 5,* 16619.

Yuksel, R., Yuksel, R. N., Sengezer, T., & Dane, S. (2016). Autonomic cardiac activity in patients with smoking and alcohol addiction by heart rate variability analysis. *Clinical and Investigative Medicine, 39,* S147–S152.

14

Drug Discovery in Psychiatry: Time for Human Genome-Guided Solutions

Andreas Papassotiropoulos[*,†,‡,§], *Dominique J.-F. de Quervain*[*,‡,¶]

[*]Transfaculty Research Platform Molecular and Cognitive Neurosciences, University of Basel, Basel, Switzerland
[†]Department of Psychology, Division of Molecular Neuroscience, University of Basel, Basel, Switzerland [‡]Psychiatric University Clinics, University of Basel, Basel, Switzerland [§]Department Biozentrum, Life Sciences Training Facility, University of Basel, Basel, Switzerland [¶]Department of Psychology, Division of Cognitive Neuroscience, University of Basel, Basel, Switzerland

I DISILLUSIONMENT IN PSYCHIATRIC PHARMACOTHERAPY

As young residents in psychiatry back in the 1990s, we were initially excited by the availability of a repertoire of different psychiatric medications. Indeed, many different compounds were available for specific diseases (e.g., amitriptyline, imipramine, and iproniazid for depression and haloperidol, chlorpromazine, and clozapine for schizophrenia), and some drugs seemed to be efficacious across disorders. In the eyes of a psychiatry novice, this broad inventory of psychoactive drugs led to the impression that the molecular paths leading to psychiatric disorders were obvious and that drugs existed that were specifically and efficiently directed toward these paths. It did not take long to realize that this was an erroneous impression. Not only was the efficacy of these drugs limited and the molecular pathways related to psychiatric disorders unclear, but also the broad repertoire of psychiatric medications could be narrowed down into less than a handful of key compounds, with most of the drugs being close relatives of one prototype. In fact the pharmacological concepts behind these prototypes were based on serendipity and were dated back to the 1950s without a significant modification since then.

Our initial disappointment with this stagnant treatment landscape was replaced by the hope that groundbreaking developments in neuroscience and the resulting gain of knowledge about molecular and neural mechanisms of cognitive and emotional processes would lead to the identification of better treatments. Now, two decades later, this expectation still remains unfulfilled (Abbott, 2011;

Hyman, 2012; Insel, 2012). In this chapter, we comment on some of the issues that, in our view, contribute to the current problematic situation and argue that human- and genome-centered research approaches (Collins, 2011; Healy et al., 2004; Hyman, 2012; Insel, 2012; Lander, 2011; Muglia, 2011; Papassotiropoulos et al., 2013; Plenge et al., 2013; Rasetti & Weinberger, 2011; Sanseau et al., 2012; Schubert et al., 2014) might help to overcome the depression in psychiatric drug discovery.

II FAILED DRUG DISCOVERY

Brain disorders are common and cause enormous emotional and economic burden to patients, relatives, caregivers, and the community. A recent comprehensive assessment of the direct and indirect financial consequences of brain disorders in Europe calculated an annual cost of 1 trillion US$, pointing out that this estimate is very likely to be conservative (Gustavsson et al., 2011). Topping the list of cost estimates are mood and anxiety disorders. Direct health-care expenses (i.e., medication, hospitalization, and visits to physicians) account for 37% of the total costs. Although the market for drugs directed against psychiatric diseases is large (i.e., 80.5 billion US$ sales in 2010) and still growing (Abbott, 2011; Smith, 2011), major pharmaceutical companies are disengaging from research and drug discovery programs related to psychiatry, as recent decades have brought no significant progress in the identification of novel and improved drugs for psychiatric diseases. In this environment, many companies have concluded that engagement in mental health drug development might be too risky (Abbott, 2011). The discrepancy between the urgent need for, and large market potential of, improved therapeutic compounds and the current lack of significant development of novel and improved drugs illustrates the importance of pursuing new strategies aimed at identifying druggable targets related to psychiatric disease.

III LLL-SUITED DISEASE MODELS

Human psychiatric disorders are human-specific conditions, characterized by the interplay of genetic, environmental, and social factors. There is growing awareness of the limitations of some widely used animal models (Anon, 2013) and of the fact that many of these models poorly reflect human disease. For example, widely used murine models of depression do not even model appropriately the therapeutic action of antidepressants (Hyman, 2012). Therefore it is time to seriously reappraise the usefulness of animal experiments claiming to model human mental disease. The questionable comparability between animals and humans is not an issue specific to psychiatry, but it seems to be inherent also to other complex disorders. A recent, large study comparing transcriptional responses to inflammatory insults in mice and humans revealed that among genes changed significantly in humans, the murine orthologs poorly match their human counterparts (Seok et al., 2013).

Despite these significant caveats, ill-suited models are still being used to make go or no-go decisions to carry drug candidates forward into clinical trials (Seok et al., 2013). The time has come, especially in psychiatry, to utilize the appropriate research tools and focus on the human situation to understand the paths leading to human-specific psychiatric disorders and by this to increase the success rates of drug discovery. Because of the high heritability rates of psychiatric disorders, human genetics represents such an appropriate, human-centered research tool.

IV PROMISING HUMAN GENOME

Improving understanding, diagnosis, and therapy of human disease was a central promise of the human genome project (Anon, 2011). This promise is being increasingly fulfilled, at least in some medical research fields. For example, cancer research has benefitted dramatically from the discoveries of the human genome project (Lander, 2011), mainly because the genomic mechanisms leading to the development of many cancers are amenable to direct observation. The situation is different for disorders in which the underlying molecular events are not easily accessible, as is the case for mental

disorders. Thus it is logical to ask whether utilizing genome information will have a significant impact on the understanding of mental disease and on the development of better therapies.

Recent advances in the development of high-throughput genotyping platforms, analytical software, and collaborative efforts have led to the identification of numerous well-validated genetic risk factors for common, complex diseases (www.genome.gov/gwastudies). Importantly, known drug targets for such complex diseases, like type 2 diabetes, hyperlipidemia, multiple sclerosis, and psoriasis, have turned up in the genome-wide association studies (GWAS) (Collins, 2011). Recent megaanalyses have also led to the robust identification of genetic risk factors for common psychiatric disorders (Schizophrenia Working Group of the Psychiatric Genomics Consortium, 2014; Sklar et al., 2011; Wang et al., 2009) and to the notion that many of these factors are shared across diagnostic categories (Cross-Disorder Group of the Psychiatric Genomics Consortium and Genetic Risk and Outcome of Psychosis Consortium, 2013). Thus the use of genetic information is also likely to provide important clues about potential drug targets for psychiatric disorders.

V ILL-SUITED PHENOTYPES FOR DRUG DISCOVERY

Notwithstanding these recent human genetics-driven discoveries, it is important to point out that the success and relevance of human genetic research stands and falls with the choice of the appropriate phenotype. In this respect, current diagnostic constructs in psychiatry, such as those used in most GWAS, are clearly suboptimal.

Imagine a patient presenting with the following symptoms in the same 2-week period: loss of interest, feelings of guilt, weight loss, insomnia, and psychomotor agitation. This patient fulfills the diagnostic criteria for major depressive disorder (MDD) (American Psychiatric Association, 2013). Now imagine another patient presenting with the following symptoms in the same 2-week period: depressed mood, fatigue, weight gain, hypersomnia, and psychomotor retardation. This patient also fulfills the diagnostic criteria for MDD, despite a different clinical picture. This example instantly highlights one of the central problems in psychiatry: the absence of biologically operationalized diagnostic criteria. Psychiatric diagnoses are still treated as constructs based on clinical phenomenology and represent a consensus list of different symptoms, mostly unrelated to underlying biology. Such diagnostic lists, although useful in clinical terms, compromise the search for biological underpinnings of disease and jeopardize the development of targeted and causal therapies.

Sole dependency on phenomenology and a list of symptoms may lead to arbitrary classifications. For example, although the current DSM version (DSM-5) offers 11 combinations of criteria to arrive at diagnostic threshold for autism spectrum disorder, the previous DSM version (DSM-IV) offered a total of 2027 different combinations (McPartland et al., 2012). There is no biological rationale for the justification of either number of combinations, yet the influence of this arbitrary classification on research and the development of targeted treatments is obvious.

VI APPROPRIATE PHENOTYPES FOR DRUG DISCOVERY

As long as psychiatric classification results in an insufficient description of the neurobiological heterogeneity of human psychopathology, the search for targeted—and hopefully more effective—therapies of disorders related to pathological cognitive or emotional states will be seriously compromised. Mental disorders are not just dichotomous categories. At least in the case of common and heterogeneous disorders, such as depression, schizophrenia, and autism, evidence suggests that mental disease represents the extremes of a normal distribution of symptoms on multiple dimensions, for example, cognitive, emotional, and behavioral dimensions (Anderson et al., 1993; Kendler & Gardner, 1998; Lundstrom et al., 2012; Quee et al., 2014; Toulopoulou et al., 2007; van Os, 2013; Whittington & Huppert, 1996; Zammit et al., 2013) (but see also David, 2010; Weinberger & Goldberg, 2014). Fortunately, this fact has received increased attention. The idea of deconstructing psychiatric categories into measurable and biologically informed dimensions has its roots in the endophenotype/intermediate phenotype concept (Gottesman & Gould, 2003; Rasetti & Weinberger, 2011), which supports the notion that genetic associations will be stronger at the level of biological substrates of a given psychiatric illness than at the level of the respective diagnostic category (Rasetti & Weinberger, 2011). Recent work, mostly related to the schizophrenia spectrum and to functional brain imaging as the relevant intermediate phenotype, has provided important empirical evidence in support of this rationale (Egan et al., 2004; Callicott et al., 2005; Esslinger et al., 2009; Bigos et al., 2010).

Along the lines of the intermediate phenotype concept, the recently launched Research Domain Criteria (RDoC) initiative aims at establishing a new classification framework for research on mental disorders by capitalizing on the current developments in neuroscientific methodology (Insel et al., 2010). However, caution needs to be exercised when deconstructing psychiatric categories and diagnoses into biologically informed dimensions. As pointed out recently, even clear and important behavioral dimensions observed in patients may be difficult to assess in healthy individuals (Weinberger & Goldberg, 2014). Specifically, it might prove erroneous to assume that the mechanisms that give rise to a given dimension in a non-diseased population are the same mechanisms associated with this dimension in patients. For example, auditory hallucinations can be observed in healthy individuals; however, it is unclear whether the same mechanisms are in place in psychiatric patients experiencing acoustic hallucinations during a psychotic episode (David, 2010). Consequently a study on the genetic underpinnings of acoustic hallucinations performed in the general population might identify biological mechanisms that are not linked to psychosis and are ill-suited for further consideration as potential drug targets.

Despite these caveats, we believe that a deconstruction of psychiatric categories and diagnoses into biologically informed domains will be key for improving drug discovery in psychiatry provided that such domains fulfill the criteria listed in Box 14.1.

VII APPROPRIATE DATA MINING FOR DRUG DISCOVERY

GWAS employing single-marker statistics have been successful in identifying trait-associated single-gene loci (Manolio, 2010). It is, however, widely accepted that single-marker-based analyses have limited power to identify the genetic basis of a given trait. For example, many loci will fail to reach stringent genome-wide significance threshold, despite the fact that they may be genuinely associated with the trait. Statistical approaches for the analysis of gene expression have recently made

---

BOX 14.1

KEY CRITERIA FOR BIOLOGICALLY INFORMED DOMAINS IN PSYCHIATRY

- *Relevance criterion*: The domains represent physiological traits known to be disturbed in neuropsychiatric diseases. Examples: working memory, episodic memory, attention, verbal fluency, and cognitive control.
- *Neural correlate criterion*: The domains have specific and testable neural correlates (e.g., as shown in human brain imaging studies). This allows for further corroboration at the neural systems level.
- *Genetic criterion*: The domains are heritable. This enables the utilization of human genetic information to identify molecules related to the domain under study.

gene-set-based analytic methods available. These methods aim to identify biologically meaningful sets of genes associated with a certain trait, rather than focusing on a single GWAS gene locus (Wang et al., 2010). By taking into account prior biological knowledge, gene-set-based approaches examine whether test statistics for a group of related genes has consistent deviation from chance (Mooney et al., 2014; Wang et al., 2010). As shown recently in studies on working memory (Heck et al., 2014), autism (Voineagu et al., 2011), bipolar disorder (Holmans et al., 2009; Sklar et al., 2011), attention deficit hyperactivity disorder (ADHD) (Stergiakouli et al., 2012), and schizophrenia (O'Dushlaine et al., 2011), such approaches can identify convergent molecular pathways relevant to neuropsychiatry. Importantly the identification of groups of functionally related genes is likely to facilitate drug discovery, because the most significant single loci from a GWAS might not be the best candidates for therapeutic intervention (Gaudet et al., 2010; Muglia, 2011). This pathway approach has been already integrated into corporate drug discovery pipelines, because the more genes for any given pathway are identified, the greater the confidence that this pathway should be prioritized over others (Schubert et al., 2014).

VIII FACILITATING DRUG DISCOVERY THROUGH HUMAN GENETICS

With the launch of multinational collaborative efforts and the initiation of large-scale GWAS, the robustness and reliability of genetic association findings on complex traits and disorders have finally reached the level of confidence required for further consideration of the trait-associated genes as starting points for drug discovery. This increase in confidence also applies to the psychiatric genetics field, which has also benefitted from the formation of large consortia (Schizophrenia Working Group of the Psychiatric Genomics Consortium, 2014). However, phenotype definition in these large studies still relies on diagnostic constructs, which, as described earlier, are not driven by biological information.

In the last few years, pharmaceutical companies have made large investments with the expectation that some of the GWAS findings will ultimately lead to novel therapeutic agents. It remains to be seen how successful the translation of such findings to novel drugs will be. Allowing enough time for this judgment will be critical, because the GWAS field is rather new and the road from target identification to regulatory approval of a new drug takes more than a decade. Nevertheless, it is fair to ask whether there is any support for this genetic approach's potential to facilitate drug discovery. Evidence comes from a recent study, which assessed the utility of GWAS in identifying alternative indications of existing drugs (Sanseau et al., 2012). By implementing a systematic and comprehensive analysis pipeline, the study demonstrated that GWAS genes are significantly more likely to be theoretically druggable or biopharmable targets than expected just by chance. Importantly the study showed that GWAS data may lead to immediate translational opportunities for drug discovery and development through successful drug repositioning (Sanseau et al., 2012). It is therefore logical to assume that GWAS have the potential of translating into novel treatment targets for psychiatric conditions. Indeed, strategies for applying human genetics to drug discovery in neuroscience are being developed (Schubert et al., 2014).

Recently, we conducted a study focusing on a physiological cognitive domain fulfilling the criteria described in Box 14.1 and addressed the question of whether the use of human genetic information would lead to the identification of compounds modulating human cognition (Papassotiropoulos et al., 2013). In a first step, we performed a multinational collaborative study that included assessment of aversive memory—a trait central to posttraumatic stress disorder—and a genetic analysis in healthy individuals. Gene-set-based analysis identified two pathways, the neuroactive ligand-receptor interaction and the long-term depression pathway that were enriched with genes associated with aversive memory. A total of 20 genes constituting these pathways were replicated in two independent cohorts and represent potential drug target genes. To provide a rapid proof of concept for human genetic-guided identification of memory-modulating drugs, we selected from these 20 candidate targets only gene products with already existing therapeutic compounds. Diphenhydramine, which targets the histamine 1 receptor (encoded by *HRH1*), a member of the neuroactive ligand-receptor interaction pathway, was given highest priority for a subsequent pharmacological intervention trial. Diphenhydramine led to significant reduction of aversive memory in healthy participants. Further studies are needed to assess whether this finding translates into the therapy of a clinical condition, such as posttraumatic stress disorder. This study demonstrated that genome information can be used as a starting point for the identification of memory-modulating compounds. Besides diphenhydramine and other already well-known drugs, it provided several novel drug targets that may serve drug development purposes.

Notwithstanding the potential of the human genetics approach for drug discovery in psychiatry, it is important to realize that this approach also comes with challenges and limitations. For example, despite the reliable identification of disease-associated genomic loci, the connection between GWAS locus and biology of the respective disorder is not always readily clear and straightforward (Plenge et al., 2013). In this context the assignment of a given locus to a gene and the identification of causal genes are important, nontrivial tasks. Even when a gene is identified as causal, the direction of the effect of this gene on disease processes is rarely known (Schubert et al., 2014). Further the genomic architecture of psychiatric disorders is complex. Human genetics alone cannot entirely capture and describe this complexity and should be therefore considered one of multiple possible sources of information within a framework of methods aimed at identifying causal biological mechanisms of disease and pharmacological remedies for targeting these mechanisms. Another issue worth mentioning is the fact that most GWAS to date are performed in populations of European ancestry. Thus it is not clear whether a drug identified through such GWAS will be efficacious in populations with distinct genetic backgrounds.

Of note the vast majority of the GWAS-derived risk variants show low effect sizes (e.g., the odds ratios of the recent large GWAS of schizophrenia cluster around 1.1). One may think of this observation as another potential limitation. However, it is not possible to extrapolate from a gene's odds ratio to biological effect size (Okada et al., 2014) and to the therapeutic potency of a drug discovered through genetic association (Plenge et al., 2013; Schubert et al., 2014). For example, the effect size of common variants of the genes encoding HMG-CoA-reductase (*HMGCR*) and peroxisome proliferator-activated receptor-γ (*PPARG*) on blood lipid levels and risk for type 2 diabetes is small, yet these genes point to potent drugs for the treatment of hyperlipidemia and type 2 diabetes, respectively (Altshuler et al., 2000; Burkhardt et al., 2008; Kathiresan et al., 2008).

IX CONCLUDING REMARKS

Drug discovery in psychiatry has been jeopardized by the use of ill-suited disease models and diagnostic constructs unrelated to underlying biological mechanisms. We anticipate that the exponential increase in knowledge about the genetic basis of complex human traits, including neuropsychiatric disorders, will be a game changer. Along with appropriate, biologically informed phenotypes and appropriate data mining methodology, such knowledge is an ideal starting point for the identification of novel drug targets. Although it remains to be seen whether such approaches will ultimately improve therapeutic outcomes, they bear considerable potential for understanding the neurobiology of human psychopathology.

References

Abbott, A. (2011). Novartis to shut brain research facility. *Nature, 480*, 161–162.

Altshuler, D., et al. (2000). The common PPARgamma Pro12Ala polymorphism is associated with decreased risk of type 2 diabetes. *Nature Genetics, 26*, 76–80.

American Psychiatric Association. (2013). *Diagnostic and statistical manual of mental disorders* (5th ed. (DSM-5)). American Psychiatric Association.

Anderson, J., et al. (1993). Normality, deviance and minor psychiatric morbidity in the community. A population-based approach to general health questionnaire data in the health and lifestyle survey. *Psychological Medicine, 23*, 475–485.

Anon. (2011). Best is yet to come. *Nature, 470*, 140.

Anon. (2013). Of men, not mice. *Nature Medicine, 19*, 379.

Bigos, K. L., et al. (2010). Genetic variation in CACNA1C affects brain circuitries related to mental illness. *Archives of General Psychiatry, 67*, 939–945.

Burkhardt, R., et al. (2008). Common SNPs in *HMGCR* in micronesians and whites associated with LDL-cholesterol levels affect alternative splicing of exon13. *Arteriosclerosis, Thrombosis, and Vascular Biology, 28*, 2078–2084.

Callicott, J. H., et al. (2005). Variation in DISC1 affects hippocampal structure and function and increases risk for schizophrenia. *Proceedings of the National Academy of Sciences of the United States of America, 102*, 8627–8632.

Collins, F. S. (2011). Reengineering translational science: the time is right. *Science Translational Medicine, 3*, 90cm17.

Cross-Disorder Group of the Psychiatric Genomics Consortium and Genetic Risk and Outcome of Psychosis Consortium. (2013). Identification of risk loci with shared effects on five major psychiatric disorders: a genome-wide analysis. *Lancet, 381*, 1371–1379.

David, A. S. (2010). Why we need more debate on whether psychotic symptoms lie on a continuum with normality. *Psychological Medicine, 40*, 1935–1942.

Egan, M. F., et al. (2004). Variation in GRM3 affects cognition, prefrontal glutamate, and risk for schizophrenia. *Proceedings of the National Academy of Sciences of the United States of America, 101*, 12604–12609.

Esslinger, C., et al. (2009). Neural mechanisms of a genome-wide supported psychosis variant. *Science, 324*, 605.

Gaudet, M. M., et al. (2010). Common genetic variants and modification of penetrance of BRCA2-associated breast cancer. *PLoS Genetics, 6*, e1001183.

Gottesman, I. I., & Gould, T. D. (2003). The endophenotype concept in psychiatry: etymology and strategic intentions. *The American Journal of Psychiatry, 160*, 636–645.

Gustavsson, A., et al. (2011). Cost of disorders of the brain in Europe 2010. *European Neuropsychopharmacology, 21*, 718–779.

Healy, D. G., et al. (2004). Population genetics for target identification. *Drug Discovery Today: Technologies, 1*, 69–74.

Heck, A., et al. (2014). Converging genetic and functional brain imaging evidence links neuronal excitability to working memory, psychiatric disease, and brain activity. *Neuron, 81*, 1203–1213.

Holmans, P., et al. (2009). Gene ontology analysis of GWA study data sets provides insights into the biology of bipolar disorder. *American Journal of Human Genetics, 85*, 13–24.

Hyman, S. E. (2012). Revolution stalled. *Science Translational Medicine, 4*, 155cm111.

Insel, T., et al. (2010). Research domain criteria (RDoC): toward a new classification framework for research on mental disorders. *The American Journal of Psychiatry, 167*, 748–751.

Insel, T. R. (2012). Next-generation treatments for mental disorders. *Science Translational Medicine, 4*, 155ps119.

Kathiresan, S., et al. (2008). Six new loci associated with blood low-density lipoprotein cholesterol, high-density lipoprotein cholesterol or triglycerides in humans. *Nature Genetics, 40*, 189–197.

Kendler, K. S., & Gardner, C. O., Jr. (1998). Boundaries of major depression: an evaluation of DSM-IV criteria. *The American Journal of Psychiatry, 155*, 172–177.

Lander, E. S. (2011). Initial impact of the sequencing of the human genome. *Nature, 470*, 187–197.

Lundstrom, S., et al. (2012). Autism spectrum disorders and autistic like traits: similar etiology in the extreme end and the normal variation. *Archives of General Psychiatry, 69*, 46–52.

Manolio, T. A. (2010). Genomewide association studies and assessment of the risk of disease. *The New England Journal of Medicine, 363*, 166–176.

McPartland, J. C., et al. (2012). Sensitivity and specificity of proposed DSM-5 diagnostic criteria for autism spectrum disorder. *Journal of the American Academy of Child and Adolescent Psychiatry, 51*, 368–383.

Mooney, M. A., et al. (2014). Functional and genomic context in pathway analysis of GWAS data. *Trends in Genetics, 30*, 390–400.

Muglia, P. (2011). From genes to therapeutic targets for psychiatric disorders—what to expect? *Current Opinion in Pharmacology, 11*, 563–571.

O'Dushlaine, C., et al. (2011). Molecular pathways involved in neuronal cell adhesion and membrane scaffolding contribute to schizophrenia and bipolar disorder susceptibility. *Molecular Psychiatry, 16*, 286–292.

Okada, Y., et al. (2014). Genetics of rheumatoid arthritis contributes to biology and drug discovery. *Nature, 506*, 376–381.

Papassotiropoulos, A., et al. (2013). Human genome-guided identification of memory-modulating drugs. *Proceedings of the National Academy of Sciences of the United States of America, 110*, E4369–E4374.

Plenge, R. M., et al. (2013). Validating therapeutic targets through human genetics. *Nature Reviews. Drug Discovery, 12*, 581–594.

Quee, P. J., et al. (2014). Cognitive subtypes in non-affected siblings of schizophrenia patients: characteristics and profile congruency with affected family members. *Psychological Medicine, 44*, 395–405.

Rasetti, R., & Weinberger, D. R. (2011). Intermediate phenotypes in psychiatric disorders. *Current Opinion in Genetics & Development, 21*, 340–348.

Sanseau, P., et al. (2012). Use of genome-wide association studies for drug repositioning. *Nature Biotechnology, 30*, 317–320.

Schizophrenia Working Group of the Psychiatric Genomics Consortium. (2014). Biological insights from 108 schizophrenia-associated genetic loci. *Nature, 511*, 421–427.

Schubert, C. R., et al. (2014). Translating human genetics into novel treatment targets for schizophrenia. *Neuron, 84*, 537–541.

Seok, J., et al. (2013). Genomic responses in mouse models poorly mimic human inflammatory diseases. *Proceedings of the National Academy of Sciences of the United States of America, 110*, 3507–3512.

Sklar, P., et al. (2011). Large-scale genome-wide association analysis of bipolar disorder identifies a new susceptibility locus near ODZ4. *Nature Genetics, 43*, 977–983.

Smith, K. (2011). Trillion-dollar brain drain. *Nature, 478*, 15.

Stergiakouli, E., et al. (2012). Investigating the contribution of common genetic variants to the risk and pathogenesis of ADHD. *The American Journal of Psychiatry, 169*, 186–194.

Toulopoulou, T., et al. (2007). Substantial genetic overlap between neurocognition and schizophrenia: genetic modeling in twin samples. *Archives of General Psychiatry, 64*, 1348–1355.

van Os, J. (2013). The dynamics of subthreshold psychopathology: implications for diagnosis and treatment. *The American Journal of Psychiatry, 170*, 695–698.

Voineagu, I., et al. (2011). Transcriptomic analysis of autistic brain reveals convergent molecular pathology. *Nature, 474*, 380–384.

Wang, K., et al. (2009). Common genetic variants on 5p14.1 associate with autism spectrum disorders. *Nature, 459*, 528–533.

Wang, K., et al. (2010). Analysing biological pathways in genome-wide association studies. *Nature Reviews. Genetics, 11*, 843–854.

Weinberger, D. R., & Goldberg, T. E. (2014). RDoCs redux. *World Psychiatry, 13*, 36–38.

Whittington, J. E., & Huppert, F. A. (1996). Changes in the prevalence of psychiatric disorder in a community are related to changes in the mean level of psychiatric symptoms. *Psychological Medicine, 26*, 1253–1260.

Zammit, S., et al. (2013). Psychotic experiences and psychotic disorders at age 18 in relation to psychotic experiences at age 12 in a longitudinal population-based cohort study. *The American Journal of Psychiatry, 170*, 742–750.

Glossary

Cohort A group of people with one or more common statistical characteristics (e.g., healthy adults, aged between 18 and 35 years).

Complex trait A quantifiable property of an organism influenced by both genetic and environmental factors and interactions between them.

Drug repositioning The use of existing drugs for new therapeutic indications. Also known as drug repurposing.

Endophenotype/intermediate phenotype A heritable, disease-related trait (e.g., disturbed working memory) that is observed in patients and their healthy relatives. Genes contributing to an endophenotype represent a subset of the genes contributing to the respective disease.

Episodic memory A memory system that enables conscious recollection of past experiences (e.g., autobiographical episodes and learned material) along with their spatial and temporal contexts.

Gene-set-based analytic methods In contrast to single-marker statistics, which focus on single variants and the corresponding main effects, gene-set-based analysis attempts to identify biologically meaningful sets of genes associated with a certain complex trait. By taking into account prior biological knowledge, gene-set-based approaches examine whether test statistics for a group of related genes have consistent deviation from chance.

Genome-wide association study (GWAS) An analysis of genetic variants (usually hundreds of thousands of variants, ideally all of the genetic variants throughout the human genome) in groups of individuals to test for statistical association of these variants with a given trait. GWAS can be performed in a case-control setting (i.e., the trait of interest is represented by a binary variable, e.g., patients with schizophrenia vs healthy controls) and/or by using a quantitative trait approach (i.e., the trait of interest is represented by a continuous variable, e.g., memory performance). In contrast to methods that specifically test one or a few genes, GWAS investigate the entire genome.

Heritability A population-based statistical value that indicates how much of the phenotypic variance is attributable to heritable factors. Heritability values range between 0 (i.e., heritable factors explain 0% of the phenotypic variance) and 1 (i.e., heritable factors explain 100% of the phenotypic variance). Heritability is specific to the population under study and does not apply to traits not showing any variability.

High-throughput genotyping platform Array- or sequencing-based technologies enabling high-throughput analysis of genetic variants.

Long-term depression pathway Genes constituting this pathway are involved in the modulation of synaptic strength between nerve cells.

Neuroactive ligand-receptor interaction pathway Genes constituting this pathway encode several neuronal receptors along with their binding partners.

Odds ratio (OR) A numerical value that describes the strength of the association between two binary variables. In genetic association studies, OR describes the strength of the association between a given genetic variant and a binary trait (e.g., disease status).

Phenotype An observable characteristic of an organism with respect to a physiological trait (e.g., blue eye color and memory performance) or disease (e.g., depression).

Single-marker statistics This type of genetic analysis tests for statistical association of a variant with a given trait independently of the association of other variants with that trait. In a genome-wide setting engaging the analysis of 1 million variants, this type of analysis yields 1 million independent test results.

Trait-associated single-gene locus A gene variant that is statistically associated with the trait under study.

Variant In genetics, a difference in DNA sequence among individuals. A common form of a genetic variant is a single nucleotide polymorphism (SNP), which occurs when a nucleotide—A, T, C, or G—differs between individuals. The human genome contains millions of SNPs.

Working memory A limited-capacity neural network capable of actively maintaining task-relevant information during the execution of a cognitive task. Working memory deficits are characteristic for many psychiatric disorders.

15

Use of Cognition to Guide Decisions About the Safety and Efficacy of Drugs in Early-Phase Clinical Trials

Paul Maruff*,†

*Florey Institute for Neuroscience and Mental Health, University of Melbourne, Parkville, VIC, Australia †Cogstate, New Haven, CT, United States

I INTRODUCTION

In early-phase clinical trials, cognition is measured to provide an index of the extent to which the experimental drug influences function of the central nervous system (CNS). Such influence may manifest as a disruption to normal thinking, indicated by a decline in performance from baseline on the cognitive tests (e.g., Fredrickson et al., 2008; Harel, Pietrzak, & Snyder, 2013; Maruff et al., 2006). Alternatively, the effects of a drug on the CNS may be facilitatory in which case they will manifest as an improvement in performance on the cognitive tests (e.g., Hilt, Safirstein, Hassman, et al., 2009a; Pietrzak, Snyder, & Maruff, 2010; Snyder, Bednar, Cromer, & Maruff, 2005). The ability of different cognitive tests to identify disruption or facilitation of CNS function depends on the aspects of cognition measured by those tests and whether these align with the biological systems influenced by any experimental drug. However, even with the appropriate aspects of cognition measured, sensitivity to both the disruptive and the facilitatory effects of experimental drugs depends on the design of the study itself and the theoretical context within which that study is conducted. Because of these issues, researchers measuring cognition to guide decisions about CNS drug effects in early-phase clinical trials should be clear on the question addressed by those tests and the audience for whom the outcome of such questions are important. Consideration of these issues will guide the selection of tests, their application in the study, and analyses and interpretation of the results. In this chapter, I will consider how such issues can influence decision-making in early-phase medicine.

II MEASURING COGNITION TO GUIDE DECISIONS ABOUT THE FACILITATORY EFFECTS OF CNS-ACTIVE DRUGS

Most early-phase clinical trials that use cognition to aid understanding about the CNS effects of their drugs are a part of a broader drug development program aimed at developing a treatment for some aspect of CNS disease, disorder, or injury. In early-phase trials, such drugs are administered to humans for the first time, albeit after substantial bench and animal research has ensured their pharmacological activity and safety (McGonigle, 2014). Thus preclinical programs deliver experimental drugs to human clinical trials in humans with a package of experimental results from preclinical studies. In addition to providing information about the absorption, distribution, metabolism, and excretion of these new drugs, such packages may also provide data to guide in the dosing necessary for the drug to cross the blood-brain barrier and influence behavior (McGonigle, 2014), for example, by showing how treatment with the drug can improve or disrupt performance on rodent cognitive models such as the Morris water maze or balance beam walking test. Consequently, when such drugs are brought to studies in humans, it is expected that they will also cross the blood-brain barrier (BBB) in sufficient amounts and thereby influence CNS function. Despite experimental drugs being developed for specific therapeutic uses, in the majority of CNS drug development programs, the first in humans, single ascending dose (SAD) and multiple ascending dose (MAD) studies are often conducted in healthy adults. Therefore cognitive tests utilized to direct understanding of drug effects at this stage of medicine are selected based on their utility in healthy adults rather than in any patient group. The ultimate context of use for an experimental drug may also influence the type of cognitive tests selected for use in SAD and MAD studies in healthy adult humans. Ideally the cognitive tests selected will therefore measure those aspects of cognition that have demonstrated validity in models of cognitive impairment in the therapeutic area of interest, appropriate for use in clinical trials and in healthy adults.

One example of these multiple constraints on cognitive test selection is seen in studies that have supported the development of drugs designed for the amelioration of cognitive impairment associated with schizophrenia (CIAS) (Nuechterlein, Green, et al., 2008; Wallace, Ballard, Pouzet, Riedel, & Wettstein, 2011). For many such drugs, early-phase studies that include assessments of cognition will begin in healthy human adults. Furthermore, in some cases, early studies in programs developing drugs for CIAS will require evidence that the drug has some beneficial CNS effects for the drug to be advanced into phase IIb or phase III clinical trials in people with CIAS (Hilt, Safirstein, Hassman, et al., 2009b). Fig. 15.1 provides an example of the logic of this translation using outcomes of cognitive testing first in human studies and studies of patients with CIAS to understand the potential therapeutic benefit of dopamine agonists in CIAS. Fig. 15.1 summarizes data from different experiments that show that psychomotor function, measured using the Cogstate detection task, is improved acutely in healthy adults following treatment with dopamine agonist dexamphetamine (20 mg, p.o.) (Pietrzak et al., 2010) (Fig. 15.1A), is impaired in patients with chronic and stable schizophrenia compared with matched controls (Maruff, Thomas, Cysique, et al., 2009) (Fig. 15.1B), and is improved in CIAS by daily dosing of oral lisdexamfetamine, where the starting dose of 70 mg p.o. was increased every 5 days (Fig. 15.1C) (Martin, Dirks, Gertsik, et al., 2014).

Together, these data illustrate how first psychomotor function can be measured reliably in healthy adults and in patients with CIAS. Furthermore, this same test can be used unchanged to provide an index of psychomotor function in patients with schizophrenia and to also detect improvement in this function following treatment with the experimental drug (e.g., Girgis, Van Snellenberg, Kegeles, et al., 2014). Thus the same test can provide a basis for decision-making in early-phase studies and a context for planning and interpreting data from drugs with a similar mechanism of action in the target therapeutic area. It is worthwhile highlighting here that impairment in psychomotor function is not sufficient for a clinical classification of CIAS (Nuechterlein et al., 2008). In clinical trials conducted to date, this clinical construct has required evidence of impairment across multiple

domains of cognition (Jarskog, Lowy, Grove, et al., 2015; Lieberman, Dunbar, Segreti, et al., 2013; Walling, Marder, Kane, et al., 2015). Decisions made in these studies measuring psychomotor function are about the extent to which the study drugs act on the CNS generally, and in CIAS specifically. With this positive evidence, drug development programs now must design larger clinical trials with cognitive and clinical endpoints that adhere to regulatory recommendations to determine whether the positive CNS effects observed in these early-phase studies are sufficient to influence the target clinical syndrome. Nonetheless, these positive early-phase data provide important information for that design process.

Another example of a translation of cognitive data from studies in healthy humans to studies in patients with CNS disease is shown in Fig. 15.2 for treatment of dementia with histamine H3 receptor antagonists. This series of experiments demonstrate the extent to which treatment based on histamine H3 receptor antagonism can improve aspects of higher cognitive function in both healthy adults (Cho, Maruff, Connell, et al., 2011) and adults with Alzheimer's disease (Grove, Harrington, Mahler, et al., 2014), diagnosed clinically. However, in the healthy adults the CNS activity of the H3 receptor antagonist was demonstrated by the extent to which treatment with this drug could ameliorate the deleterious effects of the muscarinic antagonist scopolamine.

Fig. 15.2A shows how a single acute dose of scopolamine (0.5 mg, s.c.) was associated with a deterioration in executive function, measured using the Groton Maze Learning Test (GMLT) (Cho et al., 2011). The figure shows that this disruption reached the nadir 2h after dosing after which performance begins to improve and ultimately it returns to baseline levels 8h after dosing. This deleterious effect of scopolamine on cognition is a well-reported phenomenon and has proven to be a highly predictable psychopharmacological assay for challenging the CNS effects of putative cognitive-enhancing drugs in healthy adults for many years (Fredrickson et al., 2008). In such assays, the putative beneficial effects of CNS-active drugs are determined by the extent to which they can overcome this scopolamine-related cognitive decline. For example, Fig. 15.2A shows that if the same healthy adults are treated with a H3 inverse agonist 2h prior to baseline, the deleterious effects of scopolamine at 2h postdose and beyond were reduced substantially. This indicates and suggests that the H3 inverse agonist did enter the CNS and in sufficient amount to overcome cognitive impairment arising from the scopolamine-related disruption to acetylcholine neurotransmission. In a related series of studies, an H3 receptor antagonist was given to approximately 180 patients with clinically classified AD of mild or moderate severity and who were not receiving any approved pharmacotherapies for AD (e.g., cholinesterase inhibitors or

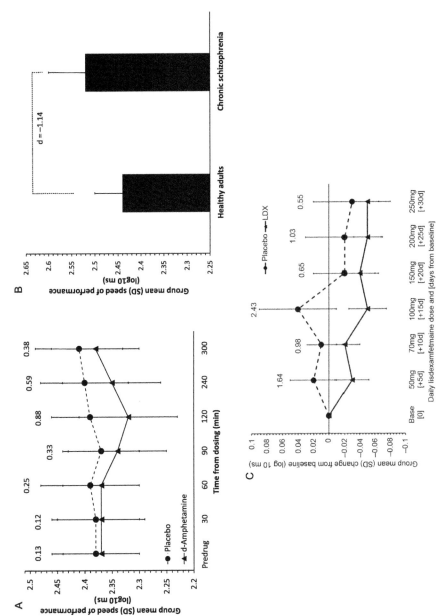

FIG. 15.1 Example of translation of effects of a dopamine agonist drug on psychomotor function from healthy adults to patients with chronic schizophrenia measured using the Cogstate detection test. The figures have been reconstructed from data presented in studies (Martin et al., 2014; Pietrzak et al., 2009, 2010). For each figure, negative values indicate better performance (i.e., a reduction in performance speed). (A) Group mean speed of performance before and every 30 min after dosing of dexamphetamine (20 mg, p.o.) (*solid line*) and placebo (*dashed line*). The values are the top of each error bar and reflect the measure of effect size for differences in change from baseline scores at that timepoint. (B) Group mean speed of performance in patients with schizophrenia and matched controls. The value at the top of the figure shows magnitude of difference (Cohen's *d*). (C) Group mean change in speed of performance under placebo (*dashed line*) and lisdexamfetamine (LDX), where the dose of LDX was increased every 5 days. Values on top of each bar reflect the measure of effect size for differences in change from baseline scores at that timepoint.

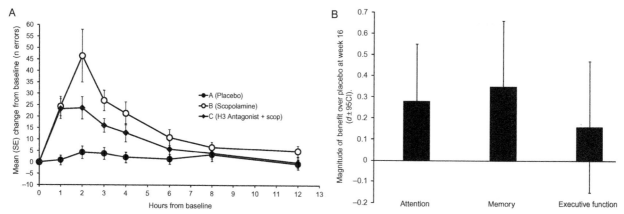

FIG. 15.2   Example of translation of effects of histamine H3 receptor antagonism on executive function from healthy adults to patients with clinically classified dementia measured using the Groton Maze Learning Test (GMLT). The figures have been reconstructed from the data presented in studies (Cho et al., 2011; Grove et al., 2014). Increasing positive values indicate worsening performance (i.e., increasing errors). (A) The group mean change from baseline in total GMLT errors after dosing with scopolamine alone *(open circles)*, scopolamine, and the histamine H3 inverse agonist MK3134 *(filled diamonds)* and with placebo *(filled circles)*. (B) The effect sizes reflecting the difference between 16 weeks treatment with the histamine H3 antagonist GSK239512 and placebo for composite measures of attention, memory, and executive function derived from performance on the Cogstate Alzheimer's disease battery. Where the 95% CIs associated with an effect size do not include zero (attention and memory) indicates that performance was significantly better under the drug.

memantine) (Grove et al., 2014; Nathan, Boardley, Scott, et al., 2013). The data in Fig. 15.2B show the results of a study where the drug or placebo was administered for 16 weeks. The results indicated that treatment with the H3 antagonist was associated with moderate improvement in attention and memory and a smaller improvement in executive function (Fig. 15.2B) (Grove et al., 2014). Together, these data show that antagonism of the histamine H3 receptor can facilitate aspects of higher cognition such as executive function and memory and that this facilitation can be observed in both healthy adults and individuals with brain disease. However, to observe the positive cognitive effects associated with modulation of the histamine H3 receptor in healthy adults, it was necessary to superimpose the drug treatment on a pharmacological challenge designed to disrupt normal CNS function. In the individuals with brain disease, the beneficial effects of histamine H3 antagonists were evident without any challenge.

In the past the extent to which experimental drugs might be beneficial in patients with AD was often investigated using scopolamine challenge models in healthy adults, predominantly because biological models considered dementia due to Alzheimer's disease (AD) to be a disorder of cholinergic neurotransmission (e.g., Snyder et al., 2005). More recent developments in AD research show that the cholinergic hypothesis described only a small part of the disease process with current models of AD biology emphasizing cellular dysfunction related to accumulation of abnormally occurring protein fragments such as amyloid and tau (Knopman, Haeberlein, Carrillo, et al., 2018; Villemagne, Burnham, Bourgeat, et al., 2013). However, even with the diminished importance of

acetylcholine in AD and the absence of a direct theoretical link between scopolamine challenges and the pathology of AD, scopolamine challenge studies still prove useful for guiding decisions about whether putative cognitive-enhancing drugs can enter the CNS in sufficient amount to influence cognition in healthy adults. Thus, in the example in Fig. 15.2, the effect of the antagonism of the histamine H3 receptor in overcoming scopolamine-related impairment in executive function in the healthy adults did not by itself predict the positive effects of the use of the same class of drug in patients with AD. However, the finding that such drugs could cross the BBB in sufficient amount to overcome scopolamine-related cognitive impairment in healthy adults did provide an important basis for guiding expectations that it could also improve higher cognitive function in the phase II study in AD.

III MEASURING COGNITION TO GUIDE DECISIONS ABOUT THE DELETERIOUS EFFECTS OF CNS-ACTIVE DRUGS

Another important reason for the measurement of cognition in early-phase trials is to determine whether treatment with an experimental drug is associated with disruption to CNS function and, if so, whether the magnitude of this is clinically important. Many classes of drug designed to act on the CNS will also disrupt cognition. Fig. 15.3 shows the magnitude of decline in performance on the Groton Maze Learning Test in healthy adults who had been treated with lorazepam 2 mg (p.o.) (Chen, Jacobs, De Kam, et al., 2014) (Fig. 15.3A) and after

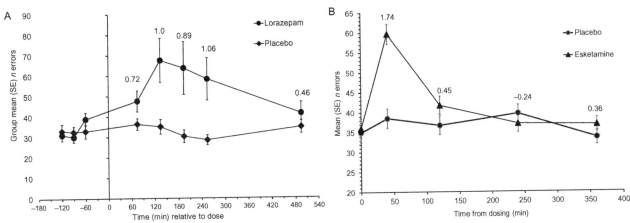

FIG. 15.3 Example of disruption to executive function in healthy adults following a single dose of lorazepam and esketamine measured using the Groton Maze Learning Test (GMLT). The figures have been reconstructed from the data presented in studies (Chen et al., 2014; Morrison et al., 2018). Increasing positive values indicate worsening performance (i.e., increasing errors). (A) Group mean errors on the GMLT prior to an after dosing with lorazepam 2 mg *(filled circles)* and placebo *(filled diamonds)* as assessments up to 9 h after dosing. (B) Group mean errors on the GMLT before and following dosing with esketamine (84 mg, intranasal) *(filled triangles)* and placebo *(filled circles)* for timepoints up to 3 h after dosing. Both figure values on top of each bar reflect the measure of effect size for differences in change from baseline scores at that timepoint.

treatment in a different group with esketamine 84 mg (intranasal) (Fig. 15.3B) (Morrison, Fedgchin, Singh, et al., 2018). In each figure, it is evident that treatment with the placebo resulted in no change in performance on the GMLT across the assessment sessions. Hence, there was no practice effect associated with the multiple reassessments using the GMLT. However, for both drugs, there was an acute increase in the number of errors made while (Hindmarch, 2009) learning the GMLT, and this increase occurred immediately after dosing. With time, however, the number of errors made on the GMLT returns to predose levels. For both drugs, the cognitive disruption immediately after dosing is considered important. For lorazepam, these adverse cognitive effects have also been observed when patients with anxiety disorders use the medication therapeutically and have resulted clinicians being made aware of potential cognitive changes (Patat, Paty, & Hindmarch, 2001). For esketamine, the effects of treatment on cognitive effects in clinical groups are still investigated.

To provide context for the effects of these different drugs, Fig. 15.3 provides a summary of data from studies where healthy older adults have been treated with a single dose of a drug known to disrupt CNS function and where the acute effects of these drugs on psychomotor function and executive function have been measured. The data shown reflect the magnitude of the difference in change from baseline performance between the drug and placebo at that timepoint at which the difference was greatest. This difference is expressed as a measure of effect size (Cohen's *d*). As a reference for the importance of these effects, the effects on the same cognitive tests are shown for two conditions that are well understood by the general population and indeed by legislators concerned with road safety, to be associated with potentially dangerous levels of cognitive impairment and intoxication with alcohol at the limit for driving and with 24 h sleep restriction (Falleti, Maruff, Collie, Darby, & McStephen, 2003).

Comparison of the data for the changes in executive function arising from single dose of lorazepam and esketamine, shown in Fig. 15.3, to the metaanalytic framework in Fig. 15.4 shows that the acute effects of treatment with both lorazepam and esketamine lead to cognitive impairments that are greater than those associated with the legal driving limit (0.05% blood alcohol content [BAC]) and greater or equal to the impairments in executive function observed after 24 h sleep deprivation. Hence, both impairments are important from a social context. Drug-related changes in executive function returns to values less than these socially relevant criteria within 2 h after dosing for esketamine and 8 h after doing for lorazepam.

IV USE OF COGNITIVE TESTS TO DEMONSTRATE CNS SAFETY

Given the sensitivity of different aspects of cognition to changes in CNS function arising from treatment with experimental drugs that cross the BBB, it is now becoming common for investigators to include cognitive assessments in their phase II and even phase III studies, as a way of showing that the use of the experimental medicine in the therapeutic area for which it is designed

FIG. 15.4 Summary of the magnitude of negative effects (Cohen's *d*) of a single dose of a CNS-active drug at the time after dosing where those effects were greatest in healthy adults. To provide a socially meaningful benchmark, these negative effects are compared with those for performance on the same task after low levels of alcohol intoxication and 24 h sleep restriction, also in healthy adults.

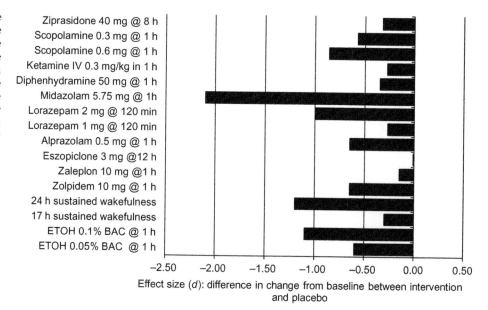

FIG. 15.5 Magnitude of the difference in group change from baseline under the fesoterodine 4 mg, fesoterodine 4 mg/8 mg, and alprazolam 1 mg conditions compared with that under placebo in the same participants. The figure has been reconstructed from the data presented in Kay, Maruff, Scholfield, et al. (2012). In this figure, negative effect sizes indicate that performance under drug became worse than under placebo.

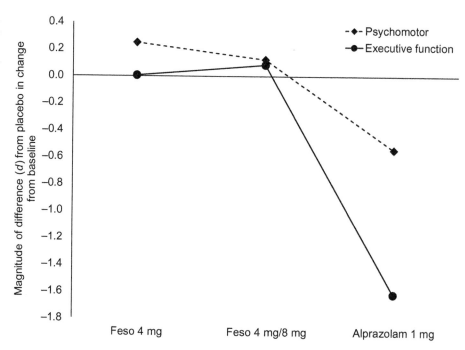

is not associated with any adverse CNS effects. For example, anticholinergic drugs are used commonly to treat overactive bladder (OAB) in older adults. However, as is shown in Fig. 15.2, anticholinergic drugs like scopolamine can cause a decline in cognition raising the possibility that chronic treatment for OAB could give rise to an abiding cognitive impairment. One solution to this issue is to develop anticholinergic drugs that do not cross the BBB. The data in Fig. 15.5 show the results from a study of the effects of an anticholinergic drug, developed to treat OAB, but with low or no BBB penetrance on psychomotor function and executive function (Kay, Maruff,

Scholfield, et al., 2011). To demonstrate the CNS safety of this compound, cognition was investigated in older adults who received 6 days of treatment with the drug at two doses: 4 mg for 6 days or 4 mg for 3 days followed by 8 mg for 3 days. The effect of these treatment regimens on cognition was compared in the same people with that under placebo and under acute challenge with alprazolam in a randomized crossover design. In this case, the alprazolam challenge can be considered an index of assay sensitivity (Kay et al., 2011).

Fig. 15.5 shows an example of the outcome for the measure of episodic memory used in this study, the Cogstate

Continuous Paired Associate Learning (CPAL) test. The figure shows that 6 days of treatment with either dose of fesoterodine did not influence episodic memory at all, whereas in the same older adults, the single dose of alprazolam induced a substantial impairment. Together, these data confirmed that treatment with fesoterodine did not disrupt CNS function in older adults (Kay et al., 2011). The approach of showing that cognition does not change following acute or chronic treatment with a drug that may cross the BBB has now become common in phase II and even phase III studies of CNS-active drugs (Goldman, Loebel, Cucchiaro, Deng, & Findling, 2017; Jacques, Harel, Schembri, et al., 2016; Stocchi, Rascol, Destee, et al., 2013). In these cases, a negative effect on cognition is not necessarily a characteristic that would result in an experimental drug program being stopped; rather the cognitive tests provide a method for understanding and quantifying some of the potential negative effects of experimental and even licensed medicines. Such data then provide a reliable basis for decisions about the costs and benefits of any treatment and a cognitive side effect profile that may itself require management.

V ADVERSE COGNITIVE OUTCOMES CAN INDICATE THE PRESENCE OF OCCULT CNS DISEASE

In the preceding sections, it has been shown that CNS-active drugs can have negative cognitive effects when administered to healthy adults. In experimental medicine contexts, such negative effects can be exploited to assist in understanding drug actions. In clinical trials, they can inform decisions about safety or adverse effects in respect of dosing. Recent experiments show that the adverse CNS effects of some CNS-active drugs can also be used as

challenges to unmask occult CNS disease in individuals who are cognitively normal but who carry risk factors for the disease of interest. In the section earlier the changing relationship between models of cholinergic neurotransmission and dementia due to Alzheimer's disease was considered. Recent models of AD now focus on the accumulation of amyloid and tau proteins and the biology of the downstream neuronal dysfunction and death associated with this (Knopman et al., 2018; Villemagne et al., 2013).

Recently, Snyder and coworkers showed that in relatively young adults (i.e., mean age 62 years), for whom PET scanning had identified abnormally high levels of amyloid, a very small dose of scopolamine (0.2 mg, s.c.) was associated with a substantial and protracted decline in executive function, measured using the GMLT (Lim, Maruff, Schindler, et al., 2015; Snyder, Lim, Schindler, et al., 2014). An example of this low-dose scopolamine interaction with amyloid is shown in Fig. 15.6 in terms of the level of increase in the number of errors made on the GMLT over the 7 h following the single 0.2-mg (s.c.) scopolamine dose. In a group of well-matched adults for whom amyloid PET showed no abnormal accumulation of amyloid, the same dose of scopolamine had only a small and transient effect on executive function. There was no difference in objectively measured or self-reported cognitive function between these two groups prior to treatment. The effect of the low dose of scopolamine is shown in Fig. 15.6.

The observation that a very small dose of scopolamine can unmask cognitive impairment in early AD shows that disruption to normal cholinergic neurotransmission remains a central aspect of the disease, even in its asymptomatic stages. Therefore, cholinesterase inhibitors approved for use in people with dementia may provide some therapeutic utility in preclinical AD. These data

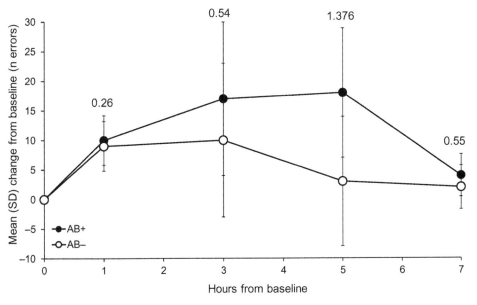

FIG. 15.6 Group mean change from baseline in executive function following a low dose of scopolamine (0.2 mg, s.c.) in healthy older adults with abnormally high levels of amyloid (*filled circles*) and normal amyloid levels (*open circles*). The figure has been reconstructed from the data presented in Lim et al. (2015). Increasing positive values indicate worsening performance (i.e., increasing errors). Positive amyloid levels were determined by PET and executive function measured using the GMLT. Values on top of each bar reflect the measure of effect size for differences in change from baseline scores at that timepoint.

also suggest that the use of a scopolamine challenge may provide a suitable challenge model for identifying individuals with preclinical AD, prior to their undergoing amyloid PET scanning.

VI CHARACTERISTICS OF COGNITIVE TESTS FOR USE IN EARLY-PHASE OR EXPERIMENTAL MEDICINE

The psychopharmacological effects discussed here are in general based on the results of studies where drugs are given for relatively short periods of time (i.e., once or over days) in either healthy adults or patients with or at risk of CNS disease. In each of these studies, the ability to make decisions about the disruptive or facilitatory CNS effects of those drugs has depended on the repeated application of the cognitive tests with inferences based on the extent to which performance on these has been changed in association with the administration of the drug. Consequently, the cognitive tests used in early-phase studies must be appropriate for repeated administration and have metric properties that make them optimal for detecting both improvement and deterioration in cognition in the study sample of interest. Translational models are also improved if the same cognitive tests can be used in studies of healthy adults and in patients with CNS diseases and disorders such as CIAS or AD, for example, as shown in Figs. 15.1 and 15.2.

These characteristics are evident in the figures. First, the tests themselves have alternative forms so that novel challenges of equivalent difficulty can be presented at each reassessment. If this is the case, then individuals who do the test repeatedly should not show any performance improvements. For the GMLT, this stability of performance in the absence of any drug treatment, even over very short retest intervals (i.e., hours), is evident in Figs. 15.2 and 15.3. This absence of practice effects is also shown for the Cogstate detection task in the trials of the different dopamine agonists in the studies of healthy adults and in schizophrenia summarized in Fig. 15.1. These data indicate that in the absence of true CNS change, performance on the cognitive tests themselves does not change. Of course, in these studies the period over which reassessment occurred in these studies is too brief (i.e., hours or days) for there to have been any true maturational, or even degenerative, CNS change to have occurred. Importantly the absence of practice effects means that true CNS improvement (i.e., as in Fig. 15.1C) or disruption (Fig. 15.2) does not need to be separated from the unintended consequences of the repeated administration of the test. Put in more basic terms, it is difficult to determine if the phenomena being measured are changing and if the measurement device itself can also change with reapplication.

A second characteristic optimal for the cognitive tests utilized in early-phase studies is that they have psychopharmacological sensitivity. This characteristic is difficult to design but becomes obvious with the use of the cognitive test in experiments where they are exposed to the effects of different CNS drugs. The studies presented here show how the detection, GMLT, and CPAL tasks each possess strong psychopharmacological sensitivity, so that both the presence and the magnitude of change, or indeed the absence of any change, in response to treatment with an experimental drug provide a sound basis for decision-making.

Finally, the tests applied should reflect the cognitive domains relevant to the mechanism of action of the drug and ideally to the clinical manifestations of cognitive dysfunction in the disease or disorder for which the experimental drug is intended. Each of the tests discussed here has been shown to be sensitive to the presence of disorders such as schizophrenia and depression and diseases such as Alzheimer's disease and Parkinson's disease.

In summary, cognitive tests can be applied in early-phase medicine studies to guide decisions about the nature and magnitude of CNS effects of drugs in development. Such decisions can be about the dosing at which drugs may disrupt or facilitate cognition; the time course of such CNS effects; the group in which it occurs; the extent to which effects seen, or not observed, in healthy adults are observed in patient groups; or the interaction of drugs with other risk markers of biomarkers of disease. For cognitive tests to be used optimally to guide these different types of decision, it is important that main experimental questions are clear from the outset of the study.

References

Chen, X., Jacobs, G., De Kam, M., et al. (2014). The central nervous system effects of the partial GABA-Aα2,3-selective receptor modulator AZD7325 in comparison with lorazepam in healthy males. *British Journal of Clinical Pharmacology*, 78(6), 1298–1314. https://doi.org/10.1111/bcp.12413.

Cho, W., Maruff, P., Connell, J., et al. (2011). Additive effects of a cholinesterase inhibitor and a histamine inverse agonist on scopolamine deficits in humans. *Psychopharmacology*. 218(3). https://doi.org/10.1007/s00213-011-2344-y.

Falleti, M. G., Maruff, P., Collie, A., Darby, D. G., & McStephen, M. (2003). Qualitative similarities in cognitive impairment associated with 24 h of sustained wakefulness and a blood alcohol concentration of 0.05%. *Journal of Sleep Research*, 12(4), 265–274. doi:363 [pii].

Fredrickson, A., Snyder, P. J., Cromer, J., Thomas, E., Lewis, M., & Maruff, P. (2008). The use of effect sizes to characterize the nature of cognitive change in psychopharmacological studies: an example with scopolamine. *Human Psychopharmacology*, 23(5), 425–436. https://doi.org/10.1002/hup.942

Girgis, R., Van Snellenberg, J., Kegeles, L., et al. (2014). A proof of concept, randomized clinical trial of DAR-0100a, a dopamine-1 receptor agonist, for cognitive enhancement in schizophrenia. *Neuropsychopharmacology*, 39, S370. https://doi.org/10.1038/npp.2014.281.

Goldman, R., Loebel, A., Cucchiaro, J., Deng, L., & Findling, R. L. (2017). Efficacy and safety of lurasidone in adolescents with schizophrenia: a 6-week, randomized placebo-controlled study. *Journal of Child and Adolescent Psychopharmacology*, 27, 516–525. https://doi.org/10.1089/cap.2016.0189.

Grove, R., Harrington, C., Mahler, A., et al. (2014). A randomized, double-blind, placebo-controlled, 16-week study of the H₃ receptor antagonist, GSK239512 as a monotherapy in subjects with mild-to-moderate Alzheimer's disease. *Current Alzheimer Research, 11*(1), 47–58. https://doi.org/10.2174/1567205010666131212110148.

Harel, B. T., Pietrzak, R. H., & Snyder, P. J. (2013). Effect of cholinergic neurotransmission modulation on visual spatial paired associate learning in healthy human adults. *Psychopharmacology, 228*, 673–683. https://doi.org/10.1007/s00213-013-3072-2.

Hilt, D., Safirstein, B., Hassman, D., et al. (2009a). EVP-6124—safety, tolerability and cognitive effects of a novel alpha-7 nicotinic receptor agonist in Alzheimer's disease patients on stable donepezil or rivastigmine therapy. *Alzheimer's & Dementia: The Journal of the Alzheimer's Association, 1*, e32.

Hilt, D., Safirstein, B., Hassman, D., et al. (2009b). EVP-6124—safety, tolerability and cognitive effects of a novel α7 nicotinic receptor agonist in Alzheimer's disease patients on stable donepezil or rivastigmine therapy. *Alzheimer's & Dementia: The Journal of the Alzheimer's Association, 5*, e32, 6124.

Hindmarch, I. (2009). Cognitive toxicity of pharmacotherapeutic agents used in social anxiety disorder. *International Journal of Clinical Practice, 63*(7), 1085–1094. https://doi.org/10.1111/j.1742-1241.2009.02085.x.

Lieberman, J. A., Dunbar, G., Segreti, A. C., et al. (2013). A randomized exploratory trial of an alpha-7 nicotinic receptor agonist (TC-5619) for cognitive enhancement in schizophrenia. *Neuropsychopharmacology, 38*(6), 968–975. https://doi.org/10.1038/npp.2012.259.

Jacques, F. H., Harel, B. T., Schembri, A. J., et al. (2016). Cognitive evolution in natalizumab-treated multiple sclerosis patients. *Multiple Sclerosis Journal—Experimental, Translational and Clinical, 2*, 0–5. https://doi.org/10.1177/2055217316657116.

Jarskog, L. F., Lowy, M. T., Grove, R. A., et al. (2015). A phase II study of a histamine H 3 receptor antagonist GSK239512 for cognitive impairment in stable schizophrenia subjects on antipsychotic therapy. *Schizophrenia Research, 164*, 136–142. https://doi.org/10.1016/j.schres.2015.01.041.

Kay, G. G., Maruff, P., Scholfield, D., et al. (2011). Evaluation of cognitive function in healthy older adults treated with fesoterodine. *Postgraduate Medicine, 124*(3), 7–15.

Kay, G. G., Maruff, P., Scholfield, D., et al. (2012). Evaluation of cognitive function in healthy older subjects treated with fesoterodine. *Postgraduate Medicine, 124*(3), 7–15.

Knopman, D. S., Haeberlein, S. B., Carrillo, M. C., et al. (2018). The National Institute on Aging and the Alzheimer's Association Research Framework for Alzheimer's disease: perspectives from the research roundtable. *Alzheimer's & Dementia, 14*(4), 563–575. https://doi.org/10.1016/j.jalz.2018.03.002.

Lim, Y. Y., Maruff, P., Schindler, R., et al. (2015). Disruption of cholinergic neurotransmission exacerbates Aβ -related cognitive impairment in preclinical Alzheimer's disease. *Neurobiology of Aging, 36*(10), 2709–2715. https://doi.org/10.1016/j.neurobiolaging.2015.07.009.

Martin, P., Dirks, B., Gertsik, L., et al. (2014). Safety and pharmacokinetics of lisdexamfetamine dimesylate in adults with clinically stable schizophrenia. *Journal of Clinical Psychopharmacology, 34*(6), 682–689. https://doi.org/10.1097/JCP.0000000000000205.

Maruff, P., Thomas, E., Cysique, L. A. J., et al. (2009). Validity of the CogState brief battery: relationship to standardized tests and sensitivity to cognitive impairment in mild traumatic brain injury, schizophrenia, and AIDS dementia complex. *Archives of Clinical Neuropsychology, 24*(2), 165–178. https://doi.org/10.1093/arclin/acp010.

Maruff, P., Werth, J., Giordani, B., Caveney, A. F., Feltner, D., & Snyder, P. J. (2006). A statistical approach for classifying change in cognitive function in individuals following pharmacologic challenge: an example with alprazolam. *Psychopharmacology, 186*(1), 7–17. https://doi.org/10.1007/s00213-006-0331-5.

McGonigle, P. (2014). Animal models of CNS disorders. *Biochemical Pharmacology, 87*(1), 140–149. https://doi.org/10.1016/j.bcp.2013.06.016.

Morrison, R. L., Fedgchin, M., Singh, J., et al. (2018). Effect of intranasal esketamine on cognitive functioning in healthy participants: a randomized, double-blind, placebo-controlled study. *Psychopharmacology, 235*(4), 1107–1119. https://doi.org/10.1007/s00213-018-4828-5.

Nathan, P. J., Boardley, R., Scott, N., et al. (2013). The safety, tolerability, pharmacokinetics and cognitive effects of GSK239512, a selective histamine H₃ receptor antagonist in patients with mild to moderate Alzheimer's disease: a preliminary investigation. *Current Alzheimer Research, 10*, 240–251.

Nuechterlein, K. H., Green, M. F., et al. (2008). The MATRICS Consensus Cognitive Battery, part 1: test selection, reliability, and validity. *The American Journal of Psychiatry, 165*, 203–213.

Patat, A., Paty, I., & Hindmarch, I. (2001). Pharmacodynamic profile of Zaleplon, a new non-benzodiazepine hypnotic agent. *Human Psychopharmacology, 16*(5), 369–392. https://doi.org/10.1002/hup.310.

Pietrzak, R. H., Olver, J., Norman, T., Piskulic, D., Maruff, P., & Snyder, P. J. (2009). A comparison of the CogState Schizophrenia Battery and the Measurement and Treatment Research to Improve Cognition in Schizophrenia (MATRICS) battery in assessing cognitive impairment in chronic schizophrenia. *Journal of Clinical and Experimental Neuropsychology, 31*, 848–859. https://doi.org/10.1080/13803390802592458.

Pietrzak, R. H., Snyder, P. J., & Maruff, P. (2010). Use of an acute challenge with d-amphetamine to model cognitive improvement in chronic schizophrenia. *Human psychopharmacology, 25*(4), 353–358. https://doi.org/10.1002/hup.1118.

Snyder, P. J., Bednar, M. M., Cromer, J. R., & Maruff, P. (2005). Reversal of scopolamine-induced deficits with a single dose of donepezil, an acetylcholinesterase inhibitor. *Alzheimer's & Dementia, 1*(2), 126–135. https://doi.org/10.1016/j.jalz.2005.09.004.

Snyder, P. J., Lim, Y. Y., Schindler, R., et al. (2014). Microdosing of scopolamine as a "cognitive stress test": rationale and test of a very low dose in an at-risk cohort of older adults. *Alzheimer's & Dementia, 10*(2), 262–267. https://doi.org/10.1016/j.jalz.2014.01.009.

Stocchi, F., Rascol, O., Destee, A., et al. (2013). AFQ056 in Parkinson patients with levodopa-induced dyskinesia: 13-week, randomized, dose-finding study. *Movement Disorders, 28*(13), 1838–1846. https://doi.org/10.1002/mds.25561.

Villemagne, V. L., Burnham, S., Bourgeat, P., et al. (2013). Amyloid β deposition, neurodegeneration, and cognitive decline in sporadic Alzheimer's disease: a prospective cohort study. *Lancet Neurology, 12*(4), 357–367. https://doi.org/10.1016/S1474-4422(13)70044-9.

Wallace, T. L., Ballard, T. M., Pouzet, B., Riedel, W. J., & Wettstein, J. G. (2011). Drug targets for cognitive enhancement in neuropsychiatric disorders. *Pharmacology, Biochemistry, and Behavior, 99*(2), 130–145. https://doi.org/10.1016/j.pbb.2011.03.022.

Walling, D., Marder, S. R., Kane, J., et al. (2015). Phase 2 trial of an alpha-7 nicotinic receptor agonist (TC-5619) in negative and cognitive symptoms of schizophrenia. *Schizophrenia Bulletin*, 1–9. https://doi.org/10.1093/schbul/sbv072.

16

Digital Biomarkers in Clinical Drug Development

Amir Kalali*, Sarah Richerson†, Emilia Ouzunova†,
Ryan Westphal‡, Bradley Miller‡

*San Diego, CA, United States †Neuroscience Center of Excellence, IQVIA, Durham, NC, United States
‡Eli Lilly and Company, Indianapolis, IN, United States

I INTRODUCTION

The term biomarker, or biological marker, generally refers to an indicator of some biological state or medical condition that can be measured objectively, accurately, and repeatedly (Strimbu & Tavel, 2001). The Biomarkers Definitions Working Group (Biomarkers Definitions Working Group, 2001) of the National Institutes of Health established the following definition a biomarker as "a characteristic that is objectively measured and evaluated as an indicator of normal biological processes, pathogenic processes, or pharmacological responses to a therapeutic intervention." A joint venture on chemical safety, the International Programme on Chemical Safety, led by the World Health Organization (WHO) and in coordination with the United Nations and the International Labor Organization, has defined a biomarker as "any substance, structure, or process that can be measured in the body or its products and influence or predict the incidence of outcome or disease" (WHO, 2001). It is also critical to consider the environmental impacts such as chemical contaminants, pollution, and climate alongside the incidence of disease and what consequences these effects may have on the treatment intervention and outcomes. According to the WHO report on the validity of biomarkers in environment risk assessment, the definition of biomarkers should also include "almost any measurement reflecting an interaction between a biological system and a potential hazard, which may be chemical, physical, or biological. The measured response may be functional and physiological, biochemical at the cellular level, or a molecular interaction" (WHO, 1993). There are many potential roles for biomarkers within pharmaceutical discovery and development (IOM, 2010).

Biomarkers primarily play a crucial role in understanding the mechanism of action of a drug, in identifying efficacy or safety signals at an early stage of development, and in identifying patients likely to respond to treatment (Jenkins et al., 2011).

In defining what a digital biomarker is and applying the same principles as those of standard biological biomarkers, those measurable indicators would be patient- or participant-generated data that are collected through connected digital tools that can be quantitatively used to explain or predict pathophysiological, behavioral, functional, and other health measures and outcomes. Connected digital tools are those such as digital medical devices, wearable devices, sensors, and mobile technologies. We are beginning to see the potential of digital biomarkers being used as primary outcome measures in clinical research (Redfield et al., 2015).

More recently a digital biomarker has been defined by Wang, Azad, and Rajan (2016) as "consumer-generated physiological and behavioral measures collected through connected digital tools that can be used to explain, influence, and/or predict health-related outcomes. This excludes patient-reported measures (e.g., survey data), genetic information, and data collected through traditional medical devices and equipment." Digital biomarkers are essentially the data output of digital health technologies such as mobile apps, connected devices, and wearables and may prove to be more useful than conventional molecular biomarkers (Wang et al., 2016).

Wireless technologies have revolutionized the way we communicate. Eric Topol said "When healthcare executives understand biology from genomics and physiology through wireless phenotyping, this will set up a unique

Drug development industry?
+ software engineers?
↓　SOCS

ability to understand each individual at the granular level, and enable prevention and optimal management in a precise and efficient manner" (Buell, 2011). Individualized treatments through technological advancements will become the new norm. Ideally a digital biomarker that can be passively and objectively measured will serve as a clinical trial outcome measure in place of a more subjective "soft" outcome measure such as a symptom rating scale.

Biomarkers to predict human behavior and psychiatric disorders have been sought for decades. Despite great efforts the etiology of many psychiatric illnesses remains unclear, and it has been difficult to categorize such disorders (Singh & Rose, 2009). It is hoped that biomarker approaches will advance the field (Laughren, 2010). This has been emphasized by many in the research community (e.g., Filiou & Turck, 2011; Li et al., 2012; Nutt & Attridge, 2014). With the advent of digital technologies, especially if they can help improve psychiatric diagnosis and predictive analytics, we are at the dawn of a new age. We might be able to predict who is at risk for a disorder and the likelihood of response to specific treatments.

II CLASSIFICATION OF BIOMARKERS AND DIGITAL BIOMARKERS

Nondigital biomarkers have been classified based on many different parameters such as imaging biomarkers (CT, PET, and MRI) or molecular- or DNA-based biomarkers (e.g., biological measurement samples such as blood samples, biopsy, and serum). Standard biomarkers have also been classified based on diagnosis, disease progression, disease response, and those that measure the clinical response to a therapeutic intervention. Digital biomarkers are those that frequently and quantitatively measure biological events or patient function through the use of a technological sensor or device. To classify digital biomarkers further, it is central to know what is actually being measured and the actionable clinical insight being derived from that measurement (Wang et al., 2016). One example of this is that of CardioMEMS, where continuous heart rate is monitored from a remote wireless device implanted much like that of a pacemaker to reduce risk of sudden death from heart failure (Koprowski, 2014). Classification of digital biomarkers can be described by the type of data collected (e.g., consumer-generated physiological or behavioral measures), the collection method (connected digital tools, medical devices, equipment, and assays), and the nature of the biomarker itself (e.g., wellness, disease, or pharmaceutical agent) (Wang et al., 2016).

In clinical drug development the interest in identifying, evaluating, and qualifying innovative biomarker technologies for use in establishing therapeutically relevant disease understanding is growing. In continuing

to further developments with the advance of digital biomarkers, it is essential to obtain or build fit-for-purpose apps/devices to inform disease understanding in clinical trial settings. The drug development industry must collaborate to further the path of improvement and reiteration in demonstrating the value of digital biomarkers. Indeed, by providing access to its trials, the drug development industry can offer a path to demonstration of value and validation of use that is otherwise inaccessible to most if not all early-stage ventures.

III TYPES OF DIGITAL BIOMARKERS

By 2020 the average person will have four personal connected devices (Mobile World Live, 2014). We are moving from a period of episodic care to continuous care. It is therefore critical that we embrace these new technologies to collect data to help us understand the fundamental drivers of disease and the health and economic burdens. It will be equally critical to understand how to process and analyze these data and transform them into actionable insights. There are several types of digital biomarkers that are collected by a variety of digital heath devices and sensors. Examples of different data types that can be monitored by a variety of collection measures include medication adherence, activity, heart rate, blood pressure, sleep, glucose, pulse oximetry, weight, breathing, temperature, nutrition, spirometry, and smoking (Validic, 2016).

IV WEARABLES

Wearable technology, wearable devices, wearables, fashion electronics, and fabric sensors all refer to electronic technologies or computers that are implanted into items of clothing and accessories that can be easily and comfortably worn on the body (Slideshare, 2014; WearableDevices.com, 2013). These technologies are gaining an increasingly prominent role in many industries and are evolving at a rapid rate in terms of both popularity, functionality, size, and real-time applications (Marr, 2016). These technologies are all good examples of the "Internet of Things" (IoT), since they are part of the network of physical devices or "things" embedded with electronics, software, sensors, and connectivity to enable objects to exchange data with a manufacturer, operator, and/or other connected devices, without requiring human intervention (Brown, 2016; ITU, 2015). Wearable devices have the potential to allow physicians to remotely supervise connected real-time patient information.

As health-care delivery shifts to more patient-centered care, a key consideration in designing the clinical trial of the future is the ability to make the experience integrated into the patient's everyday life. Currently the clinical trials process can be inconvenient and burdensome for participants, both in terms of time spent traveling to and from the trial location and the time required to record physiological and drug reactions. Wearable devices can reduce the number of times patients need to go to a clinic and can provide a better, more complete picture of physiological data needed to measure a drug's impact. Clinical trials are often criticized for not being sufficiently patient-centric. Innovating through the use of wearable devices can help address this challenge by streamlining the process and creating greater patient engagement. This could not only lead to faster recruitment of patients and completion of the clinical trial but also improve the diversity of participants in terms of gender, ethnicity, geography, and economic status. We might also give researchers a better picture of individual variation, compared with current methods (Powerfulpatients, 2015).

It is well recognized both within the health-care industry and among technology companies that the development of wearable medical devices and body sensor technologies for applications in telemedicine and mobile health are imperative and have the opportunity to considerably transform the lives of people who endure chronic illness (Glatter, 2014).

The current popularity of wearables in the consumer space is accelerating with one out of five American adults having a wearable device according to the 2014 Price-WaterhouseCoopers wearable future report (PwC US, 2014), and these devices are being used to monitor everything from fitness to location, tracking, health care, and security (Zalud, 2015). In health care, wearables have been used since the 1950s with the invention of the first wearable, the hearing aid, worn behind the ear. Since that time, there have been many advances in this field, and wearable products are now being used to detect, record, and transmit various physiological signals and to manage many chronic diseases (Dittmar, Axisa, Delhomme, & Gehin, 2004; Mazzoldi, de Rossi, Lorussi, Scilingo, & Paradiso, 2002; Pacelli, Caldani, & Paradiso, 2006).

According to the ClinicalTrials.gov website, 104 studies were completed by June 2016 using consumer grade wearable devices. Consumer wearables started to accelerate in 2013 with activity trackers such as Fitbit and Jawbone. Other recently launched wearables include the smart watches such as Apple Watch, Moto 360, and Pebble Classic. The Apple Watch consumers with diabetes are able to use a Dexcom glucose monitor together with an app to monitor their glucose levels. The app allows the user to view their own glucose level data and another app that can share that data with their physician as needed (Whitney, 2015). Another example is that of the wrist worn Parkinson's KinetiGraph (PKG) Data logger, which was developed by Global Kinetics and tracks Parkinson's movement information that

is then translated into objective graphs that the physician can use to help determine how well the patient is responding to therapy and medication.

Wearables have advanced the field in disorders such as sleep apnea and are also showing promise for chronic conditions such as diabetes, Alzheimer's disease, dementia, neuropathic pain, depression, and macular degeneration (Bizjournals.com, 2014). Heart rate monitors such as the Qardio are detecting for arrhythmia, and smart biosensors such as TempDrop are reading body temperature making fertility tracking simple and convenient. The fertility app Natural Cycles appears to show an improved effectiveness on traditional fertility awareness-based methods by tracking ovulation and period cycles and combining that data with a set algorithm to pinpoint fertility and prevent pregnancy (Berglund Scherwitzl, Gemzell Danielsson, Sellberg, & Scherwitzl, 2016). Smartwatches and activity trackers such as actigraphy devices are also being used to track physical activity and sleep.

Motion biosensors wearable devices have been used in dermatology studies. These provide objective nocturnal scratching event measurements to evaluate impact of a particular therapy or medication pertaining to atopic dermatitis (Actigraphy, n.d.).

Identification of novel motion end points based on objective analysis may be more sensitive to the effects of the treatment, thereby providing new insights. These novel endpoints are often exploratory at first, but their value may grow as their physiological meaning becomes well established with additional studies (Slideshare, 2014). While actigraphy data are usually collected in studies of sleep and sleep disorders, there are many more disease states that may benefit from such objective analysis—such as chronic obstructive pulmonary disease (COPD), Alzheimer's disease, rheumatoid arthritis, Parkinson's disease, fibromyalgia, and bipolar disorder (Actigraphy, n.d.).

The advantages to wearables is clear with many experts agreeing that the amplified connectivity from the growth of the IoT will result in the next revolution in digital technology, and significant progress will be seen between now and 2025 (Anderson & Rainie, 2014). These devices bring many current advantages including simplicity, ease of use, and improved safety by having access to real-time information. Future advances could be even more dramatic (Revesencio, 2015), for example, translating currently available miniaturized sensing technologies such as pedometers, altimeters, and accelerometers into wearable formats (Papagrigoriou, 2016). The next generation of wearables may in fact include ingestibles, for example, with the digital pill; invisibles such as skin on mini sensors; or embeddables/minichips that are inserted into muscles, skin, or nerves (Proteus, n.d.). These sensors have the capability to capture vital signs from the skin and then be either absorbed by the body or simply peeled off (Jervis, 2016). The biggest value will be the ability to transform the data into information that enables better and more actionable insights. The data generated from these devices will allow researchers to conduct more robust clinical trials and support more objective outcome-driven research.

One clear challenge to the use of wearables to generate data and be used as digital biomarkers is the issue of patient adherence. There is evidence that adherence to wearing the current generation of wearables is not high, and in the future the more unobtrusive the "wearables" become the better for patient adherence (Why nobody's wearing wearables, n.d.).

V MOBILE APPLICATIONS

Mobile technologies can be used as diagnostic support tools. These technologies are readily accessible and can provide continuous evaluation through multidimensional evaluations. Across all therapeutic areas, researchers are leveraging these smartphone application technologies to reach broader populations of people at scale. Smartphone supported research has helped detect and monitor the symptoms of patients with Parkinson's disease by capturing symptoms using a diagnostic test, which measured voice, gait, posture, finger tapping, and response time (Arora et al., 2015).

There are also a growing number of mobile phone apps available to support people in taking their medications and to improve medication adherence. One such example is that of AiCure where their clinical validated artificial intelligence platform visually confirms medication ingestion using facial recognition and motion-sensing technology to help eliminate inappropriate subjects in clinical trials. Electronic monitoring of medication adherence information is important in clinical research as access to real-time dosing data can provide more accurate pharmacokinetic, pharmacodynamic, and efficacy analyses (Agot et al., 2015; Fossler, 2015). It is important to have information on factual adherence to help eliminate unnecessary miscalculations in safety signals and effect sizes (Shiovitz et al., 2016). Other companies such as Proteus Digital Health, eTect, and Xhale (Proteus, n.d.; eTect, n.d.; Xhale, n.d.) are also confirming medical ingestion using ingestible sensors embedded into each dose. The sensors are activated on ingestion and transmit a signal to a patch worn by the patient. Using medication adherence digital biomarkers such as this does come with its own set of implementation challenges such as acceptability, ease of use, scalability, privacy, and cost, but they will help in ensuring that changes in health outcomes can be directly attributed to treatment and not an inflated, unknown, or inaccurate interpretation.

Clinical studies will be required to validate the use of digital biomarkers. It is yet unknown if the data coming from these digital biomarker devices will truly bring

about change to management of health care and chronic disease or if they will have meaningful impact on signal detection over nondigital biomarkers (Wang et al., 2016). The process of collecting the data is relatively straightforward; interpreting it and connecting it to disease outcomes will be more challenging. Companies must embrace a spirit of data sharing and transparency to effectively innovate and deliver products that have real medical value.

VI VOCAL BIOMARKERS

Not long ago, scientists discovered vocal features with distinctive characteristics. Researchers labeled these features "vocal biomarkers." These can serve as a diagnostic tool (Medicalfuturist, 2017).

Computer speech analysis that focuses on the features of our speech, such as how many pauses we take, how we articulate our words, and the quality of our voice have the potential to reveal important clues about our emotional and psychological state. Speech and language involve interactions between many areas of the brain.

Several companies are already attempting to utilize advances in computerized speech analysis to help identify mental health conditions. Changes in speech have been described as being a defining characteristic for many mental disorders, including schizophrenia, depression, autism, and bipolar disorder (Psychscenehub, 2016).

VII COMPUTERIZED SPEECH AND LANGUAGE ANALYSIS

Cogito, a Boston company, has developed a smartphone app that can decipher emotions in the human voice, an early warning system that could bring timely help to those with mental illness (Betaboston, n.d.).

VIII SENSORED PILL BOTTLES

Smart wireless pill bottles can help identify and improve medication adherence. Medication vial cap devices are currently being used by patients in the pharmaceutical and research environments to collect and store all adherence data and have become common practice in general medicine (Byerly et al., 2005; Byerly, Nakonezny, & Lescouflair, 2007; Diaz, Neuse, Sullivan, Pearsall, & Woods, 2004; Nakonezny, Byerly, & Rush, 2008; Osterberg & Blaschke, 2005). The data are also readily accessible to physicians in real time. Built-in systems automatically monitor the information coming from these connected devices and populate the data onto dashboards that can then be visualized by physicians. If doses are missed, patients can receive customizable audio or visual alerts, and early interventions can take place using automated phone calls and texts. A newly developed digital medicine system (DMS) offers an innovative approach to objectively report actual medication adherence. This technology incorporates a sensor embedded in an ingestible tablet, a wearable sensor, and a cloud-based computing system (Psychscenehub, 2017).

IX WEARABLE BIOSENSORS AND BIOELECTRIC TATTOOS

Flexible wearable sensors can be used in measuring motor-related symptoms such as balance, gait, and spasticity symptoms. Designing patient friendly sensors than can be easily placed on the hips, knees, and legs will allow the development of more digital biomarker systems that can continuously monitor patient activity and improve patient care (Papagrigoriou, 2016). The BioStamp Research Connect system marks the first in bioelectric tattoos. The system consists of ultrathin, flexible wearable sensors that easily conform to the human body and report to a cloud-based software system to offer users access to secure real-time essential physiological data (Mc10 BiostampRC, n.d.). Conducting clinical studies with wearable or remote biosensors and mobile health platforms will help us get a more detailed and real-world understanding of patient's physiology, behavior, and treatment response.

X SOCIAL MEDIA

We now have the ability to collect data utilizing social media. Content placed on social media sites is subject to behavioral analysis (PLoS, n.d.). Longitudinal device-related data can be combined with the digital footprint that many consumers are willing to share on social media sites to create a more complete clinical picture of the patient and help with health-care decisions.

XI GAMIFICATION

Gamification was first described on Google in 2010 and has quickly gained interest since that time, as a way of improving patient engagement. There are seven key elements behind gamification: status, milestones, competition, rankings, social connectedness, immersion reality, and personalization. As consumers become more engaged with their own health, they may take greater responsibility for managing their condition.

Companies such as Akili Labs are offering gamified solutions that sensitively measure neural function and intervene, offering a closed loop system of diagnosis and treatment. By using their Project, EVO platform, which is pending FDA approval currently, they are potentially able to improve cognition and disease symptoms using exclusively at home digital interactions. These games could have the potential to improve symptoms of inattention, working memory, and executive function and could have benefit in the areas of Alzheimer's disease, autism, depression, traumatic brain injury, and many other neurological diseases (FierceBiotech, 2016).

XII VIRTUAL REALITY

Although still early in development, several virtual reality and simulation "challenge tests" have been tested for use in clinical trials (Bordnick et al., 2008; Carter, Bordnick, Traylor, Day, & Paris, 2008). This approach promises to bring the empirical rigor of neurocognitive testing, focused on performance, to critical areas of focus, such as functioning in social, occupational, and everyday life virtual environments. In psychiatry, there is potential to collect rich data using virtual reality technology, for example, assessing symptoms, establishing symptom correlates, identifying predictive variables, causal factors, determining environmental predictors, and developing treatments to meet the individuals need (Medicalfuturist, 2016; Vahabzadeh, 2016).

XIII MACHINE LEARNING

Machine learning is a type of artificial intelligence (AI) that provides computers with the ability to learn without being explicitly programed. Machine learning focuses on the development of computer programs that can change when exposed to new data.

One study has shown that machine learning is up to 93% accurate in correctly classifying a suicidal person and 85% accurate in identifying a person who is suicidal and has a mental illness but is not suicidal or neither. These results provide evidence for using advanced technology as a decision-support tool to help clinicians and caregivers identify and prevent suicidal behavior, says John Pestian, PhD, professor in the Divisions of Biomedical Informatics and Psychiatry at Cincinnati Children's Hospital Medical Center and the study's lead author (Neurosciencenews, n.d.).

XIV VALUE OF DIGITAL BIOMARKERS

The value of a biomarker must be assessed before it can be used in clinical drug development. There are several types of biomarkers: diagnostic biomarkers that determine if a disease already exists, prognostic biomarkers that measure disease outcome independent of the type of treatment, and predictive biomarkers that predict the likelihood of response to a particular treatment or a class of treatments. Drug-related biomarkers indicate drug effectiveness in specific patients and how the patient's body processes it.

In 2015, in recognition of this potential, President Obama announced a nationwide research effort, the Precision Medicine Initiative (Precision Medicine Initiative, n.d.). That same year, 28% of drugs approved by the FDA were considered precision therapies, and an October 2016 report predicted the global precision medicine market will reach nearly $113 billion by 2025 (Catalyst, 2016). By definition, precision drugs target only a subset of a particular patient population. Diagnostic tests to identify responsive patients must be created alongside a new targeted therapy and appropriate biomarkers.

New digital solutions are likely to require collaboration among diverse companies (see "Collaborative Opportunities" section of this chapter).

The adoption of digital biomarkers creates new opportunities in clinical research. There are various processes throughout the clinical trial development cycle that can be potentially impacted by inclusion of technologies using digital biomarkers such as the following:

- Centralized recruitment: e-recruiting, social media, online patient support groups, crowdsourcing
- Patient retention: via mobile apps, patient communities
- Medication adherence: smart pills, smart bottles, real-time tracking via AI on mobile apps
- Virtual assessments: neuroimaging, fall detection, gamified solutions
- Cognitive monitoring: virtual reality, mobile apps measuring all functions of cognitive domain
- Patient engagement: web-based trials, telemedicine
- Biometric monitoring: measuring vitals, ECGs, temperature, physical activity, body posture, gait with mobile apps, patches, devices
- Remote data collection and real-time monitoring and tracking: eSource solutions

Having access to real-time data can help provide more accurate and objective data (e.g., medication adherence) and support early decision-making in clinical drug development (Validic, 2016). It can potentially lead to advances in many areas throughout the clinical trial process helping get drugs to market better, faster, and cheaper. Potential areas for improvement with the adoption of digital biomarkers in clinical trial design are listed in the succeeding text:

- Enriched populations selected
- More accurate and real-time quality data through targeted sensitive data collection measures
- Ability to track and monitor disease progression
- Identify and quantify treatment effects
- Monitoring of an illness and correlating symptoms

- Greater patient experience via education and communication—promotes self-management of disease using health apps, personalized treatment
- Improve patient adherence, reduce professional patients, less missing data and out of window events
- Increased patient–clinician interaction—eVisits
- Increased patient/caregiver engagement and support—work more as a team
- Overall patient wellness
- Enhanced collaboration among clinical trial sites, staff, and patients
- Reduction in study visits, enable cost-effectiveness
- Generate actionable insights through analytics
- Potential increase in study outcomes, demonstration of efficacy, and real-world value

Could there be a digital signature for mental health? Some examples of quantitative mental status examinations that could be used with digital biomarker technologies are listed in the succeeding text:

Measure	Technology	Commercial example
Appearance	Image matching	Google Cloud Vision API
Behavior	ML-based behavioral analysis	Android Wear + Google Fit
Motor	Motion-tracking software	Xbox Kinect
Affect	Face tracking	Microsoft Cognitive Cloud
Speech	Voice analysis	Cogito
Thought patterns	Natural language processing	Google Prediction, Viv.ai
Thought content	Natural language processing	IBM Watson Personality

(Vahabzadeh, 2016).

XV CHALLENGES

It is clear that there will be a lot of value obtained from digital biomarkers, but how will this value measure against burden? Will we share this digital information, how we will create standards for use, how we will design and develop clinical trial protocols using digital biomarker data as key endpoints, how will we navigate the data outputs, how will we know what are the critical data points to use, and how we will transfer these data into our EHRs and clinical trial databases? These are some of the critical questions regarding digital biomarkers that will be addressed as more companies include them in their clinical trial designs.

There are various challenges affecting digital health (including digital biomarkers) today, including regulatory compliance, digital health oversight, and overall unknown risks to the patients. Multiple government entities have interests in digital health and the value in digitally generated data, including the Food and Drug Administration (FDA), Federal Communications Commission (FCC), Federal Trade Commission (FTC), Department of Health and Human Services (DHHS), Center for Medicare & Medicaid Services (CMS), and Drug Enforcement Agency (DEA). Since digital biomarkers encompass a wide variety of technologies for use across broad populations, purposes, and data types, there needs to be a strong digital health oversight process put in place—one that provides scrutiny over design integrity, security, and privacy in much the same way standard clinical trials without digital technologies are required to follow. It is not yet clear who exactly is responsible for data protection and issuing the safeguards needed to secure the data and how the patient data should be handled. There are also risks to the patient to be considered. Perhaps, these digital biomarkers are inaccurate or unreliable due to inadequate software development practices and the lack of clinical input. There is also the chance that the data will be misinterpreted by the patient directly. Patient safety and protection is of utmost importance, but there are risks to loss of patient's personal privacy when individual digital biomarker data are released.

A key issue is determining the relationship between any given measurable biomarker and relevant clinical endpoints. When used as outcomes in clinical trials, biomarkers are considered to be surrogate endpoints; that is, they act as surrogates or substitutes for clinically meaningful end points. But not all biomarkers are surrogate endpoints, nor are they all intended to be. Surrogate endpoints are a small subset of well-characterized biomarkers with well-evaluated clinical relevance. To be considered a surrogate endpoint, there must be solid scientific evidence (e.g., epidemiological, therapeutic, and/or pathophysiological) that a biomarker consistently and accurately predicts a clinical outcome, either a benefit or harm. In this sense a surrogate endpoint is a biomarker that can be trusted to serve as a stand-in for, but not as a replacement of, a clinical endpoint. There are a number of advantages to using biomarkers as surrogate endpoints in trials. Primary clinical endpoints, such as survival, can occur so infrequently that their use in clinical trials can be highly impractical or even unethical. For many diseases, clear clinical endpoints such as survival or recurrence of, for instance, a cardiovascular event may occur only after many years of treatment. Biomarkers can provide researchers interim evidence about the safety and efficacy of such treatments, while more definitive clinical data are collected. In some cases, it may be preferable to use established biomarkers as surrogate endpoints to reduce the risk of harm to subjects: the early data provided by biomarkers can allow researchers the opportunity to stop interventions potentially harmful to subjects before the associated clinical data would be available. In other cases, biomarkers may simply allow researchers to design smaller, more efficient studies, reducing the number of subjects exposed to a given experimental treatment. By shortening the time to approval of new treatments, more efficient trials could speed the overall drug development process, allowing effective treatments to reach their target patient populations sooner, while conserving both material and human resources for other clinical trials (Strimbu & Tavel, 2001).

Another challenge in the digital biomarker arena and its application to the process of drug discovery is the fact that in some therapeutic areas, for example, psychiatry, the current clinical measures to which digital biomarkers are being evaluated and validated against are subjective and suboptimal. Scales for depression, schizophrenia, anxiety, etc. are subjective scales that historically exhibit high variability and placebo response. It is these clinical instruments that the validation of the sophisticated, data-rich, biomarkers is ultimately dependent. The design and development of novel ways to overcome this hurdle will be essential to the ultimate application of digital biomarkers to allow them to fulfill their great potential.

Biomarkers have been approved by the US Food and Drug Administration (FDA) regulation for use as surrogate endpoints (Medicalfuturist, 2017). The FDA allows provisional intervention approval with surrogate marker-defined efficacy but further requires phase IV follow-up studies that prove relevant clinical endpoint correlation exists. Some researchers have suggested that biomarkers are most effective in and best left for use as endpoints in phase I and phase II trials. Their use can help determine what potential treatments are worth the effort and resources of a large, well-powered phase III trial. This use case has the practical benefits that companies directing these trials may choose the degree to which they rely on these data for decision-making. In these early-stage clinical trials, such data would be used principally to inform on patient selection and dose setting. There is, of course, the risk that researchers may inappropriately abandon large-scale research on treatments that, although actually effective at improving clinical outcomes, do not appear to be effective on the basis of early-stage trial biomarker analysis (Food and Drug Administration, n.d.; Strimbu & Tavel, 2001).

This type of false negative has in fact already occurred in at least one case involving a trial of chronic granulomatous disease in children (The International Chronic Granulomatous Disease Cooperative Study Group, 1991). The trial, which was designed to measure both surrogate outcomes and true clinical endpoints, ultimately showed that the treatment, interferon gamma, was effective at reducing mortality, but there was no associated improvement on the surrogate outcome, production of superoxide, which was expected to increase the patients' ability to kill bacteria. The biological process that led to improved clinical outcomes, in other words, was not captured by the biological mechanism proposed and predicted by the researchers. Had the trial been designed to rely first on interim analysis of the biomarkers alone, the true and clinically relevant effects of the treatment might never have been discovered. The lesson here is the same as in the cases of false positives: treatments' effects on biomarkers used as surrogate outcomes do not necessarily predict true clinical outcomes. Without confirmatory clinical endpoint analysis, the overreliance on biomarkers, even ones previously considered validated in particular treatment contexts, presents a serious and persistent risk of producing misleading and, in some cases dangerous, erroneous conclusions.

Studies using biomarkers should always have as ultimate measures clinical outcomes, at least for retrospective analysis of biomarker correlation success. Without continual reevaluation of the relationship between surrogate endpoints and true clinical endpoints, we risk again approving whole classes of drugs that either have no additional benefit or, worse, harm patients.

Other challenges remain as outlined in Table 3, for example, the CDISC standards for how to assimilate and analyze the data collected by digital biomarker technologies have not yet been established. Easy to interpret analytic and visualization tools are needed for clinicians and patients to interpret the potentially vast amount of digital biomarker data that will be collected especially if the use of that data is to understand clinical symptoms and overall behavioral and wellness information (Jervis, 2016).

Identifying and validating clinical and digital biomarkers that meet criteria that can be used as trial outcome measures and that would also be accepted by regulatory authorities for drug approvals are in its infancy (Lesko & Atkinson Jr, 2001). This has had limited success in most indications.

Challenges with digital biomarker use in clinical drug development include the following:

- Data accuracy and standardization—selecting the right endpoint, understanding the critical elements, how to reduce noise and variability
- Data analytic limitations—arising from multiplicity of solutions
- Design and process integration—implementing the digital device into the study design and including the data in the clinical trials platform or system
- Consent—patients', sites', and caregivers' ability and willingness to use the technology
- Privacy regulations—invasion of patient privacy
- Security—patients' vulnerability to personal security breaches

- Adoption—understanding and selecting the right technology
- Regulatory requirements and ethical framework—21CRF Part II compliance, the lack of FDA guidance
- Financial costs—understanding return on investment
- Globalization—accessibility of technologies

XVI COLLABORATIVE OPPORTUNITIES

From this discussion, it is clear that (at least with regard to the pharmaceutical industry) there is a catch-22 paradox with the potential to limit development of this novel class of biomarkers. Specifically the challenges in developing and validating these biomarkers are large, requiring demonstrations of equivalence and/or superiority to currently validated clinical measures, while at the same time, the size and cost of clinical trials (especially late stage trials with highly regulatory burdens on measures and outcomes) act as a pressure against the incorporation and development of novel biomarkers. In this environment, companies developing digital biomarkers may find it difficult to generate the data needed that results in broad application to clinical research and practice.

Based upon these challenges, one approach to advance the digital biomarker field may be through a virtuous cycle of developer-pharmaceutical company collaboration. In this model, developers work closely with drug makers to facilitate testing, optimization, and validation of digital biomarkers to support drug development. The digital biomarker innovators can benefit from the experience of working in clinical trials to understand precisely the clinical endpoints and most useful details of testing frequency and depth desired by pharmaceutical companies working in various medical indications. Pharmaceutical companies, in turn, have the opportunity to invest their resources by prioritizing access to their clinical trials for such tools, enabling their refinement and validation as part of the clinical trial cycle. Through iterative approaches in optimizing the biomarker, the knowledge gap can be addressed to result in the improvement in biomarker performance that benefits both the development company and the pharmaceutical company.

XVII VALIDATION PATHWAY

Measurement validity is the degree to which a digital biomarker indicates what it purports to indicate and it is imperative to validate the methods and technology used for capturing, integrating, and analyzing the longitudinal real-world data to make it more actionable for all stakeholders. It is important to prospectively identify the dataset of interest, how it will be measured and collected and how it will be validated in clinical trials. This can only be achieved if more research is conducted using digital biomarkers data as endpoints. The quality of the data also greatly depends on patients' willingness to capture and communicate these data with clinicians (Jervis, 2016). Comparator studies using digital biomarkers against current gold measurements standards will be needed to assess digital biomarker noninferiority, equivalence, or superiority (Wang et al., 2016).

XVIII CONCLUSION

The development of digital biomarkers is at its earliest stage, and the path is unclear as yet for how digital biomarkers will be integrated in each therapeutic area. However, digital biomarkers clearly have the potential to revolutionize treatment development, delivery, and assessment.

With the mass amounts of data that can be generated, digital biomarkers are ideally situated to feed the artificial intelligence/big data environment that will further transform all aspects of health care.

If we look at the accelerating evolution and adoption of technology in general as an illustration, we can anticipate that the availability and utility of digital biomarkers will revolutionize our approach to capturing, quantitating, and interpreting patient information.

References

Actigraphy, n.d. http://www.actigraphy.respironics.com/.

Agot, K., Taylor, D., Corneli, A. L., Wang, M., Ambia, J., Kashuba, A. D., et al. (2015). Accuracy of self-report and pill-count measures of adherence in the FEM-PrEP clinical trial: implications for future HIV-prevention trials. *AIDS and Behavior*, 19, 743–751.

Anderson, J., & Rainie, L. (2014). *The internet of things will thrive by 2025. May.* http://www.pewinternet.org/2014/05/14/internet-of-things/.

Arora, S., Venkataraman, V., Zhan, A., Donohue, S., Biglan, K. M., Dorsey, E. R., et al. (2015). Detecting and monitoring the symptoms of Parkinson's disease using smartphones: a pilot study. *Parkinsonism & Related Disorders*, 21(6), 650–653.

Berglund Scherwitzl, E., Gemzell Danielsson, K., Sellberg, J. A., & Scherwitzl, R. (2016). Fertility awareness-based mobile application for contraception. *The European Journal of Contraception & Reproductive Health Care*, 21(3), 234–241.

Betaboston n.d. http://www.betaboston.com/news/2016/02/24/cogito-and-mgh-test-voice-app-to-monitor-moods/.

Biomarkers Definitions Working Group (2001). Biomarkers and surrogate endpoints: preferred definitions and conceptual framework. *Clinical Pharmacology and Therapeutics*, 69(3), 89–95.

Bizjournals.com (2014). *Wearables the cure: Five diseases wearables are tackling.* August.

Bordnick, P. S., Traylor, A., Copp, H. L., Graap, K. M., Carter, B., Ferrer, M., et al. (2008). Assessing reactivity to virtual reality alcohol based cues. *Addictive Behaviors*, 33(6), 743–756.

Brown, E. (2016). *Who needs the internet of things?* 13 September 2016. Linux.com. Retrieved 23 October 2016.

Buell, J. M. (2011). *The digital medicine revolution in healthcare.* Reprinted from Healthcare Executive. Jan/Feb https://www.ache.org/abt_ache/JF11_F3reprint.pdf.

Byerly, M., Fisher, R., Whatley, K., Holland, R., Varghese, F., Carmody, T., et al. (2005). A comparison of electronic monitoring vs. clinician rating of antipsychotic adherence in outpatients with schizophrenia. *Psychiatry Research*, 133(2–3), 129–133.

Byerly, M. J., Nakonezny, P. A., & Lescouflair, E. (2007). Antipsychotic medication adherence in schizophrenia. *The Psychiatric Clinics of North America*, 30(3), 437–452.

Carter, B. L., Bordnick, P., Traylor, A., Day, S. X., & Paris, M. (2008). Location and longing: the nicotine craving experience in virtual reality. *Drug & Alcohol Dependence*, 95(1), 73–80.

Catalyst. (2016). http://Catalyst.Phrma.Org/2015-A-Banner-Year-For-Personalized-Medicine.

Diaz, E., Neuse, E., Sullivan, M. C., Pearsall, H. R., & Woods, S. W. (2004). Adherence to conventional and atypical antipsychotics after hospital discharge. *The Journal of Clinical Psychiatry*, 65(3), 354–360.

Dittmar, A., Axisa, F., Delhomme, G., & Gehin, C. (2004). New concepts and technologies in home care and ambulatory monitoring. *Studies in Health Technology and Informatics*, 108, 9–35. Review.

eTect n.d. http://etectrx.com/.

FierceBiotech (2016). *Akili, Pfizer successfully gamify Alzheimer's.* December. http://www.fiercebiotech.com/medical-devices/can-a-videogame-detect-early-alzheimer-s-akili-pfizer-data-points-to-yes.

Filiou, M. D., & Turck, C. W. (2011). General overview: biomarkers in neuroscience research. *International Review of Neurobiology*, 101, 1–17. https://doi.org/10.1016/B978-0-12-387718-5.00001-8. Review.

Food and Drug Administration n.d. Food and Drug Administration Modernization Act of 1997, 21CFR314.

Fossler, M. J. (2015). Patient adherence: clinical pharmacology's embarrassing relative. *Journal of Clinical Pharmacology*, 55(4), 365–367.

Glatter, R. (2014). *Wearable technology and digital healthcare strategies should shift focus to chronic medical illness.* http://www.forbes.com/sites/robertglatter/2014/11/20/wearable-technology-and-digital-healthcare-strategies-should-shift-focus-to-chronic-medical-illness/#3182bf8129a7.

IOM. (2010). *Evaluation of biomarkers and surrogate endpoints in chronic disease.* Washington, DC: National Academies Press. http://www.iom.edu/Reports/2010/Evaluation-of-Biomarkers-and-Surrogate-Endpoints-in-Chronic-Disease.aspx.

ITU. (2015). *Internet of things global standards initiative.* Retrieved 26 June.

Jenkins, M., Flynn, A., Smart, T., Harbron, C., Sabin, T., Ratnayake, J., et al. (2011). A statistician's perspective on biomarkers in drug development. *Pharmaceutical Statistics*, 10, 494–507.

Jervis, S. (2016). *The future will eat itself: Digesting the next generation of wearable tech.* Dec. https://www.theguardian.com/media-network/2016/jan/13/future-eat-digesting-next-generation-wearabletech?utm_content=buffer433cf&utm_medium=social&utm_source=twitter.com&utm_campaign=buffer.

Koprowski, G. J. (2014). *Hospitals trial wireless heart monitor technology.* http://www.foxnews.com/tech/2014/10/31/hospitals-trial-wireless-heart-monitor-tech.html.

Laughren, T. P. (2010). What's next after 50 years of psychiatric drug development: an FDA perspective. *The Journal of Clinical Psychiatry*, 71(9), 1196–1204. https://doi.org/10.4088/JCP.10m06262gry.

Lesko, L. J., & Atkinson, A. J., Jr. (2001). Use of biomarkers and surrogate endpoints in drug development and regulatory decision making: criteria, validation, strategies. *Annual Review of Pharmacology and Toxicology*, 41, 347–366.

Li, B.-S., Zhao, Y.-L., Guo, G., Li, W., Zhu, E.-D., Luo, X., et al. (2012). Plasma microRNAs, miR-223, miR-21 and miR-218, as novel potential biomarkers for gastric cancer detection. *PLoS One*, 7(7), e41629. https://doi.org/10.1371/journal.pone.0041629.

Marr, B. (2016). *15 Noteworthy facts about wearables in 2016*. Forbes.

Mazzoldi, A., de Rossi, D., Lorussi, F., Scilingo, E. P., & Paradiso, R. (2002). Smart textiles for wearable motion capture systems. *AUTEX Research Journal, 2*(4).

Mc10 BiostampRC. n.d. Source: http://www.mc10inc.com/our-products/biostamprc.

Medicalfuturist. (2017). http://medicalfuturist.com/the-most-exciting-medical-technologies-of-2017/.

Medicalfuturist. (2016). http://medicalfuturist.com/5-ways-medical-vr-is-changing-healthcare/.

Mobile World Live. (2014). *Number of devices to hit 4.3 per person by 2020- Mobile World Live report*. http://www.mobileworldlive.com/featured-content/home-banner/connected-devices-to-hit-4-3-per-person-by-2020-report/.

Nakonezny, P. A., Byerly, M. J., & Rush, A. J. (2008). Electronic monitoring of antipsychotic medication adherence in outpatients with schizophrenia or schizoaffective disorder: an empirical evaluation of its reliability and predictive validity. *Psychiatry Research, 157*(1–3), 259–263.

Neurosciencenews n.d. http://neurosciencenews.com/suicide-machine-learning-5448/.

Nutt, D. J., & Attridge, J. (2014). CNS drug development in Europe—past progress and future challenges. *Neurobiology of Disease, 61*, 6–20. https://doi.org/10.1016/j.nbd.2013.05.002.

Osterberg, L., & Blaschke, T. (2005). Adherence to medication. *The New England Journal of Medicine, 353*(5), 487–497.

Pacelli, M., Caldani, L., & Paradiso, R. (2006). Textile piezoresistive sensors for biomechanical variables monitoring. In *vol. 1. Conference proceedings: annual international conference of the IEEE engineering in medicine and biology society* (pp. 5358–5361).

Papagrigoriou, V. (2016). *4 key barriers to the development of digital biomarker monitoring systems*. 01 June. http://blog.cambridgeconsultants.com/medical-technology/4-key-barriers-to-the-development-of-digital-biomarker-monitoring-systems/.

PLoS n.d. http://journals.plos.org/plosone/article?id=10.1371/journal.pone.0169693.

Powerfulpatients (2015). *Can digital wearables help in clinical trials?* https://www.powerfulpatients.org/author/marieennis/page/2.

Precision Medicine Initiative n.d. https://www.nih.gov/research-training/allofus-research-program.

Proteus n.d. Proteus http://www.proteus.com/.

Psychscenehub. (2016). http://psychscenehub.com/psychinsights/computerized-speech-analysis-psychiatry/.

Psychscenehub. (2017). http://psychscenehub.com/psychinsights/digital-medicine-systems-psychiatry/.

PwC US. (2014). *Wearable technology future is ripe for growth—Most notably among millennials*. Says http://www.pwc.com/us/en/press-releases/2014/wearable-technology-future.html.

Redfield, M. M., Anstrom, K. J., Levine, J. A., Koepp, G. A., Borlaug, B. A., Chen, H. H., et al. (2015). Isosorbide mononitrate in heart failure with preserved ejection fraction. *The New England Journal of Medicine, 373*, 2314–2324.

Revesencio, J. (2015). *Exploring the benefits of wearable technology*. http://www.huffingtonpost.com/jonha-revesencio/exploring-the-benefits-of_b_7910662.html.

Shiovitz, T. M., Bain, E. E., DJ, M. C., Skolnick, P., Laughren, T., Hanina, A., et al. (2016). Mitigating the effects of nonadherence in clinical trials. *Journal of Clinical Pharmacology, 56*(9), 1151–1164. https://doi.org/10.1002/jcph.689.

Singh, I., & Rose, N. (2009). Biomarkers in psychiatry. *Nature, 460*(7252), 202–207.

Slideshare (2014). *History of wearables*. https://www.slideshare.net/RohiniVaze/wearable-technology-report.

Strimbu, K., & Tavel, J. A. (2001). What are biomarkers? Biomarkers Definition Working Group Biomarkers and surrogate endpoints: preferred definitions and conceptual framework. *Clinical Pharmacology and Therapeutics, 69*, 89–95.

The International Chronic Granulomatous Disease Cooperative Study Group. (1991). A controlled trial of interferon gamma to prevent infection in chronic granulomatous disease. *The New England Journal of Medicine, 324*, 509–516.

Vahabzadeh, A. (2016). *The future of psychiatry and mental health*. https://www.youtube.com/watch?v=5WdkdhzpmKs.

Validic. (2016). *Insights on digital health technology survey*. http://pages.validic.com/rs/521GHL511/images/Digital_Health_Survey_Results_Pharma_2016.pdf.

Wang, T., Azad, T., & Rajan, R. (2016). *The emerging influence of digital biomarkers on healthcare*. https://rockhealth.com/reports/the-emerging-influence-of-digital-biomarkers-on-healthcare/.

WearableDevices.com (2013). *What is a wearable device?* Retrieved 29 October 2013.

Whitney, L. (2015). *Apple Watch app will track glucose levels for diabetics*. https://www.cnet.com/news/apple-watch-app-will-track-your-glucose-levels.

WHO (1993). *WHO International Programme on Chemical Safety biomarkers and risk assessment: Concepts and principles*. Retrieved from http://www.inchem.org/documents/ehc/ehc/ehc155.htm.

WHO (2001). *WHO International Programme on Chemical Safety biomarkers in risk assessment: Validity and validation*. Retrieved from http://www.inchem.org/documents/ehc/ehc/ehc222.htm.

Why nobody's wearing wearables. n.d. Bloomberg https://www.bloomberg.com/view/articles/2016-03-03/why-nobody-s-wearing-wearables.

Xhale n.d. http://xhale.com/.

Zalud, B. (2015). *The age of wearables is on us*: January (pp. 72–73). SDM.

17

Lessons Learned From Public Private Partnerships and Consortia: The ADNI Paradigm

Enchi Liu

E-Scape Bio, South San Francisco, CA, United States

I INTRODUCTION

Development of effective therapies for Alzheimer's disease (AD) has been an international priority for at least the last 4 decades (G8 Summit Report, 2013; Progress Report, 2015). Several therapies have been approved that provide modest symptomatic benefits but do not modify the underlying neurodegenerative disease processes (Masters et al., 2015). The major effort since then has been directed at developing therapies targeting the key pathological features of AD (Scheltens et al., 2016); specifically, amyloid-β aggregation and toxicity (β-amyloid (Aβ)) that is believed to initiate the cascade of events that lead to the disease (Selkoe & Hardy, 2016). More recently, efforts are underway for therapies targeting tau pathology that reflect downstream neuronal injury and neurodegeneration and that correlate more directly with functional deficits (Scheltens et al., 2016; Zetterberg, 2017).

The search for a disease-modifying therapy for AD—one that slows or halts both the underlying disease pathology and the progression of cognitive decline and functional disability—has so far been unsuccessful (Schneider et al., 2014).

Drug discovery and development for CNS diseases are in general complicated, uncertain, and lengthy, largely reflecting the nature of CNS diseases and their underlying molecular mechanisms. In AD, while Aβ and tau pathology reflect the final common pathway, the upstream etiologies, with the exception of the familial (genetic) forms of AD, are elusive and may be innumerable. Further, AD is insidious and slowly progressive with disease pathology initiating decades before overt clinical symptoms are apparent (Scheltens et al., 2016).

It is still not certain what is the optimal window or stage of disease to treat and how long to treat for to be able to observe a measurable benefit. Phase 2 or 3 clinical trials evaluating a potential disease-modifying therapy for AD are typically 18–24-month treatment protocols with additional time needed for recruitment. Thus a given development program may require up to 15 years from discovery to marketing approval (PHRMA, 2015 Research *Alzheimer's Medicines: Setbacks and Stepping Stones*). The cost for a single development program may be upward of $2.6 billion (2000–10; in constant 2013 dollars) including cost of failures. Only approximately 12% of candidate therapies that reach phase 1 clinical development obtain FDA approval (PHRMA, 2016 Profile Report). In light of this, in the past 5–7 years, several companies have drastically reduced their neuroscience programs or terminated them (http://www.reuters.com/article/us-neuroscience-pharma-idUSTRE71A2E120110211). Yet Alzheimer's disease affects 5.5 million patients in the United States in 2017 (Alzheimer's Disease Fact and Figures, 2017); there will be approximately 64,000 new cases among people aged 65–74, 173,000 new cases among people aged 75–84, and 243,000 new cases among people aged 85 and older (the "oldest old"). Worldwide estimates for the prevalence of dementia are that 46.8 million people worldwide are living with dementia in 2015. This number will almost double every 20 years, to 74.7 million in 2030 and 131.5 million in 2050 (World Alzheimer Report, 2015). Thus it is extremely important to continue the search for clinically meaningful treatments for AD patients; it is a grievous disease with high unmet medical need. Fortunately the recent contractions in neuroscience programs within large companies reflect a new thinking on and moving away from the older models of neuroscience R&D (http://cenblog.org/the-haystack/2012/03/wither-neuroscience-rd-pfizers-ehlers-doesnt-think-so/; http://neomed.ca/wp-content/uploads/2015/06/01_Sci_Transl_Med-2015-Munos-286ps12_EN.pdf).

Likewise, there has been evolving thinking that there could be greater efficiency in utilizing resources in collaborative efforts, both financial and research know-how, compared with single development programs, reducing duplication and accelerating research productivity through these collaborations in a precompetitive forum. One pioneering example of this type of collaboration is Alzheimer's Disease Neuroimaging Initiative (ADNI), launched in 2004, with public-private funding, which continues today. This chapter reviews the genesis, mission, achievements to date, and influence of ADNI on other collaborative efforts.

II ADNI

Alzheimer's Disease Neuroimaging Initiative (ADNI; NCT00106899) was launched in 2004 with the overarching goal to improve the speed and success rate of clinical trials of novel Alzheimer's disease (AD) therapeutics by accelerating the understanding and validation of biomarkers. One key innovation was the framework of collaboration for the 5-year research project among researchers from academia, industry, governmental agencies, and nonprofit organizations through a public-private partnership. Public funding was provided by the National Institute of Aging (NIA), and private funding, provided by pharmaceutical companies and a nonprofit foundation (Alzheimer's Disease Association), is facilitated and managed through the Foundation for the National Institutes of Health (FNIH), an independent, not-for-profit organization authorized by the US Congress in 1990 to support the mission of the NIH through the creation and facilitation of public-private partnerships. Another important innovation was open access to all data collected, without embargo, deposited in the database housed at the Laboratory of Neuro Imaging (LONI) (http://www.loni.usc.edu/). Over 1000 research articles based on analyses of ADNI data have been published (Weiner et al., 2013; Weiner, Veitch, Aisen, et al., 2015; Weiner, Veitch, Hayes, et al., 2015; Weiner et al., 2017). The ADNI project was renewed in 2010 with the 2-year NIH-funded Grand Opportunity (ADNI-GO) grant (NCT01078636), a 5-year grant renewal in 2011 that funded ADNI-2 (NCT01231971), and most recently with another 5-year grant renewal in 2016 to initiate ADNI-3 (NCT0285403).

At the time ADNI-1 was formed and continues to a large degree today, patient functioning and cognition are considered "gold standard" measures to determine disease state and effectiveness of a therapy. However, brain functioning may be affected by factors other than AD pathology, and evaluation of any therapy as potentially disease modifying using cognitive and clinical instruments may not provide sufficient power and may not be achievable within the time frame and constraints (e.g., sample size) of clinical trials. Use of biomarkers of AD pathology and pathophysiology measured by imaging or in cerebrospinal fluid could provide more precise measures of disease progression and the impact of therapy. Thus biomarkers could help reduce the size and duration of clinical trials that in turn would reduce the cost and time to develop an effective therapy.

The ADNI-1 protocol is an observational study comprising cross-sectional and longitudinal biomarker and clinical data collection across the disease continuum, including cognitively normal (CN), mild cognitive impairment (MCI), and Alzheimer's disease (AD) patients (Mueller et al., 2005). In contrast to small cohort studies led by one or two academic centers, ADNI-1 enrolled 800 participants by harnessing a network of ~57 US and Canadian clinical sites through Alzheimer's Disease Cooperative Study (ADCS), a research consortium funded by the NIA that conducts multicenter clinical trials. A core focus of the project has been the validation and standardization of biomarker acquisition and analysis methods for application to multicenter clinical trials. In 2010 the ADNI-GO grant enabled enrollment of a new cohort of early MCI subjects and optimization of biomarker methods (http://www.adcs.org/studies/imagineadni.aspx). The ADNI-2 renewal in 2011 allowed for continued data collection from the original ADNI-1 and ADNI-GO study participants and enrolment of an additional ~550 participants, including those in the earlier stages of the disease and predementia individuals, and with an added goal of using biomarkers to detect the earliest stages of AD pathology (https://www.nia.nih.gov/alzheimers/clinical-trials/alzheimers-disease-neuroimaging-initiative-2-adni2). The ADNI-3 renewal in 2016 allowed for the enrolment of additional elderly cognitively normal, MCI and early AD participants, annual tau PET imaging, computerized online cognitive testing, and improved CSF bioanalysis methods (https://www.nia.nih.gov/alzheimers/clinical-trials/alzheimers-disease-neuroimaging-initiative-3-adni3).

The ADNI structure is composed of eight cores under the umbrella of the Administrative Core, directed by Dr. Weiner, the principal investigator. A steering committee governs ADNI and consists of Dr. Weiner, all core leaders, all site investigators, representatives from the NIH and FDA, and representatives from each of the contributing companies/organizations as observers only. The day-to-day decisions are made by the ADNI Executive Committee, which is composed of the ADNI principal investigator, core leaders, and a representative from NIA and the current, past, and future chairs of the Private Partners Scientific Board (PPSB) and a representative from the FNIH as observers.

The eight cores are responsible for specific aspects of the ADNI protocol, each led by an expert at a specific institution within the ADCS network. The Clinical Core is responsible for the recruitment of study participants, development of an electronic data capture system at each clinical site, development of protocols and procedures, and clinical assessments of subjects. The MRI Core is responsible for all MRI procedures and developing standardized imaging methods. The PET Core is responsible for all PET procedures and developing standardized imaging methods. The Biomarker Core is responsible for the central collection and analysis of biomarkers in biofluids and for establishing a biobank for archiving biofluids. The Genetics Core is responsible for genotyping participants, genome-wide association studies (GWAS), whole-genome sequencing (WGS), and other potential projects. The Neuropathology Core is responsible for the establishment of protocols to facilitate brain autopsies of ADNI participants who die and who consent. The Biostatistics Core is responsible for the statistical analysis of data generated from the other cores. The Informatics Core is responsible for the establishment of a website to facilitate the sharing of data generated by ADNI projects. In addition, two other committees were formed to help manage other aspects of ADNI. The Data and Publications Committee developed the policy, ratified by the executive and steering committees, regarding data access and publications and conducted an administrative review of all manuscripts based on the analyses of ADNI data. The resource allocation review committee reviews proposals from researchers regarding the use of banked biofluid samples collected from ADNI participants; if approved, then requests would be sent to the Biomarker Core for the release of samples to the study investigator.

Since its inception, ADNI has emphasized a collaborative relationship between the academic researchers and the representatives from the private partner organizations that contributed to ADNI funding. The private partners' voice is included in the ADNI steering committee (two representatives from each organization) and in the ADNI executive committee (current, past, and future chairs of the PPSB). Further, private partners may join the ADNI core meetings and provide their specific expertise; for example, selected industry MRI experts will join the MRI core meetings. The Private Partner Scientific Board (PPSB) convenes, with FNIH facilitation, for member organizations to formulate input on aspects of study design, biomarker selection, clinical assessments, organization of data and relevant methodological issues, and trial management. In addition the PPSB serves as an independent, open, and precompetitive forum in which all private sector and not-for-profit partners in ADNI can collaborate, share information, and offer scientific and private sector perspectives and expertise on issues relating to the ADNI project. While now largely accepted the notion of a precompetitive consortium in which private companies that would otherwise be competitors could come together to address common methodological issues was rare earlier on (Schmidt et al., 2010). Several PPSB initiatives emerged as part of this collaborative environment, some directly impacting ADNI methodology, while others arose out of agreement on common needs of or recognition of gaps in knowledge for developing AD therapies (Liu, Luthman, et al., 2015; Liu, Schmidt, et al., 2015). As an example, while ADNI-1 collected CSF and blood for future work, funding for the work was not originally part of the ADNI-1 grant. A subset of the private partners provided funding to complete proteomic analyses in both blood and CSF, under the auspices of the FNIH Biomarker Consortium, which resulted in several publications (Craig-Shapiro et al., 2011; Hu et al., 2012; Soares et al., 2012; Spellman et al., 2015). Another example was the collaboration between private partner PET imaging experts in pharmaceutical companies and imaging core labs and academia to set out guidance and best practices for reducing technical variability in longitudinal PET scanning in the setting of multicenter clinical trials, with different scanners, image reconstruction, and processing methods (Schmidt et al., 2014).

Initially the private partner participation skewed toward large pharmaceutical companies, but as ADNI has evolved, so too has the profile of the industry and nonprofit partners. The PPSB is now composed of representatives of private, for-profit entities including pharmaceutical, biotechnology, diagnostics, imaging companies, and imaging contract research organizations, as well as nonprofit organizations. This wider group of stakeholders has broadened the perspectives, insight, and expertise the PPSB is able to provide to ADNI. A complete list of private partners that have participated in the ADNI project is provided here https://fnih.org/what-we-do/current-research-programs/adni3.

III VALUE AND IMPACT OF ADNI

There is no question of the value of ADNI; they are innumerable (Weiner, Veitch, Aisen, et al., 2015; Weiner, Veitch, Hayes, et al., 2015) and highlighted here. The ADNI mission was to validate biomarkers for clinical trial use to enable faster routes to developing effective therapies for AD. The longitudinal collection of clinical and biomarker data in a large sample ranging from cognitively normal elderly and subjective memory-complaint participants to patients represents the full spectrum of the disease: early/late MCI (predementia) and dementia patients greatly enhanced our knowledge of the natural history of the disease. ADNI holds a rich dataset; in some patients, data have been collected for more than 10 years. The open access configuration allows for individual researchers to researchers at for-profit private organizations to interrogate the data collected. Data in many forms are available for download (http://adni.loni.usc.edu/). The data collected and analyzed by the ADNI cores are posted to the LONI Image and Data Archive (LDA) for researchers to access clinical measures (demographics, physical exam, and cognitive assessments), fluid biomarker concentrations (T-tau, p-tau, and Aβ), and analyzed MR and PET imaging data. Source images for MRI and PET are posted as well; thus researchers can perform their own image processing and analyses, some to optimize/validate newly developed methods. Genotyping (apolipoprotein E, TOMM40) and GWAS data are available. Whole-genome sequencing data are available if one provides suitable hardware to copy data due to its large size (150 terabytes). Data from two dementia studies complementary to ADNI are also available through the LDA. They are the Department of Defense-ADNI study (effects of traumatic brain injury and posttraumatic stress disorder on AD in Veterans in ADNI) and Australian Imaging Biomarkers and Lifestyle Study of Aging (AIBL; Ellis et al., 2009).

Along with a few other key large datasets, including AIBL and the Dominantly Inherited Alzheimer's Disease Network (DIAN; Bateman et al., 2012; Moulder et al., 2013), ADNI greatly contributed to our understanding of natural history of the disease. The power of the ADNI dataset is that multiple biomarkers are assessed along with clinical/cognitive measures in the same individual over a long period of time. These studies provided insight regarding the relationship between biomarkers especially in predementia stages of the disease and in cognitively normal elderly and provided evidence that biomarkers of AD pathology and pathophysiology becomes abnormal decades prior to the onset of dementia (e.g., Bateman et al., 2012; Jack et al., 2010; Jack et al., 2011; Mormino et al., 2009; Shaw et al., 2009; Villemagne et al., 2013). Thus a model of a hypothetical ordering of AD biomarkers was proposed (Jack et al., 2010) in which amyloids either in CSF or imaged by PET are the earliest biomarkers to become abnormal, followed by biomarkers of tau pathology (at the time, only CSF measures of tau were available) and indication of synaptic dysfunction with FDG-PET glucose metabolism, and later followed by structural MRI abnormalities preceding clinical symptoms and eventually dementia. This was later amended (Jack et al., 2013) as new data emerged, particularly, evidence that medial temporal lobe tau pathology can occur in normal aging independent of amyloid (Braak, Thal, Ghebremedhin, & Del Tredici, 2011). Additional modeling work based on ADNI data improved our

understanding of the relationship between biomarkers and clinical measures (Donohue et al., 2014; Jedynak et al., 2012).

Our current thinking of the optimal time frame during which to intervene in the disease and to evaluate a potential therapy for AD has shifted, in part due to ADNI. Until recently, most therapeutic clinical trials recruited AD patients who have already crossed the threshold of dementia, with overt cognitive and functional deficits. But this has slowly changed as drug developers are focusing on earlier stages of the disease before extensive neurodegeneration occurred (Mattsson et al., 2015; Schneider et al., 2014) and even in secondary prevention prior to clinical symptoms appear (Bateman et al., 2011; Reiman et al., 2011; Sperling et al., 2014). Further, while anti-Aβ therapies have been the majority of those in development since the approvals of symptomatic treatments for AD, with the better understanding of relationship and timing of biomarkers relative to symptoms and the advent of in vivo brain tau imaging with PET, researchers are beginning to develop therapies against tau.

For clinical trial design, one of the greatest utilities is the ability to track change over time in clinical and biomarker measures and allow more accuracy in estimating variability and enabling sample size calculations (e.g., Beckett et al., 2010). The ADNI study design is unprecedented in its contribution to these efforts. Another innovation is to utilize an enrichment strategy not only to ensure that the right patients are enrolled but also to reduce trial sample size. Several investigators utilized ADNI data to evaluate the impact of using patient enrichment strategies on trial sample size compared with no enrichment when using an instrument such as the Clinical Dementia Rating-Sum of the Boxes (CDR-SB) as an outcome measure. As an example, Grill et al. (2013), using ADNI data, estimated that a sample size of 458 (95% CI: 334–679) per arm would be needed for a 24-month trial with MCI patients using the CDR-SB, but that number can be reduced to 258 (95% CI: 173–437) per arm if enriched for MCI patients with a pathological CSF biomarker signature (CSF ratio T-tau/Aβ). Another set of investigators (Holland, McEvoy, Desikan, & Dale, 2012), also using ADNI data, found that without enrichment, in a 24-month clinical trial evaluating MCI patients using the CDR-SB, a sample size of 583 (95% CI: 416, 894) per arm would be needed. This can be reduced to 234 (95% CI: 151, 455) if the MCI subject were enriched for abnormal MRI and CSF Aβ and p-tau, which can be further reduced to 60 (95% CI: 42, 100) per arm if entorhinal cortex atrophy is used instead of CDR-SB as the outcome measure.

ADNI provided the model that multicenter image collection of both PET and MRI not only is possible but also can produce meaningful data by the use of standardized protocols (http://adni.loni.use.edu/methods/)

that allow direct comparison of data collected at different centers. The PET core designed standardized protocols to accommodate different scanners hardware and software combinations at 16 PET centers within the ADNI participating sites and for different PET radiotracers (FDG for glucose metabolism, PiB and florbetapir for amyloid burden, and now AV-1451 for tau burden). PET image acquisition, quality control methods (including a Hoffman phantom), and preprocessing methods to prepare image data for analysis resulted in the availability of sets of PET images that investigators can download and perform image analysis themselves. Similarly, standardized MRI protocols were developed by the MRI core that also allowed the use of different scanners from three different companies (GE, Siemens, and Philips), different hardware/software combinations, and two different MRI field strengths (1.5T and 3T) collected at 59 sites. Standardized methods for image acquisition, QC methods (including a phantom to monitor scanner performance), and postacquisition corrections produced images available for download and further image analysis by investigators. ADNI MR images were used to optimize automated processing to measure brain and hippocampal volumes and atrophy (Jorge Cardoso et al., 2013; Leung et al., 2010; Leung et al., 2011). Sponsored clinical trials have incorporated the ADNI PET and MRI model and demonstrated their utility in treatment trials (e.g., Liu, Luthman, et al., 2015; Liu, Schmidt, et al., 2015; Novak et al., 2016; Salloway et al., 2014). The ADNI Biomarker Core has developed the laboratory manual for sample collection, handling, and shipment with precautions such as the collection material as some proteins ($A\beta$) will stick to the tube and confound the measurement. The Biomarker Core is responsible for the bioanalysis of the samples collected and has optimized testing methods that have been incorporated into clinical trials (e.g., Streffer et al., 2013).

Another important impact is the additional research studies conducted outside of the ADNI protocol with archived blood and CSF samples from ADNI subjects. The mandate from the ADNI Executive and Steering Committees for the use of the ADNI samples is for the validation of new technologies that already has initial evidence of utility rather than discovery of novel technologies. The Resource Allocation Review Committee (RARC), composed of academic experts and an PPSB expert member, was formed to review investigator proposals for the use of banked ADNI samples. The RARC has approved several projects that lead to publications. For example, one study showed that CSF α-synuclein is highly correlated with CSF p-tau and MMSE scores (Korff, Liu, Ginghina, Shi, & Zhang, 2013; Toledo, Korff, Shaw, Trojanowski, & Zhang, 2013). Another study evaluated the utility of plasma tau. Higher plasma tau was associated with AD dementia, higher CSF tau,

and lower CSF $A\beta_{42}$, but the correlations were weak and may not be a good tool to differentiate between cognitively normal and stages of AD. Longitudinal analysis in ADNI showed significant associations between plasma tau and worse cognition, greater atrophy, and greater hypometabolism during follow-up and may be a better utility of plasma tau (Mattsson et al., 2016).

IV ADNI-INSPIRED OTHER INITIATIVES

ADNI served as a model for neuroimaging initiatives worldwide. The Worldwide ADNI (WW-ADNI), sponsored by the Alzheimer's Disease Association, is an umbrella organization with programs in Japan, Australia, Argentina, Taiwan, China, Korea, Europe, and Italy (Carrillo, Bain, Frisoni, & Weiner, 2012). A key goal of WW-ADNI is to harmonize protocols and share standardized data across the international research community with the hope to standardize diagnosis of AD worldwide. WW-ADNI utilizes the established ADNI protocols for structural MRI; PET imaging; and collection of cognitive, blood, and genomic data. European ADNI (E-ADNI) was initiated shortly after ADNI, first as a pilot, now with a network of 50 centers. One key focus had been standardizing protocols for measuring hippocampal volume (Frisoni, 2010; Frisoni & Jack, 2011). An informatics platform (neuGRID) funded by the European Union was designed to allow E-ADNI data interface with the LONI data repository. Similarly the data from Australian ADNI, known as AIBL, established in 2006, are available through the LONI data repository. WW-ADNI demonstrates a unique example of international cooperation, with hope that collaborative research will lead to the development of novel therapies for AD.

ADNI also inspired projects in other neurodegenerative disorders. The Department of Defense (DOD) is funding a study, DOD-ADNI, to evaluate traumatic brain injury and posttraumatic stress disorder in military veterans as potential risk factors for AD utilizing ADNI methods (Weiner et al., 2014). The Parkinson's Disease Progressive Markers Initiative (PPMI), launched in 2010, with private-public funding initiated by the Michael J Fox Foundation for Parkinson's Disease Research, has a similar goal to ADNI: to identify biomarkers of Parkinson's disease (PD) progression and to further the understanding of the natural history of PD in order to increase the speed and success of clinical trials for novel PD therapies (Parkinson's Disease Progression Markers I, 2011). A 3-year pilot study, the Down Syndrome Biomarker Initiative, has begun at UC San Diego through the ADCS with pharmaceutical funding (Ness et al., 2012). Its aim is to evaluate the potential relationship between Down syndrome and AD with assessments of cognition, MRI, PET imaging, and fluid biomarkers.

V POTENTIAL FURTHER ENHANCEMENTS FOR GREATER EFFICIENCY AND ACCELERATION OF DRUG DEVELOPMENT

ADNI brought together researchers from academia, industry, governmental agency, and a nonprofit organization and created a forum for discussion, sharing of ideas, and brainstorming. Out of this came the identification of further work that could be done in the precompetitive space rather than by single entities that could benefit all AD clinical research. For example, it is recognized that more precise instruments to measure cognitive and functional deficits in earlier stages of disease are needed. Other examples include technical validation of promising new platforms to measure fluid biomarkers and novel imaging methods. However, the grants funding ADNI specifies the scope of the research through the grant application itself, the study protocol, and associated technical or laboratory manuals. Thus these new areas of research would need to be championed, and new funding would need to be identified. One option is for the PPSB, already in existence and associated with ADNI, to find existing grants that would be suitable for the new projects. An example is the CSF research mentioned earlier where funding from the NIH Biomarker Consortium was obtained for plasma and CSF multiplex biomarker identification. Another option is to include scope in a future ADNI grant renewal that would fund research related to the ADNI study, but not strictly tied to the protocol. The newly identified research could be done through collaboration that will speed our progress to discovering new effective, potentially disease-modifying therapies for AD.

VI SUMMARY

ADNI provided an exemplar model of a public-private-funded clinical research study. The innovations include bringing together funding from a novel combination of sources, the scale of the ADNI study, and open access to data collected. Its overarching goal to validation of biomarkers and standardization of assessment and analysis methods through longitudinal collection of data has impacted the designs of current clinical trials evaluating novel therapies for Alzheimer's disease. ADNI has developed standardized biomarkers for use in clinical trial subject selection and developed standardized protocols that are being used in therapeutic clinical trials. ADNI has demonstrated that a multicenter collection of imaging data, both MRI and PET, is possible. ADNI data have been utilized in over 1000 publications, leading to greater understanding of the relationship between biomarkers and AD progression and identification of novel biomarkers. ADNI has inspired other ADNI-like projects such as the WW-ADNI, projects investigating TBI and PTSD in military populations, Parkinson's disease, and Down's syndrome. Most importantly is the ADNI legacy: generation of a rich dataset accessible to AD research community so that effective, novel AD therapies can become available to patients.

References

Alzheimer's Association. (2017). *Alzheimer's fact and figures 2017.* Retrieved from https://www.alz.org/documents_custom/2017-facts-and-figures.pdf.

Alzheimer's Disease International. (2015). *World Alzheimer's report 2015.* Retrieved from https://www.alz.co.uk/research/WorldAlzheimerReport2015.pdf.

Bateman, R. J., Aisen, P. S., De Strooper, B., Fox, N. C., Lemere, C. A., Ringman, J. M., et al. (2011). Autosomal-dominant Alzheimer's disease: a review and proposal for the prevention of Alzheimer's disease. *Alzheimer's Research & Therapy, 3,* 1. https://doi.org/10.1186/alzrt59.

Bateman, R. J., Xiong, C., Benzinger, T. L., Fagan, A. M., Goate, A., Fox, N. C., et al. (2012). Clinical and biomarker changes in dominantly inherited Alzheimer's disease. *The New England Journal of Medicine, 367,* 795–804. https://doi.org/10.1056/NEJMoa1202753. Erratum in: The New England Journal of Medicine, 23 (2012), 367, 780.

Beckett, L. A., Harvey, D. J., Gamst, A., Donohue, M., Komak, J., Zhang, H., et al. (2010). The Alzheimer's Disease Neuroimaging Initiative: annual change in biomarkers and clinical outcomes. *Alzheimers Dement, 6,* 257–264.

Braak, H., Thal, D. R., Ghebremedhin, E., & Del Tredici, K. (2011). Stages of the pathologic process in Alzheimer's disease: age categories from 1 to 100 years. *Journal of Neuropathology and Experimental Neurology, 70,* 960–969. https://doi.org/10.1097/NEN.0b013e318232a379.

Carrillo, M. C., Bain, L. J., Frisoni, G. B., & Weiner, M. W. (2012). Worldwide Alzheimer's disease neuroimaging initiative. *Alzheimers Dement, 8,* 337–342.

Craig-Shapiro, R., Kuhn, M., Xiong, C., Pickering, E. H., Liu, J., Misko, T. P., et al. (2011). Multiplexed immunoassay panel identifies novel CSF biomarkers for Alzheimer's disease diagnosis and prognosis. *PLoS One, 6,* e18850.

Donohue, M. C., Jacqmin-Gadda, H., Le Goff, M., Thomas, R. G., Raman, R., Gamst, A. C., et al. (2014). Estimating long-term multivariate progression from short-term data. *Alzheimers Dement, 10,* S400–S410. https://doi.org/10.1016/j.jalz.2013.10.003.

Ellis, K. A., Bush, A. I., Darby, D., De Fazio, D., Foster, J., Hudson, P., et al. (2009). The Australian imaging, biomarkers and lifestyle (AIBL) study of aging: methodology and baseline characteristics of 1112 individuals recruited for a longitudinal study of Alzheimer's disease. *International Psychogeriatrics, 21,* 672–687. https://doi.org/10.1017/S1041610209009405.

Frisoni, G. B. (2010). Alzheimer's Disease Neuroimaging Initiative in Europe. *Alzheimers Dement, 6,* 280–285.

Frisoni, G. B., & Jack, C. R. (2011). Harmonization of magnetic resonance-based manual hippocampal segmentation: a mandatory step for wide clinical use. *Alzheimers Dement, 7,* 171–174.

Global Action Against Dementia. (2013). *G8 dementia summit declaration.* December 11. Retrieved from https://www.gov.uk/government/uploads/system/uploads/attachment_data/file/265869/2901668_G8_DementiaSummitDeclaration_acc.pdf. Accessed 20 October 2016.

Global Action Against Dementia. (2015). *Progress report December 2013-March 2015*. Retrieved from http://www.ohchr.org/Documents/Issues/OlderPersons/Dementia/GAADReport.pdf. Accessed 20 October 2016.

Grill, J. D., Di, L., Lu, P. H., Lee, C., Ringman, J., Apostolova, L. G., et al. (2013). Estimating sample sizes for pre-dementia Alzheimer's trials based on the Alzheimer's Disease Neuroimaging Initiative. *Neurobiology of Aging*, 34, 62–72. https://doi.org/10.1016/j.neurobiolaging.2012.03.006.

Holland, D., McEvoy, L. K., Desikan, R. S., & Dale, A. M. (2012). Enrichment and stratification for predementia Alzheimer disease. *PLoS One*, 7(10), e47739.

Hu, W. T., Holtzman, D. M., Fagan, A. M., Shaw, L. M., Perrin, R., Arnold, S. E., et al. (2012). Plasma multianalyte profiling in mild cognitive impairment and Alzheimer's disease. *Neurology*, 79, 897–905.

Jack, C. R., Knopman, D. S., Jagust, W. J., Petersen, R. C., Weiner, M. W., Aisen, P. S., et al. (2013). Tracking pathophysiological processes in Alzheimer's disease: an updated hypothetical model of dynamic biomarkers. *Lancet Neurology*, 12, 207–216.

Jack, C. R., Knopman, D. S., Jagust, W. J., Shaw, L. M., Aisen, P. S., Weiner, M. W., et al. (2010). Hypothetical model of dynamic biomarkers of the Alzheimer's pathological cascade. *Lancet Neurology*, 9, 119–128.

Jack, C. R., Vemuri, P., Wiste, H. J., Weigand, S. D., Aisen, P. S., Trojanowski, J. Q., et al. (2011). Evidence for ordering of Alzheimer disease biomarkers. *Archives of Neurology*, 68, 1526–1535.

Jack, C. R., Wiste, H. J., Vemuri, P., Weigand, S. D., Senjem, M. L., Zeng, G., et al. (2010). Brain beta-amyloid measures and magnetic resonance imaging atrophy both predict time-to-progression from mild cognitive impairment to Alzheimer's disease. *Brain*, 133, 3336–3348.

Jedynak, B. M., Lang, A., Liu, B., Katz, E., Zhang, Y., Wyman, B. T., et al. (2012). A computational neurodegenerative disease progression score: method and results with the Alzheimer's Disease Neuroimaging Initiative cohort. *NeuroImage*, 63, 1478–1486.

Jorge Cardoso, M., Leung, K. K., Modat, M., Keihaninejad, S., Cash, D., Barnes, J., et al. (2013). STEPS: similarity and truth estimation for propagated segmentations and its application to hippocampal segmentation and brain parcellation. *Medical Image Analysis*, 17, 671–684.

Korff, A., Liu, C., Ginghina, C., Shi, M., & Zhang, J. (2013). ADNI: alpha-synuclein in cerebrospinal fluid of Alzheimer's disease and mild cognitive impairment. *Journal of Alzheimer's Disease*, 36, 679–688.

Leung, K. K., Barnes, J., Modat, M., Ridgway, G. R., Bartlett, J. W., Fox, N. C., et al. (2011). Brain MAPS: an automated accurate and robust brain extraction technique using a template library. *NeuroImage*, 55, 1091–1108.

Leung, K. K., Barnes, J., Ridgway, G. R., Bartlett, J. W., Clarkson, M. J., Mac Donald, K., et al. (2010). Automated cross-sectional and longitudinal hippocampal volume measurement in mild cognitive impairment and Alzheimer's disease. *NeuroImage*, 51, 1345–1359.

Liu, E., Luthman, J., Cedarbaum, J. M., Schmidt, M. E., Cole, P. E., Hendrix, J., et al. (2015). Perspective: The Alzheimer's Disease Neuroimaging Initiative and the role and contributions of the Private Partner Scientific Board (PPSB). *Alzheimers Dement*, 11, 840–849.

Liu, E., Schmidt, M. E., Margolin, R., Sperling, R., Koeppe, R., Mason, N. S., et al. (2015). Amyloid-β 11C-PiB-PET imaging results from 2 randomized bapineuzumab phase 3 AD trials. *Neurology*, 85, 692–700. https://doi.org/10.1212/WNL.0000000000001877.

Masters, C., Bateman, R., Blennow, K., Rowe, C., Sperling, R., & Cummings, J. (2015). Alzheimer's disease. *Nature Reviews. Disease Primers*, 1, 15056.

Mattsson, N., Carrillo, M. C., Dean, R. A., Devous, M. D., Sr., Nikolcheva, T., Pesini, P., et al. (2015). Revolutionizing Alzheimer's disease and clinical trials through biomarkers. *Alzheimers Dement*, 1, 412–419.

Mattsson, N., Zetterberg, H., Janelidze, S., Insel, P. S., Andreasson, U., Stomrud, E., et al. (2016). Plasma tau in Alzheimer's disease. *Neurology*, 87, 1827–1835.

Mormino, E. C., Kluth, J. T., Madison, C. M., Rabinovici, G. D., Baker, S. L., Miller, B. L., et al. (2009). Episodic memory loss is related to hippocampal-mediated beta-amyloid deposition in elderly subjects. *Brain*, 132(Pt5), 1310–1323.

Moulder, K. L., Snider, B. J., Mills, S. L., Buckles, V. D., Santacruz, A. M., Bateman, R. J., et al. (2013). Dominantly inherited Alzheimer network: facilitating research and clinical trials. *Alzheimer's Research & Therapy*, 5, 48. https://doi.org/10.1186/alzrt213.

Mueller, S. G., Weiner, M. W., Thal, L. J., Petersen, R. C., Jack, C. R., Jagust, W., et al. (2005). Ways toward an early diagnosis in Alzheimer's disease: the Alzheimer's Disease Neuroimaging Initiative (ADNI). *Alzheimers Dement*, 1, 55–66.

Ness, S., Raffi, M., Aisen, P., Krams, M., Siverman, W., & Manji, H. (2012). Down's syndrome and Alzheimer's disease: toward secondary prevention. *Nature Reviews. Drug Discovery*, 11, 655–656.

Novak, G., Fox, N., Clegg, S., Nielsen, C., Einstein, S., Lu, Y., et al. (2016). Changes in brain volume with bapineuzumab in mild to moderate Alzheimer's disease. *Journal of Alzheimer's Disease*, 49, 1123–1134. https://doi.org/10.3233/JAD-150448.

Parkinson's Progression Markers Initiative. (2011). The Parkinson progression markers initiative. *Progress in Neurobiology*, 95, 629–635.

PHRMA. (2015). *Research Alzheimer's medicines: setbacks and stepping stones*. Retrieved from http://phrma-docs.phrma.org/sites/default/files/pdf/alzheimers-setbacks-and-stepping-stones.pdf.

PHRMA. (2016). *Profile report*. Retrieved from http://phrma-docs.phrma.org/sites/default/files/pdf/biopharmaceutical-industry-profile.pdf.

Reiman, E. M., Langbaum, J. B., Fleisher, A. S., Caselli, R. J., Chen, K., Ayutyanont, N., et al. (2011). Alzheimer's Prevention Initiative: a plan to accelerate the evaluation of presymptomatic treatments. *Journal of Alzheimer's Disease*, 26(Suppl 3), 321–329.

Salloway, S., Sperling, R., Fox, N. C., Blennow, K., Klunk, W., Raskind, M., et al. (2014). Two phase 3 trials of bapineuzumab in mild-to-moderate Alzheimer's disease. *The New England Journal of Medicine*, 370, 322–333. https://doi.org/10.1056/NEJMoa1304839.

Scheltens, P., Blennow, K., Breteler, M. M., de Strooper, B., Frisoni, G. B., Salloway, S., et al. (2016). Alzheimer's disease. *Lancet*, 388, 505–517.

Schmidt, M. E., Chiao, P., Klein, G., Matthews, D., Thurfjell, L., Cole, P. E., et al. (2014). The influence of biological and technical factors on quantitative analysis of amyloid PET: points to consider and recommendations for controlling variability in longitudinal data. *Alzhemers Dement*, 11, 1050–1068. https://doi.org/10.1016/j.jalz.2014.09.004.

Schmidt, M. E., Siemers, E., Snyder, P. J., Potter, W. Z., Cole, P., & Soares, H. (2010). The Alzheimer's Disease Neuroimaging Initiative: perspectives of the Industry Scientific Advisory Board. *Alzheimers Dement*, 6, 286–290.

Schneider, L., Mangialasche, F., Andreasen, N., Feldman, H., Giacobini, E., Jones, R., et al. (2014). Clinical trials and late-stage drug development for Alzheimer's disease: an appraisal from 1984 to 2014. *Journal of Internal Medicine*, 275, 251–283.

Selkoe, D. J., & Hardy, J. (2016). The amyloid hypothesis of Alzheimer's disease at 25 years. *EMBO Molecular Medicine*, 8, 595–608.

Shaw, L. M., Vanderstichele, H., Knapik-Czajka, M., Clark, C. M., Aisen, P. S., Petersen, R. C., et al. (2009). Cerebrospinal fluid biomarker signature in Alzheimer's Disease Neuroimaging Initiative subjects. *Annals of Neurology*, 65, 403–413.

Soares, H. D., Potter, W. Z., Pickering, E., Kuhn, M., Immermann, F. W., Shera, D. M., et al. (2012). Plasma biomarkers associated with the apolipoprotein E genotype and Alzheimer's disease. *Archives of Neurology*, 69, 1310–1317.

Spellman, D., Wildsmith, K. R., Honigberg, L. A., Tuefferd, M., Baker, D., Raghavan, N., et al. (2015). Development and evaluation of a multiplexed mass spectrometry-based assay for measuring candidate peptide biomarkers in ADNI. *Proteomics. Clinical Applications, 9,* 715–731. https://doi.org/10.1002/prca.201400178.

Sperling, R. A., Rentz, D. M., Johnson, K. A., Karlawish, J., Donohue, M., Salmon, D. P., et al. (2014). The A4 study: stopping AD before symptoms begin? *Science Translational Medicine, 6*(228), 228fs13.

Streffer, J. R., Blennow, K., Salloway, S., Zetterberg, H., Xu, Y. Z., Lu, Y., et al. (2013). Effect of bapineuzumab on CSF p-tau and t-tau in mild to moderate Alzheimer's disease: results from two phase 3 trials in APOE ε4 carriers and non-carriers. *Alzheimers Dement, 9,* P138.

Toledo, J. B., Korff, A., Shaw, L. M., Trojanowski, J. Q., & Zhang, J. (2013). CSF alpha-synuclein improves diagnostic and prognostic performance of CSF tau and Abeta in Alzheimer's disease. *Acta Neuropathologica, 126,* 683–697.

Villemagne, V. L., Burnham, S., Bourgeat, P., Brown, B., Ellis, K. A., Salvao, O., et al. (2013). Amyloid β deposition, neurodegeneration, and cognitive decline in sporadic Alzheimer's disease: a prospective cohort study. *Lancet Neurology, 12,* 357–367. https://doi.org/10.1016/S1474-4422(13)70044-9.

Weiner, M. W., Veitch, D. P., Aisen, P. S., Beckett, L. A., Cairns, N. J., Cedarbaum, J., et al. (2014). Effect of traumatic brain injury and posttraumatic stress disorder on Alzheimer's disease in veterans, using the Alzheimer's Diesease Neuroimaging Initiative. *Alzheimers Dement, 10,* S226–S235. https://doi.org/10.1016/j.jalz.2014.04.005.

Weiner, M. W., Veitch, D. P., Aisen, P. S., Beckett, L. A., Cairns, N. J., Cedarbaum, J., et al. (2015). Impact of the Alzheimer's Disease Neuroimaging Initiative, 2004 to 2014. *Alzheimers Dement, 11,* 865–884.

Weiner, M. W., Veitch, D. P., Aisen, P. S., Beckett, L. A., Cairns, N. J., Green, R. C., et al. (2013). The Alzheimer's Disease Neuroimaging Initiative: a review of papers published since its inception. *Alzheimers Dement, 9,* e111–e194. https://doi.org/10.1016/j.jalz.2013.05.1769.

Weiner, M. W., Veitch, D. P., Aisen, P. S., Beckett, L. A., Cairns, N. J., Green, R. C., et al. (2017). Recent publications from the Alzheimer's Disease Neuroimaging Initiative: reviewing progress toward improved AD clinical trials. *Alzheimers Dement, 13,* e1–e85. https://doi.org/10.1016/j.jalz.2016.11.007.

Weiner, M. W., Veitch, D. P., Hayes, J., Neylan, T., Grafman, J., Aisen, P. S., et al. (2015). 2014 Update of the Alzheimer's Disease Neuroimaging Initiative: a review of papers published since its inception. *Alzheimers Dement, 11,* e1–120.

Zetterberg, H. (2017). Review: tau in biofluids—relation to pathology, imaging and clinical features. *Neuropathology and Applied Neurobiology, 43,* 194–199. https://doi.org/10.1111/nan.12378.

18

Regulatory Perspectives on the Use of Biomarkers and Personalized Medicine in CNS Drug Development: The FDA Viewpoint

Mathangi Gopalakrishnan, Jogarao V.S. Gobburu

Center for Translational Medicine, School of Pharmacy, University of Maryland, Baltimore, Baltimore, MD, United States

Abbreviations

AD	Alzheimer's disease
ADAS-Cog	Alzheimer's Disease Assessment Scale-Cognitive Subscale
ADNI	Alzheimer's Disease Neuroimaging Initiative
AEDs	Antiepileptic drugs
AFA	Alzheimer's Foundation of America
APOE	Apolipoprotein E
ARR	Annualized relapse rate
BLA	Biologics license application
BQP	Biomarker qualification program
BRAIN	Brain Research through Advancing Innovative Neurotechnologies
CAMD	Coalition against major diseases
CDER	Center for Drug Evaluation and Research
CNS	Central nervous system
CPIM	Critical Path Innovation Meeting
CPP	Critical Path for Parkinson's
CYP	Cytochrome P450
DAT	Dopamine transporter
DMD	Duchenne muscular dystrophy
EEG	Electroencephalogram
FDA	Food and Drug Administration
FDASIA	Food and Drug Safety and Innovation Act
GAIN	Generating Antibiotic Incentives Now
HbA1c	Glycosylated hemoglobin
HLA	Human leucocyte antigen
IND	Investigational new drug
MRI	Magnetic resonance imaging
NDA	New drug application
NIH	National Institutes of Health
PD	Parkinson's disease
PD	Pharmacodynamics
PDUFA	Prescription Drug User Fee Act
PEACE	Pediatric Epilepsy Academic Consortium on Extrapolation
PET	Positron emission tomography
PGC	Psychiatric Genomics Consortium
PK	Pharmacokinetics
PoC	Proof of concept
PSTC	Predictive Safety Testing Consortium
QIDP	Qualified infectious disease product
RWE	Real-world evidence
SSRI	Selective serotonin reuptake inhibitor

I CNS DRUG DEVELOPMENT

A Current Status

The socioeconomic burden of neurological illnesses and mental disorders in the United States (US) costs more than $760 billion a year. Nearly 100 million Americans are affected by more than 1000 neurological and neurodegenerative diseases, such as schizophrenia, traumatic brain injury, autism, Alzheimer's disease, and Parkinson's disease (PD) (Society for Neuroscience, 2017). Stroke is the second leading cause of death worldwide, and central nervous system (CNS) disorders account for 12 of the top 20 causes of years living with disability (Vos et al., 2012). Based on the 2010 US census, the disease burden due to PD is expected to double by 2040 with the currently diagnosed cases at 630,000 in the United States (Kowal, Dall, Chakrabarti, Storm, & Jain, 2013). The incidence, prevalence, mortality, and economic burden of Alzheimer's disease (AD) are even more stark with one new case of AD expected to develop every 33 seconds by 2050 as opposed to every 67 seconds in 2015 (Alzheimer's Association, 2015). Despite the grave economic and disease impact of CNS disorders on the society, development of more effective therapies and prevention strategies are heavily lacking.

A survey of two decades of new drug development for CNS disorders (Kesselheim, Hwang, & Franklin, 2015)

demonstrated that there had been a slowdown in CNS drug development with fewer drugs entering Phase I when compared between 1990 and 2012 (11% vs 7%). In addition, the Phase III failure rates were 53% with lack of efficacy being the main reason (46%) based on trials across different CNS indications between 1990 and 2012. Another discouraging aspect for CNS drug development is that several large pharmaceutical companies have recently discontinued their neuroscience research to a significant extent. Between 2009 and 2014 the total number of CNS program portfolios has decreased by 51% (267 programs in 2009 vs 129 in 2014) in large pharma companies (Choi et al., 2014). These industry cutbacks in neuroscience research reflect the challenges in the drug development that are driven by higher cost (1.8–3.9$ billion Khanna, 2012) to discover and develop a drug but with lower overall success rates.

B Emerging Horizons

However, despite the bleak depiction of overall CNS drug development, a silver lining is emerging with varied initiatives taken up by the NIH; disease foundations such as the Michael J. Fox Foundation, the Coalition Against Major Diseases (CAMD) (Critical Path Institute, 2017a), and the Alzheimer's Foundation of America (AFA); and several US Food and Drug Administration (FDA) initiatives to propel drug development for CNS disorders and avail the opportunities to develop path-breaking therapies. President Obama's Brain Research through Advancing Innovative Neurotechnologies (BRAIN) initiative (National Institutes of Health, 2015) unveiled in 2013 offers researchers a great opportunity to understand the complexity of how the brain works and how diseases occur, which may eventually lead to a paradigm shift in developing drugs for neurodegenerative and psychiatric diseases.

It has been recognized that traditional clinical trial methods are falling short in CNS drug development as CNS disorders are chronic, slow degenerative processes with ill-defined and undefined pathophysiology, which are highly subjective, thus leading to poor signal-to-noise characteristics for the treatments tested. Because of the highly subjective nature of CNS disorders, the need of the hour would be to identify reliable biomarkers and use them as an integral part of the clinical trial design and in the decision-making process. To that extent, efforts are already underway through CAMD, a public-private partnership with one of the goals being the qualification of objective biomarkers for neurodegenerative diseases that can be intended for patient enrichment in the trials. The FDA guidance on enrichment (Guidance for Industry: Food and Drug Administration, 2012) further supports prognostic or predictive strategies to enrich

a trial using biomarkers of different characteristics (pathophysiological, genomic or proteomic, clinical, and psychological characteristics). With additional impetus from the precision medicine initiative (Collins & Varmus, 2015) and the 21st Century Cures Act (21st Century Cures, 2014), the use of biomarkers for personalized medicine to prevent, diagnose, and treat is seemingly promising with supportive regulatory (FDA) initiatives. The increased collection and use of pharmacogenomic information in the drug labels and many pre- and postmarketing assessments of pharmacogenomic markers are paving the way for personalized medicine (Zineh & Pacanowski, 2011). Physicians have been practicing personalized medicine in the clinic while treating patients, where the treatment options are chosen for the patient under consideration based on certain characteristics. The pharmacotherapeutic decision made by the physician for a patient in the clinic is typically based on prior experience and clinical judgment. However, the new interest in personalized medicine and the use of varied biomarkers during the drug development process have opened up novel opportunities for CNS drug development. In this chapter, we will discuss the regulatory (FDA) perspective on the use of biomarkers for personalized medicine in CNS drug development. The chapter is organized as follows: In Section II, introduction to the use of biomarkers in drug development and regulation is discussed with case studies in CNS drugs. Section III discusses the US FDA initiatives for use of biomarkers and personalized medicine. Section IV provides the challenges and future perspective for the use of biomarkers in CNS drug development.

II BIOMARKERS AND PERSONALIZED MEDICINE IN DRUG DEVELOPMENT AND REGULATION

The term "biomarker" is "a characteristic that is objectively measured and evaluated as an indicator of normal biological processes, pathogenic processes, or pharmacologic responses to a therapeutic intervention" (Biomarkers Definitions Working Group, 2001). Biomarkers play an important and broad role in guiding decisions regarding optimal dose selection; benefit–risk assessment; and regulatory approvals related to new formulations, populations, and/or indications for drugs with demonstrated effectiveness using clinical endpoints. There are several excellent reviews and FDA guidance on the nature and role of biomarkers in drug development and regulation (e.g., Amur, Frueh, Lesko, & Huang, 2008; Lesko & Atkinson, 2001). In the literature, there are different categorizations of biomarkers to describe the nature and the use of the biomarkers. In this section, we have described the role and use of biomarkers based on how the biomarkers could assist in answering the key questions/decisions during each stage of clinical drug development and in a drug's life cycle. Contrary to the typical representation of clinical drug development in phases, we present the drug development process in terms of key questions that are answered in early-phase and late-phase trials and postmarketing. In addition, throughout the chapter, the term biomarker is used in general and refers to both genomic and nongenomic biomarkers, which could be based on the genotypic/genomic or phenotypic expression, respectively.

A Key Questions in Clinical Drug Development

The go/no-go decisions to move the compound from one phase to another are dependent on the answers to the key questions that are posed at each stage of the drug development.

1 Tolerability and Safety

Assessment of tolerability and safety in the first-in-human studies will dictate further use of the drug compound in the clinic. The goal of the early clinical development biomarkers would be to seek answers to three questions: (i) Is the compound tolerable? (ii) Does the compound show reasonable promise of efficacy in patients who are most likely to respond? (iii) What dose range can be tested in subsequent trials? More often, at this stage, biomarkers pertaining to safety/adverse events are of importance, and evaluation of such biomarkers, typically in healthy volunteers, is applicable for all therapeutic areas including CNS drug development.

Drug-induced prolongation of QT interval is a biomarker for *torsades de pointes*, when escalating doses of the drug (supratherapeutic doses) are tested. The clinical implication of prolonged QT interval could also lead to nonapproval of a drug despite demonstrated efficacy in pivotal trials. For example, sertindole, an atypical antipsychotic, was not considered for approval in the United States due to the risk of QT prolongation despite superior efficacy in schizophrenic patients. Based on a drug concentration–QT analysis, it was predicted that coadministration of sertindole with CYP2D6 and CYP3A4 inhibitors would lead to substantial QT prolongation and hence a high-risk profile (Food and Drug Administration, 2009). Although the sertindole example demonstrates the use of the QT as a safety biomarker in late-phase trials, evaluation of the QT prolongation is typically performed in early-phase trials and pertains to assessing the tolerability of a drug.

In addition, pharmacogenomic biomarkers that explain the genetic polymorphisms in metabolism of a drug (e.g., poor metabolizers) provide information on the type of patients to be enrolled in late-phase clinical trials and doses to be studied in future clinical studies. The pharmacogenomic biomarker information obtained from animal and in vitro studies can also be prospectively applied to exclude patients at risk of excessive drug exposures.

Identification and qualification of predictive safety biomarkers based on preclinical and clinical studies have been the focus of the Predictive Safety Testing Consortium (PSTC), led by the Critical Path Institute (Critical Path Institute, 2017b). Through the consortium, there is collaborative data sharing between 18 different pharmaceutical companies to identify and qualify predictive safety biomarkers. Recently, seven additional biomarkers were qualified to monitor drug-induced kidney injury in addition to serum creatinine and blood urea nitrogen in rats. The predictive ability testing of these biomarkers in humans is undergoing.

2 Proof of Concept (PoC)

The purpose of biomarkers for efficacy during early clinical development is for supporting the two questions: (i) Does the compound show reasonable promise of efficacy in patients who are most likely to respond? (ii) What dose range can be tested in the subsequent trials? The proof-of-concept (PoC) biomarkers are perhaps most useful when the measurement of clinical endpoints may be too time-consuming (requiring long follow-up) to provide timely PoC or dose-ranging information. For example, early trials of antidiabetics can primarily focus on fasting plasma glucose levels and not necessarily glycosylated hemoglobin (HbA1c) both for deciding if the target glycemic control can be achieved with the new compound and for selecting the dose/regimen range for the next trials. Effects on fasting plasma glucose, albeit variable, can be realized in 4 weeks, whereas effects on HbA1c take up 26 weeks. The relationship between fasting plasma glucose and HbA1c has been established reliably over decades. However, with respect to CNS drug development, identifying biomarkers for target identification, validation, and engagement is the most critical piece for the new drug discovery and development pipeline. Complexity of the brain, lack of proper understanding of the disease mechanism, and paucity of animal models provide additional challenges for CNS drug development. Despite such challenges, recently, several articles have described how to accelerate neurological and psychiatric drug development, including in pediatrics (Grabb & Gobburu, 2016; Pankevich, Altevogt, Dunlop, Gage, & Hyman, 2014).

In this section, biomarkers of drug effects are grouped as follows: (i) Early signals of drug action (acute effects) are considered pharmacodynamics (PD) signals, and (ii) treatment effects as a result of chronic exposure that correlate with behavioral outcomes (what one would see in an efficacy trial) are considered outcome measures or intermediate phenotypes. PD measures assess the interaction between neuroactive drug and biological target and can provide additional information about dosing in early-stage trial designs. PD measures can be used with acute dose-range studies to determine the doses of the investigational drug that exert the CNS effects expected. (PD measures could also be built into pharmacokinetic (PK) bridging studies, as described in the succeeding text.)

Reproducible effects on electroencephalogram (EEG) recordings are considered PD biomarkers for several CNS drug classes including benzodiazepines, antidepressants, antipsychotics, opioids, and anticonvulsants, making this measure a good candidate for PD assessments in trials evaluating those drug classes. As stated earlier, target engagement is very critical to assess the activity of a neuroactive drug. Does the drug reach the target? Is the concentration sufficient? Is the residence time in the target tissue as desired? Does the drug interact with the assumed molecular target? Is there an interaction with sufficient number of target molecules per cell? Is the interaction of the drug at the molecular target of sufficient duration? Finally, is there a cellular response? Is this a cellular response of sufficient degree and duration to induce a therapeutic outcome? These are some of the critical questions that could not be answered before. But now, with positron emission tomography (PET) as an imaging biomarker (Gomez-Mancilla et al., 2005), the distribution and the kinetics of the drug in the brain can be measured. Use of a PET-microdosing study enables researchers to understand how the drug distributes through the blood-brain barrier, its relative concentration in the brain in relation to plasma concentration, and the rate of exchange between brain and plasma, thus providing an opportunity to relate brain drug concentrations to PD effects rather than plasma drug concentrations. In addition, PET imaging is being used to evaluate specific targets in neurodegenerative diseases, especially the amyloid tracers in Alzheimer's disease, and also to inform on receptor occupancy. Recently, in an early clinical study (Rabiner et al., 2001), PET was used to examine whether the dose of pindolol used to augment a selective serotonin reuptake inhibitor (SSRI) is sufficient and whether significant occupancy of the serotonin type 1A (5-HT(1A)) autoreceptor is achieved in depressed patients receiving the SSRI. The PET study showed that the autoreceptor occupancy by pindolol at the dose studied was at a modest 19%, much lower than the 75% occupancy required to exert an augmenting PD effect (based on rat studies). However, due to the complex nature of brain physiology and the pathophysiological mechanism, very low receptor occupancy may indicate poor efficacy, but the converse may not be true (Gomez-Mancilla et al., 2005).

Higher failure rates in CNS drug development are attributed to the high heterogeneity observed among patients of CNS disorders, which may obscure a signal of drug effect in early-phase trials and which may result in halting a drug development program. Enrichment and stratification strategies using diagnostic/prognostic or predictive biomarkers that could be either phenotypic or genomic in nature can improve the ability of early-phase trials to detect a drug's effect clearly. Diagnostic biomarkers ensure that the right patient is chosen for the study, whereas prognostic biomarkers help identify patients with a greater likelihood of having a disease-related endpoint event when no treatment is administered. On the other hand, predictive biomarkers assist in choosing patients who are likely to respond to drug treatments. All the aforementioned biomarkers provide a strategy to decrease heterogeneity/variability in the outcome measure and hence to provide higher statistical power to detect the drug effect. Through the use of clinical phenotyping and identification of common genetic variants, small clinical trials enriched by biomarkers could be useful to establish proof of concept and further understanding of disease pathogenesis.

An early-phase (Phase 1b) study investigated the use of amyloid PET imaging as a screening tool to identify amyloid-positive (prognostic biomarker) Alzheimer's patients as an enrichment strategy for the clinical trial (Sevigny et al., 2016). Using combined biomarkers of amyloid-positive pathology and neurodegeneration, it was demonstrated that the signal-to-noise ratio improved substantially in the clinical endpoint measures, that is, the mean 2-year change in ADAS-Cog 13 increased by 66% as compared with the unenriched population, thus reducing the sample size by 45%–60% depending on the outcome measure (Wolz et al., 2016). These studies also explained the reason for failure of two Phase III trials of antiamyloid therapy (bapineuzumab and solanezumab) as being due to not enriching the trials based on amyloid-positive biomarkers.

Recently the Critical Path for Parkinson's (CPP) consortium submission of imaging of dopamine transporter (DAT) has received a letter of support (Food and Drug Administration, 2015) from the FDA's biomarker qualification program (described later in the chapter) as a prognostic enrichment biomarker in clinical trials of Parkinson's disease for patient enrollment. It is envisaged that reduction in DAT may serve to identify patients with increased likelihood of having progression of their Parkinson's disease symptoms, thus decreasing the variability in clinical trials.

A simpler approach for trial enrichment in early clinical trials would identify placebo responders through either a phenotypic or genotypic biomarker during a run-in phase and then exclude the placebo responders from the randomization phase. Such enrichment biomarkers are common in psychiatric indications especially in migraine trials in adolescents (Evers, Marziniak, Frese, & Gralow, 2009).

Knowledge management from prior trials in a therapeutic area can help develop a biomarker/clinical outcome relationship, which could be further utilized to evaluate the proof of concept for a new drug based on biomarker data from early trials. In multiple sclerosis, early PoC trials utilize magnetic resonance imaging (MRI) lesion counts as an endpoint, whereas the registration trials employ annualized relapse rate at 24 months (ARR-24) as the clinical endpoint. The PoC trials in multiple sclerosis are typically 12 months' duration, and the go/no-go decisions for promising drug/dose selections are qualitative in nature leading to failure rates in the registration trials. In a recent publication (Gopalakrishnan, Minocha, & Gobburu, 2015), the authors developed a biomarker/clinical endpoint relationship of MRI-T2/ARR-24 using pooled data from six prior multiple sclerosis development programs and demonstrated by clinical trial simulations that at least 60% reduction in MRI-T2 counts from placebo is needed to achieve an 80% probability of technical success of a drug in the late-phase registration trial, where ARR-24 is used as the clinical endpoint. This example demonstrates the utility of MRI-T2 as a predictive biomarker for decision-making in early clinical trials for promising drug compounds and the value of prior knowledge utilization.

3 Efficacy and Safety

The goal of late clinical development is to seek responses to three questions: (i) Is the drug safe and efficacious? (ii) What is the best strategy to use the drug (dosing) in practice? (iii) What are the insights into the trial database to optimize future development? Late clinical trials are registration trials based on which drug approval decisions are made. The role of biomarkers for enrichment and stratification in registration trials to identify subgroups of patients who are most likely to respond is to assist in magnifying the treatment effect by accounting for heterogeneity. It is acceptable to design these trials to recruit patients who are most likely to respond to an intervention based on placebo run-in response or enriching a trial using biomarker-positive subjects. There are several examples of the use of biomarkers in registration trials across therapeutic areas; however, in this section, we will highlight the case studies pertaining to CNS drugs. As an example, an early-phase trial of rosiglitazone evaluated for treatment of Alzheimer's disease detected significant treatment benefit in cognition in apolipoprotein E (APOE)-e4–negative subjects (Risner et al., 2006). However, a late clinical trial enriched using APOE-e4–negative subjects did not demonstrate any efficacy benefit over placebo (Gold et al., 2010). The treatment benefit found in the early trial in the biomarker-negative subjects could not be replicated in a larger efficacy trial indicating the importance of validation of biomarkers. More discussion on the validation of biomarkers is presented in Section B.

The role of biomarkers, both phenotypic and genomic, for the safe and effective use of drugs is multifactorial. Biomarkers have been used (i) for approval under the accelerated approval pathway, (ii) to propose one or more doses in patients with risk of toxicity, and (iii) as the basis for extensions of original approval.

The FDA-accelerated approval pathway (Guidance for Industry: Food and Drug Administration, 2014a) facilitates access to important treatments for serious or life-threatening diseases with limited or no therapeutic options. The approval is based upon a determination that the product has an effect on a surrogate endpoint that is reasonably likely to predict clinical benefit or on a clinical endpoint that can be measured earlier than irreversible morbidity or mortality that is reasonably likely to predict an effect on irreversible morbidity or mortality or other clinical benefit, taking into account the severity, rarity, or prevalence of the condition and the availability or lack of alternative treatments. The new provisions for accelerated approval under FDASIA (Food & Drug Safety and Innovation Act of 2012) provide additional flexibility to consider pharmacologic or other evidence using biomarkers or other scientific methods or tools, in conjunction with other data, in determining whether an endpoint is reasonably likely to predict clinical benefit. Recently, eteplirsen was granted accelerated approval for a rare genetic disease, Duchenne muscular dystrophy (DMD) (Food and Drug Administration, 2016b). DMD is caused by an absence of dystrophin, a protein that helps keep muscle cells intact. The approval, though very controversial, considered the evidence from a biomarker, that is, dystrophin increase (minimal) in skeletal muscle observed in some drug-treated subjects. There are several examples of accelerated approval for anticancer drugs; however, the examples in CNS drugs are limited.

Plasma drug concentrations and drug metabolism-based biomarkers are commonly used to identify a subset of patients, who may be at higher risk for toxicity or therapeutic failure, and the information is included in the drug labeling. Dosing recommendations in the drug label for tetrabenazine (Clinical Pharmacology review: Food and Drug Administration, 2008), a drug indicated for the treatment of chorea associated with Huntington's disease, mandates genotyping for the drug-metabolizing enzyme CYP2D6 to determine if the patient is a poor or extensive metabolizer, when the patients require doses greater than 50 mg per day. This is an example where pharmacogenomic biomarkers are used before approval to provide dosing recommendations and personalize therapy for a subset of the population.

The use of biomarkers for product updates and extensions includes the approval of new dosing regimens and use of the drug in different population or in some cases different indications. New formulations and generic drug products are typically approved on the basis of bioequivalence studies in which the biomarker is usually plasma drug concentrations. New dosing regimens for indications that were previously approved on the bases of mortality and/or morbidity endpoints could be approved on the basis of changes in biomarkers without repeating adequate and well-controlled clinical studies (Gobburu, 2009).

There are several examples where plasma drug concentration (exposure) have been used as a biomarker and assessment of the similarity of exposure-response relationships have been used as the basis for extrapolation and for deriving dosing recommendations. Drugs approved using the animal rule use efficacy data (dose-response relationship) from animals and pharmacokinetic information from humans to derive dosing recommendations in humans. For example, pyridostigmine (Clinical Pharmacology review: Food and Drug Administration, 2003), a reversible acetylcholine esterase inhibitor in combination with atropine and 2-pralidoxime, was approved as preventative/rescue treatment against the soman nerve agent. Using the monkey dose-survival data and the human and simian pharmacokinetic information, a dose that produced an equivalent concentration in humans was obtained. The extrapolation of efficacy from monkeys to humans was possible assuming the response to treatment (survival) will be similar in monkeys and humans and assuming similar plasma drug concentrations/clinical outcome relationship, where the drug concentration can be considered as a phenotypic biomarker. Similarly, extrapolation of efficacy from adults to pediatrics also assumes similarity in plasma drug concentration/response relationship between adults and pediatrics. Recently the FDA, Pediatric Epilepsy Academic Consortium on Extrapolation (PEACE), and University of Maryland investigated accumulated evidence from clinical trials with multiple antiepileptic drugs (AEDs) to evaluate whether extrapolation of efficacy from adult to pediatric patients can be made for the treatment of partial seizures. Based on quantitative analysis indicating similarity of drug exposure/response analysis between adults and pediatrics across several AEDs, the FDA concluded that extrapolation of efficacy from adults to pediatric patients 4 years of age and older is acceptable. The policy will make it unnecessary to conduct efficacy trials in pediatrics for future AEDs for partial-onset seizures and will lead to automatic approval in pediatrics when adult data are available (Pediatric News, 2016).

4 Postmarketing

The registration trials used for drug approval provide important information on the efficacy and safety of the drug; however, information on safety of a drug continues to evolve over the months and even years during the product's life cycle in the marketplace. Hence postmarketing studies are typically requested as part of the approval process to obtain complete information of the safety of the drug, and during that time, labels are constantly updated. There are several examples where pharmacogenomic biomarkers have been used to personalize dosing that may influence drug response, tolerability, or safety of drugs and updates to labeling have been made (Center for Drug Evaluation and FDA, 2016). In a recent article the labeling updates based on pharmacogenomic biomarkers to guide pharmacotherapy of 32 neuropsychiatric drugs were presented, including drug selection/dosing for tricyclic antidepressants, carbamazepine and phenytoin (Drozda, Müller, & Bishop, 2014). It is known that several of the neuropsychiatric drugs are substrates of the CYP metabolic enzymes that may put patients to higher risk of toxicity or therapeutic failure. Based on the review, of the 32 neuropsychiatric drugs listed, 27 (84%) have CYP2D6 metabolizer status listed as an important biomarker, 3 (9%) have CYP2C19 metabolizer status as an important biomarker, and 3 (9%) pertain to other genetic markers (e.g., major histocompatibility complex human leucocyte antigen [HLA] allele HLA-B*1502 for carbamazepine and phenytoin, carbamoyl phosphate synthetase 1, and ornithine carbamoyltransferase for valproic acid). The pharmacogenomic-guided dosing instructions are available for 10 (32%) neuropsychiatric drugs including pimozide.

Recently the labeling update for pimozide recommended new CYP2D6 genotype-guided dosing recommendations to prevent the occurrence of pimozide-induced arrhythmias in patients who are CYP2D6 poor metabolizers (Rogers et al., 2012). The occurrence of pimozide-induced arrhythmias is concentration dependent, and hence by using pharmacogenomic information and pharmacokinetic simulations, the updated label provided clearer dosing, titration, and genotype testing recommendations in both adults and pediatric patients, where CYP2D6 testing is recommended only above a certain dose threshold where QT prolongation risk is higher. The case studies presented here involved labeling updates based on postmarketing information, and they were incorporated into clinical practice as technology to perform the genetic testing became available. The FDA strongly encourages drug developers to address potential genetic liabilities before marketing, when possible, to support prescribing recommendations and to collect controlled data from exploratory studies.

III US FDA INITIATIVES

In a recent commentary (Woodcock, 2016), Janet Woodcock from the FDA has provided an interesting viewpoint of how the concept of precision medicine would lead to precision drug development in the future, where population-based results obtained from conventional drug development may no longer be acceptable and might give way to developing patient-focused interventions. More quantitative, mechanism-based understanding of the disease and the ability to predict using rapidly expanding computational tools could consequently reduce the variability of human responses to drugs, thereby significantly improving the benefit/risk balance of medicines on an individual level. Currently, several drugs have been coapproved with in vitro diagnostic tests that eventually tailor the treatment to individual patients, though almost all of these tests are for oncologic agents (Pacanowski & Huang, 2016). In this changing drug development landscape focused on small subsets of patients, identification of sensitive and specific biomarkers of drug response will play a crucial role, since measuring clinical outcomes for small populations might prove difficult. It can be clearly seen that the next generation of medicines is upon us and this section will discuss several FDA initiatives in the context of personalized medicine, and examples for CNS drugs will be provided wherever available.

A PDUFA VI Research Priorities

In 2016 the US FDA published its goals and commitment letters for the reauthorization of its Prescription Drug User Fee Act (PDUFA) for fiscal year 2018–22, known as PDUFA VI (Food and Drug Administration, 2016a). The document reflects the agency's performance and procedural goals to expedite bringing safer therapies to patients. One of the top research priorities stated was advancing the use of biomarkers and pharmacogenomics and exploring the use of real-world evidence for use in regulatory decision-making. Arguably, as discussed before, because of the complexities involved in CNS drug development, advancing the use of biomarkers would enormously benefit the CNS therapeutic area. The potential utility of imaging biomarkers in the development of CNS drug candidates has been discussed in detail in the literature (Hargreaves et al., 2015; Palmer, 2014), and their use can range from patient selection and stratification, measurement of engagement of a CNS drug to a target (target engagement) to measuring the pharmacodynamic effect at the site of action (intact brain). Advances in imaging technologies have led to increased use of molecular imaging biomarkers at various phases in the drug discovery and development process. A recent FDA consumer update document (FDA, 2016) has stressed the role of biomarkers to help identify patients to be enrolled in an Alzheimer's trial in an effort to increase the likelihood of trial success. The 2013 FDA draft guidance on Alzheimer's disease (Guidance for Industry: Food and Drug Administration, 2013) encourages research and discusses the FDA's thinking about conducting new clinical trials at the very early stages of

Alzheimer's disease in patients with no obvious symptoms or even no symptoms at all. The crucial aspect of conducting such a trial would be to correctly identify patients at risk of developing Alzheimer's disease. Considerable research is underway to identify biomarkers (gene mutations), which can help identify such patients at high risk of developing the disease, and this is a paradigm shift in Alzheimer's drug development, as the previously conducted trials had predominantly involved patients who already had the onset of disease.

A recent article from the FDA (Sherman et al., 2016) on the use of real-world evidence (RWE) for regulatory decision-making, though in a cautious tone, has discussed the FDA's commitment to robust policy development for use of real-world evidence to assess the safety and effectiveness of drugs in both premarketing and postmarketing scenarios. Use of RWE in CNS drug development can help potentially conduct enriched prospective studies that may complement the traditional development strategies and could lead to more generalizable therapeutic options.

B Biomarker Qualification Program

Another important FDA initiative is the biomarker qualification program (BQP) (Food and Drug Administration, 2017) established to support the Center for Drug Evaluation and Research (CDER) work with external stakeholders to develop biomarkers that aid in the drug development process. Development of validated and reliable biomarkers is critical for trial enrichment strategies, correct patient selection, and dose selection and to assess safety and efficacy of drugs. There remain very significant challenges for identifying and developing brain-based biomarkers that identify patient subpopulations or measure target engagement or surrogate outcome and a pressing need to validate or invalidate these potential biomarkers for a well-defined context of use. The FDA accepts the use of biomarkers in the drug approval process through investigational new drug (IND), NDA (and supplementary NDA), and biologics license application (BLA) submissions, where the burden to provide evidence on the biomarker is from a single sponsor and, if accepted, the biomarker information gets embedded in drug labels. For example, the carbamazepine label has a boxed warning that "testing for HLA-B*1502 should be performed in patients with ancestry in populations in which HLA-B*1502 may be present." However, to generalize the use of validated biomarkers across different drug development programs, in that context, the BQP offers a standardized framework, where the burden of proof is with a consortium rather than a single sponsor. If the biomarker is qualified, the information on the use of the biomarker gets announced as a draft guidance. At present, through the

BQP, four biomarkers have been qualified, where three of them are response (safety) biomarkers for nephrotoxicity and cardiotoxicity and one is a diagnostic biomarker for patient selection for a fungal infection (Amur, LaVange, Zineh, Buckman-Garner, & Woodcock, 2015).

The success of the BQP relies on multiple stakeholders, including the entire scientific community and the FDA, to identify potential biomarkers for qualification and to determine the level of evidence required for a biomarker to be validated. Based on a survey requested through the Federal Register Notice, with the goal to identify potential biomarkers for qualification and describing the context of use to address areas important to drug development, the two main therapeutic areas in critical need for biomarkers were neurological and neuropsychiatric diseases and oncology (Amur, 2015). The field of oncology has made excellent progress in developing biomarkers, but due to the highly complex nature of neurological and neuropsychiatric diseases, the progress with respect to biomarkers has been slow. However, the oncology experience can help inform CNS clinical research, using analytical and clinical validation and statistical considerations as critical steps in the successful identification of a biomarker use in drug trials as outlined in the succeeding text.

1 Analytical Validation

To establish analytical validation, one needs to create performance characteristics for accurately and reliably measuring the marker or measure of interest in the laboratory, including evaluating the sensitivity, specificity, reliability, and reproducibility of the method. Conditions for collecting the biomarker need to be standardized, quality control processes need to be used to document the biomarker's performance in clinical studies (and to determine if the data are of high quality or need to be discarded), and clinical staff need to be trained on the process. It also needs to be determined how compatible different manufacturers' devices are in obtaining the same signal for the biomarker of interest.

2 Clinical Validation

Clinical validation requires the detection or prediction of an associated disorder/domain of function or an outcome from very narrowly defined patient groups. Additional clinical validation considerations include age-related differences, medication status (e.g., in pediatric patients who may already be on other medications before the start of the drug trial, dose levels, and time of exposure would also need to be considered), illness stage, main symptom intensity, areas of the brain assessed, comorbid disorders, etc. One needs to understand the sources of technical and biological variability

in the biomarker measures, including test–retest reliability (intrasubject, intersite, and between site), task-specific brain activation (signal strength), longitudinal evaluation of measurement properties (stability), data quality, and data analyses. Finally, once all analytical and clinical validations are complete, the long-term goal would be to establish consensus methods and evidentiary standards to validate biomarkers, or composite biomarkers, for their intended context of use through the FDA-BQP. This process enables the biomarker to be used in late clinical trials as the efficacy endpoint.

3 Statistical Considerations for Validation

Since specific guidelines for statistical analysis to support qualification have not been developed, general statistical principles used by CDER-BQP are outlined based on published literature (Amur et al., 2015). The choice of the statistical methods for biomarker qualification will be a function of the context of use for a biomarker. Methods relevant to prognostic and predictive biomarkers are presented here. In general, prognostic biomarkers should require less statistical rigor as compared with a predictive biomarker. In addition to choice of the statistical methods, it is important to understand the data source, general knowledge about the nature of relationship between the biomarker and the clinical outcome.

For a prognostic biomarker the first step in qualification would be to show that a relationship exists and to estimate the strength of the relationship; a statistically significant positive Pearson's correlation coefficient will suffice, given that the relationship can be assumed to be approximately linear. Unknown or more complex relationships may be ascertained by modeling the relationship. Given the biomarker/clinical outcome relationship, the specific use of prognostic markers for stratification and enrichment comes into play. For the purpose of stratification, it is necessary to define discrete values of the biomarker to group the patients into homogeneous strata, if the biomarker is a continuous variable. The stratification using the prognostic biomarker can be done at randomization and/or adjusted in the statistical analysis (ANCOVA, analysis of covariance) to improve the statistical power to detect the treatment effects in the clinical trial. Upon stratification the statistical analysis for qualification involves demonstrating a sufficiently strong biomarker/clinical outcome relationship within the strata. If the prognostic biomarker is used for enrichment, then an important factor for qualification is to know the target range of the biomarker within which a specified proportion of patients will be enrolled.

A predictive biomarker is measured prior to therapy and provides information on the likely patients who would respond to the intervention. Hence the most straightforward way to establish that a biomarker is predictive is to demonstrate a statistically significant interaction between treatment and the biomarker status (marker positive vs negative) in the analysis to demonstrate the treatment effect.

C Regulatory Pathways for Approval

The FDA offers four programs (Guidance for Industry: Food and Drug Administration, 2014b) to facilitate and expedite the development and review of new drugs to address unmet medical need in the treatment of serious or life-threatening conditions, namely, fast-track designation, breakthrough therapy designation, accelerated approval, and priority review designation. So far, it can be said that the field of oncology, antibiotics, and HIV/ AIDS and the approval of certain rare diseases have benefitted the most with the help of these approval pathways, which have potentially accelerated drug development. In the case of CNS drug development, despite advances in neuroscience, there is a large uncertainty in the drug development pipeline due to the fact that the pathophysiology of major CNS disorders is less well delineated as compared with the other therapeutic areas mentioned previously. For example, the measurements of viral load and CD4 counts serve as surrogate endpoints predictive of clinical outcome and qualify well for accelerated approval. Similarly, in the case of oncology agents, demonstrating clinical benefit using intermediate clinical endpoints, such as tumor shrinkage often using a single arm trial, has led to accelerated approvals. However, in the case of neurology and neuropsychiatry, due to the fact that the likelihood of failure may be high, these policies have not still incentivized CNS drug development, and increased knowledge about and willingness to use these pathways for CNS disorders are needed. Given the enhanced research of identifying biomarkers for, for example, Alzheimer's disease and Parkinson's disease, the accelerated approval pathway could most benefit CNS drug development in the long run. A recent white paper from the FDA (White Paper: Food and Drug Administration, 2015) has described its willingness to use accelerated approval to accelerate Alzheimer's drug development, allowing companies to rely on more easily demonstrated short-term clinical endpoints, such as cognitive improvement or stabilization, if there is clear evidence of an effect on a valid and reliable cognitive assessment. The white paper also discusses the use of unqualified biomarkers for enriched trial designs and continued collaboration with ADNI and CAMD to expedite biomarker development.

The Generating Antibiotic Incentives Now (GAIN) of 2012 served as a pull incentive for developing new antibiotic agents offering breakthrough or fast-track designation if the antibiotic agent can obtain the qualified infectious disease product (QIDP) status. It is possible that a similar policy could act as a pull incentive for pharmaceutical companies to invest in CNS research and thereby accelerate CNS drug development (Choi et al., 2014).

D Critical Path Innovation Meeting

The Critical Path Innovation Meeting (CPIM) (Guidance for Industry: Food and Drug Administration, 2015) is another FDA initiative through which the Center for Drug Evaluation and Research (CDER) and drug developers can communicate to improve efficiency and success in drug development. Through the CPIM forum, drug developers can discuss with the FDA, topics such as biomarkers in early phase of development and which are not yet ready for the BQP and innovative conceptual approach to clinical trial design and analysis. To date, there has been 33 CPIMs (Research, 2017) across different therapeutic areas and a wide range of topics. The CPIM discussions could be another venue for CNS drug developers to efficiently engage with the FDA early on in the drug development program.

IV CHALLENGES AND FUTURE PERSPECTIVES

Drug development for CNS disorders has had mixed success. The challenges to CNS drug development are numerous due to various reasons, such as lack of clear understanding of mechanisms of neuropsychiatric disorders, heterogeneity of CNS disorders, absence of traditional preclinical disease models to translate potential efficacy to early-phase trials, and/or high placebo response and dropouts. For example, several experimental compounds for modifying Alzheimer's disease failed in late clinical trials. The lack of biomarkers is compounded by the fact that we do not have any experience with disease-modifying drug effects. The difficulty in developing and commercializing new treatments has led large biopharmaceutical companies to downsize or restrategize their early-stage CNS drug development efforts despite the growing CNS disease burden. However, scientific advances in genetic sequencing, multimodality fusion imaging technologies, and other methods offer promising ways to accelerate the understanding of the systems biology of disease and to help in the development of new therapies.

Given the rapid scientific improvements, the next generation of medicines are projected to involve the development and evaluation of targeted therapies

(Pacanowski, Leptak, & Zineh, 2014). Identifying a specific subset of patients, defined by molecular and pharmacologic targets, has proven to be a viable pathway for bringing drugs to market. Since 2010 the approval of targeted therapies by the FDA constitutes about 20% of approvals. The FDA's approval of 19 in vitro companion diagnostics for 12 drugs, mostly in the oncology area, suggests the FDA's and industry's acceptance of personalized medicine strategies. Though there exist hurdles to the development of companion in vitro diagnostics for neuropsychiatric disorders (Nikolcheva, Jäger, Bush, & Vargas, 2011), the advent of molecular imaging biomarkers and next-generation genomic sequencing could spur the development of new diagnostics that could identify patients to enroll in the trial who are more likely to respond to treatment, thus enhancing the signal-to-noise ratio to detect treatment effects.

Recently, advances in genomic sequencing have led to the understanding that depression by itself may not be one disease rather an amalgamation of many diseases as the first robust genetic links to depression are being reported (Ledford, 2015). The researchers hope that as more genetic links are found, the pathways involved could be investigated as drug targets and could help in making the diagnosis of depression more definitive and objective, rather than being based on subjective endpoints. In addition, the use of molecular imaging biomarkers might be valuable to set new standards of treatment for patients in the latent phase of progressive neurodegenerative disorders, as the biomarkers could enable detection, prediction, prevention, and tracking of disease as opposed to a delayed treatment paradigm, that is, waiting for the symptoms of the disease to emerge.

Over the past decade, there has been an overload of data and information owing to the "big data" revolution. The field of neuroscience has also been characterized by the systematic collection of multimodal, longitudinal data in initiatives such as the Alzheimer's Disease Neuroimaging Initiative (ADNI Alzheimer's Disease Neuroimaging Initiative, 2013), the Coalition Against Major Disease (CAMD Critical Path Institute, 2017a), and the BRAIN initiative. Without doubt the upcoming decades are going to follow a similar or more vigorous trend of accumulating vast amounts of data. With rapid innovation in mobile health and wearable devices, the way the endpoints are collected will likely change (Capone, 2015). For example, the 6-minute walk test, used as an assessment of disease severity, in clinical trials involving cardiovascular, respiratory, and CNS diseases, might now be measured continuously from patients wearing devices. Since the device will continuously monitor and measure, the physician and clinical trial researchers will have access to much richer and more nuanced data of disease severity as opposed to the traditional 6-minute test. This rich information can be valuable to identify the right patients to enroll in the

trial and can lead to a more informed objective measure of the clinical endpoint.

Given the informatics revolution the main challenge remains as to how to turn big data into actionable knowledge that can further the understanding of brain function and identify biomarkers for prognostic and diagnostic uses and thus increase the efficiency of CNS drug development. The most important challenge is identified as the need for an open data culture (Manji, Insel, Insel, & Narayan, 2014), having mechanisms or processes for researchers to share data despite the competitive environment. The article mentioned earlier states that certain public-private partnerships such as the Biomarker's consortium and the Accelerating Medicine Partnership do exist but with limited success. The data-sharing process is limited by unresolved policy issues from the funding agencies like NIH, the largest funding agency for biomedical research. Though NIH requires researchers to submit a data-sharing plan for investigator-initiated grants above $500,000/year, the process does not seem to be effective if the clinical study duration is long.

In spite of the limitations, there exist some successful examples of efficient data sharing in NIH contract-based projects in neuroscience, such as the Human Connectome Project, a consortium funded by NIH institutes, to map the connectional brain anatomy of 1200 volunteers. All data from this study are available publicly. Similarly the National Database for Autism Research (NDAR) with data from 70,000 subjects and the Psychiatric Genomics Consortium (PGC) consisting of genomic data shared between 80 investigators from 30 countries are success stories with efficient mechanisms for sharing patient-level data. More such success stories are needed to help understand the complexity of the brain, to identify biomarkers effectively, and to allow their use for efficient trial designs (e.g., adaptive trial designs). Adaptive trials use accumulating data in an ongoing trial to make subsequent clinical study decisions, such as adaptive allocation or early stopping for futility/success. Early-phase trials can use adaptive allocation based on response to a sensitive biomarker or decide on doses to carry forward based on a pharmacodynamic response, thus leading to efficient go/no-go decisions. Adaptive trial designs have been effectively used in the field of oncology and diabetes and have led to significant reduction in late-phase attrition. Judicious use of adaptive designs could help the CNS drug development area especially by facilitating efficient go/no-go decisions in early-phase clinical trials.

V CONCLUSIONS

The current climate of rapid scientific advancements and innovations in technology to understand the intricacies of the brain holds great promise for a more successful

CNS drug development paradigm. With increased understanding of brain mechanisms, the use of valid molecular and imaging biomarkers for target identification, target engagement, and trial enrichment could lead to better decision-making during drug development and thus may reduce the late-phase failure rate. Converting the unprecedented amount of diverse data ranging from genetic to clinical phenotypes collected as part of the "big data" revolution to actionable knowledge could revolutionize development of effective treatments for neurodegenerative disorders, such as Alzheimer's disease and Parkinson's disease. It can be envisioned that the FDA would play an important role through its PDUFA VI commitments to enhance the use of biomarkers and pharmacogenomic information and also to incorporate the use of "big data" and RWE for evaluating the safety and the effectiveness of drugs in the forthcoming decades. Regulatory initiatives such as BQP and CPIM could be immensely useful for efficient CNS drug development.

References

21st Century Cures. (2014). Retrieved September 15, 2016, from https://energycommerce.house.gov/cures.

Alzheimer's Association. (2015). 2015 Alzheimer's disease facts and figures. *Alzheimer's & Dementia: The Journal of the Alzheimer's Association, 11*(3), 332–384.

Alzheimer's Disease Neuroimaging Initiative. (2013). *ADNI home*. Retrieved January 31, 2017, from http://www.adni-info.org/.

Amur, S. (2015). *FDA's effort to encourage biomarker development and qualification*. Retrieved January 4, 2017, from https://c-path.org/wp-content/uploads/2015/08/EvConsid-Symposium-20150821-I-01-SAmur-FINAL.pdf.

Amur, S., Frueh, F. W., Lesko, L. J., & Huang, S. -M. (2008). Integration and use of biomarkers in drug development, regulation and clinical practice: a US regulatory perspective. *Biomarkers in Medicine, 2*(3), 305–311. https://doi.org/10.2217/17520363.2.3.305.

Amur, S., LaVange, L., Zineh, I., Buckman-Garner, S., & Woodcock, J. (2015). Biomarker qualification: toward a multiple stakeholder framework for biomarker development, regulatory acceptance, and utilization. *Clinical Pharmacology & Therapeutics, 98*(1), 34–46.

Biomarkers Definitions Working Group (2001). Biomarkers and surrogate endpoints: preferred definitions and conceptual framework. *Clinical Pharmacology and Therapeutics, 69*(3), 89–95. https://doi.org/10.1067/mcp.2001.113989.

Capone, M. (2015). *How wearables and mobile health tech are reshaping clinical trials*. Retrieved January 31, 2017, from http://venturebeat.com/2015/04/17/how-wearables-and-mobile-health-tech-are-reshaping-clinical-trials/.

Center for Drug Evaluation and FDA (2016). *Genomics—Table of pharmacogenomic biomarkers in drug labeling [WebContent]*. Retrieved January 3, 2017, from http://www.fda.gov/drugs/scienceresearch/researchareas/pharmacogenetics/ucm083378.htm.

Centre for Research on the Epidemiology of Disasters (2017). *Drug innovation—Critical path innovation meeting (CPIM) topics held to date*. Retrieved April 13, 2017, from https://www.fda.gov/Drugs/DevelopmentApprovalProcess/DrugInnovation/ucm444165.htm.

Choi, D. W., Armitage, R., Brady, L. S., Coetzee, T., Fisher, W., Hyman, S., et al. (2014). Medicines for the mind: policy-based "pull" incentives for creating breakthrough CNS drugs. *Neuron, 84*(3), 554–563. https://doi.org/10.1016/j.neuron.2014.10.027.

Clinical Pharmacology review: Food and Drug Administration. (2003). *20-414 Pyridostigmine bromide clinical pharmacology biopharmaceutics review Part 1—20-414_Pyridostigmine Bromide_biopharmr_P1.pdf*. Retrieved April 14, 2017, from https://www.accessdata.fda.gov/drugsatfda_docs/nda/2003/20-414_Pyridostigmine%20Bromide_biopharmr_P1.pdf.

Clinical Pharmacology review: Food and Drug Administration. (2008). *Tetrabenazine clinical pharmacology review*. Retrieved October 1, 2016, from http://www.accessdata.fda.gov/drugsatfda_docs/nda/2008/021894s000_ClinPharmR_P1.pdf.

Collins, F. S., & Varmus, H. (2015). A new initiative on precision medicine. *New England Journal of Medicine, 372*(9), 793–795.

Critical Path Institute. (2017a). *CAMD | critical path institute*. Retrieved September 15, 2016, from https://c-path.org/programs/camd/.

Critical Path Institute. (2017b). *PSTC | critical path institute*. Retrieved September 28, 2016, from https://c-path.org/programs/pstc/.

Drozda, K., Müller, D. J., & Bishop, J. R. (2014). Pharmacogenomic testing for neuropsychiatric drugs: current status of drug labeling, guidelines for using genetic information, and test options. *Pharmacotherapy, 34*(2), 166–184. https://doi.org/10.1002/phar.1398.

Evers, S., Marziniak, M., Frese, A., & Gralow, I. (2009). Placebo efficacy in childhood and adolescence migraine: an analysis of double-blind and placebo-controlled studies. *Cephalalgia, 29*(4), 436–444. https://doi.org/10.1111/j.1468-2982.2008.01752.x.

FDA. (2016). *FDA consumer update: FDA facilitates research on earlier stages of Alzheimer's disease*. Retrieved January 2, 2017, from http://www.fda.gov/downloads/ForConsumers/ConsumerUpdates/UCM520398.pdf.

Food and Drug Administration. (2009). *Sertindole advisory committee meeting*. Retrieved September 28, 2016, from http://www.fda.gov/downloads/AdvisoryCommittees/CommitteesMeetingMaterials/Drugs/PsychopharmacologicDrugsAdvisoryCommittee/UCM248413.pdf.

Food and Drug Administration. (2015). *Letter of support to the critical path's institute's caolition against major diseases (CAMD)*. Retrieved September 29, 2016, from http://www.fda.gov/downloads/Drugs/DevelopmentApprovalProcess/UCM439715.pdf.

Food and Drug Administration. (2016a). *PDUFA reauthorization performance goals and procedures fiscal years 2018 through 2022*. Retrieved January 4, 2017, from http://www.fda.gov/downloads/ForIndustry/UserFees/PrescriptionDrugUserFee/UCM511438.pdf.

Food and Drug Administration. (2016b). *Press announcements—FDA grants accelerated approval to first drug for Duchenne muscular dystrophy [WebContent]*. Retrieved October 2, 2016, from http://www.fda.gov/NewsEvents/Newsroom/PressAnnouncements/ucm521263.htm.

Food and Drug Administration. (2017). *Biomarker qualification program*. Retrieved April 14, 2017, from https://www.fda.gov/Drugs/DevelopmentApprovalProcess/DrugDevelopmentToolsQualificationProgram/BiomarkerQualificationProgram/default.htm.

Gobburu, J. V. S. (2009). Biomarkers in clinical drug development. *Clinical Pharmacology and Therapeutics, 86*(1), 26–27. https://doi.org/10.1038/clpt.2009.57.

Gold, M., Alderton, C., Zvartau-Hind, M., Egginton, S., Saunders, A. M., Irizarry, M., et al. (2010). Rosiglitazone monotherapy in mild-to-moderate Alzheimer's disease: results from a randomized, double-blind, placebo-controlled phase III study. *Dementia and Geriatric Cognitive Disorders, 30*(2), 131–146. https://doi.org/10.1159/000318845.

Gomez-Mancilla, B., Marrer, E., Kehren, J., Kinnunen, A., Imbert, G., Hillebrand, R., et al. (2005). Central nervous system drug development: an integrative biomarker approach toward individualized medicine. *NeuroRX, 2*(4), 683–695. https://doi.org/10.1602/neurorx.2.4.683.

Gopalakrishnan, M., Minocha, M., & Gobburu, J. (2015). Leveraging magnetic resonance imaging—annualized relapse rate relationship to aid early decision making in multiple sclerosis clinical drug development. *Medical Research Archives, 2*(3), 15–21.

Grabb, M. C., & Gobburu, J. V. S. (2016). Challenges in developing drugs for pediatric CNS disorders: a focus on psychopharmacology. *Progress in Neurobiology*. https://doi.org/10.1016/j.pneurobio. 2016.05.003.

Guidance for Industry: Food and Drug Administration. (2012). *Enrichment strategies for clinical trials to support approval of human drugs and biological products—ucm332181.pdf*. Retrieved September 15, 2016, from http://www.fda.gov/downloads/drugs/guidancecompliance regulatoryinformation/guidances/ucm332181.pdf.

Guidance for Industry: Food and Drug Administration. (2013). *Alzheimer's disease: Developing drugs for the treatment of early stage disease—ucm338287.pdf*. Retrieved January 4, 2017, from http://www.fda. gov/downloads/drugs/guidancecomplianceregulatoryinformation/ guidances/ucm338287.pdf.

Guidance for Industry: Food and Drug Administration. (2014a). *Expedited programs for serious conditions—Drugs and biologics—UCM358301.pdf*. Retrieved October 2, 2016, from http://www.fda.gov/downloads/ Drugs/GuidanceComplianceRegulatoryInformation/Guidances/ UCM358301.pdf.

Guidance for Industry: Food and Drug Administration. (2014b). *Expedited programs for serious conditions—Drugs and biologics—ucm358301.pdf*. Retrieved January 5, 2017, from http://www.fda.gov/downloads/ drugs/guidancecomplianceregulatoryinformation/guidances/ucm 358301.pdf.

Guidance for Industry: Food and Drug Administration. (2015). *Critical path innovation meetings guidance for industry—UCM417627.pdf*. Retrieved April 13, 2017, from https://www.fda.gov/downloads/ Drugs/GuidanceComplianceRegulatoryInformation/Guidances/ UCM417627.pdf.

Hargreaves, R., Hoppin, J., Sevigny, J., Patel, S., Chiao, P., Klimas, M., et al. (2015). Optimizing central nervous system drug development using molecular imaging. *Clinical Pharmacology & Therapeutics*, 98(1), 47–60. https://doi.org/10.1002/cpt.132.

Kesselheim, A. S., Hwang, T. J., & Franklin, J. M. (2015). Two decades of new drug development for central nervous system disorders. *Nature Reviews Drug Discovery*, 14(12), 815–816. https://doi.org/10.1038/ nrd4793.

Khanna, I. (2012). Drug discovery in pharmaceutical industry: productivity challenges and trends. *Drug Discovery Today*, 17(19–20), 1088–1102. https://doi.org/10.1016/j.drudis.2012.05.007.

Kowal, S. L., Dall, T. M., Chakrabarti, R., Storm, M. V., & Jain, A. (2013). The current and projected economic burden of Parkinson's disease in the United States. *Movement Disorders*, 28(3), 311–318. https://doi. org/10.1002/mds.25292.

Ledford, H. (2015). First robust genetic links to depression emerge. *Nature News*, 523(7560), 268. https://doi.org/10.1038/523268a.

Lesko, L. J., & Atkinson, A. J. (2001). Use of biomarkers and surrogate endpoints in drug development and regulatory decision making: criteria, validation, strategies. *Annual Review of Pharmacology and Toxicology*, 41, 347–366. https://doi.org/10.1146/ annurev.pharmtox. 41.1.347.

Manji, H. K., Insel, T. R., Insel, T. W., & Narayan, V. A. (2014). Harnessing the informatics revolution for neuroscience drug R&D. *Nature Reviews. Drug Discovery*, 13(8), 561–562. https://doi.org/ 10.1038/nrd4395.

National Institutes of Health. (2015). *Brain research through advancing innovative neurotechnologies (BRAIN)—National institutes of health (NIH)*. Retrieved September 14, 2016, from https://www. braininitiative.nih.gov/about/index.htm.

Nikolcheva, T., Jäger, S., Bush, T. A., & Vargas, G. (2011). Challenges in the development of companion diagnostics for neuropsychiatric disorders. *Expert Review of Molecular Diagnostics*, 11(8), 829–837. https://doi.org/10.1586/erm.11.67.

Pacanowski, M., & Huang, S. M. (2016). Precision medicine. *Clinical Pharmacology and Therapeutics*, 99(2), 124–129. https://doi.org/ 10.1002/cpt.296.

Pacanowski, M. A., Leptak, C., & Zineh, I. (2014). Next-generation medicines: past regulatory experience and considerations for the future. *Clinical Pharmacology and Therapeutics*, 95(3), 247–249. https://doi.org/10.1038/clpt.2013.222.

Palmer, A. M. (2014). The utility of biomarkers in CNS drug development. *Drug Discovery Today*, 19(3), 201–203. https://doi. org/10.1016/j.drudis.2013.11.016.

Pankevich, D. E., Altevogt, B. M., Dunlop, J., Gage, F. H., & Hyman, S. E. (2014). Improving and accelerating drug development for nervous system disorders. *Neuron*, 84(3), 546–553. https://doi.org/10.1016/ j.neuron.2014.10.007.

Pediatric News (2016). *Journal of Pediatric Pharmacology and Therapeutics*, 21(1), 98. https://doi.org/10.5863/1551-6776-21.1.98.

Rabiner, E. A., Bhagwagar, Z., Gunn, R. N., Sargent, P. A., Bench, C. J., Cowen, P. J., et al. (2001). Pindolol augmentation of selective serotonin reuptake inhibitors: PET evidence that the dose used in clinical trials is too low. *The American Journal of Psychiatry*, 158(12), 2080–2082. https://doi.org/10.1176/appi. ajp.158.12.2080.

Risner, M. E., Saunders, A. M., Altman, J. F. B., Ormandy, G. C., Craft, S., Foley, I. M., et al. (2006). Efficacy of rosiglitazone in a genetically defined population with mild-to-moderate Alzheimer's disease. *The Pharmacogenomics Journal*, 6(4), 246–254. https://doi.org/ 10.1038/sj.tpj.6500369.

Rogers, H. L., Bhattaram, A., Zineh, I., Gobburu, J., Mathis, M., Laughren, T. P., et al. (2012). CYP2D6 genotype information to guide pimozide treatment in adult and pediatric patients: basis for the U.S. Food and Drug Administration's new dosing recommendations. *The Journal of Clinical Psychiatry*, 73(9), 1187–1190. https://doi.org/10.4088/JCP.11m07572.

Sevigny, J., Suhy, J., Chiao, P., Chen, T., Klein, G., Purcell, D., et al. (2016). Amyloid PET screening for enrichment of early-stage Alzheimer disease clinical trials: experience in a Phase 1b clinical trial. *Alzheimer Disease and Associated Disorders*, 30(1), 1–7. https:// doi.org/10.1097/WAD.0000000000000144.

Sherman, R. E., Anderson, S. A., Dal Pan, G. J., Gray, G. W., Gross, T., Hunter, N. L., et al. (2016). Real-world evidence—what is it and what can it tell us? *The New England Journal of Medicine*, 375(23), 2293–2297. https://doi.org/10.1056/NEJMsb1609216.

Society for Neuroscience. (2017). *Society for neuroscience*. Retrieved September 13, 2016, from http://www.sfn.org/Advocacy/ Neuroscience-Funding/Science-Funding-Advocacy-Tools/Making- the-Case-for-NIH-Funding.

Vos, T., Flaxman, A. D., Naghavi, M., Lozano, R., Michaud, C., Ezzati, M., et al. (2012). Years lived with disability (YLDs) for 1160 sequelae of 289 diseases and injuries 1990-2010: a systematic analysis for the global burden of disease study 2010. *Lancet*, 380(9859), 2163–2196. https://doi.org/10.1016/S0140-6736(12) 61729-2. [London, England].

White Paper: Food and Drug Administration. (2015). *Targeted drug development: Why are many diseases lagging behind?— UCM454996.pdf*. Retrieved September 20, 2016, from http://www. fda.gov/downloads/AboutFDA/ReportsManualsForms/Reports/ UCM454996.pdf.

Wolz, R., Schwarz, A. J., Gray, K. R., Yu, P., Hill, D. L. G., & Alzheimer's Disease Neuroimaging Initiative. (2016). Enrichment of clinical trials in MCI due to AD using markers of amyloid and neurodegeneration. *Neurology*, 87(12), 1235–1241. https://doi.org/ 10.1212/WNL.0000000000003126.

Woodcock, J. (2016). "Precision" drug development? *Clinical Pharmacology and Therapeutics*, 99(2), 152–154. https://doi.org/ 10.1002/cpt.255.

Zineh, I., & Pacanowski, M. A. (2011). Pharmacogenomics in the assessment of therapeutic risks versus benefits: inside the United States food and drug administration. *Pharmacotherapy*, 31(8), 729–735. https://doi.org/10.1592/phco.31.8.729.

19

Regulatory Considerations for the Use of Biomarkers and Personalized Medicine in CNS Drug Development: A European Perspective

Eamon O'Loinsigh, Anjana Bose

Synchrogenix, A Certara Company, Delaware Corporate Center, Wilmington, DE, United States

I INTRODUCTION

The lack of effective treatments for central nervous system (CNS) disorders represents an area of major unmet medical need. The growing economic burden associated with an aging population more prone to developing neurodegenerative and psychiatric diseases poses a unique challenge (GBD 2015 Neurological Disorders Collaborator Group, 2017; Millan, Goodwin, Meyer-Lindenberg, & Ögren, 2015). Neurology is an area in which few new effective treatments for severe debilitating diseases, such as stroke, Alzheimer's disease (AD), and Parkinson's disease (PD), have been successfully developed in recent decades. In psychiatry, there is a need to develop more effective therapeutic interventions with improved efficacy to manage mood disorders and reduce the rates of suicide attributable to diseases such as major depression and schizophrenia.

Regulatory agencies have recognized the urgent need for treatments of rare disorders that are serious but neglected, given the multiple challenges with development and market access (e.g., difficulties in performing clinical trials in smaller patient populations, pricing to recuperate the research, and development costs). The increased awareness was in part driven by a patient-centric movement and greater involvement of multiple patient groups and caregivers who are seeking solutions. Much progress has occurred in the area of oncology, and approaches developed for oncology are gradually spreading to other disease areas including neurological disorders that often have complex genetic roots and may benefit from targeted (or personalized) approaches to therapeutic interventions. Technological innovations coupled with novel methodologies that are supported by regulatory agencies have led to multiple new initiatives to drive new approaches to address these needs.

Personalized medicine (PM) is a new paradigm that represents a shift from the traditional medicine approach, which applies the same treatment regimen to all patients affected by a disease regardless of phenotype. Personalized medicine capitalizes on known variability in gene expression that results in differences in susceptibility to diseases and responses to medicines. Genetic data are combined with environment and lifestyle data to group patients according to their likely response to a specific intervention to better target treatment and prevention (World Health Organization, 2017).

This chapter will summarize the European regulatory system that is embracing new technologies and research approaches in support of the development and approval of PMs for CNS disorders.

II THE EUROPEAN REGULATORY NETWORK AND APPROACH TO PERSONALIZED MEDICINE

European pharmaceutical legislation facilitates the creation of a single market for pharmaceuticals in the European Union (EU). The European Medicines Agency (EMA) is a decentralized agency responsible for the scientific evaluation, supervision, and safety monitoring of medicines in the EU (EMA, 2016a, 2017a). It oversees a network of individual national Regulatory Agencies in the EU member states (e.g., Medicines Evaluation Board in the Netherlands and Medical Products Agency in Sweden).

ISSN: 1569-7339
https://doi.org/10.1016/B978-0-12-803161-2.00019-9

The EMA is a major contributor to international regulatory science (Hemmings, Germain, & Warner, 2018), and over the last two years, the EMA has initiated a new agenda focused on innovation and greater patient access, leading to increasing collaboration with other global regulatory authorities including the US Food and Drug Administration (FDA) and the Japanese Pharmaceuticals and Medical Devices Agency (PMDA) to determine new regulatory approaches for novel methodologies in drug development research including PM.

A primary responsibility of the EMA is the evaluation of marketing authorization applications (MAAs) submitted through the centralized procedure leading to a single EU-wide approval with identical approved indications and conditions of use in all member states including the prescribing information (summary of product characteristics [SmPC]). The centralized procedure is compulsory for certain types of new medicines including those used to treat neurodegenerative disorders, medicines for rare diseases (orphan drugs), medicines derived from biotechnology products, and advanced therapy medicinal products (ATMPs), such as gene therapy, somatic cell therapy, and tissue-engineered medicines. The inclusion of neurodegenerative diseases, including AD, PD, and multiple sclerosis (MS), within the mandatory scope of the centralized procedure acknowledges the major unmet medical need in this therapeutic area and ensures that any new treatments will be made available to all EU citizens upon approval. There is also increased cooperation with notified bodies and other groups responsible for the development and approval of medical devices (e.g., in vitro diagnostic devices), because many PM approaches are dependent on the codevelopment and approval of companion diagnostics (CDx) in identifying patient subgroups likely to benefit from the treatment (EMA, 2017e, 2017h, 2018b).

The EMA is committed to innovation in drug development, and its strategic plan acknowledges the need to strengthen regulatory capability across the EU regulatory network to address new areas, such as PMs, and ensure that it has the capability to adequately assess, regulate, and monitor novel products of the future as well as provide access to patients without delay (EMA, 2015i, 2017h, 2018f). The EMA's Innovation Task Force (ITF) aims to facilitate the development of innovative medicines by addressing gaps in regulatory support for products in early development. The scope of ITF activities encompasses emerging therapies (e.g., gene therapy, cellular therapy, and engineered tissues), technologies (e.g., genomics or proteomics surrogates), and new methods of defining target populations (e.g., pharmacogenomics [PGx]) (EMA, 2014c).

While there is no universally accepted definition of PM, the following definition adopted by the European Council is now widely accepted in Europe:

> A medical model using characterisation of individuals' phenotypes and genotypes (e.g., molecular profiling, medical imaging, lifestyle data) for tailoring the right therapeutic strategy for the right person at the right time, and/or to determine the predisposition to disease and/or to deliver timely and targeted prevention. Personalised medicine relates to the broader concept of patient-centred care, which takes into account that, in general, healthcare systems need to better respond to patient needs. (Council of the European Union, 2015, p. C421/3)

The EMA notes that related terms include precision medicine and stratified medicine. Precision medicine is used as a synonym for personalized medicine, although some reserve it for targeted treatment guided by biomarkers (BMs). Alternatively, stratified medicine refers more specifically to the process of using genetic and physical characteristics to identify the right therapeutic strategy for subsets of patients. These concepts differ from the concept of individualized medicine, which refers to a tailor-made medical treatment (e.g., products based on a patient's own cells) (EMA, 2017c).

In March 2017 the EMA convened a workshop on PMs to gather the perspectives of various stakeholders responsible for overseeing the development, approval, reimbursement, and delivery of health care in Europe on aspects and challenges of developing PMs (EMA, 2017b, 2017c). The various EMA groups represented included the Committee for Medicinal Products for Human Use (CHMP), Pharmacovigilance Risk Assessment Committee (PRAC), Committee for Orphan Medicinal Products (COMP), Committee for Advanced Therapies (CAT), Pediatric Committee, and Scientific Advice Working Party (SAWP). The workshop also included representatives from health technology appraisal agencies, who are responsible for assessing the clinical effectiveness and cost-effectiveness of new treatments, and patient groups in recognition of the fact that regulatory approval is only one component of providing access to innovative new treatments.

A key message from the workshop was that PM requires a major change in the way medicines are tested and evaluated and must bring together all stakeholders. Changes in the way health care is delivered and how health-care systems are structured must also occur. Widespread implementation of PM will require improving patients' health literacy to allow them to become the center of health care. Educating health-care professionals to enable interpretation of the new types of data will also be beneficial.

Certain PM approaches require the use of extensive individual patient- and population-based data, which raises challenges for the integration and communication of those data into clinical practice. Finally the growth of PMs creates additional challenges in terms of data protection and patient privacy, particularly given the increased focus on transparency and disclosure of clinical trial data (e.g., EMA Policy 0070) and increasingly stringent EU data protection requirements (e.g., the new EU General Data Protection Regulation [GDPR]) (EMA, 2014d; EU General Data Protection Regulation, 2018).

This manuscript will discuss ongoing application of PM approaches in CNS drug development (Section III). The different mechanisms and regulatory pathways by which the EMA can facilitate the development of PMs are also discussed, including PGx, genetic BMs, CDx, and rare diseases/orphan drugs (Section IV); novel methodologies and innovative clinical trial designs (Section V); development of ATMPs, including gene therapies, somatic cell therapies, and tissue-engineered medicines (Section VI); and miscellaneous other approaches including public-private research partnerships (Section VII).

III CNS DISEASES AND PERSONALIZED MEDICINE

In recent decades, there have been few major advances in the development of new medicines to treat major neurological and psychiatric diseases despite the enormous resources devoted to this area by the pharmaceutical industry and academia. However, there are multiple ongoing initiatives to improve the diagnosis and classification of CNS disorders; develop a deeper knowledge of underlying risk factors; improve the design and outcome of clinical trials; develop reliable BMs to identify patient subgroups and predict medication efficacy; and promote collaborative approaches to innovation by uniting regulators, industry, academia, and patients (Arnerić, Kern, & Stephenson, 2018; EMA, 2013a, 2014a; Gooch, Pracht, & Borenstein, 2017; Insel & Cuthbert, 2015; Millan et al., 2015).

An aging population, improved diagnostic capabilities, innovative research, and rising health-care costs have led to greater awareness of the detection and treatment of rare and chronic progressive neurological conditions. The primary focus has been on improving quality of life with improved social and occupational functioning by early detection and developing treatments that reduce the severity of the diseases, reduce the rate of progression, and prevent relapses. Regulatory agencies have opened new avenues for designing and executing studies, thus setting the stage for new treatment approvals focused on specific patient groups in rare disease areas (Arnerić et al., 2018; Council of the European Union, 2016; National Institutes of Health, n.d.).

Significant inroads have been made in oncology through precision medicine, leading to many innovative therapies targeting specific genetic markers. In many other disease areas, although considerable efforts have been made to identify specific genotypes linked to a specific disorder, establishing a direct correlation remains

an elusive goal due to the complex interplays of the hundreds of genes, each with a small contribution. As the diagnoses of many mental disorders are built on signs and symptoms derived from self-reports or observations from family members/caregivers rather than underlying biology, establishing the diagnostic validity for promising targets identified in preclinical studies and bringing them to studies in patients continue to be challenging. Research is ongoing to obtain a better understanding of brain function and to integrate the learnings with behavioral components to move from "symptom-based categories" to "data-driven categories" (Insel & Cuthbert, 2015; Schizophrenia Working Group of the Psychiatric Genomic Consortium, 2014; Venigalla et al., 2017).

The US National Institute of Mental Health (NIMH)'s "precision medicine for mental disorders project," known as the Research Domain Criteria (RDoC) initiative, is a prominent example of an attempt to rethink psychopathology by building a framework beyond basic symptomatology that includes genetic, behavioral, and self-reported aspects (Cuthbert & Insel, 2013; Insel et al., 2010; Insel & Cuthbert, 2015). This approach considers how personalized (or precision) medicine could deconstruct traditional symptom-based disease categories by studying patients with a range of mood disorders across several analytical platforms (e.g., genetic risk, brain activity, and physiology) to parse current heterogeneous symptoms into more discrete homogeneous clusters. It is noted that the RDoC initiative has served as a catalyst for other efforts to transform the diagnostic process outside the United States including the EU-funded Roadmap for Mental Health Research and the CNS projects funded by the EU Innovative Medicines Initiative (IMI) (see Section VII) to link clinical neuropsychiatry and quantitative neurobiology (Insel & Cuthbert, 2015; Millan et al., 2015). Identification of newly defined homogeneous data-driven patient clusters will require prospective replication and stratified clinical trials to validate these new clinical constructs and the ability of therapeutic interventions to improve clinical outcomes for these patient groups. It should be noted that using a variety of data sources (e.g., combining PGx signatures with brain imaging and life experiences) to identify new patient clusters and demonstrate therapeutic effects in clinical trials may raise practical challenges in identifying the same patient groups in a real-world clinical setting and may require changes to the education of health-care professionals and to the use of new diagnostic tools (e.g., beyond clinical interview for affective and other neuropsychiatric disorders).

Central nervous system disorders with a genetic basis are an area where a PM approach can be effectively leveraged through identification of patients with an underlying genetic mutation responsible for the disease (e.g., Friedreich's ataxia and Huntington's disease), genetic risk factors (e.g., the presence of apolipoprotein E4 [APOE4] in AD), or deterministic genes (e.g., presenilin-1/2 in AD and mutated dystrophin gene in Duchenne muscular dystrophy [DMD]). Other areas include developing therapies to target genetic disorders (e.g., eteplirsen for treatment of DMD and cerliponase alfa for treatment of neuronal ceroid lipofuscinosis type 2 disease [CLN2]). European regulatory considerations for the development of various PM approaches are discussed further in the following sections.

IV BIOMARKERS, PGx, CDx, AND RARE DISEASES

Biomarkers can be defined as characteristics (e.g., a molecular, histologic, radiographic, or physiologic characteristic) that are measured as indicators of normal biological processes, pathogenic processes, or responses to an exposure or intervention, including therapeutic interventions (Daniel, McClellan, Richardson, & Nosair, 2016). This definition broadly captures a variety of BMs that serve several important functions in the nonclinical and clinical settings of medical product development and clinical practice. These include utilization of BMs as measurable indicators to identify patients at risk of disease (e.g., with genetic signatures for diagnosis of genetically based diseases), for enrichment of trials (e.g., patient subgroups with a common disease risk factor or higher likelihood of response to drug therapy), or as surrogate endpoints in predicting clinical outcomes once validated during development. Biomarkers are becoming increasingly prevalent in drug developments, and both the EMA and FDA are fostering collaboration and providing new guidance, encouraging use of innovative clinical trial designs and endpoints, and offering increased opportunity for agency interactions throughout development. Genetic BMs, including PGx, are discussed in Section A, and nongenetic BMs are addressed in Section V.

A Genetic Biomarkers and Pharmacogenomics

An area where a PM approach to drug development has been very successful is the use of PGx, defined as how the variability of the expression of genes between people leads to differences in susceptibility to diseases and responses to medicines. PGx are instrumental in the development and life cycle management of targeted therapies for PM. An increasing number of large prospective randomized studies incorporating PGx-BMs have allowed the identification of closely defined patient populations for therapeutic intervention, most notably in the oncology therapy area.

The CHMP has adopted a series of guidance documents (EMA, 2012a) to assist drug developers in regard to good PGx and genomics BM practices, methodological considerations, pharmacokinetic evaluation, and pharmacovigilance (EMA, 2007, 2011a, 2015c, 2016b). The CHMP's multidisciplinary PGx Working Party (PgWP) (EMA, 2018k) oversees matters relating directly or indirectly to PGx, including preparing guidelines for the preparation and assessment of the PGx sections of regulatory submissions, providing advice to the CHMP on general and product-specific matters relating to PGx, and supporting educational efforts in this developing area. Sponsors can arrange informal (i.e., nonbinding) briefing meetings with PgWP to discuss technical, scientific, and regulatory issues that may arise due to the inclusion of PGx in the development strategy and to assess their potential implications for the regulatory process (EMA, 2006, 2018k). Such interactions allow for information exchange, can minimize development risks associated with using PGx, and can inform future scientific advice and MAA submissions for both the sponsor and the agency. There is also an avenue for joint FDA-EMA voluntary genomic data submissions to allow the agencies to develop a better understanding of new genomic data.

B Companion Diagnostics (CDx) and Use of PGx to Optimize Drug Therapy

Orchestrating an individualized therapeutic regimen based on the individual's characteristics, genetic makeup, and the mechanism of action of the treatment inherently relies on the development of BMs. The assay used to measure the BM is considered a CDx. The EMA defines a CDx as a device that is used to identify patients who are most likely to benefit from or to identify patients who are likely to be at increased risk for serious adverse reactions as a result of treatment with the corresponding medicinal product; therefore CDx is essential to the safe and effective use of the corresponding medicinal product (EMA, 2017f).

The inclusion of PGx information in a drug's labeling can contribute to an improved benefit/risk balance through optimization of the target population and dosing recommendations potentially resulting in improved efficacy and safety including minimization of adverse drug reactions. The EU SmPC guideline (European Commission, 2009) contains specific recommendations on how PGx information should be presented in the label, and the EMA has created training materials to guide incorporation of PGx information into drug labeling. A 2015 paper by the EMA summarized all PGx-related information mentioned in EU product labels and classified it according to its main effect and function on drug treatment. As reported in the paper, approximately 15% of all EMA-evaluated medicines (i.e., those that have been centrally authorized in the EU) contain PGx information in the SmPC in the indications, posology, contraindications, warnings, or clinical trial (pharmacodynamics) sections, all of which directly impact patient treatment (EMA, 2018j). The PGx-BM information was related to 48 different genes of which 14 encode for drug targets or other gene variations having predictive information regarding the drug treatment outcome, while the rest of the genes were related to drug metabolism and transport. Examples include the need to use a CDx for selection of patients as defined in the therapeutic indication (e.g., HER-2 testing for trastuzumab and EGFR testing for cetuximab), use of CDx screening prior to treatment (e.g., the presence of HLA-B*5701 allele before the use of abacavir), or potential use of CDx testing to inform dosing or metabolism (e.g., dose adjustment of clopidogrel in poor CYP2C19 metabolizers) (ERBITUX, 2017; HERCEPTIN, 2017; PLAVIX, 2018; ZIAGEN, 2017). A recent example of an FDA approval based on genetic information is Kalydeco (ivacaftor) for the treatment of cystic fibrosis (CF) in patients with at least one mutation in their CF gene, which expanded the use from 10 to 33 mutations (KALYDECO, 2017). The extended approval was based on laboratory data that identified patients with certain rare gene mutations who are likely to respond to this treatment, and this development was a close collaboration with the Cystic Fibrosis Foundation (US FDA, 2017b). Thus PGx information in the drug label has an impact on appropriate prescribing (and, by extension, the reimbursement) of the drug as well as the associated CDx.

The 2015 EMA paper (Ehmann et al., 2015) notes that several alleles act as pharmacogenetic BMs for more than one drug (e.g., the HER-2 allele responsible for encoding the human epidermal growth factor receptor 2 present in a subset of breast cancer patients) or more than one indication (e.g., HER-2 presence in certain stomach or pancreatic neoplasms). The identification of specific groups of patients who will benefit from targeted therapy, even across different indications, is a key principle of a PM approach. Analyses of the patient population noted by the therapeutic indication section of the SmPC (Section 4.1) demonstrated that 19 out of 30 products were only tested in the BM-selected patient population and 10 out of 30 products included BM-positive and BM-negative patients in their pivotal clinical trials and that some inclusions of PGx information were based on retrospective analyses as distinct from prospectively designed clinical trials. These observations indicate that there is inherent flexibility in the European regulatory approach to the assessment and approval of drugs that rely on PGx-BMs or other PM methodologies.

Additional examples of utilization of PGx markers include physiologically based pharmacokinetic modeling and simulation using the Virtual Twin technology developed by Certara. This technology enables the creation of a computer-simulated model for each patient using the patient's various attributes (e.g., age, sex, race, and genetics of drug metabolizing enzymes) that can be used to evaluate a drug's effect on the patient as well as the optimal drug-dosing regimen. Two recently completed proof-of-concept studies using this technology include prediction of olanzapine (an antipsychotic) exposure in individual patients (Polasek et al., 2018) and prediction of the likely occurrence of cardiotoxic events with citalopram (an antidepressant) (Patel, Wiśniowska, Jamei, & Polak, 2018).

C Rare Diseases

Many rare and orphan diseases are associated with heritable mutations, and therefore genetic approaches to patient characterization are featured prominently in the development of orphan drugs for these diseases. Orphan designation can be granted in Europe for drugs meeting the following criteria (EMA, 2018h):

- Intended for the treatment, prevention, or diagnosis of a disease that is life-threatening or chronically debilitating.
- The prevalence of the condition in the EU must not be more than 5 in 10,000, or it must be unlikely that marketing of the medicine would generate sufficient returns to justify the investment needed for its development.
- No satisfactory method of diagnosis, prevention, or treatment of the condition concerned can be authorized, or if such a method exists, the medicine must be of significant benefit to those affected by the condition.

The EMA has established COMP as a specialist committee to oversee the development of orphan drugs. This committee assigns orphan designation early in the drug's development and assesses whether the drug constitutes a significant benefit over existing therapies. It also performs an orphan maintenance assessment to confirm orphan designation at the time of market authorization. Drugs that qualify for orphan designation can receive incentives for development such as reduced fees for regulatory activities, MAAs, inspections, and postauthorization changes (EMA, 2018d, 2018g, 2018i). Orphan drugs authorized for marketing also benefit from 10 years of protection from market competition from other approved drugs with similar indications. This period of protection can be extended by two years for medicines that have completed agreed-upon pediatric investigation plans.

Because rare diseases are a global issue, the EMA works closely with its international counterparts, in particular, the FDA, on the designation and assessment of orphan drugs. This includes sharing information on orphan drugs, developing common procedures for applying for orphan designation, and submitting annual reports on the status of development of designated orphan drugs. In addition, the EMA and FDA set up a new "cluster" in September 2016 to work jointly on advanced treatments for patients with rare diseases with the goal of expediting the review and approval of treatments for rare diseases (EMA, 2016g). The EMA also works with organizations representing patients with rare diseases through the European Organization for Rare Diseases (EURORDIS).

Within CNS drug development the implementation of PGx approaches has been slow due to difficulty in establishing a correlation between diagnostic BMs and drug response despite considerable progress in genomic research and disease pathology. Other impediments include difficulties in implementing PGx in large randomized clinical trials. However, examples of drug approvals relying on PGx for rare CNS diseases include the following:

- Translarna (ataluren) approved in the EU only for treatment of patients aged 5 years and older with DMD who are able to walk. Translarna is for use in patients whose disease is due to the presence of certain defects (called nonsense mutations) in the dystrophin gene, which prematurely stop the production of a normal dystrophin protein, leading to a shortened dystrophin protein that does not function properly (EMA, 2017i). Translarna works in these patients by enabling the protein-making apparatus in cells to move past the defect, allowing the cells to produce a functional dystrophin protein.
- The EMA and FDA approval of Spinraza (nusinersen), a survival motor neuron-2-directed antisense oligonucleotide, indicated for the treatment of 5q spinal muscular atrophy (SMA) in children and adults. Spinal muscular atrophy is a rare and fatal genetic disease affecting muscle strength and movement (EMA, 2017d). The efficacy of Spinraza was demonstrated in a clinical trial of 121 patients with infantile-onset SMA who were less than 7 months old at the time of their first dose.
- The EMA and FDA approval of Brineura (cerliponase alfa) for treatment of a specific form of Batten disease (to slow the loss of ambulation) in symptomatic pediatric patients 3 years of age and older with late infantile neuronal CLN2, also known as tripeptidyl peptidase-1 deficiency (EMA, 2018n). Brineura is administered into the cerebrospinal fluid (CSF) by infusion via a specific surgically implanted reservoir and catheter in the head (intraventricular access device). The efficacy of Brineura was established in a nonrandomized, single-arm, dose-escalation clinical trial in 23 symptomatic pediatric patients with CLN2 disease who were at least 3 years of age and had motor or language symptoms.
- The FDA approval of EXONDYS 51 (eteplirsen) injection for treatment of DMD for patients who have a confirmed mutation of the dystrophin gene amenable to exon 51 skipping (US FDA, 2018b). The approval was based on three small clinical trials using a surrogate endpoint of dystrophin increase in skeletal muscle predictive of reasonable clinical benefit with the confirmed mutation. The EU MAA for EXONDYS 51 is currently under review by the EMA, and a decision is expected by the first half of 2018.

In addition, there are NDA and MAA applications under review for cannabidiol for Dravet syndrome (severe myoclonic epilepsy of infancy) and Lennox-Gastaut syndrome (pediatric epilepsy) as of March 2018 (GW Pharmaceuticals, 2017; GW Pharmaceuticals, 2018). There are also numerous other CNS drugs with orphan designation currently in the preauthorization development stage (European Commission, 2018).

D PGx Approaches in Nonrare Diseases

In addition to PGx-BMs being used to diagnose diseases such as DMD and Huntington's disease, there is ongoing research into the genetic bases for various psychiatric disorders through the use of genome-wide association studies including major depression and schizophrenia (Allardyce et al., 2018; Ledford, 2015; Power et al., 2017; Schizophrenia Working Group of the Psychiatric Genomic Consortium, 2014). Other approaches include PGx profiling of samples collected during previously completed drug trials to retroactively identify subgroups of patients who demonstrated an increased drug response and/or improved safety profile (Li & Lu, 2012). Such approaches may lead to improved identification of more homogeneous patient subgroups and optimized drug labeling. Using PGx to identify subsets of patients within otherwise nonrare conditions with the intent of filing an orphan application can be a challenging issue for COMP. This issue is of importance in cases when BMs redefine the classification of medical diseases or syndromes. Reclassification based on BMs necessitates the validation of the link between the BM and the condition in question and must exclude effects outside the BM-defined subset to secure orphan designation (Tsigkos et al., 2014).

Validated PGx-BMs may ultimately be able to identify new homogenous target populations and distinguish them from a broader patient group defined by a symptomatology-based classification. Specific identification of new patient populations with shared prognostic indicators or genomic signatures could be used for enrichment of clinical trial populations and may lead to differential treatment effects for these populations.

Finally the availability of new CDx capable of identifying specific patient populations may guide the choice of therapeutic interventions in CNS disorders in the future.

V NONGENETIC BMs AND NOVEL METHODOLOGIES, NEW CLINICAL TRIAL DESIGNS, PATIENT-REPORTED OUTCOMES (PROs), AND DIGITAL/ WEARABLE TECHNOLOGIES

In addition to PGx approaches, there are ongoing efforts to validate other novel methodologies and innovative approaches for drug development and PM in CNS disorders. These include the development of new (nongenetic) BMs to identify more tightly defined patient groups to enrich clinical trial populations or for diagnostic purposes (e.g., AD and PD). Other BMs under development include more sophisticated tools that capture patient-reported outcomes (PROs), which can be used to identify patients sharing common features or more prominent impairments within a disease (e.g., ataxia subtypes and tremor types [essential, cerebellar, and psychogenic]) or patients with PD with differing impairments of motor function or can provide more sensitive and/or specific measures of drug effects (e.g., new clinical outcome measures for prodromal AD). For many neurological diseases, caregiver input provides useful data to assess the environmental impacts and sociodynamics, which are hard to derive from physiological or clinical assessments and could further inform PM approaches.

New approaches to clinical trial design (e.g., basket, master, and umbrella designs) using enriched patient populations or novel methods of patient selection based on BMs (e.g., new modalities such as brain imaging, common symptom clusters, and genetics) are being used to investigate new PM approaches that embody many of the concepts proposed by RDoC and related initiatives to redefine CNS drug development (see Section III). These approaches have led to a notable shift on the part of the regulatory agencies through new guidance documents embracing the changes and establishing pathways for increased dialogue with various stakeholders to encourage the development of new tools and diagnostics, the use of alternative study designs and endpoints, and options for accelerated approval pathways.

A Qualification of BMs and Diagnostics for Medicine Development

The EMA has established a regulatory pathway for the qualification of novel methodologies and innovative approaches to drug development similar to the clinical outcome assessment qualification program implemented by the FDA (EMA, 2018m; US FDA, 2018e). The EMA offers scientific advice to support the qualification of innovative development methods for a specific intended use in the context of research and development into pharmaceuticals and has issued a guidance that addresses the essential considerations for successful qualification of novel methodologies (EMA, 2017m). The CHMP can issue a qualification opinion on the acceptability of a specific use of a method, such as use of a novel methodology or an imaging method in nonclinical or clinical trials, and its application as a novel BM. In addition, the CHMP can issue qualification advice on protocols and methods that are intended to develop a novel method with the aim of moving towards qualification. Based on qualification advice the EMA may propose a letter of support as an option when the novel methodology under evaluation cannot yet be qualified but has promising preliminary data. The letter of support is intended to encourage data sharing and to facilitate further studies aimed at eventual qualification for the novel methodology under evaluation. The EMA's commitment to regulatory convergence in drug development is demonstrated through the availability of a joint letter of intent template from the EMA and FDA intended to facilitate parallel submissions for qualification to both agencies. The intent is to promote the sharing of scientific perspectives and advice and to provide coordinated responses to applicants where feasible.

To date, BMs have been used frequently for patient enrichment of CNS trials; however, none have been qualified for use as a diagnostic tool or as an outcome measure in CNS clinical trials (Arnerić et al., 2018). Diseases with significant ongoing work in BMs include AD, amyotrophic lateral sclerosis (ALS), autism spectrum disorders (ASD), major depressive disorder, Huntington's disease, MS, PD, schizophrenia, traumatic brain injury (TBI), and rare CNS diseases. The recently published 2018 EMA guideline on the clinical investigation of medicines for the treatment of AD notes that BMs in AD clinical trials can be separated according to their potential context of use as diagnostic (for determining diagnosis), enrichment (for selecting populations), prognostic (for determining course of illness), predictive (for predicting a future clinical response to therapy and for safety assessment), and pharmacodynamic (for determination of intended or unintended activities) (EMA, 2018b). This new AD guideline clarifies that qualification of BMs for any of these uses will require testing in both BM-positive and BM-negative AD patients.

The CHMP has granted a qualification opinion for the following novel methodologies in CNS disorders:

- Cerebrospinal fluid protein levels (containing both low Aβ1-42 and high T-tau proteins) and/or positron emission tomography amyloid imaging (positive/ negative) as BMs for enrichment of clinical trial populations with mild to moderate AD who are at increased risk of having an underlying AD neuropathology. Both were qualified for trial enrichment only and not as a diagnostic tool or outcome measure (EMA, 2012b).

- Low hippocampal volume (atrophy) as measured by magnetic resonance imaging as a marker of progression to dementia in patients with cognitive deficit compatible with the predementia stage of AD for enrichment of clinical trial populations (EMA, 2011b).

- A novel data-driven model of disease progression and trial evaluation in mild and moderate AD to provide a quantitative rationale for selection of study design and inclusion criteria (EMA, 2013b).

There is also an ongoing open consultation as of March 2018 on the qualification of molecular neuroimaging of the dopamine transporter as an enrichment BM in clinical trials to identify patients with early-manifest Parkinsonism in PD (EMA, 2017l).

Both EMA and FDA encourage the use of diagnostic tests for early detection and measurement. The FDA recently granted an approval of the Banyan Brain Trauma Indicator, a blood test measuring two proteins (UCH-L1 and GFAP) that are released into the bloodstream following a head injury to evaluate mild TBI. This test means that health-care professionals can assess the need for a computed tomography scan, thereby reducing costs and risks of unnecessary scans and radiation exposure (US FDA, 2018f).

In addition to qualified methodologies, the EMA has issued many letters of support to promote the development of BMs, novel methodologies, and PRO measures evaluated directly from the patient (see also Section C). These include a patient data platform for capturing PRO measures in Dravet syndrome, which facilitates data capture and supports longitudinal tracking of patient symptoms and multiple letters of support for developments in ASD, including new methods for stratifying populations and measuring impairments or clinical outcomes (EMA, 2015d, 2015e, 2015f, 2015g, 2015h, 2016c). These ongoing research efforts are expected to facilitate successful development of new PMs for CNS disorders.

B Innovative Study Designs and New Clinical Guidelines

In recent years, increased participation by dedicated patient groups (including caregivers) has resulted in an urgency to seek and utilize new pathways to accelerate availability of treatment options for patients with serious diseases with unmet need and with rare disorders. Because of small patient populations in rare diseases, traditional randomized controlled trials may not be feasible to establish the effectiveness of the medicinal product.

Trials using newer approaches (e.g., adaptive trials, basket trials, platform trials, and observational trials) that have been used in other disease areas are being explored for use in trials focused on CNS disorders.

The EMA has adopted multiple new guidelines to assist drug developers investigating new therapies for AD (EMA, 2018o), ASD (EMA, 2017k), ALS (EMA, 2015b), DMD and Becker muscular dystrophy (EMA, 2015j), and MS (EMA, 2015a). In February 2018 the FDA released a batch of new guidance documents (US FDA, 2018d) for treatment of neurological disorders to encourage drug development for AD, DMD, ALS, migraine, and pediatric epilepsy. These guidelines redefine primary endpoints, diagnostic criteria, the use of BMs for early detection of disease, and the use of new PROs. Of particular note is the incorporation of patient input into these new FDA guidance documents, specifically the new DMD guidance (US FDA, 2018c), which was preceded by a pioneering effort from Parent Project Muscular Dystrophy, which submitted their own independent proposed draft guidance in 2014. This independently authored draft guidance provided important scientific and patient input from the DMD community and stimulated the production of the new FDA guidance (US FDA, 2014). Similarly the new draft of the ALS-related guidance (US FDA, 2018a) was influenced by a comprehensive proposed draft guidance from the ALS Association (US FDA, 2017c). In the EU the revised EMA guidelines on AD and MS were preceded by workshops involving multiple stakeholders including patient representatives (EMA, 2013a, 2014a). The development history of these guidelines highlights the increasing importance placed by the regulatory authorities on patients' perspectives in the development of CNS drugs and PMs.

Master protocols are innovative study designs used to evaluate targeted therapies in rare disorders using PGx-BMs and include the evaluation of more than one treatment within the same trial. These protocols include multiple trials that share key design components and operational aspects. Use of master protocols (e.g., umbrella or basket trials) is common in oncology and is designed to evaluate multiple therapies for a single disease or single targeted therapy for multiple disease subtypes (Woodcock & LaVange, 2017). Examples include the B2225 master protocol investigating a common BM in multiple treatment combination and the National Cancer Institute's Molecular Analysis for Therapy Choice master protocol evaluating multiple genetic markers and associated targeted therapies for cancers that carry the targeted mutation. Another innovative dynamic design concept includes platform trials, an extension of the umbrella trial type that utilizes a decision algorithm to allow therapies to enter or leave the platform. Examples include I-SPY-2, an exploratory-phase platform trial (12 therapies from nine sponsors as of March 2017), and

Lung-MAP, a phase 2–3 master protocol evaluating multiple targeted therapies independently of others based on genetic subgroups in advanced squamous non-small-cell lung cancer. The recent FDA approval of Keytruda (pembrolizumab) for all solid tumors with a specific genetic signature regardless of where in the body the cancer started was based on novel trial designs and has led to the approval of a tumor-type agnostic indication (KEYTRUDA, 2017). Early notable examples of master trials in CNS disorders in the United States sponsored by the NIH include the Clinical Antipsychotic Trials of Intervention Effectiveness (CATIE), which compared various atypical antipsychotic drugs for the treatment of schizophrenia, and the Sequenced Treatment Alternatives to Relieve Depression (STAR*D) trial, which compared various antidepressants in patients diagnosed with major depressive disorder (Lieberman et al., 2005; Rush, Lavori, Trivedi, Sackeim, et al., 2004). Examples of ongoing platform trials in CNS include the Dominantly Inherited Alzheimer Network Trial (DIAN-TU) trial, which tests multiple drugs to slow or prevent the progression of AD in autosomal dominant AD (ADAD) caused by genetic mutations (Washington University School of Medicine, n.d.; Bateman et al., 2017). The first trial in this platform was launched in 2012, a placebo-controlled, two-year BM trial involving two drugs, solanezumab and gantenerumab, and then transitioned to a phase 3 cognitive endpoint trial in 2014.

The earlier examples highlight the potential of innovative trial designs to lead the identification of new patient groups, leading ultimately to the approval of new therapies for CNS disorders. This could include testing multiple new therapies for a CNS disorder within a single trial or, alternatively, enrolment of different patient populations in a single interventional trial to demonstrate safety and efficacy across multiple indications.

C PROs, Wearable Technologies, and Real-World Data

Other avenues under consideration in the expanded approach to CNS drug development include the use of electronic health records, registries, wearable devices, and the use of PROs to ensure greater involvement of patients in the decision-making process and to obtain access to observational and real-world data that could be used for regulatory decision-making (GetReal Initiative, 2018). Technological advances have led to novel solutions such as extracting critical information from unstructured notes included in patient records for use by clinicians and researchers without conducting a clinical trial (e.g., Flatiron Health) (Flatiron, n.d.).

A PRO measures the patient's symptoms or the effect of a medical condition on a patient's functioning. These are reported by or measured directly in the patient and include observations recorded in diaries or other tracking devices (e.g., pain intensity scales, seizure episodes, and sleep diaries). The use of PROs as primary or secondary endpoints in clinical trials is gaining momentum and is encouraged by regulatory agencies (DeMuro et al., 2013; EMA, 2014b; Gnanasakthy & De Muro, 2016; Gnanasakthy, Mordin, Evans, Doward, & DeMuro, 2017; Storf, 2013; US FDA, 2017a; Venkatesan, 2016). A specific PRO measure utilized in CNS trials and often mentioned in product labeling is quality of life, which covers physical and cognitive functioning. Other measures include activities of daily living, motor function, somatic symptoms (e.g., pain intensity and degree of impairment), and assessment of seizure severity (rufinamide for seizures associated with Lennox-Gastaut syndrome) (AMPYRA, 2017; DeMuro et al., 2013; Gnanasakthy et al., 2012; NUCYNTA, 2017; SAVELLA, 2016; VYVANSE, 2017). Use of disease-specific PROs in pivotal clinical trials allows for more relevant and patient-centric treatment effects to be considered in regulatory decision-making.

Wearable devices using sensors are being introduced in clinical trials utilizing data science technologies to gain understanding of the underlying symptoms, disease progression, and drug effects. A recent 2017 paper from the Critical Path Institute's Electronic Patient-Reported Outcome Consortium has proposed recommendations regarding the selection and evaluation of wearable devices and their measurement for use in regulatory trials and to support labeling claims (Byrom et al., 2017). The data collected by wearable devices provide an alternative to traditional PRO measures, because the data can be collected without the patient or clinician actively recording the information. These devices have potential for tracking longitudinal data without long-term extension trials. Recent examples include (i) a wearable device used in a pilot study of an experimental drug in patients with PD to track data on movement and effect of medication, (ii) use of Fitbit monitors in MS studies to track patient movements in real time, and (iii) a 3-year study using wearable devices and other technology to assess potential physical changes that could be associated with cognitive decline (Evans Center Affinity Research Collaborative, n.d.; Bachlin et al., 2010; Balto, Kinnett-Hopkins, & Motl, 2016).

Other data sources potentially contributing to the development of PM include technological advances in procuring and organizing "big data" obtained from alternative data sources, bioinformatics, and large national health databases (e.g., UK Clinical Practice Research Datalink collates data from the entire UK National Health Service) (Clinical Practice Research Datalink, 2018). The extensive use of patient- and population-based data raises challenges for integration and analyses, implementation in clinical practice, and the education of health-care providers and patients themselves. These challenges, including the need to respect data protection and patient confidentiality, must be carefully considered to optimize the benefits and risks.

Data collected using wearable technologies and other nontraditional sources are expected to provide meaningful input regarding patient characteristics, lifestyle, and other environmental factors and have the potential for further development as enrichment tools or diagnostic markers. The burgeoning data arising from many different sources present both great promise and challenges. An area of relevance is the field of quantitative science, which is evolving to incorporate data from adjacent disciplines and diverse technologies. New quantitative tools that can analyze, simulate, and present data in decision-informing ways are being developed. One example of a relevant interdisciplinary quantitative science innovation is the pharmacology-to-payer platform developed in the infectious disease area, which links pharmacology, epidemiology, and health economics (Kamal et al., 2017). This platform can be applied to capture a patient journey for a new medicinal product and project health economic outcome or inform the target product profile of a new agent. These approaches could be applied to the CNS therapeutic area to provide better information regarding patient-centric decision-making.

In summary the development of novel methodologies and BMs, new clinical trial design approaches, the use of new PROs and wearable technologies, and integration of multiple data sources all have potential utility in the development of new PMs for CNS diseases.

VI ADVANCED THERAPY MEDICINAL PRODUCTS AND PERSONALIZED MEDICINE

Advanced therapy medicinal products include gene therapies, somatic cell therapies, and tissue-engineered medicines for rare diseases and other chronic disorders with limited treatment options (Council of the European Union, 2017). The EMA along with the European Commission's Directorate General for Health and Food Safety works with competent authorities in member states to support the development and authorization of high-quality, safe, and effective ATMPs. The CAT is the EMA committee responsible for assessing the quality, safety, and efficacy of ATMPs and for following scientific developments in the field (EMA, 2018a). The CAT provides scientific expertise related to the development of innovative medicines and therapies and provides scientific recommendations on the classification of ATMPs. It also contributes to the scientific advice in cooperation with the SAWP and participates in CHMP or PRAC

procedures delivering advice on the conduct of efficacy follow-up, pharmacovigilance, or risk-management systems for ATMPs (EMA, 2018c).

The CAT has produced multiple guidelines that address developers of advanced therapies on quality, nonclinical, clinical, and risk-management aspects of these innovative new therapeutic interventions (EMA, 2008a, 2008b, 2011c, 2012c, 2016d, 2017j). A multistakeholder meeting was held on May 2016 to explore ways to foster ATMP development and to expand patient access to ATMPs (EMA, 2016d, 2017j). The areas covered included optimizing regulatory processes for ATMPs; facilitating research and development; improving funding, investment, and patient access; and moving from hospital exemption to marketing authorization. At an international level, a forum for dialogue was set up in October 2017 among the EMA, FDA, Health Canada, and PMDA to share experience on ATMPs (EMA, 2017j).

To date, 18 MAAs for ATMPs have been submitted to the EMA, and 10 products have been approved. No ATMPs for CNS disorders have yet been approved. Examples of EMA-approved ATMPs include the following:

- Spherox (spherical aggregates of ex vivo expanded autologous matrix-associated chondrocytes), an implant suspension to repair defects to the cartilage in the knee (EMA, 2017g)
- Zalmoxis, genetically modified allogenic T cells for hematopoietic stem cell transplantation (EMA, 2016f)
- Strimvelis, autologous CD34+ cells transduced with a retroviral vector that encodes for the human adenosine deaminase cDNA sequence for treatment of severe combined immunodeficiency due to adenosine deaminase deficiency (EMA, 2016e)
- Alofisel (darvadstrocel), which are expanded adipose-derived stem cells for treatment of complex perianal fistulas in patients with Crohn's disease (EMA, 2018p)

The MAA review of the gene therapy Luxturna (voretigene neparvovec-rzyl) for the treatment of patients with confirmed biallelic *RPE65* mutation-associated retinal dystrophy that leads to vision loss is ongoing (Spark Therapeutics, 2017).

Ongoing developments for ATMPs for CNS disorders include research into alternative treatment options in epilepsy and evaluating stem cell transplantation therapy using either autologous or allogeneic cell types. Autologous transplant approaches avoid controversial ethical concerns, because the donor is the host and potentially has reduced safety risks compared with allogeneic transplantations where the donor is not the same as the host but is a close match. Other avenues include induced pluripotent stem cells that are derived from adult stem cells and provide the promise of autologous cell replacement and gene therapy to reprogram cells (Rao, Mashkouri, Aum, Marcet, & Borlongan, 2017). Additional projects include the recent announcement of a strategic collaboration to develop a gene therapy platform to produce anti-tau antibodies within the brain to promote cellular stability and function in AD and other neurodegenerative diseases and stem cell transplantation for treatment of PD (AbbVie, 2018; Kegel, 2017).

Making regenerative cellular therapies a reality for patients faces many challenges at all stages of development and post approval, including validation, large-scale manufacturing, and affordability (Corbett, Webster, Hawkins, & Woolacott, 2017). These therapies are expensive and require considerable support from health-care providers, and the long-term prospects are also yet to be worked out. Advanced therapies such as gene therapy and stem cells may eventually provide effective PM therapies for currently incurable neurodegenerative diseases.

VII MISCELLANEOUS

Other European initiatives of relevance to PM in CNS include public-private partnerships such as the IMI, which is jointly funded and driven by the European Commission, pharmaceutical industry, and academia (European Federation of Pharmaceutical Industries and Associations, 2014). The EMA is using IMI projects to expand and inform its future-facing strategies to regulation of these new treatment modalities (see also Section A on qualification of novel methodologies). The IMI projects where PM approaches for CNS disorders are in development include the following:

- EBiSC: European Bank for induced pluripotent stem cells including AD (Innovative Medicines Initiative, 2018a)
- EPAD: European prevention of Alzheimer's dementia consortium (Innovative Medicines Initiative, 2018b)
- EU-AIMS: European Autism Interventions—a multicenter study for developing new medications (Innovative Medicines Initiative, 2018c)
- NEWMEDS: Novel methods leading to new medications in depression and schizophrenia (Innovative Medicines Initiative, 2018d)
- PRISM: Psychiatric Ratings using Intermediate Stratified Markers: providing quantitative biological measures to facilitate the discovery and development of new treatments for social and cognitive deficits in AD, schizophrenia, and major depression (Innovative Medicines Initiative, 2018e)

The IMI projects also include the adaptive pathway program from the EMA that utilizes a life cycle approach of acquiring and assessing evidence through the product life cycle, namely, development, licensing, reimbursement, monitoring/postlicense evidence, and utilization (Eichler et al., 2015). The collaborative project Accelerated

Development of Appropriate Patients Therapies led by the EMA (ADAPT SMART, n.d.) was set up to formulate collaborative solutions to foster the development of Medicines Adaptive Pathways to Patients (MAPPs) in Europe. This is a platform to accelerate the availability of MAPPs to all health-care stakeholders and foster access to beneficial treatments to the patients in the product life span in a sustainable manner and could hold promise for development of new PMs and label expansion for new patient groups. Another relevant IMI project is the GetReal initiative, which aims to show how new methods of real-world evidence collection and synthesis could be adopted earlier in pharmaceutical research and development and the health-care decision-making process (GetReal Initiative, 2018).

A new initiative from the EMA is the Priority Medicines (PRIME) scheme (similar to the FDA Breakthrough Therapy designation process) (EMA, 2018l), which provides proactive regulatory support for accelerated assessment in disease areas with high unmet need. The key benefits of this approach include appointment of a rapporteur from the CHMP during the development phase, meetings with the CHMP/CAT rapporteur and group of experts, scientific advice at key development milestones, and a dedicated EMA contact point. As of January 2018, two products each in neurology and psychiatry have been accepted into the PRIME program. These are (i) adeno-associated viral vector serotype 9 containing the human SMN gene (AVXS-101) for treatment of pediatric patients diagnosed with SMA type 1, (ii) aducanumab for treatment of AD, (iii) allopregnanolone for treatment of postpartum depression, and (iv) rapastinel for adjunctive treatment of major depressive disorder. Products with PRIME designation can expect to be eligible for accelerated assessment at time of MAA (i.e., reduced review time by EMA, similar to FDA priority review) to further expedite approval and patient access to such innovative treatments. Products accepted into facilitated regulatory development schemes such as PRIME may also be eligible for conditional (i.e., early) marketing authorization on the basis of compelling preliminary clinical data, which could provide early access to innovative therapies for unmet clinical needs but may require more intensive risk minimization strategies due to limited clinical trial knowledge of a drug's safety profile at time of approval, in addition to conduct of postauthorization confirmatory efficacy studies (EMA, 2018e).

VIII CONCLUSIONS

As described earlier, there are multiple mechanisms through which the European regulatory network is encouraging and facilitating the development of PM approaches including those for CNS disorders. A greater focus on patient centricity and involvement of patient groups has led to increasing support from regulatory agencies in the form of new guidance documents, scientific standards, encouraging use of surrogate endpoints, and use of small nonrandomized trials. In addition to provision of enhanced regulatory support such as dedicated scientific advice resources, the EMA has also established platforms such as multistakeholder meetings to address challenges within this important emerging area of pharmaceutical medicine development. These initiatives have resulted in some early successes in the development of PM approaches for CNS disorders such as BM qualification, and it is hoped that these initiatives and the commitment of the EMA and other regulatory agencies will aid future approval of personalized CNS medicines to address this area of high unmet clinical need.

References

AbbVie. (2018). *AbbVie and voyager therapeutics announce global strategic collaboration to develop potential new treatments for Alzheimer's disease and other tau-related neurodegenerative diseases [Press release].* Retrieved from https://news.abbvie.com/news/abbvie-and-voyager-therapeutics-announce-global-strategic-collaboration-to-develop-potential-new-treatments-for-alzheimers-disease-and-other-tau-related-neurodegenerative-diseases.htm.

ADAPT SMART. (n.d.). Retrieved from http://adaptsmart.eu.

Allardyce, J., Leonenko, G., Hamshere, M., Pardiñas, A. F., Forty, L., Knott, S., et al. (2018). Association between schizophrenia-related polygenic liability and the occurrence and level of mood-incongruent psychotic symptoms in bipolar disorder. *JAMA Psychiatry, 75*(1), 28–35.

AMPYRA. (2017). *(Dalfampridine) [Prescribing information].* Ardsley, NY: Acorda Therapeutics, Inc. Retrieved from http://www.ampyra.com/prescribing-information.pdf.

Arnerić, S. P., Kern, V. D., & Stephenson, D. T. (2018). Regulatory-accepted drug development tools are needed to accelerate innovative CNS disease treatments. *Biochemical Pharmacology, 151,* 291–306. https://doi.org/10.1016/j.bcp.2018.01.043.

Bachlin, M., Plotnik, M., Roggen, D., Maidan, I., Hausdorff, J. M., Giladi, N., et al. (2010). Wearable assistant for Parkinson's disease patients with the freezing of gait symptom. *IEEE Transactions on Information Technology in Biomedicine, 14*(2), 436–446.

Balto, J. M., Kinnett-Hopkins, D. L., & Motl, R. W. (2016). Accuracy and precision of smartphone applications and commercially available motion sensors in multiple sclerosis. *Multiple Sclerosis Journal–Experimental, Translational and Clinical, 2,* 2055217316634754.

Bateman, R. J., Benzinger, T. L., Berry, S., Clifford, D. B., Duggan, C., Fagan, A. M., et al. (2017). The DIAN-TU next generation Alzheimer's prevention trial: adaptive design and disease progression model. *Alzheimers Dement, 13*(1), 8–19.

Byrom, B., Watson, C., Doll, H., Coons, S. J., Eremenco, S., Ballinger, R., et al. (2017). Selection of and evidentiary considerations for wearable devices and their measurements for use in regulatory decision making: recommendations from the ePRO Consortium. *Value in Health.* Retrieved from http://www.valueinhealthjournal.com/article/S1098-3015(17)33532-5/pdf.

Clinical Practice Research Datalink. (2018). Retrieved March 26, 2018 from https://www.cprd.com/home/.

Corbett, M. S., Webster, A., Hawkins, R., & Woolacott, N. (2017). Innovative regenerative medicines in the EU: a better future in evidence? *BMC Medicine, 15*(1), 49.

Council of the European Union. (2015). Council conclusions on personalised medicine for patients [C 421/03]. *Official Journal of the European Union, 58*(C 421), 2–5. Retrieved from http://eur-lex.europa.eu/legal-content/EN/TXT/?uri=CELEX%3A52015XG1217(01).

Council of the European Union. (2016). Regulation (EU) 2016/679 of the European parliament and of the council of 27 April 2016 on the protection of natural persons with regard to the processing of personal data and on the free movement of such data, and repealing directive 95/46/EC (general data protection regulation) [L 119/1]. *Official Journal of the European Union, 59*(L 119), 1–88. Retrieved from http://eur-lex.europa.eu/eli/reg/2016/679/oj.

Council of the European Union. (2017). Regulation (EC) No 1394/2007 of the European parliament and of the council of 13 November 2007 on advanced therapy medicinal products and amending directive 2001/83/EC and Regulation (EC) No 726/2004 [L 324/121]. *Official Journal of the European Union, 50*(L 324), 121–137. Retrieved from https://ec.europa.eu/health/sites/health/files/files/eudralex/vol-1/reg_2007_1394/reg_2007_1394_en.

Cuthbert, B. N., & Insel, T. R. (2013). Toward the future of psychiatric diagnosis: the seven pillars of RDoC. *BMC Medicine, 11*(1), 126.

Daniel, G. W., McClellan, M. B., Richardson, E., & Nosair, W. (2016). *Facilitating biomarker development: Strategies for scientific communication, pathway prioritization, data-sharing, and stakeholder collaboration.* Duke-Margolis Center for Healthy Policy. Retrieved from https://healthpolicy.duke.edu.

DeMuro, C., Clark, M., Doward, L., Evans, E., Mordin, M., & Gnanasakthy, A. (2013). Assessment of PRO label claims granted by the FDA as compared to the EMA (2006-2010). *Value in Health, 16*(8), 1150–1155. https://www.firstwordpharma.com/node/1562479.

Ehmann, F., Caneva, L., Prasad, K., Paulmichl, M., Maliepaard, M., Llerena, A., et al. (2015). Pharmacogenomic information in drug labels: European medicines agency perspective. *The Pharmacogenomics Journal, 15*(3), 201–210.

Eichler, H. G., Baird, L. G., Barker, R., Bloechl-Daum, B., Børlum-Kristensen, F., Brown, J., et al. (2015). From adaptive licensing to adaptive pathways: delivering a flexible life-span approach to bring new drugs to patients. *Clinical Pharmacology and Therapeutics, 97*(3), 234–246.

ERBITUX *(Cetuximab)* [Summary of the product characteristics]. (2017). Retrieved March 27, 2018 from http://www.ema.europa.eu/ema/index.jsp?curl=pages/medicines/human/medicines/000558/human_med_000769.jsp&mid=WC0b01ac058001d124.

EU General Data Protection Regulation. (2018). Retrieved March 27, 2018 from https://www.eugdpr.org/.

European Commission. (2009). *A guideline on summary of product characteristics (SmPC).* In *Vol. 2C. Rules governing medicinal products in the European Union. Notice to applicants.* Retrieved from https://ec.europa.eu/health/sites/health/files/files/eudralex/vol-2/c/smpc_guideline_rev2_en.pdf.

European Commission. (2018). *Pharmaceuticals community register.* Retrieved March 27, 2018 from http://ec.europa.eu/health/documents/community-register/html/alforphreg.htm.

European Federation of Pharmaceutical Industries and Associations. (2014). The world's largest public private partnership in healthcare just got over 50% BIGGER *[Press release].* Retrieved from https://www.efpia.eu/news-events/the-efpia-view/statements-press-releases/140709-the-world-s-largest-public-private-partnership-in-healthcare-just-got-over-50-bigger/.

European Medicines Agency. (2006). Guidelines on pharmacogenetics briefing meeting. *EMEA/CHMP/PGxWP/20227/2004.* London, UK: European Medicines Agency. Retrieved from http://www.ema.europa.eu/docs/en_GB/document_library/Scientific_guideline/2009/09/WC500003886.pdf.

European Medicines Agency. (2007). *ICH E15 Definitions for genomic biomarkers, pharmacogenomics, pharmacogenetics, genomic data and sample coding categories.* CHMP/ICH/437986/06. London, UK: European Medicines Agency. Retrieved from http://www.ema.europa.eu/ema/index.jsp?curl=pages/regulation/general/general_content_001298.jsp&mid=WC0b01ac058002958e.

European Medicines Agency. (2008a). Guideline on human cell-based medicinal products. *EMEA/CHMP/410869/2006.* London, UK: European Medicines Agency. Retrieved from http://www.ema.europa.eu/docs/en_GB/document_library/Scientific_guideline/2009/09/WC500003894.pdf.

European Medicines Agency. (2008b). Guideline on safety and efficacy follow-up - Risk management of advanced therapy medicinal products. *EMEA/149995/2008.* London, UK: European Medicines Agency. Retrieved from http://www.ema.europa.eu/docs/en_GB/document_library/Scientific_guideline/2009/10/WC500006326.pdf.

European Medicines Agency. (2011a). *Methodological issues with pharmacogenomic biomarkers in relation to clinical development and patient selection.* EMA/CHMP/446337/2011. London, UK: European Medicines Agency. Retrieved from http://www.ema.europa.eu/ema/index.jsp?curl=pages/regulation/general/general_content_001397.jsp&mid=WC0b01ac058002958e.

European Medicines Agency. (2011b). *Qualification opinion of low hippocampal volume (atrophy) by MRI for use in clinical trials for regulatory purpose—in pre-dementia stage of Alzheimer's disease.* EMA/CHMP/SAWP/809208/2011/. London, UK: European Medicines Agency. Retrieved from http://www.ema.europa.eu/docs/en_GB/document_library/Regulatory_and_procedural_guideline/2011/12/WC500118737.pdf.

European Medicines Agency. (2011c). *Reflection paper on design modifications of gene therapy medicinal products during development.* EMA/CAT/GTWP/44236/2009. London, UK: European Medicines Agency. Retrieved from http://www.ema.europa.eu/docs/en_GB/document_library/Scientific_guideline/2012/02/WC500122743.pdf.

European Medicines Agency. (2012a). *Use of pharmacogenetic methodologies in the pharmacokinetic evaluation of medicinal products.* EMA/CHMP/37646/2009. London, UK: European Medicines Agency. Retrieved from http://www.ema.europa.eu/ema/index.jsp?curl=pages/regulation/general/general_content_001399.jsp&mid=WC0b01ac058002958e.

European Medicines Agency. (2012b). *Qualification opinion of Alzheimer's disease novel methodologies/biomarkers for the use of CSF AB 1-42 and t-tau and/or PET-amyloid imaging (positive/negative) as biomarkers for enrichment, for use in regulatory clinical trials in mild and moderate Alzheimer's disease.* EMA/CHMP/SAWP/893622/2011. London, UK: European Medicines Agency. Retrieved from http://www.ema.europa.eu/docs/en_GB/document_library/Regulatory_and_procedural_guideline/2012/04/WC500125019.pdf.

European Medicines Agency. (2012c). *Guideline on quality, non-clinical and clinical aspects of medicinal products containing genetically modified cells.* London, UK: European Medicines Agency. EMA/CAT/GTWP/671639/2008. Retrieved from http://www.ema.europa.eu/docs/en_GB/document_library/Scientific_guideline/2012/05/WC500126836.pdf.

European Medicines Agency. (2013a). *Workshop on the clinical investigation of new medicines for the treatment of multiple sclerosis.* London, UK: European Medicines Agency. Retrieved from http://www.ema.europa.eu/ema/index.jsp?curl=pages/news_and_events/events/2013/06/event_detail_000724.jsp&mid=WC0b01ac058004d5c3.

European Medicines Agency. (2013b). *Qualification opinion of a novel data driven model of disease progression and trial evaluation in mild and moderate Alzheimer's disease.* EMA/CHMP/SAWP/567188/2013. London, UK: European Medicines Agency. Retrieved from http://www.ema.europa.eu/docs/en_GB/document_library/Regulatory_and_procedural_guideline/2013/10/WC500151309.pdf.

European Medicines Agency. (2014a). *European Medicines Agency workshop on the clinical investigation of medicines for the treatment of Alzheimer's disease.* London, UK: European Medicines Agency. Retrieved from http://www.ema.europa.eu/ema/index.jsp?curl=pages/news_and_events/events/2014/04/event_detail_000932.jsp&mid=WC0b01ac058004d5c3.

European Medicines Agency. (2014b). *Reflection paper on the use of patient reported outcome (PRO) measures in oncology studies.* EMA/CHMP/292464/2014. London, UK: European Medicines Agency. Retrieved from http://www.ema.europa.eu/docs/en_GB/document_library/Scientific_guideline/2014/06/WC500168852.pdf.

European Medicines Agency. (2014c). *Mandate of the European medicines agency innovation task force (ITF).* EMA/48440/2014. London, UK: European Medicines Agency. Retrieved from http://www.ema.europa.eu/docs/en_GB/document_library/Other/2009/10/WC500004912.pdf.

European Medicines Agency. (2014d). *European medicines agency policy on publication of clinical data for medicinal products for human use.* EMA/240810/2014. London, UK: European Medicines Agency. Retrieved from http://www.ema.europa.eu/docs/en_GB/document_library/Other/2014/10/WC500174796.pdf.

European Medicines Agency. (2015a). *Guideline on clinical investigation of medicinal products for the treatment of multiple sclerosis.* EMA/CHMP/771815/2011, Rev. 2. London, UK: European Medicines Agency. Retrieved from http://www.ema.europa.eu/docs/en_GB/document_library/Scientific_guideline/2015/03/WC500185161.pdf.

European Medicines Agency. (2015b). *Guideline on clinical investigation of medicinal products for the treatment of amyotrophic lateral sclerosis (ALS).* EMA/531686/2015, Corr. 1. London, UK: European Medicines Agency. Retrieved from http://www.ema.europa.eu/docs/en_GB/document_library/Scientific_guideline/2015/12/WC500199241.pdf.

European Medicines Agency. (2015c). *Key aspects for the use of pharmacogenomic methodologies in the pharmacovigilance evaluation of medicinal products.* EMA/CHMP/281371/2013. London, UK: European Medicines Agency. Retrieved from http://www.ema.europa.eu/ema/index.jsp?curl=pages/regulation/general/general_content_001396.jsp&mid=WC0b01ac058002958e.

European Medicines Agency. (2015d). *Letter of support for eye tracking to be used to stratify populations of people with autism spectrum disorder (ASD).* EMA/794527/2015. London, UK: European Medicines Agency. Retrieved from http://www.ema.europa.eu/docs/en_GB/document_library/Other/2015/12/WC500198349.pdf.

European Medicines Agency. (2015e). *Letter of support for measures of executive function and basic emotions to be used to stratify populations of people with autism spectrum disorder (ASD) and predict clinical outcome.* EMA/794534/2015. London, UK: European Medicines Agency. Retrieved from http://www.ema.europa.eu/docs/en_GB/document_library/Other/2015/12/WC500198350.pdf.

European Medicines Agency. (2015f). *Letter of support to explore clinical outcomes assessments utility to measure clinical symptoms in people with autism spectrum disorders.* EMA/794457/2015. London, UK: European Medicines Agency. Retrieved from http://www.ema.europa.eu/docs/en_GB/document_library/Other/2015/12/WC500198347.pdf.

European Medicines Agency. (2015g). *Letter of support to explore EEG utility to measure deficits in social recognition in people with autism spectrum disorders (ASD) and its potential to stratify patient groups.* EMA/794518/2015. London, UK: European Medicines Agency. Retrieved from http://www.ema.europa.eu/docs/en_GB/document_library/Other/2015/12/WC500198348.pdf.

European Medicines Agency. (2015h). *Letter of support to explore MRI methodology to be used to stratify populations of people with autism spectrum disorder (ASD).* EMA/794542/2015. London, UK: European Medicines Agency. Retrieved from http://www.ema.europa.eu/docs/en_GB/document_library/Other/2015/12/WC500198351.pdf.

European Medicines Agency. (2015i). *EU medicines agencies network strategy to 2020: Working together to improve health.* EMA/MB/151414/2015. London, UK: European Medicines Agency. Retrieved from http://www.ema.europa.eu/docs/en_GB/document_library/Other/2015/12/WC500199060.pdf.

European Medicines Agency. (2015j). *Guideline on the clinical investigation of medicinal products for the treatment of Duchenne and Becker muscular dystrophy.* EMA/CHMP/236981/2011, Corr.1. London, UK: European Medicines Agency. Retrieved from http://www.ema.europa.eu/docs/en_GB/document_library/Scientific_guideline/2015/12/WC500199239.pdf.

European Medicines Agency. (2016a). *The European regulatory system for medicines: A consistent approach to medicines regulation across the European Union.* EMA/716925/2016. London, UK: European Medicines Agency. Retrieved from http://www.ema.europa.eu/docs/en_GB/document_library/Leaflet/2014/08/WC500171674.pdf.

European Medicines Agency. (2016b). *Good pharmacogenomic practice.* EMA/CHMP/268544/2016. London, UK: European Medicines Agency. Retrieved from http://www.ema.europa.eu/docs/en_GB/document_library/Scientific_guideline/2016/05/WC500205758.pdf.

European Medicines Agency. (2016c). *Letter of support for patient data platform for capturing patient-reported outcome measures for Dravet syndrome.* EMA/327846/2016. London, UK: European Medicines Agency. Retrieved from http://www.ema.europa.eu/docs/en_GB/document_library/Other/2016/05/WC500206671.pdf.

European Medicines Agency. (2016d). Advanced therapy medicines: exploring solutions to foster development and expand patient access in Europe: Outcome of a multi-stakeholder meeting with experts and regulators held at European medicines agency on Friday 27 *May* 2016. EMA/345874/2016. London, UK: European Medicines Agency. Retrieved from http://www.ema.europa.eu/docs/en_GB/document_library/Report/2016/06/WC500208080.pdf.

European Medicines Agency. (2016e). *Strimvelis.* London, UK: European Medicines Agency. Retrieved from http://www.ema.europa.eu/ema/index.jsp?curl=pages/medicines/human/medicines/003854/human_med_001985.jsp&mid=WC0b01ac058001d124.

European Medicines Agency. (2016f). *Zalmoxis.* London, UK: European Medicines Agency. Retrieved from http://www.ema.europa.eu/ema/index.jsp?curl=pages/medicines/human/medicines/002801/human_med_002016.jsp&mid=WC0b01ac058001d124.

European Medicines Agency. (2016g). *EU-US collaboration to boost medicine development for rare diseases.* London, UK: European Medicines Agency. Retrieved from http://www.ema.europa.eu/ema/index.jsp?curl=pages/news_and_events/news/2016/09/news_detail_002609.jsp&mid=WC0b01ac058004d5c1.

European Medicines Agency. (2017a). *About us—European medicines agency (EMA).* London, UK: European Medicines Agency. Retrieved from http://www.ema.europa.eu/ema/index.jsp?curl=pages/about_us/document_listing/document_listing_000426.jsp&mid.

European Medicines Agency. (2017b). Personalised medicines—Report from a workshop on personalised medicines held by EMA on 14 *March* 2017a. EMA/185440/2017. London, UK: European Medicines Agency. Retrieved from http://www.ema.europa.eu/docs/en_GB/document_library/Report/2017/05/WC500227797.pdf.

European Medicines Agency. (2017c). *Patients' and consumers' working party (PCWP) and healthcare professionals' working party (HCPWP) joint workshop on personalised medicines.* London, UK: European Medicines Agency. Retrieved from http://www.ema.europa.eu/ema/index.jsp?curl=pages/news_and_events/events/2017/03/event_detail_001400.jsp&mid=WC0b01ac058004d5c3.

European Medicines Agency. (2017d). *First medicine for spinal muscular atrophy.* London, UK: European Medicines Agency. Retrieved from http://www.ema.europa.eu/ema/index.jsp?curl=pages/news_and_events/news/2017/04/news_detail_002735.jsp&mid=WC0b01ac058004d5c1.

European Medicines Agency. (2017e). Experience and opportunities for the co-development (process) of companion diagnostics and medicinal products (in the EU). EMA stakeholder platform meeting. London, UK: European Medicines Agency. Retrieved from http://www.ema.europa.eu/docs/en_GB/document_library/Presentation/2017/05/WC500227702.pdf.

European Medicines Agency. (2017f). *Concept paper on predictive biomarker-based assay development in the context of drug development and lifecycle.* EMA/CHMP/800914/2016. London, UK: European Medicines Agency. Retrieved from http://www.ema.europa.eu/docs/en_GB/document_library/Scientific_guideline/2017/07/WC500232420.pdf.

European Medicines Agency. (2017g). *Spherox.* London, UK: European Medicines Agency. Retrieved from http://www.ema.europa.eu/ema/index.jsp?curl=pages/medicines/human/medicines/002736/human_med_002138.jsp&mid=WC0b01ac058001d1.

European Medicines Agency. (2017h). Concept paper on development and lifecycle of personalised medicines and companion diagnostics [*Press release*]. London, UK: European Medicines Agency. Retrieved from http://www.ema.europa.eu/ema/index.jsp?curl=pages/news_and_events/news/2017/07/news_detail_002788.jsp&mid=WC0b01ac058004d5c1.

European Medicines Agency. (2017i). *Translarna.* EMEA/H/C/002720-II/0036. London, UK: European Medicines Agency. Retrieved from http://www.ema.europa.eu/ema/index.jsp?curl=pages/medicines/human/medicines/002720/human_med_001742.jsp&mid=WC0b01ac058001d124.

European Medicines Agency. (2017j). *New action plan to foster development of advanced therapies.* London, UK: European Medicines Agency. Retrieved from http://www.ema.europa.eu/ema/index.jsp?curl= pages/news_and_events/news/2017/10/news_detail_002831.jsp.

European Medicines Agency. (2017k). *Guideline on the clinical development of medicinal products for the treatment of autism spectrum disorder (ASD).* EMA/CHMP/598082/2013. London, UK: European Medicines Agency. Retrieved from http://www.ema.europa.eu/docs/en_GB/document_library/Scientific_guideline/2017/11/WC500238886.pdf.

European Medicines Agency. (2017l). *Draft qualification opinion on molecular neuroimaging of the dopamine transporter as biomarker to identify patients with early manifest Parkinsonism in Parkinson's disease.* EMA/765041/2017. London, UK: European Medicines Agency. Retrieved from http://www.ema.europa.eu/docs/en_GB/document_library/Regulatory_and_procedural_guideline/2018/01/WC500242219.pdf.

European Medicines Agency. (2017m). *Essential considerations for successful qualification of novel methodologies.* EMA/750178/2017. London, UK: European Medicines Agency. Retrieved from http://www.ema.europa.eu/docs/en_GB/document_library/Other/2017/12/WC500239928.pdf.

European Medicines Agency. (2018a). *Advanced therapy medicinal products.* London, UK: European Medicines Agency. Retrieved March 27, 2018 from http://www.ema.europa.eu/ema/index.jsp?curl=pages/regulation/general/general_content_000294.jsp&mid=WC0b01ac05800241e0.

European Medicines Agency. (2018b). *Authorisation of medicines.* London, UK: European Medicines Agency. Retrieved March 27, 2018 from http://www.ema.europa.eu/ema/index.jsp?curl=pages/about_us/general/general_content_000109.jsp&mid=WC0b01ac0580028a47.

European Medicines Agency. (2018c). *Committee for advanced therapies (CAT).* London, UK: European Medicines Agency. Retrieved March 27, 2018 from http://www.ema.europa.eu/ema/index.jsp?curl=pages/about_us/general/general_content_000266.jsp&mid=WC0b01ac05800292a4.

European Medicines Agency. (2018d). *Committee for orphan medicinal products (COMP).* London, UK: European Medicines Agency.

Retrieved March 27, 2018 from http://www.ema.europa.eu/ema/index.jsp?curl=pages/about_us/general/general_content_000263.jsp&mid= WC0b01ac0580028e30.

European Medicines Agency. (2018e). *Conditional marketing authorisation*. London, UK: European Medicines Agency. Retrieved March 27, 2018 from http://www.ema.europa.eu/ema/index.jsp?curl=pages/regulation/general/general_content_000925.jsp&mid=WC0b01ac05809f843b.

European Medicines Agency. (2018f). *Innovation in medicines*. London, UK: European Medicines Agency. Retrieved March 27, 2018 from http://www.ema.europa.eu/ema/index.jsp?curl=pages/regulation/general/general_content_000334.jsp&mid=WC0b01ac05800ba1d9.

European Medicines Agency. (2018g). *Legal framework: Orphan designation*. London, UK: European Medicines Agency. Retrieved March 27, 2018 from http://www.ema.europa.eu/ema/index.jsp?curl=pages/regulation/general/general_content_000552.jsp&mid=WC0b01ac058061ecb7.

European Medicines Agency. (2018h). *Orphan designation*. London, UK: European Medicines Agency. Retrieved March 27, 2018 from http://www.ema.europa.eu/ema/index.jsp?curl=pages/regulation/general/general_content_000029.jsp&mid=WC0b01ac0580b18a41.

European Medicines Agency. (2018i). *Orphan incentives*. London, UK: European Medicines Agency. Retrieved March 27, 2018 from http://www.ema.europa.eu/ema/index.jsp?curl=pages/regulation/general/general_content_000393.jsp&mid=WC0b01ac058061f017.

European Medicines Agency. (2018j). *Pharmacogenomics information in SmPC*. London, UK: European Medicines Agency. Retrieved March 27, 2018 from http://www.ema.europa.eu/docs/en_GB/document_library/Presentation/2013/01/WC500137032.pdf.

European Medicines Agency. (2018k). *Pharmacogenomics working party*. London, UK: European Medicines Agency. Retrieved March 27, 2018 from http://www.ema.europa.eu/ema/index.jsp?curl=pages/contacts/CHMP/people_listing_000018.jsp&mid=WC0b01ac0580028d91.

European Medicines Agency. (2018l). *PRIME: Priority medicines*. London, UK: European Medicines Agency. Retrieved March 27, 2018 from http://www.ema.europa.eu/ema/index.jsp?curl=pages/regulation/general/general_content_000660.jsp.

European Medicines Agency. (2018m). *Qualification of novel methodologies for medicine development*. London, UK: European Medicines Agency. Retrieved March 27, 2018 from http://www.ema.europa.eu/ema/index.jsp?curl=pages/regulation/document_listing/document_listing_000319.jsp&mid=WC0b01ac0580022bb0.

European Medicines Agency. (2018n). *Brineura*. London, UK: European Medicines Agency. Retrieved from http://www.ema.europa.eu/ema/index.jsp?curl=pages/medicines/human/medicines/004065/human_med_002111.jsp&mid=WC0b01ac058001d124.

European Medicines Agency. (2018o). *Guideline on the clinical investigation of medicines for the treatment of Alzheimer's disease*. CPMP/EWP/55395 Rev. 2. London, UK: European Medicines Agency. Retrieved from http://www.ema.europa.eu/docs/en_GB/document_library/Scientific_guideline/2018/02/WC500244609.pdf.

European Medicines Agency. (2018p). *Alofisel*. London, UK: European Medicines Agency. Retrieved from http://www.ema.europa.eu/ema/index.jsp?curl=pages/medicines/human/medicines/004258/human_med_002222.jsp&mid=WC0b01ac058001d124.

Evans Center Affinity Research Collaborative. (n.d.). Precision medicine for Alzheimer disease and related disorders. Retrieved from http://www.bumc.bu.edu/genetics/research/precision-medicine-in-alzheimer-disease-arc/.

Flatiron. (n.d.). Retrieved March 27, 2018 from https://flatiron.com/.

GBD 2015 Neurological Disorders Collaborator Group. (2017). Global, regional, and national burden of neurological disorders during 1990-2015: a systematic analysis for the Global Burden of Disease Study 2015. *Lancet Neurology, 16*(11), 877–897.

GetReal Initiative. (2018). Retrieved March 27, 2018 from http://www.imi-getreal.eu/.

Gnanasakthy, A., & De Muro, C. (2016). Outcome assessments of primary endpoints of new drugs approved by the FDA (*2011-2015*). *Poster presented at the ISPOR 21st Annual International Meeting, May 21-25, 2016, Washington, DC, United States* Retrieved from https://www.rtihs.org/sites/default/files/27533%20Gnanasakthy%202016%20Outcome%20assessments%20of%20primary%20endpoints%20of%20new%20drugs%20approved%20by%20the%20FDA%20(2011-2015).pdf.

Gnanasakthy, A., Mordin, M., Clark, M., DeMuro, C., Fehnel, S., & Copley-Merriman, C. (2012). A review of patient-reported outcome labels in the United States: 2006 to 2010. *Value in Health, 15*(3), 437–442.

Gnanasakthy, A., Mordin, M., Evans, E., Doward, L., & DeMuro, C. (2017). A review of patient-reported outcome labeling in the United States (2011-2015). *Value in Health, 20*(3), 420–429.

Gooch, C. L., Pracht, E., & Borenstein, A. R. (2017). The burden of neurological disease in the United States: a summary report and call to action. *Annals of Neurology, 81*(4), 479–484.

Hemmings, R., Germain, A., & Warner, J. (2018). Future-proofing with regulatory science. *Regulatory Rapporteur, 15*(3), 15–17.

HERCEPTIN. (2017). (Trastuzumab) *[Summary of product characteristics]*. Retrieved from http://www.ema.europa.eu/docs/en_GB/document_library/EPAR_-_Product_Information/human/000278/WC500074922.pdf.

Innovative Medicines Initiative. (2018a). *EBiSC: European Bank for induced pluripotent stem cells*. Retrieved March 27, 2018 from http://www.imi.europa.eu/projects-results/project-factsheets/ebisc.

Innovative Medicines Initiative. (2018b). *EPAD: European prevention of Alzheimer's dementia consortium*. Retrieved March 27, 2018 from http://www.imi.europa.eu/projects-results/project-factsheets/epad.

Innovative Medicines Initiative. (2018c). *EU-AIMS: European autism interventions—A multicentre study for developing new medications*. Retrieved March 27, 2018 from http://www.imi.europa.eu/projects-results/project-factsheets/eu-aims.

Innovative Medicines Initiative. (2018d). *NEWMEDS: Novel methods leading to new medications in depression and schizophrenia*. Retrieved March 27, 2018 from http://www.imi.europa.eu/projects-results/project-factsheets/newmeds.

Innovative Medicines Initiative. (2018e). *PRISM: Psychiatric ratings using intermediate stratified markers: Providing quantitative biological measures to facilitate the discovery and development of new treatments for social and cognitive deficits in AD, SZ, and MD*. Retrieved March 27, 2018 from http://www.imi.europa.eu/projects-results/project-factsheets/prism.

Insel, T. R., & Cuthbert, B. N. (2015). Medicine. brain disorders? Precisely. *Science, 348*(6234), 499–500.

Insel, T., Cuthbert, B., Garvey, M., Heinssen, R., Pine, D. S., Quinn, K., et al. (2010). Research domain criteria (RDoC): toward a new classification framework for research on mental disorders. *The American Journal of Psychiatry, 167*(7), 748–751.

KALYDECO. (2017). *(Ivacaftor) [Prescribing information]*. Boston, MA: Vertex. Retrieved from https://pi.vrtx.com/files/uspi_ivacaftor.pdf.

Kamal, M. A., Smith, P. F., Chaiyakunapruk, N., Wu, D. B., Pratoomsoot, C., Lee, K. K., et al. (2017). Interdisciplinary pharmacometrics linking oseltamivir pharmacology, influenza epidemiology and health economics to inform antiviral use in pandemics. *British Journal of Clinical Pharmacology, 83*(7), 1580–1594.

Kegel, M. (2017). *First dose group in Parkinson's stem cell trial successfully transplanted*. Parkinson's News Today. Retrieved from https://parkinsonsnewstoday.com/2017/04/26/first-dose-group-parkinsons-stem-cell-trial-successfully-transplanted/.

KEYTRUDA. (2017). *(Pembrolizumab) [Prescribing information]*. Whitehouse Station, NJ: Merck. Retrieved from https://www.merck.com/product/usa/pi_circulars/k/keytruda/keytruda_pi.pdf.

Ledford, H. (2015). First robust genetic links to depression emerge. *Nature, 523*(7560), 268–269.

Li, J., & Lu, Z. (2012). Systematic identification of pharmacogenomics information from clinical trials. *Journal of Biomedical Informatics, 45*(5), 870–878.

Lieberman, J. A., Stroup, T. S., McEvoy, J. P., Swartz, M. S., Rosenheck, R. A., Perkins, D. O., et al. (2005). Effectiveness of antipsychotic drugs in patients with chronic schizophrenia. *The New England Journal of Medicine, 353*(12), 1209–1223.

Millan, M. J., Goodwin, G. M., Meyer-Lindenberg, A., & Ögren, S. O. (2015). Learning from the past and looking to the future: emerging perspectives for improving the treatment of psychiatric disorders. *European Neuropsychopharmacology, 25*(5), 599–656.

National Institutes of Health. (n.d.). Report on NIH funding vs. global burden of disease. Bethesda, MD: National Institutes of Health. Retrieved from https://report.nih.gov/info_disease_burden.aspx.

NUCYNTA. (2017). *(Tapentadol) [Prescribing information]*. Gurabo, PR: Janssen Ortho, LLC.. Retrieved from https://www.nucynta.com/assets/pdf/nucynta-pi_0.pdf.

Patel, N., Wiśniowska, B., Jamei, M., & Polak, S. (2018). Real patient and its virtual twin: application of quantitative systems toxicology modelling in the cardiac safety assessment of citalopram. *The AAPS Journal, 20*(1), 6.

GW Pharmaceuticals. (2017). GW Pharmaceuticals announces acceptance of NDA filing for Epidiolex® (cannabidiol) in the treatment of Lennox-Gastaut syndrome and Dravet syndrome *[Press release]*. Retrieved from http://ir.gwpharm.com/news-releases/news-release-details/gw-pharmaceuticals-announces-acceptance-nda-filing-epidiolexr.

GW Pharmaceuticals. (2018). GW Pharmaceuticals announces the European Medicines Agency (EMA) accepts Epidiolex® (cannabidiol) marketing authorization application (MAA) for review *[Press release]*. Retrieved from http://ir.gwpharm.com/news-releases/news-release-details/gw-pharmaceuticals-announces-european-medicines-agency-ema.

PLAVIX. (2018). (Clopidogrel) *[Summary of product characteristics]*. Retrieved from https://www.medicines.org.uk/emc/product/5935# DOCREVISION.

Polasek, T. M., Tucker, G. T., Sorich, M. J., Wiese, M. D., Mohan, T., Rostami-Hodjegan, A., et al. (2018). Prediction of olanzapine exposure in individual patients using physiologically based pharmacokinetic modelling and simulation. *British Journal of Clinical Pharmacology, 84*(3), 462–476.

Power, R. A., Tansey, K. E., Buttenschon, H. N., Cohen-Woods, S., Bigdeli, T., Hall, L. S., et al. (2017). Genome-wide association for major depression through age at onset stratification: major depressive disorder working group of the psychiatric genomics consortium. *Biological Psychiatry, 81*(4), 325–335.

Rao, G., Mashkouri, S., Aum, D., Marcet, P., & Borlongan, C. V. (2017). Contemplating stem cell therapy for epilepsy-induced neuropsychiatric symptoms. *Neuropsychiatric Disease and Treatment, 13*, 585–596.

Rush, A. J., Lavori, P. W., Trivedi, M. H., Sackeim, H. A., … STAR*D Investigators Group. (2004). Sequenced treatment alternatives to relieve depression (STAR* D): rationale and design. *Contemporary Clinical Trials, 25*(1), 119–142.

SAVELLA. (2016). *(Milnacipran HCl) [Prescribing information]*. Irvine, CA: Allergan. Retrieved from https://www.allergan.com/assets/pdf/savella.

Schizophrenia Working Group of the Psychiatric Genomic Consortium. (2014). Biological insights from 108 schizophrenia-associated genetic loci. *Nature, 511*(7510), 421–427.

Spark Therapeutics. (2017). Spark therapeutics submits marketing authorization application to European medicines agency for investigational LUXTURNA™ (voretigene neparvovec) *[Press release]*. Retrieved from http://ir.sparktx.com/news-releases/news-release-details/spark-therapeutics-submits-marketing-authorization-application.

Storf, M. (2013). *The impact of FDA and European medicines agency guidances regarding patient reported outcomes (PRO) on the drug development and approval process*. Unpublished manuscript. Bonn, Germany: DGRA e.V. Retrieved from http://dgra.de/deutsch/studiengang/master-thesis/2013-Max-Storf-The-impact-of-FDA-and-European-Medicines-Agency-guidances-regarding-Patient-Reported-O.

Tsigkos, S., Llinares, J., Mariz, S., Aarum, S., Fregonese, L., Dembowska-Baginska, B., et al. (2014). Use of biomarkers in the context of orphan medicines designation in the European Union. *Orphanet Journal of Rare Diseases, 9*(1), 13.

U.S. Food and Drug Administration. (2014). *Parent project muscular dystrophy submission of a proposed draft guidance for industry on Duchenne muscular dystrophy developing drugs for treatment over the spectrum of disease; establishment of a public docket*. Retrieved from https://www.regulations.gov/docket?rpp=100&so=DESC&sb=docId&po=0&D=FDA-2014-D-1264.

U.S. Food and Drug Administration. (2017a). *Value and use of patient-reported outcomes (PROs) in assessing effects of medical devices: CDRH strategic priorities 2016–2017*. Retrieved from https://www.fda.gov/downloads/AboutFDA/CentersOffices/OfficeofMedicalProductsandTobacco/CDRH/CDRHVisionandMission/UCM588576.pdf.

U.S. Food and Drug Administration. (2017b). *FDA expands approved use of Kalydeco to treat additional mutations of cystic fibrosis: Laboratory evidence used to support efficacy*. Retrieved from https://www.fda.gov/NewsEvents/Newsroom/PressAnnouncements/ucm559212.htm.

U.S. Food and Drug Administration. (2017c). *Guidance for industry from the ALS association*. Retrieved from https://www.regulations.gov/document?D=FDA-2017-D-6503-0001.

U.S. Food and Drug Administration. (2018a). *Amyotrophic lateral sclerosis: Developing drugs for treatment. Guidance for industry*. Retrieved from https://www.fda.gov/downloads/Drugs/GuidanceComplianceRegulatoryInformation/Guidances/UCM596718.pdf.

U.S. Food and Drug Administration. (2018b). *Drugs@FDA: FDA approved products*. Retrieved March 27, 2018 from https://www.accessdata.fda.gov/scripts/cder/daf/index.cfm.

U.S. Food and Drug Administration. (2018c). *Duchenne muscular dystrophy and related dystrophinopathies: Developing drugs for treatment. Guidance for industry*. Retrieved from https://www.fda.gov/downloads/Drugs/GuidanceComplianceRegulatoryInformation/Guidances/UCM450229.pdf.

U.S. Food and Drug Administration. (2018d). *Statement from FDA Commissioner Scott Gottlieb, M.D. on advancing the development of novel treatments for neurological conditions; part of broader effort on modernizing FDA's new drug review programs*. Retrieved from https://www.fda.gov/NewsEvents/Newsroom/PressAnnouncements/ucm596897.htm.

U.S. Food and Drug Administration. (2018e). *Clinical outcome assessment qualification program*. Retrieved from https://www.fda.gov/Drugs/DevelopmentApprovalProcess/DrugDevelopmentToolsQualificationProgram/ucm284077.htm.

U.S. Food and Drug Administration. (2018f). *FDA authorizes marketing of first blood test to aid in the evaluation of concussion in adults: New quick testing option to help reduce need for CT scans, radiation exposure for patients*. Retrieved from https://www.fda.gov/newsevents/newsroom/pressannouncements/ucm596531.htm.

Venigalla, H., Mekala, H. M., Hassan, M., Rizwan, A., Zain, H., Dar, S., et al. (2017). An update on biomarkers in psychiatric disorders. *Mental Health in Family Medicine, 13*, 471–479.

Venkatesan, P. (2016). New European guidance on patient-reported outcomes. *The Lancet Oncology, 17*(6), 30113–30119.

VYVANSE. (2017). *(Lisdexamfetamine dimesylate)* *[Prescribing information]*. Lexington, MA: Shire. Retrieved from http://pi.shirecontent.com/PI/PDFs/Vyvanse_USA_ENG.pdf.

Washington University School of Medicine. (n.d.). *Dominantly inherited Alzheimer Network Trial: An opportunity to prevent dementia. A study of potential disease modifying treatments in individuals at risk for or with a type of early onset Alzheimer's disease caused by a genetic mutation (DIAN-TU).* ClinicalTrials.gov Identifier: NCT01760005. Retrieved from https://clinicaltrials.gov/ct2/show/NCT01760005.

Woodcock, J., & LaVange, L. M. (2017). Master protocols to study multiple therapies, multiple diseases, or both. *The New England Journal of Medicine, 377*(1), 62–70.

World Health Organization. (2017). *Fact sheet: Mental health of older adults.* Geneva, Switzerland: World Health Organization. Retrieved from http://www.who.int/mediacentre/factsheets/fs381/en/.

ZIAGEN. (2017). (Abacavir) *[Summary of product characteristics].* Retrieved from http://www.ema.europa.eu/docs/en_GB/document_library/EPAR_-_Product_Information/human/000252/WC500050343.pdf.

Regulatory Science Objectives and Biomarker Qualification Through Public-Private Partnerships Are Critical to Delivering Innovative Treatments for CNS Diseases

Diane T. Stephenson*, Stephen P. Arnerić†

*Critical Path for Parkinson's, Critical Path Institute, Tucson, AZ, United States †Critical Path for Alzheimer's Disease, Critical Path Institute, Tucson, AZ, United States

Abbreviation List

ADNI	Alzheimer's Disease Neuroimaging Initiative
AD	Alzheimer's disease
ACR	American College of Radiology
ALS	amyotrophic lateral sclerosis
ASD	autism spectrum disorder
BEST	biomarkers, endpoints and other tools
BMD	biometric monitoring device
BBB	blood-brain barrier
CDISC	Clinical Data Interchange Standards Consortium
CDER	Center for Drug Evaluation and Research
CDRH	Center of Devices and Radiological Health
CMS	Centers for Medicare and Medicaid Services
CFAST	Coalition for Accelerating Standards and Therapies
COAs	clinical outcome assessments
CNS	central nervous system
CAMD	Coalition Against Major Diseases
CHMP	Committee for Medicinal Products for Human Use
CDEs	common data elements
COU	context of use
CPAD	Critical Path for Alzheimer's Disease
CPP	Critical Path for Parkinson's
CPTR	Critical Path for Tuberculosis Research
CPIM	Critical Path Innovation Meetings
C-Path	Critical Path Institute
CSDR	Clinical Study Data Repository
CSF	cerebrospinal fluid
DBS	deep brain stimulation
DDT	drug development tool
DMD	Duchenne muscular dystrophy
EU	European Union
EMA	European Medicines Agency
IDEAS	Imaging Dementia-Evidence for Amyloid Scanning
FR	Federal Register
FNIH	Foundation for the National Institutes of Health
HD	Huntington disease
ImPACT	Immediate Postconcussion Assessment and Cognitive Testing
IVD	in vitro diagnostic
iPS	induced pluripotent stem cells
ICFs	informed consent forms
IND	investigational new drug
PMDA	Japanese Pharmaceutical Medical Devices Agency
LOS	letter of support
MDDT	Medical Device Drug Development Tool
MDIC	Medical Device Innovation Consortium
MJFF	Michael J. Fox Foundation
MND	motor neuron disease
MRI	magnetic resonance imaging
MS	multiple sclerosis
MSOAC	Multiple Sclerosis Outcome Assessment Consortium
NCATS	National Center for Advancing Translational Sciences
NINDS	National Institute of Neurological Disorders and Stroke
NIH	National Institutes of Health
NDA	new drug application
NME	new molecular entity
PD	Parkinson's disease
PPMI	Parkinson's Progression Marker Initiative
PET	positron-emission tomography
PKD	Polycystic Kidney Disease Consortium
PMA	premarket approval
SARDAA	Schizophrenia and Related Disorders Alliance of America
SAWP	Scientific Advice Working Party
SOD	superoxide dismutase
TED	Traumatic Brian Injury Endpoints Development Initiative
TAUGs	Therapeutic Area User Guides
TDI	Therapy Development Institute
TRACK-TBI	Transforming Research and Clinical Knowledge in Traumatic Brain Injury
TBI	traumatic brain injury
FDA	US Food and Drug Administration

Translational Medicine in CNS Drug Development, Volume 29
ISSN: 1569-7339
https://doi.org/10.1016/B978-0-12-803161-2.00020-5

I BACKGROUND

This section addresses three key areas that shape the scientific community's ability to deliver innovative treatments for brain diseases: (1.1) unmet needs, (1.2) current strategies, and (1.3) regulatory science considerations.

A Unmet Need

The growing economic burden of brain diseases is becoming unsustainable (Table 20.1). Factors that contribute to this increase include dramatic advances in medicine and society infrastructures that increase longevity. Many therapies are marketed for the treatment of CNS conditions, yet most treat symptoms as opposed to the underlying disease process itself, and many patients remain refractory to drugs on the market. Failure rates in late-stage clinical trials are disproportionately high for neurological and psychiatric diseases when compared with other disease areas (Paul et al., 2010). Critically, no cures exist for brain diseases. As one example, the US annual costs for treating and caring for those afflicted with dementias pose greater financial burden than for heart disease or cancer (Hurd, Martorell, & Delvande, 2013). Critically, psychiatric and neurological diseases comprise the largest category of disability within the United States (Murray, 2013). When expanded to a global perspective, the economic and emotional burden is daunting.

In recent years, national and global initiatives are being launched across specific CNS diseases to escalate visibility of unmet needs and develop nationwide plans aimed at increasing awareness, funding, and outreach. Examples of public provide partnerships devoted to CNS diseases include Alzheimer's consortia (Snyder, Kim, Bain, Egge, & Carrillo, 2014), neuropsychiatry (Brady & Potter, 2014), and neurodevelopmental disorders (Murphy, Goldman, Loth, & Spooren, 2014). These initiatives bring together nontraditional partners with a shared goal, and they also share risks and costs among all stakeholders. In fact the rapid rise in the number of

TABLE 20.1 Economic Burden of Representative CNS Diseases

Disease	Incidence/prevalence	Economic burden
Alzheimer's disease (AD)	• 5.4 million Americans are living with AD • By 2050, up to 16 million will have AD • Every 66 seconds, someone is diagnosed	• 2016 costs = $236 billion • Nearly 1 of every 5 Medicare dollars is spent on people with AD
Autism spectrum disorder (ASD)	• 1 in 68 children (1 in 42 boys and 1 in 189 girls) • More than 3.5 million Americans live with an ASD	• Autism services cost US citizens $236–262 billion annually • A majority of costs in the United States are in adult services—$175–196 billion, compared with $61–66 billion for children
Depression	• Affects more than 15 million American adults, or about 6.7% of the US population age 18 and older in a given year	• Mood disorders cost the United States more than $42 billion a year, almost one-third of the country's $148 billion total mental health bill
Multiple sclerosis (MS)	• More than 400,000 people in the United States • About 2.5 million people worldwide • About 200 new cases are diagnosed each week in the United States	• Direct and indirect health care costs range from $8,528 to $54,244 per patient per year in the United States • MS ranks second only to congestive heart failure in terms of costliness compared with other chronic conditions
Parkinson's disease (PD)	• As many as 1 million Americans live with PD, which is more than the combined number of people diagnosed with multiple sclerosis, muscular dystrophy, and Lou Gehrig's disease • Approximately 60,000 Americans are diagnosed with PD each year, and this number does not reflect the thousands of cases that go undetected • More than 10 million people worldwide are living with PD	• The combined direct and indirect cost of Parkinson's disease, including treatment, social security payments, and lost income from inability to work, is estimated to be nearly $25 billion per year in the United States alone • Medication costs for an individual person with PD average $2,500 a year, and therapeutic surgery can cost up to $100,000 dollars per patient
Schizophrenia	• Schizophrenia can be found in approximately 1.1% of the world's population, regardless of racial, ethnic, or economic background • Approximately 3.5 million people in the United States are diagnosed with schizophrenia, and it is one of the leading causes of disability	• Treatment and other economic costs due to schizophrenia are enormous, estimated between $32.5 and $65 billion annually

Alzheimer's info: http://www.alz.org/documents_custom/2016-Facts-and-Figures-Fact-Sheet.pdf (Alzheimer's Association, 2016)
Autism: https://www.autismspeaks.org/what-autism/prevalence and http://www.autism-society.org/what-is/facts-and-statistics/ (Autism Society, 2015; Autism Speaks Website, 2018)
Depression: https://www.adaa.org/about-adaa/press-room/facts-statistics (Anxiety and Depression Association of America (ADAA), 2010/2016)
Multiple sclerosis: http://www.healthline.com/health/multiple-sclerosis/facts-statistics-infographic (Healthline, Pitrangelo, & Higuera, 2015)
Parkinson's disease: http://www.parkinson.org/Understanding-Parkinsons/Causes-and-Statistics/Statistics (Parkinson's Foundation, n.d.)
Schizophrenia: http://www.sardaa.org/resources/about-schizophrenia/ (Schizophrenia and Related Disorders Alliance of America, 2008/2017)

PPPs has created a crowded landscape. *FasterCures* initiated the consortia-pedia project to better understand the breadth and scope of approaches that a wide range of PPPs have adopted as a resource (http://consortiapedia.fastercures.org) (Consortia Pedia, and FasterCures, 2017).

Despite such incremental progress for each of these CNS diseases, the resources needed to match the scope and scale are not being mobilized fast enough to augment therapeutic development.

One clear gap is the absence of efforts that coordinate common lessons learned across the specific CNS disease areas. While there are some efforts seeking common biochemical threads across diseases (Berlyand et al., 2016; Lleó et al., 2014; Trojanowski & Hampel, 2011), given the high degree of comorbidities and disabilities across diseases, it is surprising that more isn't being done to understand both the common elements and differentiating features for the observed disabilities (Table 20.2). Understanding these elements would be particularly insightful in building translational drug development tools that could be implemented across these diseases.

B Current Strategies

Across all disease areas the failure rate of drugs in development is exceptionally high in diseases of the brain. For example, the clinical trial failure rate for late-stage AD therapies from 2002 to 2012 was 99.6% (Cummings, Morstorf, & Zhong, 2014). A comprehensive study across multiple diseases reported success rates for new drugs in clinical development by measuring the likelihood of the FDA approval over 10 years (2006–16) (AMPLION REPORT/Thomas, Albert, Petersen, & Aisen, 2016). The compound probability of progressing from Phase 1 to US FDA approval revealed that only 9.6% of drug development programs successfully make it to the market (AMPLION REPORT/Thomas et al., 2016). The lowest success rate was consistently transitioning in Phase 2

(30.7%) with the second lowest phase transition success found in Phase 3 (58.1%). This is significant in that the longest and most costly trials to conduct are at these late stages of development. The clinical success evaluation reported that neurology and psychiatry represent disease areas that have the highest number of transitions while having the lowest likelihood of approval (Thomas et al., 2016). Moreover the duration of time for regulatory review and approval of drugs for neurology is the longest across all disease areas; for example, neurology drugs took on average 2 years to approve, while oncology drugs were approved in nearly half the time (1.1 yrs.). A comprehensive analysis of FDA-approved drugs for neurological disorders shows that 95% of all new molecular entities approved belong into five groupings. These include seizures (39%) followed by Parkinson's disease (23%), neuromuscular diseases (20%), Alzheimer's disease (7%), and narcolepsy (6%) (Kinch, 2015). Most of the many CNS diseases lack effective therapies. Collectively, these facts pose serious pharmacoeconomic challenges for pharmaceutical companies to continue to invest given the limited return on investment (Choi et al., 2014).

There are a multitude of reasons that contribute to the high rate of failure of new drug approvals for brain diseases. In Table 20.3 are some of the key challenges facing developers of CNS therapeutics for chronic diseases.

A central issue has been the standard pharmaceutical approach of trying to identify and modulate a single, key target within the complex pathophysiology of a given CNS disease, only to find after much time and effort that modulation of the single target in humans has no significant clinical impact.

More recently, there has been a resurgence of multitarget therapeutic strategies given the acknowledgement that complex CNS diseases such as AD are going to require combination therapeutic strategies (Das & Basu, 2017; Perry et al., 2015; Stephenson et al., 2015). The increasing recognition that brain diseases affect multiple

TABLE 20.2 Comorbid Disabilities Across CNS Diseases

Disease and disability	Impaired cognition	Impaired mobility	Aberrant sleep	Psychosis	Depression/Anxiety	Pain
Alzheimer's disease (AD)	Yes	Yes	Yes	Yes	Yes	Yes
Amyotrophic lateral sclerosis (ALS)	Sometimes	Yes	Sometimes	No	Yes	Yes
Autism spectrum disorder (ASD)	Yes	Sometimes	Yes	Rarely	Yes	Rarely
Depression	Yes	Sometimes	Yes	Sometimes	Yes	Yes
Huntington's disease (HD)	Yes	Yes	Yes	Yes	Yes	Yes
Multiple sclerosis (MS)	Yes	Yes	Yes	Rarely	Yes	Yes
Parkinson's disease (PD)	Yes	Yes	Yes	Yes	Yes	Yes
Schizophrenia	Yes	No	Yes	Yes	Sometimes	Rarely

References: refer to links in Table 20.1.

TABLE 20.3 Challenges in Developing CNS Therapeutics

Patient heterogeneity	• Heterogeneity of patient populations • Lack of accurate diagnostic criterion • Heterogeneity of comorbid symptoms
Patient selection/ study design	• Recruitment speed • Patients without disease • Underpowered studies • High dropout rates
Mechanism	• Limited understanding of the neurobiology of disease • Uncertainty of translation of preclinical models to human studies • Biomarkers for target engagement not fully developed
Treatment delivery	• Need to cross the blood-brain barrier (BBB) • Chronic disease requires long-duration treatment • Increasing rationale to treat in presymptomatic stages of the disease
Clinical outcome assessments	• Subjective, relatively insensitive outcome measures • Outcome assessments in presymptomatic stages not yet developed
Economic	• Reimbursement issues/payers; uncertainty in terms of what is needed by payers to facilitate adoption and coverage (e.g., IDEAs study) • High cost of trials due to imaging, ancillary assessments, and chronic nature of trials • Expense in running dose-response studies in Phase 3
Regulatory	• Need for qualified drug development tools to accelerate the development process • Infrequent use of CDISC standards across individual trials that would enable the ability to pool and analyze data across all trials

system circuits should reinforce the notion that a new molecular entity (NME) with a single mechanism of action is unlikely to normalize the multitude of affected brain functions.

C Regulatory Science Considerations

Over a decade ago the FDA publicly recognized the importance of improved efficiencies in drug development with the 2004 Critical Path Initiative Challenges and Opportunities report (FDA, 2004). A comprehensive list of recommendations provided an example of striking innovation and vision. The list encompasses diverse approaches to tackling drug development. Translational biomarker development was deemed as a top priority. The importance of biomarkers in drug approvals has been increasingly recognized since the Critical Path Initiative report. Biomarkers are the foundation of precision medicine, and their successful implementation in drug development has been transformational (Collins & Varmus, 2015). The 21st Century Cures Initiative has more recently highlighted the urgent need for biomarkers in catalyzing drug development (Chandra, 2017); 21st

Century Cures Act (H.R. 6) (10 Jul 2015) (https://www.congress.gov/114/bills/hr6/BILLS-114hr6rfs.pdf) (Senate of the United States, 2015).

The FDA recently reviewed the role of PPPs in catalyzing the Critical Path (Maxfield, Buckman-Garner, & Parekh, 2017)

There is growing interest in seeking regulatory endorsement of biomarkers (Amur, LaVange, et al., 2015; Amur, Sanyal, et al., 2015; Arnerić et al., 2017; Lavezzari & Womack, 2016). Efficiencies gained from qualified biomarkers can have impact across multiple disease areas and can be applied broadly, independent of individual targets or sponsors. A formal mechanism now exists for identification, evaluation, and qualification of biomarkers for use in drug development in the FDA and EMA (Amur, LaVange, et al., 2015; Amur, Sanyal, et al., 2015; Manolis, Vamvakas, & Isaac, 2011).

The Critical Path Institute (C-Path) is a nonprofit organization founded in 2005 to deliver on the vision of FDA's Critical Path Initiative. The mission of C-Path is to be a catalyst in the development of tools to advance medical innovation and regulatory science. This is achieved by leading teams that share data, knowledge, and expertise resulting in sound, consensus-based science. In its 12-year history, C-Path has created 12 consortia that focus on expediting drug development in many therapeutic areas. The consortia include over 1450 scientists from 65 global pharmaceutical companies and government agencies such as the FDA, the EMA, and the National Institutes of Health (NIH). The founding C-Path consortia are partially funded by the FDA as part of the Critical Path public-private partnerships (Woodcock & Woosley, 2008). Core competencies of C-Path include (1) development of consensus data standards, (2) acquiring and integrating patient-level data, and (3) advancing regulatory science paths for the endorsement of drug development tools. Public-private partnerships, like C-Path, can enable efficiencies in regulatory acceptance of biomarkers (Haas et al., 2015; Maxfield et al., 2017; Rouse et al., 2017; Woodcock, 2014). Examples will be highlighted of how the development of data standards and the integration of anonymized and patient-level data have advanced the regulatory science required to validate Drug Development Tools (DDTs).

II EVIDENCE THAT BIOMARKERS INCREASE SUCCESSFUL APPROVALS

Clinical symptoms using instruments that are subjective and heterogeneous in nature are still the primary outcome measure for determining efficacy of novel therapeutics for most CNS diseases. Molecular biomarkers are a cornerstone of the precision medicine initiative and enable an approach to tailor treatments to the biochemical phenotypes/fingerprints of individual patients. When approved as DDTs by regulators, they can be used by all sponsors for a defined context of use (COU). The COU for biomarkers can be viewed as analogous to the product label for new drugs as it specifies precisely the conditions for application in drug development. COUs can range in utility from preclinical safety biomarkers to patient stratification biomarkers, to surrogate endpoints that are reasonably likely to predict clinical outcomes.

A recent assessment aimed to quantify the benefit of using biomarkers for patient selection decisions in drug development across 10,000 clinical trials across multiple therapeutic areas (Thomas et al., 2016). The report evaluated data from Amplion's BiomarkerBase and BioMedTracker clinical transition records. The results demonstrated that biomarker-enabled programs for patient selection are three times more likely to reach approval. The benefit from selection biomarker use raises the likelihood of approval from Phase 1 from one in four versus less than one in 10 without biomarkers. Use of selection biomarkers increased the likelihood of approval at all four phases of transition (Ph 1-2, Ph 2-3; Ph 3-NDA/BLA; NDA/BLA approval). In addition, when biomarkers are used, patients can be better stratified, and those who were unlikely to respond are not subjected to undue adverse effects. Overall, the use of relevant patient selection biomarkers can increase statistical power, and the likelihood of observing treatment responses in clinical development. This later point is key, as it has been reported that most CNS trials are underpowered (Button et al., 2013; Halpern, Karlawish, & Berlin, 2002).

III GAPS IN BUILDING TRANSLATIONAL BIOMARKERS FOR CNS DISEASES

Four key dimensions need to be addressed to advance translational biomarkers for CNS diseases: (1) understand the biological responses in preclinical models under normal and treatment groups, (2) understand the biological responses in target human populations and following treatment interventions, (3) understand and define how these responses evolve over the course of model/disease progression, and (4) Build an integrated database with standardized data elements that relate model/disease progression with treatment responses. The following sections detail examples of how this can and should be approached.

A Bench to the Clinic

Challenges in animal model research have become more visible given the augmented attention to reproducibility of research (Lithgow, Driscoll, & Phillips, 2017;

McGonigle, 2014; Snyder et al., 2016). A high percent of animal model research cannot be replicated (Pankevich, Altevogt, Dunlop, Gage, & Hyman, 2014). There is a need to pay more attention to the scientific validity of data when decisions have impact to moving from the laboratory into clinical trials. Standardization of assay performance and reproducibility is a key gap. It has been suggested that traditional animal models for CNS may even be screening out potential effective compounds (Hyman, 2016; Pankevich et al., 2014). Publication bias exists in that negative findings are not typically published and incorrect conclusions can be made from poorly designed and executed animal models. The problems become even more augmented when making decisions to advance therapies based on data in animal models, particularly when trying to mimic complex human neurophysiology and measures such as cognition.

On the positive side, animal models have led to improved understanding of nervous system disorders. They can be used to make informed decisions or dose predictions, establish the therapeutic index of efficacy versus safety, and evaluate target engagement and test-specified hypotheses and/or mechanisms. Key learnings in terms of use of biomarkers in drug development can be highlighted by looking carefully at the field of Alzheimer's disease. An informative evaluation of drug development programs across different sponsors over many years revealed that many of the studies failed to demonstrate target engagement in humans (Gold, 2017; Karran & Hardy, 2014). Biomarkers in animal models of amyloid deposition have been very powerful in providing a way to assess mechanisms of action of drug targets in distinct compartments including plasma, cerebrospinal fluid (CSF), and the brain (Lanz, Hosley, Adams, & Merchant, 2004). True in vivo pharmacology experiments have been successfully carried out in this fashion allowing determination of target engagement and dose selection for clinical trials.

The decreasing confidence in the ability of nonclinical models to predict human efficacy of drugs, especially those with unprecedented mechanisms of action, is catalyzing the development of alternative innovative approaches using human tissues. For example, translational tools such as human induced pluripotent stem (iPS) cells are providing a bridging strategy for the divide between animal models and human clinical studies. Biomarkers measured in such models hold more promise in translating to humans given the genetic focus (genotype to phenotype concept) of direct translatability. The status and promise of human iPS cells are reviewed by (Hosoya & Czysz, 2016). Notably the FDA has highlighted how such technologies can be used in the regulatory process (Rouse et al., 2017).

B Clinic to the Bench

The overarching rationale is to focus on human disease rather than expecting that nonclinical models will recapitulate drug efficacy in chronic human brain diseases. Although the field seems to be going in this direction, true principles and examples of what has been termed "back translation" are still in the early stages of development.

One impactful example of validating back translation has been carried out in ALS. The ALS Therapy Development Institute (TDI) nonprofit organization embarked on an intensive initiative to evaluate the predictive accuracy of the most commonly used mouse model of ALS, experimental animals carrying human superoxide dismutase (SOD) mutations. The G93A genetic mouse model of motor neuron disease developed in 1994 (Gurney et al., 1994) has been used extensively to evaluate candidate therapies for ALS with no efficacy found to date from all trials despite promising results in animals. Investigators gathered in 2006 to agree on standard operating procedures for preclinical research in ALS/motor neuron disease (MND) as a way to foster translatability and consensus by applying learnings from independent investigators (Ludolph et al., 2010).

The ALS TDI investigators tested more than 100 potential drugs in the most commonly used mouse model of ALS using defined experimental conditions (Perrin, 2014). No treatment benefit was observed in the experiments carried out by ALS TDI investigators, despite the fact that many of the drugs tested included those that had been previously reported to slow disease progression in the same mouse model. Eight of these compounds ultimately failed in clinical trials. In a disease, as rare and devastating as ALS, eight trials represent a relatively large population subjected to experimental treatments with no benefit or data available from failed trials to apply to future clinical trials. One approach to discover novel therapeutic targets and biomarkers that is promising is to analyze patient-level data from human ALS patients integrated into an open source data repository (Atassi et al., 2014; Küffner et al., 2014).

C Building Foundational Understanding of Disease Progression

The clinical community has embraced the concept to create an open-access database of anonymized patient-level data as a means to understand disease progression and to model treatment responses. Prime examples of this approach include the Alzheimer's Disease Neuroimaging Initiative (ADNI) (Jones-Davis & Buckholtz, 2015) and the fit-for-purpose clinical trial simulation tool for Alzheimer's disease developed by the Coalition Against Major Diseases (CAMD) (Romero et al., 2014), arising from its creation of the first publicly available database of anonymized clinical trial data (Neville et al., 2015). Other CNS diseases that have developed impactful databases for quantifying disease progression include ALS (Atassi et al., 2014), PD (Heinzel, Lerche, Maetzler, & Berg, 2017; Lawton et al., 2016; Marek et al., 2011), and MS (LaRocca et al., 2017). Biomarkers are key in defining pathophysiologic correlates of clinically meaningful change at various stages of the disease spectrum that typifies chronic neurodegenerative diseases.

Analysis of clinical data is a powerful way to understand disease progression and onset. Analysis of individual studies is valuable, yet many important research questions cannot be answered as the number of subjects in studies is typically too small. The strategy to combine different studies for enhancing the sample size can be powerful yet is hampered by the lack of comparability of symptoms/target populations and specific assessments. This is particularly problematic for biomarkers as each has characteristic preanalytical and analytical variables to consider. Clearly, there is true benefit to agree on common strategies across studies to define and validate predictive and progression markers.

1 Consensus Data Standards are Foundational for Success

The Coalition for Accelerating Standards and Therapies (CFAST) was launched in October 2012, as a partnership between the Clinical Data Interchange Standards Consortium (CDISC) and C-Path to accelerate clinical research and medical product development by facilitating the creation and maintenance of data standards, tools, and methods for conducting research in therapeutic areas important to public health. CDISC is a nonprofit organization with approximately 300 supporting member organizations from across the global clinical research and healthcare arenas. CDISC catalyzes productive collaboration to develop industry-wide data standards enabling the harmonization of clinical data and streamlining research processes from protocol through analysis and reporting, including the use of electronic health records to facilitate the collection of high-quality research data. CDISC standards and innovations have substantially decreased the time and cost of medical research and improved quality, thus contributing to the faster development of safer and more effective medical products.

A list of Therapeutic Area User Guides (TAUGs) of CDISC standards for CNS disease is shown in Table 20.4. Many of the CDISC clinical standards for brain diseases were developed from foundational elements originating from National Institute of Neurological Disorders and Stroke (NINDS) plus common data elements (CDEs) (Grinnon et al., 2012; Stone, 2010). In the absence of having these standards in place, there will be clear gaps achieving efficient regulatory approval of

TABLE 20.4 Status of CDISC Standard Development for Key CNS Diseases

Disease TAUGs	Available	In planning	In progress	Comments
Alzheimer's disease (AD) V2.0	Yes	V3.0		Structural and fluid biomarkers integrated into V2.0; Future plans for presymptomatic stages of the disease that include biometric monitoring devices (V3.0)
Amyotrophic lateral sclerosis (ALS)	No			
Autism spectrum disorder (ASD)	No			
Depression	Yes			Biomarkers not included
Huntington's disease (HD)	Yes			Includes biomarkers across modalities
Multiple sclerosis (MS)	Yes			Contains imaging biomarkers
Parkinson's disease (PD) v1.0	Yes	Yes		
Schizophrenia	Yes			Limited fluid biomarkers
Traumatic brain injury (TBI)	Yes			Imaging and fluid biomarkers included

All CDISC Therapeutic Area User Guides can be accessed free at www.cdisc.org.

products. Development of standards for biomarkers serves as a true catalyst for enhancing reliability and reproducibility of biomarkers in multisite studies, thus enabling their efficient use in clinical decision-making.

With the aim of accelerating the review process, the FDA has issued guidance (http://www.fda.gov/downloads/drugs/guidances/ucm292334.pdf) (FDA, CDER, USDHS, 2014) that requires that all new drug applications will need to be submitted in CDISC format. Similar requirements have been set forth in Japan with PMDA as of October 2016 (http://www.pmda.go.jp/files/000206451.pdf) (PMDA, 2015). These standards also serve to enable integration of clinical trial data into a fully integrated database.

The Critical Path Institute has leveraged through its public-private partnerships two key elements required to advance regulatory science (Brumfield, 2014; Stephenson et al., 2015). It has done so through building consensus science in the areas of data standards and creating standardized, actionable databases in multiple disease areas. These areas and specific accomplishment are shown in Fig. 20.1.

Unfortunately, comparable comprehensive open-access databases for many preclinical studies for advancing CNS targets do not exist (Fig. 20.2). Some interest in the development of preclinical data standards has been highlighted in the CNS conditions of TBI (Smith et al., 2015), spinal cord injury (Lemmon et al., 2014), spinal muscular atrophy (Willmann, Dubach, & Chen, 2011), and ALS (Ludolph, 2011). However, the primary focus in big data approaches and standardization for CNS diseases has been in humans as opposed to preclinical models.

The success of C-Path's qualification of nonclinical safety biomarkers for drug-induced kidney toxicity by

FIG. 20.1 Actionable, standardized databases to advance regulatory science by the Critical Path Institute. Consortia and their area of focus, Coalition Against Major Diseases (CAMD)—Alzheimer's disease (currently known as CPAD, Critical Path for Alzheimer's Disease); Coalition for Accelerating Standards for Therapies (CFAST), multiple therapeutic areas; Critical Path for Parkinson's (CPP), Parkinson's disease; Critical Path for Tuberculosis Research (CPTR), Tuberculosis; Multiple sclerosis outcome assessment consortium (MSOAC), multiple sclerosis; and Polycystic Kidney Disease Consortium (PKD), Polycystic Kidney Disease.

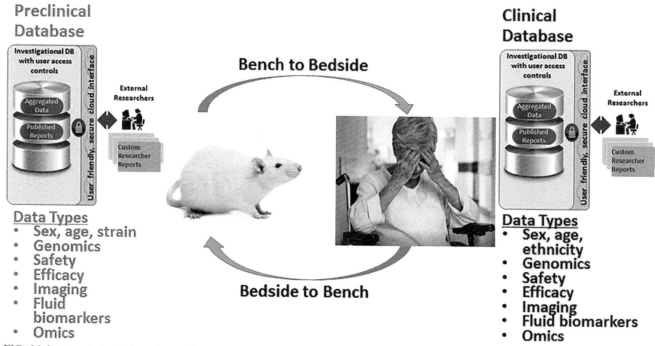

FIG. 20.2   Translational biomarkers will require open-access databases to facilitate validation and modeling of disease progression both at a preclinical and clinical level.

FDA, EMA, and PMDA was only achievable by the contribution of data from many of preclinical toxicology studies contributed by individual sponsors into a collective data repository (Dennis, Walker, Baker, & Miller, 2013; Dieterle et al., 2010; Sistare & DeGeorge, 2011). If forward and back translations of biomarkers are to become a reality, data sharing needs to become the new standard. With the launch of global initiatives aimed at advancing brain science (US BRAIN and European Human Brain Projects) and growing international efforts toward transparency and increased access to publicly funded research in the neurosciences, it is hoped that greater attention on such needs will emerge.

In summary a key gap that needs addressing are consensus data standards that are openly available and comply both with CDISC clinical data and NIH's common data elements (CDE) initiative (Smith et al., 2015; Willmann et al., 2011). For CNS diseases, this has been accomplished for Alzheimer's disease, depression, multiple sclerosis, Parkinson's disease, schizophrenia, and traumatic brain injury (https://www.cdisc.org/therapeutic) (CDISC, 2014). Aggregation of these actionable datasets into integrated databases will facilitate a greater understanding of disease progression and provide the foundation for advancing the regulatory science required to develop drug development tools that accelerate the delivery of innovative treatments. Regulatory agencies in both the United States and Europe have identified quantitative disease models as a drug development tool platform to accelerate drug development. Quantitative pharmacometric modeling is an innovative

approach that enables sponsors to analyze data from integrated multiple sources to accurately design prospective clinical trials. Model-informed drug development is highlighted as a key goal of PDUFAVI (https://www.fda.gov/forindustry/userfees/prescriptiondruguserfee/ucm446608.htm) (FDA, 2018b), and examples exist to aid in the efficiency of drug development across all stages (Rouse et al., 2017). Utilizing all of these measures would help foster improvements in biomarker utilization independent of disease area and enable translational (forward and backwards) cross species validation.

Given that drug development in neurodegenerative diseases is moving toward early intervention with the goal of treatments that slow or even prevent disease onset and progression (Gray, 2016), many trials are being designed and conducted where at-risk populations will be treated prior to onset of clinical symptoms using currently available clinical instruments or biomarkers (Jack et al., 2016; Sperling, Jack, & Aisen, 2011). Evidence suggests that many neurodegenerative diseases take several decades to manifest frank clinical signs (Trojanowski & Hampel, 2011). The importance of novel biomarkers in these early interventional strategies is even more critical as clinical signs and symptoms are not evident or readily measurable. Thus the challenge and dilemma is to validate sensitive biomarkers in the earliest, premanifest stages of the disease such that the correct patients are chosen and decades will not be required to pass before less sensitive validated clinical outcome assessments (COAs) can detect improvements in outcomes (Fig. 20.3).

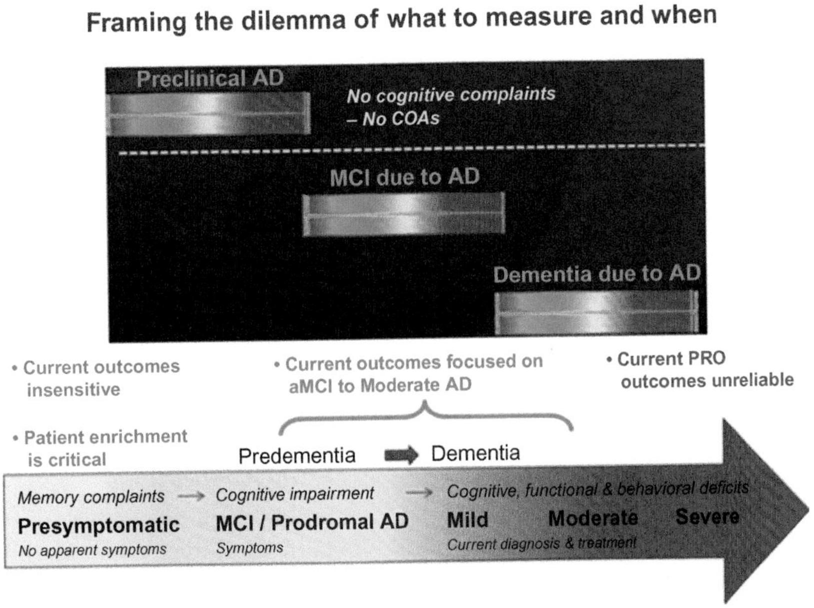

Framing the dilemma of what to measure and when

FIG. 20.3 The challenge of targeting early interventions reinforces the need for validated biomarkers and clinical outcome assessments (COAs). Which patients should be selected for trials? In the absence of very sensitive biomarkers or clinical instruments, how long must we wait before a treatment effect can be measured?

IV REGULATORY CONSIDERATIONS TO VALIDATE BIOMARKERS

Historically the field has focused on biomarkers that aid in the diagnosis of disease, assessment of target engagement, and responses to treatments. The complexity of this field has grown in recent years. Government agencies have recognized that precise, clear communication across all scientific research disciplines is crucial for more efficient translation of laboratory discoveries into new treatments for patients. To this end, NIH and the FDA recently published an open-access textbook: the Biomarkers, EndpointS and other Tools (BEST) resource (FDA-NIH Biomarker Working Group, 2016; Robb, McInnes, & Califf, 2016). This glossary is focused on clarifying important definitions; capturing the distinction between biomarkers and clinical assessments; and describing some of the hierarchical relationships, connections, and dependencies among the terms. Such an open-access resource originating from NIH and FDA is intended to ensure a consistent use of terms and a common understanding of issues.

What constitutes a validated biomarker? The response to this question depends upon what context of use (COU) the biomarker will be used for. These three COU elements are key: (1) What concept (analyte, structure, and function) is the biomarker measuring? (2) What is its purpose in drug development? (3) What clinical decision/action is taken based on the results? In 2010, FDA issued a draft guidance and road map for validation and qualification of biomarkers that was revised in 2014 (FDA, C, 2014). More recently, formal regulatory processes to advance

biomarkers for regulatory endorsement are in place to enable review and advice. Consensus data-driven strategies are now enabled to align with best practices for biomarker development and validation (Amur, LaVange, et al., 2015; Amur, Sanyal, et al., 2015; Leptak et al., 2017; McShane, 2017). The final qualified biomarkers are issued as guidance documents that apply across drug classes and mechanisms of action and can be used by all sponsors. Overall the goal is to reduce the risk of errors in decision-making based on biomarkers and avoids the need for independent sponsors to generate the levels of evidence for reliability and reproducibility of biomarkers on their own.

The FDA continues to work with the broader scientific community to define specific levels of validation appropriate to specific COUs, including fit-for-purpose biomarkers, qualified biomarkers, and validated biomarkers. In October 2016 the Biomarkers Consortium Evidentiary Standards Writing Group published the "Framework for Defining Evidentiary Criteria for Biomarker Qualification" (http://fnih.org/sites/default/files/final/pdf/Evidentiary%20Criteria%20Framework%20Final%20Version%20Oct%2020%202016.pdf) (Biomarkers Consortium Evidentiary Standards Writing Group, 2016; Leptak et al., 2017). New revisions to the biomarker qualification program are underway.

Many of the biomarkers that neuroscientists take for granted for use in clinical studies are *"fit for purpose."* That is the biomarker has "good enough" validation/qualification information to be used internally by a company/institution for internal decisions or possibly for publication. This does not require regulatory submission or review. However, the FDA has more recently created a fit-for-purpose designation for regulatory endorsement of modeling tools (http://www.fda.gov/Drugs/DevelopmentApprovalProcess/ucm505485.htm) (FDA, 2016b).

"Qualified Biomarkers" are typically those that are not assay or algorithm-specific and may require both observational and clinical trial data sets for cross validation of performance. The level of clinical decision made (i.e., the COU) with the biomarker will determine the level of evidence required to qualify the biomarker (Amur, LaVange, et al., 2015). Biomarker qualifications require regulatory review. This process has been recently updated by the FDA (https://www.fda.gov/Drugs/DevelopmentApprovalProcess/DrugDevelopmentToolsQualificationProgram/BiomarkerQualificationProgram/ucm536018.htm) (FDA, 2018a).

"Validated Biomarkers" are often done for specific assays that are to be used as companion diagnostics with a therapeutic agent. These assays are fully validated and require regulatory review of the assay methods, calibration standards, accuracy, precision, reproducibility, linearity, quality specifications, etc. These validation elements are well-known among bioanalytical, radiologic, and diagnostic laboratory professionals and are described in the FDA guidance on in vitro laboratory developed testing (http://www.fda.gov/downloads/MedicalDevices/DeviceRegulationandGuidance/GuidanceDocuments/UCM416685.pdf) (FDA, 2014).

Fig. 20.4 graphically depicts the categories of biomarkers and proposed level of evidence required for a specific context of use.

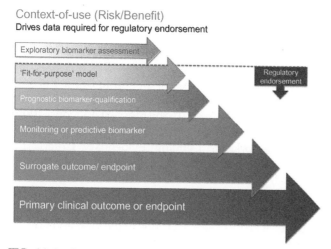

FIG. 20.4   Context of use drives the level of evidence required for biomarker validation. See BEST for definitions (https://www.ncbi.nlm.nih.gov/books/NBK338448/) (NCBI, NLM, NIH, 2016). As a conceptual framework, as the context of use has greater benefit/risk to the patient, the more evidence will be required to support the qualification of that biomarker.

A Regulatory Endorsement Pathways

There are different processes for regulatory endorsement of new biomarkers; importantly, each of these three paths is not mutually exclusive. The three paths are (1) endorsement within individual INDs, (2) approval as companion diagnostics, and (3) biomarker qualification programs. The first two paths are most common and affiliated with individual drug submissions, whereas the biomarker qualification path has implications across different therapeutic candidates and is independent of mechanism of action. Reviewed in the succeeding text are the practices of the FDA, EMA, and PMDA.

1 US Biomarker Acceptance in Individual New Drug Applications

Sponsors typically employ biomarkers in clinical development at their own risk through the conventional investigational new drug (IND)/new drug application (NDA) pathways. This is the most common approach where individual sponsors submit biomarker data for regulatory review along with their candidate therapeutic agent. This is viewed by some organizations as the most efficient path to advancing biomarkers to clinical trials. In this path the individual sponsor who is submitting the application bears the entire burden of biomarker validation and supportive evidence in its clinical and analytical validity. The review divisions own the assessment of risk/benefit and level of evidence required for accepting the biomarker as being suitable for a defined application or intended context of use. Sponsors using biomarkers should proactively discuss the intended use of the biomarker and test assay system in clinical studies with regulatory agencies to guide study design considerations and better understand the assay validation process for each specific biomarker. In Phase 3 clinical trials, the risk for successful biomarker implementation and eventual use in regulatory decision-making for a drug development program is higher, as the criteria for reproducibility and reliability are inherently greater in large multicenter global trials.

The implications of such reviews are that the biomarker pertains only to a single drug development program. Results can be applied only for that specific drug and cannot be applied to other drugs of the same or different class or drugs submitted by different drug sponsors. All regulatory reviews are confidential, and information cannot be shared beyond that one sponsor. Biomarker information is embedded in the specific drug labels and reviews with this mechanism.

2 Biomarker Approval as Companion Diagnostics—Center for Devices and Radiologic Health (CDRH)

Biomarkers can be measured in a variety of ways, such as by magnetic resonance imaging (MRI) of brain structures or quantification of analytes in body fluids and follow medical device regulations and guidance. Commonly a specific biomarker in the body fluid is measured using an in vitro diagnostic (IVD) kit.

In the United States the regulatory path for a specific biomarker IVD is clearance as a premarket notification 510(k) application or approval as premarket approval (PMA) application through CDRH at FDA and the type of application depends on the class of the device and the availability of similar devices that have already been approved (http://www.fda.gov/MedicalDevices/Device RegulationandGuidance/IVDRegulatoryAssistance/ucm 123682.htm) (FDA, 2015a).

In Europe, depending again on the class of the IVD, the manufacturer of the IVD prepares a technical file that includes both assay performance and clinical data to support the proposed IVD indication and appoints an authorized representative to manage the regulatory interactions. The technical file is audited by a European notified body, an accredited third party. Upon successful audit completion a CE-Mark certification is issued for an IVD by the notified body. The IVD must then be registered in the European Competent Authority where the authorized representative is located and may require additional notifications based on each country's requirements. It is important to note that the European Union (EU) IVD regulations are in the process of revision and are expected to be released in the near future.

Impactful examples of devices approved for use in CNS are growing with impact on patients' lives. Most notable is the approval of deep brain stimulation (DBS) for the treatment of PD (Hickey & Stacy, 2016). The FDA also recently cleared devices that can help determine the need for imaging following a head injury. The Immediate Postconcussion Assessment and Cognitive Testing (ImPACT) and ImPACT Pediatric are the first medical devices permitted for marketing that are intended to assess cognitive function following a possible concussion in adults and children, respectively. They are intended as part of the medical evaluation that physicians perform to assess signs and symptoms of a head injury. Medical products are not yet cleared for the specific diagnosis or treatment of TBI, yet efforts, such as the Traumatic Brian Injury Endpoints Development (TED) Initiative, are progressing in paths that hold promise (Manley et al., 2017). Most recently, two blood-based biomarkers for TBI received letters of support from FDA and approval by CDRH for use in diagnosis of concussion.

3 Biomarker Qualification (FDA/EMA/PMDA)

The FDA's Center for Drug Evaluation and Research's (CDER) Biomarker Qualification Program was established in 2005 to support work to develop biomarkers and provide a framework for scientific development and regulatory acceptance of biomarkers for use in drug development. Biomarker qualification allows a biomarker to be used for multiple drug development programs independent of the drug mechanism of action or sponsor submitting the drug candidate for review. Submitters for biomarker qualification are frequently public-private partnerships where sharing of the work, data, and risk among members can be enhanced (Brumfield, 2014; Woodcock, 2014). Regulatory endorsement of biomarkers through the biomarker qualification process is an iterative process that is enhanced by collaboration and sharing of data. Qualified biomarkers are announced as draft guidance to the drug development community. Qualification indicates that the FDA accepts the use of the biomarker in a drug development program with a defined context of use. Qualified biomarkers have been reviewed extensively for specific applications in drug development and do not need to undergo additional review when submitted for the same context of use in future drug applications. The FDA has published a detailed guidance highlighting the process of biomarker qualification (http://www.fda. gov/downloads/Drugs/uidanceComplianceRegulatory Information/Guidances/UCM230597.pdf) (FDA, 2014) and outlined the process in peer-reviewed publications (Amur, LaVange, et al., 2015; Amur, Sanyal, et al., 2015).

In the United States a total of 13 biomarkers, 10 nonclinical and 3 clinical biomarkers, have been qualified by the US FDA to date (https://www.fda.gov/Drugs/Development ApprovalProcess/DrugDevelopmentToolsQualification Program/BiomarkerQualificationProgram/ucm535383. htm) (FDA, 2017c). A select group of the biomarkers were qualified as a panel for nonclinical use (i.e., kidney safety biomarkers including urinary biomarkers: albumin, $\beta2$-microglobulin, clusterin, cystatin C, KIM-1, total protein, and trefoil factor-3). The clinical biomarkers include imaging and biofluid biomarkers. As of January 2018, none of these have been for CNS diseases. More than 20 biomarker submissions are in the consultation and advice stages. Learnings from the six qualification decisions for qualified biomarkers by FDA are being applied to new candidate biomarker submissions. There is a growing recognition by all stakeholders that there is a need for greater clarity on evidentiary expectations needed for qualification of biomarkers (Lavezzari & Womack, 2016).

Both the EMA and the PMDA have adopted similar drug development tool qualification initiatives for biomarkers. The Critical Path Institute's Predictive Safety Testing Consortium (PSTC) successfully advanced nonclinical kidney safety biomarkers for qualification by EMA, FDA, and PMDA (Dieterle et al., 2010).

Biomarker qualification by the EMA is a stepwise data-driven process achieved by first requesting scientific advice from the EMA to support the qualification of innovative development methods for a specific intended use in the context of research and development into pharmaceuticals. A guidance is available that outlines the procedure and steps for the qualification of novel methodologies for drug development: guidelines for applicants (http://www.ema.europa.eu/docs/en_GB/document_library/Regulatory_and_procedural_guideline/2009/10/WC500004201.pdf) (EMA, 2009). The Committee for Medicinal Products for Human Use (CHMP) can issue an opinion on the acceptability of a specific use of a method, such as the use of a novel methodology in the context of research and development. CHMP gives advice based on recommendations of the Scientific Advice Working Party (SAWP), a group of experts whose remit is to provide scientific advice and protocol assistance to applicants. The CHMP can provide qualification advice or a qualification opinion that is publicly posted and based on the assessment of data submitted to the agency. Before the final adoption of a qualification opinion, the CHMP makes its evaluation open for public consultation by the scientific community. This ensures that the CHMP shares information, as agreed with the applicant, and is open to scientific scrutiny and discussion. As of September 2017 a total of eight biomarkers have been qualified by EMA, four of which are focused on predementia stages of Alzheimer's disease http://www.ema.europa.eu/ema/index.jsp?curl=pages/regulation/document_listing/document_listing_000319.jsp (EMA, 2014).

For PMDA, qualified clinical biomarkers are lacking for CNS diseases. AD biomarkers in Japan have been highlighted in selected peer-reviewed publications (Moritoyo, 2015). The importance of using pharmacogenomics/biomarkers for enhancing recruitment into clinical trials and for evaluating the efficacy of treatments has been recently reviewed (https://www.ncbi.nlm.nih.gov/pubmed/25963998) (Otsubo, 2015). Collaborations between universities and the PMDA have been encouraged as a way to enhance development of drugs for AD in Japan (https://www.ncbi.nlm.nih.gov/pubmed/26342202) (Moritoyo, 2015).

V RECENT REGULATORY INITIATIVES TO ACCELERATE BIOMARKER QUALIFICATION AND APPROVALS

The FDA encourages the identification of novel and emerging biomarkers and is fostering new ways to enhance the role of biomarkers in drug development.

Critical Path Innovation Meetings (CPIM) represent a mechanism to meet with FDA on a case-by-case basis to discuss general challenges in drug development for specific indications.

The CPIM process enables CDER to meet with investigators from industry, academia, patient advocacy groups, and other governmental agencies in a 90-min meeting that oftentimes includes representatives across several different FDA divisions. The topics may include biomarkers, outcome measures, and clinical trial designs. More than 20 CPIM meetings took place in 2016. Submitters receive a brief high-level summary of nonbinding comments from FDA following the CPIM meeting that may include recommendations for further research. A complete guidance document outlines the process to request a CPIM (http://www.fda.gov/downloads/Drugs/GuidanceCompliance RegulatoryInformation/Guidances/UCM417627.pdf) (FDA, 2015b).

In addition to the qualification of biomarkers, CDER's Office of Translational Sciences has initiated efforts to facilitate the integration of biomarkers into the regulatory review process, encourages the identification of novel and emerging biomarkers, and reaches out to stakeholders in industry and academia to foster biomarker development. The FDA initiated a new process in 2015 called the letter of support (LOS) for biomarkers that hold promise and are in need of additional data for clinical qualification. The letter signed by FDA leadership is publicly posted to the FDA's website and is intended to express FDA's interest in qualification of the candidate biomarker, promote data collection and sharing, and stimulate additional studies that will catalyze the steps to qualification. As of September 2017 the FDA has issued a total of 14 letters of support across different therapeutic areas (http://www.fda.gov/Drugs/DevelopmentApproval Process/ucm434382.htm) (FDA, 2017a).

EMA initiated the letter of support process soon after FDA. As of September 2017, EMA has issued 14 letters of support of which six are related to CNS diseases.

FDA has taken a leadership role to enhance interactions with consortia through organizations such as the National Center for Advancing Translational Sciences (NCATS), the Foundation for the National Institutes of Health (FNIH), the Critical Path Institute, and European partners such as the Innovative Medicines Initiative (IMI). Presentations, publications, and an FDA webpage containing information for submitters, as well as information about previously qualified biomarkers, are also part of the communication strategy https://www.fda.gov/Drugs/DevelopmentApproval Process/DrugDevelopmentToolsQualificationProgram/ BiomarkerQualificationProgram/default.htm (FDA, 2017d).

The FDA continues to improve communication around the biomarker qualification process by enhanced interaction with submitters. This includes beginning context of use discussions very early in the process. For example, FDA is encouraging biomarker qualification for a limited context of use. This will allow the biomarker to be more rapidly integrated into drug development, allowing for the generation of additional data that may help to qualify the biomarker for an "expanded" context of use.

In 2015, FDA completed an external survey via a Federal Register (FR) posting (FDA, 2015b) intended to identify potential biomarkers for qualification and describing contexts of use, to address areas important to drug development. A summary of comments is posted on FDA's webpage (https://www.regulations.gov/ docket?dct=FR+PR+N+O+SR&rpp=10&po=0&D=FDA-2014-N-2187) (FDA, 2015e). Examples of promising biomarkers for CNS conditions that were included in the responses to FDA are tau positron-emission tomography (PET) and CSF neurogranin for AD, as markers of progression (http://www.regulations.gov/#!documentDetail;D=FDA-2014-N-2187-0029) (FDA, 2015c), and blood-based biomarkers and structural MRI for acute TBI (http://www.regulations.gov/#!documentDetail;D=FDA-2014-N-2187-0015) (FDA, 2015d).

In 2017, FDA has announced significant changes to the biomarker qualification process based on the 21st Century Cures legislation (https://www.fda.gov/drugs/developmentapprovalprocess/drugdevelopmenttoolsqualificationprogram/) (FDA, 2017b). The legislation is aimed at reforming the development and approval of processes for medical products. In addition to providing FDA the much-needed resources for staffing, initiatives that are prioritized include effective advancement of biomarkers for use in clinical trials, incorporating the patient perspective into the approval process and evaluation of the potential to use real-world evidence for drug development. Regarding "qualification of drug development tools," this is expected to set forth the standards and scientific approaches that will support the development of biomarkers, as well as the requirements and process for the qualification program. Changes that are already being announced include greater transparency and a defined set of requirements for novel biomarkers to be formally accepted into the qualification program.

The use of biomarkers for therapeutic trial decision-making can be highlighted in regulatory led disease-specific guidance. Guidance documents are typically prepared for FDA staff and agency stakeholders that describe the agency's interpretation of, or policy on, a regulatory issue. These documents are related to design, production, labeling, promotion, manufacturing, and testing of regulated products and the processing of content and evaluation or approval of submissions and inspection and enforcement policies. Draft guidance documents outline recommendations on the development of treatments for specific diseases. Such guidance oftentimes incorporate consensus from the research community through public meetings including outreach for comments and in this way can align regulatory science with contemporary advances in the field. The most

notable example is in AD, where both FDA and EMA have issued specific guidance drafts (FDA: http://www.fda.gov/downloads/drugs/guidancecomplianceregulatoryinformation/ guidances/ucm338287.pdf (FDA, CDER, and Kozauer, 2013). EMA: http://www.ema.europa.eu/docs/en_GB/document_library/Scientific_guideline/2016/02/WC500200830.pdf) (EMA, 2016a).

Examples of guidance in disorders of the nervous system, where biomarkers are included, are the use of dystrophin measurements for Duchenne muscular dystrophy (DMD) (http://www.fda.gov/downloads/drugs/guidancecomplianceregulatoryinformation/guidances/UCM450229.pdf) (FDA Center for Drug Evaluation, 2015), amyloid biomarkers in AD (http://www.regulations.gov/#!documentDetail;D=FDA-2014-N-2187-0029), and eye tracking/functional neuroimaging/MRS in autism http://www.ema.europa.eu/docs/en_GB/document_library/Scientific_guideline/2016/03/WC500202650.pdf (EMA, 2016c).

The FDA's Center of Devices and Radiological Health (CDRH) sponsors initiatives aimed at accelerating device and diagnostic development including biomarkers, namely, the Medical Device Innovation Consortium (MDIC) and the Medical Device Drug Development Tool (MDDT) programs. MDIC is a public-private partnership created with the objective of advancing medical device regulatory science (Kampfrath & Cotten, 2013). Neurodiagnostics and device development are also covered in specific guidance documents some of which are focused on CNS conditions, "Clinical Considerations for Investigational Device Exemption (IDE) for Neurological Devices Targeting Disease Progression and Clinical Outcomes" (FDA, 2016c) http://www.fda.gov/downloads/MedicalDevices/DeviceRegulationandGuidance/GuidanceDocuments/UCM489111.pdf?source=govdelivery&utm_medium=email&utm_source=govdelivery) (FDA, 2016c).

Finally, FDA sponsors fellows to focus their attention on specific areas of human health where regulatory science can catalyze progress. One example is the appointment of an FDA commissioner's fellow, tasked with coordinating FDA's collaborations with the TBI Endpoints Development (TED) enterprise, and the agency's efforts to engage in regulatory science with research and industry communities in the field of TBI more broadly. In 2016, FDA sponsored the first dedicated meeting on TBI biomarkers (http://www.fda.gov/MedicalDevices/NewsEvents/WorkshopsConferences/ucm483551.htm) (FDA, 2016a).

EMA has been successful at employing and expanding the qualification opinion path for driving innovation in regulatory science. Most notable, there are a total of five qualification opinions for AD, many of which are aimed at enabling treatments in early stages of the disease. In addition to biomarker qualification opinions, EMA has qualified the use of specific outcome measures for defined diseases (e.g., chronic obstructive pulmonary disease), quantitative drug development tools for enhancing the use of modeling and simulation of CAMD's AD model of disease progression and trial evaluation (http://www.ema.europa.eu/docs/en_GB/document_library/Regulatory_and_procedural_guideline/2013/07/WC500 146179.pdf) (EMA, 2013), and physiologic-based pharmacodynamic (PKPD) modeling and simulation (http://www.ema.europa.eu/docs/en_GB/document_library/Scientific_guideline/2016/07/WC500211315.pdf) (EMA, 2016d), and recently the use of digital health technologies to monitor drug compliance (http://www.ema.europa.eu/docs/en_GB/document_library/Regulatory_and_procedural_guideline/2016/02/WC500201943.pdf) (EMA, 2016b). These represent candid examples that emphasize the fact that in the majority of CNS conditions of idiopathic origins, no single biomarker will be sufficient to define a homogeneous population for clinical trials nor adequate to monitor disease progression. Most likely a strategy that employs combinatorial biomarkers, genetic factors, and modeling tools will be required to successfully enable precision medicine-based treatments for CNS conditions.

In summary, publications by regulatory agencies based on data across trials are working to reveal some of the root causes of CNS clinical trial failures (Butlen-Ducuing et al., 2016). While efficacy and safety were the most frequent cause of failure, others such as inability to target the "right" patient populations, underpowered studies, and lack of evidence of target engagement and other biomarker assessments remain critical gaps.

In summary, public-private partnerships are uniquely positioned to advance biomarker discovery and development. Initiatives such as ADNI clearly paved the way for open data sharing, consensus data standards, and collaboration. Similar initiatives that are essentially equivalent to ADNI are underway, including the Transforming Research and Clinical Knowledge in Traumatic Brain Injury (TRACK-TBI), the Transforming Research and Clinical Knowledge in Huntington's Disease (TRACK-HD), and the Parkinson's Progression Marker Initiative (PPMI) sponsored by the Michael J. Fox Foundation (MJFF). Such initiatives are paving the way to early intervention by demonstrating that the origin of pathologic hallmarks can be detected in the brain decades before the onset of symptoms.

A Pharmacoeconomic, Analytics and Informed Consent Considerations

There is increasing recognition that regulators and payers require that clinical studies demonstrate not only statistical significance of an effect but also, even more importantly, clinical and pharmacoeconomic significance (Ranganathan, Pramesh, & Buyse, 2015). Such standards also apply to biomarkers. A case example of an ongoing study that serves to emphasize the impact of this issue is a study being carried out focused on amyloid PET imaging. PET imaging is very costly and challenging to implement successfully in multisite global trials and in routine clinical care, particularly in rural communities. Despite the successful approval by regulatory agencies of multiple amyloid PET neuroimaging ligands to aid in the detection of amyloid in the living human brain, CMS will not cover the costs of PET imaging. The main concerns center around whether there is sufficient evidence showing a beneficial impact on improving patient outcomes.

The Centers for Medicare and Medicaid Services (CMS) and the Alzheimer's Association in conjunction with the American College of Radiology (ACR) are carrying out a 4-year, $100 million study called Imaging Dementia-Evidence for Amyloid Scanning (IDEAS). The IDEAS study aims to evaluate the clinical relevance of amyloid PET imaging with an ambitious plan for recruitment; from 2016 to 2017, IDEAS aims to enroll 18,488 Medicare beneficiaries aged 65 and older. This experience and its outcome will hopefully inform the scientific, regulatory, and payer communities as to how to identify clinical relevance in early phases of biomarker development and approval (Anand & Sabbagh, 2017). The scientific community has developed a set of specific appropriate use criteria recommendations for defining the types of patients and clinical circumstances for the use of amyloid PET (Johnson et al., 2013). The community anxiously awaits empirical evidence of the impact of amyloid PET imaging on clinical outcomes.

Further work is needed to engage the scientific community in defining the key evidentiary considerations required for biomarker development. Specific recommendations include the concept of the requirement by the US Senate to ask FDA to participate in public-private consortia that convene expert panels to guide and coordinate disease-specific biomarker research (Huber & Howard, 2015; http://www.manhattan-institute.org/html/how-can-policymakers-bridge-gap-between-precision-medicine-and-fda-regulation-8230.html) (Huber & Howard, 2015).

With the expansive landscape in human health care and required access to patient-level data to support big data analytics, the attentiveness to data sharing is growing at a rapid pace. Real-world examples exist among PPPs focused on data sharing, pathway-based analysis, crowdsourcing, and mechanistic modeling (Artigas et al., 2017; Atassi et al., 2014; Haas et al., 2016; Hofmann-Apitius et al., 2015; Neville et al., 2015; Romero et al., 2014) generating substantial impact for CNS diseases. Yet significant barriers still exist around privacy, consent, intellectual property, costs, enabling competitors, infrastructure, and potential inappropriate or false conclusions. Efficiencies will be gained by sharing

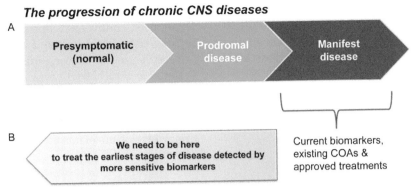

FIG. 20.5   The future of biomarker development across CNS diseases. Panel (A) depicts three major stages of disease progression. Panel (B) depicts where clinical outcome assessments are in terms of measuring disability (COAs = clinical outcome assessments). More sensitive biomarker assessments would facilitate the stratification of patients for viable preventative treatment options.

learnings across organizations that host data repositories and engaging legal representatives from various stakeholder groups to define true risk versus benefit. The field is clearly advancing yet not adequately enough to meet the needs for the patients most of who are eager to share their data (Frasier, 2016; Hake, et al., 2017; Mursaleen, Stamford, Jones, Windle, & Isaacs, 2017).

B New Technologies Enabling Potential Digital Biomarker Assessments in the Future

The general public and the collective biomedical research enterprise are increasingly embracing local and remote monitoring technologies, mobile smartphone apps, and portable and wearable devices into their everyday activities of daily living (Hird, Ghosh, & Kitano, 2016). Many view that these devices are used as the future path to provide a deeper understanding of disease processes and patients' functional status (Chiauzzi, Rodarte, & DasMahapatra, 2015). The terminology surrounding the use and clinical trial utility of these devices is evolving, with terms found in the literature such as digital biomarkers, digital measurement devices, technology-observed measures, and biometric monitoring systems. In this review, we define the term biometric monitoring device (BMD) to refer to the use of a biosensor(s) to collect objective data on a biological recognition element, for example, blood glucose, sodium, potassium, and hormone levels (Mobile Healthcare, 2016), or integrated physiological parameters (e.g., heart rate, blood pressure, electroencephalography, mobility, speech and sleep patterns, social engagement such as work/phone/computer, speed of information processing, driving and habits) (Roe, 2016). These BMDs utilize algorithms to transform these data (signal output) into a format that is interpretable as a specific measure or an aggregate functional outcome (Austin, Cross, Hayes, & Kaye, 2014; Dodge, Mattek, Austin, Hayes, & Kaye, 2012; Dodge, Zhu, Lee, Chang, & Ganguli, 2014; Espay et al., 2016; Hayes, Riley, Mattek,

Pavel, & Kaye, 2014; Kaye et al., 2014). Health platforms using biometric monitoring are providing clinical researchers real-time evidence that allows efficient collection of high-frequency data with decreased assessment time to monitor clinically meaningful parameters (Dodge et al., 2015) The challenge of successful advancement and acceptance of translational biomarkers for CNS diseases is significant and often not realized (Borsook, Becerra, & Hargreaves, 2011; Borsook, Hargreaves, & Becerra, 2011). While BMDs represent a paradigm shift in potentially developing biomarker outcome platforms for clinical trials that could span multiple CNS disease conditions (Arnerić et al., 2016; Dorsey, Papapetropoulos, Xiong, & Kieburtz, 2017), they will require an even more robust approach to integrate data collection and analysis (Fig. 20.5).

VI CONCLUSIONS AND RECOMMENDATIONS FOR THE FUTURE

The high failure rate of drugs to treat CNS diseases has led to business decisions by some large pharmaceutical companies to discontinue drug development in this space (Pankevich et al., 2014). A sense of urgency is needed to collectively work together to solve complex problems associated with drug development in this space (Hyman, 2016). Patients suffering from brain diseases are waiting and oftentimes do not survive the length of time to advance a new target or biomarker for use that will benefit them in their lifetime. Existing outcome measures may not be reflective of what is truly important to patients in their daily lives. The path to defining if biomarkers can serve to fill such gaps is underway, particularly in light of the advances in real-world evidence (Sherman et al., 2016) and new technologies (Arnerić et al., 2016; Dorsey et al., 2017).

Public-private partnerships represent a unique platform to share costs and risks. Policy-makers have become engaged themselves in raising issues and recommendations that are aimed at bridging the gap between precision medicine and FDA regulation. The total cost for biomarker-related initiatives, currently covered by industry, health insurance programs, philanthropists, and federally funded research institutions, is far less than what is needed to advance innovative treatments across CNS diseases.

To successfully advance the field and achieve regulatory qualification of biomarkers for use in clinical trials in CNS disorders, several steps are recommended to catalyze progress:

- Create incentives that enable and reward data sharing for those who contribute patient-level data to support biomarker development. Examples include crowdsourcing and reward mechanisms, as defined by scientific journals, government agencies, and consortia.
- Generate simple and concise informed consent forms (ICFs) for sponsors to employ in prospective clinical trials that will enable sharing of data beyond specific sponsored trials.
- Share bioanalytical performance data supporting reliability and reproducibility of all relevant biomarker platforms. This would include assay analytical results and imaging test/retest and algorithm performance characteristics.
- Encourage engagement and active participation of relevant diagnostic companies and manufacturers, particularly as contributors of data.
- Use CDISC biomarker and clinical data standards in ongoing and prospective clinical trials in diseases where Therapeutic Area User Guides have been developed and develop CDISC standards in those areas where they do not exist.
- Harmonize data generated by different technology platforms, as well as documentation of the commutability of the analytical system with a reference method.
- Share biomarker and clinical data from international initiatives (e.g., Innovative Medicines Initiative) to analyze biomarker performance in independent global observational cohorts.
- Continuously engage and submit data analyses and methodology to regulatory agencies to ensure efforts lead to qualification.
- Create cross consortia collaborations to share learnings and avoid duplication of effort.
- Nurture training of statisticians/pharmacometricians who have the expertise to tackling complex analysis approaches that integrate multiple sources of data collectively.
- Engage payers early in biomarker development so as to assure that approved/qualified biomarkers will be reimbursed and made available to for widespread use (IDEAS study in AD, as an example to learn from).

- Improve compliance with sharing of clinical trials that have been completed (clinical trials.gov, peer-reviewed publications) and improve communication of results of data-sharing initiatives, especially those related to clinical trials, for example, the Clinical Study Data Repository (CSDR; Rockhold, Nisen, & Freeman, 2016).
- Proactively outline strategies to determine clinical relevance of novel biomarkers early in discovery and throughout the clinical development.

Acknowledgements

The authors gratefully acknowledge the assistance of Peggy Abbott and Nicky Kuhl for their role in helping with this chapter, and we thank Anne Pedata for her review.

References

Alzheimer's Association (2016). *2016-facts-and-figures-fact-sheet—Alzheimer's Association.pdf*. Retrieved from http://www.alz.org/documents_custom/2016-Facts-and-Figures-Fact-Sheet.pdf.

Amur, S., LaVange, L., Zineh, I., Buckman-Garner, S., & Woodcock, J. (2015). Biomarker qualification: toward a multiple stakeholder framework for biomarker development, regulatory acceptance, and utilization. *Clinical Pharmacology & Therapeutics*, 98(1), 34–46. https://doi.org/10.1002/cpt.136.

Amur, S. G., Sanyal, S., Chakravarty, A. G., Noone, M. H., Kaiser, J., McCune, S., et al. (2015). Building a roadmap to biomarker qualification: challenges and opportunities. *Biomarkers in Medicine*, 9(11), 1095–1105. https://doi.org/10.2217/bmm.15.90.

Anand, K., & Sabbagh, M. (2017). Amyloid imaging: poised for integration into medical practice. *Neurotherapeutics*, 14(1), 54–61. https://doi.org/10.1007/s13311-016-0474-y.

Anxiety and Depression Association of America (ADAA). (2010/2016). *Anxiety and Depression Association of America*. Retrieved from https://adaa.org/about-adaa/press-room/facts-statistics.

Arnerić, S. P., Batrla-Utermann, R., Beckett, L., Bittner, T., Blennow, K., Carter, L., et al. (2016). Cerebrospinal fluid biomarkers for Alzheimer's disease: a view of the regulatory science qualification landscape from the coalition against major diseases CSF biomarker team. *Journal of Alzheimer's disease*, 55(1), 19–35. https://doi.org/10.3233/JAD-160573.

Arnerić, S. P., Cedarbaum, J. M., Khozin, S., Papapetropoulos, S., Hill, D., Ropacki, M., et al. (2017). Biometric monitoring devices for assessing end points in clinical trials: developing an ecosystem. *Nature Reviews. Drug Discovery*, 16, 736. https://doi.org/10.1038/nrd.2017.153.

Artigas, F., Schenker, E., Celada, P., Spedding, M., Llado-Pelfort, L., Jurado, N., et al. (2017). Defining the brain circuits involved in psychiatric disorders: IMI-NEWMEDS. *Nature Reviews. Drug Discovery*, 16(1), 1–2.

Atassi, N., Berry, J., Shui, A., Zach, N., Sherman, A., Sinani, E., et al. (2014). The PRO-ACT database design, initial analyses, and predictive features. *Neurology*, 83(19), 1719–1725.

Austin, D., Cross, R. M., Hayes, T., & Kaye, J. (2014). Regularity and predictability of human mobility in personal space. *PloS One*, 9(2), e90256.

Autism Society (2015). *Autism Society Website*. Retrieved from http://www.autism-society.org/what-is/facts-and-statistics/.

Autism Speaks Website (2018). *Autsim Speaks Website*. Retrieved from https://www.autismspeaks.org/what-autism/prevalence.

Berlyand, Y., Weintraub, D., Xie, S. X., Mellis, I. A., Doshi, J., Rick, J., et al. (2016). An Alzheimer's disease-derived biomarker signature identifies Parkinson's disease patients with dementia. *PLoS One. 11*(1), e0147319. https://doi.org/10.1371/journal.pone.0147319.

Biomarkers Consortium Evidentiary Standards Writing Group (2016). *Framework for defining evidentiary criteria for biomarker qualification.* Retrieved from http://fnih.org/sites/default/files/final/pdf/Evidentiary%20Criteria%20Framework%20Final%20Version%20Oct%2020%202016.pdf.

Borsook, D., Becerra, L., & Hargreaves, R. J. (2011). Borsook et al. Biomarkers for chronic pain 2011.pdf. *Discovery Medicine, 11*(58), 197–207.

Borsook, D., Hargreaves, R. J., & Becerra, L. (2011). Can functional magnetic resonance imaging improve success rates in central nervous system drug discovery? *Expert Opinion on Drug Discovery, 6*(6), 597–617. https://doi.org/10.1517/17460441.2011.584529.

Brady, LS., & Potter, WZ. (2014). Public-private partnerships to revitalize psychiatric drug discovery. *Expert Opinion on Drug Discovery, 9*(1), 1–8. Epub 2013 Dec 6. https://doi.org/10.1517/17460441.2014.867944.

Brumfield, M. (2014). The Critical Path Institute: transforming competitors into collaborators. *Nature Reviews Drug Discovery, 13*(11), 785–786. https://doi.org/10.1038/nrd4436.

Butlen-Ducuing, F., Pétavy, F., Guizzaro, L., Zienowicz, M., Salmonson, T., Haas, M., et al. (2016). Regulatory watch: challenges in drug development for central nervous system disorders: a European Medicines Agency perspective. *Nature Reviews Drug Discovery, 15*(12), 813–814.

Button, K. S., Ioannidis, J. P. A., Mokrysz, C., Nosek, B. A., Flint, J., Robinson, E. S. J., et al. (2013). Power failure: Why small sample size undermines the reliability of neuroscience. *Nature Reviews Neuroscience, 14*(5), 365.

CDISC. (2014). *Therapeutic areas*. Retrieved from https://www.cdisc.org/standards/therapeutic-areas.

Chandra, S. R. (2017). Alzheimer's disease: an alternative approach. *The Indian Journal of Medical Research, 145*(6), 723.

Chiauzzi, E., Rodarte, C., & DasMahapatra, P. (2015). Patient-centered activity monitoring in the self-management of chronic health conditions. *BMC Medicine, 13*(1). https://doi.org/10.1186/s12916-015-0319-2.

Choi, D. W., Armitage, R., Brady, L. S., Coetzee, T., Fisher, W., Hyman, S., et al. (2014). Medicines for the mind: policy-based "Pull" incentives for creating breakthrough CNS drugs. *Neuron, 84*(3), 554–563. https://doi.org/10.1016/j.neuron.2014.10.027.

Collins, F. S., & Varmus, H. (2015). A new initiative on precision medicine. *New England Journal of Medicine, 372*(9), 793–795.

Consortia Pedia, & FasterCures. (2017). *Consortia-pedia*. Retrieved from http://consortiapedia.fastercures.org/.

Cummings, J. L., Morstorf, T., & Zhong, K. (2014). Alzheimer's disease drug-development pipeline: few candidates, frequent failures. *Alzheimer's Research & Therapy, 6*(4), 37. https://doi.org/10.1186/alzrt269.

Das, S., & Basu, S. (2017). Multi-targeting strategies for Alzheimer's disease therapeutics: pros and cons. *Current Topics in Medicinal Chemistry, 17*(27), 3017–3061. https://doi.org/10.2174/1568026617666170707130652.

Dennis, E. H., Walker, E. G., Baker, A. F., & Miller, R. T. (2013). Opportunities and challenges of safety biomarker qualification: perspectives from the predictive safety testing consortium: value of safety biomarker qualification. *Drug Development Research, 74*(2), 112–126. https://doi.org/10.1002/ddr.21070.

Dieterle, F., Sistare, F., Goodsaid, F., Papaluca, M., Ozer, J. S., Webb, C. P., et al. (2010). Renal biomarker qualification submission: a dialog between the FDA-EMEA and predictive safety testing consortium. *Nature Biotechnology, 28*(5), 455–462. https://doi.org/10.1038/nbt.1625.

Dodge, H. H., Mattek, N. C., Austin, D., Hayes, T. L., & Kaye, J. A. (2012). In-home walking speeds and variability trajectories associated with mild cognitive impairment. *Neurology, 78*(24), 1946–1952.

Dodge, H. H., Zhu, J., Lee, C. -W., Chang, C. -C. H., & Ganguli, M. (2014). Cohort effects in age-associated cognitive trajectories. *The Journals of Gerontology Series A: Biological Sciences and Medical Sciences, 69*(6), 687–694. https://doi.org/10.1093/gerona/glt181.

Dodge, H. H., Zhu, J., Mattek, N. C., Austin, D., Kornfeld, J., & Kaye, J. A. (2015). Use of high-frequency in-home monitoring data may reduce sample sizes needed in clinical trials. *PLoS One, 10*(9) e0138095. https://doi.org/10.1371/journal.pone.0138095.

Dorsey, E. R., Papapetropoulos, S., Xiong, M., & Kieburtz, K. (2017). The First Frontier: Digital Biomarkers for Neurodegenerative Disorders. *Digital Biomarkers.* https://doi.org/10.1159/000477383.

EMA. (2009). *Qualification of novel methodologies for drug development: guidance to applicants.* European Medicines Agency. Retrieved from http://www.ema.europa.eu/docs/en_GB/document_library/Regulatory_and_procedural_guideline/2009/10/WC500004201.pdf.

EMA. (2013). *Qualification opinion of a novel data driven model of disease progression and trial evaluation in mild and moderate Alzheimer's disease.* European Medicines Agency. Retrieved from http://www.ema.europa.eu/docs/en_GB/document_library/Regulatory_and_procedural_guideline/2013/07/WC500146179.pdf.

EMA. (2014). *Qualification of novel methodologies for medicine development.* Retrieved from http://www.ema.europa.eu/ema/index.jsp?curl=pages/regulation/document_listing/document_listing_000319.jsp.

EMA. (2016a). *Draft guideline on the clinical investigation of medicines for the treatment of Alzheimer's disease and other dementias.* European Medicines Agency. Retrieved from http://www.ema.europa.eu/docs/en_GB/document_library/Scientific_guideline/2016/02/WC500200830.pdf.

EMA. (2016b). *Qualification opinion on ingestible sensor system for medication adherence as biomarker for measuring patient adherence to medication in clinical trials.* European Medicines Agency. Retrieved from http://www.ema.europa.eu/docs/en_GB/document_library/Regulatory_and_procedural_guideline/2016/02/WC500201943.pdf.

EMA. (2016c). *Guideline on the clinical development of medicinal products for the treatment of Autism Spectrum Disorder (ASD)—Draft.* European Medicines Agency. Retrieved from http://www.ema.europa.eu/docs/en_GB/document_library/Scientific_guideline/2016/03/WC500202650.pdf.

EMA. (2016d). *Guideline on the qualification and reporting of physiologically based pharmacokinetic (PBPK) modelling and simulation.* European Medicines Agency. Retrieved from http://www.ema.europa.eu/docs/en_GB/document_library/Scientific_guideline/2016/07/WC500211315.pdf.

Espay, A. J., Bonato, P., Nahab, F. B., Maetzler, W., Dean, J. M., Klucken, J., et al. (2016). Technology in Parkinson's disease: challenges and opportunities: Technology in PD. *Movement Disorders.* https://doi.org/10.1002/mds.26642.

FDA. (2004). *Innovation or stagnation: challenge and opportunity on the critical path to new medical products*: (p. 38). .Retrieved from https://wwwfda.gov/scienceresearch/specialtopics/criticalpath initiative/criticalpathopportunitiesreports/ucm077262.htm.

FDA. (2014). *Framework for regulatory oversight of laboratory developed tests (LDTs).* Retrieved from https://www.fda.gov/Drugs/DevelopmentApprovalProcess/DrugDevelopmentToolsQualificationProgram/BiomarkerQualificationProgram/ucm536018.htm.

FDA. (2015a). *Overview of IVD Regulation*. Retrieved from https://www.fda.gov/MedicalDevices/DeviceRegulationandGuidance/IVDRegulatoryAssistance/ucm123682.htm.

FDA. (2015b). *Critical path innovation meetings—guidance for industry*. Retrieved from https://www.fda.gov/downloads/Drugs/GuidanceComplianceRegulatoryInformation/Guidances/UCM417627.pdf.

FDA. (2015c). *Comment from critical path institute*. Retrieved from http://www.regulations.gov/#!documentDetail;D=FDA-2014-N-2187-0029.

FDA. (2015d). *Comments from TRACK-TBI and TED investigators*. Retrieved from https://www.regulations.gov/document?D=FDA-2014-N-2187-0015.

FDA. (2015e). *Identifyig potential biomarkers for qualification and describing contexts of use to address areas important to drug development*. Retrieved from https://www.regulations.gov/docket?dct=FR+PR+N+O+SR&rpp=10&po=0&D=FDA-2014-N-2187.

FDA. (2016a). *Public workshop—advancing the development of biomarkers in traumatic brain injury*: (p. 2016). Critical Path Institute.. March 3. Retrieved from https://www.fda.gov/MedicalDevices/NewsEvents/WorkshopsConferences/ucm483551.htm.

FDA. (2016b). *Drug development tools: fit-for-purpose initiative*. Retrieved from https://www.fda.gov/Drugs/DevelopmentApprovalProcess/ucm505485.htm.

FDA. (2016c). *Clinical considerations for investigational device exemptions (IDEs) for neurological devices targeting disease progression and clinical outcomes*. U.S. Food and Drug Administration. Retrieved from https://www.fda.gov/downloads/MedicalDevices/DeviceRegulationandGuidance/%20GuidanceDocuments/UCM489111.pdf?source=govdelivery&utm_medium=email&utm_source=govdelivery.

FDA. (2017a). *Letter of support initiative*. Retrieved from https://www.fda.gov/Drugs/DevelopmentApprovalProcess/ucm434382.htm.

FDA. (2017b). *Drug Development Tools (DDT) qualification programs*. Retrieved from https://www.fda.gov/drugs/developmentapprovalprocess/drugdevelopmenttoolsqualificationprogram/.

FDA. (2017c). *List of qualified biomarkers*. Retrieved from https://www.fda.gov/downloads/Drugs/%20GuidanceComplianceRegulatoryInformation/Guidances/UCM230597.pdf.

FDA. (2017d). *Biomarker qualification program*. Retrieved from https://www.fda.gov/Drugs/DevelopmentApprovalProcess/DrugDevelopmentToolsQualificationProgram/BiomarkerQualificationProgram/default.htm.

FDA. (2018a). *Biomarker guidances and reference materials*. Retrieved from https://www.fda.gov/Drugs/DevelopmentApprovalProcess/DrugDevelopmentToolsQualificationProgram/BiomarkerQualificationProgram/ucm536018.htm.

FDA. (2018b). *PDUFA VI: fiscal years 2018-2022*. Retrieved from https://www.fda.gov/forindustry/userfees/prescriptiondruguserfee/ucm446608.htm.

FDA, C. (2014). *FDA—Guidance for Qualification process for drug development tools 2014.pdf (Procedural)*. U.S. Department of Health and Human Services. Retrieved from http://www.fda.gov/downloads/drugs/guidancecomplianceregulatoryinformation/guidances/ucm230597.pdf.

FDA, CDER, & Kozauer, N. (2013). *Guidance for industry Alzheimer's disease: developing drugs for the treatment of early stage disease*. Retrieved from https://www.fda.gov/downloads/drugs/%20guidancecomplianceregulatoryinformation/guidances/ucm338287.pdf.

FDA, CDER, USDHS. (2014). *Providing regulatory submissions in electronic format: standardized study data*. U.S. Food and Drug Administration. Retrieved from http://www.fda.gov/downloads/drugs/guidances/ucm292334.pdf.

FDA Center for Drug Evaluation, D. (2015). *Duchenne muscular dystrophy and related dystrophinopathies: developing drugs for treatment guidance for industry*. Retrieved from https://www.fda.gov/downloads/drugs/guidancecomplianceregulatoryinformation/guidances/UCM450229.pdf.

FDA-NIH Biomarker Working Group (2016). *BEST (Biomarkers, endpointS, & other tools) resource*. Retrieved from https://www.ncbi.nlm.nih.gov/books/NBK326791/.

Frasier, M. (2016). Perspective: data sharing for discovery. *Nature, 538*(7626), S4.

Gold, M. (2017). Phase II clinical trials of anti–amyloid β antibodies: when is enough, enough? *Alzheimer's & Dementia: Translational Research & Clinical Interventions, 3*(3), 402–409. https://doi.org/10.1016/j.trci.2017.04.005.

Gray, B. B. (2016). Signs before symptoms: some clinical trials are focusing on the early stages of neurologic disease, or on people who are genetically at risk but have no known symptoms. *Neurology, 12*(3), 34–36.

Grinnon, S. T., Miller, K., Marler, J. R., Lu, Y., Stout, A., Odenkirchen, J., et al. (2012). National Institute of Neurological Disorders and Stroke Common Data Element Project—approach and methods. *Clinical Trials, 9*(3), 322–329. https://doi.org/10.1177/1740774512438980.

Gurney, M. E., Pu, H., Chiu, A. Y., Canto, M. C. D., Polchow, C. Y., Alexander, D. D., et al. (1994). Motor neuron degeneration in mice that express a human Cu,Zn superoxide dismutase mutation. *Science, 264*(5166), 1772–1775.

Haas, M., Stephenson, D., Romero, K., Gordon, M. F., Zach, N., & Geerts, H. (2016). Big data to smart data in Alzheimer's disease: real-world examples of advanced modeling and simulation. *Alzheimer's & Dementia, 12*(9), 1022–1030. https://doi.org/10.1016/j.jalz.2016.05.005.

Haas, M., Mantua, V., Haberkamp, M., Pani, L., Isaac, M., Butlen-Ducuing, F., et al. (2015). The european medicines agency's strategies to meet the challenges of Alzheimer disease. *Nature Reviews Drug Discovery, 14*(4), 221–222. https://doi.org/10.1038/nrd4585.

Hake, A. M., Dacks, P. A., Armerić, S. P., CAMD ICF working group, et al. (2017). Concise informed consent to increase data and biospecimen access may accelerate innovative Alzheimer's disease treatments. *Alzheimer's & Dementia: Translational Research & Clinical Interventions, 3*, 536–541. Retrieved from http://www.sciencedirect.com/science/article/pii/S2352873717300495.

Halpern, S. D., Karlawish, J. T., & Berlin, J. A. (2002). The continuing unethical conduct of underpowered clinical trials. *JAMA, 288*(3), 358–362. https://doi.org/10.1001/jama.288.3.358.

Hayes, T. L., Riley, T., Mattek, N., Pavel, M., & Kaye, J. A. (2014). Sleep habits in mild cognitive impairment. *Alzheimer Disease & Associated Disorders, 28*(2), 145–150. https://doi.org/10.1097/WAD.0000000000000010.

Healthline, Pitrangelo, A., & Higuera, V. (2015). *Multiple sclerosis by the numbers: facts, statistics, and you*. Retrieved from https://www.healthline.com/health/multiple-sclerosis/facts-statistics-infographic.

Heinzel, S., Lerche, S., Maetzler, W., & Berg, D. (2017). Global, yet incomplete overview of cohort studies in parkinson's disease. *Journal of Parkinson's Disease, 1*–10. https://doi.org/10.3233/JPD-171100.

Hickey, P., & Stacy, M. (2016). Deep brain stimulation: a paradigm shifting approach to treat Parkinson's disease. *Frontiers in Neuroscience, 10*, 173. https://doi.org/10.3389/fnins.2016.00173.

Hird, N., Ghosh, S., & Kitano, H. (2016). Digital health revolution: perfect storm or perfect opportunity for pharmaceutical R&D? *Drug Discovery Today, 21*(6), 900–911. https://doi.org/10.1016/j.drudis.2016.01.010.

Hofmann-Apitius, M., Ball, G., Gebel, S., Bagewadi, S., de Bono, B., Schneider, R., et al. (2015). Bioinformatics mining and modeling methods for the identification of disease mechanisms in neurodegenerative disorders. *International Journal of Molecular Sciences. 16*(12). https://doi.org/10.3390/ijms161226148.

Hosoya, M., & Czysz, K. (2016). Translational prospects and challenges in human induced pluripotent stem cell research in drug discovery. *Cells. 5*(4). https://doi.org/10.3390/cells5040046.

Huber, P., & Howard, P. (2015). *FDLI-Food and drug policy forum. Volume 5, Issue 7.* Retrieved from www.fdli.org.

Hurd, M., Martorell, P., & Delvande, A. (2013). Dementia costs top cancer, heart disease—neurology reviews.html. *Neurology Reviews, 21*(5), 47.

Hyman, S. E. (2016). Back to basics: luring industry back into neuroscience. *Nature Neuroscience, 19*(11), 1383–1384.

Jack, C. R. J., Wiste, H. J., Weigand, S. D., Therneau, T. M., Lowe, V. J., Knopman, D. S., et al. (2016). Defining imaging biomarker cut points for brain aging and Alzheimer's disease. *Alzheimer's & Dementia: The Journal of the Alzheimer's Association, 13*(3), 205–216. https://doi.org/10.1016/j.jalz.2016.08.005.

Johnson, K. A., Minoshima, S., Bohnen, N. I., Donohoe, K. J., Foster, N. L., Herscovitch, P., et al. (2013). Appropriate use criteria for amyloid PET: a report of the amyloid imaging task force, the Society of Nuclear Medicine and Molecular Imaging, and the Alzheimer's Association. *Journal of Nuclear Medicine, 54*(3), 476–490. https://doi.org/10.2967/jnumed.113.120618.

Jones-Davis, D. M., & Buckholtz, N. (2015). The impact of the Alzheimer's disease neuroimaging initiative 2: what role do public-private partnerships have in pushing the boundaries of clinical and basic science research on Alzheimer's disease? *Alzheimer's & Dementia, 11*(7), 860–864. https://doi.org/10.1016/j.jalz.2015.05.006.

Kampfrath, T., & Cotten, S. W. (2013). The new collaborative path in medical device development: The medical device innovation consortium. *Clinical Biochemistry, 46*(15), 1320–1322. https://doi.org/10.1016/j.clinbiochem.2013.03.021.

Karran, E., & Hardy, J. (2014). A critique of the drug discovery and phase 3 clinical programs targeting the amyloid hypothesis for Alzheimer disease: amyloid hypothesis for AD. *Annals of Neurology, 76*(2), 185–205. https://doi.org/10.1002/ana.24188.

Kaye, J., Mattek, N., Dodge, H. H., Campbell, I., Hayes, T., Austin, D., et al. (2014). Unobtrusive measurement of daily computer use to detect mild cognitive impairment. *Alzheimer's & Dementia, 10*(1), 10–17. https://doi.org/10.1016/j.jalz.2013.01.011.

Kinch, M. S. (2015). An analysis of FDA-approved drugs for neurological disorders. *Drug Discovery Today, 20*(9), 1040–1043. https://doi.org/10.1016/j.drudis.2015.02.003.

Küffner, R., Zach, N., Norel, R., Hawe, J., Schoenfeld, D., Wang, L., et al. (2014). Crowdsourced analysis of clinical trial data to predict amyotrophic lateral sclerosis progression. *Nature Biotechnology, 33*(1), 51–57. https://doi.org/10.1038/nbt.3051.

Lanz, T. A., Hosley, J. D., Adams, W. J., & Merchant, K. M. (2004). Studies of Abeta pharmacodynamics in the brain, cerebrospinal fluid, and plasma in young (plaque-free) Tg2576 mice using the gamma-secretase inhibitor N2-[(2S)-2-(3,5-difluorophenyl)-2-hydroxyethanoyl]-N1-[(7S)-5-methyl-6-oxo-6,7-dihydro-5H-dibenzo[b,d]azepin-7-yl]-L-alaninamide (LY-411575). *Journal of Pharmacology and Experimental Therapeutics, 309*(1), 49. https://doi.org/10.1124/jpet.103.060715.

LaRocca, N. G., Hudson, L. D., Rudick, R., Amtmann, D., Balcer, L., Benedict, R., et al. (2017). The MSOAC approach to developing performance outcomes to measure and monitor multiple sclerosis disability. *Multiple Sclerosis Journal, 24*(11), 1469–1484. https://doi.org/10.1177/1352458517723718.

Lavezzari, G., & Womack, A. (2016). Industry perspectives on biomarker qualification. *Clinical Pharmacology & Therapeutics, 99*(2), 208–213. https://doi.org/10.1002/cpt.264.

Lawton, M., Kasten, M., May, M. T., Mollenhauer, B., Schaumburg, M., Liepelt-Scarfone, I., et al. (2016). Validation of conversion between mini-mental state examination and montreal cognitive assessment. *Movement Disorders: Official Journal of the Movement Disorder Society, 31*(4), 593–596. https://doi.org/10.1002/mds.26498.

Lemmon, V. P., Ferguson, A. R., Popovich, P. G., Xu, X. -M., Snow, D. M., Igarashi, M., et al. (2014). Minimum information about a spinal cord injury experiment: a proposed reporting standard for spinal cord injury experiments. *Journal of Neurotrauma, 31*(15), 1354–1361. https://doi.org/10.1089/neu.2014.3400.

Leptak, C., Menetski, J. P., Wagner, J. A., Aubrecht, J., Brady, L., Brumfield, M., et al. (2017). What evidence do we need for biomarker qualification? *Science Translational Medicine, 9*(417), eaal4599.

Lithgow, G. J., Driscoll, M., & Phillips, P. (2017). A long journey to reproducible results. *Nature, 548*(7668), 387.

Lleó, A., Cavedo, E., Parnetti, L., Vanderstichele, H., Herukka, S. K., Andreasen, N., et al. (2014). Cerebrospinal fluid biomarkers in trials for Alzheimer and Parkinson diseases. *Nature Reviews Neurology, 11*(1), 41–55. https://doi.org/10.1038/nrneurol.2014.232.

Ludolph, A. C. (2011). Motor neuron disease: urgently needed—biomarkers for amyotrophic lateral sclerosis. *Nature Reviews Neurology, 7*(1), 13–14. https://doi.org/10.1038/nrneurol.2010.196.

Ludolph, A. C., Bendotti, C., Blaugrund, E., Chio, A., Greensmith, L., Loeffler, J. -P., et al. (2010). Guidelines for preclinical animal research in ALS/MND: a consensus meeting. *Amyotrophic Lateral Sclerosis, 11*(1–2), 38–45. https://doi.org/10.3109/17482960903545334.

Manley, G. T., Mac Donald, C. L., Markowitz, A. J., Stephenson, D., Robbins, A., Gardner, R. C., et al. (2017). The traumatic brain injury endpoints development (TED) initiative: progress on a public-private regulatory collaboration to accelerate diagnosis and treatment of traumatic brain injury. *Journal of Neurotrauma.* https://doi.org/10.1089/neu.2016.4729.

Manolis, E., Vamvakas, S., & Isaac, M. (2011). New pathway for qualification of novel methodologies in the european medicines agency. *PROTEOMICS—Clinical applications, 5*(5–6), 248–255. https://doi.org/10.1002/prca.201000130.

Marek, K., Jennings, D., Lasch, S., Siderowf, A., Tanner, C., Simuni, T., et al. (2011). The parkinson progression marker initiative (PPMI). *Progress in Neurobiology, 95*(4), 629–635. https://doi.org/10.1016/j.pneurobio.2011.09.005.

Maxfield, K. E., Buckman-Garner, S., & Parekh, A. (2017). The role of public-private partnerships in catalyzing the critical path: the role of PPPs in catalyzing the critical path. *Clinical and Translational Science, 10*(6), 431–442. https://doi.org/10.1111/cts.12488.

McGonigle, P. (2014). Animal models of CNS disorders. *Special Issue: Pharmacology in 21st Century Biomedical Research, 87*(1), 140–149. https://doi.org/10.1016/j.bcp.2013.06.016.

McShane, L. (2017). In pursuit of greater reproducibility and credibility of early clinical biomarker research. *Clinical and Translational Science, 10*(2), 58–60. https://doi.org/10.1111/cts.12449.

Mobile Healthcare (2016). *Introduction of mobile healthcare has greatly influenced medical science.* Retrieved from http://www.bio-itworld.com/Press-Release/Introduction-of-Mobile-Healthcare-has-Greatly-Influenced-Medical-Science/. Accessed 29 November 2016.

Moritoyo, T. (2015). Accelerating regulatory science initiatives for the development of drugs for Alzheimer's disease in Japan. *Clinical Therapeutics, 37*(8), 1622–1626. https://doi.org/10.1016/j.clinthera.2015.04.014.

Murphy, D. G., Goldman, M., Loth, E., & Spooren, W. (2014). Public-private partnership: a new engine for translational research in neurosciences. *Neuron, 84*(3), 533–536. Epub 2014 Nov 5. https://doi.org/10.1016/j.neuron.2014.10.006.

Murray, C. J. L. (2013). The state of US Health, 1990–2010: burden of diseases, injuries, and risk factors. *JAMA, 310*(6), 591. https://doi.org/10.1001/jama.2013.13805.

Mursaleen, L. R., Stamford, J. A., Jones, D. A., Windle, R., & Isaacs, T. (2017). Attitudes towards data collection ownership and sharing among Parkinson's disease patients. *Journal of Parkinson's Disease,* 1–9. https://doi.org/10.3233/JPD-161045.

NCBI, NLM, NIH. (2016). *BEST (Biomarkers, endpointS, and other tools) resource [Internet].* Retrieved from https://www.ncbi.nlm.nih.gov/books/NBK338448.

Neville, J., Kopko, S., Broadbent, S., Avilés, E., Stafford, R., Solinsky, C. M., et al. (2015). Development of a unified clinical trial database for

Alzheimer's disease. *Alzheimer's & Dementia*, 11(10), 1212–1221. https://doi.org/10.1016/j.jalz.2014.11.005.

Otsubo, Y. (2015). Use of Pharmacogenomics and Biomarkers in the Development of New Drugs for Alzheimer Disease in Japan. The views expressed in this article are those of the author and do not necessarily reflect the official views of ministry of health, labor, and welfare, or the pharmaceuticals medical devices agency of Japan. *Clinical Therapeutics*, 37(8), 1627–1631. https://doi.org/10.1016/j.clinthera.2015.04.010.

Pankevich, D. E., Altevogt, B. M., Dunlop, J., Gage, F. H., & Hyman, S. E. (2014). Improving and accelerating drug development for nervous system disorders. *Neuron*, 84(3), 546–553. https://doi.org/10.1016/j.neuron.2014.10.007.

Parkinson's Foundation. (n.d.). Parkinson's Foundation Statistics. Retrieved from http://www.parkinson.org/Understanding-Parkinsons/Causes-and-Statistics/Statistics

Paul, S. M., Mytelka, D. S., Dunwiddie, C. T., Persinger, C. C., Munos, B. H., Lindborg, S. R., et al. (2010). How to improve R&D productivity: the pharmaceutical industry's grand challenge. *Nature Reviews Drug Discovery*, 9(3), 203.

Perrin, S. (2014). Make mouse studies work. *Nature*, 507, 424–425.

Perry, D., Sperling, R., Katz, R., Berry, D., Dilts, D., Hanna, D., et al. (2015). Building a roadmap for developing combination therapies for Alzheimer's disease. *Expert Review of Neurotherapeutics*, 15(3), 327–333. https://doi.org/10.1586/14737175.2015.996551.

PMDA. (2015). *Notification of practical operations of electronic study data submissions*. Retrieved from http://www.pmda.go.jp/files/000206451.pdf.

Ranganathan, P., Pramesh, C. S., & Buyse, M. (2015). Common pitfalls in statistical analysis: clinical versus statistical significance. *Perspectives in Clinical Research*, 6(3), 169–170. https://doi.org/10.4103/2229-3485.159943.

Robb, M. A., McInnes, P. M., & Califf, R. M. (2016). Biomarkers and surrogate endpoints: developing common terminology and definitions. *JAMA*, 315(11), 1107–1108.

Rockhold, F., Nisen, P., & Freeman, A. (2016). Data sharing at a crossroads. *New England Journal of Medicine*, 375(12), 1112–1115. https://doi.org/10.1056/NEJMp1608351.

Roe, D. (2016). *Future market trends for wearable devices*. Retrieved from http://www.meddeviceonline.com/doc/future-market-trends-for-wearable-devices-0001. Accessed 29 November 2016.

Romero, K., Sinha, V., Allerheiligen, S., Danhof, M., Pinheiro, J., Kruhlak, N., et al. (2014). Modeling and simulation for medical product development and evaluation: highlights from the FDA-C-Path-ISOP 2013 workshop. *Journal of Pharmacokinetics and Pharmacodynamics*, 41(6), 545–552. https://doi.org/10.1007/s10928-014-9390-0.

Rouse, R., Kruhlak, N., Weaver, J., Burkhart, K., Patel, V., & Strauss, D. G. (2017). Translating new science into the drug review process: the US FDA's division of applied regulatory science. *Therapeutic Innovation & Regulatory Science*, 52, 244–255. https://doi.org/10.1177/2168479017720249.

Schizophrenia and Related Disorders Alliance of America (2008/2017). *Quick facts about schizophrenia*. Retrieved from https://sardaa.org/resources/about-schizophrenia/.

Senate of the United States (2015). *21st century cures act H.R.6.*

Sherman, R. E., Anderson, S. A., Dal Pan, G. J., Gray, G. W., Gross, T., Hunter, N. L., et al. (2016). Real-world evidence—what is it and what can it tell us? *New England Journal of Medicine*, 375(23), 2293–2297. https://doi.org/10.1056/NEJMsb1609216.

Sistare, F. D., & DeGeorge, J. J. (2011). Promise of new translational safety biomarkers in drug development and challenges to regulatory qualification. *Biomarkers in Medicine*, 5(4), 497.

Smith, D. H., Hicks, R. R., Johnson, V. E., Bergstrom, D. A., Cummings, D. M., Noble, L. J., et al. (2015). Pre-clinical traumatic brain injury common data elements: toward a common language across laboratories. *Journal of Neurotrauma*, 32(22), 1725–1735. https://doi.org/10.1089/neu.2014.3861.

Snyder, H. M., Kim, H., Bain, L. J., Egge, R., & Carrillo, M. C. (2014). Alzheimer's disease public-private partnerships: update 2014. *Alzheimers Dement*, 10(6), 873–880. Epub 2014 Sep 11. https://doi.org/10.1016/j.jalz.2014.06.014.

Snyder, H. M., Shineman, D. W., Friedman, L. G., Hendrix, J. A., Khachaturian, A., Le Guillou, I., et al. (2016). Guidelines to improve animal study design and reproducibility for Alzheimer's disease and related dementias: for funders and researchers. *Alzheimer's & Dementia: The Journal of the Alzheimer's Association*, 12(11), 1177–1185. https://doi.org/10.1016/j.jalz.2016.07.001.

Sperling, R. A., Jack, C. R., & Aisen, P. S. (2011). Testing the right target and right drug at the right stage. *Science Translational Medicine*, 3(111), 111cm33.

Stephenson, D., Perry, D., Bens, C., Bain, L. J., Berry, D., Krams, M., et al. (2015). Charting a path toward combination therapy for Alzheimer's disease. *Expert Review of Neurotherapeutics*, 15(1), 107–113. Epub 2014. https://doi.org/10.1586/14737175.2015.995168.

Stone, K. (2010). Comparative effectiveness research in neurology: healthcare reform will increase the focus on finding the most effective-and affordable-treatment. *Annals of Neurology*, 68(1), A10–A11. https://doi.org/10.1002/ana.22113.

Thomas, R. G., Albert, M., Petersen, R. C., & Aisen, P. S. (2016). Longitudinal decline in mild-to-moderate Alzheimer's disease: analyses of placebo data from clinical trials. *Alzheimer's & Dementia*, 12(5), 598–603.

Trojanowski, J. Q., & Hampel, H. (2011). Neurodegenerative disease biomarkers: guideposts for disease prevention through early diagnosis and intervention. *Biological Markers for Neurodegenerative Diseases*, 95(4), 491–495. https://doi.org/10.1016/j.pneurobio.2011.07.004.

Willmann, R., Dubach, J., & Chen, K. (2011). Developing standard procedures for pre-clinical efficacy studies in mouse models of spinal muscular atrophy. *Neuromuscular Disorders*, 21(1), 74–77. https://doi.org/10.1016/j.nmd.2010.09.014.

Woodcock, J. (2014). Paving the critical path of drug development: the CDER perspective. *Nature Reviews. Drug Discovery*, 13(11), 783–784.

Woodcock, J., & Woosley, R. (2008). The FDA critical path initiative and its influence on new drug development. *Annual Review of Medicine*, 59(1), 1–12. https://doi.org/10.1146/annurev.med.59.090506.155819.

21

The Assessment of Cognition in Translational Medicine: A Contrast Between the Approaches Used in Alzheimer's Disease and Major Depressive Disorder

John E. Harrison[*,†,‡], *Suzanne Hendrix*[§]

*Metis Cognition Ltd., Kilmington Common, Kilmington, United Kingdom [†]Alzheimer's Center, VUmc, Amsterdam, The Netherlands [‡]Institute of Psychiatry, Psychology and Neuroscience, King's College London, London, United Kingdom [§]Pentara Corporation, Salt Lake City, UT, United States

List of Abbreviations

AD	Alzheimer's disease
ADAS-cog	Alzheimer's Disease Assessment Scale-cognitive subscale
ADCS-PACC	Alzheimer's Disease Cooperative Study Pre-Alzheimer's Cognitive Composite
APCC	Alzheimer's Prevention Initiative Composite Cognitive
CANTAB	Cambridge Neuropsychological Test Automated Battery
ChEI	cholinesterase inhibitor therapy
CIAS	cognitive impairment associated with schizophrenia
COWAT	Controlled Oral Word Association Test
CRT	choice reaction time
DSST	digit symbol substitution test
EF	executive function
GRECO	Groupe de Réflexion sur les Evaluations Cognitives
IS	imperative signal
LC	Language Comprehension Scale (from the ADAS-cog)
MCCB	MATRICS Consensus Cognitive Battery
MDD	major depressive disorder
MMSE	Mini-Mental State Examination
MoCA	Montreal Cognitive Assessment
NTB	neuropsychological test battery
OFN	Object and Finger Naming test (from the ADAS-cog)
RAVLT	Rey Auditory Verbal Learning Test
SLA	Spoken Language Ability scale (from the ADAS-cog)
SRT	simple reaction time
TBRI	to-be-remembered item
WAIS	Wechsler Adult Intelligence Scale
WM	working memory
YoE	years of education

I A CRITICAL REVIEW OF COGNITIVE ASSESSMENT IN PATIENTS WITH ALZHEIMER'S DISEASE

The assessment of cognition in clinical drug trials of potential new therapies for Alzheimer's disease (AD) has traditionally been conducted using the Mini-Mental State Examination (MMSE—Folstein, Folstein, & McHugh, 1975) and variants of the Alzheimer's Disease Assessment Scale-cognitive subscale (ADAS-cog—Rosen, Mohs, & Davis, 1984). The MMSE is a portmanteau measure of very brief tests of memory, attention, praxis, and language, originally designed as a brief, bedside measure of general cognition. In AD trials, it has traditionally been employed as a means of both selecting and stratifying patient cohorts. The earliest trials of AD therapies were designed to assess treatment effects in patients labeled as having "mild-to-moderate" disease severity, typically a range of 14–26 on the MMSE. The MMSE is ranged from 0 to 30 with higher scores indicating superior performance. An informal scheme has evolved by which patients with scores in the 0–12 range are classified as "severe," 13–20 as "moderate," and 21–26 as "mild."

The ADAS-cog has for the past 20 years near universally been employed as the primary cognitive efficacy measure in AD trials (Hobart et al., 2013; Wesnes & Harrison, 2003). Like the MMSE, it is also a portmanteau measure and is similarly composed of memory, language, and praxis subtests. A major limitation of the ADAS-cog is the omission of working memory, attention and executive function (EF) tests, and a deficiency highlighted by Mohs et al. (1997). In an attempt to remedy these deficiencies, Mohs et al. (1997) recommended the inclusion of maze and number cancellation tests, as well as a delayed recall version of the original Word Recall measure. These additional measures are referred to as ADAS-cog+ tests and on occasion have been augmented by the "concentration/distractibility" element of the ADAS-noncog. These additional measures have been variously added to the original 11 ADAS-cog subtests to create new configurations, the most common of which is the ADAS-cog13. The elements and standard running order of the ADAS-cog13 are shown in Table 21.1.

The most serious deficiency of the ADAS-cog is the conspicuous lack of measures to evaluate attention, working memory, and EF. These domains are all known to be compromised early in the disease process and have repeatedly been specified for evaluation by expert groups (e.g., Ritchie et al., 2016; Vellas, Andrieu, Sampaio,

TABLE 21.1   ADAS-cog and ADAS-cog+ Subtests

No	Subtest	Range	Comments
1	Immediate Word Recall	0–10	Episodic verbal memory test
2	Naming Fingers and Objects	0–5	Confrontation naming
3	Commands	0–5	Tests comprehension and praxis
4	Delayed Word Recall[b]	0–10	Episodic verbal memory test
5	Constructional Praxis	0–5	Design copy
6	Ideational Praxis	0–5	Familiar task execution
7	Orientation	0–8	Semantic and episodic memory
8	Word Recognition	0–12	Episodic verbal memory test
9	Remembering Test Instructions	0–5	Episodic memory
10	Spoken Language Ability[c]	0–5	Language production
11	Word-finding Difficulty[c]	0–5	Semantic memory
12	Language Comprehension[c]	0–5	Language comprehension
13	Number Cancellation[b]	0–5	Attention
14	Maze[b]	0–5	Executive function (in theory)
15	Concentration/ Distractibility[c,a]	0–5	Attention

[a] An ADAS-noncog test.
[b] Denotes an addition to the original version.
[c] A subjectively rated assessment.

Coley, & Wilcock, 2008). Furthermore, in trials such as the proof-of-concept studies of PBT2 and encenicline, the use of EF tests has yielded evidence of drug efficacy (Faux et al., 2010; Hilt et al., 2009). However, in practice, these functions, particularly EF, have only rarely been assessed. This is to be regretted as evidence suggests that EF is improved by already marketed drugs. Rockwood, Black, Robillard, and Lussier (2004) inquired of health-care practitioners their experience of donepezil therapy on patients with AD. Those responding stated that the most conspicuous change was on the patients' ability to organize their thoughts, synonymous with executive skills. Rockwood comments that these skills "have not been systematically assessed in clinical drug trials." Test selection for indexing EF seems a very worthwhile endeavor, though it is important that selected measures meet best practice guidance, a topic we will address later in this chapter.

A ADAS-Cog Subtest Content

Memory assessment accounts for 45 out of 70 of the available points in the original ADAS-cog11. Word Recall and Word Recognition are tests of episodic verbal memory and account for 22 possible points. Remembering Test Instructions relates to the need for reminders while completing Word Recognition, though was originally for reminders needed across the whole ADAS-cog11. A significant element of the Commands subtest is the ability to recall the instructions. The orientation subtest requires the patient to recall the elements of orientation to time, place, and self. The worst possible score is eight, with seven points for episodic memory items (What is the date, day, month, etc.?) and just one related to semantic memory (What is your name?). Semantic memory is also assessed in the context of the Object and Finger Naming (OFN) and Word-finding Difficulty subtests. The addition of Delayed Word Recall adds a further 10 points of memory measurement. When employed, this test is inserted after the OFN and Commands subtests, which yields a delay of approximately 5 min between the immediate and delayed recall tasks.

Completion of the "overlearned" task of folding a piece of paper to fit an envelope, sealing the envelope, addressing the envelope to oneself, and indicating where the stamp should be placed constitutes the Ideational Praxis subtest. Constructional praxis is measured by having the patient copy figures of increasing complexity from a circle to a 3D cube. The formal assessment of praxis contributes a total of 10 points to the overall score, although there are elements of praxis in the Commands subtest (see Hendrix & Welsh-Bomer, 2013; Tatsuoka et al., 2013).

Language skills are assessed subjectively, largely through an opening conversation with the patient, but also during the administration of other ADAS-cog subtests. On the basis of these interactions, the patient is rated on a 0–5 scale for their Spoken Language Ability (SLA) and Language Comprehension (LC) skills. Word-finding Difficulty is sometimes classed as a language measure but as outlined earlier is better considered as a measure of semantic memory.

The same 0–5 rating scale is used when the ADAS-noncog test of "concentration/distractibility" is employed. Subjective assessments of attention are of very limited utility. The Number Cancellation subtest potentially offers accurate assessment of attentional skills. It was one of the additional subtests recommended for use as a means of filling gaps in the ADAS-cog11 with respect to the measurement of attention, working memory, and EF. Since the 1997 recommendations, the assessment of EF has been assessed using the Mazes subtest (Mohs et al., 1997).

B Variation in ADAS-Cog Administration

The ADAS-cog is typically considered to be homogeneously administered across clinical drug trials. However, there are significant sources of variation. Word Recall administration varies with respect to whether the same or different word lists are used on successive visits. Repetition of the same list runs the risk of learning effects. This is less of an issue with respect to moderate and later stage patients, who are typically unable to recall any of the 10 words read after even very short intervals, as occurs with the Delayed Word Recall subtest. An additional source of variation is whether the three iterations of the word list within a visit are repeated in the same or different order for each iteration. Using the same order for the word list allows assessment of primacy and recency effects (Shankle et al., 2005), but nearly all clinical trials use a different order for each iteration. The same issue of different or the same word list is also a consideration with respect to the Word Recognition subtest. A further source of variation of the Word Recognition subtest extends to the number of trials administered. The original ADAS-cog requires that this subtest be administered three times, necessitating the use of three different word lists. In practice, clinical drug trial versions include only one trial of Word Recognition, and rarely two, presumably because the extra time required to administer trials 2 and 3 adds unacceptably to the administration time required. The OFN task varies with respect to the items employed with substitution sometimes occurring if objects are perceived to be uncommonly encountered. For example, the tongs item is variously substituted with tweezers or a funnel.

Concerning language assessment, substantial variation exists with respect to the content and duration of the opening conversation. The length of this conversation is rarely specified, and there have been lingering concerns that the greater the duration, the higher the probability that speech errors will be made. An attempt to control for this issue was the introduction of the GRECO version of the ADAS-cog (Puel & Hugonot-Diener, 1996). In this version, a controlled 10-min conversation with a prescribed content is required. Scoring of LC, WFD, and SLA is then completed before the administration of Word Recall.

The content and administration of the OFN, Commands, Ideational Praxis, and Constructional Praxis subtests are invariant not only within a trial but also across trials. This consistency and repetition likely contribute to the tendency for these tests to exhibit ceiling effects, even across trials of relatively long duration.

C Criticisms of the ADAS-Cog

A substantial literature exists with regard to the challenges, quirks, and inadequacies of the ADAS-cog in its different versions (e.g., Wesnes & Harrison, 2003; Hobart et al., 2013). An already considered challenge with the Word Recall subtest is those occasions when the word list is unchanged across study visits. Best practice requires that when the test content for episodic memory tests can be retained between assessments, then parallel versions of the test should be used. However, crucially the different versions of the test must be demonstrably equivalent. Recent data suggest that the standard variants employed in AD drug trials may not have this attribute. Data from expeditions 1, 2, and 3 have clearly illustrated the problem. Word list 4 typically yields higher scores than the others employed (Hendrix & Ellison, 2017). The frequent use of the ADAS-cog in mild and even prodromal patients can lead to learning effects, and it is evident that the invariant content of several subtests can lead to artificially sustained levels of performance (Sevigny, Peng, Liu, & Lines, 2010).

A further issue is the lack of sensitivity of many subtests across the full range of disease severity in which the ADAS-cog is commonly used. Performance on the ADAS-cog subtests of praxis; language; and semantic memory, attention, and EF are at ceiling in more than three-quarters of mild-stage patients (Winblad et al., 2008). This gives the false impression that these cognitive skills are unimpaired in prodromal and mild-stage patients. However, there is substantial evidence that these skills are compromised, as well as working memory skills, which are not indexed by the ADAS-cog in any of its variants (Hort et al., 2014). When cognitive domains are measured by items that are at ceiling for most study participants, there is a potential risk that changes seen on these scales may reflect more noise than true signal. In mild diseases, using eight items from the ADAS-cog has been shown to be more sensitive to AD progression over time than the full 11 item ADAS-cog score (Hendrix et al., 2010).

As well as issues with content and administration, ADAS-cog scoring conventions have also received criticism. For example, performance on the three administrations of the 10-item Word Recall subtest is averaged across the three trials to yield a score of 0–10, where 10 represents worst possible performance. This requirement reduces the dynamic range of the test from 0–30 to 0–10. As the Word Recognition subtest has almost universally been reduced to a single trial, averaging performance is not a requirement for scoring this subtest. It is important to note that in those trials that require two or three administrations of Word Recognition, there is greater potential to require repetition of the test instructions. This can give to higher scores on the Remembering Test Instructions subtest. However, in spite of the potential for 22 reminders per Word Recognition trial, the ADAS-cog scoring scheme specifies a worst possible scaled score of five, which occurs when seven reminders have been given. The conversion from a scale with 22 possible outcomes to a scale with only 5 is conducted informally rather than mathematically, making the conversion somewhat subjective. The reduction of raw score values to scaled scores is also a characteristic of the OFN test. In this subtest the patient's ability to name the 5 fingers and 12 objects has the potential to yield a score with a 0–17 range. However, the ADAS-cog scoring scheme requires that this be compressed to a 0–5 range again, resulting in further lost information.

There is also a loss of information when translating an ordered list of remembered and forgotten words to a total count of words forgotten. Serial position of each word is important. Words that are near the beginning of the word list are more often remembered as are words near the end, corresponding to a frequency distribution of words remembered within an unimpaired population. This distribution also differs across the three trials based on the order of the words within the prior and current trials, unless the same word order is used for each trial. When cognition becomes impaired, the pattern of the words remembered could differ based on whether the loss of words is related to attention issues or working memory impairment. Capturing this type of information, rather than just an average number of words recalled across the three trials, could result in better sensitivity to important cognitive changes (Shankle et al., 2005). Noise may also be reduced by weighting words appropriately for the type of cognitive change that is expected.

II BEST PRACTICE GUIDANCE FOR THE ASSESSMENT OF COGNITION

The evidence reviewed in the previous section indicates that the selection of the ADAS-cog was based largely on the measure having been relatively recently published at the time of ChEI compound development in the 1990s. Despite only modest efficacy being captured on this measure, its use became habitual thereafter. Its continued use seems to be due to a perception among drug developers that its selection is mandated by regulatory agencies, in spite of clear indications to the contrary. In the previous section, we detailed concerns with the continued use of the ADAS-cog and focused on its psychometric deficiencies, insensitivity to change, and inadequate coverage of key cognitive domains. But what alternatives exist with regard to the measurement of cognition in clinical drug trials of putative therapies for AD? Advice has long been forthcoming, but in spite of the sound good sense behind various endeavors, it has been virtually entirely ignored. As far back as 1997, Ferris et al. offered advice on the selection of objective measurement tools (Ferris et al., 1997). They specified a number of characteristics that selected tests should possess, including acceptable and ideally high levels of reliability, validity, and sensitivity. In addition, the authors also suggest the use of parallel test versions to mitigate content learning effects, the selection of culture independent tests, and the use of computerized testing where possible. The authors also wisely suggested the use of simple manipulanda, such as external key pads and button boxes, to ensure that it was the cognitive skills of study participants that were being assessed and not their computer literacy.

These suggestions for best practice have been reiterated and augmented by subsequent authors (Harrison, 2016; Harrison & Caveney, 2011). More recent recommendations have included advice on the statistical characteristics exhibited by appropriate tests. These include the absence of range restrictions, such as floor and ceiling effects, appropriate data transformations, and the management of extreme values and statistical outliers (Harrison & Maruff, 2008). Advice has also been forthcoming regarding methods of standardization. The most obvious manifestation of this has been the use of a z-score transformation to correct for different methods of scaling (Harrison et al., 2007, 2014). For example, the challenge has been to compare measures of latency, which are typically reckoned in milliseconds, with word list learning that can have a dynamic range of as little as 0–10. Simple addition of the two measures has the potential for latency changes to swamp those in Word Recall. Transforming individual change scores puts these metrics on a common scale. However, substantial differences may be observed in these standardized z-score values depending on the method used for standardization.

A z-score is calculated by taking the original score and subtracting a mean and then dividing this difference by a standard deviation. The critical question is which mean and standard deviation to use. Z-score standardization of cognitive test scores is often conducted relative to age- and education-related "norms" so that the resulting z-score indicates how an individual performs relative to typical controls with similar age and education. This results in a z-score that can be interpreted as the number of standard deviations away from the mean, with negative scores corresponding to scores below the mean and positive scores above the mean. A z-score that is less than −2 would indicate that a person is below average by 2 or more standard deviations and may be impaired from an original normal state. However, in clinical trials where all of the subjects enter the study in a cognitively impaired state, the mean and standard deviation used for calculating the z-scores are often taken from the baseline distribution. A test item that has small variability at baseline, due to ceiling effects for instance, will have a small standard deviation resulting in z-scores that are potentially larger in magnitude for those who are not at ceiling. This may result in higher weighting of items that are skewed and are only changing for a few individuals. Perhaps, this is acceptable, but these assumptions need to be considered. Another possible way to calculate z-scores to allow summing of individual item scores is to use the mean and standard deviation of the change in the placebo group at the same time point in the study. It is also possible to standardize items relative to the total range of the scale, so that, for instance, every score has a range from 0 to 1, or alternatively to standardize to the total range of the scale as observed in the study. Each of these approaches may be appropriate according to circumstance.

Other sources of guidance have focused on methods of reducing error variance. Here the focus has been on environmental factors, test procedures, and managing internal sources of variable performance. Management of the testing environment includes ensuring the location at which testing will occur is quiet, well lit, and comfortable. A full list of precautions is beyond the remit of this chapter, but a basic list includes the following:

(1) telephones and pagers turned off and a "Do not disturb" sign for the door,
(2) a room that is well lit and set to a comfortable temperature,
(3) accessible seating with a table at which the patient can be sat comfortably.

Continuity is the key factor with respect to test management procedure. This precaution extends to these issues:

(1) Where possible, use the same rater for the same patient for all study visits.
(2) Conduct testing in the same room.
(3) Assessments should be made at approximately the same time of day.

Item three crosses over into internal factors. Diurnal variation in cognitive performance is a well-documented phenomenon in even healthy normal controls and if anything might be exacerbated in patients with neurological and psychiatric disorders, for example, the well-known "sundowning" effect often seen in patients with AD. A further factor is the extent to which the patient is free of distractions that might disrupt their ability to concentrate. This is part of the rationale for conducting assessments in appropriate locations, which can help mitigate the impact of any possible stress and agitation. Good test selection can help manage changes in performance due to familiarity effects, that is, improvements over time due to familiarity with test content. For many patients, the experience of being tested might be new and could induce some mild anxiety. This can be managed by exposing the patient to study testing procedures on their screening visit. Other precautions include ensuring that the patient is not distracted by discomfort. The impact of not being able to void a full bladder has been documented to have a clear negative effect on cognition (Lewis et al., 2011). Patients should also be given access to light refreshments before being tested and access to liquids during testing procedures.

A Determining Which Cognitive Domains to Assess

Preclinical study results can sometimes yield information about the cognitive safety and efficacy of novel treatments. However, these studies can have limited predictive validity for effects in human study participants. Preclinical results can certainly inform trial design for human studies, but the recommended approach for cognitive assessment is to ensure that a broad range of skills are assessed. Cognitive tests are often labeled as measures of specific domains. Various endeavors have sought to specify these domains. A key approach has been to determine whether cognitive skills can be dissociated, either neuropsychologically through the investigation of selective lesions or pharmacologically. An example of the former approach is patient HM who had bilateral hippocampal damage. HM was capable of learning a new motor task but had no recollection of doing, illustrating a dissociation between episodic memory and procedural memory (Scoville & Milner, 1957).

One approach to enumerating and characterizing cognitive domains was the endeavor of the MATRICS group

TABLE 21.2   MATRICS Cognitive Domains

	Cognitive domain
1	Information processing speed
2	Attention/vigilance
3	Working memory
4	Verbal learning
5	Visual learning
6	Reasoning and problem solving
7	Social cognition

(Nuechterlein et al., 2008). This was conducted in the context of defining domains for investigating cognitive dysfunction in patients with cognitive impairment associated with schizophrenia (CIAS). However, this approach can be applied to cognitive performance in healthy controls and other indications. The MATRICS domains are listed in Table 21.2.

Pure measures of these domains are difficult if not impossible to obtain, and so the proposed domains have the status of hypothetical constructs. There is nevertheless some consensus about the proposed division of cognition. Other commentators have argued that deficits in information processing speed are in fact due to dysfunction in other domains and often attentional deficits. It is also not clear that visual and verbal learning can be usefully dissociated and even within the MATRICS group, there was some disagreement about the status of social cognition as a discrete domain. Verbal and visual learning would include memory for events, usually referred to as episodic memory. The MATRICS domain of reasoning and problem solving is analogous to the idea of EF.

With regard to the assessment of cognition in patients with AD, the domains listed in Table 21.2 do not include either measures of language or praxis. As with information processing speed, deficits in these two domains of function can be viewed as due to dysfunction in other core domains, that is, praxis and language deficits are due to compromised output of working memory and attentional and executive dysfunction. Expert groups have previously recommended the assessment of episodic memory, working memory, and EFs. We would endorse this approach and would add the assessment of attention to yield a set of core domains to be assessed in exploratory studies of new chemical entities for the putative treatment of AD. Indexing performance on both visual and verbal memory assessments would also be of benefit. For many patients, it is changes in episodic memory that then lead to seek help from health-care professionals. However, clinical research has routinely indicated that patients exhibit deficits in other key cognitive domains. Recognition of these issues has recently led

the EPAD Scientific Advisory Group for Clinical and Cognitive Outcomes to specify measures of these domains (Ritchie et al., 2016).

Identification of the key domains is a critical step. Once they have been specified, all that is then required is the selection of measures that have acceptable levels of reliability, responsiveness, and validity. As discussed earlier, further considerations concern the characteristics of the data yielded by the selected measures. Tests that yield data free of range restrictions (e.g., floor and ceiling effects), demonstrate responsiveness to change with disease progression and have tight variance, commend themselves for use. However, it is important to recognize that while the suggested procedures and precautions can help maximize the probability of detecting treatments effects, these are based on a set of prior assumptions about testing changes in cognition. A further aspect of good test selection is what we will refer to as assay sensitivity validation. By this, we mean whether measures have been shown to be sensitive to treatment effects known to impair or enhance cognition. Experimental medicine has routinely employed scopolamine and tryptophan depletion to model cognitive deficits observed in CNS disorders. The preservation or rescue of cognition with ChEI therapy in patients with AD has been observed on composite measures (i.e., the ADAS-cog and NTB). Positive treatment effects have also been seen on individual NTB tests, such as the COWAT (Hilt et al., 2009) and the Rey Auditory Verbal Learning Test (RAVLT; Scheltens et al., 2012).

The need for cognitive assessment within the context of AD extends to identifying sensitive measures of impairment for study selection, for example, measures of cognitive change and assessments that can be employed for the purposes of population screening. Evidence-based treatment of cognition also requires the use of tools capable of detecting individual patient response to treatment. These requirements have not yet been met in the context of assessing cognition in AD, but the means by which it could be achieved have been outlined earlier. The need as we see it is for a cohesive approach based on the assessment of the appropriate cognitive domains using measures that meet best practice to:

(a) Screen for cognitive deficits cross-sectionally and longitudinally.
(b) Detect cognitive change in interventional studies.
(c) Obtain evidence-based measures of treatment effects.

In the following section, we will describe how these needs were met in detecting and monitoring cognitive change in the context of major depressive disorder (MDD). We will in this section also consider whether the same model could be successfully applied to the management of cognitive issues in AD.

III MEASURING COGNITIVE IMPAIRMENTS AND COGNITIVE CHANGE IN PATIENTS WITH DEPRESSION: A MODEL FOR AD?

Patients with MDD often suffer cognitive difficulties. It has often been supposed that these difficulties were due to a reduction in mood. The logic of this position is that alleviation of the patient's mood complaints would be accompanied by a restoration of their cognitive skills. However, research has shown that patients with MDD complain that their thinking is impaired between depressive episodes. Residual cognitive difficulties have been a major challenge in restoring patients to full functionality. A variety of interventions have been proposed to help patients with this pattern of partial remission. A pharmacological approach is among those that have been proposed, and the multimodal antidepressant Trintellix (vortioxetine) has recently been approved by regulatory agencies for the treatment of both mood and cognition.

The first procognitive effect of vortioxetine was reported from a study of elderly individuals with a history of MDD (Katona, Hansen, & Olsen, 2012). In this study, cognitive effects of treatment with duloxetine and vortioxetine were assessed using the RAVLT and the digit symbol substitution test (DSST). The results showed that treatment with both compounds improved performance on the RAVLT. However, vortioxetine also showed beneficial treatment effects on the DSST.

The DSST has a long history of use in clinical psychology and has been employed extensively in a variety of neurological and psychiatric disorders (Lezak, 1995). These disorders include AD, and based on its use in this indication, the DSST is a component of the recently proposed ADCS-PACC (Donohue et al., 2014). The DSST was independently identified as an important item for measuring early cognitive progression, based on a modeling approach used to create the API-APCC (Langbaum et al., 2014, 2015). Its role in the PACC is to function as a measure of EF, a role endorsed recently by the European Medicines Agency who describes the DSST as a "timed executive function test" (CHMP, 2014, p. 15). While true the DSST in fact measures a good deal more than EF. Analysis of the test content indicates that the functional integrity of cognitive resources such as attention and working memory are also required for successful completion of the test. It is relatively a brief task (c.2–3 min in duration) and reliable. However, the simplicity of the tests belies the fact that it makes keen demands on the cognitive skills listed. It is therefore perhaps unsurprising that it is a sensitive measure of dysfunction in a number of indications and that DSST performance can be influenced by a number of psychopharmacological interventions. This makes the test a useful measure of both cognitive deficit and change, but one that is uninformative with respect to which domains have been impacted by treatment. Therefore, a key question for the follow-up study described in the following section (McIntyre, Lophaven, & Olsen, 2014) was which cognitive domains are positively impacted by treatment with vortioxetine?

This follow-up study was designed primarily to be a replication of the Katona et al. study. Consequently, the chosen primary outcome was a composite measure of the RAVLT and DSST. The secondary question was which cognitive domains were positively impacted by treatment. As acknowledged earlier, there are no entirely pure measures of specific cognitive domains. However, the task demands of many cognitive tests have been constructed to index performance on key domains. For example, measures of episodic verbal memory, such as the RAVLT, require the recollection of single word items. Measures of attention are derived from latency data that indicate how attentive the study participant was when the "go" signal was given.

A number of measures have been employed to measure cognitive deficits in patients with MDD, and the typical pattern of impairment is usually in the order of -0.8 SD from normal in patients in the midst of a depressive episode (Douglas, Porter, Knight, & Maruff, 2011). On average, patients between episodes tend to exhibit deficits in the effect size range of -0.3 to -0.6 (Rock, Roiser, Riedel, & Blackwell, 2013). However, the detailed investigation of patient cohorts has revealed that the cognitive deficits are not necessary sequelae of MDD. Approximately, 40% fail to show clear evidence of cognitive deficit at baseline presentation (Maruff & Jaeger, 2016). While the evaluation of cognitive deficits in patients with MDD has been commonly reported, changes in the signs and symptoms of the disease have usually been focused on measures of mood.

In contrast to the prevailing inclination for picking traditional, but often not appropriate measures for AD trials, there was no legacy of specific tests for use in the FOCUS study. Instead a two-stage process of (a) specifying cognitive domains of interest and (b) selecting appropriate tests was implemented. The domains of attention, episodic memory, working memory, and EF were selected for assessment. Test selection was informed by available best practice guidance and especially with reference to requirements for reliability, validity, and responsiveness. The selected assessments, together with the domains indexed, are listed in Table 21.3.

The CogState simple and choice reaction time tasks, known as "Detection" and "Identification," respectively, were selected as our measures of attention and processing speed. The rationale for this is that latency for correct responses on these tasks represents a measure of attention. At the time of test selection, acceptable computerized versions of the remaining tasks were unavailable

TABLE 21.3   Cognitive Measures Selected for the FOCUS Study

Test	Domain
Rey Auditory Verbal Learning Test (RAVLT)	Episodic memory
Simple and Choice Reaction Time (SRT and CRT)	Attention/psychomotor speed
Trail Making Test Part A (TMT A)	Attention/psychomotor speed
Trail Making Test Part B (TMT B)	Executive function
Stroop Test	Executive function
Digit Symbol Substitution Test (DSST)	Multiple skills

and so were administered in their traditional "paper-and-pencil" (P&P) formats. The full assessment required just less than 30 min to administer. This was a deliberate policy to avoid fatiguing patients. Experience suggests that this is a sensible limitation for cognitive assessment in patients with neurological or psychiatric disorders. Assessments of this length avoid factors that impact performance such as fatigue and ennui. Significant precautions were taken to ensure tests were well understood prior to testing, as well as careful site selection, data review, and rater training.

The results of the FOCUS study showed significant improvements on the DSST and in performance across all the preselected cognitive domains after 8 weeks of treatment with vortioxetine (Harrison, Lam, Baune, & McIntyre, 2016). The DSST effect was further replicated in the later CONNECT study, together with positive treatment effects on elements of the longer cognitive assessment selected for that trial (Mahableshwarkar, Zajecka, Jacobson, Chen, & Keefe, 2015). The time needed to administer the selected measures in the CONNECT study was extended well past our recommended 30-min maximum, often taking as long as 1 h. This extended administration time might account for the smaller effect sizes observed after vortioxetine treatment in CONNECT. For example, the positive treatment effect size observed on the DSST in the FOCUS study was slightly better than 0.5. For context, this is substantially greater than the effect size improvements seen in patients with AD after ChEI initiation. A recent metaanalysis of the use of the DSST in vortioxetine studies reported a positive effect size of 0.35 (McIntyre, Harrison, Loft, Jacobson, & Olsen, 2016). The level of cognitive impairment in DSST is more comparable to MCI than to mild or moderate AD. Detecting treatment effects in predementia AD is substantially more challenging than in later disease. We suggest that the FOCUS study offers a rational, scientific solution to the issue of testing new chemical entities for cognitive efficacy. The process of domain identification followed by best practice guidance test

selection yielded a conspicuous success in characterizing the procognitive profile of vortioxetine. The FOCUS study results showed also that the selected measures are capable of detecting psychopharmacological treatment effects, a critical component of the validation process. We have recently been critical of studies reporting null treatment effects, but those that have used cognitive measures, we would consider to be not fit for purpose (Harrison, Lophaven, & Olsen, 2016). As well as validating the responsiveness to treatment effects of the selected measures, the FOCUS study also permitted validation of the z-score methodology as used in this study. The logic of this approach is to avoid oversimplifying the assessment of cognition by analyzing a single, composite measure while also avoiding the need for harsh statistical correction for family-wise error that might be considered necessary if multiple items were individually analyzed. Assessing and interpreting cognitive data by domain and then balancing the contribution of each domain to an overall composite (using z-scores) or global outcome (using ranks) seem to us the most informative means of determining treatment effects.

Interestingly, the process used to derive the API-APCC composite score utilized a similar approach but in reverse order (Ayutyanont et al., 2014). First, all responsive composite combinations available within the Rush database were considered, and then one was selected based on inclusion of measures that assessed the relevant cognitive domains. Although the weighting of the items was balanced across domains, it was based on modeling historical datasets in order to optimize responsiveness over time rather than assuming a balanced weighting by utilizing z-scores to combine items. This approach is intended to maximize the ability to detect changes within a population that shows very subtle decline—preclinical AD.

An analytic approach that is conceptually remarkably similar to the z-score composite outcome example earlier is the use of a global test statistic as has been used recently in Parkinson's disease (NET-PD, 2012, 2015). The O'Brien global test statistic approach involves ranking the scores (or change scores) across all treatment groups for each outcome of interest and then summing (or averaging) the ranks to get a global score per person. This global score is then analyzed with a t-test, ANCOVA, or other analysis to get an overall assessment of a person's disease status. The ranking of each item essentially reduces each scale to the observed range within the study and then equally weights them in the global score (unequal weighting is also possible). The philosophy behind this approach is that multiple, correlated measures may be used to measure Parkinson's disease and an intervention that affects them all similarly may result in some successful and some unsuccessful outcomes by chance. Rather than rolling the die by choosing one outcome as primary, combining the outcomes into an overall disease measure will improve power over any single test. Correction for multiple comparisons in this case substantially diminishes the power since most corrections assume independence of outcomes and overcorrect for correlated outcomes.

These approaches for combining relevant outcomes, items, or domains offer the promise of improved responsiveness to disease and to treatment effects by utilizing multiple measures of the same underlying process to stabilize estimates. With all of these approaches— z-score composites, optimized composites, and global test statistics—combining a sensitive outcome with an insensitive outcome will result in less power than the sensitive outcome alone. But when there are multiple sensitive measures of the same disease process, the combination has better power than any single component.

An important caveat to attach is that we do not consider that these combination approaches result in a single "gold standard" for measuring a specific disease. The assessment used in FOCUS should not be considered a "gold standard" for assessing cognition in patients with MDD. The FOCUS assessments have robustly demonstrated their capacity to detect treatment effects and were most likely successful because the cognitive composite was selected to include responsive measures of relevant cognitive domains and combine them in a way that reflects the expected treatment benefits. We consider that the process outlined earlier represents a suitable and scientifically defendable approach to assessing cognition in new chemical entity trials. We would further suggest that such an approach be reasonably adopted for assessing cognition or global treatment effects for any drug class and any indication. We contend that there are no tests that magically deliver evidence of efficacy or safety warnings and that the reification of tests into gold standard assessments does considerably more harm than good, as the dogmatic use of unfit for purpose measures such as the ADAS-cog admirably demonstrates.

A further benefit of using appropriate tests is that the paradigms employed can more readily be integrated into the translational medicine approach. One of the greatest indictments of the continued use of the ADAS-cog is that this measure is to the best of our knowledge never employed as a standard assessment in health-care practice such as AD centers or memory clinics. The use of appropriate measures of critical cognitive domains facilitates the process of population screening. This is particularly important in the detection of "caseness" and by extension to the recruitment of appropriate patients for assessment in clinical drug trials. Recent studies have shown that cognitive deficits are not routinely investigated during health-care visits for patients with MDD. On the rare occasions when assessments are conducted, they tend to be with poor measures, such as the MMSE, and usually only in elderly patients on the assumption

that dementia might be an issue. Various groups have initiated programs designed to raise awareness of the importance of cognitive issues in MDD. Among these has been the THINC Task Force, who has sought to raise awareness of cognitive issues in depression through educational programs. The early work of this group identified that while functional deficits and partial remission are well-known features of the condition, these problems were often not obviously connected to the cognitive deficits. Dialogue with health-care providers revealed a keenness to discuss these issues informally and to evaluate the nature of the cognitive difficulties with subjective and objective assessments (McAllister-Williams et al., 2016).

In response to this expressed need, the task force developed and specified a combination of objective cognitive measures and a brief, five-item version of the Perceived Deficits Questionnaire (McIntyre et al., 2017). Interviews with potential users revealed the need for a brief, computerized screening assessment that can be administered by nonexperts. Delivery of results was required to be presented in a rudimentary fashion and available in real time. In response to these specifications, the THINC Task Force designed an assessment with the following attributes:

(1) Brevity—ideally less than 15-min administration time
(2) Nonexpert administration—tests delivered by available staff
(3) Simple interpretation—basic reporting using a traffic light system
(4) Gamified tests—the use of visually engaging measures

The objective cognitive measures for THINC-it were drawn from paradigms that have previously shown sensitivity to impairment but which might also function successfully as measures of change. The following paradigms were selected:

(1) Choice reaction time
(2) One-back test
(3) Repeatable trails B test
(4) Symbol-coding test

In addition to these measures, the five-item version of the Perceived Deficits Questionnaire was also included (Lam et al., 2013). The chosen measures were selected as assessments of attention, working memory, EF, and broad measure of cognitive function, analogous to the use of the DSST. Thus, three of the four measures mirrored the use of paradigms included in the FOCUS study. All the THINC-it measures were validated in a study of $N = 100$ normal controls so that key psychometric data could be obtained (Harrison et al., 2018). Two facets of these data were the collection of temporal (aka test-retest) reliability and stability. Individuals were tested after a 1-week interval in order to estimate temporal reliability correlation (test-retest reliability). Study participants were assessed at three "back-to-back" assessments on their first study visit to assess stability based on a within-subject standard deviation value. These two metrics are required for the calculation of Reliable Change Index (RCI) scores that is the smallest difference that could be considered a "real" change in performance on an individual level, rather than due to the error variance or noise in the test. Various methods of calculating RCI were included into the THINC-it system to facilitate decisions about individual changes in performance.

IV CONCLUSIONS AND RECOMMENDATIONS

We began this chapter by reviewing the use of measures routinely employed in the assessment of cognition in patients with AD. Our review of the two key measures, the MMSE and ADAS-cog, indicates that there are significant difficulties attached to their continued use. Among our criticisms are the following:

- a lack of reliability caused by repeated assessment with identical measures,
- range restrictions and especially ceiling effects on tests of praxis and language,
- poor item selection for key cognitive domains,
- deficiencies with respect to the cognitive domains assessed.

These inadequacies have been evident for some time, and the dogged use of these measures is likely a factor in the lack of progress made in developing new pharmacological interventions. The acknowledged limitations have also been a barrier to the use of these measures in screening for cases of Alzheimer's dementia and providing evidence-based data for deciding about treatment efficacy.

We have offered a methodology for selecting measures that address our criticisms. This approach also provides statistical methods for detecting cases, measuring cognitive change, and providing data upon which evidenced-based decisions about health care may be based. This methodology relies on the initial selection of the appropriate domains of interest. We have further suggested that test selection must be made via the application of best practice guidance based on empirical data. Using the experience of assessing cognition in patients with MDD, we have provided an example of how this approach can be applied, not only to AD but also to any CNS indication.

The aim of translational medicine is to provide a "bench-to-bedside" approach to facilitate the discovery of new diagnostic tools and treatments. In the context of cognitive assessment, we believe that the proposed integrated approach can assist with case detection, clinical trial recruitment, issues of drug safety and efficacy, and data for making evidence-based care decisions.

References

Ayutyanont, N., Langbaum, J. B., Hendrix, S. B., Chen, K., Fleisher, A. S., Friesenhahn, M., et al. (2014). The Alzheimer's prevention initiative composite cognitive test score: sample size estimates for the evaluation of preclinical Alzheimer's disease treatments in presenilin 1 E280A mutation carriers. *Journal of Clinical Psychiatry, 75*(6), 652–660.

CHMP. (2014). *Discussion paper on the clinical investigation of medicines for the treatment of Alzheimer's disease and other dementias.* EMA/CHMP/539931/2014.

Donohue, M. C., Sperling, R. A., Salmon, D. P., Rentz, D. M., Raman, R., Thomas, R. G., et al. (2014). The preclinical Alzheimer cognitive composite: measuring amyloid-related decline. *JAMA Neurology, 71*(8), 961–970.

Douglas, K. M., Porter, R. J., Knight, R. G., & Maruff, P. (2011). Neuropsychological changes and treatment response in severe depression. *British Journal of Psychiatry, 198*(2), 115–122.

Faux, N. G., Ritchie, C. W., Gunn, A., Rembach, A., Tsatsanis, A., Bedo, J., et al. (2010). PBT2 rapidly improves cognition in Alzheimer's disease: additional phase II analyses. *Journal of Alzheimer's Disease, 20,* 509–516.

Ferris, S. H., Lucca, U., Mohs, R., Dubois, B., Wednes, K., Erzigkeit, H., et al. (1997). Objective psychometric tests in clinical trials of dementia drugs. Position paper from the International Working Group on Harmonization of Dementia Drug Guidelines. *Alzheimer's Disease Association Disorders, 11*(S3), 34–38.

Folstein, M. F., Folstein, S. E., & McHugh, P. R. (1975). "Mini-mental state": a practical method for grading the cognitive state of patients for the clinician. *Journal Psychiatric Research, 12,* 189–198.

Harrison, J., Minassian, S. L., Jenkins, L., Black, R. S., Koller, M., & Grundman, M. (2007). A neuropsychological test battery for use in Alzheimer disease clinical trials. *Archives of Neurology, 64,* 1323–1329.

Harrison, J. E., & Maruff, P. (2008). Measuring the mind: assessing cognitive change in clinical drug trials. *Expert Reviews in Clinical Pharmacology, 1*(4), 471–473.

Harrison, J., & Caveney, A. (2011). 10 years of the neuropsychological test battery (NTB). *PRO Newsletter, 46,* 21–24.

Harrison, J. E., Rentz, D. M., McLaughlin, T., Niecko, T., Gregg, K. M., Black, R. S., et al. (2014). Cognition in MCI & Alzheimer's disease: baseline data from a longitudinal study of the NTB. *The Clinical Neuropsychologist, 28*(2), 252–268.

Harrison, J. E., Lophaven, S., & Olsen, C. K. (2016). Which cognitive domains are improved by treatment with vortioxetine? *International Journal of Neuropsychopharmacology,* pyw054.

Harrison, J. E., Lam, R. W., Baune, B. T., & McIntyre, R. S. (2016). Selection of cognitive tests for trials of therapeutic agents. *The Lancet Psychiatry, 8*(8), 1–13.

Harrison, J. E. (2016). Measuring the mind: detecting cognitive deficits and measuring cognitive change in patients with depression. In R. S. McIntyre & D. Cha (Eds.), *Cognitive impairment in major depressive disorder* (pp. 229–241). Cambridge: Cambridge University Press.

Harrison, J. E., Barry, H., Baune, B. T., Best, M. W., Bowie, C. R., Cha, D. S., et al. (2018). Stability, reliability and validity of the

THINC-it screening tool for cognitive impairment in depression: a psychometric exploration in healthy volunteers. *International Journal of Methods in Psychiatric Research, 27*(3), e1736.

Hendrix, S. B., Wells, B. M., & the Alzheimer's Disease Neuroimaging Initiative (ADNI). (2010). Time course of cognitive decline in subjects with mild Alzheimer's disease based on ADAS-cog subscales and neuropsychological tests measured in ADNI. *Alzheimer's & Dementia, 6*(4), e50.

Hendrix, S., & Welsh-Bomer, K. (2013). Separation of cognitive domains to improve prediction of progression from mild cognitive impairment to Alzheimer's disease. *Alzheimer's Research & Therapy, 5*, 22.

Hendrix, S., & Ellison, N. (2017). Are Alzheimer's treatment failures due to inactive compounds or are we doing something wrong? *Alzheimer's & Dementia, 13*(7), P617.

Hilt, D., Gawryl, M., Koenig, G., & the EVP-6124 Study Group. (2009). EVP-6124: Safety, tolerability and cognitive effects of a novel A7 nicotinic receptor agonist in Alzheimer's disease patients on stable Donepezil or Rivastigmine therapy. *Alzheimer's and Dementia, 5*(4), Supp., e32, P4-348.

Hobart, J., Cano, S., Posner, H., Selnes, O., Stern, Y., Thomas, R., et al. (2013). Putting the Alzheimer's cognitive test to the test I: traditional psychometric methods. *Alzheimers & Dementia, 9*(1 Suppl), S4–S9.

Hort, J., Andel, R., Mokrisova, I., Gazova, I., Amlerova, J., Valis, M., et al. (2014). Effect of donepezil in Alzheimer disease can be measured by a computerized human analog of the morris water maze. *Neurodegenerative Disorders, 13*(2–3), 192–196.

Katona, C., Hansen, T., & Olsen, C. K. (2012). A randomized, double-blind, placebo-controlled, duloxetine-referenced, fixed-dose study comparing the efficacy and safety of Lu AA21004 in elderly patients with major depressive disorder. *International Clinical Psychopharmacology, 27*(4), 215–223.

Lam, R. W., Saragoussi, D., Danchenko, N., Rive, B., Lamy, F. X., & Brevig, T. (2013). Psychometric validation of perceived deficits questionnaire—depression (PDQ-D) in patients with major depressive disorder (MDD). *Value in Health, 16*(7), A330.

Langbaum, J. B., Hendrix, S. B., Ayutyanont, N., Chen, K., Fleisher, A. S., Shah, R. C., et al. (2014). An empirically derived composite cognitive test score with improved power to track and evaluate treatments for preclinical Alzheimer's disease. *Alzheimer's & Dementia, 10*(6), 666–674.

Langbaum, J. B., Hendrix, S., Ayutyanont, N., Bennett, D. A., Shah, R. C., Barnes, L. L., et al. (2015). Establishing composite cognitive endpoints for use in preclinical Alzheimer's disease trials. *The Journal of Prevention of Alzheimer's Disease, 2*(1), 2–3.

Lewis, M. S., Snyder, P. J., Pietrzak, R. H., Darby, D., Feldman, R. A., & Maruff, P. (2011). The effect of acute increase in urge to void on cognitive function in healthy adults. *Neurourology and Urodynamics, 30*(1), 183–187.

Lezak, M. D. (1995). *Neuropsychological assessment.* Oxford: Oxford University Press.

Mahableshwarkar, A. R., Zajecka, J., Jacobson, W., Chen, Y., & Keefe, R. (2015). A randomized, placebo-controlled, active-reference, double-blind, flexible-dose study of the efficacy of vortioxetine on cognitive function in major depressive disorder. *Neuropsychopharmacology, 40*, 2015–2037.

Maruff, P., & Jaeger, J. (2016). Understanding the importance of cognitive dysfunction and cognitive change in major depressive disorder. In R. S. McIntyre & D. Cha (Eds.), *Cognitive impairment in major depressive disorder* (pp. 15–29). Cambridge: Cambridge University Press.

McAllister-Williams, R. H., Bones, K., Goodwin, G. M., Harrison, J. E., Katona, C., Rasmussen, J., et al. (2016). Analysing UK clinicians' understanding of cognitive symptoms in major depression: a survey of primary care physicians and psychiatrists. *Journal of Affective Disorders, 207*, 346–352.

McIntyre, R. S., Lophaven, S., & Olsen, C. K. (2014). A randomized, double-blind, placebo-controlled study of vortioxetine on cognitive function in depressed adults. *International Journal of Neuropsychopharmacology, 30*, 1–11.

McIntyre, R. S., Harrison, J. E., Loft, H., Jacobson, W., & Olsen, C. K. (2016). The effects of vortioxetine on cognitive function in patients with major depressive disorder (MDD): a meta-analysis of three randomized controlled trials. *International Journal of Neuropsychopharmacology,* pyw055.

McIntyre, R. S., Barry, H., Baune, B. T., Best, M. W., Bowie, C. R., Cha, D. S., et al. (2017). The THINC-integrated tool (THINC-it) screening assessment for cognitive dysfunction: validation in patients with major depressive disorder. *Journal of Clinical Psychiatry, 78*(7), 873–881.

Mohs, R. C., Knopman, D., Petersen, R. C., Ferris, S. H., Ernesto, C., Grundman, M., et al. (1997). Development of cognitive instruments for use in clinical trials of antidementia drugs: additions to the Alzheimer's disease assessment scale that broaden its scope. The Alzheimer's disease cooperative study. *Alzheimer Disease & Associated Disorders, 11*(Suppl 2), 13–21.

Nuechterlein, K. H., Green, M. F., Kern, R. S., Baade, L. E., Barch, D. M., Cohen, J. D., et al. (2008). The MATRICS consensus cognitive battery, Part 1: test selection, reliability, and validity. *American Journal of Psychiatry, 165*, 203–213.

Puel, M., & Hugonot-Diener, I. (1996). Présentation de l'adaptation en langue française par le groupe greco, d'une échelle d'évaluation cognitive utilisée dans les démences de type Alzheimer. *Presse Médicale, 25*, 1028–1032.

Ritchie, K., Ropacki, M., Albala, B., Harrison, J. E., Kay, J., Kramer, J., et al. (2016). Recommended cognitive outcomes in pre-clinical Alzheimer's disease: consensus statement from the European Prevention of Alzheimer's Dementia (EPAD) Project. *Alzheimer's and Dementia, 13*(2), 186–195.

Rock, P. L., Roiser, J. P., Riedel, W. J., & Blackwell, A. D. (2013). Cognitive impairment in depression: a systematic review and meta-analysis. *Psychological Medicine, 29*, 1–12.

Rockwood, K., Black, S. E., Robillard, A., & Lussier, L. (2004). Potential treatment effects of donepezil not detected in Alzheimer's disease clinical trials: a physician survey. *International Journal of Geriatric Psychiatry, 19*(10), 854–860.

Rosen, W. G., Mohs, R. C., & Davis, K. L. (1984). A new rating scale for Alzheimer's disease. *American Journal of Psychiatry, 141*(11), 1356–1364.

Scheltens, P., Twisk, J. W. R., Blesa, R., Scarpini, E., von Arnim, C. A. F., Bongers, A., et al. (2012). Efficacy of souvenaid in mild Alzheimer's disease: results from a randomized, controlled trial. *Journal of Alzheimer's Disease, 31*, 225–236.

Scoville, W. B., & Milner, B. (1957). Loss of recent memory after bilateral hippocampal lesions. *Journal of Neurology, Neurosurgery & Psychiatry, 20*(1), 11–21.

Sevigny, J. J., Peng, Y., Liu, L., & Lines, C. R. (2010). Item analysis of ADAS-Cog: effect of baseline cognitive impairment in a clinical AD trial. *American Journal of Alzheimer's Disease & Other Dementias, 25*, 119–124.

Shankle, W. R., Romney, A. K., Hara, J., Fortier, D., Dick, M. B., Chen, J. M., et al. (2005). Methods to improve the detection of mild cognitive impairment. *Proceedings of the National Academy of Sciences of the United States of America, 102*, 4919–4924.

Tatsuoka, C., Tseng, H., Jaeger, J., Varadi, F., Smith, M. A., Yamada, T., et al. (2013). Modeling the heterogeneity in risk of progression to Alzheimer's disease across cognitive profiles in mild cognitive impairment. *Alzheimer's Research & Therapy, 5*, 10–14.

Vellas, B., Andrieu, S., Sampaio, C., Coley, N., & Wilcock, G. (2008). Endpoints for trials in Alzheimer's disease: a European task force consensus. *Lancet Neurology, 7*(5), 436–450.

Wesnes, K. A., & Harrison, J. E. (2003). The evaluation of cognitive function in the dementias: methodological and regulatory considerations. *Dialogues in Clinical Neuroscience, 5*(1), 77–88.

Winblad, B., Gauthier, S., Scinto, L., Feldman, H., Wilcock, G. K., Truyen, L., et al. (2008). Safety and efficacy of galantamine in subjects with mild cognitive impairment. *Neurology, 70*(22), 2024–2035.

Writing Group for the NINDS Exploratory Trials in Parkinson's Disease (NET-PD). (2012). Design innovations and baseline findings in a long-term Parkinson's trial: NET-PD LS1. *Movement Disorders, 27*(12), 1513–1521.

Writing Group for the NINDS Exploratory Trials in Parkinson's Disease (NET-PD). (2015). Effects of creatine monohydrate on clinical progression in patients with Parkinson's disease: a randomized clinical trial. *Journal of the American Medical Association, 10*(313), 584–593.

Translational Medicine Strategies in Drug Development for Neurodevelopmental Disorders

Siddharth Srivastava*, Mustafa Sahin*,†, Lisa Prock†,‡

*Department of Neurology, Boston Children's Hospital, Harvard Medical School, Boston, MA, United States
†Translational Neuroscience Center, Boston Children's Hospital, Harvard Medical School, Boston, MA, United States
‡Developmental Medicine Center, Division of Developmental Medicine, Department of Medicine,
Boston Children's Hospital, Harvard Medical School, Boston, MA, United States

I INTRODUCTION

Neurodevelopmental disorders (NDDs) refer to a large, heterogeneous group of childhood-onset conditions that result from disrupted brain development and/or functioning. NDDs can impact an affected individual in many domains including cognition, self-help, social communication, behavior, and motor control. NDDs are often diagnosed based on clinical descriptions of developmental impairments, such as autism spectrum disorder (ASD), intellectual disability (ID), language disorder, specific learning disorder, and attention-deficit hyperactivity disorder (ADHD) (American Psychiatric Association et al., 2013). A subset of NDDs are genetically defined diseases, such as Rett syndrome (RTT), tuberous sclerosis complex (TSC), and fragile X syndrome (FXS), which are commonly associated with variable developmental impairments (Sahin & Sur, 2015).

Collectively, NDDs have a tremendous financial and clinical impact on patients and their families/caretakers. In the United States alone, from 2006 to 2008, around one in six children had a diagnosis of an NDD (Boyle et al., 2011). In addition to limiting a child's ability to function in everyday life throughout their life, NDDs are known to cause significant parental distress (Craig et al., 2016). The financial costs associated with these disorders is also substantial, taking into account both direct and indirect health-care expenses (Buescher, Cidav, Knapp, & Mandell, 2014; Olesen et al., 2012; Cohen et al., 2012). For example, one study estimated the lifetime cost of raising a child with ASD and ID as $2.4 million in the United States and £1.5 million in the United Kingdom (Buescher et al., 2014). Another study suggested that the total economic cost in 2010 of "brain disorders" in Europe was €798 billion (Olesen et al., 2012).

With an increasing understanding of specific genetic causes and therefore improved insight into the possible mechanisms underlying NDDs, the potential role of effective targeted treatments continues to grow. Current treatments for NDDs, whether behavioral or pharmacological, primarily address specific symptoms rather than targeting disease-specific pathophysiology. Nonpharmacological interventions, such as applied behavioral analysis therapy, can be effective strategies for improving maladaptive behavior, but they require a deep investment of time and effort that even leads to occupational burnout (Hurt, Grist, Malesky, & McCord, 2013). Based on the prevalence of burnout for professionals choosing to work with individuals with NDDs, the burden for parents—who may live with affected individuals throughout their life span—may be overwhelming. As a result, both behavioral and psychopharmacologic interventions aim to measure clinical improvement for affected individuals and the impact on family members and caretakers.

In the realm of psychopharmacology, there are broad categories of medications—such as antidepressants, anxiolytics, stimulants, mood stabilizers, and antipsychotics—that are widely used to target certain symptom clusters prevalent in individuals with NDDs. Existing pharmacological interventions may reduce target symptoms, but they may be only partly effective. They can also have nonspecific mechanisms of action, contributing to negative side effects that might not be as prevalent with more specific and targeted treatments. Because of their side effect profile, they may require numerous empirical trials adjusting dose or family of medication before an optimal regimen becomes apparent. Even with this trial-and-error approach, there is no guarantee of success or ability to predict who will respond to which medication at what dose (Bousman & Hopwood, 2016). At best, current psychopharmacologic treatments address symptoms but not the underlying cause of specific behavioral challenges for individuals with NDDs.

Not surprisingly, within NDDs, the prospect of better understanding specific disease pathophysiology to develop specific, targeted treatments is a tantalizing one. As the genetic landscape of NDDs expands to include a growing number of pathogenic single-gene changes, chromosomal copy number variants, trinucleotide repeat expansions, and other genomic alterations (Kiser, Rivero, & Lesch, 2015), there is increasing attention toward studying NDDs through the lens of specific genetic entities—with the hope that better understanding and treatment of these disorders may have implications for currently idiopathic NDDs. The hypothesis driving investigation into single-gene disorders is the hope that disparate genetic disorders associated with neurodevelopmental dysfunction may converge on final common pathways leading to phenotypes like ID and ASD (Geschwind, 2008). Therefore treatments developed from a biological understanding of one disorder may be useful for other disorders with shared circuit dysfunction and phenotypic (but not necessarily genotypic) features (Plummer, Gordon, & Levitt, 2016; Sahin & Sur, 2015).

Hence, in this chapter, we will focus on strategies for translational medicine in the field of NDDs, ranging from preclinical and animal models to clinical trials to date. Following a discussion of challenges translating experiment data in animals to interventional trials in the clinic, we will highlight three examples of NDDs associated with ID and ASD—RTT, TSC, and FXS—where insights into the underlying genetics and neurobiology have led to clinical trials targeting cognition and behavior based on both preclinical and animal data. For each of these disorders, we will discuss critical milestones in scientific understanding that paved the way for clinical trials, starting with the discovery of gene function, followed by investigation of animal models, and culminating in hypothesis testing in patients and potential future targeted treatments. From these three examples, we will provide some overarching observations about the promises and pitfalls of translational neuroscience for NDDs as a whole.

II CONSIDERATIONS IN EXAMINING ANIMAL MODEL DATA RELEVANT TO HUMAN CLINICAL TRIALS

Drug development strategies have traditionally considered compelling animal model data as a critical prerequisite to instituting clinical trials in humans. For disorders with clear biomarkers related to gene or cellular function with a similar and biologically measurable profile in both animals and humans, this has been a productive approach. However, with animal data pertinent to individuals with a range of NDDs, generalizability from animals to humans has been more challenging. For example, outcomes such as "cognition," "behavior," and "anxiety" may rely on very different paradigms in animals in comparison with humans. However, to successfully translate basic science discoveries into clinical practice, investigators benefit from considering a range of models and collaborations with other investigators and the medical community to ensure clinical relevance of outcomes and facilitate translation of basic science findings to the clinic (Zon, 2016).

Prior to examining any mouse or rat model data for eventual use in translational research, several overarching principles are critical to consider, including **construct validity**, **reproducibility and robustness**, and **face validity**. With respect to translational research efforts, **construct validity** refers to how closely molecular underpinnings of a disease in an animal model mirror those in humans. **Reproducibility** is the idea that multiple laboratories should, in theory, generate similar results, while **robustness** pertains to the ability to extend results to different animal models and experimental frameworks, including humans. Finally, **face validity** is the notion that the phenotype in an animal model has significant overlap with the phenotype in humans (Katz et al., 2012). As outlined in the disease-specific sections in the succeeding text, current efforts in translational efforts from animals to humans have relatively successfully addressed construct validity with respect to underlying genetics; increased attention is directed toward reproducibility and robustness of results between animal models. However, outcomes in animals and face validity with human phenotypes of NDDs continue to be evasive in many cases.

In terms of **construct validity**, an animal model typically does not completely mirror the underlying genetics and clinical outcomes seen in a patient with the associated condition. For example, to study a disease caused by inactivating mutations, researchers often create an animal model that abolishes all expression of the defective protein. For certain mutation categories, such as large deletions, complete cessation of protein production is the likely outcome. However, other mutation types, including missense, nonsense, and frameshift alterations, may cause the loss of function while still preserving some degree of protein activity. Moreover, in some instances, two different mutations may have completely opposite mechanisms of action, resulting in variable phenotypes. These possibilities, plus others, underlie some of the molecular discrepancies seen between animal models and affected humans (Katz et al., 2012).

In the area of **reproducibility and robustness**, many factors come into play that may cause the findings of one laboratory working with an animal model to be different from those of another. When modeling neurogenetic disorders like RTT, TSC, or FXS in mice, how knockout models are created, when in development the knockout takes effect, and how researchers take into account the genetic background of the strain are all considerations that may make it difficult to establish an animal model of a disease that is consistent from one laboratory to the next. To complicate matters further, external and epigenetic influences—such as the quality/type of food, resting environment, and overall care provided to the offspring—can interact with the underlying genetics in sometimes unforeseen ways that may or may not be applicable to humans. Taken together, all of these factors emphasize the great need for ensuring that the phenotype under investigation for a specific model of an NDD is consistent across research settings and animal models (Katz et al., 2012).

Face validity refers to the concordance in phenotype between animal models of a disease and humans with the condition. Face validity is often considered to be an especially important—if not the most important—aspect of a model. However, one must consider certain factors that can limit the interpretation and applicability of animal model to humans. From an evolutionary standpoint, species have diverged on the basis of the effects of a behavior, rather than the behavior. A behavior in one animal (e.g., mice) may have different evolutionary and functional purposes than the same behavior in another (e.g., humans). As a result an aberrant behavior shared between two different species may not reflect the exact same pathophysiological underpinnings. In addition, even though there is potential usefulness in studying animal models that have lower biological complexity (e.g., flies), establishing behavioral correlates in these models can be a subjective endeavor (van der Staay, Arndt, & Nordquist, 2009). Moreover the concordance between animal and human models of disease is not always clear. Clinical observations of interest in humans (e.g., reduced language or social interest, decreased global cognitive abilities, and anxiety) are not well represented in animal models of NDDs, which has significant implications for considering outcomes of interest in translational efforts. For these reasons and others, establishing face validity requires careful consideration in the interpretation of animal model data relevant to NDDs.

In the following sections, we will highlight translational efforts and challenges with respect to three genetically defined NDDs: RTT, TSC, and FXS.

III RETT SYNDROME

A Overview

RTT is a rare neurodevelopmental disorder characterized by acquired microcephaly, epilepsy, regression in language and hand use, motor stereotypies, and apraxic gait (Hagberg, Aicardi, Dias, & Ramos, 1983; Rett, 1966). It has an estimated prevalence of 1 in 8500 females (Laurvick et al., 2006). In the classic form the disease is progressive and manifests in four different stages. Prior to onset, there may be subtle abnormalities in spontaneous movements (Einspieler, Kerr, & Prechtl, 2005), while the first stage is characterized by only mild developmental delay between 6 and 18 months of age (Hagberg, 2002). The second stage is characterized by a pronounced period of gradual or rapid regression in language and motor abilities, particularly spoken communication and functional hand use. Motor stereotypies emerge, including hand-wringing/squeezing behaviors, along with other autistic features, decelerating head growth, and breathing abnormalities such as hyperventilation and breath-holding. Following this period of regression is a pseudostationary third stage. Though some of the behavioral and communication concerns may improve, seizures become frequent, and breathing irregularities worsen. The fourth and final phase is the late motor deterioration stage, notable for the development of scoliosis, rigidity, spasticity, dystonia, and the loss of ambulation (Hagberg, 2002; https://www.ninds.nih.gov/Disorders/Patient-Caregiver-Education/Fact-Sheets/Rett-Syndrome-Fact-Sheet).

The most common cause of classic RTT is a de novo mutation in the X-linked gene methyl-CpG-binding protein 2 (*MECP2*) (Neul et al., 2008). *MECP2* mutations are inherited in X-linked dominant fashion, and females are almost exclusively affected. However, increasing numbers of males with *MECP2* mutations are being reported with presentations that range from neonatal encephalopathy to ID (Villard, 2007). Finally, *MECP2* duplication in males is commonly associated with ID, ASD, and anxiety, while in female carriers, it can be associated with features like anxiety, depression, and compulsions even with favorable X-inactivation skewing (Ramocki et al., 2009).

B Gene Function

MECP2 binds preferentially to methylated DNA (Ishibashi, Thambirajah, & Ausió, 2008); however, it can also interact with chromatin and unmethylated DNA (Nikitina et al., 2007; Hansen, Ghosh, & Woodcock, 2010). Interestingly, MECP2 is involved in both the repression and activation of transcriptional activity (Chahrour et al., 2008). MECP2 acts as a transcriptional repressor when it interacts with 5-methylcytosine elements at CpG dinucleotides (Guy, Cheval, Selfridge, & Bird, 2011). In contrast, it serves as a transcriptional activator when it binds to 5-hydroxymethylcytosine, which is a signature of actively expressed genes in the central nervous system (CNS) (Mellén, Ayata, Dewell, Kriaucionis, & Heintz, 2012).

MECP2 is involved in several processes tied to neuronal maturation and synaptogenesis. In neurons, expression level of *MECP2* is related to degree of maturation (Shahbazian, Antalffy, Armstrong, & Zoghbi, 2002; Balmer, Goldstine, Rao, & LaSalle, 2003). In astrocytes, MECP2 helps support normal morphology of neuronal dendrites (Maezawa, Swanberg, Harvey, LaSalle, & Jin, 2009; Ballas, Lioy, Grunseich, & Mandel, 2009). In addition to its role in neuronal maturation, MECP2 regulates synaptic connectivity, synaptic plasticity, and a normal ratio between the number of excitatory and inhibitory synapses (Feldman, Banerjee, & Sur, 2016; Boggio, Lonetti, Pizzorusso, & Giustetto, 2010). Some of the transcriptional and posttranscriptional targets of MECP2 are synaptic players as well. Among the transcriptional targets are brain-derived neurotrophic factor (BDNF) (Chen et al., 2003; Chang, Khare, Dani, Nelson, & Jaenisch, 2006) and insulin-like growth factor 1 (IGF-1) (Mellios et al., 2014), which serve as growth factors mediating neuronal survival and synaptic plasticity (Lipsky & Marini, 2007; Fernandez & Torres-Alemán, 2012; Khwaja & Sahin, 2011; Huat et al., 2014). Similarly the posttranscriptional targets of MECP2 include synaptic proteins and synaptic vesicle proteins that help support brain development after birth (Nguyen et al., 2012).

C Animal Model Data

Animal models, particularly mice, have captured several aspects of the neurobehavioral phenotype of RTT. There are investigations on a wide array of *Mecp2*-deficient mice (Leonard, Cobb, & Downs, 2016), ranging from complete knockouts (Guy, Hendrich, Holmes, Martin, & Bird, 2001) to conditional knockouts with *Mecp2* expression restricted in certain cell lines (Meng et al., 2016; Chao et al., 2010), brain regions (Gemelli et al., 2006), and time points in development (Cheval et al., 2012). Findings from these studies show that mice completely lacking *Mecp2* demonstrate many characteristics of RTT, including stereotypies, reduced movement, abnormal gait, irregular breathing, and low weight (Guy et al., 2001). Even local loss of *Mecp2* can recapitulate some of the clinical characteristics of RTT. In mice, eliminating *Mecp2* expression from glutamatergic cells results in a phenotype of early lethality, obesity, tremor, anxiety, and impaired startle response (Meng et al., 2016), while eliminating *Mecp2* expression from GABAergic neurons leads to symptoms that resemble the complete knockout model (Chao et al., 2010). *Mecp2* restriction from the mouse forebrain causes RTT-like behavioral abnormalities—such as hindlimb clasping, impaired motor coordination, anxiety, and abnormal socialization—but not deficits in locomotor activity or fear conditioning (Gemelli et al., 2006), again affirming that the anatomic distribution of

MECP2 expression plays a role in mediating at least a subset of the features of the overall RTT phenotype. From a temporal expression standpoint, inactivation of *Mecp2* in the majority of murine neurons at various postnatal age windows causes symptoms of RTT, including abnormalities in motor skills and learning and premature death (Cheval et al., 2012). Similarly, inactivation of *Mecp2* in fully mature adult mice causes manifestations similar to those in the germline knockout. Taken together, these findings suggest that normal neurological function requires ongoing *MECP2* expression, even during adulthood.

Animal models have also highlighted potential cellular deficits that may occur in RTT. *Mecp2* knockout leads to disrupted synaptic function, in particular experience-dependent synapse remodeling past the initial point of synapse formation (Noutel, Hong, Leu, Kang, & Chen, 2011). Though neurological dysfunction in RTT is primarily due to disrupted MECP2 function in neurons (Luikenhuis, Giacometti, Beard, & Jaenisch, 2004), other CNS cell types, such as glia, may be involved. In mice with global depletion of *Mecp2*, astrocyte-specific restoration of *Mecp2* rescues some of the systemic and neurological manifestations, including locomotion, anxiety, respiratory pattern, and life span (Lioy et al., 2011). Moreover, when wild-type neurons are cultured with astrocytes from a RTT mouse model, they demonstrate atypical dendritic arborization, possibly related to abnormal release of trophic factors in the surrounding milieu (Ballas et al., 2009). Dendritic toxicity in the murine model may also be the result of excessive glutamate released by microglia, and this detrimental effect improves with the administration of glutamate receptor antagonists (Maezawa & Jin, 2010). GABAergic signaling is also dysregulated in the animal model, though it may be amenable to GABAergic modulators (El-Khoury et al., 2014). Finally, there is evidence in *Mecp2*-null mice that downstream targets of MECP2 are dysregulated, leading to decreased levels of both BDNF (Wang et al., 2006; Chang et al., 2006; Li & Pozzo-Miller, 2014) and IGF-1 (Castro et al., 2014).

There are certain biomarkers studied in RTT mouse models that have correlates in affected humans. In a study on *Mecp2* heterozygous female mice and girls with RTT, both groups showed reductions in visual evoked potential amplitude, especially in relation to disease stage (LeBlanc et al., 2015). As researchers take knowledge from preclinical data and use it to develop treatments for humans, having an objective biomarker that is simultaneously applicable to animal and human subjects may be helpful for validating an intervention when it has the desired outcome or troubleshooting it when it does not.

D Therapeutic Strategies and Trials

In light of the extensive scientific discoveries surrounding the role of *MECP2* dysfunction in the pathogenesis

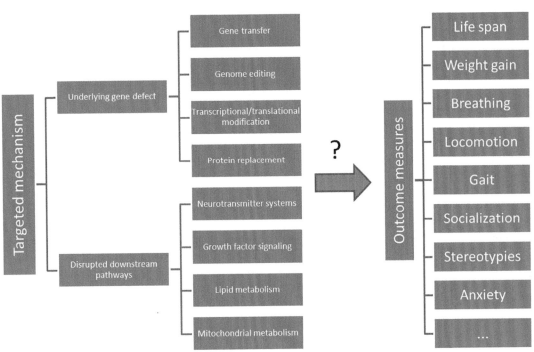

FIG. 22.1 Overview of selected translational strategies for treatment of RTT and potential outcome measures these strategies are targeting. The question denotes uncertainty of addressing each possible outcome measure. Strategies are applicable to other NDDs besides RTT. *Data from Leonard, H., Cobb, S., Downs, J. (2016). Clinical and biological progress over 50 years in Rett syndrome. Nature Reviews. Neurology, 13(1), 37–51.*

of RTT, researchers have proposed multiple routes to treatment for the disorder based on knowledge of MECP2 function (Fig. 22.1). These strategies are designed to address either the underlying gene defect or downstream pathways implicated in the disorder (Leonard et al., 2016). Potential for these approaches is based on studies that demonstrated that restoring *Mecp2* expression in knockout mice rescues many of the neurological and systemic abnormalities that mirror the phenotype of RTT (Guy, Gan, Selfridge, Cobb, & Bird, 2007; Robinson et al., 2012; Ure et al., 2016).

1 Gene Therapy and Transcriptional/Translational Modification

Directly or indirectly manipulating faulty copies of the *MECP2* gene, transcript, or protein is an appealing approach to treating RTT. Generally speaking, to restore partial or complete working copies of a gene, currently available options include gene transfer (Naldini, 2011, 2015; Tillotson et al., 2017), genome editing (Maeder & Gersbach, 2016), RNA repair (Sinnamon et al., 2017), transcriptional/translational modification (Aartsma-Rus & van Ommen, 2007), and protein replacement. Out of these approaches that may be applicable to RTT, we will discuss gene transfer and transcriptional/translational modification here.

Gene transfer involves transduction of host cells with a vector carrying wild-type DNA (Walther & Stein, 2000). Animal models of RTT have demonstrated success for this procedure using the viral vector adeno-associated virus serotype 9 (AAV9) (Gadalla et al., 2013; Garg et al., 2013). However, the challenges of gene therapy for RTT are plentiful. First the precise level of *MECP2* in cells is important, evident from the fact that *MECP2* duplication syndrome also leads to neurological impairment (Van Van Esch, 1993). Therefore, in the process of gene delivery, engineering the exact number of copies of *MECP2* needed to regain normal gene function without causing toxicity is a delicate balancing act that continues to be the subject of study. Second the delivery route for a gene vector can pose problems: *MECP2* administration through intravenous injection has the potential for peripheral toxicity; at the same time, delivery into the CNS either intrathecally or directly into the brain can be quite invasive with associated complications such as infection. Third, *MECP2* is expressed throughout the brain but preferentially in certain cell types, so even if the delivery route is completely optimal, the corrected gene must reach the full span of affected neurons and glial cells without disrupting unaffected cell types. Finally, there is a size restriction on the amount of genetic material that a vector can contain; therefore certain portions of the *MECP2* gene may be missing from the DNA construct. Ideally a version of *MECP2* reduced to its core motifs has the same function as an unstripped copy. However, practically speaking, delivery of an incomplete gene may lead to inefficient phenotypic rescue or, worse, unforeseen consequences (Gadalla et al., 2015).

Recent studies suggest that a truncated form of *Mecp2* may be sufficient to prevent or reverse neurological symptoms in a mouse model of RTT (Tillotson et al., 2017). Having larger animal or nonhuman primate models of RTT could help overcome some of the challenges associated with gene-transfer approaches.

Read-through strategies attempt to bypass the deleterious effect of certain *MECP2* mutations during the process of transcription and translation. Nonsense mutations are common causes of RTT (Laccone, Huppke, Hanefeld, & Meins, 2001), which creates the opportunity for nonsense read-through. This method entails the use of small molecules that bind to and alter the geometry of premature stop codons so that ribosomes no longer recognize these as stop codons and proceed with translation. As a result of this nonsense mutation suppression, the translational machinery produces increased amounts of full-length protein (Velho, Sperb-Ludwig, & Schwartz, 2015). Certain existing compounds, such as the aminoglycoside antibiotic gentamicin, can induce these effects (Manuvakhova, Keeling, & Bedwell, 2000). However, efficiency of this approach is limited by gentamicin toxicity, leading to the development of novel synthetic aminoglycosides, like NB54 (Nudelman et al., 2009). NB54 has been the subject of study in RTT. In ex vivo experiments involving fibroblasts from both RTT patients and mouse models with common nonsense mutations, NB54 promoted recovery of full-length MECP2 protein production with an efficiency as high as 38% for the human cells and 28% for the murine cells. This effect was mutation-specific and superior to that of gentamicin (Vecsler et al., 2011; Brendel et al., 2011). Another compound potentially useful for premature stop codon read-through is the drug molecule PTC124, which is under investigation for cystic fibrosis (Shoseyov, Cohen-Cymberknoh, & Wilschanski, 2016) and dystrophinopathies (Bushby et al., 2014). However, based on multiple reporter assays, it has failed to show efficacy for this purpose, raising doubts about its mechanism of action (McElroy et al., 2013). Efforts are ongoing to discover other, perhaps better understood, compounds that promote nonsense read-through for children with RTT due to nonsense mutations.

2 Modulation of Downstream Targets

In addition to modifying the underlying genetic defect, researchers have also focused on understanding the downstream pathways implicated in RTT in the hope that compensatory therapeutic pathways can be identified. These impacted pathways include neurotransmitter systems, growth factor signaling, mitochondrial metabolism, and lipid metabolism (Leonard et al., 2016; Katz et al., 2016).

The finding that RTT is associated with abnormalities in neural circuits involving different neurotransmitters—including glutamate (Johnston, Blue, & Naidu, 2005), GABA (Johnston et al., 2005), monoamines (Panayotis,

Ghata, Villard, & Roux, 2011), and acetylcholine (Zhang et al., 2016)—has paved the way for modulation of each of these systems (Katz, Menniti, & Mather, 2016). Clinical trials are underway evaluating the use of two different N-methyl-D-aspartate (NMDA) receptor antagonists, dextromethorphan and ketamine, to improve outcomes like epilepsy in RTT, in light of evidence of glutamate excitotoxicity in animal models (Maezawa & Jin, 2010) and affected humans (Hamberger, Gillberg, Palm, & Hagberg, 1992; Horská et al., 2009). In *Mecp2*-null mice, hippocampal neurons become damaged when subjected to conditioned medium from microglia. The most likely culprit is glutamate, because microglia from such animal models produce elevated levels of glutamate, and administration of drugs that inhibit release or action of glutamate rescues the neurotoxicity (Maezawa & Jin, 2010). In patients with RTT, CSF glutamate is elevated, and MR spectroscopy reveals increased glutamate and glutamine/Cr ratio (Hamberger et al., 1992; Horská et al., 2009).

The hypothesis that elevated CNS glutamate may cause excitotoxicity in RTT led to an open-label study of oral dextromethorphan in RTT, with the primary outcome measure focusing on seizures/EEG spike counts—but not glutamate measurement (ClinicalTrials.gov: NCT00593957). Though the primary endpoint did not improve with this study, one of the secondary outcomes, receptive speech, did improve (Johnston, Blue, & Naidu, 2015). These results culminated in a randomized, double-blind, placebo-controlled investigation of dextromethorphan for females with RTT (ClinicalTrials.gov: NCT01520363) showing improvement in seizures, receptive language, and hyperactivity (Pubmed ID: 28931647).

Respiratory dysfunction in RTT may also be amenable to neurochemical manipulation (Katz, Dutschmann, Ramirez, & Hilaire, 2009), in particular modulation of monoamine neurotransmitters (Zanella et al., 2014). Activation of the noradrenergic system through administration of desipramine led to improvement of respiratory regulation and survival of *Mecp2*-deficient mice (Roux, Dura, Moncla, Mancini, & Villard, 2007) and in 2015 a double-blind, placebo-controlled trial (ClinicalTrials.gov: NCT00990691) investigated whether this medication ameliorated breathing abnormalities (including number of apneas) in patients with RTT, but it did not show clinical efficacy (PubMed ID: 29468173). Sarizotan, a 5-HT (5-hydroxytryptamine receptor)$_{1A}$ and dopamine D$_2$-like receptor agonist, reduced apneas in three mouse models of RTT (Abdala et al., 2014); a randomized, double-blind, placebo-controlled investigation (ClinicalTrials.gov: NCT02790034) for this drug is completed (results not yet posted) . On the surface, these seem to be aligned outcome measures in terms of respiratory pattern, but further trial implementation details will be necessary to determine if a treatment failure might represent a problem in the clinical trial design and/or a failure to translate from the preclinical data.

Among neurotrophic factors that could be amenable to modulation, two compelling targets are BDNF and IGF-1. BDNF is upregulated by neuronal activity and has multiple exon-specific promoters leading to different transcripts. *Mecp2* mice knockouts demonstrate downregulated transcription of *Bdnf*, though this downregulation may be dependent on age and cell type (Sun & Wu, 2006). Essential to BDNF's role in synaptic maturation and plasticity is its activation of tyrosine receptor kinase B (TrkB) receptors, which trigger various pathways involving phosphoinositide phospholipase C γ (PLCγ), phosphoinositide 3-kinase (PI3K)/Akt, and mitogen-activated protein kinases (MAPK)/extracellular signal-regulated kinase (ERK) signaling that upregulate production of the neuronal scaffolding protein postsynaptic density protein 95 (PSD95). IGF-1, too, stimulates pathways such as PI3K/Akt signaling that increase the synthesis of PSD95 (Banerjee, Castro, & Sur, 2012). Exogenous BDNF has poor penetration into the blood-brain barrier, limiting its use. However, administration of the ampakine drug CX546, which activates glutamatergic α-amino-3-hydroxy-5-methyl-4-isoxazolepropionic acid (AMPA) receptors resulting in enhanced activity-dependent BDNF expression, led to improvement in breathing irregularities in *Mecp2*-null mice (Ogier et al., 2007). Of note, it is unclear if this improvement was exclusively due to the effects of BDNF or if enhanced AMPA activity, which may have some regulation over breathing pattern (Whitney, Ohtake, Simakajornboon, Xue, & Gozal, 2000), also contributed. Similarly, administration of the sphingosine-1-phosphate receptor modulator fingolimod, which also upregulates BDNF, increased locomotor activity and extended the life span of *Mecp2*-null mice (Deogracias et al., 2012). A phase 1 clinical study involving fingolimod in children with RTT is ongoing (NCT02061137).

Like BDNF modulators, IGF-1 is under active preclinical and clinical investigation for use in RTT. In a RTT animal model, systemic administration of recombinant human IGF-1 (rhIGF-1) restored both phenotypic features (pertaining to life span, locomotion, cardiorespiratory status, and socialization) and neurophysiological features (including synaptic amplitude and dendritic spine density) (Castro et al., 2014). In girls with *MECP2* mutations, a preliminary investigation showed that rhIGF-1 improved anxiety, mood, and an EEG marker of anxiety and depression (Khwaja et al., 2014). Double-blind, placebo-controlled trials of rhIGF-1 and NNZ-2566 (a synthetic version of the terminal tripeptide fragment of IGF-1) have recently completed (ClinicalTrials.gov: NCT01777542 and NCT01703533, respectively).

Multiple lines of evidence suggest abnormalities in mitochondrial function in RTT (Müller & Can, 2014), which may be responsive to targeted therapy. Mouse models of RTT have demonstrated enlarged mitochondria (Belichenko et al., 2009) and increased oxidative stress due to redox imbalance and mitochondrial dysfunction—biochemical changes that improved with administration of an antioxidant compound (Grosser et al., 2012). In children with RTT, analysis of muscle biopsies showed deficiencies in mitochondrial respiratory chain enzyme activity and swollen mitochondria (Coker & Melnyk, 1991; Dotti et al., 1993; Eeg-Olofsson et al., 1990). In peripheral lymphocytes from other RTT patients, the gene expression profile is notable for abnormalities in genes relevant to ATP synthesis, mitochondrial function, and proteasome degradation (Pecorelli et al., 2013). There are clinical trials for RTT for two therapies that confer protection against oxidative stress and mitochondrial dysfunction, EPI-743 (ClinicalTrials.gov NCT01822249) and UX007/triheptanoin (ClinicalTrials. gov NCT02696044), with outcomes such as seizure frequency, overall clinical severity, and oxidative stress biomarkers.

The cholesterol pathway has been the subject of study in RTT (Justice, Buchovecky, Kyle, & Djukic, 2013). In a mouse model of RTT, there were signatures of dysregulated lipid metabolism in the brain and liver, and treatment with statin drugs (3-hydroxy-3-methylglutaryl-coenzyme A reductase inhibitors), which lower levels of cholesterol, ameliorated locomotion and life span. Furthermore, in this model, the simultaneous presence of an inactivating mutation in *Sqle*, encoding an enzyme related to cholesterol biosynthesis, was associated with a milder RTT-like phenotype (Buchovecky et al., 2013). However, a later study showed that the lovastatin did not improve locomotor activity and survival in *Mecp2*-null mice (Villani et al., 2016). Possibly because of the earlier study showing more promising response to statins (Buchovecky et al., 2013), lovastatin is currently under investigation in an open-label trial for females with RTT to determine effect on gait velocity, visual attention/memory, visual pursuit, EEG, respiratory function, and quality of life (ClinicalTrials.gov: NCT02563860).

One of the major caveats to the strategy of modulating downstream targets is that it does not address the underlying genetic defect, so there is the possibility that some, but not all, of the phenotypic features may improve with the use of a pathway-specific treatment. Moreover, there should still be a testable hypothesis—generated from animal model data and other sources—regarding the pathophysiology of the disorder and how a treatment will address it.

3 Screening Drugs on Induced Pluripotent Stem Cell Models

Human-induced pluripotent stem cell (iPSC) models of RTT are undergoing increasing investigation, as they facilitate screening of vast chemical compound libraries for efficacy in restoring various defective neurophysiological parameters. iPSC technology involves the use of specific transcription factors and growing conditions to transform adult fibroblasts or other cell types into pluripotent stem cells that resemble embryonic stem cells. These pluripotent stem cells can differentiate into different cell types for further study (Yamanaka, 2012). Currently, researchers have created numerous human iPSCs from RTT patients with various *MECP2* mutations and have studied the phenotypes of neurons developed from these iPSCs (Walsh & Hochedlinger, 2010; Dajani, Koo, Sullivan, & Park, 2013; Balachandar et al., 2016). For example, in one study, neurons developed in such fashion recapitulated abnormalities seen in RTT patients, including neuronal morphological changes (reduced glutamatergic synapses, dendritic spine density, and neuronal cell body size) and functional changes (impaired calcium signaling and firing of postsynaptic currents). Furthermore, administration of IGF-1 improved the number of glutamatergic synapses, though excessively, and administration of gentamicin for one of the iPSC lines involving a nonsense *MECP2* mutation increased neuronal MECP2 amount and glutamatergic synapse number (Marchetto et al., 2010).

Such results have created the potential for developing phenotypic drug screens for testing on iPSC-derived neurons from RTT patients (Walsh & Hochedlinger, 2010; Dajani et al., 2013; Balachandar et al., 2016). However, there are certain challenges to this approach in RTT. One issue is that even if the source cell lines show random X-inactivation, highly skewed X-inactivation can occur in iPSC-derived neurons (Marchetto et al., 2010), which can affect phenotypic expression. Nonetheless, these iPSC-derived neurons could help accelerate identification of target pathways and tool compounds for therapeutic development, further expanding the pipeline of therapies to be tested in clinical trials.

IV TUBEROUS SCLEROSIS COMPLEX

A Overview

TSC is a genetic disorder associated with systemic manifestations, primarily hamartomas in multiple organ systems and prominent impairments in neurodevelopment (Curatolo, Bombardieri, & Jozwiak, 2008). The incidence of TSC is about 1 in 6000 live births (Osborne, Fryer, & Webb, 1991), and affected individuals often present with variable manifestations of ID, ASD, and epilepsy. While over half of patients (55%) have normal IQs, 14% have mild-to-severe ID, and 31% have profound cognitive impairment (Joinson et al., 2003). The prevalence of ASD in this disorder is reportedly up to 40%, based on a metaanalysis and systematic review of the literature (Richards, Jones, Groves, Moss, & Oliver, 2015). A large

percentage of patients have a history of infantile spasms, 38% in one retrospective cohort of 291 patients (Chu-Shore, Major, Camposano, Muzykewicz, & Thiele, 2010) and 32% in another cohort of 81 patients (Wilbur et al., 2017). Individuals with TSC commonly experience other neuropsychiatric impairments, including ADHD, anxiety, mood disorders, and challenging behaviors (de Vries et al., 2015; Prather & de Vries, 2004).

Developmental brain abnormalities are common in the disorder. On a macroscopic level, there are three types of CNS lesions associated with TSC: cortical tubers, subependymal nodules (SENs), and subependymal giant cell astrocytomas (SEGAs) (Mizuguchi & Takashima, 2001). Furthermore, in the disease, there is strong evidence for microstructural changes in the brain, including hypomyelination and the loss of axonal integrity, based on both histologic (Ruppe et al., 2014) and radiographic studies (Krishnan et al., 2010; Peters et al., 2012). Overall, these abnormalities convey a picture of altered neuronal connectivity in TSC that may serve as the basis for targeted therapies (Khwaja & Sahin, 2011).

B Gene Function

One of the central roles of *TSC1/TSC2*, the genes implicated in TSC, is to provide a break on the activity of the mechanistic target of rapamycin (mTOR) pathway. Within this pathway, TSC1 (also known as hamartin), tuberin (TSC2), and a third protein Tre2-Bub2-Cdc16 1 domain family, member 7 (TBC1D7) form a protein complex, which acts as a sensor of cellular growth conditions and is an essential negative regulator of mTOR complex 1 (mTORC1) (Dibble et al., 2012). TSC1 helps stabilize TSC2 (Chong-Kopera et al., 2006), and TSC2 acts as a GTPase-activating protein (GAP) for Ras homologue enriched in brain (RHEB), causing it to be inactive (Inoki, Li, Xu, & Guan, 2003). Since RHEB activates mTOR (Inoki et al., 2003), the net effect of the TSC1/TSC2/TBC1D7 complex is reduced mTOR activation. mTOR, in turn, regulates a number of different cellular processes, including cell growth, protein synthesis, and metabolism (Sarbassov, Ali, & Sabatini, 2005) in response to upstream signals related to growth factors (Jewell & Guan, 2013), nutrient status (Jewell & Guan, 2013), and hypoxia (Brugarolas et al., 2004). The downstream effects of mTOR signaling include phosphorylation of a number of targets, like eukaryotic translation initiation factor 4E-binding protein 1 (eIF-4EBP-1) and S6 kinase (S6K1), both of which are involved in regulating different aspects of protein translation (Hay & Sonenberg, 2004). The net effect from loss-of-function mutations in TSC is disinhibition of RHEB activity and overactivation of the mTOR pathway.

In the CNS the effects of mTOR signaling are widespread (Jaworski & Sheng, 2006; Switon, Kotulska, Janusz-Kaminska, Zmorzynska, & Jaworski, 2017).

mTOR is crucial for regulating neuronal morphology, playing a role in neuronal polarity (Morita & Sobue, 2009), axonal guidance (Nie et al., 2010), dendritic arborization (Jaworski, Spangler, Seeburg, Hoogenraad, & Sheng, 2005; Kumar, Zhang, Swank, Kunz, & Wu, 2005; Urbanska, Gozdz, Swiech, & Jaworski, 2012), and dendritic spine formation (Kumar et al., 2005). Multiple lines of evidence have implicated mTOR signaling in synaptic plasticity, including long-term plasticity (LTP) (Tang et al., 2002), long-term depression (LTD) (Banko, Hou, Poulin, Sonenberg, & Klann, 2006), and memory formation (Banko et al., 2005).

C Animal Model Data

Several mouse models of TSC have recapitulated a wide range of phenotypes associated with the disorder. It is worthwhile to point out that there is not one model that captures every aspect of the TSC phenotype. As noted in the succeeding text, mice heterozygous for a mutant *Tsc1* or *Tsc2* allele—which would mirror the genetics seen in affected humans—do not demonstrate seizures or structural brain lesions, two common manifestations of TSC.

Mice with conditional knockout of *Tsc1* in neural stem/progenitor cells of the subventricular zone develop lesions that in some ways look similar to SENs or SEGAs (Zhou et al., 2011). Models involving the generation of cortical tubers are sparse, but one study involved mice with both a conditional *Tsc1* and a mutant *Tsc1* allele. The conditional allele was subject to knockout in certain progenitors at a specific point in time through in utero electroporation, resulting in offspring with structures resembling tubers. However, these structures did not demonstrate gliosis, as would be expected (Feliciano, Su, Lopez, Platel, & Bordey, 2011).

Spontaneous and provoked early-onset seizures occur along with delayed myelination and enlarged neurons in mice with neuron-specific *Tsc1* deletion (Meikle et al., 2007). Likewise, mice with astrocyte-specific *Tsc1* deletion demonstrate seizures and astrocytosis and neuronal disorganization in the hippocampus (Uhlmann et al., 2002). Treatment with rapamycin, an mTOR inhibitor, in this model not only helped prevent seizures but also rescued some of the neuropathological changes, including astrogliosis and neuronal disorganization (Zeng, Xu, Gutmann, & Wong, 2008).

Some TSC animal models demonstrate primarily cognitive deficits. For example, $Tsc2^{+/-}$ (Ehninger et al., 2008) and $Tsc1^{+/-}$ mice (Goorden, van Woerden, van der Weerd, Cheadle, & Elgersma, 2007) have impairments in learning, memory, and social behavior without seizures or brain lesions. In the former, treatment with rapamycin, an mTOR inhibitor, improved performance on learning tasks.

Recent studies have highlighted the role of Purkinje cell loss in TSC as it relates to social behavioral deficits consistent with autistic behaviors in humans. Loss of *Tsc1* in mouse cerebellar Purkinje cells leads to autistic-like behaviors including abnormal social interaction and repetitive behavior and vocalizations (Tsai et al., 2012). Loss of *Tsc1* also correlates with morphological changes in Purkinje cells. Similarly, the loss of *Tsc2* in murine Purkinje cells contributes to a behavioral phenotype characterized by autistic-like social deficits, including the lack of preference between a mouse and an inanimate object and the lack of preference between a familiar and a stranger mouse (Reith et al., 2013). These autistic behaviors and social functioning deficits improved with rapamycin therapy (Reith et al., 2013; Tsai et al., 2012), which has provided motivation for further translational trials.

D Therapeutic Strategies and Trials

The hypothesis that overactive mTOR signaling in TSC may be amenable to mTOR inhibitors has led to research studies involving the use of this class of medications for SEGAs. Examples of mTOR inhibitors under clinical investigation include rapamycin and everolimus, and others are in development (de de Vries, 2010). In one early, small cases series, patients with TSC and SEGAs who received oral rapamycin experienced regression in tumor growth (Franz et al., 2006). A randomized, placebo-controlled phase 3 trial (EXIST-1) in TSC patients showed that everolimus was effective in reducing SEGA volume (Franz et al., 2013), and a 2-year open-label extension of this study demonstrated continued efficacy (Franz et al., 2014).

Like SEGAs, epilepsy can cause significant neurological impairment in TSC. Over time the seizures associated with TSC can become refractory to traditional anticonvulsants. Furthermore, the development of epilepsy in TSC is not necessarily related to structural disruption from cortical tubers; rather, defective control of the mTOR pathway may play a role in epileptogenesis (Curatolo, 2015). Accordingly, in a multicenter, open-label, phase 1/2 clinical trial, individuals with TSC who had refractory epilepsy improved with respect to seizure control after receiving everolimus (Krueger et al., 2013). A follow-up phase 3 trial involving 366 TSC patients continued to show that adjunctive everolimus was effective for refractory focal epilepsy (French et al., 2016).

Cortical tubers, SEGAs, and epilepsy may not be sufficient to explain the cognitive, psychiatric, and behavioral characteristics of the disorder; rather, in TSC, overactive mTOR signaling may underlie neurocognitive deficits through several possible mechanisms. Disinhibited mTOR signaling can lead to increased neuronal protein synthesis that prematurely consolidates synaptic plasticity, affecting learning and memory. Moreover, in addition to upregulating various proteins, mTOR downregulates the translation of neuronal players such as Kv1.1, which could affect neuronal excitability, learning, and memory. Finally, neuronal plasticity depends on some of the upstream and downstream targets of mTOR signaling (Ehninger, de Vries, & Silva, 2009).

Thus the use of mTOR inhibitors has the potential of improving cognition and behavior in TSC, but relevant clinical data are still emerging. Case reports have described clinical improvement in autistic and other maladaptive behaviors in TSC patients treated with everolimus (Hwang et al., 2016; Ishii et al., 2015). During a phase 2, nonrandomized, open-label trial to detect safety and efficacy of serolimus for renal angiomyolipoma in adults with TSC or lymphangioleiomyomatosis, seven out of eight TSC patients who underwent neurocognitive testing demonstrated improvement in immediate recall memory scores after treatment (Davies et al., 2011). In children a few clinical trials have attempted to evaluate the effects of mTOR inhibition on neurocognition. One such trial (ClinicalTrials.gov NCT01289912) was a randomized, double-blind, placebo-controlled phase 2 trial using everolimus in individuals with TSC ages 6–21 years. The study of children with baseline IQ $\geq$ 60 (and thus capable of completing comprehensive neuropsychological testing), showed trends of improvement in some standardized parent ratings of behavior (such as Social Cognition on the Social Responsiveness Scale) following everolimus treatment for 6 months. However, most neuropsychological measures failed to demonstrate a significant improvement when compared to placebo (PMID: 29296616). There are two other similarly designed trials, one for TSC patients 4–15 years old (ClinicalTrials.gov NCT01730209) and the other for TSC patients 16–60 years old (ClinicalTrials.gov NCT01954693). Both are investigating the effects of everolimus on outcome measures such as IQ, autism ratings, working memory and attention, visual motor integration, and executive functioning.

Certain caveats are noteworthy about the use of mTOR inhibitors in TSC clinical trials for neurodevelopmental symptoms. Currently, there are no clinical trials involving this class of medications for young infants with TSC (Capal & Franz, 2016). During the first 2 years of life, there is a period of rapid growth in postnatal brain development (Knickmeyer et al., 2008). Therefore, in theory, a disease mechanism–based intervention that is supposed to improve neurodevelopmental outcomes may be more effective if administered relatively early in life, though this construct, as it relates to TSC specifically, requires validation. Other questions remain. For example, it is unclear for how long a patient must receive everolimus or another mTOR inhibitor to achieve permanent neurocognitive benefit. It is not yet clear if there is a critical window during which mTOR inhibitor treatment will be effective. It will be important to predict treatment response stratified by a patient's baseline clinical severity and to continually redefine treatment response to include MRI, electrophysiological, and other novel biomarkers (de de Vries, 2010). As with disorders like RTT, researchers are developing TSC patient-derived iPSC lines differentiated into neurons to refine understanding of disorder-specific molecular characteristics that can serve as additional targets of therapy (Ebrahimi-Fakhari & Sahin, 2015). Initial studies using stem cell–derived neuronal cultures have been published (Ercan et al., 2017; Ebrahimi-Fakhari et al., 2016; Costa et al., 2016). Such iPSC lines will also be useful for large drug screens including, but not limited to, mTOR inhibitors.

V FRAGILE X SYNDROME

A Overview

FXS is an X-linked, trinucleotide repeat expansion disorder involving fragile X mental retardation 1 (*FMR1*), which results in a range of symptoms depending on an affected individual's mutation state (full mutation versus premutation carrier defined by the number of trinucleotide repeats), degree of *FMR1* methylation, and gender (Turner, Webb, Wake, & Robinson, 1996). Full mutation FXS has an estimated prevalence between 1:4000 and 1:7000 in males and two-thirds to one-half of these numbers in females (Hunter et al., 2014). FXS is widely recognized as the most common cause of inherited ID—primarily manifesting in males with the full mutation and clinically identified by phenotypic characteristics such as craniofacial dysmorphisms (long and narrow face, large ears, and macrocephaly) and autistic behaviors (reduced eye contact, limited communication, and stereotypical behaviors) (Tsai, Pickler, Tartaglia, & Hagerman, 2009). However, premutation carriers of any gender may also be affected with a range of symptoms including primary ovarian insufficiency (POI) in women, fragile X-associated tremor-ataxia syndrome (FXTAS) later in life, and a range of cognitive, behavioral, and emotional concerns that are under increasing scrutiny (Hagerman et al., 2009). Although cognitive abilities are uniformly impacted in full mutation males, the behavioral phenotype can be quite variable, with presentations ranging from symptoms of anxiety to impulsivity or social avoidance (autistic features) (McConkie-Rosell et al., 2005). Boys with full mutation FXS have an increased prevalence of seizures (10%–20%), which is lower (5%) in affected girls with the full mutation (Hagerman et al., 2009). Seizures may be of many types, but they are typically not severe, easily managed with monotherapy and often resolved by adulthood (Hagerman et al., 2009).

It is worth highlighting that given the X-linked nature of FXS, phenotypic presentations vary broadly based on gender and across individuals. For example, males with full mutation FXS are typically the most impacted with cognitive, adaptive, and communication limitations (generally in the ID range) and a range of behavioral and emotional comorbidities including anxiety, impulsivity, hyperactivity, and autistic features (Tsai et al., 2009). With increasing age, clinical features typically evolve: hyperactivity generally improves with age, and anxiety symptoms may increase with age. In contrast the phenotype of girls with full mutation FXS is much more variable given individual differences in X chromosome inactivation (and therefore the state of the *FMR1*). As a result, approximately 50% of girls with the full mutation have typical cognitive abilities, with the remainder presenting with a range of learning, emotional, and behavioral profiles (Hagerman et al., 2009).

B Gene Function

FXS is a trinucleotide repeat disorder impacting *FMR1* with downstream effects on fragile X mental retardation protein (FMRP) (Rooms & Kooy, 2011; Hagerman et al., 2009). In more than 99% of cases, the loss of function of *FMR1* is related to unstable expansion of a trinucleotide (cytosine-guanine-guanine; CGG) repeat in the 5'untranslated region of the gene. In addition, deletions, point mutations, and missense mutations in *FMR1* may also contribute to FXS symptoms (Hagerman et al., 2009). Clinical symptoms in full mutation or premutation carriers may be related to either reduced level of FMRP or increased *FMR1* mRNA levels. While classic FXS features are purportedly related to the absence of functional FMRP, symptoms in carriers may result from increased levels of *FMR1* mRNA due to increased transcription coupled with decreased FMRP production (Lozano, Rosero, & Hagerman, 2014).

In general the number of CGG repeats in *FMR1* influences clinical status. A typical number of CGG repeats in the gene in unaffected individuals are <50, while premutation states result from 50 to 200 repeats, and full mutations are due to >200 repeats. Although clinical outcomes have been described based on trinucleotide repeat status, significant heterogeneity within each mutation status (normal, premutation, and full mutation) is the norm. With premutation carriers, *FMR1* remains transcriptionally active with an increase in mRNA production, leading in turn to RNA gain-of-function toxicity. With full mutations, methylation coupling typically leads to silencing of *FMR1* and the absence of FMRP, resulting in full

mutation phenotypes that are correlated with gender. In FXS, as with many NDDs, despite an individual's genetic profile as assessed by peripheral blood analysis of mutation status, phenotypic heterogeneity is the norm and perhaps related in part to differential FMR1 protein expression in neurological tissues as opposed to peripheral blood-derived cells (Hagerman et al., 2009; Lozano et al., 2014).

FMRP, an mRNA-binding protein, has multiple important cellular roles, particularly in regulation of protein synthesis in the CNS. FMRP inhibits dendritic protein translation in response to synaptic activation by group 1 metabotropic glutamate receptors (mGluR1 and mGluR5), muscarinic (M1) acetylcholine receptors, and probably multiple Gq-linked receptors (Berry-Kravis, Knox, & Hervey, 2011; Huber, Gallagher, Warren, & Bear, 2002). These receptors activate signaling pathways, which lead to the loss of FMRP repressor function at the ribosome and a resultant induction of de novo protein synthesis. FMRP mediates the translation of multiple synaptic proteins (including striatal-enriched protein tyrosine phosphatase [STEP] and Arc), which are linked to AMPA receptor internalization (Deng et al., 2013). Based on mouse data, it is hypothesized that FMRP regulates the activity of ion channels (big potassium [BK] and slack) through direct protein-protein interactions (Berry-Kravis, 2014). Additional evidence in mice suggests that FMRP's regulatory functions appear to be critical for synaptic maturation and strength, as the absence of FMRP leads to elevated levels of synaptic proteins controlled by FMRP and the following: immature elongated dendritic spines (Grossman, Aldridge, Weiler, & Greenough, 2006); abnormal spine density; abnormal synaptic plasticity including enhanced mGluR-activated hippocampal and cerebellar LTD and impaired LTP in the hippocampus, cortex, and amygdala; and abnormal epileptiform discharges (Berry-Kravis, 2014; Berry-Kravis et al., 2011).

C Animal Model Data

Our vast understanding of the neurobiology and synaptic mechanisms underlying absent FMRP in FXS has paved the way toward future targeted treatments in FXS via extensive animal experience in both mice and fruit flies. The *Fmr1* knockout mouse model of FXS (which leads to animals with no functional FMRP) has led to understanding of the role of FMRP in neurons, helped to identify cellular targets for potential treatment, and provided a platform to explore potential treatments for FXS. Similar modeling with the *dfxr* fly has provided parallel evidence to support our current understanding of the mechanisms and provided insight into potential targeted treatments of FXS. Although levels of FMRP in animal models of FXS mirror levels in humans with the disorder, for the most part, such animal models are not associated with robust behavioral phenotypes that allow for comparison between mouse and human behavior—limiting the translatability of animal behavioral measures with respect to human outcome measures.

Largely as a result of the abnormalities noted in the mouse model of FXS, a variety of potential treatment targets have been identified and evaluated preclinically. Treatment has been successful in reversing abnormal phenotypes in the *Fmr1* knockout mouse and the *dfxr* fly via six different pathways (Table 22.1). The most extensive and promising animal work has examined the use of mGluR5 negative allosteric modulators (NAMs) and genetic receptor reduction to reduce excessive translational pathway signaling through mGluR5 receptors, as FXS phenotypes were largely reversible even in adult animals (Dölen et al., 2007).

Reports of success with mGluR5 NAMs to reduce mGluR5 receptor activity in the knockout mouse model have resulted in numerous publications documenting correction of abnormalities in dendritic spine morphology, synaptic plasticity (LTP and LTD), motor learning, open-field hyperactivity, marble burying, social behavior, prepulse inhibition (PPI), epileptiform bursts, and audiogenic seizures. While some adult phenotypes of FXS animal models are amenable to rescue, there is some evidence to suggest that treatments are most effective in young animals (Michalon et al., 2012; Meredith, de Jong, & Mansvelder, 2011; Su et al., 2011).

D Therapeutic Strategies and Trials

Building on a solid foundation of preclinical and animal work in FXS, investigators have conducted a number of clinical trials in individuals with FXS. However,

TABLE 22.1 Target Pathways for Potential Clinical Development That Have Reversed FXS Phenotypes in Preclinical Models (*Fmr1* Knockout Mouse and *dfxr* fly) (Bilousova et al., 2009; Henderson et al., 2012; Yuskaitis et al., 2010; McBride et al., 2005)

1. Reduction of excess activity in signal transduction pathways connecting group 1 mGluRs or other Gq-linked receptors to the dendritic translational machinery

2. Reduction of excessive activity of proteins normally regulated by FMRP

3. Augmentation of expression and activation of surface AMPA receptors

4. Modification of activity of GABA and other receptors/proteins regulating glutamate signaling or translational signaling pathways

5. Use of miRNAs to block excessive translation of mRNAs typically regulated by FMRP

6. Correction of abnormal ion channel activities typically directly regulated by FMRP

despite the tremendous understanding of the basic neuroscience of synaptic dysfunction and the significant success of treatment of FXS animal models (both mice and fruit flies), treatment of humans with FXS has not been as successful to date. It is worth noting that some of the outcomes in animals (e.g., audiogenic seizures) have very little relevance to human outcomes, as humans do not have audiogenic seizures. Several examples of trials in humans with FXS along with lessons learned are outlined in the succeeding text.

1 mGluR5 Antagonists

Given significant evidence that FXS pathophysiology involves upregulation of mGluR5-mediated processes (Bear, Huber, & Warren, 2004) and findings in animal models suggesting the correlation between *Fmr1* activity and mGluR5 expression (Dölen et al., 2007), a number of clinical translation efforts have emerged testing mGluR5 antagonists.

An initial open-label, single-dose trial of fenobam (an mGluR5 antagonist) in 12 adults with FXS (6 females and 6 males) revealed that at least half of patients showed reversal of abnormal sensorimotor gating, with most showing clinical improvement on reports of hyperactivity and anxiety (Berry-Kravis et al., 2009).

A subsequent double-blind, placebo-controlled crossover trial of AFQ056 (mavoglurant) in 30 adult males with FXS revealed no significant improvements in primary and secondary endpoints (Jacquemont et al., 2011). However, a small subgroup of participants with fully methylated *FMR1* promoter regions demonstrated significant behavioral improvements as demonstrated by the Aberrant Behavior Checklist-Community Edition (ABC-C), specifically in the areas of stereotypic behaviors, hyperactivity, and inappropriate speech. This finding encouraged subsequent larger, multinational, double-blind, placebo-controlled, and parallel group trials of mavoglurant (phase 2 in adults and phase 3 in adolescents), which failed to show a significant improvement on any test measures regardless of patient's dose or methylation status. However, many families reported an improvement in behavior and cognition during follow-up visits in an open-label extension, but these changes were not captured via outcome measures (Berry-Kravis et al., 2016). It has also been hypothesized that expecting cognitive outcomes to change in adolescents and adults over a several month clinical trial period may not be a reasonable outcome, given limited cognitive plasticity by this in time.

As a result, driven by lessons learned from previous trials with mGluR5 antagonists, a multicenter controlled trial of mavoglurant for children with FXS ages 2½–6 years is ongoing (Clinicaltrials.gov: NCT02920892). This trial will examine outcome measures, including language, for all participants after an intensive learning program provided to all participants and psychopharmacologic intervention provided only to some (Ligsay & Hagerman, 2016). This trial was designed specifically to target children of younger ages, as they are expected to have more neuroplasticity and therefore more potential to demonstrate change with cognitive and behavioral outcomes.

2 Minocycline

Minocycline is a tetracycline antibiotic that has been considered as a potential treatment for FXS given its hypothesized downstream effects on the mGluR5 and GABA pathways impacted by FMRP. In full mutation FXS individuals, decreased FMRP levels lead to upregulation of the protein matrix metalloproteinase-9 (MMP-9), which is associated with immature dendritic spine morphology (Bilousova et al., 2009). Initial open-label studies of minocycline in both children and adults suggested a positive impact on language, attention, hyperactivity, and overall status (Utari et al., 2010). A subsequent randomized, crossover trial demonstrated modest mood and anxiety improvements with minimal side effects, but ongoing studies are warranted to examine potential positive effects of this medication (Ligsay & Hagerman, 2016).

3 Lovastatin

As noted previously, lovastatin is an HMG-CoA reductase inhibitor used to treat hyperlipidemia and hypercholesterolemia; it also inhibits RAS-MAPK-ERK1/2 activation (Xu, McGuire, Blaskovich, Sebti, & Romero, 1996). Many of the proteins upregulated in FXS are thought to be downstream consequences of increased ERK1/2 activity. Trials of lovastatin in FXS mouse models have suggested improvements in cellular functioning paralleling a reduction in audiogenic seizures (Osterweil, Krueger, Reinhold, & Bear, 2010). Based on this murine data, a single 12-week open-label trial of lovastatin in patients with FXS (n = 15/16 subjects completing the study) suggested significant behavioral benefit based on parental report (e.g., ABC scores) (Çaku, Pellerin, Bouvier, Riou, & Corbin, 2014). Additional studies examining lovastatin paired with parent-implemented language intervention and minocycline compared with lovastatin are currently underway.

VI CHALLENGES ASSOCIATED WITH TRANSITIONING TO HUMAN CLINICAL TRIALS

A Selecting the Right Patients

Clinical trial design must take into account not only an understanding of underlying cellular mechanisms of a specific NDD, associated potential targets, and correct pharmacological treatment with the right medication but also the appropriate patients to evaluate and treat. Heterogeneity of clinical populations with a specific NDD (defined either genetically or behaviorally) is common, and clinical trials often require controlling for that variability to some extent through inclusion and exclusion criteria. Although many variables must be considered for design and participation in all clinical trials, common considerations of critical importance for individuals with NDDs are related to the known underlying etiology of each disorder. In addition to disease-specific inclusion and exclusion criteria, general considerations for trials with individuals with NDDs include age, developmental status (including cognitive and language level), and gender. We will discuss each of these criteria in detail in the succeeding text.

1 Age

When considering individuals with NDDs, both research and clinical experience suggests that intervention as soon as possible will allow for the most optimal outcomes over the life span (Rogers & Vismara, 2008). In contrast a long history of clinical research ethics has contributed to a culture of patient protection in research studies, which makes research involving adults (individuals who can themselves consent to research involvement) routinely more feasible than research involving vulnerable populations including children and individuals with ID. This dichotomy has contributed to a culture of clinical research often starting with adult patients and moving into pediatric populations based on safety considerations rather than beginning with those who may be most likely to benefit from or show a response to an intervention. Current efforts in the field are now targeting younger children who may have greater potential for modification of their disease trajectory and improved outcome over time (Damiano, Mazefsky, White, & Dichter, 2014).

2 Developmental Status

Individuals with an NDD have a range of cognitive, language, and adaptive capabilities, and many present with significant challenges across any or all of these domains. Evaluating baseline status and outcome measures requires assessment with measures that may not span the full capacity of individual profiles or capture small changes over time. For example, individuals may demonstrate cognitive improvement or decline between baseline and subsequent cognitive assessments, but cognitive measures may not be sensitive to these changes for individuals with extremely low cognitive skills. Similarly, capturing language or adaptive changes that are meaningful can be challenging for relatively low functioning individuals (Campbell, Brown, Cavanagh, Vess, & Segall, 2008).

3 Gender

Consideration of both participants and clinical outcomes in all translational trials needs to respect the impact of gender-related effects, but this is particularly important when examining outcomes for individuals with NDDs. First the prevalence of specific NDDs in some disorders may be extremely gender skewed. For example, RTT is far more prevalent in females, FXS is far more prevalent in males, and TSC is gender-neutral. Second the profile of various genetic disorders may be extremely variable based on gender and X chromosome-related gene inactivation (e.g., clinical profiles of full mutation males versus premutation females in FXS). As a result, translational trial planning needs to consider the advantages and disadvantages of both homogeneity and heterogeneity of participants in clinical trials. Some translational efforts focused on individuals with the greatest challenges (e.g., seizures in TSC) may have the most optimal response to intervention, while in other trials, individuals with the most modest impact from the disorder may demonstrate greater response to intervention (e.g., level of cognition in FXS). Finally, there is male predominance for most NDDs, including ASD, ADHD, and a range of other learning- and language-based disorders. This factor must be considered when there is inclusion of both genders in NDD trials (Boyle et al., 2011).

B Accounting for a Strong Placebo Response

The placebo response in a clinical trial can be quite powerful, enough to weaken, by comparison, the treatment advantage (Walach, Sadaghiani, Dehm, & Bierman, 2005). Clinical trials in psychiatry, for example, commonly have high associated placebo response rates (Fava, Evins, Dorer, & Schoenfeld, 2003), such as 35%–40% for antidepressant trials (Furukawa et al., 2016). Among studies of NDDs, one metaanalysis of randomized controlled trials (RCTs) enrolling patients with ID syndromes found a significant placebo response for both subjective and objective outcome measures, especially for the younger individuals (Curie et al., 2015). Similarly a metaanalysis of RCTs involving medication and dietary interventions in pediatric ASD demonstrated a moderate placebo response (Masi, Lampit, Glozier, Hickie, & Guastella, 2015).

Two factors that may mediate a placebo response are expectation of benefit from a therapeutic intervention and levels of psychological distress surrounding a condition (Brown, 2015; Kradin, 2011). In clinical trials enrolling children with special needs, parents/caretakers may be especially hopeful for a cure or treatment, possibly dissatisfied with responses to existing therapies. In addition, they often have higher levels of stress compared with other parents with typically developing children (Craig et al., 2016). Therefore it is reasonable to imagine that a strong desire for clinical benefit, coupled with potential relief of parental stress, may be enough to drive a robust placebo response in NDD trials.

This notion—that NDD clinical trials may be subject to particularly robust placebo responses due in part to some of the circumstances of raising a child with special needs—argues further for the use of more objective outcome measures in these trials. In contrast to evaluations that rely purely on parental report, biomarkers or other quantitative assessments that closely reflect disease pathophysiology specifically targeted by the medication may be less susceptible to these kinds of placebo effects (Jeste & Geschwind, 2016; Sahin & Sur, 2015). Of course, placebo neurobiology can be quite complex (Kaptchuk & Miller, 2015; Finniss, Kaptchuk, Miller, & Benedetti, 2010), so this underlying premise deserves further rigorous scientific investigation.

C Choosing the Right Outcome Measures

Choosing the right outcome measure for a clinical trial is critical to its success, especially when the focus is an NDD with complex cognitive and behavioral manifestations, which often evolve with age. The ideal outcome measure will be clinically significant, sensitive to change over the time of a clinical trial, and acceptable for the purpose of registration of a compound for use in clinical practice. Experience to date suggests a need for increased attention to the development and selection of outcome measures in clinical trials. Commonly selected outcome measures in clinical trials for NDDs often fall under one of three categories—cognition, behavior, and biomarkers related to disease status—each of which warrants specific considerations.

1 Cognition

In NDDs, one of the desired goals for interventional clinical trials is to improve cognition—a very complex concept incorporating both genetic and environmental factors and with features evolving over time as children develop into adolescents and then adults. In addition, cognition as defined in animal models preclinically is often challenging to translate into an analogous human measure. Conceptually, cognition is appealing as an outcome measure, because it is both objectively measurable as standardized scores by a trained examiner and meaningful in real-world contexts. However, delineating this outcome with a single cognitive measure has certain disadvantages. First the degree of cognitive improvement expected from clinical trials, as assessed by a single measure over the relatively brief duration of most clinical trials, may be modest or even immeasurable. Second, disease-specific pharmacotherapies often target key pathways implicated in the disorder (e.g., the mGluR5 pathway in FXS), and because these pathways may impact some, but not all, aspects of cognition and behavior, a single cognitive measure may not capture the full extent of impact of a targeted therapy on specific cognitive subsystems. Third, psychometric properties of one chosen instrument may not be valid for all age groups (Berry-Kravis et al., 2013), and therefore careful attention is required when comparing results across studies. Fourth, animal models of NDDs may recapitulate only a subset of the features of the disorder, and as a result, treatments targeting cognition with success in the animal model may not necessarily have the same results in affected humans (Cope, Powell, & Young, 2016).

2 Behavior

For many individuals with NDDs, behavioral change in the real world is an extremely meaningful outcome and therefore an area that has been often targeted with translational intervention trials. Behavioral outcomes are typically described based on an observer's report of an individual's "behavior" or as characterized via the administration of a highly standardized but potentially artificial direct assessment. In addition, in all settings, measure of a behavior or behavioral change is related to a range of factors including but not limited to an individual's underlying capacity to perform a behavior, opportunity to demonstrate behavior, and potentially a differential response to an intervention. While the use of many behavioral measures may be beneficial when considering outcome measure, it is critical to consider the potential challenges in using these measures for individuals with NDDs in a translational trial setting (Berry-Kravis et al., 2013).

Parents and other observers often provide subjective reports of an individual's behaviors (e.g., via the ABC or Vineland Adaptive Behavior Scale [VABS]) over a finite period of time as part of the standard of care related to clinical diagnosis and treatment of NDDs. Given that reports of behaviors may be considered a proxy for both underlying biological determinants of behavior and environmentally induced behavioral changes, they have also been adopted as potential outcomes in trials with NDDs. When considering observer-based outcomes of reports via behavioral measures, it is critical to consider the range of behavioral changes that are measurable, the timeframe over which meaningful change can be measured, and the potential for a proxy placebo response.

Specialized observation and documentation of behaviors in a highly standardized setting (e.g., Autism Diagnostic Observation Schedule [ADOS]) may be useful in describing behavioral profiles and outcomes. However, an individual's behavioral response may be highly determined by other factors, including comfort and anxiety level in a particular testing setting, comfort with the examiner, and distraction by other environmental factors. When using behavioral assessments as an outcome measure, careful consideration of both the behavioral construct and the assessment setting is mandatory.

3 Biomarkers

The identification of biomarkers allowing for assessment of meaningful change following interventions (pharmacological or behavioral) both preclinically and clinically continues to be an evasive target with many translational trials examining individuals with NDDs. Examination of biomarkers in NDD translational trials has included consideration of both disorder-specific contexts (e.g., FMRP levels suggestive of intervention response at a cellular level) and global contexts (e.g., functional MRI changes consistent with observer-reported functional outcomes). In contrast to many disorders with well-characterized biological determinants of disease (e.g., many oncological disorders), NDDs are defined clinically more often than biologically. As a result, examination of appropriate biomarkers for use in NDD translational efforts in heterogeneous populations is still under a great deal of investigation.

Ideally, trials with interventions for NDDs would examine outcomes proximate to the brain area impacted. However, obtaining histologic samples (e.g., brain biopsies) or even biological samples close to the CNS (e.g., cerebrospinal fluid) can be challenging, and doing so is not currently the standard of clinical care. Electrophysiological biomarkers including electroencephalography (EEG) and event-related potentials (ERPs), which have the advantage of directly measuring brain activity, hold promise with respect to translational studies in patients with NDDs such as ADHD and ASD (Modi & Sahin, 2017). However, there is need for an improved mechanistic understanding of electrophysiological measures in these disorders and approaches that accommodate diagnostic heterogeneity and individual differences rather than attempting to focus on common traits within a population (Jeste, Frohlich, & Loo, 2015).

VII OPPORTUNITIES

Some of the disadvantages that come with translational research for NDDs open the window to certain opportunities.

A Collaborations

In clinical trials for NDDs, especially those involving rare neurogenetic diseases, adequate recruitment of interested and appropriate individuals at any single site can be a significant challenge; however, this setback can serve as an opportunity for creating alliances between patients, clinicians, and researchers across a variety of geographic sites, united by a shared interest in a specific disorder, in addition to paving the pathway for alternative clinical trial models. The ideal design for a clinical trial may involve narrow inclusion criteria for various reasons, such as the desire to study clinically or genetically homogenous populations, or the need to investigate the effects of an intervention early on in the disease course. Unfortunately, placing stringent restrictions on eligibility may prevent sufficient enrollment when the overall prevalence of a disorder is low. Furthermore, families of affected individuals are often geographically scattered, and traveling to research centers can be cumbersome, not only because of distance but also due to some of the challenges of transporting a child who may have significant maladaptive behaviors and physical limitations.

In light of these obstacles, one of the ways to increase participation is to establish coalitions between patient-related organizations (including but not limited to disease foundations, support groups, and patient registries) and clinical trial researchers. These connections foster national and international collaborations, linking together people who may not have interfaced previously and providing a spark for future research ideas. At the same time, they provide feedback about which outcome measures deserve emphasis for study in clinical trials, given that the overall patient pool may be limited and not every patient can or will participate in repeated trials (Augustine, Adams, & Mink, 2013).

B Natural History Studies

One particular type of clinical trial for NDDs that greatly benefits from increased recruitment is a natural history study. Natural history studies are essential for many reasons. They elucidate the length of time needed to establish whether an intervention is effective for a particular outcome measure. They also highlight certain behavioral, cognitive, and neurophysiological biomarkers, which may evolve with age. Not only do these biomarkers serve as an objective measure of treatment response, but also they help predict disease severity and clinical outcomes, such as whether a child will develop ASD. Although the National Institute of Mental Health has proposed using the Research Domain Criteria (RDoC) to classify mental disorders, most clinical trials with NDD populations continue to focus on clinically observable or traditional biochemical outcomes. However, incorporation of RDoC criteria in further natural history studies may allow for a paradigm shift in future translational trials. Further research examining

cellular, physiological, neuroimaging, and behavioral features of NDDs correlated with clinical outcomes is clearly needed. In essence the design of an interventional trial for an NDD relies on data generated from high-quality natural history studies, which often depend on the concerted efforts of multiple research sites, sometimes spread across different countries, working together with patient advocacy organizations (Augustine et al., 2013).

C Novel Trial Designs

Recently, novel clinical trial designs have started to emerge in the field of NDDs, offering a unique set of advantages. In 2014, researchers published their findings from an Internet-based randomized, controlled clinical trial testing the efficacy of omega-3 fatty acids for hyperactivity in ASD. Nearly 60 children from 28 states enrolled in the study, which completed in 3 months. Outcome measurements, obtained from both parents and teachers, were complete for all participants. While the study did not show improvement in outcomes with the use of the therapy, it did demonstrate the feasibility of conducting an exclusively online investigational trial. The benefits of this approach are numerous: quick study completion, low overhead costs, high rates of enrollment, and the lack of need for travel adding to overall convenience (Bent et al., 2014). However, it is worthwhile to note that the medication studied, omega-3 fatty acid, is a benign supplement, which is not well regulated and may be provided in different concentrations by different manufacturers, and there can be mounting logistical challenges with the use of a pharmacological intervention that is not over the counter. Limitations aside, this kind of online platform for translational therapeutics holds promise for a wide range of NDDs given the clear advantages for participant engagement.

Another novel clinical trial design, at least in the realm of NDDs, is the simultaneous use of pharmacological and behavioral intervention. Such an approach has existed in adults for treatment of various medical issues, including obesity (Phelan & Wadden, 2002), smoking cessation (Tseng et al., 2016), and eating disorders (Grilo, Reas, & Mitchell, 2016). In children, including those with NDDs, this approach parallels clinical practice. For example, it is common for a child with ADHD to receive behavioral support, in the form of direct therapy and school accommodations, combined with stimulants or other medications. In cases of severe behavioral disturbance, an effective treatment program may in fact *depend* on both psychopharmacology plus intensive behavioral therapy, as demonstrated by one adolescent with TSC (Gipson et al., 2013). Therefore evaluation of clinical improvement must take into account pharmacological and nonpharmacological interventions. Recognizing this concept, researchers are developing clinical trial designs involving multimodal treatments for NDDs. One such investigation, a randomized, double-blind trial, combines a parent-implemented learning intervention with lovastatin or placebo for FXS (ClinicalTrials. gov: NCT02642653). Other similarly designed trials are on the horizon for NDDs.

Increasing use of smartphones and wearable electronic devices by the general public has created opportunities for health tracking that could be useful in clinical trials for NDDs. For example, biometric data collected from electronic smart watches (such as heart rate or accelerometer data) could enable seizure detection in children with epilepsy or quantification of repetitive movements in children with ASD. Furthermore, applications developed for smartphones could facilitate real-time data entry for parents characterizing maladaptive behaviors in their children, which may be an improvement over retrospective recollections. These are but a few of the possible creative applications of technology that will continue to push the boundaries of clinical trial design and implementation for the NDD population.

VIII CONCLUSIONS

There are ample promises and equally frequent challenges when considering translational medicine strategies for NDDs. Based upon preclinical data and animal data, there is increased understanding of the pathophysiology underlying neurological, cognitive, and behavioral deficits seen with different genetic disorders associated with NDDs; and these models provide hope that we can extrapolate these findings to NDDs more generally in the future. This improved understanding has opened the door to clinical trials involving therapeutics that target disrupted signaling pathways and molecular mechanisms known to contribute to NDDs. However, sometimes, there can be a translational gap—a mismatch between preclinical data and what can be demonstrated and evaluated in human clinical trials. Performing a clinical trial in a population with as many challenges as the NDD population is difficult enough; taking into account, this translational gap adds a further layer of complexity. Going forward, perhaps one of the most helpful strategies will be to design *preclinical* studies with *clinical* studies in mind. This may entail, for example, picking objective measures in an animal model that have tight corollaries, if not exact replicas, in humans. The future of this field is promising, though, because as we perform more and more clinical trials in the NDD population, we will continue to learn lessons that will provide feedback to further the translational research efforts for individuals impacted by NDDs.

References

Aartsma-Rus, A., & van Ommen, G. -J. B. (2007). Antisense-mediated exon skipping: a versatile tool with therapeutic and research applications. *RNA, 13*, 1609–1624.

Abdala, A. P., Lioy, D. T., Garg, S. K., Knopp, S. J., Paton, J. F. R., & Bissonnette, J. M. (2014). Effect of Sarizotan, a 5-HT1a and D2-like receptor agonist, on respiration in three mouse models of Rett syndrome. *American Journal of Respiratory Cell and Molecular Biology, 50*, 1031–1039.

American Psychiatric Association, American Psychiatric Association, DSM-5 Task Force. (2013). *Diagnostic and statistical manual of mental disorders: DSM-5.* Arlington, VA: American Psychiatric Association.

Augustine, E. F., Adams, H. R., & Mink, J. W. (2013). Clinical trials in rare disease: challenges and opportunities. *Journal of Child Neurology, 28*, 1142–1150.

Balachandar, V., Dhivya, V., Gomathi, M., Mohanadevi, S., Venkatesh, B., & Geetha, B. (2016). A review of Rett syndrome (RTT) with induced pluripotent stem cells. *Stem Cell Investigation, 3*, 52.

Ballas, N., Lioy, D. T., Grunseich, C., & Mandel, G. (2009). Non-cell autonomous influence of MeCP2-deficient glia on neuronal dendritic morphology. *Nature Neuroscience, 12*, 311–317.

Balmer, D., Goldstine, J., Rao, Y. M., & LaSalle, J. M. (2003). Elevated methyl-CpG-binding protein 2 expression is acquired during postnatal human brain development and is correlated with alternative polyadenylation. *Journal of Molecular Medicine, 81*, 61–68.

Banerjee, A., Castro, J., & Sur, M. (2012). Rett syndrome: genes, synapses, circuits, and therapeutics. *Frontiers in Psychiatry, 3*, 34.

Banko, J. L., Hou, L., Poulin, F., Sonenberg, N., & Klann, E. (2006). Regulation of eukaryotic initiation factor 4E by converging signaling pathways during metabotropic glutamate receptor-dependent long-term depression. *The Journal of Neuroscience, 26*, 2167–2173.

Banko, J. L., Poulin, F., Hou, L., DeMaria, C. T., Sonenberg, N., & Klann, E. (2005). The translation repressor 4E-BP2 is critical for eIF4F complex formation, synaptic plasticity, and memory in the hippocampus. *The Journal of Neuroscience, 25*, 9581–9590.

Bear, M. F., Huber, K. M., & Warren, S. T. (2004). The mGluR theory of fragile X mental retardation. *Trends in Neurosciences, 27*, 370–377.

Belichenko, P. V., Wright, E. E., Belichenko, N. P., Masliah, E., Li, H. H., Mobley, W. C., et al. (2009). Widespread changes in dendritic and axonal morphology in Mecp2-mutant mouse models of Rett syndrome: evidence for disruption of neuronal networks. *The Journal of Comparative Neurology, 514*, 240–258.

Bent, S., Hendren, R. L., Zandi, T., Law, K., Choi, J. -E., Widjaja, F., et al. (2014). Internet-based, randomized, controlled trial of omega-3 fatty acids for hyperactivity in Autism. *Journal of the American Academy of Child & Adolescent Psychiatry, 53*, 658–666.

Berry-Kravis, E. (2014). Mechanism-based treatments in neurodevelopmental disorders: fragile X syndrome. *Pediatric Neurology, 50*, 297–302.

Berry-Kravis, E., Des Portes, V., Hagerman, R., Jacquemont, S., Charles, P., Visootsak, J., et al. (2016). Mavoglurant in fragile X syndrome: results of two randomized, double-blind, placebo-controlled trials. *Science Translational Medicine, 8*, 321ra5.

Berry-Kravis, E., Hessl, D., Abbeduto, L., Reiss, A. L., Beckel-Mitchener, A., & Urv, T. K. (2013). Outcome measures for clinical trials in fragile X syndrome. *Journal of Developmental and Behavioral Pediatrics, 34*, 508–522.

Berry-Kravis, E., Hessl, D., Coffey, S., Hervey, C., Schneider, A., Yuhas, J., et al. (2009). A pilot open label, single dose trial of fenobam in adults with fragile X syndrome. *Journal of Medical Genetics, 46*, 266–271.

Berry-Kravis, E., Knox, A., & Hervey, C. (2011). Targeted treatments for fragile X syndrome. *Journal of Neurodevelopmental Disorders, 3*, 193–210.

Bilousova, T. V., Dansie, L., Ngo, M., Aye, J., Charles, J. R., Ethell, D. W., et al. (2009). Minocycline promotes dendritic spine maturation and improves behavioural performance in the fragile X mouse model. *Journal of Medical Genetics, 46,* 94–102.

Boggio, E. M., Lonetti, G., Pizzorusso, T., & Giustetto, M. (2010). Synaptic determinants of rett syndrome. *Frontiers in Synaptic Neuroscience, 2,* 28.

Bousman, C. A., & Hopwood, M. (2016). Commercial pharmacogenetic-based decision-support tools in psychiatry. *Lancet Psychiatry, 3,* 585–590.

Boyle, C. A., Boulet, S., Schieve, L. A., Cohen, R. A., Blumberg, S. J., Yeargin-Allsopp, M., et al. (2011). Trends in the prevalence of developmental disabilities in US children, 1997-2008. *Pediatrics, 127,* 1034–1042.

Brendel, C., Belakhov, V., Werner, H., Wegener, E., Gärtner, J., Nudelman, I., et al. (2011). Readthrough of nonsense mutations in Rett syndrome: evaluation of novel aminoglycosides and generation of a new mouse model. *Journal of Molecular Medicine (Berlin, Germany), 89,* 389–398.

Brown, W. A. (2015). Expectation, the placebo effect and the response to treatment. *R.I. Medical Journal (2013), 98,* 19–21.

Brugarolas, J., Lei, K., Hurley, R. L., Manning, B. D., Reiling, J. H., Hafen, E., et al. (2004). Regulation of mTOR function in response to hypoxia by REDD1 and the TSC1/TSC2 tumor suppressor complex. *Genes & Development, 18,* 2893–2904.

Buchovecky, C. M., Turley, S. D., Brown, H. M., Kyle, S. M., McDonald, J. G., Liu, B., et al. (2013). A suppressor screen in Mecp2 mutant mice implicates cholesterol metabolism in Rett syndrome. *Nature Genetics, 45,* 1013–1020.

Buescher, A. V. S., Cidav, Z., Knapp, M., & Mandell, D. S. (2014). Costs of autism spectrum disorders in the United Kingdom and the United States. *JAMA Pediatrics, 168,* 721–728.

Bushby, K., Finkel, R., Wong, B., Barohn, R., Campbell, C., Comi, G. P., et al. (2014). Ataluren treatment of patients with nonsense mutation dystrophinopathy. *Muscle & Nerve, 50,* 477–487.

Çaku, A., Pellerin, D., Bouvier, P., Riou, E., & Corbin, F. (2014). Effect of lovastatin on behavior in children and adults with fragile X syndrome: an open-label study. *American Journal of Medical Genetics. Part A, 164A,* 2834–2842.

Campbell, J. M., Brown, R. T., Cavanagh, S. E., Vess, S. F., & Segall, M. J. (2008). Evidence-based assessment of cognitive functioning in pediatric psychology. *Journal of Pediatric Psychology, 33,* 999–1014, discussion 1015-1020.

Capal, J. K., & Franz, D. N. (2016). Profile of everolimus in the treatment of tuberous sclerosis complex: an evidence-based review of its place in therapy. *Neuropsychiatric Disease and Treatment, 12,* 2165–2172.

Castro, J., Garcia, R. I., Kwok, S., Banerjee, A., Petravicz, J., Woodson, J., et al. (2014). Functional recovery with recombinant human IGF1 treatment in a mouse model of Rett Syndrome. *Proceedings of the National Academy of Sciences of the United States of America, 111,* 9941–9946.

Chahrour, M., Jung, S. Y., Shaw, C., Zhou, X., Wong, S. T. C., Qin, J., et al. (2008). MeCP2, a key contributor to neurological disease, activates and represses transcription. *Science, 320,* 1224–1229.

Chang, Q., Khare, G., Dani, V., Nelson, S., & Jaenisch, R. (2006). The disease progression of Mecp2 mutant mice is affected by the level of BDNF expression. *Neuron, 49,* 341–348.

Chao, H. -T., Chen, H., Samaco, R. C., Xue, M., Chahrour, M., Yoo, J., et al. (2010). Dysfunction in GABA signalling mediates autism-like stereotypies and Rett syndrome phenotypes. *Nature, 468,* 263–269.

Chen, W. G1., Chang, Q., Lin, Y., Meissner, A., West, A. E., Griffith, E. C., et al. (2003). Derepression of BDNF transcription involves calcium-dependent phosphorylation of MeCP2. *Science, 302,* 885–889.

Cheval, H., Guy, J., Merusi, C., De Sousa, D., Selfridge, J., & Bird, A. (2012). Postnatal inactivation reveals enhanced requirement for MeCP2 at distinct age windows. *Human Molecular Genetics, 21,* 3806–3814.

Chong-Kopera, H., Inoki, K., Li, Y., Zhu, T., Garcia-Gonzalo, F. R., Rosa, J. L., et al. (2006). TSC1 stabilizes TSC2 by inhibiting the interaction between TSC2 and the HERC1 ubiquitin ligase. *The Journal of Biological Chemistry, 281,* 8313–8316.

Chu-Shore, C. J., Major, P., Camposano, S., Muzykewicz, D., & Thiele, E. A. (2010). The natural history of epilepsy in tuberous sclerosis complex. *Epilepsia, 51,* 1236–1241.

Cohen, E., Berry, J. G., Camacho, X., Anderson, G., Wodchis, W., & Guttmann, A. (2012). Patterns and costs of health care use of children with medical complexity. *Pediatrics, 130,* e1463–e1470.

Coker, S. B., & Melnyk, A. R. (1991). Rett syndrome and mitochondrial enzyme deficiencies. *Journal of Child Neurology, 6,* 164–166.

Cope, Z. A., Powell, S. B., & Young, J. W. (2016). Modeling neurodevelopmental cognitive deficits in tasks with cross-species translational validity. *Genes, Brain and Behavior, 15,* 27–44.

Costa, V., Aigner, S., Vukcevic, M., Sauter, E., Behr, K., Ebeling, M., et al. (2016). mTORC1 Inhibition corrects neurodevelopmental and synaptic alterations in a human stem cell model of tuberous sclerosis. *Cell Reports, 15,* 86–95.

Craig, F., Operto, F. F., De Giacomo, A., Margari, L., Frolli, A., Conson, M., et al. (2016). Parenting stress among parents of children with neurodevelopmental disorders. *Psychiatry Research, 242,* 121–129.

Curatolo, P. (2015). Mechanistic target of rapamycin (mTOR) in tuberous sclerosis complex-associated epilepsy. *Pediatric Neurology, 52,* 281–289.

Curatolo, P., Bombardieri, R., & Jozwiak, S. (2008). Tuberous sclerosis. *Lancet, 372,* 657–668.

Curie, A., Yang, K., Kirsch, I., Gollub, R. L., Portes, V. d., Kaptchuk, T. J., et al. (2015). Placebo responses in genetically determined intellectual disability: a meta-analysis. *PLoS ONE, 10,* e0133316.

Dajani, R., Koo, S. -E., Sullivan, G. J., & Park, I. -H. (2013). Investigation of Rett syndrome using pluripotent stem cells. *Journal of Cellular Biochemistry, 114,* 2446–2453.

Damiano, C. R., Mazefsky, C. A., White, S. W., & Dichter, G. S. (2014). Future directions for research in autism spectrum disorders. *Journal of Clinical Child and Adolescent Psychology, 43,* 828–843.

Davies, D. M., de Vries, P. J., Johnson, S. R., McCartney, D. L., Cox, J. A., Serra, A. L., et al. (2011). Sirolimus therapy for angiomyolipoma in tuberous sclerosis and sporadic lymphangioleiomyomatosis: a phase 2 trial. *Clinical Cancer Research, 17,* 4071–4081.

de Vries, P. J. (2010). Targeted treatments for cognitive and neurodevelopmental disorders in tuberous sclerosis complex. *Neurotherapeutics, 7,* 275–282.

de Vries, P. J., Whittemore, V. H., Leclezio, L., Byars, A. W., Dunn, D., Ess, K. C., et al. (2015). Tuberous sclerosis associated neuropsychiatric disorders (TAND) and the TAND Checklist. *Pediatric Neurology, 52,* 25–35.

Deng, P. -Y., Rotman, Z., Blundon, J. A., Cho, Y., Cui, J., Cavalli, V., et al. (2013). FMRP regulates neurotransmitter release and synaptic information transmission by modulating action potential duration via BK channels. *Neuron, 77,* 696–711.

Deogracias, R., Yazdani, M., Dekkers, M. P. J., Guy, J., Ionescu, M. C. S., Vogt, K. E., et al. (2012). Fingolimod, a sphingosine-1 phosphate receptor modulator, increases BDNF levels and improves symptoms of a mouse model of Rett syndrome. *Proceedings of the National Academy of Sciences of the United States of America, 109,* 14230–14235.

Dibble, C. C., Elis, W., Menon, S., Qin, W., Klekota, J., Asara, J. M., et al. (2012). TBC1D7 is a third subunit of the TSC1-TSC2 complex upstream of mTORC1. *Molecular Cell, 47,* 535–546.

Dölen, G., Osterweil, E., Rao, B. S. S., Smith, G. B., Auerbach, B. D., Chattarji, S., et al. (2007). Correction of fragile X syndrome in mice. *Neuron, 56,* 955–962.

Dotti, M. T., Manneschi, L., Malandrini, A., De Stefano, N., Caznerale, F., & Federico, A. (1993). Mitochondrial dysfunction in Rett syndrome. An ultrastructural and biochemical study. *Brain and Development, 15,* 103–106.

Ebrahimi-Fakhari, D., Saffari, A., Wahlster, L., Di Nardo, A., Turner, D., Lewis, T. L., et al. (2016). Impaired mitochondrial dynamics and mitophagy in neuronal models of tuberous sclerosis complex. *Cell Reports, 17*, 1053–1070.

Ebrahimi-Fakhari, D., & Sahin, M. (2015). Autism and the synapse: emerging mechanisms and mechanism-based therapies. *Current Opinion in Neurology, 28*, 91–102.

Eeg-Olofsson, O., Al-Zuhair, A. G. H., Teebi, A. S., Daoud, A. S., Zaki, M., Besisso, M. S., et al. (1990). Rett syndrome: a mitochondrial disease? *Journal of Child Neurology, 5*, 210–214.

Ehninger, D., de Vries, P. J., & Silva, A. J. (2009). From mTOR to cognition: molecular and cellular mechanisms of cognitive impairments in tuberous sclerosis. *Journal of Intellectual Disability Research, 53*, 838–851.

Ehninger, D., Han, S., Shilyansky, C., Zhou, Y., Li, W., Kwiatkowski, D. J., et al. (2008). Reversal of learning deficits in a Tsc2+/- mouse model of tuberous sclerosis. *Nature Medicine, 14*, 843–848.

Einspieler, C., Kerr, A. M., & Prechtl, H. F. R. (2005). Is the early development of girls with Rett disorder really normal? *Pediatric Research, 57*, 696–700.

El-Khoury, R., Panayotis, N., Matagne, V., Ghata, A., Villard, L., & Roux, J. -C. (2014). GABA and glutamate pathways are spatially and developmentally affected in the brain of Mecp2-deficient mice. *PLoS ONE, 9*, e92169.

Ercan, E., Han, J. M., Di Nardo, A., Winden, K., Han, M. -J., Hoyo, L., et al. (2017). Neuronal CTGF/CCN2 negatively regulates myelination in a mouse model of tuberous sclerosis complex. *The Journal of Experimental Medicine, 214*, 681–697.

Fava, M., Evins, A. E., Dorer, D. J., & Schoenfeld, D. A. (2003). The problem of the placebo response in clinical trials for psychiatric disorders: culprits, possible remedies, and a novel study design approach. *Psychotherapy and Psychosomatics, 72*, 115–127.

Feldman, D., Banerjee, A., & Sur, M. (2016). Developmental dynamics of rett syndrome. *Neural Plasticity, 2016*, e6154080.

Feliciano, D. M., Su, T., Lopez, J., Platel, J. -C., & Bordey, A. (2011). Single-cell Tsc1 knockout during corticogenesis generates tuber-like lesions and reduces seizure threshold in mice. *The Journal of Clinical Investigation, 121*, 1596.

Fernandez, A. M., & Torres-Alemán, I. (2012). The many faces of insulin-like peptide signalling in the brain. *Nature Reviews. Neuroscience, 13*, 225–239.

Finniss, D. G., Kaptchuk, T. J., Miller, F., & Benedetti, F. (2010). Biological, clinical, and ethical advances of placebo effects. *Lancet, 375*, 686–695.

Franz, D. N., Belousova, E., Sparagana, S., Bebin, E. M., Frost, M., Kuperman, R., et al. (2014). Everolimus for subependymal giant cell astrocytoma in patients with tuberous sclerosis complex: 2-year open-label extension of the randomised EXIST-1 study. *The Lancet Oncology, 15*, 1513–1520.

Franz, D. N., Belousova, E., Sparagana, S., Bebin, E. M., Frost, M., Kuperman, R., et al. (2013). Efficacy and safety of everolimus for subependymal giant cell astrocytomas associated with tuberous sclerosis complex (EXIST-1): a multicentre, randomised, placebo-controlled phase 3 trial. *Lancet, 381*, 125–132.

Franz, D. N., Leonard, J., Tudor, C., Chuck, G., Care, M., Sethuraman, G., et al. (2006). Rapamycin causes regression of astrocytomas in tuberous sclerosis complex. *Annals of Neurology, 59*, 490–498.

French, J. A., Lawson, J. A., Yapici, Z., Ikeda, H., Polster, T., Nabbout, R., et al. (2016). Adjunctive everolimus therapy for treatment-resistant focal-onset seizures associated with tuberous sclerosis (EXIST-3): a phase 3, randomised, double-blind, placebo-controlled study. *Lancet, 388*, 2153–2163.

Furukawa, T. A., Cipriani, A., Atkinson, L. Z., Leucht, S., Ogawa, Y., Takeshima, N., et al. (2016). Placebo response rates in antidepressant trials: a systematic review of published and unpublished double-blind randomised controlled studies. *Lancet Psychiatry, 3*, 1059–1066.

Gadalla, K. K., Bailey, M. E., Spike, R. C., Ross, P. D., Woodard, K. T., Kalburgi, S. N., et al. (2013). Improved survival and reduced phenotypic severity following AAV9/MECP2 gene transfer to neonatal and juvenile male Mecp2 knockout mice. *Molecular Therapy, 21*, 18–30.

Gadalla, K. K., Ross, P. D., Hector, R. D., Bahey, N. G., Bailey, M. E., & Cobb, S. R. (2015). Gene therapy for Rett syndrome: prospects and challenges. *Future Neurology, 10*, 467–484.

Garg, S. K., Lioy, D. T., Cheval, H., McGann, J. C., Bissonnette, J. M., Murtha, M. J., et al. (2013). Systemic delivery of MeCP2 rescues behavioral and cellular deficits in female mouse models of Rett syndrome. *The Journal of Neuroscience, 33*, 13612–13620.

Gemelli, T., Berton, O., Nelson, E. D., Perrotti, L. I., Jaenisch, R., & Monteggia, L. M. (2006). Postnatal loss of methyl-CpG binding protein 2 in the forebrain is sufficient to mediate behavioral aspects of Rett syndrome in mice. *Biological Psychiatry, 59*, 468–476.

Geschwind, D. H. (2008). Autism: many genes, common pathways? *Cell, 135*, 391–395.

Gipson, T. T., Jennett, H., Wachtel, L., Gregory, M., Poretti, A., & Johnston, M. V. (2013). Everolimus and intensive behavioral therapy in an adolescent with tuberous sclerosis complex and severe behavior. *Epilepsy & Behaviour Case Report, 1*, 122–125.

Goorden, S. M. I., van Woerden, G. M., van der Weerd, L., Cheadle, J. P., & Elgersma, Y. (2007). Cognitive deficits in Tsc1+/- mice in the absence of cerebral lesions and seizures. *Annals of Neurology, 62*, 648–655.

Grilo, C. M., Reas, D. L., & Mitchell, J. E. (2016). Combining pharmacological and psychological treatments for binge eating disorder: current status, limitations, and future directions. *Current Psychiatry Reports, 18*, 55.

Grosser, E., Hirt, U., Janc, O. A., Menzfeld, C., Fischer, M., Kempkes, B., et al. (2012). Oxidative burden and mitochondrial dysfunction in a mouse model of Rett syndrome. *Neurobiology of Disease, 48*, 102–114.

Grossman, A. W., Aldridge, G. M., Weiler, I. J., & Greenough, W. T. (2006). Local protein synthesis and spine morphogenesis: fragile X syndrome and beyond. *The Journal of Neuroscience, 26*, 7151–7155.

Guy, J., Cheval, H., Selfridge, J., & Bird, A. (2011). The role of MeCP2 in the brain. *Annual Review of Cell and Developmental Biology, 27*, 631–652.

Guy, J., Gan, J., Selfridge, J., Cobb, S., & Bird, A. (2007). Reversal of neurological defects in a mouse model of Rett syndrome. *Science, 315*, 1143–1147.

Guy, J., Hendrich, B., Holmes, M., Martin, J. E., & Bird, A. (2001). A mouse Mecp2-null mutation causes neurological symptoms that mimic Rett syndrome. *Nature Genetics, 27*, 322–326.

Hagberg, B. (2002). Clinical manifestations and stages of Rett syndrome. *Mental Retardation and Developmental Disabilities Research Reviews, 8*, 61–65.

Hagberg, B., Aicardi, J., Dias, K., & Ramos, O. (1983). A progressive syndrome of autism, dementia, ataxia, and loss of purposeful hand use in girls: Rett's syndrome: report of 35 cases. *Annals of Neurology, 14*, 471–479.

Hagerman, R. J., Berry-Kravis, E., Kaufmann, W. E., Ono, M. Y., Tartaglia, N., Lachiewicz, A., et al. (2009). Advances in the treatment of fragile X syndrome. *Pediatrics, 123*, 378–390.

Hamberger, A., Gillberg, C., Palm, A., & Hagberg, B. (1992). Elevated CSF glutamate in Rett syndrome. *Neuropediatrics, 23*, 212–213.

Hansen, J. C., Ghosh, R. P., & Woodcock, C. L. (2010). Binding of the Rett syndrome protein, MeCP2, to methylated and unmethylated dna and chromatin. *IUBMB Life, 62*, 732–738.

Hay, N., & Sonenberg, N. (2004). Upstream and downstream of mTOR. *Genes & Development, 18*, 1926–1945.

Henderson, C., Wijetunge, L., Kinoshita, M. N., Shumway, M., Hammond, R. S., Postma, F. R., et al. (2012). Reversal of disease-related pathologies in the fragile X mouse model by selective activation of GABAB receptors with arbaclofen. *Science Translational Medicine, 4,* 152ra128.

Horská, A., Farage, L., Bibat, G., Nagae, L. M., Kaufmann, W. E., Barker, P. B., et al. (2009). Brain metabolism in Rett syndrome: age, clinical, and genotype correlations. *Annals of Neurology, 65,* 90–97.

Huat, T. J., Khan, A. A., Pati, S., Mustafa, Z., Abdullah, J. M., & Jaafar, H. (2014). IGF-1 enhances cell proliferation and survival during early differentiation of mesenchymal stem cells to neural progenitor-like cells. *BMC Neuroscience, 15,* 91.

Huber, K. M., Gallagher, S. M., Warren, S. T., & Bear, M. F. (2002). Altered synaptic plasticity in a mouse model of fragile X mental retardation. *Proceedings of the National Academy of Sciences of the United States of America, 99,* 7746–7750.

Hunter, J., Rivero-Arias, O., Angelov, A., Kim, E., Fotheringham, I., & Leal, J. (2014). Epidemiology of fragile X syndrome: a systematic review and meta-analysis. *American Journal of Medical Genetics. Part A, 164A,* 1648–1658.

Hurt, A. A., Grist, C. L., Malesky, L. A., & McCord, D. M. (2013). Personality traits associated with occupational "burnout" in ABA therapists. *Journal of Applied Research in Intellectual Disabilities, 26,* 299–308.

Hwang, S. -K., Lee, J. -H., Yang, J., Lim, C. -S., Lee, J. -A., Lee, Y. -S., et al. (2016). Everolimus improves neuropsychiatric symptoms in a patient with tuberous sclerosis carrying a novel TSC2 mutation. *Molecular Brain, 9,* 56.

Inoki, K., Li, Y., Xu, T., & Guan, K. -L. (2003). Rheb GTPase is a direct target of TSC2 GAP activity and regulates mTOR signaling. *Genes & Development, 17,* 1829–1834.

Ishibashi, T., Thambirajah, A. A., & Ausió, J. (2008). MeCP2 preferentially binds to methylated linker DNA in the absence of the terminal tail of histone H3 and independently of histone acetylation. *FEBS Letters, 582,* 1157–1162.

Ishii, R., Wataya-Kaneda, M., Canuet, L., Nonomura, N., Nakai, Y., & Takeda, M. (2015). Everolimus improves behavioral deficits in a patient with autism associated with tuberous sclerosis: a case report. *Neuropsychiatric Electrophysiology, 1,* 6.

Jacquemont, S., Curie, A., des Portes, V., Torrioli, M. G., Berry-Kravis, E., Hagerman, R. J., et al. (2011). Epigenetic modification of the FMR1 gene in fragile X syndrome is associated with differential response to the mGluR5 antagonist AFQ056. *Science Translational Medicine, 3,* 64ra1.

Jaworski, J., & Sheng, M. (2006). The growing role of mTOR in neuronal development and plasticity. *Molecular Neurobiology, 34,* 205–219.

Jaworski, J., Spangler, S., Seeburg, D. P., Hoogenraad, C. C., & Sheng, M. (2005). Control of dendritic arborization by the phosphoinositide-3′-kinase-Akt-mammalian target of rapamycin pathway. *The Journal of Neuroscience, 25,* 11300–11312.

Jeste, S. S., Frohlich, J., & Loo, S. K. (2015). Electrophysiological biomarkers of diagnosis and outcome in neurodevelopmental disorders. *Current Opinion in Neurology, 28,* 110–116.

Jeste, S. S., & Geschwind, D. H. (2016). Clinical trials for neurodevelopmental disorders: at a therapeutic frontier. *Science Translational Medicine, 8,* 321fs1.

Jewell, J. L., & Guan, K. -L. (2013). Nutrient signaling to mTOR and cell growth. *Trends in Biochemical Sciences, 38,* 233–242.

Johnston, M., Blue, M. E., & Naidu, S. (2015). Recent advances in understanding synaptic abnormalities in Rett syndrome. *F1000Research. 4,* https://doi.org/10.12688/f1000research.6987.1.

Johnston, M. V., Blue, M. E., & Naidu, S. (2005). Rett syndrome and neuronal development. *Journal of Child Neurology, 20,* 759–763.

Joinson, C., O'Callaghan, F. J., Osborne, J. P., Martyn, C., Harris, T., & Bolton, P. F. (2003). Learning disability and epilepsy in an epidemiological sample of individuals with tuberous sclerosis complex. *Psychological Medicine, 33,* 335–344.

Justice, M. J., Buchovecky, C. M., Kyle, S. M., & Djukic, A. (2013). A role for metabolism in Rett syndrome pathogenesis. *Rare Diseases, 1,* e27265.

Kaptchuk, T. J., & Miller, F. G. (2015). Placebo effects in medicine. *New England Journal of Medicine, 373,* 8–9.

Katz, D. M., Berger-Sweeney, J. E., Eubanks, J. H., Justice, M. J., Neul, J. L., Pozzo-Miller, L., et al. (2012). Preclinical research in Rett syndrome: setting the foundation for translational success. *Disease Models & Mechanisms, 5,* 733–745.

Katz, D. M., Bird, A., Coenraads, M., Gray, S. J., Menon, D. U., Philpot, B. D., et al. (2016). Rett syndrome: crossing the threshold to clinical translation. *Trends in Neurosciences, 39,* 100–113.

Katz, D. M., Dutschmann, M., Ramirez, J. -M., & Hilaire, G. (2009). Breathing disorders in Rett syndrome: progressive neurochemical dysfunction in the respiratory network after birth. *Respiratory Physiology & Neurobiology, 168,* 101–108.

Katz, D. M., Menniti, F. S., & Mather, R. J. (2016). N-methyl-D-aspartate receptors, ketamine, and rett syndrome: something special on the road to treatments? *Biological Psychiatry, 79,* 710–712.

Khwaja, O. S., Ho, E., Barnes, K. V., O'Leary, H. M., Pereira, L. M., Finkelstein, Y., et al. (2014). Safety, pharmacokinetics, and preliminary assessment of efficacy of mecasermin (recombinant human IGF-1) for the treatment of Rett syndrome. *Proceedings of the National Academy of Sciences of the United States of America, 111,* 4596–4601.

Khwaja, O. S., & Sahin, M. (2011). Translational research: Rett syndrome and tuberous sclerosis complex. *Current Opinion in Pediatrics, 23,* 633–639.

Kiser, D. P., Rivero, O., & Lesch, K. -P. (2015). Annual research review: the (epi)genetics of neurodevelopmental disorders in the era of whole-genome sequencing—unveiling the dark matter. *Journal of Child Psychology and Psychiatry, 56,* 278–295.

Knickmeyer, R. C., Gouttard, S., Kang, C., Evans, D., Wilber, K., Smith, J. K., et al. (2008). A structural MRI study of human brain development from birth to 2 years. *The Journal of Neuroscience, 28,* 12176–12182.

Kradin, R. (2011). The placebo response: an attachment strategy that counteracts the effects of stress-related dysfunction. *Perspectives in Biology and Medicine, 54,* 438–454.

Krishnan, M. L., Commowick, O., Jeste, S. S., Weisenfeld, N., Hans, A., Gregas, M. C., et al. (2010). Diffusion features of white matter in tuberous sclerosis with tractography. *Pediatric Neurology, 42,* 101–106.

Krueger, D. A., Wilfong, A. A., Holland-Bouley, K., Anderson, A. E., Agricola, K., Tudor, C., et al. (2013). Everolimus treatment of refractory epilepsy in tuberous sclerosis complex. *Annals of Neurology, 74,* 679–687.

Kumar, V., Zhang, M. -X., Swank, M. W., Kunz, J., & Wu, G. -Y. (2005). Regulation of dendritic morphogenesis by Ras-PI3K-Akt-mTOR and Ras-MAPK signaling pathways. *The Journal of Neuroscience, 25,* 11288–11299.

Laccone, F., Huppke, P., Hanefeld, F., & Meins, M. (2001). Mutation spectrum in patients with Rett syndrome in the German population: evidence of hot spot regions. *Human Mutation, 17,* 183–190.

Laurvick, C. L., de Klerk, N., Bower, C., Christodoulou, J., Ravine, D., Ellaway, C., et al. (2006). Rett syndrome in Australia: a review of the epidemiology. *The Journal of Pediatrics, 148,* 347–352.

LeBlanc, J. J., DeGregorio, G., Centofante, E., Vogel-Farley, V. K., Barnes, K., Kaufmann, W. E., et al. (2015). Visual evoked potentials detect cortical processing deficits in Rett syndrome. *Annals of Neurology, 78,* 775–786.

Leonard, H., Cobb, S., & Downs, J. (2016). Clinical and biological progress over 50 years in Rett syndrome. *Nature Reviews. Neurology, 13*(1), 37–51.

Li, W., & Pozzo-Miller, L. (2014). BDNF deregulation in Rett syndrome. *Neuropharmacology, 76*(Pt C), 737–746.

Ligsay, A., & Hagerman, R. J. (2016). Review of targeted treatments in fragile X syndrome. *Intractable Rare Disease Research, 5*, 158–167.

Lioy, D. T., Garg, S. K., Monaghan, C. E., Raber, J., Foust, K. D., Kaspar, B. K., et al. (2011). A role for glia in the progression of Rett's syndrome. *Nature, 475*, 497–500.

Lipsky, R. H., & Marini, A. M. (2007). Brain-derived neurotrophic factor in neuronal survival and behavior-related plasticity. *Annals of the New York Academy of Sciences, 1122*, 130–143.

Lozano, R., Rosero, C. A., & Hagerman, R. J. (2014). Fragile X spectrum disorders. *Intractable Rare Disease Research, 3*, 134–146.

Luikenhuis, S., Giacometti, E., Beard, C. F., & Jaenisch, R. (2004). Expression of MeCP2 in postmitotic neurons rescues Rett syndrome in mice. *Proceedings of the National Academy of Sciences of the United States of America, 101*, 6033–6038.

Maeder, M. L., & Gersbach, C. A. (2016). Genome-editing technologies for gene and cell therapy. *Molecular Therapy, 24*, 430–446.

Maezawa, I., & Jin, L. -W. (2010). Rett syndrome microglia damage dendrites and synapses by the elevated release of glutamate. *The Journal of Neuroscience, 30*, 5346–5356.

Maezawa, I., Swanberg, S., Harvey, D., LaSalle, J. M., & Jin, L. -W. (2009). Rett syndrome astrocytes are abnormal and spread MeCP2 deficiency through gap junctions. *The Journal of Neuroscience, 29*, 5051–5061.

Manuvakhova, M., Keeling, K., & Bedwell, D. M. (2000). Aminoglycoside antibiotics mediate context-dependent suppression of termination codons in a mammalian translation system. *RNA, 6*, 1044–1055.

Marchetto, M. C. N., Carromeu, C., Acab, A., Yu, D., Yeo, G. W., Mu, Y., et al. (2010). A model for neural development and treatment of Rett syndrome using human induced pluripotent stem cells. *Cell, 143*, 527–539.

Masi, A., Lampit, A., Glozier, N., Hickie, I. B., & Guastella, A. J. (2015). Predictors of placebo response in pharmacological and dietary supplement treatment trials in pediatric autism spectrum disorder: a meta-analysis. *Translational Psychiatry, 5*, e640.

McBride, S. M. J., Choi, C. H., Wang, Y., Liebelt, D., Braunstein, E., Ferreiro, D., et al. (2005). Pharmacological rescue of synaptic plasticity, courtship behavior, and mushroom body defects in a Drosophila model of fragile X syndrome. *Neuron, 45*, 753–764.

McConkie-Rosell, A., Finucane, B., Cronister, A., Abrams, L., Bennett, R. L., & Pettersen, B. J. (2005). Genetic counseling for fragile x syndrome: updated recommendations of the national society of genetic counselors. *Journal of Genetic Counseling, 14*, 249–270.

McElroy, S. P., Nomura, T., Torrie, L. S., Warbrick, E., Gartner, U., Wood, G., et al. (2013). A lack of premature termination codon read-through efficacy of PTC124 (Ataluren) in a diverse array of reporter assays. *PLoS Biology, 11*, e1001593.

Meikle, L., Talos, D. M., Onda, H., Pollizzi, K., Rotenberg, A., Sahin, M., et al. (2007). A mouse model of tuberous sclerosis: neuronal loss of Tsc1 causes dysplastic and ectopic neurons, reduced myelination, seizure activity, and limited survival. *The Journal of Neuroscience, 27*, 5546–5558.

Mellén, M., Ayata, P., Dewell, S., Kriaucionis, S., & Heintz, N. (2012). MeCP2 binds to 5hmC enriched within active genes and accessible chromatin in the nervous system. *Cell, 151*, 1417–1430.

Mellios, N., Woodson, J., Garcia, R. I., Crawford, B., Sharma, J., Sheridan, S. D., et al. (2014). β2-Adrenergic receptor agonist ameliorates phenotypes and corrects microRNA-mediated IGF1 deficits in a mouse model of Rett syndrome. *Proceedings of the National Academy of Sciences of the United States of America, 111*, 9947–9952.

Meng, X., Wang, W., Lu, H., He, L. -J., Chen, W., Chao, E. S., et al. (2016). Manipulations of MeCP2 in glutamatergic neurons highlight their contributions to Rett and other neurological disorders. *eLife, 5*. https://doi.org/10.7554/eLife.14199.

Meredith, R. M., de Jong, R., & Mansvelder, H. D. (2011). Functional rescue of excitatory synaptic transmission in the developing hippocampus in Fmr1-KO mouse. *Neurobiology of Disease, 41*, 104–110.

Michalon, A., Sidorov, M., Ballard, T. M., Ozmen, L., Spooren, W., Wettstein, J. G., et al. (2012). Chronic pharmacological mGlu5 inhibition corrects fragile X in adult mice. *Neuron, 74*, 49–56.

Mizuguchi, M., & Takashima, S. (2001). Neuropathology of tuberous sclerosis. *Brain & Development, 23*, 508–515.

Modi, M. E., & Sahin, M. (2017). Translational use of event-related potentials to assess circuit integrity in ASD. *Nature Reviews. Neurology, 13*, 160–170.

Morita, T., & Sobue, K. (2009). Specification of neuronal polarity regulated by local translation of CRMP2 and Tau via the mTOR-p70S6K pathway. *The Journal of Biological Chemistry, 284*, 27734–27745.

Müller, M., & Can, K. (2014). Aberrant redox homoeostasis and mitochondrial dysfunction in Rett syndrome. *Biochemical Society Transactions, 42*, 959–964.

Naldini, L. (2011). Ex vivo gene transfer and correction for cell-based therapies. *Nature Reviews. Genetics, 12*, 301–315.

Naldini, L. (2015). Gene therapy returns to centre stage. *Nature, 526*, 351–360.

Neul, J. L., Fang, P., Barrish, J., Lane, J., Caeg, E. B., Smith, E. O., et al. (2008). Specific mutations in methyl-CpG-binding protein 2 confer different severity in Rett syndrome. *Neurology, 70*, 1313–1321.

Nguyen, M. V. C., Du, F., Felice, C. A., Shan, X., Nigam, A., Mandel, G., et al. (2012). MeCP2 is critical for maintaining mature neuronal networks and global brain anatomy during late stages of postnatal brain development and in the mature adult brain. *The Journal of Neuroscience, 32*, 10021–10034.

Nie, D., Di Nardo, A., Han, J. M., Baharanyi, H., Kramvis, I., Huynh, T., et al. (2010). Tsc2-Rheb signaling regulates EphA-mediated axon guidance. *Nature Neuroscience, 13*, 163–172.

Nikitina, T., Shi, X., Ghosh, R. P., Horowitz-Scherer, R. A., Hansen, J. C., & Woodcock, C. L. (2007). Multiple modes of interaction between the methylated DNA binding protein MeCP2 and chromatin. *Molecular and Cellular Biology, 27*, 864–877.

Noutel, J., Hong, Y. K., Leu, B., Kang, E., & Chen, C. (2011). Experience-dependent retinogeniculate synapse remodeling is abnormal in MeCP2-deficient mice. *Neuron, 70*, 35–42.

Nudelman, I., Rebibo-Sabbah, A., Cherniavsky, M., Belakhov, V., Hainrichson, M., Chen, F., et al. (2009). Development of novel aminoglycoside (NB54) with reduced toxicity and enhanced suppression of disease-causing premature stop mutations. *Journal of Medicinal Chemistry, 52*, 2836–2845.

Ogier, M., Wang, H., Hong, E., Wang, Q., Greenberg, M. E., & Katz, D. M. (2007). Brain-derived neurotrophic factor expression and respiratory function improve after ampakine treatment in a mouse model of Rett syndrome. *The Journal of Neuroscience, 27*, 10912–10917.

Olesen, J., Gustavsson, A., Svensson, M., Wittchen, H. -U., Jönsson, B., & CDBE2010 study group, European Brain Council. (2012). The economic cost of brain disorders in Europe. *European Journal of Neurology, 19*, 155–162.

Osborne, J. P., Fryer, A., & Webb, D. (1991). Epidemiology of tuberous sclerosis. *Annals of the New York Academy of Sciences, 615*, 125–127.

Osterweil, E. K., Krueger, D. D., Reinhold, K., & Bear, M. F. (2010). Hypersensitivity to mGluR5 and ERK1/2 leads to excessive protein synthesis in the hippocampus of a mouse model of fragile X syndrome. *The Journal of Neuroscience, 30*, 15616–15627.

Panayotis, N., Ghata, A., Villard, L., & Roux, J. -C. (2011). Biogenic amines and their metabolites are differentially affected in the Mecp2-deficient mouse brain. *BMC Neuroscience, 12*, 47.

Pecorelli, A., Leoni, G., Cervellati, F., Canali, R., Signorini, C., Leoncini, S., et al. (2013). Genes related to mitochondrial functions, protein degradation, and chromatin folding are differentially expressed in lymphomonocytes of Rett syndrome patients. *Mediators of Inflammation, 2013*, 137629. https://doi.org/10.1155/2013/137629.

Peters, J. M., Sahin, M., Vogel-Farley, V. K., Jeste, S. S., Nelson, C. A., Gregas, M. C., et al. (2012). Loss of white matter microstructural integrity is associated with adverse neurological outcome in tuberous sclerosis complex. *Academic Radiology, 19*, 17–25.

Phelan, S., & Wadden, T. A. (2002). Combining behavioral and pharmacological treatments for obesity. *Obesity Research, 10*, 560–574.

Plummer, J. T., Gordon, A. J., & Levitt, P. (2016). The genetic intersection of neurodevelopmental disorders and shared medical comorbidities—relations that translate from bench to bedside. *Frontiers in Psychiatry, 7*, 142.

Prather, P., & de Vries, P. J. (2004). Behavioral and cognitive aspects of tuberous sclerosis complex. *Journal of Child Neurology, 19*, 666–674.

Ramocki, M. B., Peters, S. U., Tavyev, Y. J., Zhang, F., Carvalho, C. M. B., Schaaf, C. P., et al. (2009). Autism and other neuropsychiatric symptoms are prevalent in individuals with MeCP2 duplication syndrome. *Annals of Neurology, 66*, 771–782.

Reith, R. M., McKenna, J., Wu, H., Hashmi, S. S., Cho, S. -H., Dash, P. K., et al. (2013). Loss of Tsc2 in Purkinje cells is associated with autistic-like behavior in a mouse model of tuberous sclerosis complex. *Neurobiology of Disease, 51*, 93–103.

Rett, A. (1966). On a unusual brain atrophy syndrome in hyperammonemia in childhood. *Wiener Medizinische Wochenschrift (1946), 116*, 723–726.

Richards, C., Jones, C., Groves, L., Moss, J., & Oliver, C. (2015). Prevalence of autism spectrum disorder phenomenology in genetic disorders: a systematic review and meta-analysis. *Lancet Psychiatry, 2*, 909–916.

Robinson, L., Guy, J., McKay, L., Brockett, E., Spike, R. C., Selfridge, J., et al. (2012). Morphological and functional reversal of phenotypes in a mouse model of Rett syndrome. *Brain, 135*, 2699–2710.

Rogers, S. J., & Vismara, L. A. (2008). Evidence-based comprehensive treatments for early autism. *Journal of Clinical Child and Adolescent Psychology, 37*, 8–38.

Rooms, L., & Kooy, R. F. (2011). Advances in understanding fragile X syndrome and related disorders. *Current Opinion in Pediatrics, 23*, 601–606.

Roux, J. -C., Dura, E., Moncla, A., Mancini, J., & Villard, L. (2007). Treatment with desipramine improves breathing and survival in a mouse model for Rett syndrome. *The European Journal of Neuroscience, 25*, 1915–1922.

Ruppe, V., Dilsiz, P., Reiss, C. S., Carlson, C., Devinsky, O., Zagzag, D., et al. (2014). Developmental brain abnormalities in tuberous sclerosis complex: a comparative tissue analysis of cortical tubers and perituberal cortex. *Epilepsia, 55*, 539–550.

Sahin, M., & Sur, M. (2015). Genes, circuits, and precision therapies for autism and related neurodevelopmental disorders. *Science. 350*, https://doi.org/10.1126/science.aab3897.

Sarbassov, D. D., Ali, S. M., & Sabatini, D. M. (2005). Growing roles for the mTOR pathway. *Current Opinion in Cell Biology, 17*, 596–603.

Shahbazian, M. D., Antalffy, B., Armstrong, D. L., & Zoghbi, H. Y. (2002). Insight into Rett syndrome: MeCP2 levels display tissue- and cell-specific differences and correlate with neuronal maturation. *Human Molecular Genetics, 11*, 115–124.

Shoseyov, D., Cohen-Cymberknoh, M., & Wilschanski, M. (2016). Ataluren for the treatment of cystic fibrosis. *Expert Review of Respiratory Medicine*, 1–5.

Sinnamon, J. R., Kim, S. Y., Corson, G. M., Song, Z., Nakai, H., Adelman, J. P., et al. (2017). Site-directed RNA repair of endogenous Mecp2 RNA in neurons. *Proceedings of the National Academy of Sciences of the United States of America, 114*, E9395–E9402.

Su, T., Fan, H. -X., Jiang, T., Sun, W. -W., Den, W. -Y., Gao, M. -M., et al. (2011). Early continuous inhibition of group 1 mGlu signaling partially rescues dendritic spine abnormalities in the Fmr1 knockout mouse model for fragile X syndrome. *Psychopharmacology, 215*, 291–300.

Sun, Y. E., & Wu, H. (2006). The ups and downs of BDNF in Rett syndrome. *Neuron, 49*, 321–323.

Switon, K., Kotulska, K., Janusz-Kaminska, A., Zmorzynska, J., & Jaworski, J. (2017). Molecular neurobiology of mTOR. *Neuroscience, 341*, 112–153.

Tang, S. J., Reis, G., Kang, H., Gingras, A. -C., Sonenberg, N., & Schuman, E. M. (2002). A rapamycin-sensitive signaling pathway contributes to long-term synaptic plasticity in the hippocampus. *Proceedings of the National Academy of Sciences of the United States of America, 99*, 467–472.

Tillotson, R., Selfridge, J., Koerner, M. V., Gadalla, K. K. E3., Guy, J., De Sousa, D., et al. (2017). Radically truncated MeCP2 rescues Rett syndrome-like neurological defects. *Nature, 550*, 398–401.

Tsai, A., Pickler, L., Tartaglia, N., & Hagerman, R. (2009). Chromosomal disorders and fragile X syndrome. In W. B. Carey, A. Crocker, & W. L. Coleman (Eds.), *Developmental-behavioral pediatrics*. (4th ed., p. 224). Philadelphia: Saunder Elsevier.

Tsai, P. T., Hull, C., Chu, Y., Greene-Colozzi, E., Sadowski, A. R., Leech, J. M., et al. (2012). Autistic-like behaviour and cerebellar dysfunction in Purkinje cell Tsc1 mutant mice. *Nature, 488*, 647–651.

Tseng, T. -Y., Krebs, P., Schoenthaler, A., Wong, S., Sherman, S., Gonzalez, M., et al. (2016). Combining text messaging and telephone counseling to increase varenicline adherence and smoking abstinence among cigarette smokers living with HIV: a randomized controlled study. *AIDS and Behavior, 21*, 1964–1974.

Turner, G., Webb, T., Wake, S., & Robinson, H. (1996). Prevalence of fragile X syndrome. *American Journal of Medical Genetics, 64*, 196–197.

Uhlmann, E. J., Wong, M., Baldwin, R. L., Bajenaru, M. L., Onda, H., Kwiatkowski, D. J., et al. (2002). Astrocyte-specific TSC1 conditional knockout mice exhibit abnormal neuronal organization and seizures. *Annals of Neurology, 52*, 285–296.

Urbanska, M., Gozdz, A., Swiech, L. J., & Jaworski, J. (2012). Mammalian target of rapamycin complex 1 (mTORC1) and 2 (mTORC2) control the dendritic arbor morphology of hippocampal neurons. *The Journal of Biological Chemistry, 287*, 30240–30256.

Ure, K., Lu, H., Wang, W., Ito-Ishida, A., Wu, Z., He, L. -J., et al. (2016). Restoration of Mecp2 expression in GABAergic neurons is sufficient to rescue multiple disease features in a mouse model of Rett syndrome. *eLife, 5*.

Utari, A., Chonchaiya, W., Rivera, S. M., Schneider, A., Hagerman, R. J., Faradz, S. M. H., et al. (2010). Side effects of minocycline treatment in patients with fragile X syndrome and exploration of outcome measures. *American Journal on Intellectual and Developmental Disabilities, 115*, 433–443.

van der Staay, F. J., Arndt, S. S., & Nordquist, R. E. (2009). Evaluation of animal models of neurobehavioral disorders. *Behavioral and Brain Functions, 5*, 11.

Van Esch, H. (1993). MECP2 duplication syndrome. In R. A. Pagon, M. P. Adam, H. H. Ardinger, S. E. Wallace, A. Amemiya, L. J. Bean, T. D. Bird, C. -T. Fong, H. C. Mefford, R. J. Smith, & K. Stephens (Eds.), *GeneReviews(®)*. Seattle, WA: University of Washington, Seattle.

Vecsler, M., Ben Zeev, B., Nudelman, I., Anikster, Y., Simon, A. J., Amariglio, N., et al. (2011). Ex vivo treatment with a novel synthetic aminoglycoside NB54 in primary fibroblasts from Rett syndrome patients suppresses MECP2 nonsense mutations. *PLoS One, 6*(6), e20733.

Velho, R. V., Sperb-Ludwig, F., & Schwartz, I. V. D. (2015). New approaches to the treatment of orphan genetic disorders: mitigating molecular pathologies using chemicals. *Anais da Academia Brasileira de Ciências, 87*, 1375–1388.

Villani, C., Sacchetti, G., Bagnati, R., Passoni, A., Fusco, F., Carli, M., et al. (2016). Lovastatin fails to improve motor performance and survival in methyl-CpG-binding protein2-null mice. *eLife, 5*, e22409.

Villard, L. (2007). MECP2 mutations in males. *Journal of Medical Genetics, 44*, 417–423.

Walach, H., Sadaghiani, C., Dehm, C., & Bierman, D. (2005). The therapeutic effect of clinical trials: understanding placebo response rates in clinical trials—a secondary analysis. *BMC Medical Research Methodology, 5*, 26.

Walsh, R. M., & Hochedlinger, K. (2010). Modeling Rett syndrome with stem cells. *Cell, 143*, 499–500.

Walther, W., & Stein, U. (2000). Viral vectors for gene transfer: a review of their use in the treatment of human diseases. *Drugs, 60*, 249–271.

Wang, H., Chan, S., Ogier, M., Hellard, D., Wang, Q., Smith, C., et al. (2006). Dysregulation of brain-derived neurotrophic factor expression and neurosecretory function in Mecp2 null mice. *The Journal of Neuroscience, 26*, 10911–10915.

Whitney, G. M., Ohtake, P. J., Simakajornboon, N., Xue, Y. D., & Gozal, D. (2000). AMPA glutamate receptors and respiratory control in the developing rat: anatomic and pharmacological aspects. *American Journal of Physiology. Regulatory, Integrative and Comparative Physiology, 278*, R520–R528.

Wilbur, C., Sanguansermsri, C., Chable, H., Anghelina, M., Peinhof, S., Anderson, K., et al. (2017). Manifestations of tuberous sclerosis complex: the experience of a provincial clinic. *The Canadian Journal of Neurological Sciences, 44*, 35–43.

Xu, X. Q., McGuire, T. F., Blaskovich, M. A., Sebti, S. M., & Romero, G. (1996). Lovastatin inhibits the stimulation of mitogen-activated protein kinase by insulin in HIRcB fibroblasts. *Archives of Biochemistry and Biophysics, 326*, 233–237.

Yamanaka, S. (2012). Induced pluripotent stem cells: past, present, and future. *Cell Stem Cell, 10*, 678–684.

Yuskaitis, C. J., Mines, M. A., King, M. K., Sweatt, J. D., Miller, C. A., & Jope, R. S. (2010). Lithium ameliorates altered glycogen synthase kinase-3 and behavior in a mouse model of fragile X syndrome. *Biochemical Pharmacology, 79*, 632–646.

Zanella, S., Doi, A., Garcia, A. J., Elsen, F., Kirsch, S., Wei, A. D., et al. (2014). When norepinephrine becomes a driver of breathing irregularities: how intermittent hypoxia fundamentally alters the modulatory response of the respiratory network. *The Journal of Neuroscience, 34*, 36–50.

Zeng, L. -H., Xu, L., Gutmann, D. H., & Wong, M. (2008). Rapamycin prevents epilepsy in a mouse model of tuberous sclerosis complex. *Annals of Neurology, 63*, 444–453.

Zhang, Y., Cao, S. -X., Sun, P., He, H. -Y., Yang, C. -H., Chen, X. -J., et al. (2016). Loss of MeCP2 in cholinergic neurons causes part of RTT-like phenotypes via α7 receptor in hippocampus. *Cell Research, 26*, 728–742.

Zhou, J., Shrikhande, G., Xu, J., McKay, R. M., Burns, D. K., Johnson, J. E., et al. (2011). Tsc1 mutant neural stem/progenitor cells exhibit migration deficits and give rise to subependymal lesions in the lateral ventricle. *Genes & Development, 25*, 1595.

Zon, L. (2016). Modeling human diseases: an education in interactions and interdisciplinary approaches. *Disease Models & Mechanisms, 9*(6), 597–600. https://doi.org/10.1242/dmm.025882.

Translational Medicine Strategies in Drug Development for Mood Disorders

Zihang Pan[*,†], *Radu C. Grovu*[*], *Roger S. McIntyre*[*,†,‡,§,¶]

[*]Mood Disorders Psychopharmacology Unit (MDPU), Toronto Western Hospital, University Health Network, Toronto, ON, Canada [†]Institute of Medical Science, University of Toronto, Toronto, ON, Canada [‡]Department of Pharmacology, University of Toronto, Toronto, ON, Canada [§]Department of Psychiatry, University of Toronto, Toronto, ON, Canada [¶]Brain and Cognition Discovery Foundation, Toronto, ON, Canada

I INTRODUCTION

Affective (mood) disorders, such as major depressive disorder (MDD), are significant contributors to global disability and functional impairment. MDD affects approximately 350 million individuals worldwide with a significant lifetime prevalence of up to 15% in the general North American population (Marcus, Yasamy, Ommeren, Chisholm, & Saxena, 2012). The Diagnostic and Statistical Manual of Mental Disorders, 5th Edition (DSM-5), and the International Statistical Classification of Diseases, 10th Revision (ICD-10), characterize mood disorders, such as MDD, as a syndrome consisting of low mood, anhedonia (the loss of interest and/or pleasure), negative thought patterns, decreased energy, and impaired cognition (Gaillard, Gourion, & Llorca, 2013). World Health Organization (WHO) projections identify that mood disorders will become the leading cause of disease burden worldwide by 2030 (Collins et al., 2011). Estimated economic costs of MDD due to decreased workplace productivity result in losses of $43 billion in the United States and Canada annually (Jacobs, Ohinmaa, Escober-Doran, Patterson, & Slomp, 2009; Lépine & Briley, 2011). Depressive mood disorders significantly impair functions in multiple domains, including, but not limited to, workplace, social, and cognitive functions. Despite advances in pharmacotherapy, a significant proportion of patients with MDD and bipolar disorder (BD) remain symptomatic despite optimal treatment with approved first-line agents. The persistence of affective symptoms in MDD is associated with functional impairment, chronicity, and elevated rates of recurrence.

Major depressive episodes (MDE) are characteristic of MDD symptomatology and disease progression. Both prolonged periods of suppressed mood and anhedonia are important diagnostic criteria of an MDE. In addition, cognitive functions (i.e., learning and memory, executive function, processing speed, attention, and concentration) may become impaired in patients presenting with an MDE. These, combined with other progressive clinical observations such as anhedonia and depressed mood, are thought to be key factors subserving disability and impairment associated with mood disorders (Carvalho et al., 2014; Carvalho, Berk, Hyphantis, & McIntyre, 2014). MDD has been conceptualized as a neuroprogressive illness (Rosenblat & McIntyre, 2015). Neuroinflammatory pathways are known to play a role in the pathoetiology of MDD while contributing to mood-related neuroprogressive challenges. There is compelling evidence that proinflammatory cytokines are involved in the pathobiology of MDD, offering opportunities to identify biomarkers and treatment targets with the use of antiinflammatory agents.

Therefore, it is increasingly relevant to refine our knowledge and develop translational tools, as it relates to the prevention, neurobiology, treatment, and management of mood disorders. The following chapter highlights some of the underlying mechanisms, targets, challenges, and unmet needs that may be informative for translational approaches in treatment-resistant affective disorders.

II COGNITIVE DEFICITS IN MDD

The diagnosis of an MDE as defined by the DSM-5 combines a polythetic list of diagnostic criteria, whereby no single criterion item fully captures the diagnosis of an

ISSN: 1569-7339
https://doi.org/10.1016/B978-0-12-803161-2.00023-0

MDE (American Psychiatric Association, 2013). To determine the roles of complex phenomenological features, a domain-based approach is needed to understand cognition in MDD. For example, disturbances in concentration, thinking, and/or memory/learning are part of the DSM-5 criterion items as it relates to cognitive deficits in MDD. Domains that involve disparate neurobiological systems (i.e., impulsivity, reward/arousal, suicidality, anhedonia, psychomotor functioning, energy, and fatigue) also contribute to cognitive dysfunctions in MDD (McIntyre et al., 2016; McIntyre, Harrison, Loft, Jacobson, & Olsen, 2016).

Various typologies have been proposed to define and operationalize cognitive constructs. The conventional typology classifies cognitive functions into the subdomains of attention and concentration, processing speed, executive function, and learning and memory (Harrison, Lam, Baune, & McIntyre, 2016). Extent literature views these subdomains as interconnected and dissociable phenomena with unique, but overlapping, neurobiological substrates. For example, executive functions involve regulatory systems of planning, initiation, execution, and inhibition of thought and connect functionally to systems of working memory, learning, and psychomotor and behavioral responses (McIntyre et al., 2015).

The conceptualization of a "cold" and "hot" cognition is an important distinction in evaluating cognitive dysfunction in MDD (Roiser & Sahakian, 2013). "Cold" cognition refers to cognition that is uncoupled from emotion (e.g., some aspects of executive function and working memory). Conversely, "hot" cognition refers to cognitive processes that are intricately linked with emotion, also known as "emotionally valenced" cognition. An example of "hot" cognition is rumination, whereby thoughts and memories informed by emotion-based systems play a significant role in the decisional process (Hamilton et al., 2011). Disturbances in both "hot" and "cold" cognition may lead to clinically significant symptoms of MDD (i.e., anhedonia, psychomotor retardation, learning, and memory impairments).

Disturbances in neural substrates informed by function and reciprocity comprise the foundation for disease modeling in affective mood disorders. Brain functions and their corresponding neurobiological substrates are both integrated and segregated and may rely upon a system of reciprocity between integrated and segregated structures. Anticorrelation, or the selective activation of certain regions and deactivation of others, is critical for proper cognitive functioning (Hamilton et al., 2011). Dysregulation of normal reciprocity between nodal structures within the default mode networks, therefore, contributes to cognitive impairments in MDD (Cha et al., 2014).

Additionally, cognitive impairments in individuals with MDD may be due to an increase in neural effort.

Harvey et al. (2005) examined cerebral activity and cognitive performance in adults with MDD ($n = 10$) compared with healthy controls ($n = 10$) via the n-back test. The n-back test is a validated measure of working memory, whereby the subject is asked to determine whether previous stimuli have been observed. Performance on the n-back test did not significantly differ between the MDD subjects and the healthy controls. However, differences were observed in the activation and deactivation of nodal substrates between the MDD and non-MDD subjects. Specifically, the depressed subjects exhibited greater activation of the working memory network as a function of greater n-back complexity relative to healthy controls (Harvey et al., 2005). These findings imply depressed individuals require greater effort to achieve the same level of cognitive performance as healthy controls. There was also abnormal activity in the medial prefrontal cortex in individuals with MDD (Harvey et al., 2005), which suggests abnormal reciprocity and/or anticorrelation may play a significant role in cognitive impairments and reduced cognitive efficiency in adults with MDD.

Progress in translational research and neuroimaging studies lend additional evidence to support the notion that neurocircuits underlying emotional and cognitive deficits are overlapping and discrete (Naismith, Longley, Scott, & Hickie, 2007). Therefore, any disruptions in the frontosubcortical circuitry can directly or indirectly contribute to depressive and/or cognitive symptoms as observed in MDD. Circuits of the orbitofrontal cortex (OFC), dorsolateral prefrontal cortex (DLPFC), and anterior cingulate (ACC) are particularly relevant to the pathophysiology of MDD. The dorsal ACC, working in concert with the hippocampus and the DLPFC, contributes to the formation of dorsal "cognitive" networks, which have been postulated to be of particular importance to executive functions and emotional responses (Kheirbek & Hen, 2011). The ventral "affective" networks, on the other hand, may involve the perigenual ACC, the amygdala, the OFC, and hippocampus (Kheirbek & Hen, 2011). These networks work together to assess cognitively relevant stimuli and output-appropriate responses. Dysregulation of these networks, either via neurochemical abnormalities or neuroinflammatory insults, can lead to functional impairments, anhedonia, and cognitive dysfunctions characteristic of chronic, progressive, and treatment-resistant variants of MDD.

III ANHEDONIA IN MDD

Clinical presentations of MDD implicate brain systems involved in the regulation of mood, reward processing, and motivation (Carvalho, Berk, et al., 2014; Carvalho, Miskowiak, et al., 2014). Research Domain Criteria

(RDoC) from the National Institute of Mental Health (NIMH) is an integrative and domain-based approach that combines neurocircuitry, neurobiological systems, and behavioral responses with psychiatric disorders (Bedwell, Gooding, Chan, & Trachik, 2014). Anhedonia is classified within the RDoC matrix as a dysregulation of negative valence systems (Dillon et al., 2014). There is also an emphasis on neuroimmune dysregulations that could lead to disruptions in reward circuitry. Disturbances in negative valence systems can contribute to anxiety and fear, but disturbances in positive valence systems can also phenomenologically overlap with DSM-defined domains of anhedonia, reward, and motivational processing. The RDoC highly interweaves emotional valence constructs with cognitive phenomenology and overlapping substrates (Keedwell, Andrew, Williams, Brammer, & Phillips, 2005). Therefore, there is a reciprocal role between cognition (i.e., executive function, working memory, processing, attention, and concentration) and anhedonia.

The DSM operationally defines anhedonia as "diminished interest and/or pleasure to previously rewarding stimuli during the euthymic state" (Prien, Carpenter, & Kupfer, 1991). There is a distinction between an inability to experience pleasure and a decreased motivational drive for rewards. Categorizing anhedonia is difficult as these symptoms may represent divergent pathophysiological processes (Wiborg, 2013). Unfortunately, clinical diagnosis does not currently discriminate between these subtypes, which necessitate further research and study.

Research suggests neurotransmitters (e.g., dopamine) play a significant role in the onset of anhedonia and depressed mood within MDD (Delgado, 2000). Past studies using reserpine, a chemical agent that ablates central stores of monoamine neurotransmitters, showed an induction of depressive-like symptoms and motor retardation in animal models (Wise, 2008). Agents that elevate synaptic monoamine levels via inhibition of monoamine oxidases (e.g., iproniazid) and selective serotonin reuptake inhibitors (SSRIs) produced antidepressant effects (Nutt, 2007; Wise, 2008). Research on consummatory anhedonia (subsensitivity to rewarding stimuli) has focused on the role of dopamine-mediated responses. The attenuating effects of reserpine and chlorpromazine on intracranial self-stimulation and the potentiating effects from amphetamines both seem to have dopaminergic sources, chiefly increases and decreases in dopamine neurotransmission, respectively (Olds & Travis, 1960; Stein & Himwich, 1962).

Both the lack of experienced pleasure and a deficit in other reward-related processes may be interpreted as anhedonia (Charney & Drevets, 2012; Wiborg, 2013); therefore, these processes implicate the involvement of mesolimbic dopaminergic reward arousal and motivational pathways,

respectively (Nestler & Carlezon, 2006). Additionally, disturbances of key subcortical dopaminergic modulatory pathways, such as those found in the basal ganglia (Charney & Drevets, 2012; Nestler & Carlezon, 2006; Wiborg, 2013), may be relevant for the observed interaction and association between anhedonia, psychomotor retardation, and fatigue.

Various dopaminergic circuits may be implicated in anhedonia and MDD. Mesocortical circuits modulate hedonic drive, motivation, and cognition. Mesolimbic circuits are involved in hedonic drive and perception (Wise, 1978). Nigrostriatal circuits are involved in voluntary movement (Nestler & Carlezon, 2006; Wiborg, 2013; Wise, 1978), which may subserve observed psychomotor symptoms. Tuberoinfundibular circuits are responsible for dopamine-mediated regulation of prolactin release (Wise, 1978), which may have clinical implications for dopaminergic agonists and antagonists in treatment. Mechanisms addressing dopaminergic disturbances are currently lacking. There is a paucity of evidence connecting neurochemical disturbances with the onset of anhedonia. Recent evidence suggests proinflammatory cytokines play important roles in mediating dopaminergic effects, which in turn may contribute to the pathoetiology of anhedonia within MDD (Wise, 2008).

Anhedonia can be categorized into consummatory, motivational, anticipatory and decisional domains, encompassing subsensitivity to rewarding stimuli, lack of motivational drive, decreased reward anticipation, and impaired reward arousal and decision-making, respectively (Chentsova-Dutton & Hanley, 2010; Treadway & Zald, 2011). A dysregulation in any of the aforementioned domains may lead to clinical observations of anhedonic behavior (Treadway & Zald, 2011). Despite advances in anhedonia research, corresponding neurocircuitry has not yet been elucidated.

Available evidence suggests consummatory anhedonia involves the dysregulation of mesolimbic dopaminergic projections, chiefly the projections from the ventral tegmental area (VTA) in the midbrain to the nucleus accumbens in the ventral striatum (Chentsova-Dutton & Hanley, 2010; Horger & Roth, 1996). Rats exposed to chronic mild stress become anhedonic-like and are less likely to respond to natural and drug rewards (Espejo & Minano, 1999; Jensen et al., 2003; Lim, Huang, Grueter, Rothwell, & Malenka, 2012). Anhedonia is accompanied by a reduction in dopamine release from the nucleus accumbens, in addition to increased reward response thresholds (Jensen et al., 2003). Anticipatory, decisional, and motivational anhedonia implicate various brain regions among which are mesocortical dopaminergic projections from the VTA to the prefrontal cortex (Chentsova-Dutton & Hanley, 2010; Treadway & Zald, 2011). The reward anticipatory response, therefore, is communicated from the VTA to the nucleus accumbens and then to the DLPFC

(Treadway & Zald, 2011; Wiborg, 2013). A reduction in the activity of DLPFC may underlie motivational anhedonic processes and cognitive impairments comorbid with impaired reward behavior (Lim et al., 2012; Wiborg, 2013).

IV INFLAMMATION AND ANHEDONIA

There has been increasing recognition that proinflammatory cytokines play important roles in neuronal integrity (Haroon, Raison, & Miller, 2012; Maes, 2008), with profound effects exerted on neurocircuitry and neurotransmitter systems (Felger & Miller, 2012). Accordingly, there has been mounting interest in the study of the effects of cytokines on the development and progression of neuropsychiatric diseases.

Numerous studies have reported elevated cytokine levels in MDD and BD subjects (Dowlati et al., 2010; Maes, 1999; Maes et al., 1991; Sluzewska, 1999; Yirmiya, 2000 ; Yirmiya et al., 1999). Hepatitis C patients undergoing proinflammatory pegylated interferon (PegIFN) treatments are more likely to develop treatment-induced depression (TID); not surprisingly, treatment with PegIFN elevated levels of proinflammatory cytokines such as tumor necrosis factor alpha (TNF-α) and interleukin 1-beta (IL-1β) (Falasca et al., 2009; Krueger et al., 2011; Papafragkakis, Rao, Moehlen, Dhillon, & Martin, 2012; Sims, Whalen, Nackerud, & Bride, 2014). Moreover, chronic administration of proinflammatory cytokines such as IFN-α and IL-1β in certain cancer treatments produced behavioral alterations similar to anhedonia, fatigue, and psychomotor slowing (Yirmiya, 2000; Yirmiya et al., 1999). Converging evidence implicates a causal relationship between proinflammatory cytokines and behavioral disturbances characteristic of anhedonic symptoms in MDD (Yirmiya & Goshen, 2011).

Cytokine receptors are ubiquitous in the central nervous systems (CNS) and are likely expressed at low levels in the brain during noninflammatory states (Haas & Schauenstein, 1997; Rothwell, Luheshi, & Toulmond, 1996). Once cytokines gain entry to the CNS (either via peripheral pathways or activated neuroimmune cells), a rapid and profound response is activated. Acute responses to proinflammatory signals include changes to neurotransmitter metabolism (i.e., decreases in brain monoamines such as serotonin and norepinephrine), activation of the hypothalamic-pituitary-adrenal (HPA) axis, induction of fever, and promotion of behavioral alterations, such as reduced locomotor activity and amotivation (collectively referred to as "sickness behavior") (Dillon et al., 2014). Therefore, acute responses to cytokine signals from the periphery serve to inform the body of injury and to help protect the organism against sickness (Chen et al., 2012; Dantzer & Kelley, 2007). Acute response to cytokines may be beneficial behaviorally, as decreased motivation and activity would allow for the conservation of energy and preferential utilization of resources to clear the infection (Dantzer & Kelley, 2007). Conversely, under conditions of chronic inflammation, often as a consequence of chronic disease states, changes to CNS circuits promote anhedonia, fatigue, and psychomotor retardation that are disadvantageous toward functioning and mental health (Felger & Miller, 2012).

V TETRAHYDROBIOPTERIN IN THE PATHOPHYSIOLOGY OF ANHEDONIA

Tetrahydrobiopterin (BH4) is a crucial coenzyme involved in dopamine (DA) synthesis. DA synthesis begins with the conversion of phenylalanine to tyrosine by phenylalanine hydroxylase (PAH). Tyrosine undergoes a rate-limiting reaction to convert to L-3, 4-dihydroxyphenylalanine (L-DOPA) by the enzyme tyrosine hydroxylase (TH). L-DOPA undergoes a final conversion via dopamine decarboxylase (DDC) to become the final product dopamine (Cunnington & Channon, 2010). Both PAH and TH require BH4 as an essential coenzyme in order to initiate the conversion reactions (Cunnington & Channon, 2010; Haroon et al., 2012). A decrease in BH4 availability will consequently decrease both tyrosine and L-DOPA availability, hindering the production of dopamine for normal neuronal functioning.

BH4 is a highly redox-sensitive coenzyme and is also a participant in the crucial conversion reaction of arginine to nitric oxide (NO) via inducible nitric oxide synthase (iNOS) (Dumitrescu et al., 2007). Free BH4 can also be reversibly converted to dihydrobiopterin (BH2) and irreversibly converted to dihydroxanthopterin (XPH2) via oxidation (Cunnington & Channon, 2010; Dumitrescu et al., 2007). Increases in reactive oxygen and nitrogen species (ROS and RON) associated with inflammation have been correlated with decreased intracellular BH4 and reduced capacity of dopamine synthesis (Neurauter et al., 2008). Experimental data suggest that proinflammatory cytokines (i.e., IL-1, IL-6, and TNF-α) increase both ROS activity and the amount of inducible NOS (Neurauter et al., 2008). Increased ROS participate in the oxidation of dopaminergic neurons that leads to oxidative damage and the inactivation of intracellular BH4 (via oxidation of BH4 to XPH2 and BH2, respectively) (Dumitrescu et al., 2007). Increased inducible NOS activity results in NOS uncoupling from BH4 and the preferential generation of free radicals (Cunnington & Channon, 2010; Xia, Tsai, Berka, & Zweier, 1998). Increased concentrations of free radicals further contribute to the oxidation of BH4; decreased BH4 coenzyme availability; and overall decreased dopamine synthesis, availability, and function (Cunnington & Channon, 2010; Felger & Miller, 2012).

VI KYNURENINE IN THE PATHOPHYSIOLOGY OF ANHEDONIA

Quinolinic acid (QUIN) is a cytokine metabolite that increases ROS activity (Behan, McDonald, Darlington, & Stone, 1999; Campbell, Charych, Lee, & Möller, 2014; Santamaría et al., 2003; Zitron, Kamson, Kiousis, Juhász, & Mittal, 2013). QUIN is a by-product of the metabolism of tryptophan in the kynurenine system under proinflammatory cytokine influence (Dantzer & Kelley, 2007). Tryptophan is an amino acid precursor to serotonin, which is implicated in the pathophysiology of MDD (Wirleitner, Neurauter, Schrocksnadel, Frick, & Fuchs, 2003). Majority of dietary tryptophan is oxidatively metabolized in the liver to nicotinamide adenine dinucleotide (NAD) by tryptophan dioxygenase (TDO) via the kynurenine pathway (KP) (Wirleitner et al., 2003). In addition, tryptophan can also be extrahepatically oxidized by indoleamine 2,3-dioxygenase (IDO) (Dantzer & Kelley, 2007). Under physiological conditions, the conversion of tryptophan via IDO is negligible (Wirleitner et al., 2003).

Proinflammatory cytokines activate IDO expressed in neuroimmune cells (i.e., microglia/macrophages). Both INF-γ and TNF-α increase IDO expression (Dantzer & Kelley, 2007; Wirleitner et al., 2003). Upon IDO induction, tryptophan is quickly converted to kynurenine (KYN) (Campbell et al., 2014). One branch of the pathway triggers a cascade of reactions that convert KYN to various neurotoxic metabolites including 3-hydroxykynurenine (3-HK), 3-hydroxyanthranilic acid (3-HAA), and QUIN (Gabbay, Ely, Babb, & Liebes, 2012; Goldstein et al., 2000). IDO-induced increases in KYN in depression studies have been positively correlated with increases in 3-HK, 3-HAA, and QUIN (Campbell et al., 2014; Capuron et al., 2003; Dantzer & Kelley, 2007; Gabbay et al., 2012). The metabolism of KYN to QUIN constitutes the neurotoxic branch of the KP (Gabbay et al., 2012; Santamaría et al., 2003). In addition to the aforementioned ROS effects of QUIN, QUIN can also act as a NMDA receptor agonist and participate in glutamate (Glu) neurotransmission (Raison et al., 2010). KYN can be converted to kynurenic acid (KA) as part of the neurotropic branch of KP (Gabbay et al., 2012; Zitron et al., 2013). An increase in IDO-mediated KYN is associated with a significant decrease of KA in astrocytes (McNally, Bhagwagar, & Hannestad, 2008). KA acts as an NMDA receptor antagonist (Dantzer & Kelley, 2007; Felger & Miller, 2012; Gabbay et al., 2012). The combinations of increased glutamatergic input from QUIN and decreased glutamatergic antagonism from lowered KA result in elevated glutamate transmission and excitotoxicity of dopaminergic neurons (Gabbay et al., 2012).

Taken together, KP shows much promise as a therapeutic target in the mitigation of neuropsychiatric symptoms. However, it is quite clear that a balance of these metabolites is key in finding efficacious treatment modalities; too much or too little may both lead to pathological changes eliciting psychiatric illness (Fig. 23.1).

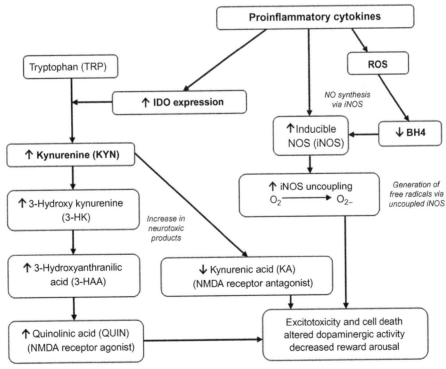

FIG. 23.1 Effect of proinflammatory cytokines on kynurenine pathway and BH4 activity.

VII INFLAMMATION AND COGNITION

The well-established abnormalities in monoamine systems in MDD likely indicate some degree of aberrant cellular signaling and neurocircuit dysfunction (Stahl, 2010). Indeed, neuroimaging studies provide preliminary evidence supporting a relationship between functional abnormalities and elevated levels of circulating inflammatory markers. Extensive literature indicates MDD is associated with elevated levels of serum immune mediators, including, but not limited to, TNF-α, c-reactive protein (CRP), IL-1, IL-6, and various chemokines (Dowlati et al., 2010; Stuart & Baune, 2014). Pathophysiological abnormalities (i.e., neuroinflammation, mitochondrial dysfunction, and oxidative stress) of the CNS are also prevalent in MDD. Smaller hippocampal volumes have also been observed in patients with MDD and have been associated with increased levels of IL-6 and CRP (Frodl & Amico, 2014). Despite a myriad of available antidepressants, the primary mechanism of action is through the inhibition of monoamine reuptake. Unfortunately, cognitive dysfunction is suboptimally treated by conventional antidepressants, where a clinical improvement in mood may not equate to recovery of function (Hasselbalch, Knorr, & Kessing, 2011). Therefore, a compelling need exists for the development of novel neuroprotective and procognitive agents (Carvalho, Berk, et al., 2014; Carvalho, Miskowiak, et al., 2014).

Peripheral inflammation and systemic release of proinflammatory cytokines can have profound effects on the brain with corresponding effects on cognition and behavior. Elevated inflammation has been highly correlated with disease states (e.g., obesity, diabetes mellitus, and autoimmune disorders). Depression, cognitive disturbances, anhedonia, and fatigue have also been well characterized in populations with inflammatory and metabolic disorders (Felger & Miller, 2012; Lam, Kennedy, McIntyre, & Khullar, 2014). These correlations are likely due to global effects on neurotransmitter metabolism and neurocircuitry. Cytokines from the periphery can enter and affect the CNS after systemic inflammation. Leaky regions of the blood-brain barrier (BBB) at the level of the choroid plexus and circumventricular organs are particularly susceptible to cytokine entry (Pan, Rosenblat, Swardfager, & McIntyre, 2017). Additionally, endothelial cells and perivascular macrophages in the cerebral vasculature can be activated to produce local inflammatory mediators such as prostaglandins, nitric oxide, and various chemokines (Cao, Matsumura, Yamagata, & Watanabe, 1997). Inflammatory signals [e.g., pathogen-associated molecular patterns (PAMPs) and damage-associated molecular patterns (DAMPs)] can lead to the recruitment and activation of immune cells (i.e., monocytes/macrophages, neutrophils, and T cells) and induce local cytokine production

(Shaftel et al., 2007). Lymphatic vessels in the dural sinuses of the CNS connect to deep cervical lymph nodes and allow for the movement of fluids and immune cells in the cerebrospinal fluid (Wood, 2015). Activation of peripheral nerve afferents (i.e., trigeminal and vagus) relay cytokine signals to relevant brain regions, such as the nucleus of the solitary tract (NTS) and the hypothalamus, allowing for communication of inflammatory signals to reach higher brain regions (Ericsson, Kovacs, & Sawchenko, 1994).

VIII TNF-α IN THE PATHOPHYSIOLOGY OF MDD

TNF-α is a pleotropic cytokine that has been increasingly recognized as a central, but not exclusive, mediator of CNS function. Elevated levels of TNF-α, along with IL-1 and IL-6, are among the most consistently identified proinflammatory cytokine abnormalities in MDD (Clark, Alleva, & Vissel, 2010). The soluble form of TNF binds to its corresponding receptors TNF-α receptor 1 (TNF-R1) and TNF-α receptor 2 (TNF-R2). Receptor-mediated effects of TNF and TNF receptor binding lead to the activation of transcription factors such as nuclear factor kappa B (NF-κB) and/or to the initiation of apoptosis via the activation of caspase-3 and caspase-8 complexes (Tracey, Klareskog, Sasso, Salfeld, & Tak, 2008). In addition, binding of TNF receptors can result in the production of cytokines, chemokines, and cell adhesion molecules (Wong, Ziring, & Korin, 2008).

TNF-α is an important mediator of cell death, immune functions, and host defense mechanisms against infections (Tracey et al., 2008). Under inflammatory conditions, proinflammatory cytokines (e.g. IL-1, IL-17, and IL-2) can induce the production of TNF-α. In turn, TNF-α induces the production of additional cytokines (e.g. IL-1, IL-6, and IL-8), which contribute to the proliferation of immune cells and activation of inflammatory and apoptotic cascades (Zhou, Wang, Yang, & Wang, 2013). Under proinflammatory conditions, peripheral production of TNF-α by monocytes results in a positive feedback in the production of TNF-α from activated microglia (Qin, Wu, & Block, 2007). These activated microglia, in turn, become the primary source of TNF-α in the CNS, which further exacerbate the inflammatory cascade with continuing increases in the production of TNF-α and other proinflammatory cytokines (Kerfoot, D'Mello, & Nguyen, 2006). Furthermore, cross talk between activated microglia and other glial cells, chiefly astrocytes and oligodendrocytes, leads to amplified inflammatory responses that may have detrimental effects on cognitive, neural, and behavioral functions (Najjar, Pearlman, Alper, Najjar, & Devinsky, 2013). Astrocytes play a significant role in the reuptake and metabolic conversion of glutamate. Microglia and astrocytes are the main producers of TNF-α in the CNS. When released, it can activate the apoptotic pathways (i.e., caspase-3 and caspase-8 pathways). TNF-α contributes to excitotoxicity by upregulating glutaminase and impairing glutamine reuptake and glutamine synthase. High levels of glutamate increase Ca^{2+} levels via the activation of Ca^{2+} permeable NMDA receptors that in turn stimulate enzymes, such as proteases (e.g., calpain), phospholipases, and endonucleases, leading to damaged cell structures and apoptosis. Excess calcium also promotes caspase processing and the opening of mitochondrial permeability transition pores, resulting in elevated cytoplasmic levels of cytochrome c, reduction of ATP levels, and oxidative stress, further exacerbating apoptosis and neuronal cell death associated with depressive symptomatology (Bortolato, Carvalho, Soczynska, Perini, & McIntyre, 2015).

IX TREATMENT STRATEGIES TARGETING ANHEDONIA

Current SSRIs have shown to be ineffective against anhedonia (Ballard et al., 2011; Capuron et al., 2002; Capuron, Hauser, Hinze-Selch, Miller, & Neveu, 2002); this supports the crucial role of dopaminergic systems in anhedonia. Further strengthening this link is the finding that anhedonia has been associated with cancer patients presenting with increased dopamine precursors and decreased dopamine availability in the brain (Zitron et al., 2013). Lowered dopamine levels are also linked with inflammation, known to damage dopaminergic neurons and be closely associated with anhedonia. Low dopamine levels and increased phenylalanine levels have been found in patients suffering from increased inflammation, including those affected by sepsis, trauma, cancer, hepatitis C, and HIV (Zitron et al., 2013), all of whom present with higher levels of inflammatory markers (e.g., IL-2 receptor, IL-6 receptor, and TNF-α receptor 2) (Falasca et al., 2009; Krueger et al., 2011; Papafragkakis et al., 2012; Sims et al., 2014; Zitron et al., 2013).

Current treatments for anhedonia focus on increasing dopamine levels in the brain to improve patient outcomes. However, stimulants such as amphetamines and dopamine reuptake inhibitors have been proven to have limited efficacy in patients suffering with anhedonia (Butler Jr. et al., 2007; Moraska et al., 2010). Even so, the norepinephrine/dopamine reuptake inhibitor bupropion has proven efficacious against anhedonia in schizophrenia comorbid with MDD (Argyropoulos & Nutt, 2013; Lipina, Fletcher, Lee, Wong, & Roder, 2013). However, disagreement in clinical efficacy against anhedonia in MDD compared with other antidepressives (Rush et al., 2011) requires more research; the increase in endogenous dopamine that results from treatment may require close monitoring to avoid adverse side effects.

Augmentation therapy with atypical antipsychotics, many of which are partial dopamine agonists, for refractory MDD has also shown capacity in improving anhedonia outcomes. Aripiprazole, in a randomized controlled trial of 101 patients with MDD, demonstrated that augmentation was superior to antidepressant switching in individuals with a current depressive episode (Greenberg, Fournier, Sisitsky, Pike, & Kessler, 2015; McIntyre & O'Donovan, 2004). Aripiprazole augmentation also showed efficacy where other conventional antidepressants had no effect or worsened patient outcomes (Nelson, Rahman, Laubmeier, et al., 2014; Nelson, Thase, Bellocchio, et al., 2012). Adjunctive aripiprazole, olanzapine, and quetiapine have statistically significant response and remission rates compared with adjunctive placebo controls (Kato & Chang, 2013). The FDA has also approved augmentation treatment strategies using, aripiprazole in combination with quetiapine, olanzapine/fluoxetine and, more recently, brexpiprazole. Brexpiprazole is a D2 receptor partial agonist with a high affinity for 5-HT1A and 5-HT2A receptors and several adrenergic receptors and is the newest FDA-approved atypical antipsychotic augmentation agent (Citrome, 2015).

The optimal duration of add-on treatment with atypical agents is currently not known; therefore it would be prudent for individuals to undergo ongoing reassessment for a minimum of 6–12 months even if the treatments are well tolerated and effective.

Stimulants increase dopamine release in the short term; however, long-term effects of increased dopamine levels are limited. A more effective approach may involve an increase in endogenous dopamine synthesis via the BH4 pathway. For example, dopamine synthesis via TH can be improved by increasing BH4 availability. A comparable therapy for phenylketonuria using sapropterin (Kuvan), a synthetic form of BH4, has shown to be effective in reducing blood phenylalanine levels and the incidence of fatigue in patients tested (Burton, Bausell, Katz, Laduca, & Sullivan, 2010). Future research will need to determine its efficacy against anhedonic symptoms in MDD as it is unclear if the improvement in symptoms is due to a decrease in phenylalanine or improved dopamine synthesis.

Inflammation can affect dopaminergic neuronal function and can therefore be associated with anhedonia. A combination of downstream cytokine-induced metabolites contributes to the development of neuropsychiatric symptoms, including fatigue, anhedonia, and psychomotor slowing (Dantzer, O'Connor, Freund, Johnson, & Kelley, 2008; Felger & Miller, 2012). It is theorized that by treating inflammation, dopamine levels can be rebalanced and anhedonic symptoms reduced.

Inhibition of the IDO-KP pathway may offer an antiinflammatory-based approach to anhedonia treatment. Treatments of LPS mice with an IDO antagonist,

1-methyl-tryptophan (1-MT), have been shown to reverse the effects of LPS-induced depressive-like behavior in early studies (O'Connor et al., 2009, 2009). Other components of the KP could also be implemented in treatment. A recent metaanalysis of antiinflammatory drugs has shown that nonsteroidal antiinflammatory drugs (NSAIDs) and cytokine inhibitors are efficacious against depressive-like symptoms compared with placebo (Kohler et al., 2014). The cyclooxygenase-2 (COX2) inhibitor celecoxib showed promising preclinical results in this regard (McCormack, 2011). Studies in mice using celecoxib significantly reduced chronic stress, depressive-like, and anhedonic-like symptoms (Guo et al., 2009; Santiago et al., 2014). Human trials have yet to be done in order to develop feasible treatment protocols.

Elevated TNF-α is a consistent inflammatory finding in MDD and is a significant upstream contributor to inflammation in the dopaminergic system (Bob et al., 2010; Kauer-Sant'anna et al., 2008). Infliximab, a monoclonal antibody that binds TNF-α and inhibits binding with its receptor, has recently been shown to improve depressive symptoms in individuals (Tracey et al., 2008). Infliximab is currently approved as a treatment for rheumatoid arthritis in Canada, and recent evidence has shown that TNF antagonists have been associated with lower rates of MDD and anxiety disorders in individuals with rheumatoid arthritis (Loftus et al., 2008). The drug has also demonstrated its efficacy in a subgroup of individuals with treatment-resistant depression (Persoons et al., 2005) with treatments having been shown to improve fatigue and depressive subscores (Papakostas et al., 2013; Raison et al., 2013; Tookman, Jones, Dewitte, & Lodge, 2008; Watanabe et al., 2015). Efficacy of adjunctive infliximab on anhedonic symptoms of MDD has not been determined and is currently being investigated by our Toronto team via a 12-week, double-blind, placebo-controlled clinical trial. Translational medicine is an interdisciplinary field. Drug discovery and implementation will be increasingly focused on the ability to integrate compatible information from disparate domains. The future of treating anhedonia in depression will rely on translating knowledge from the domains of inflammation, metabolism, and mood. The clinical investigations on infliximab demonstrate an apt example of translating knowledge from inception to implementation.

X TREATMENT STRATEGIES TARGETING COGNITIVE DYSFUNCTIONS IN MDD

The treatment of cognitive dysfunction in MDD should begin with modifying, contributing, and/or exacerbating factors, for example, management of medical and psychiatric comorbidity and identification of iatrogenic/medication confounds. Modifiable determinants of MDD present opportunities for therapeutic intervention in the domain of treating cognitive dysfunction. Current and novel treatment approaches on cognitive function will be discussed.

Standardized psychosocial approaches to the treatment of cognitive dysfunction offer a good starting point. Cognitive remediation (CR) enhances cognitive activation via the implementation of a computerized task to improve learning and stimulate neuroplasticity. These dynamic tasks keep the players/patients at relatively high success rates despite changes in performance over time to elicit sustained motivation. CR continues to enhance cognitive activation via strategy development, monitoring, and pruning. Through the support of therapists and peers, those affected by MDD improve strategies to cognitively challenging tasks over time, leading to the pruning of ineffective strategies. "Far transfer" is the final step in CR, whereby improvements in cognition and problem solving are applied to everyday environments (Medalia, Revheim, & Herlands, 2009). To date, only the first step of CR, cognitive activation, has been applied to MDD. Preliminary results also suggest that adjunctive CR could improve cognitive outcomes in adults with MDD when combined with pharmacotherapy (Bowie et al., 2013; Bowie, McGurk, Mausbach, Patterson, & Harvey, 2012; Porter, Bowie, Jordan, & Malhi, 2013).

Computerized working memory tasks have also been shown to improve cognitive functioning [e.g., the Paced Auditory Serial Addition Test (PASAT), which requires patients to add sequentially presented digits]. Siegle, Thompson, Carter, Steinhauer, and Thase (2007) reported that the use of the PASAT increased metacognitive skills, decreased maladaptive thought patterns (i.e., rumination), and improved depressive symptom severity in an MDD patient cohort. Computer programs focused on sequencing mental arithmetic problems have been shown to improve full scale intelligence quotient (IQ) in those with MDD (Cha & McIntyre, 2016). This mode of therapy has also been shown to improve psychosocial and occupational functioning, as well as hippocampal and frontotemporal activation (Deckersbach, Kaur, & Hansen, 2015). However, the specific efficacy in clinical populations is still unknown. Implementation of such computational networks are also proving difficult whereby emerging big data approaches and blockchain technologies could provide feasible next steps in the development of such computational tasks.

Manual-based psychotherapies [e.g., cognitive behavioral therapy (CBT) and mindfulness-based therapies] are highly effective in the acute treatment of MDEs and have been proven effective in the maintenance of stable mood in MDD (Lam et al., 2013; Parikh et al., 2016). Psychotherapies primarily target "hot" cognition; effectiveness targeting "cold" cognition still requires further investigation. Of the aforementioned therapies, CBT is effective for ADHD, which exhibits significant disturbances in executive function and attention (Young et al., 2015). Alternatively, brain stimulation techniques such as repetitive transcranial magnetic stimulation (rTMS) have demonstrated procognitive effects in studies in depressive cohorts (Serafini et al., 2015). The improved patient tolerability, minimal propensity for cognitive impairment, and higher rates of remission against treatment-resistant variants of MDD offer a promising approach to target cognitive dysfunction comorbid with treatment-resistant depression. Indeed, brain imagining studies in MDD patients suggest rTMS could be a viable procognitive neuromodulatory strategy.

Aerobic exercise and/or resistance training present another adjunctive therapeutic opportunity that is both cost-effective and accessible. Aerobic exercise improves cognition, has negligible side effects, and has potential to be scaled as a population-level health intervention (Smith et al., 2010). A recent review by Stanton and Reaburn (2014) found that aerobic exercise regimes should occur at an interval of 3–4 times a week for 9 weeks at moderate intensity to alleviate depressive symptoms. The SMILE study showed lower depression relapse rates for those partaking in an exercise regimen when compared with those taking sertraline, a SSRI (Deckersbach et al., 2015). Both acute and regular aerobic exercises confer improvements in memory; moreover, acute physical activity has greater impact on short- and long-term memory in comparison with chronic physical activity. Additionally, individuals with mild cognitive impairment also see greater improvements in memory compared with cognitively normal individuals when engaging in regular exercise suggesting a staggered procognitive effect due to exercise that confers greater benefits to populations with cognitive difficulties due to depressive mood disorders (Heyn, Abreu, & Ottenbacher, 2004; Smith et al., 2010; Stanton and Reaburn, 2014). Improvements in psychomotor speed, attention, visual and working memory, spatial planning, and executive control have been observed in a dose-dependent manner after exercise augmentation (Greer, Grannemann, Chansard, Karim, & Trivedi, 2015).

The catecholamines, norepinephrine and dopamine, are important neuromodulators of memory, executive function, and attention. These foregoing neuromodulators increase peripherally during and after exercise, but they are impermeable to the blood-brain barrier, providing challenges in developing efficacious approaches to target cognitive dysfunctions in MDD. Moreover, catecholamines have reciprocal relationships with glutamatergic, GABAergic, and other synaptic neurons to increase arousal in relevant brain areas (Kubesch et al., 2003; McMorris, 2009). Increased unbound tryptophan in the blood can cross the blood-brain barrier and be converted to serotonin in the brain. The dopaminergic system has been implicated in anhedonia and impaired reward arousal mechanisms in depression (Pan et al., 2017). Increased hippocampal and whole-brain serotonin concentrations, from disparate therapeutic mechanisms of action, have been linked to neurogenesis, synaptic growth, and improved connectivity between the hippocampus, prefrontal cortex and anterior cingulate gyrus (Cha & McIntyre, 2016; Ntlekofer et al., 2013). Brain-derived neurotrophic growth factor (BDNF), vascular endothelial growth factor (VEGF), and insulin-like growth factor-1 (IGF-1) are associated with exercise, neurogenesis, and long-term potentiation (Cotman, Berchtold, & Christie, 2007). These brain correlates are discussed in greater detail elsewhere and may prove useful in the development of treatment targets (Millan et al., 2012).

Available evidence concerning conventional antidepressants indicates that improvements in measures of cognitive function in adults ages 18–65 are consistent with improvements in conventional depressive outcomes. Whether most conventional antidepressants exert direct, independent, and clinically significant effects on cognitive functions in adults with MDD is currently unknown. There are no current FDA-approved agents or interventions that have proved efficacious and/or tolerable that specifically target cognitive dysfunction associated with MDD. Cha et al. (2016) examined the efficacy of several antidepressant agents in improving cognitive function in individuals with MDD. Bupropion XL and escitalopram have been reported to improve verbal memory and delayed free recall (Soczynska et al., 2014). Similarly, sertraline has been associated with improved psychomotor performance (Constant et al., 2005; Schrijvers et al., 2009). With regard to SNRIs, duloxetine has been found to significantly improve cognitive functions when compared with placebo where patients with greater severity reported greater procognitive effect after drug administration (Raskin et al., 2007). Specifically, duloxetine improved verbal learning and memory while concurrently decreasing general symptoms of depression. Raskin et al. (2012) found that duloxetine improved direct composite cognition scores by 90.0% on the geriatric depression rating scale and 81.3% on the Hamilton Depression Rating Scale-17 Items (HAM-D 17). Both duloxetine and the multimodal antidepressant, vortioxetine, have been shown to improve scores on the Rey Auditory Verbal Learning Test (RAVLT), with significant improvements in acquisition time and delayed recall. Vortioxetine demonstrated a higher direct effect than duloxetine in both RAVLT recall and acquisition scores, despite both improving depressive mood-related symptom severity (McIntyre, Lophaven, & Olsen, 2014). Additionally, duloxetine is limited to improving measures of learning and memory, while vortioxetine improves a broader range of cognitive functions (i.e., executive function, learning, memory, processing speed, and concentration). Vortioxetine has also been shown to significantly improve scores on cognitive measures after 8 weeks of treatment using either 10- or 20-mg doses when compared with placebo (Cha & McIntyre, 2016; Mahableshwarkar, Zajecka, Jacobson, Chen, & Keefe, 2015; McIntyre, Harrison, et al., 2016; McIntyre, Woldeyohannes, et al., 2016). Currently vortioxetine is the only approved treatment agent that targets cognitive impairments.

Evidence demonstrating efficacy in the treatment of depressive symptoms in MDD patients using psychostimulants is inconsistent. Lisdexamfetamine, however, has demonstrated an ability to specifically target executive dysfunctions in MDD, particularly among individuals with milder symptomatology comorbid with executive function deficits (Madhoo et al., 2014). However, in some cases with full or partial remission of depressive symptoms, lisdexamfetamine has been shown to improve executive function, and some evidence suggests an ability for lisdexamfetamine to target executive dysfunctions in MDD (Madhoo et al., 2014).

Ketamine has demonstrated efficacy in the treatment of recurrent MDD (Venero, 2014). Available evidence indicates that cognitive function may serve as a predictor of response to ketamine treatment. An evidence-based approach is still needed to improve and optimize ketamine therapy (Price et al., 2014; Zanos et al., 2016). It has been hypothesized that the antisuicide effects of ketamine may be mediated by improvements in executive functions (Lee et al., 2016). A review written by Lee et al. (2016) discusses procognitive effects of ketamine in greater detail. Translational research strategies played a significant role in supporting ketamine as a novel and efficacious therapeutic intervention in the treatment of cognitive dysfunction in depression.

Preliminary data also support the hypothesis that incretins may improve cognitive function. Incretins are involved in gastric motility and act as insulin secretion analogues. Exogenously administered glucagon-like peptide 1 (GLP-1) agonists (e.g., liraglutide) are FDA-indicated for adults with type 2 diabetes mellitus. GLP-1 is also synthesized in the nucleus tractus solitarius, and its receptors are distributed throughout the brain, with topographical distribution represented in cognitive control substrates. Preliminary evidence indicates that liraglutide administered at a dose of 1.8 mg was able to improve depressive and cognitive measures in adults with a current MDE (Mansur et al., 2017). The foregoing results validate previous findings that liraglutide has neuroprotective and neurotrophic properties. Indeed, liraglutide and similar incretin agonists provide interesting new avenues for pharmaceutical discovery. Metabolic regulators present unique opportunities as treatment targets in specific patient subpopulations with elevated metabolic and inflammatory disorder risks. Intranasal insulin

is an example of a promising new approach to improving cognition in primary MDD patients with a history of diabetes and insulin dysregulation. In the brain, insulin inhibits proapoptotic pathways and is critical for neuroplasticity, neurogenesis, and neuronal growth/survival. Insulin receptors are found throughout neural circuits involved in cognitive and emotional processing (Cha & McIntyre, 2016). Treatment strategies involving brain insulin may prove efficacious targeting cognitive dysfunctions.

Elevated levels of circulating proinflammatory cytokines have been consistently reported in depressed individuals and have been linked to cognitive impairment. TNF-α is among the most consistently identified proinflammatory cytokine abnormalities in MDD (Craft et al., 2012; Kauer-Sant'anna et al., 2008; Maes et al., 2009; McIntyre et al., 2015; Raison et al., 2013; Vevera, Uhrova, Benakova, & Zima, 2010). Infliximab, a TNF antagonist, binds to soluble and transmembrane forms of TNF-α and inhibits binding of TNF-α with its receptors (Loftus et al., 2008; Raison et al., 2013). Treatment with TNF antagonists has been associated with lower rates of MDD in subpopulations with elevated inflammation (Raison et al., 2013) and in individuals with rheumatoid arthritis (Loftus et al., 2008). Evidence also demonstrates its efficacy in subgroups with treatment-resistant depression (Persoons et al., 2005; Tookman et al., 2008). Efficacy of adjunctive infliximab on cognitive symptoms of depressive disorders has not been determined and is currently being investigated in a clinical trial (NCT02363738). The focus on translational techniques provides exciting opportunities for the discovery of novel treatment interventions for cognitive dysfunction in depression. Promising concept-linking domains traditionally distant to psychiatry are now being integrated and investigated. Translational approaches in clinical trials looking at infliximab and other antiinflammatory and/or metabolic agents allowed clinician–scientists to target chronically underserved domains of depression. Connecting multidisciplinary knowledge from bench to bedside is the future of drug discovery and development.

XI CONCLUSION

The prevailing disease models implicate cognition as a core psychopathological disturbance in MDD. The pertinence of cognitive dysfunction is that cognition is a principal mediator of psychosocial and workplace functional outcomes. The current clinical paradigms in the treatment and assessment of depressive symptoms have insufficiently addressed cognition with existing multimodal depression treatments. Several factors modify cognitive functions in MDD, providing opportunities for clinical intervention. Clinicians are encouraged to screen and assess cognitive functions in adults with MDD and to track performance on cognitive assessments to be fully informed of remission outcomes. The interest in the development of domain-based approaches to treatment provides a pragmatic impetus for therapeutic discovery. Among the treatment options in MDD, no single modality or agent has proven to be the gold standard in targeting cognition. The path toward functional recovery for many persons with MDD includes targeting impairment of cognitive function, hitherto a therapeutic target that has not been prioritized.

An inflammatory approach to anhedonia seems promising. Further steps are necessary to elucidate a mechanism involving proinflammatory metabolites in the pathogenesis of anhedonia in the context of MDD. Causal relationships still need to be established between the aforementioned metabolite pathways (chiefly BH4 and KP) with disturbances in dopaminergic systems responsible for anhedonia. Additionally, modulating influences from various other established theories should also be considered (e.g., environmental influences, activation of the HPA axis, and genetic and/or congenital defects). Biomarkers for MDD using proinflammatory cytokines are promising, which may imply the significant relationship between inflammation and MDD. Efficacy of proposed treatment methods could also be further explored to test the validity of the inflammatory hypothesis of anhedonia.

Conflict of Interest Disclosures

Roger S. McIntyre is a consultant to speak on behalf of and/or has received research support from Allergan, AstraZeneca, Bayer, Bristol-Myers, Squibb, Janssen-Ortho, Eli Lilly, Lundbeck, Merck, Otsuka, Pfizer, Sunovion, and Takeda.
Zihang Pan has received research support from the Canadian Institutes of Health Research (CIHR).
Radu C. Grovu has no conflicts of interest to disclose.

References

American Psychiatric Association. (2013). *Diagnostic and statistical manual of mental disorders (DSM-5)*. American Psychiatric Pub.

Argyropoulos, S. V., & Nutt, D. J. (2013). Anhedonia revisited: is there a role for dopamine-targeting drugs for depression? *Journal of Psychopharmacology, 27*(10), 869–877.

Ballard, I. C., Murty, V. P., Carter, R. M., MacInnes, J. J., Huettel, S. A., & Adcock, R. A. (2011). Dorsolateral prefrontal cortex drives mesolimbic dopaminergic regions to initiate motivated behavior. *The Journal of Neuroscience, 31*(28), 10340–10346.

Bedwell, J. S., Gooding, D. C., Chan, C. C., & Trachik, B. J. (2014). Anhedonia in the age of RDoC. *Schizophrenia Research, 160*(1-3), 226–227.

Behan, W. M., McDonald, M., Darlington, L. G., & Stone, T. W. (1999). Oxidative stress as a mechanism for quinolinic acid-induced hippocampal damage: protection by melatonin and deprenyl. *British Journal of Pharmacology, 128*, 1754–1760.

Bob, P., Raboch, J., Maes, M., Susta, M., Pavlat, J., Jasova, D., et al. (2010). Depression, traumatic stress and interleukin-6. *Journal of Affective Disorders, 120*(1–3), 231–234.

Bortolato, B., Carvalho, A. F., Soczynska, J. K., Perini, G. I., & McIntyre, R. S. (2015). The involvement of TNF-α in cognitive dysfunction associated with major depressive disorder: an opportunity for domain specific treatments. *Current Neuropharmacology, 13*, 558–576.

Bowie, C. R., Gupta, M., Holshausen, K., Jokic, R., Best, M., & Milev, R. (2013). Cognitive remediation for treatment-resistant depression: effects on cognition and functioning and the role of online homework. *The Journal of Nervous and Mental Disease, 201*(8), 680–685.

Bowie, C. R., McGurk, S. R., Mausbach, B., Patterson, T. L., & Harvey, P. (2012). Combined cognitive remediation and functional skills training for schizophrenia: effects on cognition, functional competence, and real-world behavior. *The American Journal of Psychiatry, 169*(7), 710–718.

Burton, B. K., Bausell, H., Katz, R., Laduca, H., & Sullivan, C. (2010). Sapropterin therapy increases stability of blood phenylalanine levels in patients with BH4-responsive phenylketonuria (PKU). *Molecular Genetics and Metabolism, 101*, 110–114.

Butler, J. M., Jr., Case, L. D., Atkins, J., Frizzell, B., Sanders, G., Griffin, P., et al. (2007). A phase III, double-blind, placebo-controlled prospective randomized clinical trial of D-threo methylphenidate HCl in brain tumor patients receiving radiation therapy. *International Journal of Radiation Oncology, Biology, Physics, 69*, 1496–1501.

Campbell, B. M., Charych, E., Lee, A. W., & Möller, T. (2014). Kynurenines in CNS disease: regulation by inflammatory cytokines. *Frontiers in Neuroscience, 8*, 12.

Cao, C., Matsumura, K., Yamagata, K., & Watanabe, Y. (1997). Involvement of cyclooxygenase-2 in LPS-induced fever and regulation of its mRNA by LPS in the rat brain. *The American Journal of Physiology, 272*, R1712–R1725.

Capuron, L., Gumnick, J. F., Musselman, D. L., Lawson, D. H., Reemsnyder, A., Nemeroff, C. B., et al. (2002). Neurobehavioral effects of interferon-alpha in cancer patients: phenomenology and paroxetine responsiveness of symptom dimensions. *Neuropsychopharmacology, 26*, 643–652.

Capuron, L., Hauser, P., Hinze-Selch, D., Miller, A. H., & Neveu, P. J. (2002). Treatment of cytokine-induced depression. *Brain, Behavior, and Immunity, 16*, 575–580.

Capuron, L., Neurauter, G., Musselman, D. L., Lawson, D. H., Nemeroff, C. B., Fuchs, D., et al. (2003). Interferon-alpha-induced changes in tryptophan metabolism: relationship to depression and paroxetine treatment. *Biological Psychiatry, 54*(9), 906–914.

Carvalho, A. F., Berk, M., Hyphantis, T. N., & McIntyre, R. S. (2014). The integrative management of treatment-resistant depression: a comprehensive review and perspectives. *Psychotherapy and Psychosomatics, 83*, 70–88.

Carvalho, A. F., Miskowiak, K., Hyphantis, T. N., Kohler, C. A., Alves, G. S., Bortolato, B., et al. (2014). Cognitive dysfunction in depression-pathophysiology and novel targets. *CNS & Neurological Disorders—Drug Targets, 13*(10), 1819–1835.

Cha, D., De Michele, F., Soczynska, J. K., Woldeyohannes, H. O., Kaidanovich-Beilin, O., Carvalho, A. F., et al. (2014). The putative impact of metabolic health on default mode network activity and functional connectivity in neuropsychiatric disorders. *CNS & Neurological Disorders—Drug Targets, 13*(10), 1750–1758.

Cha, D. S., & McIntyre, R. S. (2016). *Cognitive impairment in major depressive disorder: Clinical relevance, biological substrates, and treatment opportunities*. Cambridge University Press.

Charney, D. S., & Drevets, W. C. (2012). Neurobiological basis of anxiety disorders. In *Neuropsychopharmacology: The fifth generation of progress* (pp. 119–132). New York: Lippincott Williams and Wilkins.

Chen, Z., Jalabi, W., Shpargel, K. B., Farabaugh, K. T., Dutta, R., Yin, X., et al. (2012). Lipopolysaccharide-induced microglial activation and neuroprotection against experimental brain injury is independent of hematogenous TLR4. *Journal of Neuroscience, 32*(34), 11706–11715.

Chentsova-Dutton, Y., & Hanley, K. (2010). The effects of anhedonia and depression on hedonic responses. *Psychiatry Research, 179*(2), 176–180.

Citrome, L. (2015). Brexpiprazole: a new dopamine D2 receptor partial agonist for the treatment of schizophrenia and major depressive disorder. *Drugs of Today (Barcelona, Spain), 51*(7), 397–414.

Clark, I. A., Alleva, L. M., & Vissel, B. (2010). The roles of TNF in brain dysfunction and disease. *Pharmacology & Therapeutics, 128*(3), 519–548.

Collins, P. Y., Patel, V., Joestl, S. S., March, D., Insel, T. R., Daar, A. S., et al. (2011). Grand challenges in global mental health. *Nature, 475*(7354), 27–30.

Constant, E. L., Adam, S., Gillain, B., Seron, X., Bruyer, R., & Seghers, A. (2005). Effects of sertraline on depressive symptoms and attentional and executive functions in major depression. *Depression and Anxiety, 21*(2), 78–89.

Cotman, C. W., Berchtold, N. C., & Christie, L. A. (2007). Exercise builds brain health: key roles of growth factor cascades and inflammation. *Trends in Neurosciences, 30*(9), 464–472.

Craft, S., Baker, L. D., Montine, T. J., Minoshima, S., Watson, G. S., Claxton, A., et al. (2012). Intranasal insulin therapy for Alzheimer disease and amnestic mild cognitive impairment: a pilot clinical trial. *Archives of Neurology, 69*(1), 29–38.

Cunnington, C., & Channon, K. M. (2010). Tetrahydrobiopterin: pleiotropic roles in cardiovascular pathophysiology. *Heart, 96*, 1872–1877.

Dantzer, R., & Kelley, K. W. (2007). Twenty years of research on cytokine-induced sickness behavior. *Brain, Behavior, and Immunity, 21*(2), 153–160.

Dantzer, R., O'Connor, J. C., Freund, G. G., Johnson, R. W., & Kelley, K. W. (2008). From inflammation to sickness and depression: when the immune system subjugates the brain. *Nature Reviews. Neuroscience, 9*(1), 46–56.

Deckersbach, T., Kaur, N., & Hansen, N. S. (2015). A neurocognitive perspective. In *Symptom to synapse: A neurocognitive perspective on clinical psychology* (p. 278): CRC Press.

Delgado, P. L. (2000). Depression: the case for a monoamine deficiency. *The Journal of Clinical Psychiatry, 61*, 7–11.

Dillon, D. G., Rosso, I. M., Pechtel, P., Killgore, W. D., Rauch, S. L., & Pizzagalli, D. A. (2014). Peril and pleasure: an rdoc-inspired examination of threat responses and reward processing in anxiety and depression. *Depression and Anxiety, 31*(3), 233–249.

Dowlati, Y., Herrmann, N., Swardfager, W., Liu, H., Sham, L., Reim, E. K., et al. (2010). A meta-analysis of cytokines in major depression. *Biological Psychiatry, 67*(5), 446–457.

Dumitrescu, C., Biondi, R., Xia, Y., Cardounel, A. J., Druhan, L. J., Ambrosio, G., et al. (2007). Myocardial ischemia results in tetrahydrobiopterin (BH4) oxidation with impaired endothelial function ameliorated by BH4. *Proceedings of the National Academy of Sciences, 104*(38), 15081–15086.

Espejo, E. F., & Minano, F. J. (1999). Prefrontocortical dopamine depletion induces antidepressant-like effects in rats and alters the profile of desipramine during Porsolt's test. *Neuroscience, 88*(2), 609–615.

Ericsson, A., Kovacs, K. J., & Sawchenko, P. E. (1994). A functional anatomical analysis of central pathways subserving the effects of interleukin-1 on stress-related neuroendocrine neurons. *Journal of Neuroscience, 14*(2), 897–913.

Falasca, K., Mancino, P., Ucciferri, C., Dalessandro, M., Manzoli, L., Pizzigallo, E., et al. (2009). Quality of life, depression, and cytokine patterns in patients with chronic hepatitis C treated with antiviral therapy. *Clinical and Investigative Medicine, 32*(3), 212–218.

Felger, J. C., & Miller, A. H. (2012). Cytokine effects on the basal ganglia and dopamine function: the subcortical source of inflammatory malaise. *Frontiers in Neuroendocrinology, 33*(3), 315–327.

Frodl, T., & Amico, F. (2014). Is there an association between peripheral immune markers and structural/functional neuroimaging findings? *Progress in Neuro-Psychopharmacology & Biological Psychiatry, 48*, 295–303.

Gabbay, V., Ely, B. A., Babb, J., & Liebes, L. (2012). The possible role of the kynurenine pathway in anhedonia in adolescents. *Journal of Neural Transmission, 119*(2), 253–260.

Gaillard, R., Gourion, D., & Llorca, P. M. (2013). L'anhédonie dans la dépression. *L'Encéphale, 39*(4), 296–305.

Goldstein, L. E., Leopold, M. C., Huang, X., Atwood, C. S., Saunders, A. J., Hartshorn, M., et al. (2000). 3-Hydroxykynurenine and 3-hydroxyanthranilic acid generate hydrogen peroxide and promote α-crystallin cross-linking by metal ion reduction. *Biochemistry, 39*(24), 7266–7275.

Greenberg, P. E., Fournier, A. A., Sisitsky, T., Pike, C. T., & Kessler, R. C. (2015). The economic burden of adults with major depressive disorder in the United States (2005 and 2010). *The Journal of Clinical Psychiatry, 76*(2), 155–162.

Greer, T. L., Grannemann, B. D., Chansard, M., Karim, A. I., & Trivedi, M. H. (2015). Dose-dependent changes in cognitive function with exercise augmentation for major depression: results from the TREAD study. *European Neuropsychopharmacology, 25*(2), 248–256.

Guo, J. Y., Li, C. Y., Ruan, Y. P., Sun, M., Qi, X. L., Zhao, B. S., et al. (2009). Chronic treatment with celecoxib reverses chronic unpredictable stress-induced depressive-like behaviour via reducing cyclooxygenase-2 expression in rat brain. *European Journal of Pharmacology, 612*(1-3), 54–60.

Haas, H. S., & Schauenstein, K. (1997). Neuroimmunomodulation via limbic structures—the neuroanatomy of psychoimmunology. *Progress in Neurobiology, 51*(2), 195–222.

Hamilton, J. P., Furman, D. J., Chang, C., Thomason, M. E., Dennis, E., & Gotlib, I. H. (2011). Default-mode and task-positive network activity in major depressive disorder: implications for adaptive and maladaptive rumination. *Biological Psychiatry, 70*(4), 327–333.

Haroon, E., Raison, C. L., & Miller, A. H. (2012). Psychoneuroimmunology meets neuropsychopharmacology: translational implications of the impact of inflammation on behavior. *Neuropsychopharmacology, 37*(1), 137–162.

Harrison, J. E., Lam, R. W., Baune, B. T., & McIntyre, R. S. (2016). Selection of cognitive tests for trials of therapeutic agents. *The Lancet Psychiatry, 3*(6), 499.

Harvey, P. O., Fossati, P., Pochon, J. B., Levy, R., LeBastard, G., Lehéricy, S., et al. (2005). Cognitive control and brain resources in major depression: an fMRI study using the n-back task. *NeuroImage, 26*(3), 860–869.

Hasselbalch, B. J., Knorr, U., & Kessing, L. V. (2011). Cognitive impairment in the remitted state of unipolar depressive disorder: a systematic review. *Journal of Affective Disorders, 134*(1-3), 20–31.

Heyn, P., Abreu, B. C., & Ottenbacher, K. J. (2004). The effects of exercise training on elderly persons with cognitive impairment and dementia: a meta-analysis. *Archives of Physical Medicine and Rehabilitation, 85*(10), 1694–1704.

Horger, B. A., & Roth, R. H. (1996). The role of mesoprefrontal dopamine neurons in stress. *Critical Reviews in Neurobiology, 10*(3-4), 395–418.

Jacobs, P., Ohinmaa, A., Escober-Doran, C., Patterson, S., & Slomp, M. (2009). PMH31 measuring the economic burden of depression using patient records. *Value in Health, 12*(7), A356.

Jensen, J., McIntosh, A. R., Crawley, A. P., Mikulis, D. J., Remington, G., & Kapur, S. (2003). Direct activation of the ventral striatum in anticipation of aversive stimuli. *Neuron, 40*(6), 1251–1257.

Kato, M., & Chang, C. M. (2013). Augmentation treatments with second-generation antipsychotics to antidepressants in treatment-resistant depression. *CNS Drugs, 27*(Suppl1), S11–S19.

Kauer Sant'anna, M., Kapczinski, F., Andreazza, A. C., Bond, D. J., Lam, R. W., Trevor, Y. L., et al. (2008). Brain-derived neurotrophic factor and inflammatory markers in patients with early- vs. late-stage bipolar disorder. *The International Journal of Neuropsychopharmacology, 4*, 1–12.

Kerfoot, S. M., D'Mello, C., & Nguyen, H. (2006). TNF-alpha-secreting monocytes are recruited into the brain of cholestatic mice. *Hepatology, 43*(1), 154–162.

Keedwell, P. A., Andrew, C., Williams, S. C., Brammer, M. J., & Phillips, M. L. (2005). The neural correlates of anhedonia in major depressive disorder. *Biological Psychiatry, 58*(11), 843–853.

Kheirbek, M. A., & Hen, R. (2011). Dorsal vs ventral hippocampal neurogenesis: implications for cognition and mood. *Neuropsychopharmacology, 36*(1), 373–374. https://doi.org/10.1038/npp.2010.148.

Kohler, O., Benros, M. E., Nordentoft, M., Farkouh, M. E., Iyengar, R. L., Mors, O., et al. (2014). Effect of anti-inflammatory treatment on depression, depressive symptoms, and adverse effects: a systematic review and meta-analysis of randomized clinical trials. *JAMA Psychiatry, 71*(12), 1381–1391.

Krueger, C., Hawkins, K., Wong, S., Enns, M. W., Minuk, G., & Rempel, J. D. (2011). Persistent pro-inflammatory cytokines following the initiation of pegylated IFN therapy in hepatitis C infection is associated with treatment-induced depression. *Journal of Viral Hepatitis, 18*(7), e284–e291.

Kubesch, S., Bretschneider, V., Freudenmann, R., Weidenhammer, N., Lehmann, M., Spitzer, M., et al. (2003). Aerobic endurance exercise improves executive functions in depressed patients. *Journal of Clinical Psychiatry, 64*(9), 1005–1012.

Lam, R. W., Kennedy, S. H., McIntyre, R. S., & Khullar, A. (2014). Cognitive dysfunction in major depressive disorder: effects on psychosocial functioning and implications for treatment. *The Canadian Journal of Psychiatry, 59*(12), 649–654.

Lam, R. W., Parikh, S. V., Ramasubbu, R., Michalak, E. E., Tam, E. M., Axler, A., et al. (2013). Effects of combined pharmacotherapy and psychotherapy for improving work functioning in major depressive disorder 1. *The British Journal of Psychiatry : The Journal of Mental Science, 203*(1472–1465 (Electronic)), 358–365.

Lee, Y., Syeda, K., Maruschak, N. A., Cha, D. S., Mansur, R. B., Wium-Andersen, I. K., et al. (2016). A new perspective on the anti-suicide effects with ketamine treatment: a procognitive effect. *Journal of Clinical Psychopharmacology, 36*(1), 50–56.

Lépine, J. P., & Briley, M. (2011). The increasing burden of depression. *Neuropsychiatric Disease and Treatment, 7*(Suppl 1), 3.

Lim, B. K., Huang, K. W., Grueter, B. A., Rothwell, P. E., & Malenka, R. C. (2012). Anhedonia requires MC4R-mediated synaptic adaptations in nucleus accumbens. *Nature, 487*(7406), 183–189.

Lipina, T. V., Fletcher, P. J., Lee, F. H., Wong, A. H., & Roder, J. C. (2013). Disrupted-in-schizophrenia-1 Gln31Leu polymorphism results in social anhedonia associated with monoaminergic imbalance and reduction of CREB and β-arrestin-1,2 in the nucleus accumbens in a mouse model of depression. *Neuropsychopharmacology, 38*(3), 423–436.

Loftus, E. V., Feagan, B. G., Colombel, J. F., Rubin, D. T., Wu, E. Q., Yu, A. P., et al. (2008). Effects of adalimumab maintenance therapy on health-related quality of life of patients with Crohn's disease: patient-reported outcomes of the CHARM trial. *The American Journal of Gastroenterology, 103*(12), 3132–3141.

Madhoo, M., Keefe, R. S. E., Roth, R. M., Sambunaris, A., Wu, J., Trivedi, M. H., et al. (2014). Lisdexamfetamine dimesylate augmentation in adults with persistent executive dysfunction after partial or full remission of major depressive disorder. *Neuropsychopharmacology, 39*(6), 1388–1398.

Maes, M., Bosmans, E., Suy, E., Vandervorst, C., DeJonckheere, C., & Raus, J. (1991). Depression-related disturbances in mitogen-induced lymphocyte responses and interleukin-1β and soluble interleukin-2 receptor production. *Acta Psychiatrica Scandinavica, 84*(4), 379–386.

Maes, M. (1999). Major depression and activation of the inflammatory response system. In *Cytokines, stress, and depression* (pp. 25–46): Springer.

Maes, M. (2008). The cytokine hypothesis of depression: inflammation, oxidative & nitrosative stress (IO&NS) and leaky gut as new targets for adjunctive treatments in depression. *Neuro Endocrinology Letters, 29*(3), 287–291.

Maes, M., Yirmyia, R., Noraberg, J., Brene, S., Hibbeln, J., Perini, G., et al. (2009). The inflammatory & neurodegenerative (I&ND) hypothesis of depression: leads for future research and new drug developments in depression. *Metabolic Brain Disease, 24*(1), 27–53.

Mahableshwarkar, A. R., Zajecka, J., Jacobson, W., Chen, Y., & Keefe, R. S. (2015). A randomized, placebo-controlled, active-reference, double-blind, flexible-dose study of the efficacy of vortioxetine oncognitive function in major depressive disorder. *Neuropsychopharmacology, 40*(8), 2025–2037.

Mansur, R. B., Ahmed, J., Cha, D. S., Woldeyohannes, H. O., Subramaniapillai, M., Lovshin, J., et al. (2017). Liraglutide promotes improvements in objective measures of cognitive dysfunction in individuals with mood disorders: a pilot, open-label study. *Journal of Affective Disorders, 207*, 114–120.

Marcus, M., Yasamy, M. T., Ommeren, M. V., Chisholm, D., & Saxena, S. (2012). *Depression: A global public health concern.* 1 (pp. 6–8). WHO Department of Mental Health and Substance Abuse.

Medalia, A., Revheim, N., & Herlands, T. (2009). *Cognitive remediation for psychological disorders: Therapist guide.* Oxford University Press.

McIntyre, R. S., Harrison, J., Loft, H., Jacobson, W., & Olsen, C. K. (2016). The effects of vortioxetine on cognitive function in patients with major depressive disorder: a meta-analysis of three randomized controlled trials. *The International Journal of Neuropsychopharmacology, 19*, pyw055.

McIntyre, R. S., Lophaven, S., & Olsen, C. K. (2014). A randomized, double-blind, placebo-controlled study of vortioxetine on cognitive function in depressed adults. *International Journal of Neuropsychopharmacology, 17*(10), 1557–1567.

McIntyre, R. S., & O'Donovan, C. (2004). The human cost of not achieving full remission in depression. *Canadian Journal of Psychiatry, 49*(3 Suppl. 1), 10S–16S.

McCormack, P. L. (2011). Celecoxib. *Drugs, 71*(18), 2457.

McIntyre, R. S., Woldeyohannes, H. O., Soczynska, J. K., Maruschak, N. A., Wium-Andersen, I. K., Vinberg, M., et al. (2016). Anhedonia and cognitive function in adults with MDD: results from the International Mood Disorders Collaborative Project. *CNS Spectrums, 21*, 362–366.

McIntyre, R. S., Xiao, H. X., Syeda, K., Vinberg, M., Carvalho, A. F., Mansur, R. B., et al. (2015). The prevalence, measurement, and treatment of the cognitive dimension/domain in major depressive disorder. *CNS Drugs, 29*(7), 577–589.

McMorris, T. (2009). Exercise and cognitive function: a neuroendocrinological explanation. In *Exercise and cognitive function* (pp. 41–68): Wiley Online Library.

McNally, L., Bhagwagar, Z., & Hannestad, J. (2008). Inflammation, glutamate, and glia in depression: a literature review. *CNS Spectrums, 13*(6), 501–510.

Millan, M. J., Agid, Y., Brune, M., Bullmore, E. T., Carter, C. S., Clayton, N. S., et al. (2012). Cognitive dysfunction in psychiatric disorders: characteristics, causes and the quest for improved therapy. *Nature Reviews. Drug Discovery, 11*(2), 141–168.

Moraska, A. R., Sood, A., Dakhil, S. R., Sloan, J. A., Barton, D., Atherton, P. J., et al. (2010). Phase III, randomized, double-blind, placebo-controlled study of long acting methylphenidate for cancer-related fatigue: North Central Cancer Treatment Group NCCTG-N05C7 Trial. *Journal of Clinical Oncology: Official Journal of the American Society of Clinical Oncology, 28*, 3673–3679.

Najjar, S., Pearlman, D. M., Alper, K., Najjar, A., & Devinsky, O. (2013). Neuroinflammation and psychiatric illness. *Journal of Neuroinflammation, 10*, 43.

Naismith, S. L., Longley, W. A., Scott, E. M., & Hickie, I. B. (2007). Disability in major depression related to self-rated and objectively-measured cognitive deficits: a preliminary study. *BMC Psychiatry*, *7*(1), 32.

Nelson, J. C., Rahman, Z., Laubmeier, K. K., et al. (2014). Efficacy of adjunctive aripiprazole in patients with major depressive disorder whose symptoms worsened with antidepressant monotherapy. *CNS Spectrums*, *19*(6), 528–534.

Nelson, J. C., Thase, M. E., Bellocchio, E. E., et al. (2012). Efficacy of adjunctive aripiprazole in patients with major depressive disorder who showed minimal response to initial antidepressant therapy. *International Clinical Psychopharmacology*, *27*(3), 125–133.

Nestler, E. J., & Carlezon, W. A. (2006). The mesolimbic dopamine reward circuit in depression. *Biological Psychiatry*, *59*(12), 1151–1159.

Neurauter, G., Schrocksnadel, K., Scholl-Burgi, S., Sperner-Unterweger, B., Schubert, C., Ledochowski, M., et al. (2008). Chronic immune stimulation correlates with reduced phenylalanine turnover. *Current Drug Metabolism*, *9*(7), 622–627.

Ntlekofer, K. A., Berchtold, N. C., Malvaez, M., Carlos, A. J., McQuown, S. C., Cunningham, M. J., et al. (2013). Exercise and sodium butyrate transform a subthreshold learning event into long-term memory via a brain-derived neurotrophic factor-dependent mechanism. *Neuropsychopharmacology*, *38*(10), 2027–2034.

Nutt, D. J. (2007). Relationship of neurotransmitters to the symptoms of major depressive disorder. *The Journal of Clinical Psychiatry*, *69*, 4–7.

O'Connor, J. C., Lawson, M. A., Andre, C., Briley, E. M., Szegedi, S. S., et al. (2009). Induction of IDO by bacille Calmette-Guerin is responsible for development of murine depressive-like behavior. *Journal of Immunology*, *182*, 3202–3212.

O'Connor, J. C., Lawson, M. A., Andre, C., Moreau, M., Lestage, J., Castanon, N., et al. (2009). Lipopolysaccharide-induced depressive-like behavior is mediated by indoleamine 2,3-dioxygenase activation in mice. *Molecular Psychiatry*, *14*, 511–522.

Olds, J., & Travis, R. P. (1960). Effects of chlorpromazine, meprobamate, pentobarbital and morphine on self-stimulation. *Journal of Pharmacology and Experimental Therapeutics*, *128*(4), 397–404.

Pan, Z., Rosenblat, J. D., Swardfager, W., & McIntyre, R. S. (2017). Role of proinflammatory cytokines in dopaminergic system disturbances, implications for anhedonia features of MDD. *Current Pharmaceutical Design*, *23*(14), 2065–2072.

Papafragkakis, H., Rao, M., Moehlen, M., Dhillon, S., & Martin, P. (2012). Depression and pegylated interferon-based hepatitis C treatment. *International Journal of Interferon, Cytokine and Mediator Research*, *4*, 25–35.

Papakostas, G. I., Shelton, R. C., Kinrys, G., Henry, M. E., Bakow, B. R., Lipkin, S. H., et al. (2013). Assessment of a multi-assay, serum-based biological diagnostic test for major depressive disorder: a pilot and replication study. *Molecular Psychiatry*, *18*, 332–339.

Parikh, S. V., Quilty, L. C., Ravitz, P., Rosenbluth, M., Pavlova, B., Grigoriadis, S., et al. (2016). Canadian Network for Mood and Anxiety Treatments (CANMAT) 2016 clinical guidelines for the management of adults with Major depressive disorder: section 2. Psychological treatments. *Canadian Journal of Psychiatry*, *61*(9), 524–539.

Persoons, P., Vermeire, S., Demyttenaere, K., Fischler, B., Vandenberghe, J., Van, O. L., et al. (2005). The impact of major depressive disorder on the short- and long-term outcome of Crohn's disease treatment with infliximab. *Alimentary Pharmacology & Therapeutics*, *22*(2), 101–110.

Porter, R. J., Bowie, C. R., Jordan, J., & Malhi, G. S. (2013). Cognitive remediation as a treatment for major depression: a rationale, review of evidence and recommendations for future research. *The Australian and New Zealand Journal of Psychiatry*, *47*(12), 1165–1175.

Price, R. B., Iosifescu, D. V., Murrough, J. W., Chang, L. C., Al Jurdi, R. K., Iqbal, S. Z., et al. (2014). Effects of ketamine on explicit and implicit suicidal cognition: a randomized controlled trial in treatment-resistant depression. *Depression and Anxiety*, *31*(4), 335–343.

Prien, R. F., Carpenter, L. L., & Kupfer, D. J. (1991). The definition and operational criteria for treatment outcome of major depressive disorder: a review of the current research literature. *Archives of General Psychiatry*, *48*(9), 796–800.

Qin, L., Wu, X., & Block, M. L. (2007). Systemic LPS causes chronic neuroinflammation and progressive neurodegeneration. *Glia*, *55*(5), 453–462.

Raison, C. L., Dantzer, R., Kelley, K. W., Lawson, M. A., Woolwine, B. J., Vogt, G., et al. (2010). CSF concentrations of brain tryptophan and kynurenines during immune stimulation with IFN-α: relationship to CNS immune responses and depression. *Molecular Psychiatry*, *15*(4), 393–403.

Raison, C. L., Rutherford, R. E., Woolwine, B. J., Shuo, C., Schettler, P., Drake, D. F., et al. (2013). A randomized controlled trial of the tumor necrosis factor antagonist infliximab for treatment-resistant depression: the role of baseline inflammatory biomarkers. *JAMA Psychiatry*, *70*(1), 31–41.

Raskin, J., George, T., Granger, R. E., Hussain, N., Zhao, G. W., & Marangell, L. B. (2012). Apathy in currently nondepressed patients treated with a SSRI for a major depressive episode following randomized switch to either duloxetine or escitalopram. *Journal of Psychiatric Research*, *46*(5), 667–674.

Raskin, J., Wiltse, C. G., Siegal, A., Sheikh, J., Xu, J., Dinkel, J. J., et al. (2007). Efficacy of duloxetine on cognition, depression, and pain in elderly patients with major depressive disorder: an 8-week, double-blind, placebo-controlled trial. *American Journal of Psychiatry*, *164*, 900–909.

Rothwell, N. J., Luheshi, G., & Toulmond, S. (1996). Cytokines and their receptors in the central nervous system: physiology, pharmacology, and pathology. *Pharmacology & Therapeutics*, *69*(2), 85–95.

Roiser, J. P., & Sahakian, B. J. (2013). Hot and cold cognition in depression. *CNS Spectrums*, *18*(03), 139–149.

Rosenblat, J. D., & McIntyre, R. S. (2015). Are medical comorbid conditions of bipolar disorder due to immune dysfunction? *Acta Psychiatrica Scandinavica*, *132*(3), 180–191.

Rush, J. A., Trivedi, M. H., Stewart, J. W., Nierenberg, A. A., Fava, M., Kurian, B. T., et al. (2011). Combining medications to enhance depression outcomes (CO-MED): acute and long-term outcomes of a single-blind randomized study. *The American Journal of Psychiatry*, *168*(7), 689–701.

Santamaría, A., Flores-Escartín, A., Martínez, J. C., Osorio, L., Galván-Arzate, S., Chaverrí, J. P., et al. (2003). Copper blocks quinolinic acid neurotoxicity in rats: contribution of antioxidant systems. *Free Radical Biology and Medicine*, *35*(4), 418–427.

Santiago, R. M., Barbiero, J., Martynhak, B. J., Boschen, S. L., da Silva, L. M., Wener, M. F., et al. (2014). Antidepressant-like effect of celecoxib piroxicam in rat models of depression. *Journal of Neural Transmission*, *121*(6), 671–682.

Schrijvers, D., Maas, Y. J., Pier, M. P. B. I., Madani, Y., Hulstijn, W., & Sabbe, B. G. C. (2009). Psychomotor changes in major depressive disorder during sertraline treatment. *Neuropsychobiology*, *59*(1), 34–42.

Serafini, G., Pompili, M., Belvederi Murri, M., Respino, M., Ghio, L., Girardi, P., et al. (2015). The effects of repetitive transcranial magnetic stimulation on cognitive performance in treatment-resistant depression. A systematic review. *Neuropsychobiology*, *71*(3), 125–139.

Shaftel, S. S., Carlson, T. J., Olschowka, J. A., Kyrkanides, S., Matousek, S. B., & O'Banion, M. K. (2007). Chronic interleukin-1β expression in mouse brain leads to leukocyte infiltration and neutrophil-independent blood-brain barrier permeability without overt neurodegeneration. *Journal of Neuroscience*, *27*(35), 9301–9309.

Siegle, G. J., Thompson, W., Carter, C. S., Steinhauer, S. R., & Thase, M. E. (2007). Increased amygdala and decreased dorsolateral prefrontal BOLD responses in unipolar depression: related and independent features. *Biological Psychiatry, 61*(2), 198–209.

Sims, O. T., Whalen, C. C., Nackerud, L. G., & Bride, B. E. (2014). Longitudinal effects of selective serotonin reuptake inhibitor therapy and cytokine-related depression on hepatitis C viral logs during antiviral therapy. *Journal of Clinical Psychopharmacology, 34*(1), 80–84.

Sluzewska, A. (1999). Indicators of immune activation in depressed patients. In *Cytokines, stress, and depression* (pp. 59–73): Springer.

Smith, P. J., Blumenthal, J. A., Hoffman, B. M., Cooper, H., Strauman, T. A., Welsh-Bohmer, K., et al. (2010). Aerobic exercise and neurocognitive performance: a meta-analytic review of randomized controlled trials. *Psychosomatic Medicine, 72*(3), 239.

Soczynska, J. K., Ravindran, L. N., Styra, R., McIntyre, R. S., Cyriac, A., Manierka, M. S., et al. (2014). The effect of bupropion xl and escitalopram on memory and functional outcomes in adults with major depressive disorder: results from a randomized controlled trial. *Psychiatry Research, 220*(1), 245–250.

Stahl, S. M. (2010). Enhancing outcomes from major depression: using antidepressant combination therapies with multifunctional pharmacologic mechanisms from the initiation of treatment. *CNS Spectrums, 15*(2), 79–94.

Stanton, R., & Reaburn, P. (2014). Exercise and the treatment of depression: a review of the exercise program variables. *Journal of Science and Medicine in Sport, 17*(2), 177–182.

Stein, L., & Himwich, H. E. (1962). Effects and interactions of imipramine, chlorpromazine, reserpine and amphetamine on self-stimulation: possible neurophysiological basis of depression. In *Recent advances in biological psychiatry* (pp. 288–309). Springer.

Stuart, M. J., & Baune, B. T. (2014). Chemokines and chemokine receptors in mood disorders, schizophrenia, and cognitive impairment: a systematic review of biomarker studies. *Neuroscience and Biobehavioral Reviews, 42C*, 93–115.

Tookman, A. J., Jones, C. L., Dewitte, M., & Lodge, P. J. (2008). Fatigue in patients with advanced cancer: a pilot study of an intervention with infliximab. *Support Care Cancer, 16*, 1131–1140.

Tracey, D., Klareskog, L., Sasso, E. H., Salfeld, J. G., & Tak, P. P. (2008). Tumor necrosis factor antagonist mechanisms of action: a comprehensive review. *Pharmacology & Therapeutics, 117*(2), 244–279.

Treadway, M. T., & Zald, D. H. (2011). Reconsidering anhedonia in depression: lessons from translational neuroscience. *Neuroscience & Biobehavioral Reviews, 35*(3), 537–555.

Venero, C. (2014). Pharmacological treatment of cognitive dysfunction in neuropsychiatric disorders. In *Cognitive enhancement: Pharmacologic, environmental and genetic factors* (p. 233): Elsevier.

Vevera, J., Uhrova, J., Benakova, H., & Zima, T. (2010). Depression, traumatic stress and interleukin-6. *Journal of Affective Disorders, 120*(1-3), 231–4.6.

Watanabe, S. Y., Iga, J., Ishii, K., Numata, S., Shimodera, S., Fujita, H., et al. (2015). Biological tests for major depressive disorder that involve leukocyte gene expression assays. *Journal of Psychiatric Research, 66-67*, 1–6.

Wiborg, O. (2013). Chronic mild stress for modeling anhedonia. *Cell and Tissue Research, 354*(1), 155–169.

Wirleitner, B., Neurauter, G., Schrocksnadel, K., Frick, B., & Fuchs, D. (2003). Interferon-γ-induced conversion of tryptophan: immunologic and neuropsychiatric aspects. *Current Medicinal Chemistry, 10*(16), 1581–1591.

Wise, R. A. (1978). Catecholamine theories of reward: a critical review. *Brain Research, 152*(2), 215–247.

Wise, R. A. (2008). Dopamine and reward: the anhedonia hypothesis 30 years on. *Neurotoxicity Research, 14*(2-3), 169–183.

Wong, M., Ziring, D., & Korin, Y. (2008). TNFalpha blockade in human diseases: mechanisms and future directions. *Clinical Immunology: The Official Journal of the Clinical Immunology Society, 126*(2), 121–136.

Wood, H. (2015). Neuroimmunology: uncovering the secrets of the "brain drain": the CNS lymphatic system is finally revealed. *Nature Reviews Neurology, 11*(7), 367.

Xia, Y., Tsai, A. L., Berka, V., & Zweier, J. L. (1998). Superoxide generation from endothelial nitric-oxide synthase a Ca^{2+}/calmodulin-dependent and tetrahydrobiopterin regulatory process. *Journal of Biological Chemistry, 273*(40), 25804–25808.

Yirmiya, R. (2000). Depression in medical illness: the role of the immune system. *The Western Journal of Medicine, 173*, 333–336.

Yirmiya, R., Weidenfeld, J., Pollak, Y., Morag, M., Morag, A., Avitsur, R., et al. (1999). Cytokines, "depression due to a general medical condition," and antidepressant drugs. In *Vol. 461. Cytokines, stress, and depression* (pp. 283–316): Springer.

Yirmiya, R., & Goshen, I. (2011). Immune modulation of learning, memory, neural plasticity and neurogenesis. *Brain, Behavior, and Immunity, 25*(2), 181–213.

Young, S., Khondoker, M., Emilsson, B., Sigurdsson, J. F., Philipp-Wiegmann, F., Baldursson, G., et al. (2015). Cognitive-behavioural therapy in medication-treated adults with attention-deficit/hyperactivity disorder and co-morbid psychopathology: a randomized controlled trial using multi-level analysis. *Psychological Medicine, 45*(13), 2793–2804.

Zanos, P., Moaddel, R., Morris, P. J., Georgiou, P., Fischell, J., Elmer, G. I., et al. (2016). NMDAR inhibition-independent antidepressant actions of ketamine metabolites. *Nature, 533*(7604), 481–486.

Zhou, M., Wang, C. M., Yang, W. L., & Wang, P. (2013). Microglial CD14 activated by iNOS contributes to neuroinflammation in cerebral ischemia. *Brain Research, 1506*, 105–114.

Zitron, I. M., Kamson, D. O., Kiousis, S., Juhász, C., & Mittal, S. (2013). In vivo metabolism of tryptophan in meningiomas is mediated by indoleamine 2,3-dioxygenase 1. *Cancer Biology & Therapy, 14*(4), 333–339.

24

Translational Medicine Strategies in Alzheimer's Disease Drug Development

Veronika Logovinsky

Regeneron Pharmaceuticals Inc., Tarrytown, NY, United States

I INTRODUCTION

Alzheimer's disease (AD) is a slowly progressive neurodegenerative disease that leads to profound disability and, ultimately, to death. It is the most common cause of dementia, accounting for an estimated 60%–80% of all dementia cases (Alzheimer's Association, 2017). As the population of the United States and the world at large continues to grow older, the prevalence of AD in the United States alone is expected to increase to 13.8 million. Current estimates indicate that in 2017, 700,000 Americans aged ≥65 years will have AD at the time of their death and many of these deaths will be the result of complications of AD. By 2050 a new case of AD will develop every 33 s with nearly 1 million new cases per year (Alzheimer's Association, 2017).

At present, there are five drugs approved for the treatment of AD, including four cholinesterase inhibitors and one N-methyl-D-aspartate (NMDA) receptor antagonist. An independent review by Cummings et al. (2014) has found that 244 drugs for AD were tested in clinical trials between the years of 2002 and 2012. Unfortunately, throughout this period, AD trials have been characterized by a very high attrition rate with an overall success of approximately 0.4% (Cummings et al., 2014). While AD continues to be one of the most active drug development fields with more candidates populating the pipeline, the overall failure rate remains close to 100%. The AD field represents a field of very high unmet medical need, where development of more efficacious medications and, preferably, those that are capable of modifying the course of neurodegeneration associated with the disease, the so-called disease-modifying therapies, has become an urgent matter.

At the same time, over the past two decades, drug development has become substantially less productive across all therapeutic areas, including neurology and, particularly, AD (Mathematical Sciences & University of Reading, 2014; Pammolli et al., 2011). It can be argued that our limited understanding of the pathophysiology in the areas that concentrate research and development (R&D) resources represents the principal cause for the persistent failure of drug development efforts. While our understanding of the biology of AD has improved substantially, it remains relatively limited, making the selection of druggable targets that can relieve the disease symptoms and modify the course of the disease very challenging. At the same time, in the field of neurodegeneration, AD represents one of the most mature therapeutic fields with several identified genetic risk factors, well-characterized biomarkers, and formal diagnostic criteria for different stages of the disease. Our expanding understanding of the AD pathology has allowed us to accumulate a number of new targets for putative therapeutic agents that are progressing to clinical trials. In this environment of high medical need and increasing number of potential targets, it is particularly important to optimize the efficiency of drug development, so that putative therapeutic agents can be assessed and either eliminated or progressed to Phase 3 trials as quickly as possible. This task requires new trial designs that allow for the use of information as it is being accumulated in the course of a trial and new outcome measures that allow to resolve even relatively small but clinically meaningful treatment effects. The challenge is further complicated by the fact that while the AD field has developed a number of biomarkers, which are helpful with diagnosis, surrogate biomarkers that can predict clinical efficacy are still to be determined. Researchers must rely on clinical outcome measures to determine if a drug is efficacious, which requires large and long trials, particularly if one is studying a putative disease-modifying agent in a population at early stages of AD.

In this chapter, we review both innovative trial designs and new outcome measures that are being implemented in the new generation of AD trials. We particularly focus on Phase 2b and Phase 3 stages of development. We also review advances in trial designs in oncology, a much more successful therapeutic area over the past decade. In oncology, progress in the understanding of basic pathophysiology of different cancers has allowed for the development of targeted therapies tested using innovative trial designs that may provide useful lessons for AD and other areas of CNS drug development.

II OVERVIEW OF ADAPTIVE TRIAL DESIGNS

Later phases of drug development (Phases 2b and 3) conventionally use parallel arm, double-blind placebo-controlled studies with fixed sample sizes. In such designs, study parameters, including the number of patients to be recruited, outcome parameter, and treatment size effect, are calculated in advance of a trial, and data are collected on all of the patients before any analyses are performed. This statistical approach was originally developed for agricultural trials where all measurements occur simultaneously, at the time of harvest (Mathematical Sciences & University of Reading, 2014). In clinical trials, patients are recruited over a period of time, and measurements accumulate as a trial progresses over a period of weeks, months, or even years. It is this difference in data accumulation that calls for a different statistical approach to trial designs, specifically the use of accumulating data to adjust trial parameters, while at the same time maintaining trial integrity. One of such earlier approaches is that of "sequential trials or group sequential trials." In such trials, available patient data are analyzed at one or more interim points in the trial to see if there are sufficient data to determine conclusions on the drug's efficacy and safety (Mathematical Sciences & University of Reading, 2014).

As the research on such trial designs progressed, it became apparent that other parameters, not just decisions on efficacy/futility, can be modified in the course of a trial. This work has led to the development of a field known as "adaptive trial designs." In 2005 an adaptive design working group was formed to "foster and facilitate wider usage and regulatory acceptance of adaptive designs and to enhance clinical development, through fact-based evaluation of the benefits and challenges associated with these designs" (Gallo et al., 2006). The group was originally sponsored by the Pharmaceutical Research and Manufacturers of America (PhRMA) and, later, by the Drug Information Administration. It defined an adaptive design as "a clinical study design that uses accumulating data to decide how to modify aspects of the study

as it continues, without undermining the validity and integrity of the trial." The group particularly stressed that to maintain the trial integrity, it was important to prespecify changes prior to the start of the trial and not to introduce changes ad hoc (Gallo et al., 2006). In 2010 the Food and Drug Administration (FDA) produced a draft guidance on adaptive trial designs in which it broadly defined as adaptive, "a study that includes a prospectively planned opportunity for modification of one or more specified aspects of the study design and hypotheses based on analysis of data (usually interim data) from subjects in the study. Analyses of the accumulating study data are performed at prospectively planned time points within the study, can be performed in a fully blinded manner or in an unblinded manner, and can occur with or without formal statistical hypothesis testing" (U.S. Food and Drug Administration, 2010). In this guidance, FDA started the discussion on examining both blinded and unblinded study data while also expressing concerns about potential increases in Type I error. These broad definitions of adaptive study designs allow for a wide range of possible adaptations, including changes in the maximum sample size, study duration, treatment group allocation, dosing, number of treatment arms, or study endpoints (Kairalla et al., 2012).

At present, there are adaptive designs that are applicable to every stage of drug development. For Phase 2 exploratory trials, the emphasis is on finding safe and effective doses or dose-response modeling. Adaptive approaches that have been utilized for Phase 2 proof-of-concept studies include early stopping for either success or futility, sample-size reestimation, selection of doses at interim analysis, and dose-ranging designs with frequent Bayesian interim analyses and response-adaptive randomization ratios that allow accumulating data to guide dose assignment to favor doses that show maximum treatment response and away from doses that show no or minimal treatment response (Bhatt & Mehta, 2016). Among these the later, adaptive dose allocation approach uses Bayesian modeling to identify an appropriate dose for new participants based on previous responses. It has been argued to be particularly useful for Phase 2 proof-of-concept studies (Berry, Mueller, Griever, & Smith, 2002). If both toxicity and efficacy need to be considered, a utility index can be created to govern adaptive dose allocation using Bayesian modeling. Compared with traditional, fixed sample and fixed randomization designs, it provides greater precision for dose-response and treatment effect estimates for the same sample size. However, it is more complex to implement with more lead time needed to set up a trial and extensive simulations that use specialized software to determine the trial's operating characteristics (Bhatt & Mehta, 2016). In the field of neurology, this design has been utilized in the Acute Stroke Therapy by Inhibition of Neutrophils (ASTIN) study (Grieve & Krams, 2005; Krams et al., 2003). A similar design for a currently ongoing AD study is described in greater detail in the succeeding text.

Seamless Phase 2–3 designs cover the transition from exploratory Phase 2 to confirmatory Phase 3 development. These can be divided into two types, namely, operationally seamless and inferentially seamless (Bhatt & Mehta, 2016). Operationally seamless designs use conventional final analysis methodology with conventional estimates of dose-response and treatment effect but provide efficiency by eliminating time between Phase 2 and Phase 3. Conventional analysis at the end of the Phase 2 portion allow for the sponsor to be involved in dose selection for the Phase 3 portion and are easier to design. The drawback of this approach is that the final analysis is based only on data from the Phase 3 portion (Bhatt & Mehta, 2016). The efficiency of the inferentially seamless design is based on the fact that the final analysis combines data from both Phase 2 and Phase 3 parts of the study. The drawback of this design is that the sponsor is not involved in the dose selection for the Phase 3 portion and the dose selection algorithm must be prespecified, making the design more complex and allowing for a possibility of a suboptimal dose-response model at the end of the Phase 2 part. It is also a more complex design with a more complicated final analysis that uses nonconventional estimates of trial parameters.

Confirmatory Phase 3 trials use several different types of adaptive designs, including group sequential designs, sample-size reestimation, and population enrichment designs (Bhatt & Mehta, 2016). Different elements of these designs can be combined in a single Phase 3 trial. Group sequential designs are considered to be well understood and are the most widely used adaptive designs in confirmatory trials (Kairalla et al., 2012). These can be either of the classic type or the adaptive type. A classic group sequential design derives efficiency from allowing for early trial stop based on efficacy, futility, or toxicity, flexible alpha spending functions and allows changing maximum sample size based on blinded accumulated data. At the same time, these designs do not allow for unblinded analyses of accumulated data, are more difficult for data and safety monitoring committees as they need to review all available data before terminating a trial prematurely, and have a risk of losing statistical significance if a trial overruns (Bhatt & Mehta, 2016). Adaptive group sequential designs have all the advantages of classic group sequential designs. In addition, they allow for various adaptations, including changes in maximum sample size, in number and timing of interim analyses, and in alpha spending function and switching of endpoint from noninferiority to superiority. All of these can be done using unblinded data accumulated

in the study. While these designs allow for a number of increased efficiencies, they are more complex and are burdened with the complications that result from the use of unblinded data, including the need for very rigid firewalls to maintain trial integrity, potentially misleading interim estimates of treatment effect, thorough upfront planning with a longer lead-time period, potential for operational bias if the investigators' behavior is influenced by unwarranted inferences, need for a nonconventional final analysis and greater regulatory hurdles (Bhatt & Mehta, 2016).

Designs that include sample-size reestimation can also be based on blinded data or unblinded data. When based on blinded data, these designs allow for conventional final analyses and have fewer regulatory hurdles. However, they are significantly limited by performing sample-size adjustments only on the basis of unknown variance, which becomes better defined as trial data accumulate. Sample-size reestimation designs that use unblinded data have all the additional complications that are associated with group sequential designs that use unblinded data as described earlier (Bhatt & Mehta, 2016).

The final major class of confirmatory Phase 3 designs is population-enrichment designs that are useful if treatment is expected to show efficacy only in selected subgroups. In these designs, subgroups without any evidence of efficacy can be eliminated at interim analyses. However, these designs are characterized by their own weaknesses, including the need to identify target subgroups prospectively, the requirement for clear biomarker cutoffs if subgroup characterization is dependent on such biomarkers, the loss of power as some subgroups are eliminated, and the risk of inappropriately or prematurely eliminating subgroups (Bhatt & Mehta, 2016).

Phase 3 studies may also utilize adaptive randomization methodologies that are similar to those described for exploratory Phase 2 studies with objectives to model dose responses (see in the preceding text). Such methodologies are described in much greater detail using the example of a recently completed AD study (see below).

III USE OF ADAPTIVE DESIGNS IN ALZHEIMER'S DISEASE DRUG DEVELOPMENT

Traditional drug development follows a logical progression from successful Phase 2 studies, which establish proof of concept and define a dose response to confirmatory Phase 3 studies that use outcome measures and doses selected on the basis of Phase 2 study results. This process of drug development has proven to be particularly difficult for AD trials (Schneider et al., 2014).

As discussed earlier, our advances in the understanding of AD pathology and the urgent medical need for efficacious treatments that can slow down or arrest the neurodegeneration process in AD have shifted the field toward the development of putative disease-modifying agents that can influence the underlying AD pathology at early disease stages when neuronal damage is still relatively limited and only a relatively mild clinical status of AD patients can be preserved for longer periods of time. Development of these putative disease-modifying agents faces significant challenges in AD trials. Such trials need to enroll heterogeneous populations that are clinically mild and are more difficult to diagnose as being at well-defined disease stages of similar pathophysiological characteristics and disease severity. Although diagnostic research criteria have been offered across the AD spectrum, beginning with biomarker positive and presymptomatic stages and including MCI due to AD/prodromal AD and mild AD dementia, these diagnostic criteria are difficult to operationalize without a full alignment on the use of diagnostic clinical tests and biomarkers that have fully validated and standardized assays and well-defined cutoff points for such assays (Albert et al., 2011; Logovinsky et al., 2014; Sperling et al., 2011). Given the slow progressive nature of AD, particularly at the early stages of the disease, these difficulties with heterogeneous populations, suboptimal clinical outcomes for early disease stages that lack sensitivity and are characterized by high variability (Logovinsky et al., 2014; Riepe et al., 2011), the absence of surrogate biomarkers that are robustly correlated with clinical outcomes, proof-of-concept Phase 2b trials that seek to establish clinical efficacy, and a robust dose response become very large and long. The result is the tendency to move from small biomarker-based Phase 2a studies to confirmatory Phase 3 trials, where the high level of failure has taken place (Gray et al., 2015; Greenberg et al., 2013). Under these circumstances, adaptive Phase 2 proof-of-concept designs that offer efficient use of accumulating data while a clinical trial is ongoing can offer a very significant advantage to a program by allowing for a proper Phase 3 development that serves to confirm Phase 2 results.

Given the slow neurodegenerative process of AD, particularly at the early clinical stages, one of the principle concerns about the use of adaptive trial designs has been the ability to assess the efficacy outcomes sufficiently early in the course of a trial. If the time required to identify a treatment effect is on the order of the time period required for randomization, adaptive design efficiencies are difficult to materialize, since the data that inform such adaptations will take too long to generate. Here, we present an example of a Bayesian adaptive Phase 2 proof-of-concept trial of a humanized, monoclonal antibody, BAN2401, which targets Aβ protofibrils (Satlin et al., 2016). In this design, Bayesian statistical methodology is employed to analyze all available accumulating data, while missing endpoint data are imputed by using a longitudinal model, so that response-adaptive randomization and detection of signals predictive of success or futility at the primary analysis that uses all accumulating data to estimate probabilities of a signal of success or futility as defined by the operating characteristics of the study can take place at an interim analysis before all subjects reach the end of treatment (Satlin et al., 2016). This design represents an adaptive proof-of-concept dose-ranging design with frequent Bayesian adaptation of randomization ratios as presented earlier (Berry, 2011; Berry et al., 2010; Bhatt & Mehta, 2016). The longitudinal model is updated at every prespecified interim analysis using accruing data and correlations between observations at different time points.

Bayes' theorem represents a way of obtaining the probability of an event, based on prior knowledge of conditions that might be related to the event. It is named after Rev. Thomas Bayes, who authored an equation to allow prior knowledge to be updated by new observations (Jeffreys, 1973). In the Bayesian approach, uncertainty is measured by probability, and anything that is unknown is described by a probability distribution. All probability distributions are calculated as conditional on known values (Berry et al., 2002). When designing clinical trials, one starts with a "prior" probability, which is the initial probability of a trial parameter, p. As the probability distribution of p is updated using accumulating data, a "posterior" distribution of p is derived, which is conditioned on the events that have been observed. The final distribution at the end of a trial is a posterior distribution, while the distribution of p that reflects the probability that the next observation will be a success is a "predictive" distribution (Berry et al., 2002).

The principle parameters used in a Bayesian dose-ranging design include the maximum effective dose ($dMax$) and the effective dose 90 ($dED90$). The maximum effective dose ($dMax$) is the dose with the greatest treatment effect (greatest difference from control in mean change from baseline to end of treatment). The $dED90$ is the dose that achieves at least 90% of the treatment effect achieved by $dMax$. The two key probabilities are the probability of being superior to control and the probability of being superior to control by at least the clinically significant difference (CSD), both of which are calculated for each dose at the end of the treatment period or at the time of the primary study analysis. These parameters are estimated at each interim analysis conducted in a prespecified manner as trial data accrue and decisions about stopping early for futility or success and changing randomization ratios are made.

When designing the BAN2401 study, we defined $dED90$ or just ED_{90} as the simplest dose, that is, the smallest dose with the lowest frequency of administration, which achieves at least 90% of the treatment effect associated with the maximum effective dose, *dMax*. In this study, efficacy is measured by a novel assessment tool, Alzheimer's Disease Composite Score (ADCOMS) used as the primary study outcome (Wang et al., 2016). The primary objective of the study is to establish ED_{90} on ADCOMS at 12 months of treatment in subjects with early AD, defined as a population consisting of patients with MCI due to AD—intermediate likelihood and mild AD dementia. Five doses/dose regiments were selected for the study based on the preclinical results and the results of Phase 1 development for BAN2401, including 2.5, 5, and 10 mg/kg biweekly and 5 and 10 mg/kg monthly. While the primary analysis takes place at 12 months of treatment, the treatment in the study continues for a total of 18 months, which allows investigating whether the putative treatment effect persists and, if so, whether this treatment effect increases over time as may be expected for a disease-modifying agent.

In designing the study, we scheduled efficacy assessments at every 50 subjects recruited into the study and, then, every 3 months after all subjects had been recruited until all subjects completed 12 months of treatment. We also assumed CSD, which reflects a clinically significant effect, to be a 25% reduction in the rate of decline by 52 weeks of treatment. Other assumed parameters were a dropout rate of 20% by 52 weeks and a maximal recruitment rate of 32 subjects per month. Key operating characteristics for this study include the Type I error; the probabilities of futility, early success, overall success, and of a decision to proceed to Phase 3; and the timing of interim analyses. Key design components include the definitions of futility and success boundaries, the maximum study sample size, the allocation rules for different treatment arms, and the choice of past data sets that can be used in developing this study design. We constructed a dose-frequency response model that for each treatment arm considered the mean change from baseline in ADCOMS

at 52 weeks of treatment. This dose-frequency response model is updated at each interim analysis until all 52-week treatment data became available for all subjects. A linear regression longitudinal model based on correlations between the 6-week, 13-week, 27-week, and 39-week ADCOMS measurements is used for these updates. This regression model is adjusted as data from the ongoing trial become available.

A Bayesian adaptive design trial requires numerous simulations of the trial design to optimize the key operating characteristics for a specific set of the design features. A wide range of possible dose-frequency response scenarios were simulated for a given set of design features. Results of these simulations allow for the optimization of the operating characteristics such as balancing between the probability of stopping for futility and stopping for early success. The simulations were run for every dose-frequency response scenario and under the assumption of three response patterns, including a linear response with continuously increasing treatment effect, a symptomatic response with most of the treatment effect developing within the first 12 weeks, and a late-onset response with a delay in the appearance of treatment. Design features were repeatedly adjusted, simulations were rerun, and resulting operating characteristics were analyzed to arrive at the final optimal study design (Fig. 24.1).

The final study design uses a maximum sample size of 800 subjects and has an average overall probability of success of approximately 80% if at least one dose shows a robust, clinically significant treatment effect (Satlin et al., 2012, 2016). Simulations of the various outcomes indicate that this trial has been designed with Type I error (one sided) of approximately 10%. Bayesian interim analyses begin after a fixed randomization period of 196 subjects being enrolled in the study and subsequently take place when 250 and, then, every additional 50 subjects are enrolled. After the maximum sample size of 800 subjects

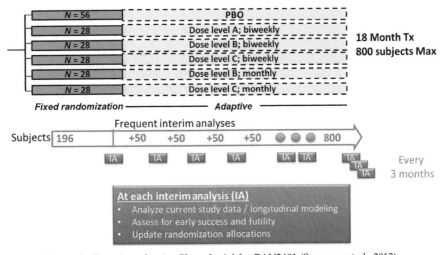

FIG. 24.1 Schematic representation of a Bayesian adaptive Phase 2 trial for BAN2401 (Swanson et al., 2013).

has been enrolled and if neither the success nor the futility boundary has been crossed, the Bayesian interim analyses take place every 3 months of treatment until the final analysis at 12 months of treatment. All available longitudinal data are used, and for subjects that have not reached 12 months of treatment, the 52-week ADCOMS measurement, which drives all of the analyses, is imputed using the longitudinal model described earlier and adjusted for all the date accumulated in the study. At each interim analysis the posterior distribution is calculated for the dose-frequency response model and guides the determination of the probability of each treatment arm to be the maximum effective dose and the ED_{90} (Fig. 24.2).

Two projects, Innovative Medicines Initiative-European Prevention of Alzheimer's Dementia (IMI-EPAD) and Global Alzheimer's Platform (GAP) are developing platforms for adaptive proof-of-concept trials of AD therapies (Ritchie et al., 2014; Vradenburg et al., 2014). While the BAN2401 study represents the first experiment in proof-of-concept Bayesian adaptive designs within the framework of a single therapeutic development program, these two initiatives are likely to expand the use of adaptive trial designs to the development of multiple agents and to provide a wide ranging area for testing the efficiency of such designs in AD.

IV CAN DEVELOPMENTS IN ONCOLOGY TRIAL DESIGNS INFORM AD TRIALS?

Over the last several decades, the focus in oncology has shifted from cytotoxic agents to so-called targeted drugs that are designed to ameliorate consequences of specific genetic alterations in a tumor. These developments in understanding molecular level of pathophysiology of various cancers have required modernization of

trial design. At the present time, there is a need to define patient populations not only at the level of tumor types or tumor histology but also at the level of specific molecular defects associated with a tumor. Once such patient subgroups are defined, one needs to develop new trial strategies to enroll these subgroups and to study putative therapeutic effects of new agents in the most efficient manner (ASCO Daily News, 2015). In this context, three new trial designs have emerged, enrichment or target, umbrella, and basket designs.

Enrichment designs are most widely used in Phase 3 trials when a new agent needs to be developed simultaneously with companion diagnostics. The classic type requires subjects to be positive on a specified diagnostic test to be eligible for a study, and the diagnostic test becomes a part of the screening procedure. This design protects biomarker negative patients from exposure to an agent that is unlikely to be beneficial for them and can result in larger effect sizes and smaller sample-size requirements. In a more complicated version of an enrichment design, when the biomarker cut point has not been determined, an adaptive element can be used. With this approach, interim analyses are prespecified to investigate accumulating data and to use a predetermined algorithm to adjust the cut point or the biomarker status.

These classical enrichment approaches are widely used in the AD field as well, where in trials of antiamyloid agent patients are screened for amyloid pathology positivity, using amyloid PET imaging or CSF sampling.

Umbrella trials enroll patients with a single tumor type or histopathology. The molecular pathology of each patient is characterized in the beginning of a trial to allow for multiple subtrials that target specific molecular pathology within the tumor type. Once a patient is characterized, this patient is assigned to the specific subtrial driven by the molecular pathology or biomarker characteristics. The advantage of such umbrella trials is that

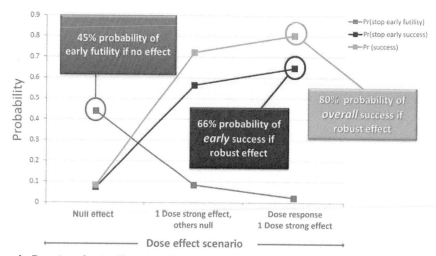

FIG. 24.2   Performance of a Bayesian adaptive Phase 2 study for BAN2401 across different dose-frequency responses (Swanson et al., 2013).

they use a centralized platform to facilitate screening of subjects by using a common genomic screening and data flow methodology and can test different drugs on different genetic mutations in a single type of cancer. Umbrella trials can have a flexible design to add, modify, or drop subtrials based on accumulating data for an agent that is targeting a specific subtrial (Azvolinsky, 2015).

Today, our knowledge of the AD pathophysiology does not reach the level of sophistication that has been obtained in oncology in terms of identifying specific genetic mutations and molecular mechanism characteristic of different tumor types. At the same time, while umbrella trials have not been implemented in the AD field and are likely to result in smaller efficiency gains as compared with oncology umbrella trials with multiple subtrials designed on the basis of specific molecular pathology, proposals for such trials have been put forward. Combination studies that put together several putative AD agents, such amyloid-directed antibodies and beta-secretase inhibitors can be designed using the framework of an umbrella trial.

Basket trials use specific genetic aberrations to include populations with these aberrations in their tumors independent of the type of tumor (Azvolinsky, 2015; Galbraith, 2014). Subjects may be randomized to different arms, depending on the specific genetic aberration and not on the tumor type or histopathology. As with umbrella trials a complicated infrastructure is required to support genetic testing and other biomarker analyses. These trials are capable of testing a variety of agents, each focused on a particular molecular/genetic feature. This means that maximum trial efficiency is obtained when industry and academia collaborate in selecting drug candidates to be included in such trials. Basket trials represent an efficient way for selecting experimental agents in earlier stages of development across multiple patient populations with different tumors.

Similar basket trial designs can be envisioned in AD. For example, it is likely that different tauopathies including AD will require different agents targeting production and clearance of toxic tau species, overall tau metabolism, and tau aggregation and spread. A basket trial may enroll patients with tau brain pathologies underlying different tauopathies, including AD, behavioral variant of frontotemporal dementia (bvFTD), nonfluent and logopenic variants of primary progressive aphasia (PPA), corticobasal degeneration (CBD), and progressive supranuclear palsy (PSP). Patients may be randomized according to the specific tauopathy with which they have been diagnosed, and multiple antitau agents can be tested in these baskets of the different tauopathies using the infrastructure of a single trial.

We can conclude the discussion of novel trial designs with the hope that as the molecular pathophysiology of AD becomes better understood and new molecular targets can be pursued, novel trial designs that are driven by both advanced statistical methodologies and biological considerations will be adopted by the AD field.

V MODERNIZATION OF CLINICAL TOOLS FOR EARLY AD

The predementia stage and the very mild stage of AD dementia are characterized by relatively minor symptoms that result from mild cognitive and functional changes. These changes are difficult to measure with well-established clinical tools that have been widely used in the dementia stages. The use of these insensitive tools in predementia and very mild dementia contributes to the very high rate of failure, particularly at the late development stages, by requiring large sample sizes and very long treatment periods. The FDA has indicated that a single composite outcome, which measures both cognitive and functional changes, may be appropriate for presymptomatic and prodromal AD (pAD) and/or MCI due to AD trials (U.S. Food and Drug Administration, 2013). In their guidance the FDA used Clinical Dementia Rating Scale-Sum of Boxes, CDR-SB, as an example of one such composite outcome. Unfortunately, CDR-SB also has limited sensitivity in early AD and represents a suboptimal approach to clinical assessments.

While the traditional, widely used clinical tools may lack overall sensitivity, certain items within these scales are more responsive to clinical decline and potentially treatment effects in MCI due to AD/pAD (Logovinsky et al., 2013; Wang et al., 2016). Several pharmaceutical companies undertook the development of novel endpoints by combining individual items from traditional scales in composite scores. The items were selected on the basis of their sensitivity to clinical decline at early stages of AD using statistical methodologies and neuropsychological characterizations of the disease (Ard et al., 2015; Raghavan et al., 2014; Wang et al., 2016). Table 24.1 summarizes four of these composites along with their constituent items. Two of these composites, ADCOMS and ADCCS, contain both cognitive and functional items, while the other two, TriAD and ProADAS, represent purely cognitive composites. Recently, another cognitive and functional composite has been created and a trail planned to validate this composite (Jutten et al., 2017).

In the area of presymptomatic the so-called biomarker positive AD or in patients that are at high risk of developing AD, the need for new clinical tools is even more urgent. Similar approaches for creating composite scores that utilize items from traditional clinical tools have been taken for such preclinical stages, including a composite used in the Alzheimer's Prevention Initiative (API) and the ADCS Preclinical Alzheimer

TABLE 24.1   Cognitive, Cognitive/Functional Composites, and Their Individual Components

Composite	ADAS-Cog items	CDR	MMSE
ADCOMS (Eisai) cognitive/functional	Delayed word recall, orientation, word recognition, word finding	CDR-SB	Orientation constructional praxis
TriAD (Janssen) cognitive	Word recall, delayed word recall, orientation	CDR-SB-Cog (memory, judgment and problem solving, and orientation)	
ADCCS (Janssen) cognitive/functional	Word recall, delayed word recall, orientation, word recognition	CDR-SB	
ProADAS (AstraZeneca) cognitive	word recall, delayed word recall, orientation, word Finding, number cancellation		

Abbreviations: ADAS-Cog, Alzheimer's disease assessment scale-cognitive subscale; ADCCS, Alzheimer's disease clinical composite score; ADCOMS, Alzheimer's disease composite score; CDR-SB, clinical dementia rating scale-sum of boxes; CDR-SB-Cog, CDR-cognitive component; MMSE, mini-mental state examination; ProADAS, prodromal Alzheimer's disease assessment scale; TriAD, tri-domain cognitive composite for Alzheimer's disease.
Adapted from Raghavan, N., et al. (2014). Validation of novel composite outcome measures for pre-dementia Alzheimer's disease. Alzheimer's & Dementia, 10 (4 Suppl), P244.

Cognitive Composite (ADCS-PACC) for the Antiamyloid Treatment in Asymptomatic Alzheimer's (A4) study (Donohue et al., 2014; Langbaum et al., 2014, 2015).

ADCOMS is currently being used prospectively as the primary clinical outcome in the Bayesian response-adaptive randomization proof-of-concept trial for BAN2401, which is described earlier. ADCOMS was created by analyzing data from placebo or untreated amnestic MCI (aMCI) arms in four studies, including the aMCI subgroup from the Alzheimer's Disease Neuroimaging Initiative (ADNI-1); the placebo group from the Alzheimer's Disease Cooperative Study (ADCS), a randomized, double-blind, placebo-controlled trial to evaluate the safety and efficacy of vitamin E and donepezil HCl (Aricept) to delay clinical progression from MCI to AD (ADCS-MCI); the placebo group of a 1-year,multicenter, randomized, double-blind, placebo-controlled evaluation of the efficacy and safety of donepezil hydrochloride in patients with MCI; and the placebo group from the *Hippocampus Study: Comparative Effect of Donepezil 10 mg/day and Placebo on Clinical and Radiological Markers.* The pooled data set from these trials was composed of 1160 subjects (Wang et al., 2016). We defined two enriched populations within this large aMCI data set. The first enriched subset consisted of aMCI subjects with amyloid beta brain pathology as measured by their cerebrospinal fluid (CSF) Aβ1-42 levels and, therefore, equivalent to a group of MCI due to AD with intermediate likelihood subjects and almost equivalent to pAD subjects. The second enriched subset consisted of aMCI subjects, who were known to be carriers of an apolipoprotein E (ApoE) ε4 allele. In addition to this large pooled data set that was used for the derivation of ADCOMS, we used data from patients in three mild AD dementia studies to determine if ADCOMS maintained its sensitivity to clinical deterioration at the stage of disease adjacent to aMCI or MCI due to AD. Patients in the

pooled aMCI data set had data from the well-established AD scales, including ADAS-Cog 14, MMSE, and CDR-SB. Additional item level data included results from Functional Activities Questionnaire (FAQ) in ADNI, ADCS-Activities of Daily Living (ADL) in ADCS-MCI, and neuropsychological tests from Neuropsychiatric Inventory (NPI) in ADNI and ADCS-MCI.

We constructed a partial least square (PLS) regression with a longitudinal clinical decline model to analyze the various item level data and to identify a weighted combination of items with the highest sensitivity to decline over time. The final composite score, ADCOMS, was a weighted linear combination of the 12 items as derived by the model (Fig. 24.3).

The mean to standard deviation ratio (MSDR) of change from baseline over a specified time period was used to assess the sensitivity of ADCOMS to clinical progression. In addition, we calculated study sample sizes required to detect a 25% reduction in clinical decline relative to placebo at 12 months of treatment for pooled aMCI, MCI due to AD with intermediate likelihood, which we treated as equivalent to pAD, ApoE ε4 aMCI carriers, and mild AD dementia patients. A two-sample t-test with a two-sided α of 0.05% and 80% statistical power was used for these calculations. We also investigated the ability of ADCOMS to detect treatment effect associated with donepezil, a known efficacious agent for AD, using results from two different studies, ADCS-MCI and 30-week, multicenter, randomized, double-blind, placebo-controlled evaluation of the safety and efficacy of E2020 (donepezil) in patients with Alzheimer's disease, abbreviated as "Eisai-MCI-302" study (Wang et al., 2016).

ADCOMS showed a robust statistically significant improvement in sensitivity over the ADAS-Cog total score and the MMSE total score and moderate but statistically significant improvement in sensitivity over the CDR-SB for the pooled aMCI population.

	Item	PLS coefficient	Maximum scaled to 100
ADAS	Delayed word recall	0.009	4.3
	Orientation	0.017	6.9
	Word recognition	0.004	2.3
	Word finding difficulty	0.016	4.1
MMSE	Orientation time	0.042	10.6
	Constructional praxis	0.038	1.9
CDR	Personal care	0.054	8.3
	Community affairs	0.109	16.6
	Home and hobbies	0.089	13.6
	Judgment and problem solving	0.070	10.6
	Memory	0.059	8.9
	Orientation	0.078	11.9
	Total		100

FIG. 24.3   Items selected by the PLS model and their corresponding PLS coefficients (Hendrix et al., 2012).

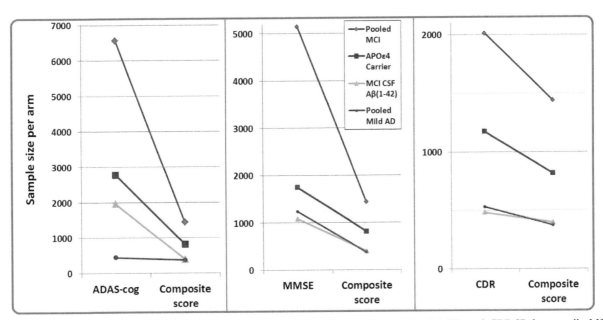

FIG. 24.4   ADCOMS significantly improves sample-size requirements relative to ADAS-Cog, MMSE, and CDR-SB for overall aMCI and two enriched subpopulations, ApoE ε4 carriers, and MCI due to AD with intermediate likelihood based on brain amyloid pathology (Wang et al., 2016). Our results also indicate that ADCOMS maintains its sensitivity in the mild AD population, with sample-size requirements being similar for ADCOMS, ADAS-Cog, MMSE, and CDR-SB.

Similar improvements were seen for ADCOMS versus ADAS-Cog, MMSE, and CDR for the two enriched populations. Corresponding improvements were identified in the study sample-size requirements for ADCOMS versus the three traditional instruments. Specifically, for ApoE ε4 carriers and pAD/MCI due to AD with intermediate likelihood, sample-size requirements were 3.7 (95% CI 2.0–6.8) and 6.3 (95% CI 2.2–17.1) times larger for ADAS-Cog, respectively; 2.3 (95% CI 1.5–3.5) and 3.0 (95% CI 1.6–5.4) times larger for MMSE, respectively; and 1.5 (95% CI 1.2–1.8) and 1.2 (95% CI 0.97–1.6) times larger for CDR-SB, respectively (Fig. 24.4).

In addition to significantly improved sensitivity to clinical progression, our investigation showed that ADCOMS was able to capture treatment effect as well or better than the three traditional measures (Table 24.2). When the results on these instruments are consistent and, at least, trend in the same direction, ADCOMS is able to measure a treatment effect more robustly. In the ADCS-MCI trial, CDR-SB showed relatively high sensitivity to clinical decline in aMCI but was less sensitive to treatment effect as compared with ADCOMS, which shows a high level of statistical significance. The same is true for the mild AD dementia population in the Eisai-AD-302 study, where ADCOMS shows the highest level of statistical significance for the well-established and most efficacious treatment with 10 mg of donepezil.

Results of the recently completed proof-of-concept study for BAN2401, which has been prospectively designed to use ADCOMS as the primary clinical outcome measure, are required for more definitive conclusions on the usefulness of ADCOMS and the approach of creating composite scores based solely on empirical, statistically rigorous analyses of available data. While it is likely that these approaches will prove to be useful and will result in smaller sample-size requirements and more manageable studies, their usefulness is limited by both the sensitivity of the contributing items and the fact that these do not represent stand-alone instruments. ADCOMS and other similar scores require full administration of the instruments that contribute their items to a score. While multiple clinical tools are very frequently administered in clinical studies, a single sensitive tool that is reasonably brief and measures well-defined cognitive and functional domains and is more easily interpretable will be preferred by drug developers, regulators, and health agencies and will carry fewer burdens for patients.

VI DEVELOPMENT OF MORE SENSITIVE CLINICAL TOOLS FOR EARLY AD PERFORMANCE-BASED FUNCTIONAL MEASURES

Traditionally, AD has been considered to follow a clinical course during which cognitive impairments that are predominantly memory-related emerge prior to functional deficits and the development of functional

TABLE 24.2   Assessment of Treatment Effect of Donepezil and Vitamin E Compared With Placebo (Wang et al., 2016)

		ADCS-MCI (12 months)			Eisai-AD-302 (6 months)		
Endpoint	Statistics	Placebo (208)	Donepezil 10 mg (182)	Vitamin E (208)	Placebo (71)	Donepezil 5 mg (74)	Donepezil 10 mg (56)
ADCOMS	LSMean (SE)	0.061 (0.0080)	0.034 (0.0085)	0.060 (0.0080)	0.053 (0.014)	−0.005 (0.014)	−0.033 (0.016)
	Difference (SE)[a]		0.027 (0.0117)	0.002 (0.0113)		0.057 (0.020)	0.085 (0.022)
	95% CI		0.005, 0.050	−0.021, 0.024		0.018, 0.096	0.043, 0.128
	P value		0.019	0.885		0.0041	<0.0001
ADAS-cog	LSMean (SE)	0.79 (0.309)	0.08 (0.328)	0.93 (0.311)	1.6 (0.552)	−0.67 (0.535)	−1.23 (0.627)
	Difference (SE)[a]		0.71 (0.451)	−0.14 (0.439)		2.27 (0.769)	2.83 (0.835)
	95% CI		−0.18, 1.59	−1.00, 0.72		0.76, 3.78	1.19, 4.47
	P value		0.116	0.755		0.0033	0.0008
MMSE	LSMean (SE)	−0.71 (0.152)	−0.20 (0.163)	−0.60 (0.154)	−0.75 (0.313)	0.83 (0.301)	0.49 (0.358)
	Difference (SE)[a]		−0.51 (0.223)	−0.12 (0.217)		−1.58 (0.434)	−1.24 (0.476)
	95% CI		−0.94, −0.07	−0.54, 0.31		−2.43, −0.72	−2.17, −0.30
	P value		0.024	0.594		0.0003	0.0095
CDR-SB	LSMean (SE)	0.38 (0.068)	0.22 (0.073)	0.46 (0.068)	0.24 (0.117)	−0.10 (0.113)	−0.19 (0.134)
	Difference (SE)[a]		0.16 (0.099)	0.08 (0.096)		0.33 (0.163)	0.43 (0.178)
	95% CI		−0.03, 0.36	−0.27, 0.11		0.01, 0.65	0.08, 0.78
	P value		0.107	0.419		0.0414	0.0173

Abbreviations: AD, Alzheimer's dementia; *ADAS-cog*, Alzheimer's disease assessment scale-cognitive subscale; *ADCOMS*, Alzheimer's disease clinical composite score; *CDR-SB*, clinical dementia rating-sum of boxes; *CI*, confidence interval; *LSMean*, least-squares means; *aMCI*, amnestic mild cognitive impairment; *MMSE*, mini-mental state exam; *SE*, standard error.

Note: Mean, treatment difference, SE, and CI are based on LSMean.

[a] *Difference = placebo − drug.*

impairments signals the onset of AD dementia. It has become increasingly clear that patients experience mild functional impairments at predementia stages, which often consist of increased numbers of errors and slower processing speed more than an overall loss of ability to perform a task (Belchior, 2015; Marshall et al., 2012; Marson, 2015). Sensitive cognitive and functional tools are required at early AD stages by the subtle nature of deficits. At the same time, for therapeutic agents that are currently in development and that target presymptomatic and MCI stages, are required by health agencies to show an effect on outcome measures that translate into life-impacting tasks, such as an ability to continue one's employment for younger patients or an ability to manage one's finances and social obligations to remain independent and to maintain good quality of life. Such outcome measures will require focused and sensitive functional assessments, which correlate with greater functional deficits at later stages and which have not been established and validated today. Commonly used functional assessment tools are inventories, which both patients and their informants complete and which require subjective input and are crude and lack accuracy and sensitivity in early AD (Harvey et al., 2017).

Over the past decade researchers have been developing new types of functional measures, so-called performance-based assessments, that hold the potential for greater sensitivity and objectivity in early AD. While a substantial number of these tools have been developed, none have been validated in large, multicenter clinical trials. The Harvard Automated Phone Task (APT), the UCSD Performance-Based Skills Assessment (UPSA), and the Financial Capacity Instrument-Short Form (FCI-SF) are some of the more advanced performance-based functional measures, which have collected initial validation data with the Harvard APT and the FCI-SF being implemented in large early AD trials (Goldberg et al., 2010; Marshal et al., 2015; Marson, 2015; Marson et al., 2016).

The Harvard APT requires one to use an interactive voice response system (IVRS) to perform three tasks:

- to refill a prescription (APT-Script),
- to select a new primary care physician (APT-PCP),
- to make a bank account transfer and a payment (APT-Bank) (Marshall et al., 2016).

Preliminary results of investigating the performance of the Harvard APT indicate that this tool is capable of discriminating clinically normal young, clinically normal elderly, and MCI subjects. It showed good psychometric properties. In clinically normal subjects the Harvard APT had a significant correlation with executive function and processing speed. Different subtests appeared to tap into different domains with APT-Script correlating better with processing speed and APT-PCP and APT-Bank correlating better with executive function. A very interesting MRI finding was that in all subjects, APT-PCP was associated with

inferior temporal cortical thinning, thus providing some hope that the results of this test can be mapped onto areas of atrophy involved in aging and in early AD. This test is being used in the A4 secondary prevention study in presymptomatic subjects with evidence of amyloid brain pathology. The hope is that the results of this study will prospectively assess the utility of this instrument, including its ability to discriminate between amyloid-positive and amyloid-negative subjects that appear to be normal on traditional neuropsychological tests and subjects that are on their path to developing MCI.

The UPSA uses standardized role-play situations to assess performance in five functional domains:

- household chores
- communication
- finance
- transportation
- planning recreational activities (Patterson et al., 2001)

It has been used extensively in schizophrenia research and has been recommended as a coprimary assessment of cognition in the Measurement and Treatment Research to Improve Cognition in Schizophrenia (MATRICS) Project (Harvey et al., 2013; Mantovani et al., 2015). The UPSA has also shown some promise in the AD research, where it was able to discriminate between healthy control, MCI, and mild AD dementia subjects (Harvey et al., 2017). In a study conducted by Goldberg et al. (2010), the UPSA identified functional deficits in MCI subjects, which were not measurable on one of the widely used traditional functional assessment scales, ADSC-ADL scale (Goldberg et al., 2010). A short version of UPSA has been developed for use in MCI (Gomar et al., 2011) and is likely to be much easier to implement in large, multicenter trials. This short version of UPSA was also able to discriminate between normal controls, MCI, and mild AD patients. However, to our knowledge, while the original UPSA has been used in numerous psychiatric trials and its psychometric properties have been well established, neither version has been used in large, multicenter AD trials. Data to show that UPSA can reliably distinguish individuals with performance-based functional deficits while these individuals remain clinically presymptomatic on traditional AD assessments but have brain amyloid pathology are yet to be collected.

Financial capacity represents a critical component for one's ability to live independently. This perspective was used by University of Alabama at Birmingham (UAB) to develop a functional tool that specifically measures financial capacity, the Financial Capacity Instrument (FCI; Marson et al., 2000). The long form, FCI-LF, was developed by Dr. Marson and his team at UAB as a tool for a comprehensive assessment of financial skills in patients with AD. Similarly to other instruments with shorter versions, the short form of FCI, FCI-SF, was derived from FCI-LF. The FCI-SF includes items from the FCI-LF, which are sensitive to progression of a cohort of well-characterized aMCI subjects to AD dementia. A complex checkbook task was added to the FCI-SF for increased sensitivity. The FCI-SF has been tested in the Mayo Clinic Study of Aging (MCSA) and at UAB and has been shown to have good psychometric properties (private communications).

The FCI-SF consists of 37 items designed to evaluate four constructs:

- monetary calculation,
- conceptual knowledge that relates to financial capacity,
- ability to use a checkbook,
- ability to use a bank statement.

It also measures time to complete four of the tasks:

- medical deductible calculation,
- simple income tax calculation,
- single-item checkbook/register task,
- three-item checkbook/register task (Marson et al., 2016).

When analyzing data from the FCI-SF, one obtains a total score, four scores for the individual constructs, and the four timing indexes. Importantly, normative data adjusted for age and education are available for the FC-SF (Gerstenecker et al., 2016).

Analyses of cross-sectional data indicate that the total score, the four component scores, and most of the time indexes robustly discriminate normal controls, aMCI, and mild AD dementia patients (D. Marson, private communications). In the MCSA study, 186 older, community dwelling subjects, who tested normal on traditional clinical tools, underwent a PET scan with the ^{11}C-Pittsburg compound B (PiB) for evidence of brain amyloid pathology and an evaluation on the FCI-SF. There was a statistically significant deficit in the performance of amyloid-positive subjects relative to their amyloid-negative counterparts on a complex coin/currency calculation, complex checkbook calculation, and a bank statement scanning task (Marson, 2015). Differences between amyloid-positive and amyloid-negative subjects on the Mental Calculations and Bank Statement Management component scores did not reach statistical significance, but indicated a strong trend. In addition, amyloid-positive subjects required more time to complete a simple income tax calculation and both single-item and three-item checkbook tasks. Timing indexes for these tasks and a composite of all timed tasks showed statistically significant slowing in the presence of amyloid brain pathology in these seemingly normal subjects (Marson, 2015). These findings raise the possibility that even when individuals appear to be cognitively normal on our rather crude widely used AD tools, the presence of amyloid pathology alone results

in subtle impairments that can be identified by very sensitive performance-based functional assessments. On the basis of these findings, the FCI-SF has been incorporated into the recently initiated third stage of ADNI (ADNI-3). As the ADNI-3 data on FCI-SF and other clinical, imaging, and fluid biomarker measures become available, our understanding of the disease may widen to incorporate both a more complex clinical framework that may be critical for the development of new AD therapeutic agents.

Rapid development of computer technologies have allowed for the advancement of computerized tools, including those that use virtual reality and video technology to simulate everyday situations. One such test is the Virtual Reality Functional Capacity Assessment (VRFCAT). VRFCAT uses virtual reality computer technology to engage subjects in a game-playing manner in four tasks:

- exploring a kitchen to find specific items,
- catching a bus to a grocery store,
- finding/purchasing food in a grocery store,
- returning home by a bus (Atkins et al., 2015).

The test has been used in schizophrenia and to identify age-related deficits. When comparing healthy younger (18–30 years of age) and healthy older (55–70 years of age) adults, VRFCAT showed a significant deficit between the two groups, with older adults taking on average 3 min longer and making on average two more mistakes. Future investigations of the use of VRFCAT and similar tests are required to show the utility and the feasibility of implementing such tests in AD trials.

VII NEW COGNITIVE CLINICAL MEASURES IN EARLY AD

Along with new performance-based functional tools, the AD field is investigating new types of cognitive tools, which use cognitive paradigms not traditionally implemented in AD. It is not possible for us to present an exhaustive review of such new measures. We use the Loewenstein-Acevedo Scale for Semantic Interference and Learning (LASSI-L) as an example of such test, which interrogates proactive semantic interference (PSI), retroactive semantic interference, and the ability to recover from the effects of PSI (Loewenstein et al., 2016). Preliminary results indicate that this test is sensitive to the presence of amyloid brain pathology in individuals that are otherwise normal on traditional neuropsychological tests. Specifically the unusual feature of the test, the measurement of recovery from the effects of PSI, was particularly correlated with total and regional amyloid load as measured by amyloid PET imaging.

This chapter will not review a wide range of digital wearable devices that open new methodological horizons for cognitive and functional assessments. Devices like actigraphy watches, which measure characteristics of sleep and wakefulness, and BioStamp nPoint, which measures vital signs, movement, activity and sleep, are expected to be used among many others to explore correlations between these biometric parameters and cognition and function in presymptomatic and early symptomatic AD patients. We hope that these devices will give us an opportunity to obtain objective and nuanced measurements in a patient's natural environment.

VIII CONCLUSION

In this chapter, we reviewed some of the new methodologies that the field of AD clinical trials is testing to increase the efficiency of development programs, particularly those for disease-modifying agents that are being studied in presymptomatic, biomarker positive and early symptomatic populations, including MCI due to AD and very mild AD dementia. Accumulating evidence indicates that even at what has been traditionally defined as presymptomatic stages of AD, the presence of amyloid brain pathology may result in subtle cognitive and functional deficits, which increase in severity as individuals develop MCI and mild AD dementia. However, the presence of such mild deficits, the heterogeneous nature of the population, and the slow rate of disease progression make trials that are aimed to investigate clinical efficacy, that is, proof-of-concept Phase 2 trials, or to confirm clinical efficacy, Phase 3 trials, very large and long, requiring huge human and financial resources. While the unmet medical need in AD is increasing, this therapeutic field has become extremely risky, in spite of the high level of preclinical and clinical research activities. The absence of surrogate biomarkers that correlate with clinical efficacy has pushed the field into abandoning Phase 2 stage of development and progressing from small Phase 1/2a trials with very limited information to large and expensive Phase 3 programs, where the level of failure has been close to 100%. We reviewed a number of novel clinical designs and more sensitive clinical tools that are either being investigated by ongoing studies or are being prepared for such investigations. Results from these trials will show if significant efficiency can be derived from the use of new trial designs and new clinical tools. What we can conclude with certainty is that there exists an urgent need for such innovations in AD drug development and that the search for these new trial approaches will have to persist if we are to maintain a chance at developing effective AD treatments.

References

Albert, M. S., et al. (2011). The diagnosis of mild cognitive impairment due to Alzheimer's disease: recommendations from the National Institute on Aging-Alzheimer's Association working groups on diagnostic guidelines for Alzheimer's disease. *Alzheimer's & Dementia, 7*, 280–292.

Alzheimer's Association. (2017). Alzheimer's Association Report. 2017 Alzheimer's disease facts and figures. *Alzheimer's & Dementia, 13*, 325–373.

Ard, M. C., et al. (2015). Optimal composite scores for longitudinal clinical trials under the linear mixed effects model. *Pharmaceutical Statistics, 14*, 418–426.

ASCO Daily News. (2015). Clinical trial designs for studying targeted therapies. A report on the "improving clinical trial efficacy: thinking outside the box" session. In *ASCO annual meeting, May 31.*

Atkins, A. S., et al. (2015). Assessment of age-related differences in functional capacity using the virtual reality functional capacity assessment tool (VRFCAT). *The Journal of Prevention of Alzheimer's Disease, 2*(2), 121–127.

Azvolinsky, A. (2015). Understanding umbrella and basket trial designs in lung cancer. *OncoTherapy Network.* http://www.oncotherapynetwork.com.

Belchior, P. C. (2015). Performance-based tools for assessing functional performance in individuals with mild cognitive impairment. *The Open Journal of Occupational Therapy, 3*, 3. article 3.

Berry, D. A. (2011). Adaptive clinical trials in oncology. *Nature Reviews. Clinical Oncology, 9*, 199–207.

Berry, D. A., Mueller, P., Griever, A. P., & Smith, M. (2002). Bayesian designs in dose-ranging drug trials. In C. Gatson, R. E. Kass, B. Carlin, A. Carriquiry, A. Gelman, I. Verdinelli, & M. West (Eds.), *Vol. 5. Case studies in Bayesian statistics* (pp. 99–181). New York: Springer.

Berry, S. M., et al. (2010). A Bayesian dose-finding trial with adaptive dose expansion to flexibly assess efficacy and safety of an investigational drug. *Clinical Trials, 7*, 121–135.

Bhatt, D. L., & Mehta, C. (2016). Adaptive designs for clinical trials. *The New England Journal of Medicine, 375*, 65–74.

Cummings, J. L., et al. (2014). Alzheimer's disease drug development pipeline: few candidates, frequent failures. *Alzheimer's Research & Therapy, 6*, 37.

Donohue, M. C., et al. (2014). The preclinical Alzheimer cognitive composite: measuring amyloid-related decline. *JAMA Neurology, 71*(8), 961–970.

Galbraith, S. (2014). The changing world of oncology drug development—a global pharmaceutical company's perspective. *Chinese Clinical Oncology, 3*(2), 40.

Gallo, P., et al. (2006). Adaptive designs in clinical drug development: an executive summary of the PhRMA working group. *Journal of Biopharmaceutical Statistics, 16*, 275–283.

Gerstenecker, A., et al. (2016). Age and education corrected older adult normative data for a short form version of the financial capacity instrument. *Psychological Assessment, 28*(6), 737–749.

Goldberg, T. E., et al. (2010). Performance-based measures of everyday function in mild cognitive impairment. *The American Journal of Psychiatry, 167*(7), 845–853.

Gomar, J. J., et al. (2011). Development and cross-validation of the UPSA short form for the performance-based functional assessment of patients with mild cognitive impairment and Alzheimer disease. *The American Journal of Geriatric Psychiatry, 19*, 915–922.

Gray, J. A., et al. (2015). The need for thorough phase 2 studies in medicines development for Alzheimer's disease. *Alzheimer's Research & Therapy, 7*, 67–69.

Greenberg, B. D., et al. (2013). Improving Alzheimer's disease phase 2 clinical trials. *Alzheimer's & Dementia, 9*, 39–49.

Grieve, A. P., & Krams, M. (2005). ASTIN: a Bayesian adaptive dose-response trial in acute stroke. *Clinical Trials, 2*, 340–351.

Harvey, P. D., et al. (2013). Hospitalization and psychosis: influences on the course of cognition and everyday functioning in people with schizophrenia. *Neurobiology of Disease, 53*, 18–25.

Harvey, P. D., et al. (2017). Performance-based and observational assessments in clinical trials across the Alzheimer's disease spectrum. *Innovations in Clinical Neuroscience, 14*(1–2), 30–39.

Hendrix, S., et al. (2012). Introducing a new tool for optimizing responsiveness to decline in early Alzheimer's disease. *Alzheimer's & Dementia, 8*(4 Suppl), S764–S765.

Jeffreys, H. (1973). *Scientific inference* (3rd ed., p. 31). Cambridge University Press.

Jutten, R. J., et al. (2017). A composite measure of cognitive and functional progression in Alzheimer's disease: design of the capturing changes in cognition study. *Alzheimer's & Dementia: Translational Research & Clinical Interventions, 3*, 130–138.

Kairalla, J. A., et al. (2012). Adaptive trial designs: a review of barriers and opportunities. *Trials, 13*, 145.

Krams, M., Lees, K. R., Hacke, W., Grieve, A. P., Orgogozo, J., & Ford, G. A. (2003). ASTIN: an adaptive dose-response study of UK-279,276 in acute ischemic stroke. *Stroke, 34*, 2543–2549.

Langbaum, J. B., et al. (2014). An empirically derived composite cognitive test score with improved power to track and evaluate treatments for preclinical Alzheimer's disease. *Alzheimer's & Dementia, 10*(6), 666–674.

Langbaum, J. B., et al. (2015). Establishing composite cognitive endpoints for use in preclinical Alzheimer's disease trials. *The Journal of Prevention of Alzheimer's Disease, 2*(1), 2–3.

Loewenstein, D. A., et al. (2016). A novel cognitive stress test for the detection of preclinical Alzheimer disease: discriminative properties and relation to amyloid load. *The American Journal of Geriatric Psychiatry, 24*, 804–813.

Logovinsky, V., et al. (2013). New composite score demonstrates sensitivity to disease progression and treatment effects. In *Presented at the 11th international conference on Alzheimer's and Parkinson's diseases, March 6–10, Florence, Italy*.

Logovinsky, V., et al. (2014). Population characterization in the early stages of Alzheimer's disease. *Alzheimer's & Dementia, 10*(4 Suppl), P243.

Mantovani, L. M., et al. (2015). UCSD performance-based skills assessment (UPSA): validation of a Brazilian version in patients with schizophrenia. *Schizophrenia Research: Cognition, 2*, 20–25.

Marshal, G. A., et al. (2015). The Harvard Automated Phone Task: new performance-based activities of daily living tests for early Alzheimer's disease. *The Journal of Prevention of Alzheimer's Disease, 2*(4), 242–253.

Marshall, G. A., et al. (2012). Activities of daily living: where do they fit in the diagnosis of Alzheimer's disease? *Neurodegenerative Disease Management, 2*(5), 483–491.

Marshall, G. A., et al. (2016). The Harvard Automated Phone Task (APT): a novel performance-based ADL instrument for early Alzheimer's disease. *Alzheimer's & Dementia, 12*(7 Suppl), P373.

Marson, D. (2015). Investigating functional impairment in preclinical Alzheimer's disease: potential measure characteristics and methodology. *The Journal of Prevention of Alzheimer's Disease, 2*(1), 4–6.

Marson, D. C., et al. (2000). Assessing financial capacity in patients with Alzheimer disease: a conceptual model and prototype instrument. *Archives of Neurology, 57*(6), 877–884.

Marson, D. C., et al. (2016). Detecting functional impairment in preclinical Alzheimer's disease using a brief performance measure of financial skills. *Alzheimer's & Dementia, 12*(7 Suppl), P373.

Mathematical Sciences, & University of Reading. (2014). *Development of novel adaptive designs to improve efficiency in clinical trials. Research excellence framework*. Impact case study, downloaded from https://impact.ref.ac.uk/CaseStudies/CaseStudy.aspx?Id=37454.

Pammolli, F., et al. (2011). The productivity crisis in pharmaceutical R&D. *Nature Reviews. Drug Discovery, 10*, 428–438.

Patterson, T. L., et al. (2001). UCSD performance-based skills assessment: development of a new measure of everyday functioning for severely mentally ill adults. *Schizophrenia Bulletin, 27*(2), 235–245.

Raghavan, N., et al. (2014). Validation of novel composite outcome measures for pre-dementia Alzheimer's disease. *Alzheimer's & Dementia, 10*(4 Suppl), P244.

Riepe, M. W., et al. (2011). Additive scales in degenerative disease—calculation of effect sizes and clinical judgment. *BMC Medical Research Methodology, 11*, 169.

Ritchie, C. W., et al. (2014). European prevention of Alzheimer's dementia (EPAD) project: an international platform to deliver proof-of-concept studies for secondary prevention of dementia. *The Journal of Prevention of Alzheimer's Disease, 1*, 221. (on behalf of the EPAD project partners) [Abstract OC7].

Satlin, A., et al. (2012). Bayesian adaptive trial design: a new approach for phase 2 clinical trials in Alzheimer's disease. In *Presented at clinical trials on Alzheimer's disease conference*.

Satlin, A., et al. (2016). Design of a Bayesian adaptive phase 2 proof-of-concept trial for BAN2401, a putative disease-modifying monoclonal antibody for the treatment of Alzheimer's. *Alzheimer's & Dementia: Translational Research & Clinical Interventions, 2*, 1–12.

Schneider, L. S., et al. (2014). Clinical trials and late-stage drug development for Alzheimer's disease: an appraisal from 1984 to 2014. *Journal of Internal Medicine, 275*, 251–283.

Sperling, R. A., et al. (2011). Toward defining the preclinical stages of Alzheimer's disease: recommendations from the National Institute on Aging-Alzheimer's Association working groups on diagnostic guidelines for Alzheimer's disease. *Alzheimer's & Dementia, 7*, 270–279.

Swanson, C. J., et al. (2013). Critical role for simulations in adaptive trial designs. In *Presentation at the 11th international conference on Alzheimer's and Parkinson's diseases, AD/PD*.

U.S. Food and Drug Administration. (2010). *Draft guidance for industry: Adaptive design clinical trials for drugs and biologics.* .

U.S. Food and Drug Administration. (2013). *Guidance for industry. Alzheimer's disease: Developing drugs for the treatment of early stage disease*. Washington, DC: U.S. Food and Drug Administration.

Vradenburg, G., et al. (2014). The global Alzheimer's platform: a building block for accelerating clinical R&D in AD. *The Journal of Prevention of Alzheimer's Disease, 1*, 221 [Abstract P2-15].

Wang, J., et al. (2016). ADCOMS: a composite clinical outcome for prodromal Alzheimer's disease trials. *Journal of Neurology, Neurosurgery, and Psychiatry, 87*, 993–999.

Experimental Medicine Models in Generalized Anxiety Disorder and Social Anxiety Disorder

David S. Baldwin*,†,‡, Ayman Abou-Aisha‡

*Clinical and Experimental Sciences, Faculty of Medicine, University of Southampton, Southampton, United Kingdom
†University Department of Psychiatry and Mental Health, University of Cape Town, Cape Town, South Africa ‡Mood and Anxiety Disorders Service, Southern Health NHS Foundation Trust, Southampton, United Kingdom

I EXPERIMENTAL MEDICINE APPROACHES IN ANXIETY DISORDERS

Evidence-based treatment guidelines for anxiety disorders synthesize knowledge of the efficacy, tolerability, and therapeutic role of currently available pharmacological and psychological treatments. Existing treatments are far from ideal, and there is much room for developing novel interventions with an earlier onset of clinical effect, greater overall effectiveness, or enhanced effectiveness in specific patient subgroups while avoiding unwanted effects such as sedation, weight gain, emotional indifference, sexual dysfunction, and the risks of tolerance and dependence (Baldwin & Brandish, 2014). However, ascertaining the potential benefit of novel treatments in pivotal large randomized controlled trials is necessarily time-consuming and costly, and novel psychotropic drug development is often regarded as being "high risk." There are typically prolonged delays before empirical innovations translate into adoption in clinical practice (Hanney et al., 2015), and many biotechnology and pharmaceutical companies have reduced their investment in neuroscience (Insel, 2015; Insel et al., 2013). Successful development of new anxiolytic medications not only could be eased through improved animal models of anxiety disorders and enhanced methods for establishing the efficacy of novel compounds (Haller, Aliczki, & Pelczer, 2013) but also may hinge upon identifying multifaceted biomarkers that combine genetic, cognitive, and neuroimaging measures (Bandelow et al., 2016, 2017).

The term "experimental medicine" connotes investigations undertaken in humans, designed to model systems, to identify mechanisms of pathophysiology or disease, or to demonstrate proof-of-concept evidence of the validity and importance of new discoveries or treatments. Experimental medicine studies can therefore provide preliminary evidence that helps to determine whether to proceed to larger, more costly efficacy studies. Table 25.1 lists suggested criteria (Bailey et al., 2011a; Guttmacher, Murphy, & Insel, 1983) for an experimental medicine model. Because experimental approaches in anxiety disorders often focus on the physiological, pharmacological, or psychological induction of anxiety symptoms, they have some potential drawbacks: for example, the effects of physiological or psychological challenges can be affected by individual resilience to challenge, and the effects of novel compounds in attenuating pharmacologically induced anxiety can be complicated by drug-drug interactions.

Despite these limitations, a range of models have been used to support investigation and treatment development in anxiety disorders, including lactate infusion and cholecystokinin challenge in panic disorder; challenge during the Trier social stress test in social anxiety disorder (SAD); oxytocin administration and attachment priming in separation anxiety disorder; threat of unpredictable shock; and carbon dioxide inhalation, in panic disorder and generalized anxiety disorder (GAD).

II GENERALIZED ANXIETY DISORDER AND SOCIAL ANXIETY DISORDER

Generalized anxiety disorder (GAD) is a common and impairing anxiety disorder, with an estimated 12-month prevalence of 1.7%–3.4% (being more prevalent in older

TABLE 25.1 Requirements for an Experimental Medicine Model in Psychiatry

Safe	For participants and investigators, ideally noninvasive
Acceptable	To participants, ethics committees, regulatory bodies
Reliable	Interperformer and repeat-performer replicability
Valid	Effects attenuated by clinically effective treatments
Translational	From lab to clinic and back again, across species
Feasible	Ease of performance in practice
Repeatable	No attenuation of response if performed again
Subjective	Measurable psychological effects
Objective	Measurable physiological effects
Inexpensive	Can be supported by academia and industry

adults) (Wittchen et al., 2011). Diagnosis of GAD requires the presence of psychological and physical anxiety symptoms for at least 6 months, these symptoms not being "understandable" as deriving from another condition. ICD-10 criteria (World Health Organization, 1993) emphasize symptoms of tension, worrying and apprehension, whereas DSM-5 criteria (American Psychiatric Association, 2013) emphasize multiple, distressing, and uncontrollable worries. Earlier versions of DSM diagnostic criteria were based on the presence of symptoms for at least 1 month and had low interrater reliability (Di Nardo et al., 1983; Mannuzza et al., 1989), but the subsequent stipulations of a 6-month duration and perception of uncontrollable worry have enhanced the overall reliability of diagnosis (Brown et al., 2001). However, there are persisting concerns about diagnostic validity, such as the distinction from major depression and the threshold for symptom severity (Brown et al., 2001; Brown, Chorpita, & Barlow, 1998), and a more dimensional approach based on measuring worry, distress, and other symptoms might help delineate the condition (Gordon & Heimberg, 2011; Rutter & Brown, 2015). Symptom severity is usually assessed through observer-rated scales, such as the Hamilton Rating Scale for Anxiety (HAMA) (Hamilton, 1959): a HAMA score of less than 9 may correspond to symptom remission, whereas a score of 24 or more indicates anxiety symptoms of at least moderate intensity (Bandelow et al., 2006), HAMA scores being strongly correlated with self-rated impairments (Stein et al., 2009).

Selective serotonin reuptake inhibitor (SSRI) prescription is the usual first-line pharmacological treatment in GAD, based on efficacy, tolerability, and safety in randomized controlled trials. Response rates are often high, but only approximately one-half of patients enter symptom remission after 2–3 months of treatment (Baldwin, Huusom, & Mæhlum, 2006). The intensity of coexisting depressive symptoms often reduces with treatment, even with anxiolytic drugs that have no efficacy in major depressive episodes (Baldwin et al., 2015; Stein et al., 2008). Response is more likely in patients whose symptoms have reduced in intensity within the initial 2 weeks of treatment (Baldwin et al., 2009, 2012a). The comparative efficacy and tolerability of differing pharmacological treatments are uncertain (Baldwin et al., 2011). Sustained treatment is usually needed as GAD is often episodic, or waxes and wanes, over long periods (Angst et al., 2008). Relapse prevention studies have demonstrated the long-term efficacy of many pharmacological treatments, including some SSRIs (escitalopram and paroxetine) (Allgulander, Florea, & Trap Huusom, 2006; Stocchi et al., 2003) and the serotonin-norepinephrine reuptake inhibitors (SNRI) duloxetine and venlafaxine (Davidson et al., 2008; Hackett, White, & Salinas, 2000; Rickels et al., 2010) and also pregabalin (Feltner et al., 2008), quetiapine (Katzman et al., 2011), and agomelatine (Stein et al., 2012). Preclinical studies with the novel "multimodal" antidepressant vortioxetine suggested potential anxiolytic effects (Baldwin & Hanumanthaiah, 2015), and it reduces anxiety symptoms in depressed patients (Baldwin et al., 2016a): however, randomized placebo-controlled trials of acute treatment of GAD have produced inconsistent findings (Pae et al., 2015), despite its efficacy in preventing relapse (Baldwin, Loft, & Florea, 2012). The place of benzodiazepine anxiolytics is much debated although benzodiazepines are often prescribed in routine practice when treating patients with GAD (Baldwin et al., 2012b). There may be a role for careful prescribing of benzodiazepines in some patients, including in short-term treatment (up to 4 weeks) while waiting for an SSRI to become effective. There may also be a place for longer-term treatment, when patients have not responded to a series of interventions, including psychological approaches and sequential treatment courses with an SSRI, SNRI, pregabalin, and the 5-HT$_{1A}$ partial agonist buspirone (Baldwin et al., 2013).

Social anxiety disorder (SAD), also known as social phobia, has an estimated 12-month prevalence of 2.3% (Wittchen et al., 2011) and typically has an onset by adolescence and runs a prolonged course. The condition is characterized by marked, persistent, and unreasonable fear of being observed or evaluated negatively by other people, in social or performance situations, with accompanying physical and psychological anxiety symptoms (American Psychiatric Association, 2013; World Health Organization, 1993). Feared situations (such as speaking to unfamiliar people or eating in public) are avoided or endured with significant distress. It is often not recognized in primary medical care, but detection in psychologically distressed patients can be enhanced with screening questionnaires. The condition is often misconstrued as mere "shyness" but can be distinguished from uncomplicated shyness by the higher levels of personal distress, severe symptoms, and greater impairment. The generalized subtype (where anxiety is associated with many situations) is more disabling and shows greater comorbidity, although patients with the nongeneralized subtype (where anxiety is focused on a limited number of situations) can also be substantially impaired. Patients with SAD often present with symptoms arising from comorbid conditions (especially depression), rather than with anxiety symptoms and avoidance of social and performance situations. There are strong and possibly two-way associations between SAD and substance use disorders.

Medications with proven efficacy in acute treatment of SAD include most SSRIs (escitalopram, fluoxetine, fluvoxamine, paroxetine, and sertraline), venlafaxine, the monoamine oxidase inhibitors phenelzine and moclobemide, some benzodiazepines (bromazepam and clonazepam), some anticonvulsants (gabapentin and pregabalin), and olanzapine. Cognitive behavioral therapy (CBT) is the most well-established psychological treatment. Network metaanalysis of the comparative efficacy of pharmacological and psychological treatments indicates that SSRIs and venlafaxine are superior to pill placebo and CBT superior to "psychological placebo" (i.e., nonspecific psychological intervention) (Mayo-Wilson et al., 2014). As with GAD, prescription of an SSRI is usually regarded as first-line pharmacological treatment, based on efficacy, tolerability, and safety in randomized controlled trials (Baldwin et al., 2014). There is rather little evidence of a dose-response relationship in acute treatment (Baldwin et al., 2016b). Response is more likely in patients whose symptoms have reduced in intensity within the initial 2 weeks of treatment (Baldwin et al., 2009). Sustained treatment is usually needed, as SAD is typically a chronic condition, but relapse prevention studies demonstrating the long-term efficacy of pharmacological treatments (escitalopram, paroxetine, sertraline, and pregabalin) are limited (Blanco et al., 2013). Patients who make only a limited response to CBT may benefit if continued CBT is combined with medication, and those who respond only partly to pharmacological treatment may benefit if it is combined with CBT (Canton, Scott, & Glue, 2012). Combination of an SSRI (sertraline) with a benzodiazepine (clonazepam) has been found superior to continuing with sertraline monotherapy or switching to venlafaxine (Pollack et al., 2014).

III ETIOLOGY OF GENERALIZED ANXIETY DISORDER AND SOCIAL ANXIETY DISORDER

The etiology of GAD is not fully established: a range of cognitive theoretical models have been elaborated to explain how excessive worrying develops and is maintained, and genetic, familial, environmental, and neurobiological factors have all been described (Stevens et al., 2014). An *avoidance theory of worry* posits that excessive worrying in GAD represents a cognitive avoidance of somatic and affective experiences through which catastrophic images are replaced by verbal-linguistic activities associated with lessened physiological arousal (Borkovec, Alcaine, & Behar, 2004). The *intolerance of uncertainty model* suggests that people with GAD find perceptions of ambiguous or uncertain information to be especially distressing and also believe that worrying either reduces the likelihood of feared outcomes or increases the ability to cope with them (Dugas, Buhr, & Ladouceur, 2004). A *metacognitive model of GAD* proposes that primary (Type I) worries relate to noncognitive and largely external anxiety-inducing situations, whereas secondarily developing (Type II) worries (metaworries) focus on the potentially adverse consequences of excessive worrying (Wells, 2005). A four-component *emotion dysregulation model* suggests individuals with GAD experience emotions more quickly and intensely, but with poorer understanding, leading to a sense of being overwhelmed by emotions, with consequent maladaptive strategies to minimize their impact (Mennin et al., 2005). An *acceptance-based model* stresses problematic relationships with internal experiences, with initial difficulties in noticing and accepting transient negative reactions and subsequent development of self-critical beliefs and avoidant behavioral restrictions (Roemer & Orsillo, 2005). *Information processing models* in GAD emphasize empirically demonstrated persistent attentional biases, with preferential attention to threatening over neutral stimuli (MacLeod & Rutherford, 2004). Other important etiological factors include genetic influences on generalized anxiety and "neuroticism" (Hettema, Prescott, & Kendler, 2004); adverse early life experiences and disordered attachment (Cassidy et al., 2009); enhanced emotion-related amygdala activation (Etkin et al., 2010); and disturbances in the hypothalamo-pituitary-adrenal (HPA) axis and in serotonergic, noradrenergic, and GABAergic neurotransmission (Gray & McEwen, 2014). Despite the wide range of cognitive theoretical models of GAD, most of which are open to pharmacological manipulation, there have been few experimental medicine studies of the effects of psychotropic medications on avoidance of worrying, intolerance of uncertainty, metacognitions, emotion dysregulation, acceptance, or information processing in healthy volunteers.

As with GAD the multifactorial etiology of social anxiety disorder is not fully established. Genetic factors are important in both "shyness" (Daniels & Plomin, 1985) and behavioral inhibition (a temperamental style characterized by fearfulness, timidity, and avoidance) (Schwartz, Snidman, & Kagan, 1999); early influences include harsh and controlling parenting styles

(Greco & Morris, 2002); parental negative information leading to negative expectancies (Mineka & Sutton, 2006); deficits in social skills (though the evidence is inconsistent) (Stravynski & Amado, 2001); attentional and interpretative cognitive biases (Heinrichs & Hofmann, 2001); sensitivity to facial expressions (Öhman, 1986); enhanced vigilance for social threat cues (Heinrichs & Hofmann, 2001); and disturbances in the HPA axis and in serotonergic, noradrenergic, and GABAergic neurotransmission (Gray & McEwen, 2014).

Social anxiety may be maintained by a perceptual bias to misread facial expressions as either disapproving (disgusted) or threatening (angry), with consequent inappropriate elicited social responses and social avoidance (Rapee & Heimberg, 1997; Schlenker & Leary, 1982). As such "retraining" individuals with high social anxiety to process faces in a less threatening manner have potential clinical benefits. However, studies that have employed "morphing" techniques to generate faces expressing varying intensities of emotion have found no evidence of a lower threshold for decoding threat emotions such as disgust (Philippot & Douilliez, 2005; Schofield, Coles, & Gibb, 2007). Furthermore, Winton et al. (1995) (Winton, Clark, & Edelmann, 1995) found no evidence for differences between cases and controls in the ability to label facial expressions as "negative" or "neutral." In addition, the studies that reported differences in emotion decoding have been inconsistent in presenting the emotions associated with social anxiety (Bell et al., 2011; Garner et al., 2009; Hunter, Buckner, & Schmidt, 2009; Silvia et al., 2006).

IV CARBON DIOXIDE INHALATION AS AN EXPERIMENTAL MEDICINE MODEL IN ANXIETY DISORDERS

Inhalation of air "enriched" with an increased proportion of carbon dioxide (CO_2) is a commonly used experimental approach in the investigation of induced anxiety. Brief inhalation of air with higher than normal concentrations of CO_2 (such as occurs with single vital capacity inhalations of 35% CO_2) is associated with the emergence of acute and severe anxiety and may induce a panic attack. By contrast, inhalation of 7.0%–7.5% CO_2 over 20 min can elicit subjective, autonomic, and neurocognitive changes that resemble the features of generalized anxiety: increases in heart rate and systolic blood pressure are detected consistently, though increased diastolic blood pressure is less frequent, and induction of panic is uncommon.

The mechanisms that underlie the provocation of anxiety by CO_2 challenge are not fully clarified (Leibold et al., 2015; Vollmer, Strawn, & Sah, 2015). It is uncertain whether anxiety is triggered by relative hypoxia or

hypercapnia, or whether both disturbances are needed: although hypercapnia and hypoxia may be equally important in driving "air hunger" in healthy subjects (Moosavi et al., 2003) and patients with panic disorder have increased sensitivity to both conditions, when compared with healthy controls (Beck, Ohtake, & Shipherd, 1999). Genetic factors may be important in hypersensitivity to CO_2 (Battaglia et al., 2007, 2008). Although inhalation of air enriched with a high proportion (35%) of CO_2 is associated with increased cortisol secretion (Argyropoulos et al., 2002; Kaye et al., 2004), it is uncertain whether this response is specific to the CO_2 challenge per se or whether it occurs as part of a general response to other study procedures (Leibold et al., 2015). The role of disturbed respiratory physiology in induction of panic following CO_2 inhalation is not certain (Schenberg, 2016), but experimentally induced panic attacks are associated with both low end-tidal CO_2 and high ventilation variance at baseline (Papp et al., 1997).

Serotonergic mechanisms influence the panic response to CO_2 challenge: tryptophan depletion alone does not induce panic (Goddard et al., 1994), but it does enhance the panic response to CO_2 inhalation (Schruers et al., 2000). Administration of the 5-HT precursor L-5-hydroxytryptophan reduces the panic response (Schruers et al., 2002). The increase in subjective anxiety, heart rate, and blood pressure after 35% CO_2 challenge suggests a shared noradrenergic-mediated mechanism underlying CO_2 sensitivity (Bailey et al., 2003). Changes in CO_2 saturation may act upon pH- or CO_2-dependent chemoreceptors in the locus coeruleus (LC) to increase noradrenaline (NA) release, as 5% CO_2 increases LC neuronal firing rate in rat brain slices (Martin et al., 2009). CO_2-induced NA release may mediate subjective and autonomic features of anxiety via projections to centers involved in cardiovascular control and to the limbic system. The cortisol response may be mediated through altered noradrenergic input into the paraventricular nucleus, enhancing corticotrophin-releasing factor release, so triggering cortisol secretion. However, although NA is important in mediating anxiety provoked by 35% CO_2 challenge, additional mechanisms must be involved as drugs that affect noradrenergic function have little effect on subjective responses to CO_2 (Pinkney et al., 2014).

CO_2 reactivity in mice is linked to chemosensors within the amygdala (Ziemann et al., 2009). The most well-characterized chemosensor is the acid-sensing ion channel 1 (ASIC-1a), which is a voltage-insensitive H^+-gated cation channel, expressed in the amygdala, dentate gyrus, cortex, striatum, and nucleus accumbens (Wemmie, 2011). ASICs can detect small reductions in brain pH (acidosis), including that arising from inhalation of an acidic gas (such as CO_2) (Sherwood, Frey, & Askwith, 2012). In the presence of fully functioning ASIC1a chemosensors, inhalation of 2%–20% CO_2 elicits normal mouse fear behavior, but pharmacological blockade (or elimination of ASIC1a in knockout mice) impairs fear responses to CO$_2$, whereas amygdala-localized reexpression restores fear behavior. Other chemosensory structures include orexin neurones in the hypothalamus, serotonergic neurones in the medullary raphe (Lin, Sun, & Chen, 2015), and hypoxia-sensitive chemosensory neurones in the periaqueductal gray (Wang, Pizzonia, & Richerson, 1998). However, perturbed chemosensor activity may not in itself fully explain the physiological effects of CO$_2$ challenge that are not understandable solely in terms of CO$_2$-provoked alterations in noradrenergic activity, and additional mechanisms may be important.

V "LOW-DOSE" CO$_2$ INHALATION AS AN EXPERIMENTAL MEDICINE MODEL OF GAD

Low-dose (7.5%) prolonged (20 min) CO$_2$ inhalation was found to induce anxiety in a double-blind, placebo-controlled trial involving healthy subjects: when compared with normal air inhalation, CO$_2$ inhalation increased heart rate and blood pressure and raised subjective anxiety (Bailey et al., 2005). In a subsequent single-blind, placebo-controlled healthy subject study, 7% CO$_2$ inhalation increased respiratory rate, minute volume, end-tidal CO$_2$, skin conductance and subjective anxiety: and participants who experienced marked anxiety underwent a second identical inhalation, with good test-retest repeatability. The study findings reveal some limitations in the model, as 30% of participants were "non-responders" and 10% experienced significant anxiety during normal air inhalation (Poma et al., 2005). However, 7.5% CO$_2$ challenge can reliably induce dysfunction in some of the neuropsychological features that characterize populations with high trait anxiety and patients with GAD, including hypervigilance and increased alertness (Garner et al., 2012), poor attention control, and selective processing of environmental threat (Garner et al., 2011). Although inhalation challenges involving less than 15% CO$_2$ provoke significantly more panic attacks in patients with panic disorder than in healthy controls (Bailey et al., 2011a), it is uncertain whether GAD is also associated with altered sensitivity to "low-dose" CO$_2$ inhalation. In a small single-blind, randomized, crossover study in medication-free GAD patients, which involved repeated inhalation of 7.5% CO$_2$ over 20 min, qualitative assessment of participants' experiences found they resembled their previous symptoms, most closely for the "physiological" symptoms (Seddon et al., 2011).

The influence of anxiolytic and antidepressant medication in attenuating CO$_2$-evoked anxiety has been assessed: in general terms, administration of a benzodiazepine reduces subjective CO$_2$-provoked anxiety but has little impact on physiological measures, whereas the effects of antidepressants are less clearly defined. When compared with placebo, lorazepam (2 mg) administration attenuated subjective anxiety with no accompanying change in autonomic measures in healthy participants undergoing 20 min of 7.5% CO$_2$ inhalation (Bailey et al., 2007). Similar effects were observed when lorazepam was employed as a control in studies using the same inhalation procedure to assess the effects of novel anxiolytic compounds (Bailey et al., 2011b; de Oliveira et al., 2012). Alprazolam (1 mg) and the GABA-A agonist zolpidem (5 mg) both attenuated subjective anxiety in healthy volunteers after 20 min of 7.5% CO$_2$ inhalation (Bailey et al., 2009). But not all evidence is consistent, as a subsequent double-blind, placebo-controlled crossover study that investigated dose-response relationships with lorazepam, using the same methodology, found no attenuation of either subjective or autonomic responses (Diaper et al., 2012).

Administration of various SSRIs, venlafaxine, tricyclic antidepressants, or the monoamine oxidase inhibitor toloxatone can attenuate the panic response to CO$_2$ challenge (Bailey et al., 2011a). SSRIs often take 4 weeks to exert sustained therapeutic effects in patients with anxiety disorders, so prolonged drug administration may be needed to generate valid findings in experimental medicine studies involving healthy subjects. A study involving 3 min of 5% CO$_2$ in individuals "at high risk of panic disorder" found that prior administration of the SSRI escitalopram had no effect on subjective or autonomic anxiety (Coryell & Rickels, 2009). By contrast, investigations in patients with established panic disorder found that 12 weeks of SSRI or SNRI treatment reduced subjective anxiety following 5% and 7% CO$_2$ challenge, when compared with the effects of baseline inhalation, before treatment (Gorman et al., 2004). Studies involving SSRI, SNRI, or beta-blocker administration in healthy volunteers using a 20-min 7.5% CO$_2$ challenge have generated variable findings, though as is the case with benzodiazepines, SSRI or SNRI administration has only limited effects on physiological responses. Placebo-controlled administration of the SSRI paroxetine for 21 days (10 mg titrated to 20 mg after day 8) reduced subjective anxiety (Bailey et al., 2007). A placebo-controlled investigation of 3-week administration of either venlafaxine (150 mg) or the anxiolytic pregabalin (200 mg) found no significant effect on ratings of subjective anxiety or autonomic response in either group (Diaper et al., 2013). The beta-blocker propranolol (40 mg) had no effect on self-reported anxiety in healthy volunteers undergoing 20 min of 7.5% CO$_2$ inhalation (Papadopoulos et al., 2010), which is consistent with its lack of efficacy in reducing psychological symptoms in patients with anxiety disorders (Steenen et al., 2016).

The CO$_2$ inhalation experimental model may be useful in determining the potential anxiolytic efficacy of novel agents in "proof-of-concept" studies and has been employed in investigations of the CRF1 receptor antagonist R317573 (Seddon et al., 2011), the NK1 receptor antagonists vestipitant and vofopitant (Poma et al., 2014), and in psychological interventions (Ainsworth et al., 2015). Positive studies with compounds that affect chemosensory mechanisms could guide the development of novel anxiolytics or "repurposed" medications—for example, with the ASIC ion channel antagonist amiloride, which has been found to have neuroprotective effects (Arun et al., 2013); with orexin receptor antagonists, which can attenuate anxiety-like responses to CO_2 challenge in rats (Johnson et al., 2012); and with the carbonic anhydrase inhibitor acetazolamide, which blocks conversion of CO_2 to carbonic acid and thence to hydrogen and bicarbonate ions (Vollmer et al., 2015).

VI EXPERIMENTAL APPROACHES TO SUPPORT TREATMENT DEVELOPMENT IN SOCIAL ANXIETY DISORDER

Animal models of social anxiety have marked limitations as they cannot determine key features such as the abhorrence of scrutiny or critical self-appraisal in social and performance situations. A range of experimental medicine approaches in healthy volunteers have therefore been undertaken to explore novel treatments for SAD, including investigations based on oxytocin, testosterone, opioid, and cannabinoid administration, sometimes coupled with human functional magnetic resonance imaging (fMRI). Oxytocin and testosterone have well-established roles in social-emotional behavior: with direct or indirect effects on the orbitofrontal cortex, anterior cingulate cortex, amygdala, ventral striatum, and hypothalamus (Neumann & Slattery, 2016; Crespi, 2016).

A Studies Involving Oxytocin Administration

A series of placebo-controlled investigations have indicated that oxytocin plays a role in modulating socially negative cues and in enhancing the integration of social responses, suggesting a potential role for oxytocin in reducing social anxiety. In a double-blind placebo-controlled within-subject fMRI investigation of the effects of single-dose intranasal oxytocin (24 IU) administration on cortical activation to facial expressions (sad, happy, and neutral) in individuals with generalized social anxiety disorder ($n=18$) and in healthy controls ($n=18$), socially anxious individuals showed heightened activity to sad faces in the medial prefrontal cortex (mPFC) and anterior cingulate cortex (ACC). However, following oxytocin administration, activation within the mPFC and ACC regions was at levels similar to that seen in controls (Labuschagne et al., 2012). In a double-blind placebo-

controlled crossover study in healthy male controls ($n=26$) and male individuals with high social anxiety ($n=16$), single-dose oxytocin (24 IU) administration was found to attenuate an attentional bias for emotional faces in the socially anxious volunteers, "normalizing" it to levels seen in the controls at baseline (Clark-Elford et al., 2015). Another randomized, placebo-controlled within-subject investigation found that intranasal oxytocin (24 IU) administration enhanced functional connectivity between the amygdala, insula, and cingulate gyrus, during the processing of fearful faces in individuals meeting criteria for generalized social anxiety disorder (but did not do so in healthy controls) (Gorka et al., 2015a).

Eckstein et al. (2015) investigated the effects of oxytocin administration on physiological and neuropsychological markers of conditioned fear in male healthy volunteers: participants were first exposed to a Pavlovian fear conditioning paradigm, then administered intranasal synthetic oxytocin (24 IU) using a randomized placebo-controlled double-blind parallel-group design. Oxytocin increased electrodermal responses and prefrontal cortex signals (determined by fMRI) to conditioned fear in the early phase of extinction but enhanced the decline of skin conductance responses in the late phase of extinction and evoked an unspecific inhibition of amygdala responses in both phases. These findings therefore suggest that oxytocin administration could be used to diminish the amygdala response to conditioned fear during cognitive behavioral therapeutic interventions.

Kirsch and colleagues (Chen et al., 2016; Kirsch et al., 2005) examined the effects of intranasal oxytocin (27 IU) on correlates of anxiety in male healthy volunteers, using a randomized placebo-controlled double-blind crossover design. In one task, simultaneous presentation of angry or fearful faces was matched with an identical target face of identical expression; in another task, participants matched one of two simultaneously presented fearful/threatening scenes with an identical target scene, with matching of simple shapes as a control intervention. When compared with placebo, oxytocin administration reduced amygdala activation and coupling of the amygdala to brain stem regions implicated in the autonomic and behavioral manifestations of fear, thereby suggesting that oxytocin could exert therapeutic effects in conditions characterized by fearful social interactions such as social anxiety disorder and autism.

Xu Chen and colleagues (Chen et al., 2016; du Plooy et al., 2014) evaluated the effects of intranasal administration of oxytocin (24 IU) or vasopressin (20 IU) in male and female healthy volunteers within a double-blind, randomized placebo-controlled pharmaco-fMRI study. Imaging was undertaken while participants played the iterated "prisoner's dilemma" game (a model of social cooperativeness, based on reciprocal altruism), being told they were playing with either same-sex human or computer partners (in reality, subjects always played against a preprogramed computer algorithm). In males, oxytocin attenuated fMRI responses to unreciprocated cooperation (negative social interaction) within the amygdala and anterior insula with human but not computer partners, and similar effects were seen with vasopressin. Conversely, in female participants, oxytocin attenuated amygdala and anterior insula responses to unreciprocated cooperation in computer but not human partners, and vasopressin did not significantly modulate amygdala or insula responses. These findings suggest oxytocin may decrease the stressful aspects of negative social interactions in men but not in women and support further examination of the effects of vasopressin antagonists as potential anxiolytic drugs.

Lynn et al. (2014)) included 40 healthy volunteers (40% women, 75% Caucasian) within a randomized double-blind placebo-controlled investigation of the effects of intranasal oxytocin (10 sprays, 3IU oxytocin per spray) on facial emotion ("angry" and "not angry") perception. The paradigm combined perceptual uncertainty with behavioral economics in a signal detection framework, using the probability of encountering angry faces and the cost of misidentifying them as "not angry" to create a risky environment wherein bias to categorize faces as angry would maximize point earnings. Consistent with the underestimation of factors creating risk (encounter rate and cost), men who were administered oxytocin exhibited a worse (i.e., less liberal) response bias than men who received placebo, whereas oxytocin did not significantly influence performance in women. These results suggest oxytocin may impair the ability to adapt to changes in risk and uncertainty when introduced to novel or changing social environments, in men. This is an intriguing finding as oxytocin-based pharmacotherapy could therefore have some unintended consequences (e.g., increased risky decision-making) while enhancing social interactions.

B Studies Involving Other Challenges

In a randomized placebo-controlled crossover study in healthy women ($n=24$), single-dose (0.5 mg) testosterone administration significantly diminished the avoidance of computerized representations of angry faces (but not the avoidance of happy faces) (Enter, Spinhoven, & Roelofs, 2014): a finding that concurs with other studies that have demonstrated the dominance-enhancing effects of testosterone administration. In another randomized placebo-controlled, crossover study in healthy women ($n=19$) and medication-free women with social anxiety disorder ($n=18$), single-dose (0.5 mg) testosterone administration was found to significantly alleviate gaze avoidance in the women with social anxiety disorder, by increasing initial gaze towards the eyes in angry facial expressions

(Enter et al., 2016). Taken together, these findings support the role of testosterone in dominance-enhancing behavior and suggest a potential role for testosterone administration in augmenting psychological and pharmacological interventions in social anxiety disorder.

Meier et al. (2016) investigated the modulating effects of administration of the opioid antagonist naltrexone (50 mg) on emotional mimicry (which is known to promote social affiliation) in female healthy volunteers, using a double-blind, randomized placebo-controlled between-subject study design. A passive viewing task with dynamic facial expressions was employed, and mimicry was measured with electromyographic recordings over three facial muscles: the *corrugator supercilii*, which is involved in frowning; the depressor jaw muscle, which pulls the corners of the mouth down to produce a sad expression; and the *zygomaticus major,* which lifts the mouth corners, producing a perky expression. There was an increase in negatively valenced facial responses to happy facial expressions after naltrexone administration when compared with placebo, consistent with lowered interest in social interaction or affiliation. This suggests a potential role for opioid agonists in modulating automatic behavioral responses to social interaction and cues of reward, which would represent a novel approach to pharmacotherapy in social anxiety.

Bergamaschi et al. (2011) compared the effects of a simulated public speaking test (SPST) in healthy controls and in treatment-naive SAD patients ($n=24$) who either received a single 600-mg dose of cannabidiol (CBD) or placebo, in a double-blind randomized manner 90 min prior to challenge. Subjective ratings (Visual Analogue Mood Scale [VAMS] and Negative Self-Statement Public Speaking scale [SSPS-N]) and physiological measures (blood pressure, heart rate, and skin conductance) were measured at six points during the SPST. After baseline measurements were obtained, pretest measurements were made 80 min after study medication had been ingested. Immediately thereafter, subjects received instructions and had 2 min to prepare a 4-min speech about "the public transportation system of your city." Participants were also told their speech would be recorded on videotape and later analyzed by a psychologist. Anticipatory speech measurements were taken before the subject started speaking. Participants spoke before a camera, viewing their own image on a TV screen. The speech was interrupted in the middle, and speech performance measurements were again taken. The speech was recorded for a further 2 min. Posttest measurements were made 15 and 35 min after the end of the speech. CBD administration was found to significantly reduce anxiety, cognitive impairment, and discomfort in speech performance. The placebo group had higher anxiety, cognitive impairment, and discomfort (assessed by VAMS) when compared with the control group, and SSPS-N scores increased significantly with placebo, but not with CBD administration.

No significant differences were observed between patients administered CBD and healthy controls in SSPS-N scores or in the cognitive impairment, discomfort, and alert factors of VAMS. The performance of SAD patients administered CBD prior to a SPST is similar to that seen in healthy controls, which suggests that CBD could have rapid therapeutic effects either as monotherapy or as an augmentation approach designed to enhance the effects of exposure-based psychological interventions.

This suggestion of potential therapeutic effects of cannabidiol is supported by the findings of a randomized placebo-controlled within-subjects investigation of $\Delta(9)$-Tetrahydrocannabinol (Δ^9THC) administration (7.5 mg) in healthy volunteers, familiar with marijuana but neither daily users nor with a lifetime history of substance use or other mental disorders. In an FMRI task designed to probe amygdala response to social threat, Δ^9THC enhanced basolateral and superficial amygdala connectivity to the rostral anterior cingulate and medial prefrontal cortex, suggesting that cannabinoids can reduce threat perception and enhance socioemotional regulation (Gorka et al., 2015b).

It is possible that the effects of psychological challenges based on observed social performance can be enhanced through combination with physiological challenges. For example, du Plooy et al. (2014) recruited healthy university students (45 men and 45 women, aged 18–27 years) to participate in a randomized double-blind placebo-controlled investigation of combining a psychological stress test (the Trier social stress test, TSST) (Kirschbaum, Pirke, & Hellhammer, 1993) with a physical stress test (the cold pressor test, CPT) (Hines & Brown, 1932) on saliva cortisol levels, heart rate, and subjective anxiety levels. Participants were pseudorandomly assigned into one of three experimental groups: the composite "fear-factor stress test" (FFST; $n=30$) (which combines the TSST and CPT, the TSST [$n=30$], or control group [$n=30$]). Outcomes were measured at baseline and at 5 and 35 min after challenge. At 5 min after challenge, cortisol levels had risen in both groups, but did not differ significantly between groups, but at 35 min, cortisol levels in the TSST group had returned to baseline values, whereas those in the FFST group continued to rise. These findings indicate that the FFST induces a more robust and sustained cortisol response, making it more suitable than the TSST for investigations of HPA function in studies involving social interaction challenges.

VII CONCLUSIONS

Despite the availability of a range of evidence-based pharmacological and psychological treatments for generalized anxiety disorder and social anxiety disorder, many patients do not experience a significant reduction in symptoms, and others will relapse despite continuing with treatment that had been effective. There is thus much scope for the development of innovative treatment approaches that have greater overall effectiveness or improved tolerability in clinical practice. However, drug development in neuroscience is time-consuming and very costly, and "late-stage" failures are not uncommon, which has led to some drawing back by pharmaceutical companies from anxiolytic drug development. Experimental medicine studies conducted in healthy volunteers can provide insights into the neurobiological mechanisms that underlie anxiety symptoms and disorders and anxiolytic effects and can provide useful information for novel targets in drug development and on the likelihood of demonstrating beneficial effects in clinical populations.

Experimental medicine studies in social anxiety disorder involving the administration of oxytocin, testosterone, or cannabinoids have all provided useful insights into the underlying neurobiological correlates of social anxiety, including the substrates of fear and avoidance and of the cognitive biases in emotion processing. In generalized anxiety disorder, studies involving "low-dose" carbon dioxide inhalation have indicated potential disturbances in noradrenergic neurotransmission and alterations in central chemosensory function and have shown beneficial effects for some current interventions in attenuating the subjective anxiety response and altered neurocognitive function associated with carbon dioxide challenge. Further studies are needed to establish whether the encouraging signals seen in these and others "proof-of-concept" studies have more than mere conceptual significance: that is, to establish whether the anxiolytic effects observable in healthy volunteers subject to experimental challenges can translate reliably into robustly beneficial effects in treating patients with anxiety disorders.

Acknowledgments

Thanks to Dr Mary Houston for help in accessing and formatting references.

References

Ainsworth, B., et al. (2015). Evaluating psychological interventions in a novel experimental human model of anxiety. *Journal of Psychiatric Research, 63*(Suppl C), 117–122.

Allgulander, C., Florea, I., & Trap Huusom, A. K. (2006). Prevention of relapse in generalized anxiety disorder by escitalopram treatment. *International Journal of Neuropsychopharmacology, 9*, 495–505.

American Psychiatric Association. (2013). *Desk reference to the diagnostic criteria from DSM-5™.* Arlington, VA: American Psychiatric Publishing, Inc.

Angst, J., et al. (2008). The generalized anxiety spectrum: prevalence, onset, course and outcome. *European Archives of Psychiatry and Clinical Neuroscience, 259*(1), 37.

Argyropoulos, S. V., et al. (2002). Inhalation of 35% CO2 results in activation of the HPA axis in healthy volunteers. *Psychoneuroendocrinology, 27*(6), 715–729.

Arun, T., et al. (2013). Targeting ASIC1 in primary progressive multiple sclerosis: evidence of neuroprotection with amiloride. *Brain, 136*(1), 106–115.

Bailey, J. E., et al. (2003). Does the brain noradrenaline network mediate the effects of the CO2 challenge? *Journal of Psychopharmacology, 17*(3), 252–259.

Bailey, J. E., et al. (2005). Behavioral and cardiovascular effects of 7.5% CO2 in human volunteers. *Depression and Anxiety, 21*(1), 18–25.

Bailey, J. E., et al. (2007). A validation of the 7.5% CO2 model of GAD using paroxetine and lorazepam in healthy volunteers. *Journal of Psychopharmacology, 21*(1), 42–49.

Bailey, J. E., et al. (2009). A comparison of the effects of a subtype selective and non-selective benzodiazepine receptor agonist in two CO2 models of experimental human anxiety. *Journal of Psychopharmacology, 23*(2), 117–122.

Bailey, J. E., et al. (2011a). Validating the inhalation of 7.5% CO2 in healthy volunteers as a human experimental medicine: a model of generalized anxiety disorder (GAD). *Journal of Psychopharmacology, 25*(9), 1192–1198.

Bailey, J. E., et al. (2011b). Preliminary evidence of anxiolytic effects of the CRF1 receptor antagonist R317573 in the 7.5% CO2 proof-of-concept experimental model of human anxiety. *Journal of Psychopharmacology, 25*(9), 1199–1206.

Baldwin, D. S., & Brandish, E. K. (2014). Pharmacological treatment of anxiety disorders. In *The Wiley handbook of anxiety disorders* (pp. 865–882): John Wiley & Sons, Ltd.

Baldwin, D. S., & Hanumanthaiah, V. B. (2015). Vortioxetine in the treatment of major depressive disorder. *Future Neurology, 10*(2), 79–89.

Baldwin, D. S., Huusom, A. K. T., & Mæhlum, E. (2006). Escitalopram and paroxetine in the treatment of generalised anxiety disorder. Randomised, placebo-controlled, double-blind study. *The British Journal of Psychiatry, 189*(3), 264–272.

Baldwin, D. S., Loft, H., & Florea, I. (2012). Lu AA21004, a multimodal psychotropic agent, in the prevention of relapse in adult patients with generalized anxiety disorder. *International Clinical Psychopharmacology, 27*(4), 197–207.

Baldwin, D. S., et al. (2009). How long should a trial of escitalopram treatment be in patients with major depressive disorder, generalised anxiety disorder or social anxiety disorder? An exploration of the randomised controlled trial database. *Human Psychopharmacology: Clinical and Experimental, 24*(4), 269–275.

Baldwin, D., et al. (2011). Efficacy of drug treatments for generalised anxiety disorder: systematic review and meta-analysis. *BMJ, 342.*

Baldwin, D. S., et al. (2012a). Does early improvement predict endpoint response in patients with generalized anxiety disorder (GAD) treated with pregabalin or venlafaxine XR? *European Neuropsychopharmacology, 22*(2), 137–142.

Baldwin, D. S., et al. (2012b). An international survey of reported prescribing practice in the treatment of patients with generalised anxiety disorder. *World Journal of Biological Psychiatry, 13*(7), 510–516.

Baldwin, D. S., et al. (2013). Benzodiazepines: risks and benefits. A reconsideration. *Journal of Psychopharmacology, 27*(11), 967–971.

Baldwin, D. S., et al. (2014). Evidence-based pharmacological treatment of anxiety disorders, post-traumatic stress disorder and obsessive-compulsive disorder: a revision of the 2005 guidelines from the British Association for Psychopharmacology. *Journal of Psychopharmacology, 28*(5), 403–439.

Baldwin, D. S., et al. (2015). Efficacy and safety of pregabalin in generalised anxiety disorder: a critical review of the literature. *Journal of Psychopharmacology, 29*(10), 1047–1060.

Baldwin, D. S., et al. (2016a). A meta-analysis of the efficacy of vortioxetine in patients with major depressive disorder (MDD) and high levels of anxiety symptoms. *Journal of Affective Disorders, 206*(Suppl C), 140–150.

Baldwin, D. S., et al. (2016b). Efficacy of escitalopram in the treatment of social anxiety disorder: a meta-analysis versus placebo. *European Neuropsychopharmacology, 26*(6), 1062–1069.

Bandelow, B., et al. (2006). What is the threshold for symptomatic response and remission for major depressive disorder, panic disorder, social anxiety disorder, and generalized anxiety disorder? *The Journal of Clinical Psychiatry, 67*(9), 1428–1434.

Bandelow, B., et al. (2016). Biological markers for anxiety disorders, OCD and PTSD—a consensus statement. Part I: neuroimaging and genetics. *The World Journal of Biological Psychiatry, 17*(5), 321–365.

Bandelow, B., et al. (2017). Biological markers for anxiety disorders, OCD and PTSD: a consensus statement. Part II: neurochemistry, neurophysiology and neurocognition. *The World Journal of Biological Psychiatry, 18*(3), 162–214.

Battaglia, M., et al. (2007). A genetic study of the acute anxious response to carbon dioxide stimulation in man. *Journal of Psychiatric Research, 41*(11), 906–917.

Battaglia, M., et al. (2008). A twin study of the common vulnerability between heightened sensitivity to hypercapnia and panic disorder. *American Journal of Medical Genetics Part B: Neuropsychiatric Genetics, 147B*(5), 586–593.

Beck, J. G., Ohtake, P. J., & Shipherd, J. C. (1999). Exaggerated anxiety is not unique to CO2 in panic disorder: a comparison of hypercapnic and hypoxic challenges. *Journal of Abnormal Psychology, 108*(3), 473–482.

Bell, C., et al. (2011). The misclassification of facial expressions in generalised social phobia. *Journal of Anxiety Disorders, 25*(2), 278–283.

Bergamaschi, M. M., et al. (2011). Cannabidiol reduces the anxiety induced by simulated public speaking in treatment-naïve social phobia patients. *Neuropsychopharmacology, 36*, 1219.

Blanco, C., et al. (2013). The evidence-based pharmacotherapy of social anxiety disorder. *International Journal of Neuropsychopharmacology, 16*(1), 235–249.

Borkovec, T. D., Alcaine, O. M., & Behar, E. (2004). Avoidance theory of worry and generalized anxiety disorder. In *Generalized anxiety disorder: Advances in research and practice* (pp. 77–108). New York, NY: Guilford Press.

Brown, T. A., Chorpita, B. F., & Barlow, D. H. (1998). Structural relationships among dimensions of the DSM-IV anxiety and mood disorders and dimensions of negative affect, positive affect, and autonomic arousal. *Journal of Abnormal Psychology, 107*(2), 179–192.

Brown, T. A., et al. (2001). Reliability of DSM-IV anxiety and mood disorders: implications for the classification of emotional disorders. *Journal of Abnormal Psychology, 110*(1), 49–58.

Canton, J., Scott, K. M., & Glue, P. (2012). Optimal treatment of social phobia: systematic review and meta-analysis. *Neuropsychiatric Disease and Treatment, 8*, 203–215.

Cassidy, J., et al. (2009). Generalized anxiety disorder: connections with self-reported attachment. *Behavior Therapy, 40*(1), 23–38.

Chen, X., et al. (2016). Effects of oxytocin and vasopressin on the neural response to unreciprocated cooperation within brain regions involved in stress and anxiety in men and women. *Brain Imaging and Behavior, 10*(2), 581–593.

Clark-Elford, R., et al. (2015). Effects of oxytocin on attention to emotional faces in healthy volunteers and highly socially anxious males. *International Journal of Neuropsychopharmacology, 18*(2), pyu012.

Coryell, W., & Rickels, H. (2009). Effects of escitalopram on anxiety and respiratory responses to carbon dioxide inhalation in subjects at high risk for panic disorder: a placebo-controlled, crossover study. *Journal of Clinical Psychopharmacology, 29*(2), 174–178.

Crespi, B. J. (2016). Oxytocin, testosterone, and human social cognition. *Biological Reviews, 91*(2), 390–408.

Daniels, D., & Plomin, R. (1985). Origins of individual differences in infant shyness. *Developmental Psychology, 21*(1), 118.

Davidson, J. R. T., et al. (2008). Duloxetine treatment for relapse prevention in adults with generalized anxiety disorder: a double-blind placebo-controlled trial. *European Neuropsychopharmacology, 18*(9), 673–681.

de Oliveira, D. C. G., et al. (2012). Oxytocin interference in the effects induced by inhalation of 7.5% CO2 in healthy volunteers. *Human Psychopharmacology: Clinical and Experimental, 27*(4), 378–385.

Di Nardo, P. A., et al. (1983). Reliability of dsm-iii anxiety disorder categories using a new structured interview. *Archives of General Psychiatry, 40*(10), 1070–1074.

Diaper, A., et al. (2012). The effect of a clinically effective and non-effective dose of lorazepam on 7.5% CO2-induced anxiety. *Human Psychopharmacology: Clinical and Experimental, 27*(6), 540–548.

Diaper, A., et al. (2013). Evaluation of the effects of venlafaxine and pregabalin on the carbon dioxide inhalation models of generalised anxiety disorder and panic. *Journal of Psychopharmacology, 27*(2), 135–145.

du Plooy, C., et al. (2014). The fear-factor stress test: an ethical, non-invasive laboratory method that produces consistent and sustained cortisol responding in men and women. *Metabolic Brain Disease, 29*(2), 385–394.

Dugas, M. J., Buhr, K., & Ladouceur, R. (2004). The role of intolerance of uncertainty in etiology and maintenance. In *Generalized anxiety disorder: Advances in research and practice* (pp. 143–163). New York, NY: Guilford Press.

Eckstein, M., et al. (2015). Oxytocin facilitates the extinction of conditioned fear in humans. *Biological Psychiatry, 78*(3), 194–202.

Enter, D., Spinhoven, P., & Roelofs, K. (2014). Alleviating social avoidance: effects of single dose testosterone administration on approach–avoidance action. *Hormones and Behavior, 65*(4), 351–354.

Enter, D., et al. (2016). Single dose testosterone administration alleviates gaze avoidance in women with social anxiety disorder. *Psychoneuroendocrinology, 63*, 26–33.

Etkin, A., et al. (2010). Failure of anterior cingulate activation and connectivity with the amygdala during implicit regulation of emotional processing in generalized anxiety disorder. *American Journal of Psychiatry, 167*(5), 545–554.

Feltner, D., et al. (2008). Long-term efficacy of pregabalin in generalized anxiety disorder. *International Clinical Psychopharmacology, 23*(1), 18–28.

Garner, M., et al. (2009). Impaired identification of fearful faces in generalised social phobia. *Journal of Affective Disorders, 115*(3), 460–465.

Garner, M., et al. (2011). Inhalation of 7.5% carbon dioxide increases threat processing in humans. *Neuropsychopharmacology, 36*, 1557.

Garner, M., et al. (2012). Inhalation of 7.5% carbon dioxide increases alerting and orienting attention network function. *Psychopharmacology, 223*(1), 67–73.

Goddard, A. W., et al. (1994). Effects of tryptophan depletion in panic disorder. *Biological Psychiatry, 36*(11), 775–777.

Gordon, D., & Heimberg, R. G. (2011). Reliability and validity of DSM-IV generalized anxiety disorder features. *Journal of Anxiety Disorders, 25*(6), 813–821.

Gorka, S. M., et al. (2015a). Oxytocin modulation of amygdala functional connectivity to fearful faces in generalized social anxiety disorder. *Neuropsychopharmacology, 40*(2), 278–286.

Gorka, S. M., et al. (2015b). Cannabinoid modulation of amygdala subregion functional connectivity to social signals of threat. *International Journal of Neuropsychopharmacology, 18*(3), pyu104.

Gorman, J. M., et al. (2004). The effect of successful treatment on the emotional and physiological response to carbon dioxide inhalation in patients with panic disorder. *Biological Psychiatry, 56*(11), 862–867.

Gray, J., & McEwen, B. (2014). Neuroendocrinology and neurotransmitters. In *The Wiley handbook of anxiety disorders* (pp. 254–273): John Wiley & Sons, Ltd.

Greco, L. A., & Morris, T. L. (2002). Paternal child-rearing style and child social anxiety: investigation of child perceptions and actual father behavior. *Journal of Psychopathology and Behavioral Assessment, 24*(4), 259–267.

Guttmacher, L. B., Murphy, D. L., & Insel, T. R. (1983). Pharmacologic models of anxiety. *Comprehensive Psychiatry, 24*(4), 312–326.

Hackett, D., White, C., & Salinas, E. (2000). Relapse prevention in patients with generalised anxiety disorder (GAD) by treatment with venlafaxine. In *Poster presented at 1st international forum on mood and anxiety disorders, Monte Carlo*.

Haller, J., Aliczki, M., & Pelczer, K. G. (2013). Classical and novel approaches to the preclinical testing of anxiolytics: a critical evaluation. *Neuroscience & Biobehavioral Reviews, 37*(10, Part 1), 2318–2330.

Hamilton, M. A. X. (1959). The assessment of anxiety states by rating. *British Journal of Medical Psychology, 32*(1), 50–55.

Hanney, S., et al. (2015). How long does biomedical research take? Studying the time taken between biomedical and health research and its translation into products, policy, and practice. *Health Research Policy and Systems, 13*(1).

Heinrichs, N., & Hofmann, S. G. (2001). Information processing in social phobia: a critical review. *Clinical Psychology Review, 21*(5), 751–770.

Hettema, J. M., Prescott, C. A., & Kendler, K. S. (2004). Genetic and environmental sources of covariation between generalized anxiety disorder and neuroticism. *American Journal of Psychiatry, 161*(9), 1581–1587.

Hines, E. A., & Brown, G. (1932). A standard stimulus for measuring vasomotor reactions: its application in the study of hypertension. In *Mayo Clinic Proceedings*.

Hunter, L. R., Buckner, J. D., & Schmidt, N. B. (2009). Interpreting facial expressions: the influence of social anxiety, emotional valence, and race. *Journal of Anxiety Disorders, 23*(4), 482–488.

Insel, T. R. (2015). The NIMH experimental medicine initiative. *World Psychiatry, 14*(2), 151–153.

Insel, T. R., et al. (2013). Innovative solutions to novel drug development in mental health. *Neuroscience & Biobehavioral Reviews, 37*(10, Part 1), 2438–2444.

Johnson, P. L., et al. (2012). Activation of the orexin 1 receptor is a critical component of CO2-mediated anxiety and hypertension but not bradycardia. *Neuropsychopharmacology, 37*, 1911.

Katzman, M. A., et al. (2011). Extended release quetiapine fumarate (quetiapine XR) monotherapy as maintenance treatment for generalized anxiety disorder: a long-term, randomized, placebo-controlled trial. *International Clinical Psychopharmacology, 26*(1), 11–24.

Kaye, J., et al. (2004). Acute carbon dioxide exposure in healthy adults: evaluation of a novel means of investigating the stress response. *Journal of Neuroendocrinology, 16*(3), 256–264.

Kirsch, P., et al. (2005). Oxytocin modulates neural circuitry for social cognition and fear in humans. *The Journal of Neuroscience, 25*(49), 11489–11493.

Kirschbaum, C., Pirke, K. M., & Hellhammer, D. H. (1993). The 'Trier Social Stress Test'—a tool for investigating psychobiological stress responses in a laboratory setting. *Neuropsychobiology, 28*(1–2), 76–81.

Labuschagne, I., et al. (2012). Medial frontal hyperactivity to sad faces in generalized social anxiety disorder and modulation by oxytocin. *International Journal of Neuropsychopharmacology, 15*(7), 883–896.

Leibold, N. K., et al. (2015). The brain acid–base homeostasis and serotonin: a perspective on the use of carbon dioxide as human and rodent experimental model of panic. *Progress in Neurobiology, 129*(Suppl C), 58–78.

Lin, S. -H., Sun, W. -H., & Chen, C. -C. (2015). Genetic exploration of the role of acid-sensing ion channels. *Neuropharmacology, 94*(Suppl C), 99–118.

Lynn, S. K., et al. (2014). Gender differences in oxytocin-associated disruption of decision bias during emotion perception. *Psychiatry Research, 219*(1), 198–203.

MacLeod, C., & Rutherford, E. (2004). Information-processing approaches: assessing the selective functioning of attention, interpretation, and retrieval. In *Generalized anxiety disorder: Advances in research and practice* (pp. 109–142). New York, NY: Guilford Press.

Mannuzza, S., et al. (1989). Reliability of anxiety assessment: I. Diagnostic agreement. *Archives of General Psychiatry, 46*(12), 1093–1101.

Martin, E. I., et al. (2009). The neurobiology of anxiety disorders: brain imaging, genetics, and psychoneuroendocrinology. *Psychiatric Clinics of North America, 32*(3), 549–575.

Mayo-Wilson, E., et al. (2014). Psychological and pharmacological interventions for social anxiety disorder in adults: a systematic review and network meta-analysis. *The Lancet Psychiatry, 1*(5), 368–376.

Meier, I. M., et al. (2016). Naltrexone increases negatively-valenced facial responses to happy faces in female participants. *Psychoneuroendocrinology, 74*(Suppl C), 65–68.

Mennin, D. S., et al. (2005). Preliminary evidence for an emotion dysregulation model of generalized anxiety disorder. *Behaviour Research and Therapy, 43*(10), 1281–1310.

Mineka, S., & Sutton, J. (2006). Contemporary learning theory perspectives on the etiology of fears and phobias. In *Fear and learning: From basic processes to clinical implications* (pp. 75–97). Washington, DC: American Psychological Association.

Moosavi, S. H., et al. (2003). Hypoxic and hypercapnic drives to breathe generate equivalent levels of air hunger in humans. *Journal of Applied Physiology, 94*(1), 141–154.

Neumann, I. D., & Slattery, D. A. (2016). Oxytocin in general anxiety and social fear: a translational approach. *Biological Psychiatry, 79*(3), 213–221.

ÖHman, A. (1986). Face the beast and fear the face: animal and social fears as prototypes for evolutionary analyses of emotion. *Psychophysiology, 23*(2), 123–145.

Pae, C. -U., et al. (2015). Vortioxetine, a multimodal antidepressant for generalized anxiety disorder: a systematic review and meta-analysis. *Journal of Psychiatric Research, 64*(Suppl C), 88–98.

Papadopoulos, A., et al. (2010). The effects of single dose anxiolytic medication on the CO2 models of anxiety: differentiation of subjective and objective measures. *Journal of Psychopharmacology, 24*(5), 649–656.

Papp, L. A., et al. (1997). Respiratory psychophysiology of panic disorder: three respiratory challenges in 98 subjects. *American Journal of Psychiatry, 154*(11), 1557–1565.

Philippot, P., & Douilliez, C. (2005). Social phobics do not misinterpret facial expression of emotion. *Behaviour Research and Therapy, 43*(5), 639–652.

Pinkney, V., et al. (2014). The effects of duloxetine on subjective, autonomic and neurocognitive response to 7.5% carbon dioxide challenge. *European Neuropsychopharmacology, 24*, S72–S73.

Pollack, M. H., et al. (2014). A double-blind randomized controlled trial of augmentation and switch strategies for refractory social anxiety disorder. *American Journal of Psychiatry, 171*(1), 44–53.

Poma, S. Z., et al. (2005). Characterization of a 7% carbon dioxide (CO2) inhalation paradigm to evoke anxiety symptoms in healthy subjects. *Journal of Psychopharmacology, 19*(5), 494–503.

Poma, S. Z., et al. (2014). Anxiolytic effects of vestipitant in a sub-group of healthy volunteers known to be sensitive to CO2 challenge. *Journal of Psychopharmacology, 28*(5), 491–497.

Rapee, R. M., & Heimberg, R. G. (1997). A cognitive-behavioral model of anxiety in social phobia. *Behaviour Research and Therapy, 35*(8), 741–756.

Rickels, K., et al. (2010). Time to relapse after 6 and 12 months' treatment of generalized anxiety disorder with venlafaxine extended release. *Archives of General Psychiatry, 67*(12), 1274–1281.

Roemer, L., & Orsillo, S. M. (2005). An acceptance-based behavior therapy for generalized anxiety disorder. In *Acceptance and mindfulness-based approaches to anxiety* (pp. 213–240): Springer.

Rutter, L. A., & Brown, T. A. (2015). Reliability and validity of the dimensional features of generalized anxiety disorder. *Journal of Anxiety Disorders, 29*(Suppl C), 1–6.

Schenberg, L. C. (2016). A neural systems approach to the study of the respiratory-type panic disorder. In A. E. Nardi & R. C. R. Freire (Eds.), *Panic disorder: Neurobiological and treatment aspects* (pp. 9–77). Cham: Springer International Publishing.

Schlenker, B. R., & Leary, M. R. (1982). Social anxiety and self-presentation: a conceptualization model. *Psychological Bulletin, 92*(3), 641–669.

Schofield, C. A., Coles, M. E., & Gibb, B. E. (2007). Social anxiety and interpretation biases for facial displays of emotion: emotion detection and ratings of social cost. *Behaviour Research and Therapy, 45*(12), 2950–2963.

Schruers, K., et al. (2000). Effects of tryptophan depletion on carbon dioxide provoked panic in panic disorder patients. *Psychiatry Research, 93*(3), 179–187.

Schruers, K., et al. (2002). Acute l-5-hydroxytryptophan administration inhibits carbon dioxide-induced panic in panic disorder patients. *Psychiatry Research, 113*(3), 237–243.

Schwartz, C. E., Snidman, N., & Kagan, J. (1999). Adolescent social anxiety as an outcome of inhibited temperament in childhood. *Journal of the American Academy of Child & Adolescent Psychiatry, 38*(8), 1008–1015.

Seddon, K., et al. (2011). Effects of 7.5% CO2 challenge in generalized anxiety disorder. *Journal of Psychopharmacology, 25*(1), 43–51.

Sherwood, T. W., Frey, E. N., & Askwith, C. C. (2012). Structure and activity of the acid-sensing ion channels. *American Physiological Journal, Cell Physiology, 303*(7), C699–C710.

Silvia, P. J., et al. (2006). Biased recognition of happy facial expressions in social anxiety. *Journal of Social and Clinical Psychology, 25*(6), 585–602.

Steenen, S. A., et al. (2016). Propranolol for the treatment of anxiety disorders: systematic review and meta-analysis. *Journal of Psychopharmacology, 30*(2), 128–139.

Stein, D. J., et al. (2008). Efficacy of pregabalin in depressive symptoms associated with generalized anxiety disorder: A pooled analysis of 6 studies. *European Neuropsychopharmacology, 18*(6), 422–430.

Stein, D. J., et al. (2009). Anxiety symptom severity and functional recovery or relapse. *Annals of Clinical Psychiatry, 21*(2), 81–88.

Stein, D. J., et al. (2012). Agomelatine prevents relapse in generalized anxiety disorder: a 6-month randomized, double-blind, placebo-controlled discontinuation study. *Journal of Clinical Psychiatry, 73*(7), 1002–1008.

Stevens, E. S., et al. (2014). Generalized anxiety disorder. In *The Wiley handbook of anxiety disorders* (pp. 378–423).

Stocchi, F., et al. (2003). Efficacy and tolerability of paroxetine for the long-term treatment of generalized anxiety disorder. *Journal of Clinical Psychiatry, 64*(3), 250–258.

Stravynski, A., & Amado, D. (2001). *From social anxiety to social phobia: Multiple perspectives.* Needham Heights: Allyn & Bacon.

Vollmer, L. L., Strawn, J. R., & Sah, R. (2015). Acid–base dysregulation and chemosensory mechanisms in panic disorder: a translational update. *Translational Psychiatry, 5*, e572.

Wang, W., Pizzonia, J. H., & Richerson, G. B. (1998). Chemosensitivity of rat medullary raphe neurones in primary tissue culture. *The Journal of Physiology, 511*(2), 433–450.

Wells, A. (2005). The metacognitive model of GAD: assessment of meta-worry and relationship with DSM-IV generalized anxiety disorder. *Cognitive Therapy and Research, 29*(1), 107–121.

Wemmie, J. A. (2011). Neurobiology of panic and pH chemosensation in the brain. *Dialogues in Clinical Neuroscience, 13*(4), 475–483.

Winton, E. C., Clark, D. M., & Edelmann, R. J. (1995). Social anxiety, fear of negative evaluation and the detection of negative emotion in others. *Behaviour Research and Therapy, 33*(2), 193–196.

Wittchen, H. U., et al. (2011). The size and burden of mental disorders and other disorders of the brain in Europe 2010. *European Neuropsychopharmacology, 21*(9), 655–679.

World Health Organization. (1993). *The ICD-10 classification of mental and behavioural disorders: diagnostic criteria for research*. Geneva: World Health Organization.

Ziemann, A. E., et al. (2009). The amygdala is a chemosensor that detects carbon dioxide and acidosis to elicit fear behavior. *Cell, 139*(5), 1012–1021.

26

Translational Medicine Strategies in PTSD Drug Development

Dan J. Stein*, Willie Daniels[†], Brian H. Harvey[‡]

*Department of Psychiatry, University of Cape Town, Cape Town, South Africa [†]Department of Physiology, University of the Witwatersrand, Johannesburg, South Africa [‡]Division of Pharmacology, School of Pharmacy and Centre of Excellence for Pharmaceutical Sciences, North-West University, Potchefstroom, South Africa

I INTRODUCTION

Posttraumatic stress disorder (PTSD) was introduced into the official psychiatric nomenclature in the 3rd edition of the Diagnostic and Statistical Manual of Mental Disorders, at a time when there was growing recognition of this condition in veterans in the United States (Stein, Seedat, Iversen, & Wessely, 2007). This formal recognition and operationalization gave additional impetus to a range of research on this condition, including investigation of its phenomenology, psychobiology, pharmacotherapy, and psychotherapy (Stein, Cloitre, Nemeroff, et al., 2009). Early work established the efficacy of antidepressants in the treatment of PTSD, and a number of selective serotonin reuptake inhibitors (SSRIs) were subsequently registered for this indication (Baldwin, Anderson, Nutt, et al., 2014). While this was certainly a significant advance, given the relatively high burden of disease due to PTSD and the relatively low effect size of SSRI pharmacotherapy, there is a clear need for novel treatment approaches (Kessler, Ruscio, Shear, & Wittchen, 2010).

In view of the availability of rigorous fear conditioning and extinction models in animals and the established use of fear extinction in exposure-based therapy of anxiety disorders, there has been considerable interest in translational approaches to PTSD, with the hope that findings can be taken from the bench to the bedside (Morrison & Ressler, 2014). Fear conditioning is a form of learning in which an aversive stimulus (the unconditioned stimulus) is repeatedly paired with a neutral cue or context, with subsequent expression of fear responses when exposed to this cue or context (the conditioned stimulus). Fear extinction refers to the decline in expression of fear responses when the conditioned stimulus is repeatedly presented without the aversive stimulus. Translational models have employed animal paradigms to explore the neurobiology of fear conditioning and extinction and then to extend this work into human populations (Thomas & Stein, 2017).

A seminal series of investigations established, for example, that D-cycloserine, a partial glutamate agonist, was able to enhance fear extinction in the laboratory and that it appeared efficacious in augmenting exposure-based psychotherapy in early clinical trials undertaken in patients with anxiety disorders (Singewald, Schmuckermair, Whittle, Holmes, & Ressler, 2015). Conversely, work on the neuroimaging, neurogenetics, and neuropharmacology of PTSD has given impetus to a range of research questions that warrant further exploration in the laboratory (Garner, Mohler, Stein, Mueggler, & Baldwin, 2009). Here, we review current work on the translational neuroscience of PTSD, and its attempts to find new novel approaches to the prevention and treatment of this condition. Several aspects of the clinical phenomenology and community epidemiology of PTSD are relevant to understanding efforts in translational neuroscience, and we begin by briefly considering this area of work. We then go on to address the translational neuroscience of PTSD, discussing cognitive-affective features, changes in neuronal circuitry, and alterations in neurochemistry in this condition. Finally, we consider current and future pharmacotherapies for PTSD.

II PHENOMENOLOGY AND EPIDEMIOLOGY

Early authors conceptualized PTSD as an understandable reaction to a highly unusual traumatic event (Yehuda & McFarlane, 1995). Epidemiological data, however, have emphasized that exposure to traumatic events is common around the world, while PTSD is a relatively uncommon sequela of such exposure (Benjet, Bromet, Karam, et al., 2015). Such data are partially consistent with an early literature that humans are well adapted to responding to stressors (with such exposures providing opportunities for learning and for growth) but that chronic uncontrollable stressors may lead to psychopathology (with corresponding alterations in psychobiology), although the literature on PTSD has emphasized that the neurobiology of this condition differs from that of the normal stress response (Selye, 1976).

Epidemiological data have shown that certain kinds of traumatic events are more likely to lead to trauma; for example, the conditional probability of PTSD after exposure to rape or combat is much higher than that after exposure to a motor vehicle accident or loss of a loved one (Atwoli, Stein, Koenen, & McLaughlin, 2015). However, as traumatic events such as motor vehicle accidents or loss of a loved one are so prevalent, at the population level, the burden of PTSD due to such events is higher than that due to less prevalent traumatic events such as rape or combat (Atwoli et al., 2015). While few relationships have been established between traumatic event type and psychobiology, early exposure to traumatic stressors may well be associated with more robust and enduring neurobiological changes (Nemeroff, 2004).

In the aftermath of exposure to traumatic events, a range of symptoms and disorders may be seen. An acute stress response is characterized by intrusion symptoms (e.g., recurrent distressing memories), negative mood, dissociative symptoms (e.g., inability to remember an important aspect of the trauma), avoidance symptoms (e.g., efforts to avoid reminders of the traumatic event), and arousal symptoms (e.g., exaggerated startle response) (American Psychiatric Association, 2013). Typically, such symptoms diminish over time; when they persist beyond a few weeks and are associated with clinically significant distress and impairment, then PTSD can be diagnosed (American Psychiatric Association, 2013). It is also noteworthy, however, that traumatic events may precipitate mental disorders other than PTSD and PTSD may co-occur together with a range of disorders, including major depression, anxiety disorders, and substance use disorders (Kessler et al., 2010).

A substantial literature has focused on risk factors for the development of PTSD. These include those that predate the trauma (e.g., female gender), peritraumatic factors (e.g., severity of the trauma and dissociation during the trauma), and posttraumatic factors (e.g., absence of social support) (Brewin, Andrews, & Valentine, 2000). A key risk factor for the development of PTSD is prior trauma; this finding is consistent with a sensitization hypothesis, where repeated exposure to stress becomes associated with more entrenched fear conditioning, increased generalization of threat, and failure to extinguish fear. Although not all data are consistent, it has been suggested that various neurobiological changes such as lowered cortisol and increased heart rate after trauma may predict PTSD, consistent with hypothalamic-adrenal-pituitary (HPA) axis/noradrenergic alterations (Morris, Hellman, Abelson, & Rao, 2016). Indeed, the risk factor literature potentially provides a number of insights into the psychobiology of PTSD.

DSM-5 has emphasized that PTSD comprises four sets of symptoms (American Psychiatric Association, 2013). Intrusive symptoms include reexperiencing the trauma in memories, dreams, or flashbacks. Avoidant symptoms include avoiding memories about the trauma and avoiding external reminders of the traumatic event. Arousal and reactivity symptoms include hypervigilance, exaggerated startle response, irritable and angry behavior, and sleep disturbance. Finally, changes in cognition and mood associated with the traumatic event may occur; these include inability to remember an important aspect of the traumatic event, persistent negative beliefs about oneself or others, and a persistent negative emotional state. It has long been suggested that such sets of PTSD symptoms are underpinned by different neuronal circuitry and different neurochemical systems, although, to date, it seems that medications that are effective for PTSD, even if acting predominantly on one neurotransmitter system, lead to a decrease in a wide range of symptoms (Charney, Deutch, Krystal, Southwick, & Davis, 1993).

III TRANSLATIONAL NEUROSCIENCE OF PTSD

PTSD may be characterized by a broad range of changes in cognitive-affective processes associated with the fear, memory, and arousal symptoms that are characteristic of this condition. A good deal of work has focused on the psychobiology of the stress response in general and on the hypothesis that PTSD is characterized by specific changes in fear conditioning, expression, and extinction in particular. Consistent with a sensitization hypothesis of PTSD, repeated exposure to stressors in the context of a range of other risk factors for the disorder (including genetic susceptibility and lack of social support) may lead to consolidation of fear conditioning (i.e., with stabilization in memory of the exposure to trauma and of associated cues and contexts), overgeneralization of fear responses (i.e., with expression of fear responses to stimuli resembling those present during trauma exposure), and diminished fear extinction (i.e., with failure of fear responses to diminish even with repeated exposure to associated cues and contexts in the absence of trauma exposure).

Laboratory investigations of stress responses, and of fear conditioning and extinction, have delineated the neurocircuitry and neurochemistry of these processes (Bowers & Ressler, 2015; Singewald et al., 2015; Uys, Stein, Daniels, & Harvey, 2003; VanElzakker, Dahlgren, Davis, Dubois, & Shin, 2014). These range from systems level alterations in communication and connectivity between brain regions to molecular and cellular level events that alter synaptic efficacy. The amygdala and related limbic structures play an important role in cued fear conditioning, with the hippocampus playing a key role in contextual fear conditioning. Fear extinction is underpinned by various brain structures, in particular the amygdala, hippocampus, medial prefrontal cortex (mPFC), periaqueductal gray, and bed nucleus of the stria terminalis (Greco & Liberzon, 2016). More recently, optogenetic studies have provided further insights into the specific neurocircuitry underlying fear and anxiety (Gordon, 2016).

Monoamine neurotransmitter systems, as well as the glutamatergic and GABAergic neurotransmitter systems, play a key role in fear-related processes, both in the brain and in peripheral organs (e.g., noradrenaline is released by chromaffin cells in the adrenal gland) (Harvey, Brand, Jeeva, & Stein, 2006; Harvey, Oosthuizen, Brand, Wegener, & Stein, 2004; Harvey & Shahid, 2012). Stress-induced release of noradrenaline into the basolateral nucleus of the amygdala (BLA) is important in emotional memory consolidation, and adrenergic signaling is involved in memory reconsolidation. Administration of an alpha-1-adrenergic receptor antagonist facilitates fear acquisition, while the beta-blocker propranolol impairs fear acquisition and blocks reconsolidation (Morrison & Ressler, 2014). The novel α_{2C} antagonist, ORM-10921, has notable cognitive effects in animal models and deserves further study in fear conditioning/extinction paradigms (Uys, Shahid, Sallinen, & Harvey, 2017). The serotonergic system also plays a role in fear conditioning/extinction, with research implicating the 5-HT_{1A}, 5-HT_2, and 5-HT_3 receptors (Singewald et al., 2015). With acute SSRI administration, there is increased cued fear conditioning and expression, while chronic SSRIs may increase fear extinction (Morrison & Ressler, 2014). The dopaminergic and cholinergic systems also play a role in fear conditioning/extinction (Brand, Groenewald, Stein, Wegener, & Harvey, 2008).

Research has addressed a range of glutamatergic receptors, and particular attention has been paid to the findings that administration of NMDA receptor antagonists into the BLA or mPFC produces extinction deficits and that allosteric modulation of the L-glycine binding site on the GluN1 subunit by D-cycloserine augments fear extinction (Singewald et al., 2015). Fear extinction is associated with upregulation of GABAergic markers in the amygdala, and the use of benzodiazepines, which are $GABA_A$ receptor allosteric modulators, is widespread in patients with anxiety disorders. However, the anxiolytic effects of benzodiazepines may interfere with fear extinction, perhaps via decreased release of key neurochemicals (e.g., noradrenaline) and stress hormones (e.g., cortisol) (Singewald et al., 2015).

A range of other neurotransmitter, neuroendocrine, and neuropeptide systems are also intimately involved in the stress response (again involving both the brain and peripheral organs). One hypothesis, for example, is that at mild levels of stress, cortisol released by the adrenal gland is preferentially bound to brain mineralocorticoid receptors, but when higher levels of stress lead to increased release of cortisol, then brain glucocorticoid receptors are also bound. This in turn leads to a range of changes in gene expression, with alterations in synaptic structure and function. Such alterations may include cellular atrophy in hippocampal (e.g., CA3) areas involved in memory and in prefrontal regions that play a role in regulating amygdala-mediated fear behavior and hypothalamus-mediated stress responses (de Kloet, 2014). Glucocorticoid administration leads to PTSD-like behaviors, while administration of glucocorticoids and other neuroactive steroids after exposure to stress can prevent these behaviors (de Quervain, Schwabe, & Roozendaal, 2017; Rasmusson, Marx, Pineles, et al., 2017). SSRI administration may alter glucocorticoid receptor function and so reverse cellular alterations (Uys et al., 2006). At the same time, animal models of PTSD indicate that the role of both glucocorticoid and serotonin systems in this disorder is complex (Harvey, Naciti, et al., 2004; Harvey, Oosthuizen, et al., 2004).

Mu-opioid antagonists increase conditioned fear by enhancing fear acquisition or blocking fear extinction, kappa-opioid receptor antagonists and the opioid receptor agonist morphine block conditioned fear acquisition, and the NOP/orphanin FQ receptor agonist administration blocks fear consolidation (Morrison & Ressler, 2014). Opioid signaling in ventrolateral periaqueductal gray matter is thought to regulate conditioned fear via activation of the mPFC and the basolateral nucleus of the amygdala (Morrison & Ressler, 2014). Oxytocin receptor antagonists impair fear extinction, while neuropeptide Y facilitates fear extinction (Singewald et al., 2015). The cholecystokinin system is another neuropeptide system that may work together with the cannabinoid system to modulate fear extinction (Bowers, Choi, & Ressler, 2012).

The endocannabinoid system has been implicated in stress sensitization, with CB_1 receptors modulating glutamatergic and GABAergic signaling in the amygdala and mPFC (Morrison & Ressler, 2014). Antagonism of cannabinoid CB_1 receptors increases anxiety-like behavior, while administration of a CB_1 agonist immediately after exposure to stressors prevents PTSD-like symptoms in rodents (Morrison & Ressler, 2014). Fibroblast growth factor-2 (FGF2) is a growth factor that plays a role in brain development and in learning; it turns out that FGF2 facilitates fear extinction (Singewald et al., 2015). Brain-derived neurotrophic factor (BDNF) has also been implicated in fear conditioning/extinction and indeed may comprise an important common signaling pathway for various systems involved in such processes; agonists of TrkB, the endogenous BDNF receptor, such as 7,8-dihydroxyflavone, may facilitate fear extinction (Andero & Ressler, 2012). Nitric oxide and associated modulatory molecules including NFκβ may also play a role in fear conditioning/extinction (Harvey, Bothma, Nel, Wegener, & Stein, 2005). There is also a growing literature on the potential relationship between stress-related and inflammatory pathways (Furtado & Katzman, 2015).

A human-based literature on the neurocircuitry and neurochemistry of fear conditioning, expression, and extinction in healthy volunteers is largely consistent with the animal literature. Functional brain imaging in healthy volunteers, for example, has confirmed that the amygdala and mPFC play a key role in fear conditioning and extinction (VanElzakker et al., 2014). Acute SSRI administration increases fear acquisition and expression, propranolol blocks primary consolidation and reconsolidation of long-term emotional memory, corticosterone administration appears to increase fear extinction, opioid agonists inhibit and opioid antagonists promote fear acquisition, delta-9-tetrahydrocannabinol (Δ9-THC) administration may facilitate extinction of conditioned fear, and oxytocin increases initial fear expression during extinction training (Bowers & Ressler, 2015; Morrison & Ressler, 2014; Singewald et al., 2015). Genetic variants in neuropeptide S and BDNF in humans are associated with alterations in fear conditioning/extinction (Singewald et al., 2015).

In recent years, a range of structural and functional neuroimaging studies of PTSD have been undertaken and have further contributed to our understanding of the neurobiology of this condition. Metaanalysis of neuroimaging studies in PTSD indicates involvement of the anterior cingulate, amygdala, and insula neurocircuitry, as well as decreases in hippocampal volume (O'Doherty, Chitty, Saddiqui, Bennett, & Lagopoulos, 2015). Much further work is needed, however; for example, it may be that hippocampal involvement is not specific to PTSD or predates the onset of this disorder (Gilbertson, Shenton, Ciszewski, et al., 2002). It has also been suggested that dissociative PTSD involves a distinctive set of neuronal circuits (Lanius, Vermetten, Loewenstein, et al., 2010),

although further translational work is needed to support or refute this hypothesis.

Studies of pharmacological challenges, pharmacotherapy dissection, and molecular imaging have again emphasized the role of the glutamatergic, GABAergic, and monoamine neurotransmitter systems and of a range of neuroendocrine and neuropeptide systems in PTSD (Kelmendi et al., 2016). Neurogenetic studies of PTSD have tentatively pointed to the role of a number of particular gene variants in PTSD, including those involved in monoamine systems (e.g., 5-HTTP), in HPA axis function (e.g., FKBP5), in opioid function (e.g., OPRM1), in cannabinoid function (e.g., CB1), and others (e.g., BDNF, PACAP, DICER, and ANKRD55) (Skelton, Ressler, Norrholm, Jovanovic, & Bradley-Davino, 2012). Genome-wide association studies (GWAS) of PTSD are at an early stage but may ultimately contribute novel insights (Koenen, Duncan, Liberzon, & Ressler, 2013).

IV CURRENT PHARMACOTHERAPY OF PTSD

Early work indicated the efficacy of tricyclic antidepressants and monoamine oxidase inhibitors in the pharmacotherapy of PTSD (Stein, Ipser, & Seedat, 2006). A subsequent set of studies, with larger sample sizes and more rigorous designs, provided convincing evidence of the efficacy of the serotonin-selective reuptake inhibitors and of the serotonin and noradrenergic reuptake inhibitor (SNRI) venlafaxine in this condition (Stein et al., 2006). Metaanalyses have not only confirmed the efficacy of these agents but also emphasized that the effect size of treatment is relatively low (Stein et al., 2006). Evidence-based treatment guidelines for PTSD list the SSRIs as the first-line treatment of choice for PTSD (Baldwin et al., 2014).

A range of other classes of medication have been investigated in PTSD. Atypical antipsychotic agents have been found efficacious both as monotherapies in PTSD and as augmenting agents (Ipser et al., 2006). A range of anticonvulsant agents, including lamotrigine, have indicated efficacy, albeit in small trials (Ipser & Stein, 2012). Prazosin, an alpha-1-adrenergic antagonist, was found efficacious for nightmares in PTSD, but subsequent work has also demonstrated efficacy in a broad range of symptoms although not all data are consistent (Ipser & Stein, 2012). In contrast, benzodiazepines do not appear efficacious in PTSD and may even exacerbate symptoms (Ipser & Stein, 2012). Other negative trials in PTSD include those with D-cycloserine as stand-alone therapy and as pharmacotherapy augmentor and with inositol (Ipser & Stein, 2012).

As alluded to earlier, D-cycloserine is effective in enhancing fear extinction in a number of laboratory models (including in fear-potentiated startle, in prolonged exposure to stressors, and in a genetic model of PTSD), where it is thought to interfere with memory consolidation (Morrison & Ressler, 2014). Early work indicated that D-cycloserine was efficacious in the augmentation of exposure-based treatment of anxiety disorders. However, subsequent data on whether D-cycloserine reduces conditioned fear in humans have been less consistent (Ori et al., 2015). In PTSD, not all studies are consistent, but D-cycloserine may be effective when administered along with virtual reality exposure-based psychotherapy, and metaanalyses emphasize the potential value of additional work in this area (Mataix-Cols, Fernandez de la Cruz, Monzani, et al., 2017).

Attention has also been paid to the use of pharmacotherapies to prevent PTSD, by administrating medication immediately after exposure to traumatic events. As alluded to earlier, a range of work has implicated alterations in resting levels or reactivity of neuroactive steroids in PTSD (Rasmusson et al., 2017). Given the evidence of increased sensitivity of the negative feedback loop between corticotropin-releasing factor (CRF) release from the hypothalamus and cortisol release from the adrenal glands, a number of studies have focused on hydrocortisone administration in the aftermath of trauma exposure, with promising results (Amos, Stein, & Ipser, 2014). There is also evidence that administration of morphine to injured patients is associated with lower rates of subsequent PTSD (Amos et al., 2014). SSRI administration after exposure to traumatic events has not, however, consistently been found to be of benefit (Suliman et al., 2015). Propranolol data in the prevention of PTSD are unfortunately also not consistent (Amos et al., 2014). A number of other agents have been prosed in this context, including anticonvulsants and circadian rhythm regulators, but much additional work remains needed (Koresh et al., 2012).

V FUTURE PHARMACOTHERAPIES FOR PTSD

Although the dopamine system plays a role in fear conditioning and extinction, further work with receptor-selective agonists and antagonists is needed in order to provide a foundation for clinical trials with these agents (Singewald et al., 2015). Nevertheless, there is some preliminary evidence that dopaminergic manipulations may ultimately be useful in the pharmacotherapy for PTSD. It is notable, for example, that L-DOPA, which enhances dopamine turnover in the frontal cortex, augments extinction-related neural activity in the mPFC of rodents and alters mPFC functional connectivity in humans (Singewald et al., 2015). Another dopamine-enhancing agent, MDMA (or "ecstasy"), was found to improve efficacy of psychotherapy in treatment-refractory PTSD patients, although this agent also acts on systems other than DA (Mithoefer, Wagner, Mithoefer, Jerome, & Doblin, 2011).

The substantial range of work on the glutamatergic system in fear conditioning/extinction was noted earlier. Ketamine is a glutamatergic agent that has received attention in animal model of PTSD, as well as in human studies of treatment-resistant depression (Averill et al., 2016). Early case series suggested that anesthetic doses of ketamine may exacerbate PTSD symptoms if administered shortly after the event, but may reduce PTSD symptoms if given well after the event (Kelmendi et al., 2016). In a randomized controlled trial of ketamine and midazolam for PTSD, both medications led to rapid reductions in symptoms, with ketamine superior to midazolam (Feder & Murrough, 2015). There is ongoing interest in a range of other translational targets for PTSD, including the cannabinoid and cholinergic systems, oxytocin, and neuroinflammation (Bowers & Ressler, 2015; Kelmendi et al., 2016; Rezaei Ardani, Hosseini, Fayyazi Bordbar, Talaei, & Mostafavi Toroghi, 2017; Thomas & Stein, 2017), with ongoing attention to the development of blood biomarkers.

In line with evidence that glucocorticoids are involved in memory consolidation in laboratory models, recent evidence suggests that glucocorticoid administration may be useful in enhancing exposure-based psychotherapy in a number of anxiety disorders (Hofmann, Mundy, & Curtiss, 2015). Early work suggested that there were short-term beneficial effects for adjunctive hydrocortisone administration and traumatic memory reactivation for patients with PTSD (Suris, North, Adinoff, Powell, & Greene, 2010). Subsequent work indicated that hydrocortisone administered in combination with traumatic memory reactivation led to greater patient retention (Yehuda, Bierer, Pratchett, et al., 2015). A placebo-controlled trial found that propranolol administered together with brief trauma reactivation sessions significantly improved PTSD symptoms (Brunet, Poundja, Tremblay, et al., 2011). Other noradrenergic agents, such as yohimbine, are also being studied in the context of enhancing exposure-based psychotherapy (Marin, Lonak, & Milad, 2015).

Earlier, it was noted that administration of a CB_1 agonist immediately after exposure to stressors prevents PTSD-like symptoms in rodents (Morrison & Ressler, 2014). Preliminary clinical data suggest that cannabinoid receptor agonists, including nabilone and Δ9-THC, may improve various symptoms, including sleep disruption and chronic pain in PTSD (Bowers & Ressler, 2015; Fraser, 2009; Roitman, Mechoulam, Cooper-Kazaz, & Shalev, 2014). Further studies are needed to assess whether such agents are useful in the treatment of PTSD, whether alone or in combination with psychotherapy.

In rodents, administration of methylene blue, an inhibitor of nitric oxide synthase (NOS), enhances memory, including fear extinction, and increases metabolic activity in mPFC (Singewald et al., 2015). A placebo-controlled trial of methylene blue added to exposure-based intervention for claustrophobia found that this agent enhanced memory and retention of fear extinction when administered after a successful exposure session but may have had a deleterious effect on extinction when administered after an unsuccessful exposure session (Telch, Bruchey, Rosenfield, et al., 2014). This finding is reminiscent of work on D-cycloserine in the augmentation of exposure-based psychotherapy for PTSD, where it has been suggested that improved findings would be obtained if the medication were given only after successful exposure sessions (Singewald et al., 2015).

A range of other drugs have been proposed as potentially useful for the augmentation of exposure-based psychotherapy of PTSD on the basis of preclinical data (Marin et al., 2015). These include histamine H2 receptor agonists, L-type calcium channel (LTCCs) modulators, and histone deacetylase (HDAC) inhibitors (Singewald et al., 2015). Growing evidence that DNA methylation is involved in fear conditioning, together with the fact that DNA methylation is reversible, suggests another target for enhancing extinction. In particular, it has been pointed out that methylcytosine dioxygenase promotes DNA demethylation (Singewald et al., 2015). Study of genomic aspects of fear conditioning/extinction has led to recognition of the role that microRNAs play in these processes, and such work may ultimately also lead to novel pharmacological targets (Wingo, Almli, Stevens, et al., 2015).

Although not immediately related to a consideration of pharmacotherapy, it is relevant to note that increased understanding of the neurocircuitry of PTSD may also give impetus to the use of neurostimulation-focused modalities for this condition. Repetitive transcranial stimulation (rTMS), deep brain stimulation (DBS), vagus nerve stimulation, and transcranial direct current stimulation (tDCS) have all been considered as potential interventions for PTSD (Bowers & Ressler, 2015). rTMS of mPFC administered together with exposure-based psychotherapy was found to decrease PTSD symptoms (Bowers & Ressler, 2015). DBS of ventral striatum enhances cued fear extinction, and DBS of amygdala leads to decreased PTSD-like symptoms and cued fear expression in rats (Bowers & Ressler, 2015). Based on these studies, DBS of the amygdala has been proposed for the treatment of PTSD.

to enhancing fear extinction in patients with PTSD, a translational approach to developing new pharmacotherapies was reasonable and feasible. Indeed, while early pharmacotherapies for PTSD were reliant on the introduction of serendipitously discovered antidepressant medications, the pioneering translation of D-cycloserine findings from bench to bedside provided the field with significant impetus. We subsequently went on to review a range of promising directions in the field, and we noted that several promising pharmacotherapy approaches to PTSD are under active investigation.

At the same time, it is important to recognize the several barriers faced by the field. PTSD is a heterogenous disorder and may involve several subtypes (Lanius et al., 2010). In addition, the psychobiology of PTSD may well change over time, as the disorder moves from acute stress disorder, to acute PTSD, to chronic PTSD. Although personalized medicine has been put forward as an important aspiration for psychiatry, genomic work on PTSD and the development of blood-based biomarkers for PTSD are at an early stage (Bandelow, Baldwin, Abelli, et al., 2017; Daskalakis, Cohen, Nievergelt, et al., 2016), and the possibility that many genes of small effect are involved in this disorder will likely make it difficult to use genetic variants to predict pharmacotherapy outcome. An approach informed by the research domain criteria (RDoC) may ultimately lead to a better understanding of intermediate phenotypes relevant to PTSD, but as this chapter has indicated, productive translational work has already been under way for some time in this area.

Given the exciting possibilities and the potential pitfalls that currently characterize the field of translational neuroscience, our view is that it is appropriate to maintain a position that allows hope but that at the same time avoids hype. In the past few decades, there have been significant advances in the neurocircuitry and neurochemistry of PTSD, as well as in the basic neuroscience of key cognitive-affective processes that appear highly relevant to this disorder. At the same time, the field remains relatively young, and a great deal remains to be learned. A broad range of potential avenues have been opened up for investigation. At the same time, the road from proof-of-concept trial to registered medication is an arduous and expensive one and succeeds in only a small proportion of cases. Given the tremendous burden of disease associated with PTSD, we eagerly look forward to the possibility of future progress.

VI CONCLUSION

This chapter began with the premise that given the availability of rigorous animal models of fear conditioning/extinction, given advances in the psychobiology of PTSD, and given the use of exposure-based approaches

References

American Psychiatric Association (Ed.), (2013). *Diagnostic and statistical manual of mental disorders.* (5th ed.). Arlington, VA: American Psychiatric Association.

Amos, T., Stein, D. J., & Ipser, J. C. (2014). Pharmacological interventions for preventing post-traumatic stress disorder (PTSD). *The Cochrane Database of Systematic Reviews, 7,* CD006239.

Andero, R., & Ressler, K. J. (2012). Fear extinction and BDNF: translating animal models of PTSD to the clinic. *Genes, Brain, and Behavior, 11*(5), 503–512.

Atwoli, L., Stein, D. J., Koenen, K. C., & McLaughlin, K. A. (2015). Epidemiology of posttraumatic stress disorder: prevalence, correlates and consequences. *Current Opinion in Psychiatry, 28*(4), 307–311.

Averill, L. A., Purohit, P., Averill, C. L., Boesl, M. A., Krystal, J. H., & Abdallah, C. G. (2016). Glutamate dysregulation and glutamatergic therapeutics for PTSD: evidence from human studies. *Neuroscience Letters 649*, 147–155.

Baldwin, D. S., Anderson, I. M., Nutt, D. J., et al. (2014). Evidence-based pharmacological treatment of anxiety disorders. posttraumatic stress disorder and obsessive-compulsive disorder: a revision of the 2005 guidelines from the British Association for Psychopharmacology. *Journal of Psychopharmacology, 28*(5), 403–439.

Bandelow, B., Baldwin, D., Abelli, M., et al. (2017). Biological markers for anxiety disorders, OCD and PTSD: a consensus statement. Part II: neurochemistry, neurophysiology and neurocognition. *The World Journal of Biological Psychiatry, 18*(3), 162–214.

Benjet, C., Bromet, E., Karam, E. G., et al. (2015). The epidemiology of traumatic event exposure worldwide: results from the World Mental Health Survey Consortium. *Psychological Medicine,* 1–17.

Bowers, M. E., Choi, D. C., & Ressler, K. J. (2012). Neuropeptide regulation of fear and anxiety: implications of cholecystokinin, endogenous opioids, and neuropeptide Y. *Physiology & Behavior, 107*(5), 699–710.

Bowers, M. E., & Ressler, K. J. (2015). An overview of translationally informed treatments for posttraumatic stress disorder: animal models of Pavlovian fear conditioning to human clinical trials. *Biological Psychiatry, 78*(5), E15–E27.

Brand, L., Groenewald, I., Stein, D. J., Wegener, G., & Harvey, B. H. (2008). Stress and re-stress increases conditioned taste aversion learning in rats: possible frontal cortical and hippocampal muscarinic receptor involvement. *European Journal of Pharmacology, 586*(1–3), 205–211.

Brewin, C. R., Andrews, B., & Valentine, J. D. (2000). Meta-analysis of risk factors for posttraumatic stress disorder in trauma-exposed adults. *Journal of Consulting and Clinical Psychology, 68*(5), 748–766.

Brunet, A., Poundja, J., Tremblay, J., et al. (2011). Trauma reactivation under the influence of propranolol decreases posttraumatic stress symptoms and disorder: 3 open-label trials. *Journal of Clinical Psychopharmacology, 31*(4), 547–550.

Charney, D. S., Deutch, A. Y., Krystal, J. H., Southwick, S. M., & Davis, M. (1993). Psychobiologic mechanisms of posttraumatic stress disorder. *Archives of General Psychiatry, 50*(4), 295–305.

Daskalakis, N. P., Cohen, H., Nievergelt, C. M., et al. (2016). New translational perspectives for blood-based biomarkers of PTSD: From glucocorticoid to immune mediators of stress susceptibility. *Experimental Neurology, 284*(Pt B), 133–140.

de Kloet, E. R. (2014). From receptor balance to rational glucocorticoid therapy. *Endocrinology, 155*(8), 2754–2769.

de Quervain, D., Schwabe, L., & Roozendaal, B. (2017). Stress, glucocorticoids and memory: implications for treating fear-related disorders. *Nature Reviews. Neuroscience, 18*(1), 7–19.

Feder, A., & Murrough, J. W. (2015). Ketamine for posttraumatic stress disorder–reply. *JAMA Psychiatry, 72*(1), 95–96.

Fraser, G. A. (2009). The use of a synthetic cannabinoid in the management of treatment-resistant nightmares in posttraumatic stress disorder (PTSD). *CNS Neuroscience & Therapeutics, 15*(1), 84–88.

Furtado, M., & Katzman, M. A. (2015). Neuroinflammatory pathways in anxiety, posttraumatic stress, and obsessive compulsive disorders. *Psychiatry Research, 229*(1–2), 37–48.

Garner, M., Mohler, H., Stein, D. J., Mueggler, T., & Baldwin, D. S. (2009). Research in anxiety disorders: from the bench to the bedside. *European Neuropsychopharmacology, 19*(6), 381–390.

Gilbertson, M. W., Shenton, M. E., Ciszewski, A., et al. (2002). Smaller hippocampal volume predicts pathologic vulnerability to psychological trauma. *Nature Neuroscience, 5*(11), 1242–1247.

Gordon, J. A. (2016). On being a circuit psychiatrist. *Nature Neuroscience, 19*(11), 1385–1386.

Greco, J. A., & Liberzon, I. (2016). Neuroimaging of fear-associated learning. *Neuropsychopharmacology, 41*(1), 320–334.

Harvey, B. H., Bothma, T., Nel, A., Wegener, G., & Stein, D. J. (2005). Involvement of the NMDA receptor, NO-cyclic GMP and nuclear factor K-beta in an animal model of repeated trauma. *Human Psychopharmacology, 20*(5), 367–373.

Harvey, B. H., Brand, L., Jeeva, Z., & Stein, D. J. (2006). Cortical/hippocampal monoamines, HPA-axis changes and aversive behavior following stress and restress in an animal model of post-traumatic stress disorder. *Physiology & Behavior, 87*(5), 881–890.

Harvey, B., Naciti, C., Brand, L., et al. (2004). Serotonin and stress: protective or malevolent actions in the biobehavioral response to repeated trauma? *Biobehavioral Stress Response: Protective and Damaging Effects, 1032*, 267–272.

Harvey, B. H., Oosthuizen, F., Brand, L., Wegener, G., & Stein, D. J. (2004). Stress-restress evokes sustained iNOS activity and altered GABA levels and NMDA receptors in rat hippocampus. *Psychopharmacology, 175*(4), 494–502.

Harvey, B. H., & Shahid, M. (2012). Metabotropic and ionotropic glutamate receptors as neurobiological targets in anxiety and stress-related disorders: focus on pharmacology and preclinical translational models. *Pharmacology, Biochemistry, and Behavior, 100*(4), 775–800.

Hofmann, S. G., Mundy, E. A., & Curtiss, J. (2015). Neuroenhancement of exposure therapy in anxiety disorders. *AIMS Neuroscience, 2*(3), 123–138.

Ipser, J. C., Carey, P., Dhansay, Y., Fakier, N., Seedat, S., & Stein, D. J. (2006). Pharmacotherapy augmentation strategies in treatment-resistant anxiety disorders. *The Cochrane Database of Systematic Reviews, 4*, CD005473.

Ipser, J. C., & Stein, D. J. (2012). Evidence-based pharmacotherapy of post-traumatic stress disorder (PTSD). *The International Journal of Neuropsychopharmacology, 15*(6), 825–840.

Kelmendi, B., Adams, T. G., Yarnell, S., Southwick, S., Abdallah, C. G., & Krystal, J. H. (2016). PTSD: from neurobiology to pharmacological treatments. *European Journal of Psychotraumatology, 7*, 31858.

Kessler, R. C., Ruscio, A. M., Shear, K., & Wittchen, H. U. (2010). Epidemiology of anxiety disorders. *Current Topics in Behavioral Neurosciences, 2*, 21–35.

Koenen, K. C., Duncan, L. E., Liberzon, I., & Ressler, K. J. (2013). From candidate genes to genome-wide association: the challenges and promise of posttraumatic stress disorder genetic studies. *Biological Psychiatry, 74*(9), 634–636.

Koresh, O., Kozlovsky, N., Kaplan, Z., Zohar, J., Matar, M. A., & Cohen, H. (2012). The long-term abnormalities in circadian expression of Period 1 and Period 2 genes in response to stress is normalized by agomelatine administered immediately after exposure. *European Neuropsychopharmacology, 22*(3), 205–221.

Lanius, R. A., Vermetten, E., Loewenstein, R. J., et al. (2010). Emotion modulation in PTSD: clinical and neurobiological evidence for a dissociative subtype. *The American Journal of Psychiatry, 167*(6), 640–647.

Marin, M. F., Lonak, S. F., & Milad, M. R. (2015). Augmentation of evidence-based psychotherapy for PTSD with cognitive enhancers. *Current Psychiatry Reports, 17*(6), 39.

Mataix-Cols, D., Fernandez de la Cruz, L., Monzani, B., et al. (2017). D-cycloserine augmentation of exposure-based cognitive behavior therapy for anxiety, obsessive-compulsive, and posttraumatic stress disorders: a systematic review and meta-analysis. *JAMA Psychiatry, 74*, 501–510.

Mithoefer, M. C., Wagner, M. T., Mithoefer, A. T., Jerome, L., & Doblin, R. (2011). The safety and efficacy of {+/-}3, 4-methylenedioxymethamphetamine-assisted psychotherapy in subjects with chronic, treatment-resistant posttraumatic stress disorder: the first randomized controlled pilot study. *Journal of Psychopharmacology, 25*(4), 439–452.

Morris, M. C., Hellman, N., Abelson, J. L., & Rao, U. (2016). Cortisol, heart rate, and blood pressure as early markers of PTSD risk: a systematic review and meta-analysis. *Clinical Psychology Review, 49,* 79–91.

Morrison, F. G., & Ressler, K. J. (2014). From the neurobiology of extinction to improved clinical treatments. *Depression and Anxiety, 31*(4), 279–290.

Nemeroff, C. B. (2004). Neurobiological consequences of childhood trauma. *The Journal of Clinical Psychiatry, 65*(Suppl 1), 18–28.

O'Doherty, D. C., Chitty, K. M., Saddiqui, S., Bennett, M. R., & Lagopoulos, J. (2015). A systematic review and meta-analysis of magnetic resonance imaging measurement of structural volumes in posttraumatic stress disorder. *Psychiatry Research, 232*(1), 1–33.

Ori, R., Amos, T., Bergman, H., Soares-Weiser, K., Ipser, J. C., & Stein, D. J. (2015). Augmentation of cognitive and behavioural therapies (CBT) with d-cycloserine for anxiety and related disorders. *The Cochrane Database of Systematic Reviews, 5,* CD007803.

Rasmusson, A. M., Marx, C. E., Pineles, S. L., et al. (2017). Neuroactive steroids and PTSD treatment. *Neuroscience Letters, 649,* 156–163.

Rezaei Ardani, A., Hosseini, G., Fayyazi Bordbar, M. R., Talaei, A., & Mostafavi Toroghi, H. (2017). Effect of rivastigmine augmentation in treatment of male patients with combat-related chronic posttraumatic stress disorder: a randomized controlled trial. *Journal of Clinical Psychopharmacology, 37*(1), 54–60.

Roitman, P., Mechoulam, R., Cooper-Kazaz, R., & Shalev, A. (2014). Preliminary, open-label, pilot study of add-on oral Delta9-tetrahydrocannabinol in chronic post-traumatic stress disorder. *Clinical Drug Investigation, 34*(8), 587–591.

Selye, H. (1976). Forty years of stress research: principal remaining problems and misconceptions. *Canadian Medical Association Journal, 115*(1), 53–56.

Singewald, N., Schmuckermair, C., Whittle, N., Holmes, A., & Ressler, K. J. (2015). Pharmacology of cognitive enhancers for exposure-based therapy of fear, anxiety and trauma-related disorders. *Pharmacology & Therapeutics, 149,* 150–190.

Skelton, K., Ressler, K. J., Norrholm, S. D., Jovanovic, T., & Bradley-Davino, B. (2012). PTSD and gene variants: new pathways and new thinking. *Neuropharmacology, 62*(2), 628–637.

Stein, D. J., Cloitre, M., Nemeroff, C. B., et al. (2009). Cape Town consensus on posttraumatic stress disorder. *CNS Spectrums, 14*(1 Suppl 1), 52–58.

Stein, D. J., Ipser, J. C., & Seedat, S. (2006). Pharmacotherapy for post traumatic stress disorder (PTSD). *The Cochrane Database of Systematic Reviews, 1,* CD002795.

Stein, D. J., Seedat, S., Iversen, A., & Wessely, S. (2007). Post-traumatic stress disorder: medicine and politics. *Lancet, 369*(9556), 139–144.

Suliman, S., Seedat, S., Pingo, J., Sutherland, T., Zohar, J., & Stein, D. J. (2015). Escitalopram in the prevention of posttraumatic stress disorder: a pilot randomized controlled trial. *BMC Psychiatry, 15,* 24.

Suris, A., North, C., Adinoff, B., Powell, C. M., & Greene, R. (2010). Effects of exogenous glucocorticoid on combat-related PTSD symptoms. *Annals of Clinical Psychiatry, 22*(4), 274–279.

Telch, M. J., Bruchey, A. K., Rosenfield, D., et al. (2014). Effects of post-session administration of methylene blue on fear extinction and contextual memory in adults with claustrophobia. *The American Journal of Psychiatry, 171*(10), 1091–1098.

Thomas, E., & Stein, D. J. (2017). Novel pharmacological treatment strategies for posttraumatic stress disorder. *Expert Review of Clinical Pharmacology, 10*(2), 167–177.

Uys, J. D., Muller, C. J., Marais, L., Harvey, B. H., Stein, D. J., & Daniels, W. M. (2006). Early life trauma decreases glucocorticoid receptors in rat dentate gyrus upon adult re-stress: reversal by escitalopram. *Neuroscience, 137*(2), 619–625.

Uys, M. M., Shahid, M., Sallinen, J., & Harvey, B. H. (2017). The alpha2C-adrenoceptor antagonist, ORM-10921, exerts antidepressant-like effects in the Flinders Sensitive Line rat. *Behavioural Pharmacology, 28*(1), 9–18.

Uys, J. D., Stein, D. J., Daniels, W. M., & Harvey, B. H. (2003). Animal models of anxiety disorders. *Current Psychiatry Reports, 5*(4), 274–281.

VanElzakker, M. B., Dahlgren, M. K., Davis, F. C., Dubois, S., & Shin, L. M. (2014). From Pavlov to PTSD: the extinction of conditioned fear in rodents, humans, and anxiety disorders. *Neurobiology of Learning and Memory, 113,* 3–18.

Wingo, A. P., Almli, L. M., Stevens, J. S., et al. (2015). DICER1 and microRNA regulation in post-traumatic stress disorder with comorbid depression. *Nature Communications, 6,* 10106.

Yehuda, R., Bierer, L. M., Pratchett, L. C., et al. (2015). Cortisol augmentation of a psychological treatment for warfighters with posttraumatic stress disorder: randomized trial showing improved treatment retention and outcome. *Psychoneuroendocrinology, 51,* 589–597.

Yehuda, R., & McFarlane, A. C. (1995). Conflict between current knowledge about posttraumatic stress disorder and its original conceptual basis. *The American Journal of Psychiatry, 152*(12), 1705–1713.

27

Unmet Medical Needs in the Treatment of Depression and the Clinical Development of a Differentiated Antidepressant: A Translational Line of Evidence

George G. Nomikos

Biogen, Clinical Development, Cambridge, MA, United States

I INTRODUCTION

Depression affects millions of people worldwide and is a major contributor to global disability, according to the World Health Organization (2017). A constellation of symptoms, physical, emotional, and cognitive, contribute to the clinical picture and the diagnosis of major depressive disorder (MDD; American Psychiatric Association, 2013). These symptoms include depressed mood, the loss of interest or pleasure, diminished ability to think/concentrate or indecisiveness, significant change in weight or appetite, insomnia or hypersomnia, psychomotor agitation or retardation, fatigue or the loss of energy, feelings of worthlessness or excessive guilt, and suicidal ideation. The essential feature of MDD is a clinical course that is characterized by one or more major depressive episodes (MDEs) without a history of manic, mixed, or hypomanic episodes. Of the nine symptoms, five or more must be present for at least 2 weeks and cause significant distress or functional impairment; at least one of the symptoms is either (1) depressed mood or (2) the loss of interest or pleasure. The course of an MDE over time from euthymia to depression and then to response, remission, and recovery or relapse and recurrence of a new MDE is depicted in Fig. 27.1.

As mentioned by Kurian, Greer, and Trivedi (2009), remission, that is, complete relief from an MDE, rather than response (merely substantial improvement) should be the goal of treatment, as it is associated with a better prognosis and better function. In this regard, first-line treatments for MDD result in remission for only approximately a third of treated patients. Failure to achieve remission, that is, the presence of residual symptoms, results in poorer prognosis for depressed individuals by increasing the likelihood of relapse, contributing to continuing or even worsening functional impairments and causing a more chronic course of illness. The residual symptoms in depression, during and between MDEs, consist primarily of cognitive problems, the lack of energy, sleeping problems, diminished interest, and feelings of worthlessness or guilt (see, e.g., Conradi, Ormel, & de Jonge, 2011); on average, two residual symptoms are present during remission, and symptoms may not disappear between MDEs, that is, during nondepressive periods. Strategies that can enhance the therapeutic efficacy of antidepressants include augmentation, switching, or combination treatments, targeting residual symptoms. The personalization of treatment may be accomplished by targeting specific symptoms or symptom clusters that remain untreated (i.e., residual) to help patients achieve full functional recovery. This "personalized," targeted approach has empirically been used in selecting antidepressants, such as the selective serotonin reuptake inhibitors (SSRIs), for the treatment of certain symptoms in MDD (e.g., psychomotor agitation or retardation) beyond depressed mood, based on whether the SSRIs are considered sedating, activating, or neutral in a continuum of clinical responses (Fig. 27.2). The same principle, that is, treating residual symptoms that constitutes a high unmet medical need, has been used the last 40 years to drive discovery and development of new drugs with different mechanisms of action from SSRIs and serotonin-norepinephrine reuptake inhibitors (SNRIs) to multimodal and fast-acting antidepressants (Fig. 27.3).

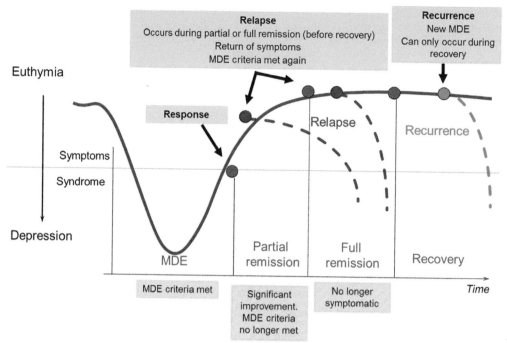

FIG. 27.1   Course of a major depressive episode (MDE). *Modified from Kupfer, D.J. (1991).*

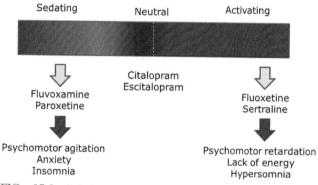

FIG. 27.2   Sedating versus activating SSRIs: A "personalized therapeutic" approach for depressive symptoms.

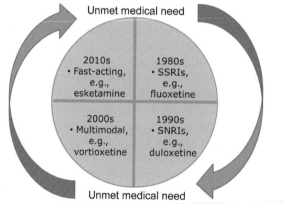

FIG. 27.3   New antidepressants with different mechanisms of action that target residual symptoms, a high unmet medical need, have been discovered and developed over the past 40 years.

II COGNITIVE IMPAIRMENT IN MDD

Cognitive dysfunction in depressed patients has been reported in a range of cognitive domains, including executive functioning, attention, memory, and processing speed, as assessed by objective neuropsychological tests (see, e.g., Rock et al., 2014; Rund et al., 2006). In particular, difficulty in concentrating and indecisiveness are reported as among the most troubling symptoms in MDD and may limit functional recovery. Cognitive deficits in memory and decision-making are present early in the course of MDD and may be accompanied by structural and functional abnormalities in the hippocampus and prefrontal cortex involved in cognitive functions. Although resolution of cognitive symptoms of depression lags behind recovery from mood symptoms in many patients, there is evidence suggesting that they not only may improve with antidepressant therapy but also can persist residually. New strategies that target cognitive symptoms of depression in addition to mood symptoms are needed to improve long-term outcomes, particularly functional recovery (Trivedi & Greer, 2014).

Many antidepressants indirectly improve cognitive function by improving mood and depressive symptoms. Previously, no conventional antidepressant had been sufficiently studied and/or had demonstrated robust procognitive effects in depression (McIntyre et al., 2013). In patients with depression, improvement in cognitive symptoms on antidepressant therapy is often incomplete, even among patients considered to be in remission (Fava et al., 2006; Trivedi, Hollander, Nutt, & Blier, 2008). Some antidepressants may even worsen cognitive dysfunction (Millan et al., 2012).

Depression is considered to be multifactorial with multiple symptom dimensions of MDD that are thought to arise from a complex interplay between dysfunction in multiple regions of the brain and alterations in multiple neurotransmitter systems, including serotonin, norepinephrine, dopamine, and glutamate. Thus, for a successful therapeutic management of symptoms of depression beyond mood, discovery and development of new antidepressants that affect multiple neurotransmitter systems in multiple regions of the brain are warranted. These new chemical entities need to be studied thoroughly in clinical trials in MDD, focusing on, for example, cognitive impairment as the primary efficacy outcome with evidence in support of functional recovery.

III VORTIOXETINE (TRINTELLIX): A MULTIMODAL DIFFERENTIATED ANTIDEPRESSANT

Vortioxetine is a new antidepressant drug that was approved in 2013 for the treatment of MDD in adults in the United States and MDEs in adults in Europe. It is an antagonist for 5-HT3, 5-HT7, and 5-HT1D receptors; an agonist for 5-HT1A receptors; a partial agonist for 5-HT1B receptors; and an inhibitor for 5-HT transporters; it is thus considered a multimodal antidepressant affecting not only the serotonin transporters, as SSRIs and SNRIs do, but also multiple serotonin receptors (see Bang-Andersen et al., 2011). Efficacy was demonstrated in six short-term studies and one relapse prevention study with a good short- and long-term safety/tolerability profile and no dose adjustments for drug-drug interactions, age, sex or ethnicity, and renal or hepatic impairment (see, e.g., TRINTELLIX (vortioxetine) prescribing information, Takeda Pharmaceuticals).

Vortioxetine's direct effects on 5-HT receptors and transporters seem to contribute to distinct indirect effects on multiple neurotransmitters, resulting in an enhancement of extracellular concentrations of 5-HT, acetylcholine, dopamine, norepinephrine, histamine, and glutamate (Pehrson & Sanchez, 2014); this multimodal pharmacological profile potentially plays a central role in its differentiated clinical actions that consist of increase in mood and cognition and decrease in sexual dysfunction, anxiety, and insomnia. In particular, vortioxetine exhibits positive effects in behavioral tests of cognition, such as cognitive flexibility, attention, and memory, or in potentially cognition-relevant mechanistic assays, such as

electroencephalography, in vivo microdialysis, in vivo or in vitro electrophysiology, and molecular assays related to neurogenesis or synaptic sprouting (see Pehrson et al., 2015). For example, vortioxetine restores recognition memory in rats with low 5-HT, while the SSRI escitalopram and the SNRI duloxetine are inactive (Pehrson & Sanchez, 2014).

IV VORTIOXETINE SHOWS A DISTINCT PROFILE IN COGNITIVE PERFORMANCE IN DEPRESSION

Vortioxetine was shown to be efficacious and well tolerated in elderly patients ($\geq$65 years old) with depression; it also improved cognitive performance, showing superiority to placebo in cognition tests of speed of processing, verbal learning, and memory (Katona, Hansen, & Olsen, 2012). The cognition endpoints were secondary in this clinical study, and interestingly, vortioxetine showed a significant separation from placebo in the digit-symbol substitution test (DSST), an objective test of speed of processing, attention, executive function, and working memory, whereas the SNRI duloxetine had no effect. In a follow-on clinical trial (FOCUS), vortioxetine improved cognitive performance in adult patients with depression across domains (McIntyre, Lophaven, & Olsen, 2014). The cognition endpoints were both primary and secondary in this clinical trial; the effects on DSST were replicated; and vortioxetine also showed antidepressant efficacy, although its positive effects on cognitive performance were largely independent of its effect on improving depressive symptoms.

In an experimental medicine fMRI study in remitted depressed subjects performing a working memory task, vortioxetine reduced the blood oxygen level-dependent (BOLD) signal within the right dorsolateral prefrontal cortex and left hippocampus (Smith et al., 2018). Acute and remitted major depression has previously been associated with increased activity within these regions, suggesting that treatment with vortioxetine may reverse the effects of the disorder within these neurocognitive systems. This raises the possibility that the cognitive effects of vortioxetine in patients with MDD may be mediated, at least in part, by its effects on neural systems supporting working memory and executive function.

Ultimately, the short-term efficacy and safety of vortioxetine on cognitive function in adults diagnosed with MDD who self-reported cognitive dysfunction were evaluated in a multicenter, randomized, double-blind, placebo-controlled, active-referenced (duloxetine), parallel-group study (CONNECT) (Mahableshwarkar, Zajecka, Jacobson, Chen, & Keefe, 2015). Efficacy was evaluated using the DSST—number of correct symbols—as the prespecified primary endpoint. Additional predefined endpoints included the objective performance-based University of San Diego performance-based skills assessment (UPSA) to measure functionality (functional capacity) and a prespecified multiple regression analysis (path analysis) to calculate direct versus indirect effects of vortioxetine on cognitive function. Vortioxetine was statistically superior to placebo on the DSST and UPSA, and path analysis indicated that vortioxetine's cognitive and functional benefits were primarily a direct treatment effect rather than due to alleviation of depressive symptoms. Duloxetine was not significantly different from placebo on the DSST or UPSA, while both duloxetine and vortioxetine improved depressive symptoms. In this study of MDD adults who self-reported cognitive dysfunction, vortioxetine significantly improved cognitive function, depression, and functionality and was generally well tolerated.

On 2 May 2018, the US Food and Drug Administration (FDA) approved a supplemental new drug application for vortioxetine. The clinical trial section of the US label now includes data from the largest replicated clinical studies on an important aspect of cognitive function in MDD. The FOCUS and CONNECT studies showed that vortioxetine has a positive effect on processing speed, an important aspect of cognitive function observed in some patients with MDD. This is the first FDA-approved treatment for MDD to have data in the US prescribing information showing a positive effect on processing speed; similarly, vortioxetine became the first antidepressant approved in Europe to improve cognitive function in patients with depression in 2015.

V CONCLUSION

This chapter provided the translational line of evidence in support of vortioxetine's unique effects on cognitive impairment in depression, a residual symptom, and a high unmet medical need. Vortioxetine shows multimodal pharmacology, affecting both 5-HT receptors and transporters and enhancing multiple neurotransmitter systems in the brain. Its procognitive effects were first shown in nonclinical tests of cognitive function and related mechanistic assays, supporting a differentiated pharmacological profile compared with established antidepressants. In the clinic, vortioxetine proved to be an efficacious, safe, and well-tolerated antidepressant with unique features, such as lower rate of sexual dysfunction and a positive effect on cognitive function in an elderly study population. An experimental medicine study in remitted MDD patients performing a cognitive task revealed brain activity in response to vortioxetine that was opposite to what has been found in depressed patients (acute and remitted), providing a neurobiological evidence for its procognitive effects in depression. Two large clinical trials in depression provided proof that vortioxetine improves cognitive function (FOCUS and CONNECT) and functional capacity (CONNECT), as assessed by objective tests, and in a manner that is independent of its shown antidepressant effects. The totality of the evidence provided and the ability to translate vortioxetine's distinct pharmacological activity in the clinic throughout its discovery and development path ultimately led to its approval by the regulatory authorities, as the first antidepressant that showed improvement in cognitive performance in MDD (Fig. 27.4).

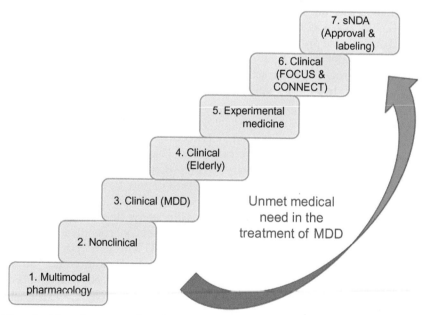

FIG. 27.4   Translational line of evidence in support of vortioxetine's distinct effect on cognitive function in depression.

References

American Psychiatric Association (2013). *Diagnostic and statistical manual of mental disorders.* (5th ed.).

Bang-Andersen, B., Ruhland, T., Jørgensen, M., Smith, G., Frederiksen, K., Jensen, K. G., et al. (2011). Discovery of 1-[2-(2,4-dimethylphenylsulfanyl)phenyl]piperazine (Lu AA21004): a novel multimodal compound for the treatment of major depressive disorder. *Journal of Medicinal Chemistry, 54*(9), 3206–3221. https://doi.org/10.1021/jm101459g.

Conradi, H. J., Ormel, J., & de Jonge, P. (2011). Presence of individual (residual) symptoms during depressive episodes and periods of remission: a 3-year prospective study. *Psychological Medicine, 41*(6), 1165–1174. https://doi.org/10.1017/S0033291710001911.

Fava, M., Graves, L. M., Benazzi, F., Scalia, M. J., Iosifescu, D. V., Alpert, J. E., et al. (2006). A cross-sectional study of the prevalence of cognitive and physical symptoms during long-term antidepressant treatment. *The Journal of Clinical Psychiatry, 67*(11), 1754–1759. https://doi.org/10.4088/JCP.v67n1113.

Katona, C., Hansen, T., & Olsen, C. K. (2012). A randomized, double-blind, placebo-controlled, duloxetine-referenced, fixed-dose study comparing the efficacy and safety of Lu AA21004 in elderly patients with major depressive disorder. *International Clinical Psychopharmacology, 27*(4), 215–223. https://doi.org/10.1097/YIC.0b013e328354245.

Kupfer, D. J. (1991). Long-term treatment of depression. *The Journal of Clinical Psychiatry, 52*(Suppl), 28–34.

Kurian, B. T., Greer, T. L., & Trivedi, M. H. (2009). Strategies to enhance the therapeutic efficacy of antidepressants: targeting residual symptoms. *Expert Review of Neurotherapeutics, 9*(7), 975–984. https://doi.org/10.1586/ern.09.53.

Mahableshwarkar, A. R., Zajecka, J., Jacobson, W., Chen, Y., & Keefe, R. S. (2015). A randomized, placebo-controlled, active-reference, double-blind, flexible-dose study of the efficacy of vortioxetine on cognitive function in major depressive disorder. *Neuropsychopharmacology: Official Publication of the American College of Neuropsychopharmacology, 40*(8), 2025–2037. https://doi.org/10.1038/npp.2015.52.

McIntyre, R. S., Cha, D. S., Soczynska, J. K., Woldeyohannes, H. O., Gallaugher, L. A., Kudlow, P., et al. (2013). Cognitive deficits and functional outcomes in major depressive disorder: determinants, substrates, and treatment interventions. *Depression and Anxiety, 30*(6), 515–527. https://doi.org/10.1002/da.22063.

McIntyre, R. S., Lophaven, S., & Olsen, C. K. (2014). A randomized, double-blind, placebo-controlled study of vortioxetine on cognitive function in depressed adults. *The International Journal of Neuropsychopharmacology, 17*(10), 1557–1567. https://doi.org/10.1017/S1461145714000546.

Millan, M. J., Agid, Y., Brüne, M., Bullmore, E. T., Carter, C. S., Clayton, N. S., et al. (2012). Cognitive dysfunction in psychiatric disorders: characteristics, causes and the quest for improved therapy. *Nature Reviews Drug Discovery, 11*(2), 141–168. https://doi.org/10.1038/nrd3628.

Pehrson, A. L., Leiser, S. C., Gulinello, M., Dale, E., Li, Y., Waller, J. A., et al. (2015). Treatment of cognitive dysfunction in major depressive disorder—a review of the preclinical evidence for efficacy of selective serotonin reuptake inhibitors, serotonin-norepinephrine reuptake inhibitors and the multimodal-acting antidepressant vortioxetine. *European Journal of Pharmacology, 753*, 19–31. https://doi.org/10.1016/j.ejphar.2014.07.044.

Pehrson, A. L., & Sanchez, C. (2014). Serotonergic modulation of glutamate neurotransmission as a strategy for treating depression and cognitive dysfunction. *CNS Spectrums, 19*(2), 121–133. https://doi.org/10.1017/S1092852913000540.

Rock, P. L., Roiser, J. P., Riedel, W. J., & Blackwell, A. D. (2014). Cognitive impairment in depression: a systematic review and meta-analysis. *Psychological Medicine, 44*, 2029–2040.

Rund, B. R., Sundet, K., Asbjornsen, A., Egeland, J., Landro, N. I., Lund, A., et al. (2006). Neuropsychological test profiles in schizophrenia and non-psychotic depression. *Acta Psychiatrica Scandinavica, 113*(4), 350–359. https://doi.org/10.1111/j.1600-0447.2005.00626.x.

Smith, J., Browning, M., Conen, S., Smallman, R., Buchbjerg, J., Larsen, K. G., et al. (2018). Vortioxetine reduces BOLD signal during performance of the N-back working memory task: a randomised neuroimaging trial in remitted depressed patients and healthy controls. *Molecular Psychiatry, 23*(5), 1127–1133. https://doi.org/10.1038/mp.2017.104.

Trivedi, M. H., & Greer, T. L. (2014). Cognitive dysfunction in unipolar depression: implications for treatment. *Journal of Affective Disorders, 152–154*, 19–27. https://doi.org/10.1016/j.jad.2013.09.012.

Trivedi, M. H., Hollander, E., Nutt, D., & Blier, P. (2008). Clinical evidence and potential neurobiological underpinnings of unresolved symptoms of depression. *The Journal of Clinical Psychiatry, 69*(2), 246–258. https://doi.org/10.4088/JCP.v69n0211.

World Health Organization (2017). *Depression and other common mental disorders: Global health estimates.* Geneva: World Health Organization.

28

Translating Neurobiology into Practice in Tobacco, Alcohol, Drug, and Behavioral Addictions

A. Benjamin Srivastava*,†, Mark S. Gold*

*Department of Psychiatry, Washington University School of Medicine, Saint Louis, MO, United States †Department of Psychiatry, Columbia University Medical Center/New York State Psychiatric Institute, New York, NY, United States

I INTRODUCTION

In 2016 the Office of the Surgeon General of the United States issued the first-of-its-kind, comprehensive report on substance use disorders as the primacy of current, public health issues (Murthy, 2016; Office of the Surgeon General, 2016). Substance use disorders take an enormous toll on American society, with recent data showing that in 2015 nearly 67 million people reported a past-month history of binge drinking and over 27 million people were current users of illicit drugs or misusers of prescription medications (Center for Behavioral Health Statistics and Quality, 2016). Furthermore, over 21 million, or approximately 10% of Americans, suffer from a substance use disorder (Center for Behavioral Health Statistics and Quality, 2016). Substance use, misuse, and addiction have led to significant morbidity and mortality, with alcohol contributing to 88,000 deaths yearly and other drugs of abuse resulting in 47,000 overdose-related deaths in 2014, over 28,000 of which were due to opioids (Rudd, Aleshire, Zibbell, & Gladden, 2016; Stahre, Roeber, Kanny, Brewer, & Zhang, 2014). Indeed these striking findings prompted the Centers for Disease Control (CDC) to label the current opioid crisis as an epidemic (Kolodny et al., 2015; Rudd et al., 2016). Substance use disorders are also financially burdensome, costing the United States over $400 billion per year from expenses related to crime, health and hospitalization, and lost productivity (National Drug Intelligence Center, 2014; Sacks, Gonzales, Bouchery, Tomedi, & Brewer, 2015).

Given the enormous public health costs, the surgeon general's report has called for implementation and dissemination of existing treatments and further research for the development of new therapeutic modalities (Office of the Surgeon General, 2016). The public health burdens incurred and perpetuation of the problems notwithstanding significant advances have been made in neuroscience over the past half-century that have fostered a more sophisticated understanding of substance use, misuse, and use disorders, to which we will hereafter refer as addictive disorders. Specifically, translational research, using animal models, neuroimaging, and genetics, has provided rational, neuroscience-informed targets that have been successfully exploited for pharmacotherapeutic innovation. However, as we will describe, current, available pharmacotherapy is rather limited in scope, necessitating a more sophisticated and comprehensive neuroscience-informed approach to drug development (Tables 28.1 and 28.2).

II GENERAL PRINCIPLES

Translational neuroscience models that guide drug development for addictive disorders must be grounded in the underlying neuroscience of addiction. Though various models of the neurobiology of addiction have been described and proposed, the Koob/Volkow "three stage" model increasingly has become the putative schematic (Koob & Volkow, 2010; Koob & Volkow, 2016;

Volkow, Koob, & McLellan, 2016) and is indeed featured in the surgeon general's recent report (Office of the Surgeon General, 2016). Koob and Volkow describe a heuristic model of three stages of binge/intoxication, withdrawal/ negative affect, and preoccupation/anticipation/craving, each with distinct neurobiological correlates and phenomenology that ultimately manifests in a self-perpetuating, downward spiral. Here we summarize the model. See Koob and Volkow (2010), Volkow et al. (2016), Koob and Volkow (2016) for full review.

A Binge/Intoxication

All drugs with abuse or addictive potential result in a phasic release of dopamine (DA) in the mesolimbic system, from the ventral tegmental area (VTA) to the nucleus accumbens (NAc), resulting in the "reward" phenomenon, which through Pavlovian conditioning becomes associated with environmental cues in which the drug is taken. Notably, this sense of "reward," which confers evolutionary fitness, is stronger than that produced by natural, survival-oriented stimuli (e.g., food and sex). This conditioning is reflective of synaptic strengthening mediated by the glutamatergic system, with neuroplasticity changes in brain areas thought to mediate drug-taking behavior, including the amygdala (stress and anxiety), hippocampus (memory), and dorsal striatum (routine motor movements).

Over time, with repeated administration, NAc dopamine receptors desensitize, leading to a decreased sense of reward (though still present nonetheless from the drug itself), but due to Pavlovian conditioning as described earlier, the contextual cues (the conditioned stimulus) result in a surge of dopamine release, manifesting as the subjective experience of craving. Several real-world examples include an individual with alcohol dependence developing a sudden craving for a drink when walking by a bar or an opioid-dependent individual experiencing euphoria when injecting water intravenously.

B Withdrawal/Negative Affect

As mentioned earlier, repeated drug administration over time results in decreased DA release into the NAc, resulting in decreased sense of euphoria compared with that previously experienced. Correspondingly, natural stimuli (e.g., food, sex, and other previously pleasurable activities) become less enjoyable, resulting in a profound state of anhedonia. This phenomenon is mediated through neuroplasticity changes in the extended amygdala vis-à-vis the so-called antireward system involving stress response hormones, including corticotrophin-releasing hormone (CRH) and dynorphin. Thus, in a period of abstinence (i.e., a temporary cessation of intake), the antireward

system and the extended amygdala, anatomically and histologically defined as the posterior shell of the NAc, central nucleus of the amygdala (CeA), and bed nucleus of the striae terminalis (BNST), become overactive, contributing to the phenotype of extreme anxiety, dysphoria, and anhedonia. This stage reveals one of the astonishing properties of addiction; the act of drug taking transitions from being *impulsive* (i.e., pleasure seeking without afterthought) to *compulsive* (undertaken to relieve stress, tension, or physical signs such as pain).

C Preoccupation/Anticipation/Cravings

So far, we have discussed the deep brain structures, namely, connections between the projections from the VTA to the NAc, dorsal striatum, hippocampus, and extended amygdala. Just as dopaminergic tone is decreased with chronic substance use in the mesolimbic (i.e., VTA to NAc) system, it is likewise dampened in the prefrontal cortical (PFC) regions, which regulate executive function, decision-making, planning and initiation, and attribution of salience, and the anterior cingulate cortex, which partially regulates error monitoring. Taken collectively, decreased dopaminergic tone in these areas and consequently impairment in associated functioning manifest as inability to resist temptation (i.e., when cued or stressed), failure to cease using despite predictable consequences, and continued use despite a sincere desire to do otherwise.

Collectively, these three stages represent the phenomenological manifestation of a rich network of neurocircuits, selected for self-preservation, that becomes hijacked with prolonged substance use. What begins as experimental or impulsive behavior driven by increased phasic dopamine release in the mesolimbic system eventually becomes compulsive behavior—an attempt at alleviating dysphoria, anhedonia, and anxiety that is mediated through the extended amygdala. While the euphoric effects of the drugs themselves wear off over time, contextual cues still produce a dopamine surge in key structures, and with aberrantly functioning prefrontal areas crucial for self-regulation, appropriation of stimuli, and learning from previous behavior, continued use becomes a foregone conclusion, perpetuating the continuous and downward spiral of addiction. Substance use over time progresses, manifesting in self-stimulating, repetitive behaviors, ultimately producing what we have called a reward deficiency syndrome, characterized by prominent depression, anhedonia, and dysphoria (Febo et al., 2017)

The seemingly deterministic and fatalistic process we have described notwithstanding, an understanding of the fundamental neurocircuitry has provided mechanisms exploitable for therapeutic intervention. As such, throughout this chapter, we will continuously refer back to this model.

TABLE 28.1 FDA-Approved Treatments for Addictive Disorders

Disorder	Drug	Mechanism	Main effects	Additional effects	References/comments
Alcohol use disorder	Acamprosate	Derivative of taurine, indirect GABA agonist, and NMDA antagonist (mGluR5)	1) Reduction of anxiety and seizures in rats 2) Increase time to relapse in humans	Inconsistent effects on craving and reduction of heavy drinking days	(Jonas, Amick, Feltner, et al., 2014; Kalk & Lingford-Hughes, 2014)
	Disulfiram	Competitive inhibitor of acetaldehyde dehydrogenase	1) Improved short-term abstinence 2) Increased time to relapse 3) Reduction in number of heavy drinking days	–	(Hempel, von Bahr-Lindstrom, & Jornvall, 1984; Jorgensen, Pedersen, & Tonnesen, 2011) • Patient should be highly motivated • Should be administered under supervision given significant, adverse effects
	Naltrexone	Mu-opioid receptor (MOR) antagonist; possibly blocks euphoric effects of alcohol mediated through the endogenous opioid system	1) Reduction in alcohol consumption in rate 2) Improved abstinence in humans 3) Reduction in heavy drinking days in humans	• Possible use as a harm avoidance tool (reduction in amount of alcohol consumed)	(Garbutt, Kampov-Polevoy, Kalka-Juhl, & Gallop, 2016; Jonas et al., 2014; Lee et al., 2005; Oslin, Leong, Lynch, et al., 2015; Schacht et al., 2013; Self & Nestler, 1995) • Effects are largely inconsistent; possible gene-gene interaction effects with naltrexone resulting in reduced striatal activation • Can be given in a monthly injectable form to improve compliance
Opioid use disorder	Buprenorphine	MOR partial agonist and kappa opioid receptor (KOR) antagonist; combined with naloxone to deter abuse	1) Improved abstinence 2) Higher treatment retention rates 3) Reduced transmission of HIV and hepatitis C	• Possible antidepressant effects through KOR antagonism • Significant relapse rates upon discontinuation	(Connery, 2015; Karp et al., 2014) • Prescribers must obtain a waiver • Prescribers are "capped" on the number of patients they are allowed to treat
	Methadone	MOR agonist	1) Improved abstinence 2) Higher treatment retention rates (even compared with buprenorphine) 3) Reduced transmission of HIV and hepatitis C	• Significant relapse rates upon discontinuation	(Connery, 2015) • Significant abuse potential • Must be dispensed at a licensed clinic • Can cause fatal respiratory depression when combined with GABA-ergic agent (e.g., benzodiazepines)
	Naltrexone	MOR antagonist (both PO and IM)	1) PO naltrexone has poor compliance with severe relapses 2) Similar efficacy to buprenorphine; though failure to maintain abstinence prior to induction may cause difficulties		(Connery, 2015; Krupitsky et al., 2011; Krupitsky et al., 2013; Lee et al., 2016) • No abuse potential • Ongoing trial comparing IM naltrexone with buprenorphine
Tobacco use disorder	Bupropion	Norepinephrine (NE) and dopamine (DA) reuptake inhibition leading to repletion of mesolimbic dopamine absent during abrupt cessation	1) Promotes abstinence 2) Reduces cravings 3) Relieves withdrawal symptoms		(Anthenelli et al., 2016; Cahill et al., 2013; Cahill, Stevens, Perera, & Lancaster, 2013; Hughes, Stead, Hartmann-Boyce, Cahill, & Lancaster, 2014; Hurt et al., 1997)
	Nicotine replacement therapy (NRT)	Nicotinic cholinergic receptor full agonist	1) Promotes abstinence 2) Reduces cravings 3) Mitigates withdrawal symptoms	Useful for harm reduction (smoking reduction)	(Anthenelli et al., 2016; Cahill et al., 2013; Chen et al., 2015; Stead et al., 2012) • Possible pharmacogenomics influence
	Varenicline	$\alpha 4\beta 2$ nicotinic cholinergic receptor partial agonist	1) Promotes abstinence 2) Reduces cravings 3) Attenuates withdrawal symptoms	Useful for harm reduction (smoking reduction)	(Anthenelli et al., 2016; Cahill et al., 2013; Cahill, Lindson-Hawley, Thomas, Fanshawe, & Lancaster, 2016; Chen et al., 2015; Lassi et al., 2016; Picciotto et al., 1998; Tapper et al., 2004) • More efficacious than bupropion or NRT • No pharmacogenomic response identified to date

GABA, gamma-aminobutyric acid; NMDA, N-methyl-d-aspartate; PO, per os or peri oral; IM, intramuscular; $\alpha 4\beta 2$, alpha 4 beta 2.

TABLE 28.2   Examples of Advances in Translational Neuroscience That May Unveil Future, Therapeutic Targets

Drug	Translational neuroscience models and targets	Potential clinical utility	References
Alcohol	Animal and postmortem studies have demonstrated dynamic regulation of D1 dopamine receptor and dopamine transporter (DAT) during prolonged abstinence	Targeting the hyperdopaminergic state in prolonged abstinence may be an avenue for preventing relapse	(Hirth et al., 2016)
	Increased histone deacetylase (HDAC) activity is correlated with increased anxiety and alcohol withdrawal behavior in rodent models, which can be reversed with HDAC inhibition	HDAC inhibition may provide a useful therapeutic target for attenuate prolonged alcohol withdrawal and thus reducing likelihood of relapse	(Moonat, Sakharkar, Zhang, Tang, & Pandey, 2013; You, Zhang, Sakharkar, Teppen, & Pandey, 2014)
	Decreased expression of histone methyltransferase Prdm2 found in alcohol-dependent rats; Prdm2 knockdown rats exhibit similar, alcoholic behaviors	Targeting PRDM2 expression may provide a novel, therapeutic target for the treatment of alcoholism, in both reduction of alcohol consumption and reducing likelihood of relapse	(Barbier et al., 2016)
Cocaine	Late positive potential encephalographic tracings, a marker of cocaine cravings, remain increased up to 6 months of abstinence from cocaine	Provides biomarker for persistence of craving that may have clinical utility in relapse prevention	(Parvaz, Moeller, & Goldstein, 2016)
	Parallel models in rhesus monkeys and humans using choice paradigms between cocaine and an alternate reward demonstrated similar	High-fidelity, translational model for testing efficacy of new therapeutic agents	(Johnson et al., 2016; Lile et al., 2016)
Cannabis	Ro 61-8048, an inhibitor of kynurenic acid, a negative allosteric modulator of α7-nicotinic acetylcholine receptors (α7 nAChR), which modulate effects of THC, reduced the rewarding effects of THC and blocked drug-induced and cue-induced relapse in primate models	α7 nAChRs in primate models may provide a neuroscience-informed therapeutic target	(Justinova et al., 2013)
Gambling disorder	Parallel models of the Iowa gaming task in rats implicate role of serotonin and dopamine in pathological gambling	Modulation of serotonin and dopamine may provide avenues for new therapeutic targets for pathological gambling	
	Human subjects with gambling disorder show decreased activation of the ventral striatum and ventromedial prefrontal cortex (vmPFC)	Modulation of reward pathways (e.g., with MOR antagonist naltrexone) or therapies targeted at reducing impulse control may prove beneficial in gambling disorder	(Potenza, 2013b; Reuter et al., 2005)
Food addiction	1) Rats exhibit binge-like tendencies when exposed to high sucrose solutions following a fast and demonstrate an opioid-like withdrawal syndrome when solution is removed or given naloxone 2) Intake of refined sugars and fats increases endogenous opioid-related gene expression in the ventral striatum and hypothalamus as well has D2R downregulation in striatum	1) Naltrexone and the GABA-B receptor agonist baclofen have been shown to reduce hedonic eating in murine models 2) Bupropion and naltrexone have shown efficacy in weight reduction and are FDA approved for obesity	(Avena, Bocarsly, Murray, & Gold, 2014; Smith & Robbins, 2013; Wang et al., 2014)

III ALCOHOL

Alcohol drinking is an immensely complex, human behavior, and thus animal models pose obvious limitations when studying alcoholism, namely, the lack of face and construct validity. However, McBride and colleagues have circumvented this issue in (McBride, Rodd, Bell, Lumeng, & Li, 2014) developing two similar strains of alcoholic rats, the alcohol-preferring (P) rats and the high-alcohol-drinking (HAD) rats that, when compared with the DSM-5 diagnostic criteria for alcohol use disorder (American Psychiatric Association, 2013), show

evidence of tolerance, the loss of control (rats will readily drink to intoxication), a great deal of time spent drinking (will binge when access is scheduled periodically throughout the day), cravings (manifested by robust seeking behavior after prolonged rest following extinction), and physical stigmata of withdrawal providing excellent face validity. Correspondingly, chronic or binge-like drinking in P rats has been shown to reduce D2 autoreceptor function, elevate extracellular DA levels, and increase DA reuptake, thus providing construct validity for a model of chronic alcoholism (Bell et al., 2016). Changes parallel to those in the dopaminergic system are also observed in the noradrenergic, glutamatergic, opioid, and corticotropin releasing factor (CRF) systems.

Animal models like McBride's have elucidated neurobiological substrates for therapeutic targets in alcoholism, and exploitation of these neurotransmitter systems and neurocircuitry continues to be of capital importance in the development of new pharmacotherapy options. Currently, three Food and Drug Administration (FDA) medications are approved for alcohol use disorder: naltrexone, acamprosate, and disulfiram.

Disulfiram is a competitive inhibitor of the enzyme acetaldehyde dehydrogenase. After ingestion, alcohol is converted to acetaldehyde via alcohol dehydrogenase, which is then converted into acetic acid, which can then be excreted (Hempel et al., 1984) Inhibition of acetaldehyde dehydrogenase results in an accumulation of acetaldehyde, resulting in a characteristic clinical syndrome of vital sign instability, nausea, vomiting, flushing, malaise, and possibly cardiovascular effects. Clinical data indicate that disulfiram improves short-term abstinence, increases days until relapse, and reduces number of heavy drinking days. However, given issues with compliance and its potentially serious side effects, it should only be given to a motivated patient under close supervision (Jorgensen et al., 2011).

Naltrexone is a mu-opioid receptor antagonist that has been employed, in both its oral and long-acting injectable form as a treatment of alcoholism. The endogenous opioid system has long been implicated as a crucial mediator of euphoria from alcohol consumption, putatively through the increase in phasic dopamine release in the ventral striatum, which has been shown in both human and animal models (Self & Nestler, 1995). Accordingly, naltrexone administration has been shown in animal models to reduce drinking, possibly mediated through reduction in ethanol-induced tyrosine hydroxylase (the enzyme that catalyzes the committed step of dopamine synthesis) mRNA expression (Lee et al., 2005). Clinical trials have shown varying results with a number of outcomes, but a consistent finding is delay in return to drinking after abstinence (i.e., relapse) and to a lesser extent reduction in heavy drinking days (Jonas et al., 2014). Reasons for variability in study response have been proposed but have been largely inconsistent in prospective studies (Garbutt et al., 2016). For example, retrospective data suggested that the functional polymorphism rs1799971 in the mu-opioid receptor gene *OPRM1* was associated with reduced risk to relapse, but this was not confirmed prospectively (Oslin et al., 2015). However, the rs1799971 SNP (single nucleotide polymorphism) may interact with the DAT1 SNP rs28363170 with naltrexone vis-à-vis reduced cue-mediated ventral striatal activation, affecting dopaminergic neurotransmission (Schacht et al., 2013).

Acamprosate, a derivative of the amino acid taurine, is structurally similar to gamma-aminobutyric acid (GABA). Chronic alcohol use results in downregulation of GABA-A receptors and upregulation of NMDA glutamate receptors, resulting in an excitotoxic state that may manifest as a protracted withdrawal syndrome. Acamprosate is thought to indirectly upregulate GABA and likewise indirectly downregulate NMDA glutamate neurotransmission, possibly through antagonism of the metabotropic glutamate receptor type 5 (mGluR5). In animal models of alcohol withdrawal, acamprosate has been shown to reduce anxiety and seizure-like symptoms. Like naltrexone, acamprosate's efficacy in clinical trials is varied, though its most pronounced effects appear to be, like naltrexone, extension of time to relapse. Effects on cravings and reduction in heavy drinking days have been inconsistent (Jonas et al., 2014; Kalk & Lingford-Hughes, 2014).

Future Directions

Pathological drinking leading to alcoholism is a very complex set of behaviors mediated by an even more complex set of neurobiological underpinnings. Referring back to the three-stage model of addiction, FDA-approved pharmacotherapies appear to target binge/intoxication phase (naltrexone) and preoccupation/anticipation/craving phase (naltrexone and questionably acamprosate), whereas any effect on withdrawal/negative affect appears negligible (though could be theoretically targeted by acamprosate). Emerging evidence suggests that the withdrawal/negative affect (especially during protracted abstinence) stage may involve the dopaminergic system in a rather complex manner. While human PET studies have shown that chronic alcohol use results in low extracellular dopamine levels and chronic D2 receptor downregulation, recent postmortem and animal data suggest that prolonged abstinence is characterized by dynamic regulation of D1 dopamine receptors and the dopamine transporter (DAT), resulting in a hyperdopaminergic state that may predict vulnerability to relapse (Hirth et al., 2016), which could be a further avenue for therapeutic intervention.

Lastly, epigenetics has unveiled an entirely new set of pharmacotherapeutic targets in animal models. Recent evidence implicates the role of increased histone deacetylase 2 (HDAC)-mediated histone modifications in anxiety-like and alcohol withdrawal behaviors in animal models (Moonat et al., 2013), the behavioral and neuropathologic manifestations of which can be reversed with HDAC inhibition (You et al., 2014). More recently, decreased expression of histone methyltransferase PR domain zinc finger protein 2 (Prdm2) was found in alcohol-dependent rats, and Prdm2 knockdown in non-alcohol-dependent rats produced a similar phenotype including increased aversion-resistant alcohol intake, increased self-administration, and enhanced stress-induced relapse (Barbier et al., 2016). Taken together, exploitation of epigenetic mechanisms may provide more refined, therapeutic targets for each of the three phases of addiction.

IV OPIOIDS

Crucial to the understanding of the current opioid epidemic and rational treatment is a sophisticated view of the endogenous opioid system. *Opioids* comprise both naturally occurring and synthetic compounds that bind to and activate any of the three known opioid receptors: mu (MOR), delta (DOR), and kappa (KOR). *Opiates* are structural derivatives of the opium plant *Papaver somniferum* and include codeine and morphine. Though perhaps the most known and therapeutic exploitation of the opioid receptors is analgesia, the downstream effects of opioid receptor agonism have profound effects on mood and behavior (Chartoff & Connery, 2014).

The mu-opioid receptor is most commonly implicated in addiction and is distributed throughout the brain both presynaptically, inhibiting neuronal transmission, or postsynaptically, regulating signal transduction pathways. Animal models have demonstrated that MORs are heavily involved in the regulation of NAc GABA-ergic neurons, VTA DA neurons, and prefrontal cortical (PFC) glutamatergic neurons. For example, rats will self-administer morphine directly into the VTA and NAc, with acute morphine causing a dopaminergic surge in the NAc. Additionally, naloxone, a MOR antagonist, will precipitate withdrawal most robustly in the locus coeruleus and periaqueductal gray (PAG) and cause conditioned place aversions when injected into the NAc (Chartoff & Connery, 2014; Koob, Maldonado, & Stinus, 1992). Increasingly, cross talk between MOR-G-coupled protein signaling and glutamatergic neurotransmission is being understood to provide causal mechanisms explaining the phenomena of opioid use, specifically correlating to the "three stage" model addressed earlier. Binge usage and the "rush" of IV use may be reflective of MOR's rapid effects on cellular activity. The transition from impulsive use to compulsive use despite negative consequences may be mediated by a delay in MOR-G-coupled proteins modulating the neurocircuitry of reward. Finally, MOR-G-coupled protein facilitation of synaptic plasticity, maintaining drug-related memories, may explain the phenomenon of contextual cues leading to relapse (Chartoff & Connery, 2014).

Given its central role in opioid addiction, the MOR is the principal target for pharmacotherapy maintenance treatment. Current FDA-approved medications for opioid use disorder include methadone, a full MOR agonist; buprenorphine, a MOR partial agonist; and naltrexone (in both oral and intramuscular forms), a MOR antagonist. Both methadone and buprenorphine have been shown to improve abstinence, retention in treatment, and reduce transmission of HIV and hepatitis C. When compared directly with buprenorphine, methadone had higher treatment retention rates. Multiple studies demonstrate that when either methadone or buprenorphine is stopped, however, relapse rates dramatically increase (Connery, 2015).

Both medications are tightly regulated; buprenorphine prescribers must obtain a waiver and are capped at the number of patients they can carry, and methadone must be dispensed at licensed clinics. While buprenorphine, due to its partial agonist properties, has a "ceiling effect," making overdoses less common, methadone, as a full agonist, has significant abuse and overdose potential and, given its long half-life, may pose a lethal threat when combined with GABA-ergic drugs such as benzodiazepines. Additionally the kappa antagonist properties of buprenorphine may provide an antidepressant effect (Karp et al., 2014). Buprenorphine can also be combined with naloxone, which, when given sublingually, is biologically inactive but when used inappropriately (i.e., intravenously) becomes active and serves as an abuse deterrent. Both methadone and buprenorphine are considered safe in pregnancy: buprenorphine causes less severe fetal-iatrogenic effects (i.e., reduced duration of neonatal abstinence syndrome), yet treatment retention is greater with methadone (Connery, 2015). In 2016 the FDA approved a subdermal implant formulation of buprenorphine in which drug concentration is maintained for 6 months, which may reduce tendency towards abuse and diversion and likewise expand access to care (American Medical Association, 2016; Sigmon & Bigelow, 2016).

Oral naltrexone has been shown to have poor compliance rates with severe relapse resulting in increased morbidity and mortality. Intramuscular naltrexone has relatively weaker evidence than agonist/partial agonist therapy, with FDA approval granted following the publications of one randomized clinical trial (Krupitsky et al., 2011) with an open-label extension phase (Krupitsky et al., 2013) demonstrating efficacy. Recently, two randomized controlled trials have demonstrated similar efficacy between buprenorphine and naltrexone; however, issues with naltrexone initiation (i.e., the patient must be abstinent from opioids 7–10 days before the injection is given) accounted for substantial recidivism in subjects receiving naltrexone (Lee et al., 2017; Tanum, Solli, Latif, et al., 2017). Nevertheless, this induction period may be shortened and facilitated with single-day buprenorphine dosing and a naltrexone bridge (occurring over 7 days with other supportive medications including clonidine, clonazepam, zolpidem, prochlorperazine, and trazodone) (Sullivan et al., 2017) In contrast to agonist treatment, naltrexone has no abuse potential and does not decrease respiratory drive (Connery, 2015). A trial comparing intramuscular naltrexone with buprenorphine is ongoing (Lee et al., 2016).

Referring back to the "three stage" model, one can easily see that available pharmacotherapy targets the binge/intoxication stage (i.e., reduces cravings with substitution therapy and limits euphoria with partial agonist/antagonist therapy) and preoccupation stages (through MOR direct and partial agonism). However, aside from possibly buprenorphine's kappa antagonist effects, absent is any pharmacotherapy that directly targets withdrawal/negative affect, which has been shown to be under top-down (i.e., cortical) control (Koob & Le Moal, 2001). Powerful opioid drugs repeatedly target the brain's euphoria and motivational centers causing dysphoria and anhedonia. The idea that all of the drug's effects on the brain and receptor systems are limited to the presence of the offending drug in the body seems misguided; drugs of abuse may have effects more akin to traumatic brain injury or concussion (Gold et al., 2009). Additionally, depression often is cooccurring and may be missed due to treatment focus on drug misuse/dependence (Gold, 2007). Thus that numerous controlled trials for buprenorphine (Weiss et al., 2011; Woody et al., 2008) and methadone (Magura & Rosenblum, 2001; Masson et al., 2004) have resulted in high relapse rates with medication discontinuation that is not surprising—it is indeed a foregone conclusion. Thus novel development of novel pharmacotherapeutic tools based on translational neuroscience that target alternate mechanisms, for example, the cortical and subcortical structures involved in prolonged withdrawal and negative affect, is prudent.

V PSYCHOSTIMULANTS: COCAINE AND AMPHETAMINES

While no pharmacologic treatments are currently FDA approved for cocaine or methamphetamine addiction, classical clinical and preclinical work from the past 30 years may provide insights for pharmacotherapeutic targets. In 1985 Gold and Dackis famously described the dopamine depletion hypothesis as an underlying mechanism for dysphoria and anhedonia that occurs with abrupt cessation of cocaine use (Dackis & Gold, 1985), further extended by Volkow et al. in classic PET imaging studies showing downregulation of postsynaptic D2 receptors (D2R) following the acute withdrawal phase in chronic cocaine addicts (Volkow et al., 1990). Nevertheless, this period was characteristically short lived, as evidenced by return of postsynaptic D2R levels following one-month abstinence (Volkow et al., 1990). Additionally, in 1986, Gawin and Kleber described, based on clinical interviews, that in cocaine addicts, cue-induced craving increased during initial abstinence from cocaine

and would remain high even during periods of prolonged abstinence (up to 28 weeks) (Gawin & Kleber, 1986), which was later replicated in animal models (Li, Venniro, & Shaham, 2016). This phenomenon is called the "incubation of craving" (Li et al., 2016). Excitingly, in 2016, these findings were extended using human electrophysiological measures, demonstrating that late positive potential (LPP) of electroencephalographic (EEG) tracings, a reliable and valid marker of motivated attention to salient stimuli, increased in the initial week of abstinence and peaks at 6 months, irrespective of subjective reports of craving (Parvaz et al., 2016). Thus translational neuroscience-informed treatment would dictate particular attention to continued intervention (e.g., for at least 6 months) following the acute withdrawal period, given changes persistent in underlying neurobiology, resulting in vulnerability to relapse.

Another issue in the development of pharmacotherapeutics for cocaine addiction, as is the case in all other addictions, is the need for high-fidelity, translational neuroscience-based screening models. Conferred in this need is a paradigm involving nonhuman primates given phylogenic similarities and translatable experimental protocols that minimize variability in procedures and variables. Recently, parallel models in rhesus monkeys and humans were developed using a choice paradigm and positive ratio schedule with cocaine and an alternate reward (food for rhesus monkeys and monetary compensation for humans), resulting in similar patterns of cocaine usage between species (Johnson et al., 2016; Lile et al., 2016). This paradigm of cross species validation provides an indispensable translational tool for the evaluation of efficacy of new therapeutic agents.

VI MARIJUANA

While no pharmacologic treatment for marijuana addiction is FDA approved, animal models have been created to explore underlying neurobiology and mechanism of reward; however, limitations exist regarding route of delivery and substance itself—most animal models use intravenous delta-9-tetrahydrocannabinol (THC), the principal psychoactive medication in marijuana rather than a smoked form of the compound itself. Some of these issues are circumvented in primate models using (1) a clear THC solution that is similar in dose compared to marijuana smoked by humans, (2) a rapid rate of infusion that parallels rapid delivery through smoke inhalation, and (3) a variable dosing schedule that encourages changes in sensitivity to dose. This final point is crucial for a model with construct, face, and predictive validity because it is necessary when evaluating reward, or even potential pharmacotherapeutic agents, to show response when the value of reward (i.e., dose) is altered

(Panlilio, Justinova, & Goldberg, 2010). In addition to IV infusion models, THC smoke exposure models have been developed that have face and predictive validity regarding locomotor activity in open-field (increased and then decreased) testing and dependence (as evidenced by rimonabant precipitated withdrawal—see succeeding text) (Bruijnzeel et al., 2016).

One class of agents that has been explored is the cannabinoid receptor antagonist class. Rimonabant, a CB1 receptor antagonist/inverse agonist, has been show in animal models to block the rewarding effect of THC, lessen the salience of environmental cues associated with drug abuse, and block both drug-induced and cue-induced relapse in animal models. Rimonabant actually entered clinical trials as an antiobesity agent but was never FDA approved due to serious, depression-like side effects (Le Foll, Gorelick, & Goldberg, 2009). More recently, high-fidelity models have been used to explore alternate and/or related circuitry. For example, the α7-nicotinic acetylcholine receptors (α7 nAChRs) are known to modulate the effects of THC, and recently, Ro 61-8048, an inhibitor of α7 nAChR-negative allosteric modulator kynurenic acid, reduced the rewarding effects of THC and blocked drug-induced and cue-induced relapse in primate models (Justinova et al., 2013). Taken together, these results suggest that high-fidelity, translational neuroscience models, though imperfect in their replication of smoking behavior, may be useful models for targeting cannabinoid reward and relapse.

VII TOBACCO

Nicotine, the putative, addictive component of tobacco smoke, has been well studied in a variety of translational models. Using intravenous drug administration models in operant conditioning paradigms, animals and humans will consistently respond, self-administering nicotine over placebo. Conditioned place preference (CPP) is another paradigm, using classical conditioning, in which the drug and placebo are paired with distinctive environments, allowing the animals to choose the environment they enter. Dopamine-mediated learning (incentive salience) is thought to reflect preference for the area paired with the drug, even when the drug is removed, a phenomenon well characterized in rodents exposed to nicotine. Finally, nicotine withdrawal, a major predictor of relapse, can be modeled in animals through either cessation of continuous delivery or the administration of a nicotinic acetylcholine receptor antagonist. Nicotine withdrawal signs in animals (bruxism, psychomotor overactivation, avoidance behavior, and reduced forced swim time [FST]) are paralleled in humans (specifically, anxiety, irritability, and depression) (Le Foll & Goldberg, 2009).

Currently, three FDA-approved treatments are available for nicotine/tobacco addiction, nicotine replacement therapy (NRT), bupropion, and varenicline (Cahill et al., 2013; Cahill et al., 2016; Hughes et al., 2014; Stead et al., 2012). In the recent multicenter EAGLES trial, all three were found to be safe and efficacious, with varenicline showing greater efficacy than bupropion or NRT, both of which were more efficacious than placebo (Anthenelli et al., 2016). Of these, varenicline is the result of neuroscience-informed drug development. Varenicline is an α4β2 nicotinic cholinergic receptor partial agonist, and translational models had previously implicated the nicotinic cholinergic receptor's α4 and β2 subunits, particularly in the VTA, as indispensable for the reinforcing effects of nicotine (Picciotto et al., 1998; Tapper et al., 2004). Evidence for bupropion, an antidepressant with the mechanism of blocking reuptake of norepinephrine (NE) and DA, as a smoking cessation agent emerged during a multitude of studies linking smoking with major depressive disorder (MDD), with the proposed mechanism of action being a repletion of mesolimbic dopamine that is absent following abrupt nicotine cessation (Hurt et al., 1997). Nicotine replacement delivers nicotine during times of abstinence to mitigate physical and psychological withdrawal symptoms, increasing the chance of remaining abstinent. Though different forms of NRT (e.g., patch, lozenge, and inhaler) exist with varying rates of delivery and ease of use, they all seem to be efficacious in promoting smoking abstinence (Stead et al., 2012).

Recently, advances in whole genome sequencing and genome-wide association studies (GWAS) have identified further avenues for the development of new pharmacotherapeutics and personalized medicine. The CHRNA5-A3-B4 gene cluster, which codes for nicotinic receptor subunits α3, α5, and β4, at single-nucleotide polymorphism (SNP) rs10730, which is in high linkage disequilibrium with rs16969968, is associated with nicotine dependence and smoking heaviness (Lassi et al., 2016). In a recent trial the rs16969968 SNP was associated with improved response to NRT, whereas in the same study, although varenicline increased abstinence, its effects did not vary by genotype (Chen et al., 2015); however, differences in experimental conditions and placebos used for each matched control may have confounded results. Nevertheless, use of sophisticated sequencing tools in collaboration with high-fidelity translational neuroscience models could be most beneficial in developing new therapeutics for tobacco addiction.

VIII NMDA ANTAGONISTS: KETAMINE AND PCP

Both phencyclidine (PCP) and ketamine are antagonists of the N-methyl-D-aspartate (NMDA) glutamate receptor and can produce a severe toxidrome consisting of autonomic hyperactivity, agitation, psychosis mimicking schizophrenia, delirium, catatonia, seizures, and coma (Dinwiddie & Farber, 1995). While these are not among the commonest drugs abused, their mechanisms are well characterized and have important implications in current neuroscience and clinical research given the NMDA hypofunction hypothesis of psychosis (Farber, 2003) and the treatment of major depressive disorder (MDD) (Caddy et al., 2015). Using murine models, Farber and Olney have described the downstream consequences of NMDA receptor hypofunction. Glutamate, via agonism of the NMDA receptors on GABA-ergic neurons, maintains tonic inhibition over three excitatory posterior cingulate (PC)/retrosplenial (RS) neurons: muscarinic cholinergic, glutamatergic, and a pathway involving neuropeptide Y (NPY) modulation of the sigma receptor. Additionally, glutamate directly modulates noradrenergic neurons that release norepinephrine at the α2 receptor on muscarinic cholinergic neurons that project to the PC/RS cortices (Olney & Farber, 1995). Thus exploiting these mechanisms, using GABA-ergic, antimuscarinic, and alpha-2-adrenergic agents, might prove fruitful for relieving the intoxicating and psychotomimetic effects of NMDA receptor antagonist ingestion (Farber, 2003; Farber, Foster, Duhan, & Olney, 1995). We must reiterate, however, that these neuroscience-informed therapeutic agents solely relieve the effects of intoxication, which can indeed be life-threatening, but do not address the other stages of addiction, specifically withdrawal/negative affect, that truly drive relapse.

IX BEHAVIORAL AND PROCESS ADDICTIONS

Process or behavioral addictions describe a series of impulsive turned compulsive behaviors that result in subjective effects of reward that continue despite negative consequences. Some examples are addiction to gambling, hedonic eating, sex and sexual behavior, shopping, exercise, and various forms of internet use (Sussman, Lisha, & Griffiths, 2011). While the validity of the conceptualization of these behaviors as addictions is not without controversy (Karim & Chaudhri, 2012), the DSM-5 marked an epoch in the annals of the field with an overwhelming acceptance of gambling disorder (GD) as a "substance-related and addictive disorder" (American Psychiatric Association, 2013), whereas in DSM-IV-TR, it was classified as "impulse-control disorder not otherwise specified (NOS)" (American Psychiatric Association, 2000). In this section, we will discuss translational models of pathological gambling and food addiction given the exquisitely delineated neurobiology and implications for treatment.

A Gambling Disorder

Several murine models of GD exist, but perhaps the model with the best face, predictive, and possibly construct validity is a modified Iowa gaming task (IGT). The IGT was developed for humans to explore decision-making: the subject is tasked with avoiding short-term, immediate high gain with unpredictable penalties in favor of smaller, immediate gains that maximize long-term benefit with minimal penalties in the form of card-based tasks. This model has been extended to rats and is thought to be related to reward hypersensitivity, vis-à-vis interindividual differences among rats (Rivalan, Ahmed, & Dellu-Hagedorn, 2009). Further work on this model has shown that manipulating the serotonergic and dopaminergic tone can change performance: Dopamine blockade can improve performance, and serotonergic (5-HT1a) agonism and dopamine reuptake inhibition may worsen performance, possibly elucidating neurobiological substrates of pathological gambling in murine models.

Several findings from the human literature are also worth discussing. In humans, emerging evidence suggests the central role of dopamine in GD largely based on a strong association between dopamine agonist treatment and pathological gambling in patients with Parkinson's disease. However, PET imaging results have been inconsistent, and manipulation of different dopamine receptors other than D2 or D3 (e.g., D4) might have differential effects on GD (Potenza, 2013a). Interestingly, fMRI work has shown decreased activation of the ventral striatum and ventromedial prefrontal cortex (vmPFC, associated with impulse control) in human subjects who exhibited pathological gambling (Reuter et al., 2005). Curiously the opioid antagonist naltrexone has been shown in numerous, randomized clinical trials to reduce gambling behaviors, suggesting a possible role of the endogenous opioid system in GD. Incidentally, success with naltrexone in GD has been linked with a positive family history of alcoholism and strong urges to gamble at the incipiency of treatment, suggesting that important genetic and individual (possibly temperamental) differences may affect treatment response (Potenza, 2013b).

B Food Addiction

While not explicitly indexed as a disorder in DSM-5, food addiction is a seminal construct in the process addiction literature that has been validated using the Yale Food Addiction Scale (YFAS) (Gearhardt, Corbin, & Brownell, 2016), the current version of which reflects DSM-5 criteria (American Psychiatric Association, 2013). Human phenomenology aside, murine models have elucidated some of the key behavioral and neurobiological constructs that define food addiction. For example, when certain rats are exposed to high sucrose solution, they exhibit binge-like tendencies at an intermittent schedule particularly following a fast, and when either the solution is removed or the rats are given the mu-opioid receptor antagonist naloxone, a characteristic, opioid-like withdrawal syndrome is produced. Additionally, self-administration protocols, well validated in animal models of drug abuse, have shown similar patterns regarding sucrose; rats will compulsively press a lever to obtain sucrose after a period of devaluation (Smith & Robbins, 2013).

Neurobiological correlates that inextricably link food addiction with substance use disorders have become increasingly clarified. First, intake of highly palatable foods (refined sugars and fats) increases hypothalamic endorphin release and enkephalin-related gene expression in the ventral striatum, clearly marking the role of the endogenous opioid system in hedonic eating. Additionally, obese individuals tend to show decreased striatal D2 receptors, mirroring well-established findings in substance use disorders (Smith & Robbins, 2013). Exploitation of these findings has led to preliminary work demonstrating that the combination of naltrexone and baclofen, a GABA-B receptor agonist that has shown efficacy in some substance use disorders and binge intake of fat, can reduce overconsumption of highly palatable foods (Avena et al., 2014). Furthermore, a bupropion/naltrexone combination has shown efficacy in weight reduction, possibly due to a bupropion-mediated attenuated hypothalamic response to food cues and enhanced activation in areas associated with inhibitory control (the anterior cingulate cortex), internal awareness (the superior frontal gyrus, insular cortex, and superior parietal lobe), and memory (the hippocampus) regions (Wang et al., 2014). Incidentally the naltrexone-bupropion combination is an FDA-approved treatment for obesity (Heymsfield & Wadden, 2017).

X CONCLUSION

In this chapter, we have shown how translational neuroscience can inform rationally designed therapeutics for addictive disorders, but in describing current progress in the context of the Koob-Volkow "three stage" model, we identify significant shortcomings. Specifically, most available therapeutics (excepting potentially, albeit weakly, varenicline, buprenorphine, and acamprosate) marginally, if at all, target the postacute withdrawal and negative affect phases. George Koob describes this negative emotional state that persists during prolonged abstinence as the "dark side of addiction" that "provides a powerful source of negative reinforcement that defines compulsive drug-seeking behavior and addiction and contributes to relapse" (Koob, 2016), a theme that we have hoped to emphasize throughout this chapter.

In the 1990s Markou and Koob developed antidepressant screening models based on cocaine (and later amphetamine) withdrawal syndromes (Lin, Koob, & Markou, 1999; Markou, Hauger, & Koob, 1992; Markou & Koob, 1991). Hence, the notion that the psychostimulant withdrawal state is characterized by depression, anhedonia, and suicidality should not be of surprise. In the new Addictions Neuroclinical Assessment (ANA) that proposes a novel, neuroscience-informed framework for conceptualizing addiction, the negative emotionality domain implicates hypohedonia as a marquee factor in relapse (Kwako, Momenan, Litten, Koob, & Goldman, 2016). Additionally, underlying negative emotionality involves an overactivity of the brain "stress system" involving neurotransmitter systems including CRF, dynorphin, norepinephrine, hypocretin, substance P, and vasopressin and an under-activity of the brain "antistress" systems including neuropeptide Y, nociception, endocannabinoids, and oxytocin (Kwako et al., 2016). Clearly many systems are disturbed by substance use, abuse, and dependence, but few patients, other than physicians with addictive disorders who enter 5-year monitoring contracts with state physician health programs (PHPs) (DuPont, McLellan, White, Merlo, & Gold, 2009), are studied sufficiently in a rigorous, longitudinal manner. Given advances in neuroscience since Markou and Koob's protracted withdrawal models in the 1990s and the reconceptualization of addictive disorders, both phenomenologically and neurobiologically, as described in the ANA, we propose that to develop more effective treatments to address America's number one public health problem, high-fidelity animal models should be developed that address treatment needs, with corresponding, adequately powered, longitudinal human studies, specifically related to protracted abstinence, anhedonia, and negative emotionality.

The Use of Translational Medicine Approaches in Drug Development for the Treatment of Addictive Disorders

References

American Medical Association. (2016). Buprenorphine implants (probuphine) for opioid dependence. *JAMA*, 316(17), 1820–1821.

American Psychiatric Association. (2000). *Diagnostic and statistical manual of mental disorders* (4th ed. text revision). Washington, DC: American Psychiatric Publishing.

American Psychiatric Association. (2013). *Diagnostic and statistical manual for of mental disorders 5th edition*. Arlington, VA: American Psychiatric Publishing.

Anthenelli, R. M., Benowitz, N. L., West, R., St Aubin, L., McRae, T., Lawrence, D., et al. (2016). Neuropsychiatric safety and efficacy of varenicline, bupropion, and nicotine patch in smokers with and without psychiatric disorders (EAGLES): a double-blind, randomised, placebo-controlled clinical trial. *The Lancet*, 387(10037), 2507–2520.

Avena, N. M., Bocarsly, M. E., Murray, S., & Gold, M. S. (2014). Effects of baclofen and naltrexone, alone and in combination, on the consumption of palatable food in male rats. *Experimental and Clinical Psychopharmacology*, 22(5), 460–467.

Barbier, E., Johnstone, A. L., Khomtchouk, B. B., Tapocik, J. D., Pitcairn, C., Rehman, F., et al. (2016). Dependence-induced increase of alcohol self-administration and compulsive drinking mediated by the histone methyltransferase PRDM2. *Molecular Psychiatry*, 22, 1746–1758.

Bell, R. L., Hauser, S., Rodd, Z. A., Liang, T., Sari, Y., McClintick, J., et al. (2016). Chapter seven—a genetic animal model of alcoholism for screening medications to treat addiction. In L. B. Richard & R. Shafiqur (Eds.), *Vol. 126. International review of neurobiology* (pp. 179–261): Academic Press.

Bruijnzeel, A. W., Qi, X., Guzhva, L. V., Wall, S., Deng, J. V., Gold, M. S., et al. (2016). Behavioral characterization of the effects of cannabis smoke and anandamide in rats. *PLoS One*, 11(4), e0153327.

Caddy, C., Amit, B. H., TL, M. C., Rendell, J. M., Furukawa, T. A., R, M. S., et al. (2015). Ketamine and other glutamate receptor modulators for depression in adults. *The Cochrane Database of Systematic Reviews*, 9, Cd011612.

Cahill, K., Lindson-Hawley, N., Thomas, K. H., Fanshawe, T. R., & Lancaster, T. (2016). Nicotine receptor partial agonists for smoking cessation. *The Cochrane Database of Systematic Reviews*, 5, Cd006103.

Cahill, K., Stevens, S., Perera, R., & Lancaster, T. (2013). Pharmacological interventions for smoking cessation: an overview and network meta-analysis. *The Cochrane Database of Systematic Reviews*, 5, Cd009329.

Center for Behavioral Health Statistics and Quality. (2016). *Results from the 2015 National survey on drug use and health: Detailed tables*. Rockville, MD: Substance Abuse and Mental Health Services Administration.

Chartoff, E. H., & Connery, H. S. (2014). It's MORe exciting than mu: crosstalk between mu opioid receptors and glutamatergic transmission in the mesolimbic dopamine system. *Frontiers in Pharmacology*, 5, 116.

Chen, L. S., Baker, T. B., Jorenby, D., Piper, M., Saccone, N., Johnson, E., et al. (2015). Genetic variation (CHRNA5), medication (combination nicotine replacement therapy vs. varenicline), and smoking cessation. *Drug and Alcohol Dependence*, 154, 278–282.

Connery, H. S. (2015). Medication-assisted treatment of opioid use disorder: review of the evidence and future directions. *Harvard Review of Psychiatry*, 23(2), 63–75.

Dackis, C. A., & Gold, M. S. (1985). New concepts in cocaine addiction: the dopamine depletion hypothesis. *Neuroscience & Biobehavioral Reviews*, 9(3), 469–477.

Dinwiddie, S. H., & Farber, N. B. (1995). Pharmacological therapies of cannabis, hallucinaogens, phencyclidine, and volatile solvent addictions. In N. S. Miller & M. S. Gold (Eds.), *Pharmacological therapies for drug & alcohol addictions*. New York: Marcel Dekkker, Inc.

DuPont, R. L., McLellan, A. T., White, W. L., Merlo, L. J., & Gold, M. S. (2009). Setting the standard for recovery: physicians' health programs. *Journal of Substance Abuse Treatment*, 36(2), 159–171.

Farber, N. B. (2003). The NMDA receptor hypofunction model of psychosis. *Annals of the New York Academy of Sciences*, 1003, 119–130.

Farber, N. B., Foster, J., Duhan, N. L., & Olney, J. W. (1995). alpha 2 adrenergic agonists prevent MK-801 neurotoxicity. *Neuropsychopharmacology: Official Publication of the American College of Neuropsychopharmacology*, 12(4), 347–349.

Febo, M., Blum, K., Badgaiyan, R. D., Baron, D., Thanos, P. K., Colon-Perez, L. M., et al. (2017). Dopamine homeostasis: brain functional connectivity in reward deficiency syndrome. *Frontiers in Bioscience (Landmark Edition)*, 22, 669–691.

Garbutt, J. C., Kampov-Polevoy, A. B., Kalka-Juhl, L. S., & Gallop, R. J. (2016). Association of the sweet-liking phenotype and craving for alcohol with the response to naltrexone treatment in alcohol dependence: a randomized clinical trial. *JAMA Psychiatry*, 73(10), 1056–1063.

Gawin, F. H., & Kleber, H. D. (1986). Abstinence symptomatology and psychiatric diagnosis in cocaine abusers. Clinical observations. *Archives of General Psychiatry*, 43(2), 107–113.

Gearhardt, A. N., Corbin, W. R., & Brownell, K. D. (2016). Development of the Yale food addiction scale version 2.0. *Psychology of Addictive Behaviors: Journal of the Society of Psychologists in Addictive Behaviors*, 30(1), 113–121.

Gold, M. S. (2007). Dual disorders: nosology, diagnosis, & treatment confusion – chicken or egg? Introduction. *Journal of Addictive Diseases*, 26(Suppl 1), 1–3.

Gold, M. S., Kobeissy, F. H., Wang, K. K., Merlo, L. J., Bruijnzeel, A. W., Krasnova, I. N., et al. (2009). Methamphetamine- and trauma-induced brain injuries: comparative cellular and molecular neurobiological substrates. *Biological Psychiatry*, 66(2), 118–127.

Hempel, J., von Bahr-Lindstrom, H., & Jornvall, H. (1984). Aldehyde dehydrogenase from human liver. Primary structure of the cytoplasmic isoenzyme. *European Journal of Biochemistry*, 141(1), 21–35.

Heymsfield, S. B., & Wadden, T. A. (2017). Mechanisms, pathophysiology, and management of obesity. *New England Journal of Medicine*, 376(3), 254–266.

Hirth, N., Meinhardt, M. W., Noori, H. R., Salgado, H., Torres-Ramirez, O., Uhrig, S., et al. (2016). Convergent evidence from alcohol-dependent humans and rats for a hyperdopaminergic state in protracted abstinence. *Proceedings of the National Academy of Sciences of the United States of America*, 113(11), 3024–3029.

Hughes, J. R., Stead, L. F., Hartmann-Boyce, J., Cahill, K., & Lancaster, T. (2014). Antidepressants for smoking cessation. *The Cochrane Database of Systematic Reviews*, 1, Cd000031.

Hurt, R. D., Sachs, D. P. L., Glover, E. D., Offord, K. P., Johnston, J. A., Dale, L. C., et al. (1997). A comparison of sustained-release bupropion and placebo for smoking cessation. *New England Journal of Medicine*, 337(17), 1195–1202.

Johnson, A. R., Banks, M. L., Blough, B. E., Lile, J. A., Nicholson, K. L., & Negus, S. S. (2016). Development of a translational model to screen medications for cocaine use disorder I: choice between cocaine and food in rhesus monkeys. *Drug and Alcohol Dependence*, 165, 103–110.

Jonas, D. E., Amick, H. R., Feltner, C., et al. (2014). Pharmacotherapy for adults with alcohol use disorders in outpatient settings: a systematic review and meta-analysis. *Journal of the American Medical Association*, 311(18), 1889–1900.

Jorgensen, C. H., Pedersen, B., & Tonnesen, H. (2011). The efficacy of disulfiram for the treatment of alcohol use disorder. *Alcoholism, Clinical and Experimental Research*, 35(10), 1749–1758.

Justinova, Z., Mascia, P., Wu, H. Q., Secci, M. E., Redhi, G. H., Panlilio, L. V., et al. (2013). Reducing cannabinoid abuse and preventing relapse by enhancing endogenous brain levels of kynurenic acid. *Nature Neuroscience*, 16(11), 1652–1661.

Kalk, N. J., & Lingford-Hughes, A. R. (2014). The clinical pharmacology of acamprosate. *British Journal of Clinical Pharmacology*, 77(2), 315–323.

Karim, R., & Chaudhri, P. (2012). Behavioral addictions: an overview. *Journal of Psychoactive Drugs*, 44(1), 5–17.

Karp, J. F., Butters, M. A., Begley, A. E., Miller, M. D., Lenze, E. J., Blumberger, D. M., et al. (2014). Safety, tolerability, and clinical effect of low-dose buprenorphine for treatment-resistant depression in midlife and older adults. *The Journal of Clinical Psychiatry*, 75(8), e785–e793.

Kolodny, A., Courtwright, D. T., Hwang, C. S., Kreiner, P., Eadie, J. L., Clark, T. W., et al. (2015). The prescription opioid and heroin crisis: a public health approach to an epidemic of addiction. *Annual Review of Public Health*, 36, 559–574.

Koob, G. F. (2016). The dark side of addiction: the Horsley Gantt to Joseph Brady connection. *The Journal of Nervous and Mental Disease*, 205, 270–272.

Koob, G. F., & Le Moal, M. (2001). Drug addiction, dysregulation of reward, and allostasis. *Neuropsychopharmacology: Official Publication of the American College of Neuropsychopharmacology*, 24(2), 97–129.

Koob, G. F., Maldonado, R., & Stinus, L. (1992). Neural substrates of opiate withdrawal. *Trends in Neurosciences, 15*(5), 186–191.

Koob, G. F., & Volkow, N. D. (2010). Neurocircuitry of addiction. *Neuropsychopharmacology: Official Publication of the American College of Neuropsychopharmacology, 35*(1), 217–238.

Koob, G. F., & Volkow, N. D. (2016). Neurobiology of addiction: a neurocircuitry analysis. *The Lancet Psychiatry, 3*(8), 760–773.

Krupitsky, E., Nunes, E. V., Ling, W., Gastfriend, D. R., Memisoglu, A., & Silverman, B. L. (2013). Injectable extended-release naltrexone (XR-NTX) for opioid dependence: long-term safety and effectiveness. *Addiction (Abingdon, England), 108*(9), 1628–1637.

Krupitsky, E., Nunes, E. V., Ling, W., Illeperuma, A., Gastfriend, D. R., & Silverman, B. L. (2011). Injectable extended-release naltrexone for opioid dependence: a double-blind, placebo-controlled, multicentre randomised trial. *Lancet (London, England), 377*(9776), 1506–1513.

Kwako, L. E., Momenan, R., Litten, R. Z., Koob, G. F., & Goldman, D. (2016). Addictions neuroclinical assessment: a neuroscience-based framework for addictive disorders. *Biological Psychiatry, 80*(3), 179–189.

Lassi, G., Taylor, A. E., Timpson, N. J., Kenny, P. J., Mather, R. J., Eisen, T., et al. (2016). The CHRNA5-A3-B4 gene cluster and smoking: from discovery to therapeutics. *Trends in Neurosciences, 39*(12), 851–861.

Le Foll, B., & Goldberg, S. R. (2009). Effects of nicotine in experimental animals and humans: an update on addictive properties. *Handbook of Experimental Pharmacology, 192*, 335–367.

Le Foll, B., Gorelick, D. A., & Goldberg, S. R. (2009). The future of endocannabinoid-oriented clinical research after CB1 antagonists. *Psychopharmacology, 205*(1), 171–174.

Lee, J. D., Nunes, E. V., Mpa, P. N., Bailey, G. L., Brigham, G. S., Cohen, A. J., et al. (2016). NIDA clinical trials network CTN-0051, extended-release naltrexone vs. buprenorphine for opioid treatment (X:BOT): study design and rationale. *Contemporary Clinical Trials, 50*, 253–264.

Lee, J. D., Nunes, E. V., Jr., Novo, P., Bachrach, K., Bailey, G. L., Bhatt, S., et al. (2017). Comparative effectiveness of extended-release naltrexone versus buprenorphine-naloxone for opioid relapse prevention (X:BOT): a multicentre, open-label, randomised controlled trial. *Lancet (London, England), 391*, P309–P318.

Lee, Y. K., Park, S. W., Kim, Y. K., Kim, D. J., Jeong, J., Myrick, H., et al. (2005). Effects of naltrexone on the ethanol-induced changes in the rat central dopaminergic system. *Alcohol and Alcoholism (Oxford, Oxfordshire), 40*(4), 297–301.

Li, X., Venniro, M., & Shaham, Y. (2016). Translational research on incubation of cocaine craving. *JAMA Psychiatry, 73*(11), 1115–1116.

Lile, J. A., Stoops, W. W., Rush, C. R., Negus, S. S., Glaser, P. E., Hatton, K. W., et al. (2016). Development of a translational model to screen medications for cocaine use disorder II: choice between intravenous cocaine and money in humans. *Drug and Alcohol Dependence, 165*, 111–119.

Lin, D., Koob, G. F., & Markou, A. (1999). Differential effects of withdrawal from chronic amphetamine or fluoxetine administration on brain stimulation reward in the rat – interactions between the two drugs. *Psychopharmacology, 145*(3), 283–294.

Magura, S., & Rosenblum, A. (2001). Leaving methadone treatment: lessons learned, lessons forgotten, lessons ignored. *The Mount Sinai Journal of Medicine, New York, 68*(1), 62–74.

Markou, A., Hauger, R. L., & Koob, G. F. (1992). Desmethylimipramine attenuates cocaine withdrawal in rats. *Psychopharmacology, 109*(3), 305–314.

Markou, A., & Koob, G. F. (1991). Postcocaine anhedonia. An animal model of cocaine withdrawal. *Neuropsychopharmacology. Official Publication of the American College of Neuropsychopharmacology, 4*(1), 17–26.

Masson, C. L., Barnett, P. G., Sees, K. L., Delucchi, K. L., Rosen, A., Wong, W., et al. (2004). Cost and cost-effectiveness of standard methadone maintenance treatment compared to enriched 180-day methadone detoxification. *Addiction (Abingdon, England), 99*(6), 718–726.

McBride, W. J., Rodd, Z. A., Bell, R. L., Lumeng, L., & Li, T. -K. (2014). The alcohol-preferring (P) and high-alcohol-drinking (HAD) rats—animal models of alcoholism. *Alcohol, 48*(3), 209–215.

Moonat, S., Sakharkar, A. J., Zhang, H., Tang, L., & Pandey, S. C. (2013). Aberrant histone deacetylase2-mediated histone modifications and synaptic plasticity in the amygdala predisposes to anxiety and alcoholism. *Biological Psychiatry, 73*(8), 763–773.

Murthy, V. H. (2016). Surgeon general's report on alcohol, drugs, and health. *JAMA, 317*, 133–134.

National Drug Intelligence Center. (2014). *National drug threat assessment*. Washington, DC: U.S. Department of Justice.

Office of the Surgeon General. (2016). *Facing addiction in America: The surgeon general's report on alcohol, drugs, and health*. Washington, DC: US Department of Health and Human Services. November 2016. Report No.

Olney, J. W., & Farber, N. B. (1995). NMDA antagonists as neurotherapeutic drugs, psychotogens, neurotoxins, and research tools for studying schizophrenia. *Neuropsychopharmacology: Official Publication of the American College of Neuropsychopharmacology, 13*(4), 335–345.

Oslin, D. W., Leong, S. H., Lynch, K. G., et al. (2015). Naltrexone vs placebo for the treatment of alcohol dependence: a randomized clinical trial. *JAMA Psychiatry, 72*(5), 430–437.

Panlilio, L. V., Justinova, Z., & Goldberg, S. R. (2010). Animal models of cannabinoid reward. *British Journal of Pharmacology, 160*(3), 499–510.

Parvaz, M. A., Moeller, S. J., & Goldstein, R. Z. (2016). Incubation of cue-induced craving in adults addicted to cocaine measured by electroencephalography. *JAMA Psychiatry, 73*(11), 1127–1134.

Picciotto, M. R., Zoli, M., Rimondini, R., Lena, C., Marubio, L. M., Pich, E. M., et al. (1998). Acetylcholine receptors containing the beta2 subunit are involved in the reinforcing properties of nicotine. *Nature, 391*(6663), 173–177.

Potenza, M. (2013a). How central is dopamine to pathological gambling or gambling disorder? *Frontiers in Behavioral Neuroscience, 7*, 206.

Potenza, M. N. (2013b). Neurobiology of gambling behaviors. *Current Opinion in Neurobiology, 23*(4), 660–667.

Reuter, J., Raedler, T., Rose, M., Hand, I., Glascher, J., & Buchel, C. (2005). Pathological gambling is linked to reduced activation of the mesolimbic reward system. *Nature Neuroscience, 8*(2), 147–148.

Rivalan, M., Ahmed, S. H., & Dellu-Hagedorn, F. (2009). Risk-prone individuals prefer the wrong options on a rat version of the Iowa Gambling Task. *Biological Psychiatry, 66*(8), 743–749.

Rudd, R. A., Aleshire, N., Zibbell, J. E., & Gladden, R. M. (2016). Increases in drug and opioid overdose deaths – United States, 2000–2014. *MMWR. Morbidity and Mortality Weekly Report, 64*(50-51), 1378–1382.

Sacks, J. J., Gonzales, K. R., Bouchery, E. E., Tomedi, L. E., & Brewer, R. D. (2015). 2010 National and state costs of excessive alcohol consumption. *American Journal of Preventive Medicine, 49*(5), e73–e79.

Schacht, J. P., Anton, R. F., Voronin, K. E., Randall, P. K., Li, X., Henderson, S., et al. (2013). Interacting effects of naltrexone and OPRM1 and DAT1 variation on the neural response to alcohol cues. *Neuropsychopharmacology: Official Publication of the American College of Neuropsychopharmacology, 38*(3), 414–422.

Self, D. W., & Nestler, E. J. (1995). Molecular mechanisms of drug reinforcement and addiction. *Annual Review of Neuroscience, 18*(1), 463–495.

Sigmon, S. C., & Bigelow, G. E. (2016). Food and drug administration approval of sustained-release buprenorphine for treatment of opioid dependence: realizing its potential. *Addiction (Abingdon, England), 112*, 386–387.

Smith, D. G., & Robbins, T. W. (2013). The neurobiological underpinnings of obesity and binge eating: a rationale for adopting the food addiction model. *Biological Psychiatry, 73*(9), 804–810.

Stahre, M., Roeber, J., Kanny, D., Brewer, R. D., & Zhang, X. (2014). Contribution of excessive alcohol consumption to deaths and years of potential life lost in the United States. *Preventing Chronic Disease, 11,* E109.

Stead, L. F., Perera, R., Bullen, C., Mant, D., Hartmann-Boyce, J., Cahill, K., et al. (2012). Nicotine replacement therapy for smoking cessation. *The Cochrane Database of Systematic Reviews, 11,* Cd000146.

Sullivan, M., Bisaga, A., Pavlicova, M., Choi, C. J., Mishlen, K., Carpenter, K. M., et al. (2017). Long-acting injectable naltrexone induction: a randomized trial of outpatient opioid detoxification with naltrexone versus buprenorphine. *The American Journal of Psychiatry, 174*(5), 459–467.

Sussman, S., Lisha, N., & Griffiths, M. (2011). Prevalence of the addictions: a problem of the majority or the minority? *Evaluation & the Health Professions, 34*(1), 3–56.

Tanum, L., Solli, K., Latif, Z., et al. (2017). Effectiveness of injectable extended-release naltrexone vs daily buprenorphine-naloxone for opioid dependence: a randomized clinical noninferiority trial. *JAMA Psychiatry, 74*(12), 1197–1205.

Tapper, A. R., McKinney, S. L., Nashmi, R., Schwarz, J., Deshpande, P., Labarca, C., et al. (2004). Nicotine activation of alpha4* receptors: sufficient for reward, tolerance, and sensitization. *Science (New York, NY), 306*(5698), 1029–1032.

Volkow, N. D., Fowler, J. S., Wolf, A. P., Schlyer, D., Shiue, C. Y., Alpert, R., et al. (1990). Effects of chronic cocaine abuse on postsynaptic dopamine receptors. *The American Journal of Psychiatry, 147*(6), 719–724.

Volkow, N. D., Koob, G. F., & McLellan, A. T. (2016). Neurobiologic advances from the brain disease model of addiction. *The New England Journal of Medicine, 374*(4), 363–371.

Wang, G. J., Tomasi, D., Volkow, N. D., Wang, R., Telang, F., Caparelli, E. C., et al. (2014). Effect of combined naltrexone and bupropion therapy on the brain's reactivity to food cues. *International Journal of Obesity, 38*(5), 682–688.

Weiss, R. D., Potter, J. S., Fiellin, D. A., Byrne, M., Connery, H. S., Dickinson, W., et al. (2011). Adjunctive counseling during brief and extended buprenorphine-naloxone treatment for prescription opioid dependence: a 2-phase randomized controlled trial. *Archives of General Psychiatry, 68*(12), 1238–1246.

Woody, G. E., Poole, S. A., Subramaniam, G., Dugosh, K., Bogenschutz, M., Abbott, P., et al. (2008). Extended vs short-term buprenorphine-naloxone for treatment of opioid-addicted youth: a randomized trial. *Journal of the American Medical Association, 300*(17), 2003–2011.

You, C., Zhang, H., Sakharkar, A. J., Teppen, T., & Pandey, S. C. (2014). Reversal of deficits in dendritic spines, BDNF and Arc expression in the amygdala during alcohol dependence by HDAC inhibitor treatment. *The International Journal of Neuropsychopharmacology, 17*(2), 313–322.

29

Translational Medicine Strategies for Drug Development for Impulsive Aggression

Emil F. Coccaro*, Royce Lee*, Neal G. Simon†

*Clinical Neuroscience and Psychopharmacology Research Unit, Department of Psychiatry, Pritzker School of Medicine, University of Chicago, Chicago, IL, United States †Department of Biological Sciences, Lehigh University, Bethlehem, PA, United States

Impulsive aggressive behavior in humans typically occurs during social interactions in which the "aggressor" perceives a threat to self that results in a swift and angry response to the "other." The swiftness of response is often characterized as "impulsive," while the anger in the response is often characterized as "affective" (Coccaro, 2012). In reality, these two features occur simultaneously, and that is why these kinds of behaviors are referred, interchangeably, as "impulsive" or "affective" aggression. That said, individuals who display impulsive/angry aggressive behavior are not always impulsive or angry between outbursts, even if they score higher on trait measures of impulsivity/anger compared with nonaggressive individuals. Ultimately, the key point is that the aggressive behavior, in these cases, is characterized by its impulsivity and anger in response to a social threat.

Treatment approaches to impulsive aggressive behavior must be developed in the context of what is known empirically about these behaviors. This includes cognitive and systems neuroscience (Coccaro, Sripada, Yanowitch, & Phan, 2011) and neurobiology (Yanowitch & Coccaro, 2011). The following sections discuss the background of impulsive aggression in the context of these areas and, then, potential strategies to test drugs for antiaggressive effects.

I COGNITIVE/SYSTEM NEUROSCIENCE AND NEUROBIOLOGICAL SUBSTRATES OF IMPULSIVE AGGRESSIVE BEHAVIOR IN HUMANS

A Cognitive Neuroscience

The most parsimonious model of impulsive aggression posits that a balance of inhibitory and excitatory neurotransmitter/modulators sets a "threshold" for impulsive aggressive responding to a perceived social threat and that impulsive aggressive behavior occurs when an external (or even internal) stimulus reaches and exceeds this threshold. In this model, individuals with prominent histories of impulsive aggressive (e.g., intermittent explosive disorder, IED; Coccaro, 2012) have a lower threshold for "exploding" than those without this history. This is not the complete model, however, because the potential social threat must first be processed along with cognitive and emotional overlays to that threat. These processes may be referred to as social-emotional information processing (SEIP; Coccaro, Fanning, Fisher, Couture, & Lee, 2017; Coccaro, Fanning, Keedy, & Lee, 2016; Coccaro, Noblett, & McCloskey, 2009), a model based on an earlier conceptual framework referred to as social information processing (SIP; Crick & Dodge, 1996; Dodge, 1991; Dodge, Pettit, Bates, & Valente, 1995; Fontaine, Burks, & Dodge, 2002).

Accordingly, interpersonal conflict may be explained within a multistage sequence of SEIP such that impulsive aggressive behavioral responses are linked to one or more of six (6) stages referred to as (1) encoding (attention to relevant social information), (2) attribution (of the intent of the behavior of the other participant in the social interaction), (3) emotional response whether negative or positive, (4) evaluation of potential responses, (5) decision-making regarding the chosen response, and (6) enactment of chosen response. Research confirms that impulsive aggressive children and adolescents (particularly those with a history of abuse) demonstrate a reduction in the encoding (or identification) of socially relevant information and the presence of a hostile attribution bias (the other person is intending harm to the subject), especially in cases of an ambiguous social interaction (Crick & Dodge, 1996; Dodge et al., 1995). Over the past several years, we extended this work to adults, specifically adults with IED (Coccaro, Fanning, Fisher, et al., 2017; Coccaro, Fanning, & Lee, 2017; Coccaro, Noblett, & McCloskey, 2009). In a series of studies, we documented that individuals with IED have reduced encoding of social cues (Coccaro, Fanning, Fisher, et al., 2017), heightened hostile attribution (Coccaro, Fanning, et al., 2016, Coccaro, Fitzgerald, Lee, McCloskey, & Phan, 2016, Coccaro, Lee, & Kavoussi, 2009), negative emotional response (Coccaro, Fanning, et al., 2016; Coccaro, Fitzgerald, et al., 2016; Coccaro, Lee, & Kavoussi, 2009), and a bias to choose directly or relationally aggressive responses to socially ambiguous cues (Coccaro, Fanning, et al., 2016; Coccaro, Fanning, Fisher, et al., 2017; Coccaro, Fanning, & Lee, 2017; Coccaro, Fitzgerald, et al., 2016). Simultaneous analysis of each of these SEIP steps reveals the critical importance of negative emotional response to socially ambiguous cues and the bias to choose aggressive responses to such cues (Coccaro, Fanning, et al., 2016; Coccaro, Fitzgerald, et al., 2016). Notably, aspects of SEIP are reported as abnormal in individuals with focal lesions of orbitofrontal cortex (Wilson, Scalaidhe, & Goldman-Rakic, 1993) and/or amygdala (Adolphs et al., 1999; George et al., 1993) brain areas implicated in the perception (Wilson et al., 1993) and correct identification of facial expression (George et al., 1993; Lane, Fink, Chau, & Dolan, 1997; Rolls, Cahusac, Feigenbaum, & Miyashita, 1993). In addition, PET studies in normal individuals show that recognition of fearful and sad faces (for example) is associated with the activation of the corticolimbic areas (Blair, Morris, Frith, Perrett, & Dolan, 1999). Individuals with lesions in these areas (who are often aggressive) often attribute a negative bias even to neutral facial expressions (Hornak, Rolls, & Wade, 1996). It is possible that deficits in SEIP are associated with aggressive behavior because the misperception of emotional stimuli and deficits in social information processing leads to inappropriate behavioral responses in impulsive aggressive individuals (Rolls et al., 1993). Work in our laboratory and of others also suggest important IED-control differences in SEIP in corticolimbic circuits (see in the succeeding text).

B Systems Neuroscience

Whether we are considering the role of neurobiology (i.e., threshold to explode) or cognitive neuroscience (processes that lead to exceeding this threshold), the generation of impulsive aggressive behavior goes through corticolimbic brain circuits (Coccaro et al., 2011). Among the more important regions in this regard are the orbital prefrontal cortex (OFC), the amygdala, and the temporoparietal junction (TPJ).

The OFC is mapped across the orbital surface of the frontal lobe and receives visual, gustatory, olfactory, auditory, and somatosensory information from primary and secondary association cortices (Rolls, 2004). In addition, the OFC is extensively interconnected with the amygdala (Ongur & Price, 2000), which processes sensory information to create emotionally valenced or conditioned object memories (Cahill, 1995), which bias decision-making processes of the OFC (Bechara, Damasio, Damasio, & Lee, 1999). The OFC plays an important role in three key aspects of impulsive aggression. These include the processing of mental reward and punishment representations: emotion, social information processing, and impulsivity. In the case of emotion, these are mental representations of internal reinforcers such as states of anger. In the case of SEIP, these are mental representations of the reinforcing or punishing attributes of social interactions, which in a social species such as man could have life-or-death consequences. In the case of impulsivity, these are mental representations of time-delayed reinforcers and punishments, such as the cost and benefits of using a resource immediately versus saving it for more beneficial use in the future.

Both medial and lateral OFC are relevant to aggression. The medial OFC is involved in the mental representation of object reward associations (reviewed in Elliott, Dolan, & Frith, 2000), especially when reward associations change or reverse over time (Elliott et al., 2000), a function dependent on interconnections with the amygdala (Ongur & Price, 2000). As with other prefrontal cortex structures, so-called delay neurons of the OFC play a working memory-like task in the sustained representation of anticipated reward during a time delay between stimulus perception and a behavioral response (Hikosaka & Watanabe, 2000). These medial OFC processes would be expected to be relevant to the processing of emotional and social information (Iversen & Mishkin, 1970). Indeed, lesions of the medial OFC are associated with deficits in processing the reward value of stimuli and disinhibited social behaviors (Damasio, 1994). Evidence from human studies indicates that the role of the medial OFC (and related structures) in processing reward associations is relevant to complex mental representations, including the subjective experience of emotions such as sadness, anger, happiness, and fear (Damasio et al., 2000). Because anger is an emotional mental operation, the medial OFC is involved in the processing of anger. In fact, experimentally evoked anger results in metabolic activation of the medial OFC (Kimbrell et al., 1999). The OFC may play a regulating role in the outward expression of anger, as imagined unrestrained physical aggression against another human is associated with decreased medial OFC metabolic activity (Pietrini, Guazzelli, Basso, Jaffe, & Grafman, 2000). These findings are consistent with reports of the effect of OFC lesions on increased behavioral aggression (Grafman et al., 1996; Zald & Sim, 1996). Thus the medial OFC may play a role in the regulation of emotional states such as anger, subject experience of anger, control of behavioral aggression, evaluation of the social milieu, and cognitive impulsivity.

The lateral OFC is involved in reversal learning and response inhibition. While some reports suggest that the lateral OFC is involved in processing of emotion valence-specific stimuli, such as negative versus positive emotions (Northoff et al., 2000), anger induction (Dougherty et al., 1999), and angry faces (Blair et al., 1999), the valence specificity of these may be due to the invocation of reversal learning and response inhibition functions. For example, angry faces may signal the punishment value of continuing a provocative behavior. This may be true of other facial expressions, however, as evidenced by a functional magnetic resonance imaging (fMRI) study using a visual reversal learning task (Kringelbach & Rolls, 2003). In addition to a role in processing emotionally salient stimuli, the lateral OFC also plays a role in the cortical inhibition of motoric impulsivity, which may serve to suppress impulsive aggressive behavior in response to provocation (Bechara, Damasio, & Damasio, 2000). In summary, both the medial and lateral sections of the OFC play key roles in processing of anger and behavioral impulsivity.

In addition to the OFC, models of human aggression also implicate the amygdala as an important structure in emotion processing and aggression (Davidson, Putnam, & Larson, 2000). This is based on data from stimulation and lesion studies in animals, nonhuman primates, and humans. In animals, stimulation of the amygdala promotes aggressive responding (Adamec, 1991), and its damage changes the nature of social interactions (Amaral et al., 2003; Kluver & Bucy, 1939). Since the amygdala and OFC are anatomically and functionally connected (Amaral & Price, 1984), their effective interactions are critical for decoding emotionally salient information and guiding goal-directed behaviors (Saddoris, Gallagher, & Schoenbaum, 2005), both of which are relevant in the control of aggression. Moreover, the OFC is hypothesized to play a key role in modulating limbic reactivity to threat (Davidson et al., 2000).

Finally, the TPJ is a recent addition to the "default mode network" and is directly involved in social-cognitive processes, including attentional state and thinking about the mental states of others (Krall et al., 2016; Koster-Hale & Saxe, 2013; Molenberghs, Johnson, Henry, & Mattingley, 2016; Price & Drevets, 2012; Saxe & Wexler, 2005). While the precise neuroanatomy and substructures of the TPJ are the subject of intense investigation, the consensus based on available results is that the right TPJ (rTPJ) is extensively involved in attentional processes and

interpretation of mental state (Mars et al., 2012; Saxe & Wexler, 2005), with at least the anterior region engaging in both functions based on an activation likelihood estimation (ALE) metaanalysis (Krall et al., 2016). Given the preceding discussion on SEIP, the rTPJ may have a critical role in triggering impulsive aggressive outbursts in impulsive aggressive individuals.

To date, there have been only two experimental medicine studies relevant to the TPJ and emotional processing. The first was a study in healthy volunteers (Lee et al., 2013) that examined fMRI blood oxygen level-dependent imaging (BOLD) signal in response to angry faces after intranasal vasopressin (AVP) with or without treatment with a novel vasopressin 1a (V1a) receptor antagonist. AVP, which is strongly associated with increased aggressive behavior in both animals and humans (see in the succeeding text), resulted in significantly elevated fMRI BOLD signal in the rTPJ and in precuneus, anterior cingulate, and putamen, when given intranasally 60 min prior to the fMRI session. As expected, the elevated fMRI BOLD signals to AVP were blocked in the presence of the V1a receptor antagonist indicting that the rTPJ (as well as other relevant corticolimbic) site is involved in anger-related social-emotional information processing. While we do not know what would be observed in impulsively aggressive individuals under the same study conditions, we have found that acute administration of the specific serotonin-reuptake inhibitor (SSRI), escitalopram, also increased the fMRI BOLD signal in the TPJ in impulsively aggressive subjects (Cremers, Lee, Keedy, Phan, & Coccaro, 2016) compared with that in healthy volunteers. Because these studies are not comparable, it is difficult to say more than that these results indicate that the TPJ may play a prominent contributory role in the manifestation of impulsive aggression.

Not surprisingly, several studies in impulsive aggressive subjects have observed anomalies in brain structure. Studies of impulsive aggressive individuals with borderline and/or antisocial personality disorders report reduced grey-matter volumes in various regions of the frontolimbic system compared with controls (Brambilla et al., 2004), findings that overlap those in individuals with IED. Specifically, individuals with IED have reduced grey-matter volume in frontolimbic areas including the orbital prefrontal cortex, ventromedial prefrontal cortex, anterior cingulate cortex, amygdala, insula, and uncus (Coccaro, Fitzgerald, et al., 2016). In addition, measures of aggression correlate directly with grey-matter volume in these areas. The shape of the amygdala is also abnormal in individuals with IED, with significantly more areas of inward deformation of the amygdala compared with healthy controls (Coccaro, Lee, McCloskey, Csernansky, & Wang, 2015). Diffusion tensor imaging (DTI) also reveals lower fractional anisotropy in two clusters located in the superior longitudinal fasciculus when compared with psychiatric and healthy controls (Lee et al., 2016), suggesting lower white-matter integrity in long-range connections between the frontal and TPJ regions of the brain and likely problems with connectivity between these brain regions.

These structural brain anomalies likely underlie results reported in functional magnetic resonance imaging studies in individuals with IED. For example, individuals with IED display greater amygdala response to exposure to angry faces compared with healthy controls whether the stimuli are presented implicitly (Coccaro, McCloskey, Fitzgerald, & Phan, 2007) or explicitly (McCloskey et al., 2016). In addition, life history of aggression measures correlate directly with amygdala response to angry faces, and the connectivity between prefrontal cortex and amygdala appears disrupted in IED compared with healthy controls. Notably, we have found evidence for a functional normalization in amygdala and in prefrontal areas in individuals with IED 12 weeks after antiaggressive pharmacotherapy (Coccaro, Fanning, Phan, & Lee, 2015). Further studies note that acute activation of 5-HT receptors by a single dose of citalopram is associated with an enhanced fMRI signal response to angry faces in the left temporal parietal junction of IED compared with healthy control individuals (Cremers et al., 2016).

C Neurobiology

The complexity of aggression reflects the role of multiple neurotransmitter interactions in regulation of the behavior. In this section, several neurotransmitter systems that have been implicated as potential therapeutic targets are briefly considered. Each is listed separately because an effort to cover the interactions is beyond the scope of this review.

An important consideration in attempting to elucidate the underlying neurobiology and neurochemistry of impulsive aggression is the generally poor translation of preclinical findings into new pharmacologic treatments (major depression is perhaps the best example). In the present case, models of aggressive behaviors utilized in laboratory settings, including resident–intruder aggression, maternal aggression, and defensive aggression, emulate productive behaviors that, in the wild, enhance access to resources and reproductive success. In this context, the regulatory systems for these behaviors almost certainly differ markedly from those governing impulsive or pathological aggression. Until better translational paradigms can be developed, results from any single preclinical model are best viewed cautiously.

1 Serotonin (5-HT)

Studies of impulsive aggression in humans have been ongoing since the late 1970s when lithium was used to treat aggressive behavior in prison inmates (Sheard, Marini, Bridges, & Wagner, 1976). It was already known that lithium had antiaggressive properties in (Sheard, 1970) and could increase 5-HT activity (Perez-Cruet, Tagliamonte, Tagliamonte, & Gessa, 1971) and that 5-HT activity correlated inversely with aggressive behavior (Valzelli & Garattini, 1968), in rodents. In addition to demonstrating an antiaggressive effect in human subjects, lithium reduced impulsive, but not premeditated, aggressive behavior (Sheard et al., 1976). Soon after, cerebrospinal fluid (CSF) studies of the 5-HT metabolite, 5-hydroxyindoleacetic acid (5-HIAA), reported consistent inverse correlations between measures of aggression and CSF 5-HIAA levels (Brown et al., 1982; Brown, Goodwin, Ballenger, Goyer, & Major, 1979). Subsequent studies reported lower levels of CSF 5-HIAA in impulsive violent offenders (Virkkunen, Nuutila, Goodwin, & Linnoila, 1987), indicating that 5-HT may be specifically related to impulsive aggression. Studies using other indices of 5-HT function largely confirmed and extended these findings (Coccaro et al., 1989). That said, a recent metaanalysis of 5-HT studies in aggression suggests that this relationship may be more modest than previously thought. Duke, Bègue, Bell, and Eisenlohr-Moul (2013) analyzed 171 studies on the serotonin-aggression relationship that employed the following: (a) CSF 5-HIAA, (b) acute tryptophan depletion (ATD), (c) pharmacochallenge, and (d) endocrine challenge methods (Duke et al., 2013). The authors found a small ($r = -0.12$), significant inverse relation between measures of 5-HT functioning and aggression overall with pharmacochallenge studies yielded the largest effect size ($r = -0.21$). These results demonstrate that the relationship between 5-HT and aggressive behavior is far more complex than previously appreciated. This should be expected given the complexity of the central 5-HT system, which manifests multiple types of receptors distributed at both pre- and postsynaptic sites, which may also exert unique and perhaps opposing effects on aggression. One must also consider the recognized role of other neurotransmitters (see in the succeeding text) and their interactions in the regulation of complex behaviors such as impulsive aggression. This is especially the case given that central 5-HT function may only account for about 4%–9% of the variance in aggression scores. Thus other neurotransmitters and neuromodulators must also play a role in the neurobiology of impulsive aggression either by inhibiting or facilitating impulsive aggressive responding.

2 Catecholamines (DA and NE)

Compared with 5-HT, much less is known about the role of catecholamines and aggression. Preclinical studies point to hyperactivity of the dopaminergic (DA) system in the mesocorticolimbic pathway during and after a provocative aggressive encounter, possibly reflecting motivational aspects of aggressive behavior

(Miczek, Fish, De Bold, & De Almeida, 2002). In humans, however, CSF homovanillic acid (HVA), the major metabolite of dopamine, has been studied with mixed results. Some studies reported an inverse relationship between CSF HVA concentration (Coccaro & Lee, 2010; Limson et al., 1991; Linnoila et al., 1983; Virkkunen, De Jong, Bartko, Goodwin, & Linnoila, 1989) or DA storage capacity in the striatum/midbrain (Schlüter et al., 2013) and aggression, but this has not been a consistent finding. Because central NE is implicated in orienting to novel stimuli, focusing attention, and enacting behavioral responses (Berridge & Waterhouse, 2003), NE has long been thought to play a role in aggression. Despite this, empirical support for this possibility in humans has been limited and mixed (Oquendo & Mann, 2000).

3 Glutamate (GLU)

GLU, the primary excitatory neurotransmitter in the CNS, is thought to play a facilitative role in aggressive behavior based on studies in cats and rodents reporting that defensive aggressive behavior is induced by GLU (and inhibited by GABA and 5-HT) in the hypothalamus (Haller, 2013). Conversely, administration of N-methyl-D-aspartate (NMDA) receptor antagonists or inhibition of GLU synthesis reduces aggression in mice. Consistent with these findings, CSF GLU concentrations correlate positively with measures of both aggression and impulsivity in personality-disordered and healthy control study participants (Coccaro, Lee, & Vezina, 2013). In humans, treatment with memantine, an uncompetitive antagonist at glutamatergic NMDA receptors, reduced agitation and aggression in individuals with Alzheimer's disease (Wilcock, Ballard, Cooper, & Loft, 2011), though no work has been performed in individuals with primary impulsive aggression.

4 Υ-Aminubutyric acid (GABA)

GABA is the primary inhibitory neurotransmitter in the brain with receptors heavily expressed in areas of frontal-limbic regions at both inhibitory-inhibitory and inhibitory-excitatory synapses. Preclinical studies have shown that aggressive animals have reduced brain GABA levels and of glutamic acid decarboxylase (GAD), the enzyme that catalyzes glutamate into GABA. In humans, Lee, Petty, and Coccaro (2009) found an inverse relationship between trait impulsivity (but not aggression) and CSF GABA levels in individuals with personality disorder and healthy control subjects. Despite the null finding for aggression, GABA levels were higher in individuals with a history of suicide attempt, a feature related to aggression. Drugs that enhance GABAergic effects (including the antipsychotic drug clozapine, anticonvulsants topiramate and valproate, and the mood stabilizer lithium) reduce aggression (Comai, Tau, & Gobbi, 2012), suicide and suicide attempts (lithium; Baldessarini, Tondo, & Hennen, 2003),

and behavioral dysregulation (carbamazepine; Cowdry & Gardner, 1988). Valproate, an agent that increases GABA, reduced aggression in individuals with IED and Cluster B personality disorder (Hollander et al., 2003). While these studies suggest an inhibitory relationship between GABA and aggression, others suggest a more complex relationship. For example, certain allosteric modulators of $GABA_A$ receptors show a bidirectional relationship with GABA in which case these agents enhance aggression at low doses and reduce aggression at high doses.

D Neuropeptides

1 Vasopressin (VASO)

VASO exerts its physiological and behavioral effects by binding to specific G-protein-coupled receptors (GPCRs) in the central nervous system and certain peripheral tissues/sites (Ring, 2005; Serradeil-Le Gal et al., 2002). Three distinct AVP receptor subtypes have been identified –V1a, V1b, and V2. V1a is the predominant VASO receptor found in the limbic system and cortex; V1b receptor is located in limbic system and pituitary gland, although it is less widespread than V1a. The V2 receptor is localized in the kidney where it mediates the antidiuretic effects of vasopressin. It is not generally thought to be expressed in the nervous systems of adult animals or humans.

Offensive and defensive aggressive behaviors are modulated by vasopressin (reviewed in Albers, 2012). Elevated VASO is associated with heightened aggression and exaggerated responses to perceived threats in both animals and humans (Ferris, 2005; Simon, 2002). We recently showed that conditioned fear responses that resulted in a hyperarousal pattern (BOLD signal activation) in the amygdala, hippocampus, and other parts of the Papez circuit in rats were attenuated by treatment with a novel V1a receptor antagonist, AVN576. Intraventricular administration of d(CH2)5Tyr(Me)AVP (Manning compound), a linear V1a antagonist that does not cross the blood-brain barrier and is used for research purposes only, blocks aggression in rodents (Ferris, 2005). Two new V1a antagonist compounds developed by Azevan Pharmaceuticals, SRX246 and SRX251, significantly reduce aggression in rats when given orally or by intraperitoneal (IP) injection (Fabio et al., 2010; Ferris et al., 2006, 2008; Simon et al., 2008). Imaging results in awake animals showed that rats treated with either of these compounds had significantly reduced fMRI BOLD activation in circuits known to mediate responses to threat and drive aggression (Ferris et al., 2008).

In humans, genetic variation of the vasopressin V1a receptor has been linked to differences in metabolic reactivity of the amygdala to emotional stimuli as measured by fMRI BOLD (Meyer-Lindenberg et al., 2009). Experimental manipulation of central vasopressin has provided proof of a causal role for vasopressin signaling in human behavior. When administered intranasally, vasopressin enhances attention to negative emotional facial expressions (Guastella, Kenyon, Alvares, Carson, & Hickie, 2010; Lee et al., 2013; Thompson, Gupta, Miller, Mills, & Orr, 2004; Thompson, George, Walton, Orr, & Benson, 2006; Zink, Stein, Kempf, Hakimi, & Meyer-Lindenberg, 2010). This effect is entirely consistent with the known relevance of vasopressin to social behavior, expression of vasopressin 1a receptors in the amygdala, and the well-known role of the amygdala in the decoding of visual information pertaining to emotional facial expressions (Lee et al., 2013; Zink et al., 2010). Intranasal vasopressin enhances TPJ and anterior cingulate metabolic activity during processing of social stimuli (Brunnlieb, Münte, Krämer, Tempelmann, & Heldmann, 2013; Rilling et al., 2012) and superior temporal sulcus activity during the expectation of punishment in the Taylor Aggression Paradigm (Brunnlieb et al., 2013). Measurements of central levels of VASO in humans by lumbar puncture were positively correlated with dimensional measures of impulsive aggression (Coccaro, Kavoussi, Cooper, & Hauger, 1998; Coccaro, Kavoussi, Hauger, Cooper, & Ferris, 1998) life history of aggressive behavior. fMRI studies show that VASO activates neural structures involved in fear regulation and social/emotional information processing (Lee et al., 2013; Zink et al., 2010; Zink et al., 2011). This finding is relevant because impulsive aggressive individuals have anomalies in social cognition and are particularly sensitive to social threat (Coccaro, Fanning, et al., 2016).

2 Oxytocin (OXY)

Like VASO, OXY plays a role in regulating social behavior, although these two neuropeptides often display opposing effects. While CSF VASO levels correlate positively, CSF OXY correlates inversely with aggression (Lee, Ferris, Van de Kar, & Coccaro, 2009). Oxytocin also reduced laboratory-assessed aggressive behavior among women with high state anxiety, suggesting an aggression-reducing anxiolytic effect of OXY (Campbell & Hausmann, 2013). However, OXY has also been shown to increase negative emotions such as envy and schadenfreude (Shamay-Tsoory et al., 2009) and to increase noncooperation toward members of out-groups (see De Dreu, 2012, for a review), suggesting that its effect on behavior may not always be positive or prosocial. Given the importance of amygdala hyperactivation in impulsively aggressive individuals (Coccaro et al., 2007; McCloskey et al., 2016), it is of note that OXY reduced the enhanced amygdala activation during exposure to angry/fearful faces in females with BPD, suggesting that OXY may reduce sensitivity to social threat in women with this disorder (Bertsch et al., 2013).

3 Substance P

The endogenous receptor for substance P is neurokinin-1 (NK1), and this receptor is widely distributed in the CNS, especially in limbic system regions (Yip & Chahl, 2001). Boin close association with 5-HT- and NE-containing neurons (Gobbi et al., 2007). A modulatory role for substance P in aggressive behavior is suggested by the presence of high concentrations of substance P in brain regions relevant to mammalian aggression (e.g., amygdala and periaqueductal grey; Smith et al., 1994). Studies in lower mammals show that substance P promotes aggressive behavior by activating hypothalamic NK1 receptors and induces rage and aggression (Barbeau, Rondeau, & Jolicoeur, 1980; Beyer, Caba, Banas, & Komisaruk, 1991; Bhatt, Gregg, & Siegel, 2003; Elliott & Iversen, 1986; Gregg & Siegel, 2001; Han, Shaikh, & Siegel, 1996; Shaikh, Steinberg, & Siegel, 1993). Conversely, NK1 receptor antagonists reduce defensive aggression in cats (Shaikh et al., 1993). A recent neurochemical study from our group reported a positive correlation between CSF levels of substance P and measures of aggression in personality disorder and control subjects, suggesting that substance P concentrations may facilitate aggression in humans (Coccaro, Lee, Owens, Kinkead, & Nemeroff, 2012). While an antagonist for NK1 receptors is available for human use (aprepitant), no study using this agent has been published to test the hypothesis that blocking NK1 receptors can reduce aggression in human subjects.

4 Neuropeptide Y

Animal models of aggression suggest that NPY can increase aggressive behavior (Karl et al., 2004; Kask & Harro, 2000; Rutkoski, Lerant, Nolte, Westberry, & Levenson, 2002). Our group reported a positive correlation between CSF NPY and measures of aggression (Coccaro, Lee, Liu, & Mathé, 2012), suggesting that NPY concentrations may facilitate aggression in humans as well. Unfortunately, to date, no antagonists for NPY receptors are available for human use so an experimental study in human subjects is not currently possible.

5 Inflammatory Cytokines

Inflammatory cytokines such as interleukin-1β (IL-1β; Hassanain, Bhatt, Zalcman, & Siegel, 2005; Hassanain, Zalcman, Bhatt, & Siegel, 2003) and interleukin-2 (IL-2; Bhatt & Siegel, 2006) modulate aggressive behavior in animals (i.e., defensive-rage aggression model in cat). In addition, IL-2 levels are higher in mice bred for high aggression versus low aggression (Petitto, Lysle, Gariepy, & Lewis, 1994) and knockout of TNF-α receptors eliminates aggressive behavior in mice (Patel et al., 1999). In human studies, circulating levels of C-reactive protein (CRP; Suarez, 2004, Marsland, Prather, Petersen, Cohen, & Manuck, 2008), a marker of inflammation, and IL-6

(Marsland et al., 2008; Suarez, 2003) also correlate positively with self-assessed hostility and tendency toward aggression in healthy adult subjects. Our group reported similar findings in personality-disordered individuals (Coccaro, 2006) and in individuals with recurrent, problematic, impulsive aggressive behavior with intermittent explosive disorder [IED; (Coccaro, Fanning, et al., 2016; Coccaro, Fitzgerald, et al., 2016; Coccaro, Lee, & Coussons-Read, 2014)]. In these studies, we found that plasma CRP (Coccaro, 2006; Coccaro et al., 2014), plasma IL-6 (Coccaro et al., 2014), and plasma sIL-1RII protein (Coccaro et al., 2016) are lowest in individuals with IED compared with psychiatric and healthy controls and that each inflammatory marker correlates directly with measures of aggression. Neurochemical studies from our group report a similar correlation with measures of aggression with CSF CRP (Coccaro, Lee, & Coussons-Read, 2015a) and with soluble CSF sIL-1RII receptor protein (Coccaro, Lee, & Coussons-Read, 2015b). While inflammatory states may reduce synaptic 5-HT by shunting tryptophan catabolism away from 5-HT to kynurenine, study of tryptophan metabolites in IED subjects reveals no change in kynurenine levels, suggesting a direct effect of these inflammatory mediators on aggressive behavior (Coccaro, Lee, et al., 2016). While studies of antiinflammatory agents have been ongoing in depression (Köhler et al., 2014), such studies have not yet been published in impulsive aggressive individuals.

II TESTING THE ANTIAGGRESSIVE EFFICACY OF PSYCHOPHARMACOLOGIC AGENTS

The classic psychopharmacologic approach to testing potential psychotropic therapeutics involves double-blind, placebo-controlled, clinical trials to determine safety and efficacy. Impulsive aggressive behavior, however, is highly variable in the short term (weeks), and such trials will likely show reductions in aggressive behavior simply due to natural ebb and flow of these behaviors during the treatment period. This is because impulsive aggressive behaviors are a function of the stable tendency to respond aggressively to social threat and the variable presence of social threats in the environment, making the use of double-blind placebo-controlled trials for impulsive aggression critical. This feature of impulsive aggression introduces an element of complexity in measuring efficacy because it increases the potential for a strong placebo response.

A Who to Study?

Typically, clinical trials are performed in individuals with a particular disease (e.g., internal medicine) or

disorder (e.g., psychiatry). However, developing pharmaceuticals for impulsive aggression is complicated by the fact that impulsive aggression occurs along a dimension. That said, research in the past two decades has determined that one can define a disorder of impulsive aggression by using DSM-5 criteria for IED (Coccaro, 2012). The validity of IED is supported by a number of data sets from family (Coccaro, Lee, & Kavoussi, 2010), neurobiological (Coccaro, Lee, & Kavoussi, 2009), neuroimaging (Coccaro et al., 2007; Cremers et al., 2016; McCloskey et al., 2016), and treatment (Coccaro, Lee, & Kavoussi, 2009) studies. Perhaps most important are data suggesting that IED represents a taxon (Ahmed, Green, McCloskey, & Berman, 2010) and not simply the presence of elevated levels of impulsive aggression, even though aggression, itself, is a dimensional construct. This is likely because IED represents impulsive aggression with a specific set of features that includes clinically significant distress and/or impairment and not simply a high aggression score. While the prevalence of IED by DSM-5 criteria is not known at this time, the prevalence of IED by DSM-IV criteria was conservatively estimated at 5.4% lifetime and 2.7% for the past year (Kessler et al., 2006). Reanalysis of these data using integrated research criteria (on which the DSM-5 criteria were based) estimates the lifetime prevalence at 3.6% and the past year prevalence at 2.2% (Coccaro, Fanning, & Lee, 2017). Since the survey data (Kessler et al., 2006) generating these estimates did not include assessment of high-frequency, low-intensity verbal/nondestructive outbursts, which can be present in the absence of low-frequency, high-intensity destructive/assaultive outbursts that were the hallmark of DSM-IV IED (Coccaro et al., 2014), both sets of numbers represent an underestimate of the prevalence of IED. Thus, DSM-5 IED represents a sizable group of potential patients for which antiaggressive pharmaceuticals could be developed. In addition to those meeting DSM-5 criteria for IED, one could also consider developing antiaggressive pharmaceuticals for individuals with impulsive aggression due to secondary causes (e.g., schizophrenia, bipolar disorder, and traumatic brain injury). In these cases the DSM-5 IED criteria can still be used to identify such individuals as long as all subjects in the trial meet all but the exclusionary criteria and belong to an identifiable group of subjects.

B Development of Clinical Trials Testing Antiaggressive Efficacy

Reviewing the evolution of the clinical psychopharmacology of impulsive aggressive behavior, the first notable study was a double-blind, placebo-controlled trial of lithium in prison inmates (Sheard et al., 1976). The rationale was based on preclinical data suggesting that lithium had antiaggressive effects in rodents (Sheard, 1970). Work in clinical pharmacology then languished until the availability of the 5-HT selective reuptake inhibitor (SSRI), fluoxetine. At that time, the idea that a 5-HT agent could have antiaggressive efficacy was predicated on substantial and consistent psychobiological data suggesting central 5-HT system hypofunction was associated with impulsive aggressive behavior (Brown et al., 1979; Coccaro et al., 1989; Linnoila et al., 1983). Soon after, an open label study with fluoxetine was published in patients with DSM-IIIR borderline personality disorder (Norden, 1989). This was followed by a small placebo-controlled, double-blind study of the effect of fluoxetine on anger in the same types of patients. Each paper suggested antiaggressive effects of fluoxetine, which led to the first double-blind, placebo-controlled study of fluoxetine in individuals with prominent histories of impulsive aggressive behavior (now defined as IED). This study was preceded by work to determine the optimal structure of antiaggressive trial (Coccaro, Harvey, Kupsaw-Lawrence, Herbert, & Bernstein, 1991). Most importantly, this work involved the development of outcome measures and led to the modification of the overt aggression scale (OAS; Yudofsky, Silver, Jackson, Endicott, & Williams, 1986), which was developed for inpatient settings, for outpatient trials (OAS-M; Coccaro et al., 1991). In addition, issues related to how to enter appropriate subjects into such studies and study duration needed to be addressed. For this first study (Coccaro & Kavoussi, 1997), the investigators selected personality-disordered subjects who scored above a relevant threshold on a measure of aggression (later, this approach was replaced by a current diagnosis of IED and a sufficiently high score on the OAS-M). To limit the number of potential "placebo responders," threshold scores on the OAS-M were required during a single-blind, 2-week, placebo lead-in phase prior to randomization. To allow a long enough period to see reliable antiaggressive effects, the investigators set the randomization phase at 12 weeks. Overall, antiaggressive effects in the fluoxetine group appeared within the first few weeks and continued through to the end of the trial in completers and in all subjects in a last observation carried forward analysis (LOCF; Coccaro & Kavoussi, 1997; Coccaro, Lee, & Kavoussi, 2009). While encouraging, attrition in both groups was about 50% by the end of trial, suggesting that impulsive aggressive subjects constitute a difficult group for an extended clinical trial. Two later trials confirmed the antiaggressive efficacy of fluoxetine (George et al., 2011; Silva et al., 2010). In addition, other studies have added further insight into the antiaggressive efficacy of fluoxetine. First, a small study reported a positive correlation between prolactin responses to pharmacochallenge with d-fenfluramine (PRL[d-FEN]) before study and the antiaggressive response manifest by OAS-M scores at the end of the trial (Coccaro, Kavoussi, & Hauger, 1997). This finding suggested that the more dysfunctional the 5-HT system (as reflected by PRL[d-FEN]), the lower the antiaggressive response was to fluoxetine. Given that the magnitude of the PRL[d-FEN] response is related to stores of newly synthesized 5-HT and to the sensitivity of postsynaptic 5-HT receptors (Coccaro, Kavoussi, Cooper, & Hauger, 1998), it is reasonable to posit that the more dysfunctional the 5-HT synapse, the less effective an SSRI will be as an antiaggressive agent. A second study (Silva et al., 2010) reported that subjects carrying the ss allele for the 5-HT transporter protein (5-HTT; a genotype associated with the production of a limited number of 5-HTT) responded less well to fluoxetine. This is not surprising because fluoxetine binds to 5-HTT and, thus, the fewer 5-HTT proteins, the lower degree of entry of 5-HTT blockade and the lower the SSRI enhancement of synaptic 5-HT. In addition, there is an inverse relationship between platelet 5-HTT number and aggression scores, indicating that a lower number of 5-HTT proteins are associated with both aggression (Coccaro, Lee, & Kavoussi, 2009) and a weaker clinical response to an SSRI intervention (Silva et al., 2010). A third study reported that antiaggressive responses to fluoxetine (a placebo arm was not included) were associated with increased activation of prefrontal cortical regions with the PET ligand fludeoxyglucose (FDG; New et al., 2004). Consistent with this finding are preliminary data suggesting that, compared with placebo, putative antiaggressive agents (fluoxetine and divalproex) reduce the fMRI BOLD response to angry faces in the amygdala and tend to rebalance the relative activation to angry faces between the amygdala (lower) and prefrontal cortex (Coccaro, Fanning, Phan, & Lee, 2015).

These data provide a number of insights into issues of relevance for the design of antiaggressive clinical trials. First, it is advisable to test agents that act on neuronal systems shown to have an influence in aggressive responding, as was exemplified by the first clinical trials of fluoxetine. Specifically, since SSRIs attach to the 5-HTT, the possibility that more aggressive individuals will have fewer 5-HTT proteins than less aggressive individuals means that SSRIs may not be very efficacious in those with moderate to severe aggression histories. Add to this the likelihood that postsynaptic 5-HT receptors are increasingly less responsive to increases in synaptic 5-HT the more aggressive the individual, the more likely that other, non-SSRI, interventions need to be investigated. This could mean selective 5-HT receptor agonists, such as lorcaserin, a 5-HT2$_c$ receptor agonist developed for weight management (Greenway, Shanahan, Fain, Ma, & Rubino, 2016). This could also mean other agents targeting other neurotransmitter/modulator systems, such as vasopressin, oxytocin, glutamate, GABA, and inflammatory cytokines.

III EXPERIMENTAL MEDICINE APPROACHES TO THE DEVELOPMENT OF ANTIAGGRESSIVE AGENTS

The risk and cost of developing or even repurposing potential antiaggressive agents, like other psychiatric drugs, are very high. The success rate for psychiatric drugs once an IND is obtained is 6.2%, which is lower than the overall 9.6% approval rate across all therapeutic classes (Thomas et al., 2016). The cost of new drug development is controversial, with recent estimates for the pharmaceutical industry reaching $2.6 billion (DiMasi, Grabowski, & Hansen, 2016). This figure has been challenged because it includes the "cost of failures" and the "cost of capital." The latter was calculated at 10.5% per year and accounts for $1.163 billion, or 44%, of the $2.6 billion figure (Avorn, 2015). Estimates for the cost of a single successful compound that allow for the cost of failure are slightly less than $200 million, although this figure does not include in-kind contributions from academic and philanthropic groups (DNDi, 2013). Regardless, the cost is very high, years of effort are required, and the risk/reward ratio is poor. This raises the question of whether experimental medication approaches can improve the development process. First, whether an agent is likely to exhibit antiaggressive properties quickly or only after a period of adaptation, like many psychotropics in the thymoleptic space, should be determined. If the agent is expected to rapidly exhibit benefit, a simple experimental study in which one dose of the agent is given and its effect on an aggression measure can be assessed at its peak exposure level, Cmax. These studies would be strengthened with a pharmacodynamic approach that incorporates assessments over a specified time period based on pharmacokinetic data. This is particularly important for CNS agents because PK based on circulating levels of a drug in blood and brain PK may differ. Studies in animals are critical in this context given the significant cost of a human brain PK study (if such work can be undertaken).

If the agent is expected to work over time, then the study should include a pretreatment phase of at least 2 weeks after which the aggression measure is used. While agents that should work over weeks (e.g., SSRIs, lithium, divalproex, and related thymoleptics) might require subchronic treatment, SSRIs typically suppress aggression acutely in analog laboratory paradigms (see in the succeeding text). Agents that might work immediately include those that stimulate or block the actions of amine (e.g., GABA) or peptide (e.g., VASO) neurotransmitter/neuromodulators. It is also important to note that such studies must include a placebo control. When assessing the acute antiaggressive effect of an agent, a within-subject design provides for the most power. This is more difficult, of course, when assessing the subacute effects. In this case, investigators are likely to be constrained to a parallel group design, which requires more subjects and resources compared with a study testing for an acute antiaggressive effect.

Second, what measure of aggression should be used in experimental medicine studies? Here, it is important to note that it is nearly impossible to observe impulsive aggressive behavior in a controlled setting. Even the most impulsively aggressive individuals can keep their behavior "under control" in a laboratory environment unless exposed to a sufficiently provocative social threat. This latter situation is fraught with difficulty due to not only issues of research ethics (i.e., deliberately provoking aggressive responses leads to intense scrutiny by Institutional Review Boards) but also the limitation that provocation scenarios are difficult to standardize across subjects. If frank provocation is not allowed, it is very unlikely that a change in aggressive behavior will be observed in the course of short-term exposure to a putative antiaggressive agent. Thus, other kinds of assessments are needed when designing an experimental medicine design to assess antiaggressive efficacy.

Measures of aggression fall into three basic categories: (a) paper and pencil assessments, (b) clinical rating assessments, and (c) analog laboratory assessments. Before detailing the ways in which aggression can be assessed, we should reiterate that it is very difficult to observe impulsive aggressive behavior in a controlled setting. Paper and pencil assessments are simple to administer but are typically trait measures of behavior not likely to manifest sufficient change for use in experimental medicine studies. Altering the time scale of a trait measure (e.g., from "in general" to the "past week"), however, does not typically improve the measure's performance for clinical trials because the statements subjects rate themselves on are inherently trait in nature. This is not necessarily true for measures of emotional states such as "anger" because these are designed to demonstrate change over short time intervals. That said, "state anger" is not the same as "aggression," even if these measures are strongly correlated.

Clinical rating assessments for aggression are few in number and have their own limitations. The assessment used in most clinical trials is the OAS-M (Coccaro et al., 1991). The assessment scale was modified to provide a frequency/severity weighted assessment of overt aggressive behavior across verbal aggression, aggression against objects, others, and self. The OAS-M Aggression score has good internal consistency ($\alpha = 0.78$; unpublished data) and excellent interrater reliability (kappa > 0.95; (Coccaro et al., 1991; Endicott, Tracy, Burt, Olson, & Coccaro, 2002) and has been used in several clinical trials of impulsive aggression (Coccaro & Kavoussi, 1997; Coccaro, Lee, & Kavoussi, 2009; George et al., 2011; Hollander et al., 2003; McCloskey, Noblett,

Deffenbacher, Gollan, & Coccaro, 2008; Silva et al., 2010). Despite this, OAS-M Aggression scores can vary substantially when assessed weekly. This is because impulsive aggressive behaviors vary with the number of social threat stimuli individuals are exposed to over short time periods. This reduces the power of the OAS-M Aggression score as an outcome measure for experimental medicine studies investigating acute antiaggressive effects. On the other hand, OAS-M Aggression scores represent a good outcome measure in studies designed to assess the antiaggressive effects of a drug over weeks to months.

There are currently only two analog laboratory measures of aggression, both of which were developed 20 or more years ago. The first is the Taylor Aggression Paradigm (TAP; Taylor, 1967), and the second is the point-subtraction-aggression paradigm (PSAP; Cherek, 1981). In both paradigms, provocation is standardized and not specific to the subject, which can be the case in anger provocation scenarios (Spoont, Kuskowski, & Pardo, 2010).

In the TAP, the provocation is a mild electric shock given to the subject when he/she loses a reaction time trial. The shock levels are individually tailored and never set above 90% of the shock level the subject deems "uncomfortable." Because electric shock is a physical stimulus, it is considered analogous to a physical threat to the subject. Because this occurs in the context of a social setting, it represents a physical social threat. The "cover story" given to the subject is that the strength of the shock was selected by the person they are paired with. While the subject never meets this person, the activity occurring around the subject supports the cover story they are paired with a real person. During the TAP, the strength of the shocks are increased over time, and subjects can increase, decrease, or not change the strength of shock they select to be delivered to the other person when that other person loses a reaction time trial. Generally speaking, higher shock levels from the other person lead to higher shock levels selected by the subject. We have found that impulsively aggressive subjects with IED set significantly higher shocks compared with healthy volunteers or nonaggressive psychiatric controls, clearly supporting the idea that mean shock levels set against the other person during the TAP reflect aggressive responding (Giancola & Parrott, 2008). Most important for the purpose of experimental medicine studies, the TAP has been used extensively to test the effect of various agents on aggression. All such TAP studies have been done as parallel group designs due to a desire to administer the TAP to subjects only once even though this increases the number of subjects needed compared to a within-subject design. In response to this concern, we have repeated the TAP on four occasions over 3 months and have found that aggressive responding during the TAP is quite stable. Accordingly, our data suggest

that the TAP can be used effectively in a within-subject design. To date, the TAP has been utilized to examine antiaggressive effects of various agents including the SSRI, paroxetine (Berman, McCloskey, Fanning, Schumacher, & Coccaro, 2009).

In the PSAP, the provocation is the loss of earned points worth a minimal amount of money (e.g., 10 or 25 cents). Again, subjects are paired with another person. Both people are set in front of monitor and a "button box" and are tasked with pressing button "A" as many times as they can during a 25-minute session. One hundred presses of the "A" button earn the subject a point (exchangeable for money at the end of the study). At the same time the "other person" can press Button "B" ten (10) times and take a point from the other subject who pressed100 times. This provocation is experienced as a frustration/obstacle social threat rather than as a physical social threat. In turn, the subject can do three things: (a) ignore the point loss and continue to press the "A" button, (b) press the "B" button ten times and cause the other person to lose a point (aggressive option), or (c) press the "C" button ten times and be protected from further point loss for a variable period of time ("escape option"). Selecting either button "B" or "C" protects the subjects from further point loss of a time. The number of "B" presses is the index of aggressive responding (Cherek, Moeller, Schnapp, & Dougherty, 1997), though we have had clearer results when using the ratio of "B" button responses to "All" button responses. Unlike the TAP, the PSAP has been used in within-subject designs when exploring the effect of pharmaceuticals on aggression. To date, the PSAP has been utilized to examine antiaggressive effects of various agents such as d-fenfluramine (Cherek & Lane, 2001) and paroxetine (Cherek, Lane, Pietras, & Steinberg, 2002), among others. While we have found that the TAP and PSAP correlate with other measures of aggression, these correlations are modest. In addition, TAP and PSAP are weakly correlated, which may be due to the likelihood that they reflect different aspects of aggression; that is, aggression in response to a physical social threat versus a frustrating/obstacle social threat.

To date, TAP and PSAP protocols have largely been used to assess the role of drugs of potential abuse on aggression (e.g., alcohol, opiates, and cocaine). That said, some studies have been conducted using agents of interest in the treatment of aggression including agents related to 5-HT—paroxetine (Berman et al., 2009), fenfluramine (Cherek & Lane, 2001), and ipsapirone (Moeller et al., 1998)—and to GABA: baclofen (Cherek, Lane, Pietras, Sharon, & Steinberg, 2002), gabapentin (Cherek, Tcheremissine, Lane, & Pietras, 2004), and tiagabine (Lieving, Cherek, Lane, Tcheremissine, & Nouvion, 2008), with each showing evidence of acute antiaggressive effects. Except for the SSRIs, however, none of these agents have advanced to Phase II clinical trials. While similar studies involving intranasal administration of vasopressin (Brunnlieb et al., 2013) or oxytocin (Alcorn 3rd., Rathnayaka, Swann, Moeller, & Lane, 2015) have yet to demonstrate evidence of antiaggressive effects, these studies were performed in nonaggressive healthy individuals in whom it would be difficult to observe pro- or antiaggressive effects.

Recently, we began pilot work using the TAP and PSAP in the context of experimental medicine studies of potential antiaggressive agents based on our own neurotransmitter-based studies (e.g., aprepitant for substance P and lorcaserin for 5-HT2$_c$). In order to create a protocol that is feasible in a single day, we administer the drug in question in the morning, wait until the drug is presumed to reach Cmax, and then ask the subject to complete a TAP session and a PSAP session each of which takes about 45 minutes start to finish. While much work with the PSAP has involved multiple sessions a day over several days, a single session is sufficient for an experimental medicine study (Golomb, Cortez-Perez, Jaworski, Mednick, & Dimsdale, 2007).

Another approach to drug development is to design a program that aligns preclinical and experimental medicine studies. The strength of this paradigm is in its potential to allow extensive neurobiological studies in preclinical models that can then inform the experimental medicine translational study design. The opportunity to use this approach to derisk potential antiaggressive agents and thus impact development time and costs has been enhanced with the now routine availability of imaging technologies that can be used in awake nonhuman species and in humans.

Two studies that are part of the development program for SRX246, a new V1a receptor antagonist that is being tested for the treatment of anger, aggression, and irritability in multiple indications (see clinicaltrials.gov: NCT02055638 for the treatment of intermittent explosive disorder, NCT02507284 for the treatment of Huntington's disease, and NCT02733614 for the treatment of PTSD), exemplify the potential utility of aligning preclinical and experimental medicine studies. SRX246 and a highly similar compound, SRX251, significantly reduced aggression in rats when given orally or by IP injection (reviewed in Simon et al., 2008; Fabio et al., 2010). In a subsequent fMRI study in awake animals, rats receiving these compounds had significantly reduced BOLD activation in circuits known to mediate responses to threat and drive aggression (Ferris et al., 2008). After an IND for SRX246 was received and Phase I clinical trial data established safety, an experimental medicine fMRI study was designed based on the preclinical findings (Lee et al., 2013). The results showed that while intranasal arginine vasopressin (AVP) administration led to an exaggerated BOLD response to angry faces in regions recognized for

their involvement in responses to fear and negatively valanced emotional stimuli, treatment with oral SRX246 (120 mg BID in capsules) significantly attenuated BOLD activation seen in response to VASO. Among the major sites where SRX246 treatment blunted the change in BOLD signal seen in response to AVP were the amygdala, anterior cingulate cortex, and temporoparietal junction, all regions that are part of a neural circuit implicated in excessive aggression, fear, and the processing of emotional stimuli (Coccaro, Kavoussi, Cooper, & Hauger, 1998; Coccaro, Kavoussi, Hauger, et al., 1998; Hayes, Vanelzakker, & Shin, 2012; Price & Drevets, 2012). Studies in the rhesus monkey brain and postmortem human brain have shown that these regions, as well as other parts of the limbic system and cortex, are enriched in V1a receptor (Lu, Simon, Palkovits, & Brownstein, 2013; Young, Toloczko, & Insel, 1999). The experimental medicine fMRI study provided a well-powered, lower-cost, more rapid test of potential therapeutic utility compared with a small POC Phase II trial. The results also provided validation in terms of circuitry where the compound exerted its effects, which provided important mechanism of action data. These observations not only reduce the development risk for the compound but also provided a strong basis for further development of V1a receptor antagonists as a novel treatment for irritability/aggression.

In addition to the importance of assessing the behavioral effects of potential antiaggressive agents, the Lee et al. (2013) study demonstrates that experimental medicine studies also can be enhanced by assessing the effects of the agents in question on functional brain outcome measures. We suggest that after TAP/PSAP sessions are completed, subjects receive MRI scanning to assess changes in corticolimbic activation to relevant stimuli (e.g., anger faces and videos of ambiguous social interactions that result in an aggressive encounter, such as the V-SEIP) associated with administration of the agent in question. For example, we have found that chronic exposure to antiaggressive agents suppresses amygdala responses to angry faces (Coccaro, Fanning, Phan, & Lee, 2015) and that acute exposure to an SSRI (s-citalopram) is associated to a reduction in fMRI BOLD activation in the TPJ (Cremers et al., 2016). While the latter study was not coupled with a simultaneous laboratory assessment of aggression, experimental medicine studies should include some form of functional neuroimaging to more robustly link potential behavioral effects to the agent being tested. Thus, antiaggressive effects appearing in experimental medicine studies for a particular agent, particularly if they are coupled with the expected changes in corticolimbic function (e.g., blunting of amygdala and enhancement of prefrontal responses to emotional stimuli), should provide the rationale to conduct a Phase II clinical trial of that agent.

IV CONCLUSION

As the scientific understanding of impulsive aggression has evolved, the opportunity has arisen to apply a translational medicine approach to the development of treatments of impulsive aggression. A neural circuit model, combined with the psychological model of social and emotional information processing, allows for the targeting of neural circuits involved in the impulsive aggressive responding with biological interventions. Past experience with this population has led to the recognition that a sophisticated approach is required in designing clinical trials in this population. Critical issues regarding measurement of impulsive aggressive behavior in human research volunteers must be addressed in the design of such clinical trials. Some of the difficulties inherent in measuring the change of aggressive behavior over time may be addressed with innovative approaches, such as the measurement of laboratory analogues and functional markers of neural circuit activation.

Acknowledgments

Dr. Coccaro reports being the recipient of funding from the National Institute of Mental Health and the Pritzker-Pucker Family Foundation and being on the Scientific Advisory Board of Azevan Pharmaceuticals, Inc.; Dr. Lee reports being the recipient of a research grant from Azevan Pharmaceuticals, Inc.; and Dr. Simon reports that he is the CEO of Azevan Pharmaceuticals, Inc.

References

Adamec, R. E. (1991). Individual differences in temporal lobe sensory processing of threatening stimuli in the cat, Physiology and Behavior, 49(3), 455–464.

Adolphs, R., Tranel, D., Hamann, S., Young, A. W., Calder, A. J., Phelps, E. A., et al. (1999). Recognition ofs facial emotion in nine individuals with bilateral amygdala damage. Neuropsychologia, 37(10), 1111–1117.

Ahmed, A. O., Green, B. A., McCloskey, M. S., & Berman, M. E. (2010). Latent structure of intermittent explosive disorder in an epidemiological sample. Journal of Psychiatric Research, 44(10), 663–672.

Albers, H. E. (2012). The regulation of social recognition, social communication and aggression: vasopressin in the social behavior neural network. Hormones and Behavior, 61(3), 283–292.

Alcorn, J. L., 3rd., Rathnayaka, N., Swann, A. C., Moeller, F. G., & Lane, S. D. (2015). Effects of intranasal oxytocin on aggressive responding in antisocial personality disorder. Psychological Record, 65(4), 691–703.

Amaral, D. G., Bauman, M. D., Capitanio, J. P., Lavenex, P., Mason, W. A., Mauldin-Jourdain, M. L., et al. (2003). The amygdala: is it an essential component of the neural network for social cognition? Neuropsychologia, 41(4), 517–522.

Amaral, D. G., & Price, J. L. (1984). Amygdalo-cortical projections in the monkey (Macaca fascicularis). Journal of Comparative Neurology, 230(4), 465–496.

Avorn, J. (2015). The $2.6 billion pill—methodologic and policy considerations. The New England Journal of Medicine, 372, 1877–1879.

Baldessarini, R. J., Tondo, L., & Hennen, J. (2003). Lithium treatment and suicide risk in major affective disorders: update and new findings. Journal of Clinical Psychiatry, 64(Suppl 5), 44–52.

Barbeau, A., Rondeau, D. B., & Jolicoeur, F. B. (1980). Behavioral effects of substance P in rats. International Journal of Neurology, 14, 239–252.

Bechara, A., Damasio, H., & Damasio, A. R. (2000). Emotion, decision making, and the orbitofrontal cortex. Cerebral Cortex, 10, 295–307.

Bechara, A., Damasio, H., Damasio, A. R., & Lee, G. P. (1999). Different contributions of the human amygdala and ventromedial prefrontal cortex to decision-making. Journal of Neuroscience, 19, 5473–5481.

Berman, M. E., McCloskey, M. S., Fanning, J. R., Schumacher, J. A., & Coccaro, E. F. (2009). Serotonin augmentation reduces response to attack in aggressive individuals. Psychological Science, 20(6), 714–720.

Berridge, C. W., & Waterhouse, B. D. (2003). The locus coeruleus-noradrenergic system: modulation of behavioral state and state-dependent cognitive processes. Brain Research. Brain Research Reviews, 42(1), 33–84.

Bertsch, K., Gamer, M., Schmidt, B., Schmidinger, I., Walther, S., Kästel, T., et al. (2013). Oxytocin and reduction of social threat hypersensitivity in women with borderline personality disorder. The American Journal of Psychiatry, 170, 1169–1177.

Beyer, C., Caba, M., Banas, C., & Komisaruk, B. R. (1991). Vasoactive intestinal polypeptide (VIP) potentiates the behavioral effect of substance P intrathecal administration. Pharmacology, Biochemistry, and Behavior, 39, 695–698.

Bhatt, S., Gregg, T. R., & Siegel, A. (2003). NK1 receptors in the medial hypothalamus potentiate defensive rage behavior elicited from the midbrain periaqueductal gray of the cat. Brain Research, 966, 54–64.

Bhatt, S., & Siegel, A. (2006). Potentiating role of interleukin 2 (IL-2) receptors in the midbrain periaqueductal gray (PAG) upon defensive rage behavior in the cat: role of neurokinin NK(1) receptors. Behavioural Brain Research, 167(2), 251–260.

Blair, R. J., Morris, J. S., Frith, C. D., Perrett, D. I., & Dolan, R. J. (1999). Dissociable neural responses to facial expressions of sadness and anger. Brain, 122(5), 883–893.

Brambilla, P., Soloff, P. H., Sala, M., Nicoletti, M. A., Keshavan, M. S., & Soares, J. C. (2004). Anatomical MRI study of borderline personality disorder patients. Psychiatry Research, 131(2), 125–133.

Brown, G. L., Ebert, M. H., Goyer, P. F., Jimerson, D. C., Klein, W. J., Bunney, W. E., et al. (1982). Aggression, suicide, and serotonin: relationships to CSF amine metabolites. The American Journal of Psychiatry, 139(6), 741–746.

Brown, G. L., Goodwin, F. K., Ballenger, J. C., Goyer, P. F., & Major, L. F. (1979). Aggression in humans correlates with cerebrospinal fluid amine metabolites. Psychiatry Research, 1(2), 131–139.

Brunnlieb, C., Münte, T. F., Krämer, U., Tempelmann, C., & Heldmann, M. (2013). Vasopressin modulates neural responses during human reactive aggression. Social Neuroscience, 8(2), 148–164.

Cahill, L. (1995). The amygdala and emotional memory. Nature, 377, 295–296.

Campbell, A., & Hausmann, M. (2013). Effects of oxytocin on women's aggression depend on state anxiety. Aggressive Behavior, 39, 316–322.

Cherek, D. R. (1981). Effects of smoking different doses of nicotine on human aggressive behavior. Psychopharmacology, 75(4), 339–345.

Cherek, D. R., & Lane, S. D. (2001). Acute effects of D-fenfluramine on simultaneous measures of aggressive escape and impulsive responses of adult males with and without a history of conduct disorder. Psychopharmacology, 157(3), 221–227.

Cherek, D. R., Lane, S. D., Pietras, C. J., Sharon, J., & Steinberg, J. L. (2002). Acute effects of baclofen, a gamma-aminobutyric acid-B agonist, on laboratory measures of aggressive and escape responses of adult male parolees with and without a history of conduct disorder. Psychopharmacology, 164(2), 160–167.

Cherek, D. R., Lane, S. D., Pietras, C. J., & Steinberg, J. L. (2002). Effects of chronic paroxetine administration on measures of aggressive and impulsive responses of adult males with a history of conduct disorder. Psychopharmacology, 59(3), 266–274.

Cherek, D. R., Moeller, F. G., Schnapp, W., & Dougherty, D. M. (1997). Studies of violent and nonviolent male parolees: I. Laboratory and psychometric measurements of aggression. *Biological Psychiatry*, *41*(5), 514–522.

Cherek, D. R., Tcheremissine, O. V., Lane, S. D., & Pietras, C. J. (2004). Acute effects of gabapentin on laboratory measures of aggressive and escape responses of adult parolees with and without a history of conduct disorder. *Psychopharmacology*, *171*(4), 405–412.

Coccaro, E. F. (2006). Association of C-reactive protein elevation with trait aggression and hostility in personality disordered subjects: a pilot study. *Journal of Psychiatric Research*, *40*(5), 460–465.

Coccaro, E. F. (2012). Intermittent explosive disorder as a disorder of impulsive aggression for DSM-5. *The American Journal of Psychiatry*, *169*(6), 577–588.

Coccaro, E. F., Fanning, J. R., Fisher, E., Couture, L., & Lee, R. J. (2017). Social emotional information processing in adults: development and psychometrics of a computerized video assessment in healthy controls and aggressive individuals. *Psychiatry Research*, *248*, 40–47.

Coccaro, E. F., Fanning, J. R., Keedy, S. K., & Lee, R. J. (2016). Social cognition in Intermittent Explosive Disorder and aggression. *Journal of Psychiatric Research*, *83*, 140–150.

Coccaro, E. F., Fanning, J. R., & Lee, R. (2017). Intermittent explosive disorder and substance use disorder: analysis of the national comorbidity survey replication sample. *The Journal of Clinical Psychiatry*, *78*, 697–702.

Coccaro, E. F., Fanning, J. R., Phan, K. L., & Lee, R. (2015). Serotonin and impulsive aggression. *CNS Spectrums*, *20*(3), 295–302.

Coccaro, E. F., Fitzgerald, D. A., Lee, R., McCloskey, M. S., & Phan, K. L. (2016). Fronto-limbic morphomentric abnormalities in intermittent explosive disorder and aggression. *Biological Psychiatry: Cognitive Neuroscience and Neuroimaging*, *1*, 32–38.

Coccaro, E. F., Harvey, P. D., Kupsaw-Lawrence, E., Herbert, J. L., & Bernstein, D. P. (1991). Development of neuropharmacologically based behavioral assessments of impulsive aggressive behavior. *The Journal of Neuropsychiatry and Clinical Neurosciences*, *3*(2), S44–S51.

Coccaro, E. F., & Kavoussi, R. J. (1997). Fluoxetine and impulsive aggressive behavior in personality-disordered subjects. *Archives of General Psychiatry*, *54*(12), 1081–1088.

Coccaro, E. F., Kavoussi, R. J., Cooper, T. B., & Hauger, R. (1998). Acute tryptophan depletion attenuates the prolactin response to d-fenfluramine challenge in healthy human subjects. *Psychopharmacology*, *138*(1), 9–15.

Coccaro, E. F., Kavoussi, R. J., & Hauger, R. L. (1997). Serotonin function and antiaggressive response to fluoxetine: a pilot study. *Biological Psychiatry*, *42*(7), 546–552.

Coccaro, E. F., Kavoussi, R. J., Hauger, R. L., Cooper, T. B., & Ferris, C. F. (1998). Cerebrospinal fluid vasopressin levels: correlates with aggression and serotonin function in personality-disordered subjects. *Archives of General Psychiatry*, *55*(8), 708–714.

Coccaro, E. F., & Lee, R. (2010). Cerebrospinal fluid 5-hydroxyindolacetic acid and homovanillic acid: reciprocal relationships with impulsive aggression in human subjects. *Journal of Neural Transmission (Vienna)*, *117*(2), 241–248.

Coccaro, E. F., Lee, R., & Coussons-Read, M. (2014). Elevated plasma inflammatory markers in individuals with intermittent explosive disorder and correlation with aggression in humans. *JAMA Psychiatry*, *71*(2), 158–165.

Coccaro, E. F., Lee, R., & Coussons-Read, M. (2015a). Cerebrospinal fluid and plasma C-reactive protein and aggression in personality-disordered subjects: a pilot study. *Journal of Neural Transmission (Vienna)*, *122*(2), 321–326.

Coccaro, E. F., Lee, R., & Coussons-Read, M. (2015b). Cerebrospinal fluid inflammatory cytokines and aggression in personality disordered subjects. *The International Journal of Neuropsychopharmacology*, *18*(7), pvy001.

Coccaro, E. F., Lee, R., Fanning, J. R., Fuchs, D., Goiny, M., Erhardt, S., et al. (2016). Tryptophan, kynurenine, and kynurenine metabolites:

relationship to lifetime aggression and inflammatory markers in human subjects. *Psychoneuroendocrinology*, *71*, 189–196.

Coccaro, E. F., Lee, R., & Kavoussi, R. J. (2010). Inverse relationship between numbers of 5-HT transporter binding sites and life history of aggression and intermittent explosive disorder. *Journal of Psychiatric Research*, *44*(3), 137–142.

Coccaro, E. F., Lee, R., Liu, T., & Mathé, A. A. (2012). Cerebrospinal fluid neuropeptide Y-like immunoreactivity correlates with impulsive aggression in human subjects. *Biological Psychiatry*, *72*(12), 997–1003.

Coccaro, E. F., Lee, R., McCloskey, M., Csernansky, J. G., & Wang, L. (2015). Morphometric analysis of amygdla and hippocampus shape in impulsively aggressive and healthy control subjects. *Journal of Psychiatric Research*, *69*, 80–86.

Coccaro, E. F., Lee, R., Owens, M. J., Kinkead, B., & Nemeroff, C. B. (2012). Cerebrospinal fluid substance P-like immunoreactivity correlates with aggression in personality disordered subjects. *Biological Psychiatry*, *72*(3), 238–243.

Coccaro, E. F., Lee, R., & Vezina, P. (2013). Cerebrospinal fluid glutamte concentration correlates with impulsive aggression in human subjects. *Journal of Psychiatric Research*, *47*(9), 1247–1253.

Coccaro, E. F., Lee, R. J., & Kavoussi, R. J. (2009). A double-blind, randomized, placebo-controlled trial of fluoxetine in patients with intermittent explosive disorder. *The Journal of Clinical Psychiatry*, *70*, 653–662.

Coccaro, E. F., McCloskey, M. S., Fitzgerald, D. A., & Phan, K. L. (2007). Amygdala and orbitofrontal reactivity to social threat in individuals with impulsive aggression. *Biological Psychiatry*, *62*(2), 168–178.

Coccaro, E. F., Noblett, K. L., & McCloskey, M. S. (2009). Attributional and emotional responses to socially ambiguous cues: validation of a new assessment of social/emotional information processing in healthy adults and impulsive aggressive patients. *Journal of Psychiatric Research*, *43*, 915–925.

Coccaro, E. F., Siever, L. J., Klar, H. M., Maurer, G., Cochrane, K., Cooper, T. B., et al. (1989). Serotonergic studies in patients with affective and personality disorders. Correlates with suicidal and impulsive aggressive behavior. *Archives of General Psychiatry*, *46*(7), 587–599.

Coccaro, E. F., Sripada, C. S., Yanowitch, R. N., & Phan, K. L. (2011). Corticolimbic function in impulsive aggressive behavior. *Biological Psychiatry*, *69*(12), 1153–1159.

Comai, S., Tau, M., & Gobbi, G. (2012). The psychopharmacology of aggressive behavior: a translational approach: part 1: neurobiology. *Journal of Clinical Psychopharmacology*, *32*, 83–94.

Cowdry, R. W., & Gardner, D. L. (1988). Pharmacotherapy of borderline personality disorder. *Archives of General Psychiatry*, *45*, 111–119.

Cremers, H., Lee, R., Keedy, S., Phan, K. L., & Coccaro, E. F. (2016). Effects of escitalopram administration on face processing in intermittent explosive disorder: an fMRI study. *Neuropsychopharmacology*, *41*(2), 590–597.

Crick, N. R., & Dodge, K. A. (1996). Social information-processing mechanisms in reactive and proactive aggression. *Child Development*, *67*(3), 993–1002.

Damasio, A. R. (1994). *Descartes' error*. New York: Putnam.

Damasio, A. R., Brabowski, T., Bechara, A., Damasio, H., Ponto, L. L. B., Parvazi, J., et al. (2000). Subcortical and cortical brain activity during the feeling of self-generated emotions. *Nature Neuroscience*, *10*, 1049–1056.

Davidson, R. J., Putnam, K. M., & Larson, C. L. (2000). Dysfunction in the neural circuitry of emotion regulation—a possible prelude to violence. *Science*, *289*(5479), 591–594.

De Dreu, C. K. (2012). Oxytocin modulates cooperation within and competition between groups: an integrative review and research agenda. *Hormones and Behavior*, *61*, 419–428.

DiMasi, J. A., Grabowski, H. G., & Hansen, R. W. (2016). Innovation in the pharmaceutical industry: new estimates of R&D costs. *Journal of Health Economics*, *47*, 20–33.

DNDi: Drugs for Neglected Diseases Initiative. (2013). *Annual report.* New York, NY: DNDi.

Dodge, K. A. (1991). The structure and function of reactive and proactive aggression. In D. Peppler & K. Rubin (Eds.), *The development and treatment of childhood aggression* (pp. 201–218). Hillsdale, NJ: Erlbaum.

Dodge, K. A., Pettit, G. S., Bates, J. E., & Valente, E. (1995). Social information-processing patterns partially mediate the effect of early physical abuse on later conduct problems. *Journal of Abnormal Psychology, 104*(4), 632–643.

Dougherty, D. D., Shin, L. M., Alpert, N. M., Pitman, R. K., Orr, S. P., Lasko, M., et al. (1999). Anger in healthy men: a PET study using script-driven imagery. *Biological Psychiatry, 46*(4), 466–472.

Duke, A. A., Bègue, L., Bell, R., & Eisenlohr-Moul, T. (2013). Revisiting the serotonin-aggression relation in humans: a meta-analysis. *Psychological Bulletin, 139*(5), 1148–1172.

Elliott, P. J., & Iversen, S. D. (1986). Behavioural effects of tachykinins and related peptides. *Brain Research, 381*, 68–76.

Elliott, R., Dolan, R. J., & Frith, C. D. (2000). Dissociable functions in the medial and lateral orbitofrontal cortex: evidence from human neuroimaging studies. *Cerebral Cortex, 10*, 308–317.

Endicott, J., Tracy, K., Burt, D., Olson, E., & Coccaro, E. F. (2002). A novel approach to assess inter-rater reliability in the use of the Overt Aggression Scale-Modified. *Psychiatry Research, 112*(2), 153–159.

Fabio, K., Guillon, C., Lu, S., Heindel, N., Miller, M., Ferris, C., et al. (2010). Vasopressin antagonists as anxiolytics and antidepressants: recent developments. *Front CNS Drug Discovery, 1*, 156–183.

Ferris, C. F. (2005). Vasopressin/oxytocin and aggression. *Novartis Foundation Symposium, 268*, 190–198.

Ferris, C. F., Lu, S. -F., Messenger, T., Guillon, C. D., Heindel, N., Miller, M., et al. (2006). Orally active vasopressin V1a receptor antagonist, SRX251, selectively blocks aggressive behavior. *Pharmacology, Biochemistry, and Behavior, 83*, 169–174.

Ferris, C. F., Stolberg, T., Kulkarni, P., Murugavel, M., Blanchard, R., Blanchard, D. C., et al. (2008). Imaging the neural circuitry and chemical control of aggressive motivation. *BMC Neuroscience, 9*, 111.

Fontaine, R. G., Burks, V. S., & Dodge, K. A. (2002). Response decision processes and externalizing behavior problems in adolescents. *Development and Psychopathology, 14*(1), 107122.

George, D. T., Phillips, M. J., Lifshitz, M., Lionetti, T. A., Spero, D. E., Ghassemzedeh, N., et al. (2011). Fluoxetine treatment of alcoholic perpetrators of domestic violence: a 12-week, double-blind, randomized, placebo-controlled intervention study. *The Journal of Clinical Psychiatry, 72*(1), 60–65.

George, M. S., Ketter, T. A., Gill, D. S., Haxby, J. V., Ungerleider, L. G., Herscovitch, P., et al. (1993). Brain regions involved in recognizing facial emotion or identity: an oxygen-15 PET study. *The Journal of Neuropsychiatry and Clinical Neurosciences, 5*(4), 384–394.

Giancola, P. R., & Parrott, D. J. (2008). Further evidence for the validity of the Taylor Aggression Paradigm. *Aggressive Behavior, 34*(2), 214–229.

Gobbi, G., Cassano, T., Radja, F., Morgese, M. G., Cuomo, V., Santarelli, L., et al. (2007). Neurokinin 1 receptor antagonism requires norepinephrine to increase serotonin function. *European Neuropsychopharmacology, 17*, 328–338.

Golomb, B. A., Cortez-Perez, M., Jaworski, B. A., Mednick, S., & Dimsdale, J. (2007). Point subtraction aggression paradigm: validity of a brief schedule of use. *Violence and Victims, 22*(1), 95–103.

Grafman, J., Schwab, K., Warden, D., Pridgen, A., Brown, H. R., & Salazar, A. M. (1996). Frontal lobe injuries, violence, and aggression: a report of the Vietnam head injury study. *Neurology, 46*, 1231–1238.

Greenway, F. L., Shanahan, W., Fain, R., Ma, T., & Rubino, D. (2016). Safety and tolerability review of lorcaserin in clinical trials. *Clinical Obesity, 6*(5), 285–295.

Gregg, T. R., & Siegel, A. (2001). Brain structures and neurotransmitters regulating aggression in cats: implications for human aggression. *Progress in Neuro-Psychopharmacology & Biological Psychiatry, 25*, 91–140.

Guastella, A. J., Kenyon, A. R., Alvares, G. A., Carson, D. S., & Hickie, I. B. (2010). Intranasal arginine vasopressin enhances the encoding of happy and angry faces in humans. *Biological Psychiatry, 67*(12), 1220–1222.

Haller, J. (2013). The neurobiology of abnormal manifestations of aggression: a review of hypothalamic mechanisms in cats, rodents, and humans. *Brain Research Bulletin, 93*, 97–109.

Han, Y., Shaikh, M. B., & Siegel, A. (1996). Medial amygdaloid suppression of predatory attack behavior in the cat: I. Role of a substance P pathway from the medial amygdala to the medial hypothalamus. *Brain Research, 716*, 59–71.

Hassanain, M., Bhatt, S., Zalcman, S., & Siegel, A. (2005). Potentiating role of interleukin-1beta (IL-1beta) and IL-1beta type 1 receptors in the medial hypothalamus in defensive rage behavior in the cat. *Brain Research, 1048*(1-2), 1–11.

Hassanain, M., Zalcman, S., Bhatt, S., & Siegel, A. (2003). Interleukin-1 beta in the hypothalamus potentiates feline defensive rage: role of serotonin-2 receptors. *Neuroscience, 120*(1), 227–233.

Hayes, J. P., Vanelzakker, M. B., & Shin, L. M. (2012). Emotion and cognition interactions in PTSD: a review of neurocognitive and neuroimaging studies. *Frontiers in Integrative Neuroscience, 9*(6), 89.

Hikosaka, K., & Watanabe, M. (2000). Delay activity of orbital and lateral prefrontal neurons of the monkey varying with different rewards. *Cerebral Cortex, 10*, 263–271.

Hollander, E., Tracy, K. A., Swann, A. C., Coccaro, E. F., McElroy, S. L., Wozniak, P., et al. (2003). Divalproex in the treatment of impulsive aggression: efficacy in cluster B personality disorders. *Neuropsychopharmacology, 28*(6), 1186–1197.

Hornak, J., Rolls, E. T., & Wade, D. (1996). Face and voice expression identification in patients with emotional and behavioural changes following ventral frontal lobe damage. *Neuropsychologia, 34*(4), 247–261.

Iversen, S. D., & Mishkin, M. (1970). Perseverative interference in monkeys following selective lesions of the inferior prefrontal convexity. *Experimental Brain Research, 11*(4), 376–386.

Karl, T., Lin, S., Schwarzer, C., Sainsbury, A., Couzens, M., Wittmann, W., et al. (2004). Y1 receptors regulate aggressive behavior by modulating serotonin pathways. *Proceedings of the National Academy of Sciences of the United States of America, 101*(34), 12742–12747.

Kask, A., & Harro, J. (2000). Inhibition of amphetamine- and apomorphine-induced behavioural effects by neuropeptide Y Y(1) receptor antagonist BIBO 3304. *Neuropharmacology, 39*(7), 1292–1302.

Kessler, R. C., Coccaro, E. F., Fava, M., Jaeger, S., Jin, R., & Walters, E. (2006). The prevalence and correlates of DSM-IV intermittent explosive disorder in the National Comorbidity Survey Replication. *Archives of General Psychiatry, 63*(6), 669–678.

Kimbrell, T. A., George, M. S., Parekh, P. I., Ketter, T. A., Podell, D. M., Danielson, A. L., et al. (1999). Regional brain activity during transient self-induced anxiety and anger in healthy adults. *Biological Psychiatry, 46*, 454–465.

Kluver, H., & Bcy, P. C. (1939). Preliminary analysis of functions of the temproal lobes in monkees. *Archives of Neurology and Psychiatry, 42*(6), 979–1000.

Köhler, O., Benros, M. E., Nordentoft, M., Farkouh, M. E., Iyengar, R. L., Mors, O., et al. (2014). Effect of anti-inflammatory treatment on depression, depressive symptoms, and adverse effects: a systematic review and meta-analysis of randomized clinical trials. *JAMA Psychiatry, 71*(12), 1381–1391.

Koster-Hale, J., & Saxe, R. (2013). Theory of mind: a neural prediction problem. *Neuron, 79*(5), 836–848.

Krall, S. C., Volz, L. J., Oberwelland, E., Grefkes, C., Fink, G. R., & Konrad, K. (2016). The right temporoparietal junction in attention and social interaction: a transcranial magnetic stimulation study. *Human Brain Mapping, 37*(2), 796–807.

Kringelbach, M. L., & Rolls, E. T. (2003). Neural correlates of rapid reversal learning in a simple model of human social interaction. *NeuroImage, 20*(2), 1371–1383.

Lane, R. D., Fink, G. R., Chau, P. M., & Dolan, R. J. (1997). Neural activation during selective attention to subjective emotional responses. *Neuroreport, 8*(18), 3969–3972.

Lee, R., Arfanakis, K., Evia, A. M., Fanning, J., Keedy, S., & Coccaro, E. F. (2016). White matter integrity reductions in intermittent explosive disorder. *Neuropsychopharmacology, 41*(11), 2697.

Lee, R., Coccaro, E. F., Cremers, H., McCarron, R., Lu, S. F., Brownstein, M. J., et al. (2013). A novel V1a receptor antagonist blocks vasopressin-induced changes in the CNS response to emotional stimuli: an fMRI study. *Frontiers in Systems Neuroscience, 7*, 1–11.

Lee, R., Ferris, C., Van de Kar, L. D., & Coccaro, E. F. (2009). Cerebrospinal fluid oxytocin, life history of aggression, and personality disorder. *Psychoneuroendocrinology, 34*(10), 1567–1573.

Lee, R., Petty, F., & Coccaro, E. F. (2009). Cerebrospinal fluid GABA concentration: relationship with impulsivity and history of suicidal behavior, but not aggression, in human subjects. *Journal of Psychiatric Research, 43*(4), 353–359.

Lieving, L. M., Cherek, D. R., Lane, S. D., Tcheremissine, O. V., & Nouvion, S. O. (2008). Effects of acute tiagabine administration on aggressive responses of adult male parolees. *Journal of Psychopharmacology, 22*(2), 144–152.

Limson, R., Goldman, D., Roy, A., Lamparski, D., Ravitz, B., Adinoff, B., et al. (1991). Personality and cerebrospinal fluid monoamine metabolites in alcoholics and controls. *Archives of General Psychiatry, 48*(5), 437–441.

Linnoila, M., Virkkunen, M., Scheinin, M., Nuutila, A., Rimon, R., & Goodwin, F. K. (1983). Low cerebrospinal fluid 5-hydroxyindoleacetic acid concentration differentiates impulsive from nonimpulsive violent behavior. *Life Sciences, 33*(26), 2609–2614.

Lu, S., Simon, N. G., Palkovits, M., & Brownstein, M. J. (2013). *Identification and distribution of vasopressin 1a receptor in human and monkey brain.* San Diego, CA: Society for Neuroscience. 673.11.

Mars, R. B., Sallet, J., Schüffelgen, U., Jbabdi, S., Toni, I., & Rushworth, M. F. (2012). Connectivity-based subdivisions of the human right "temporoparietal junction area": evidence for different areas participating in different cortical networks. *Cerebral Cortex, 22*(8), 1894–1903.

Marsland, A. L., Prather, A. A., Petersen, K. L., Cohen, S., & Manuck, S. B. (2008). Antagonistic characteristics are positively associated with inflammatory markers independently of trait negative emotionality. *Brain, Behavior, and Immunity, 22*, 753–761.

McCloskey, M. S., Noblett, K. L., Deffenbacher, J. L., Gollan, J. K., & Coccaro, E. F. (2008). Cognitive-behavioral therapy for intermittent explosive disorder: a pilot randomized clinical trial. *Journal of Consulting and Clinical Psychology, 76*(5), 876–886.

McCloskey, M. S., Phan, K. L., Angstadt, M., Fettich, K. C., Keedy, S., & Coccaro, E. F. (2016). Amygdala hyperactivation to angry faces in intermittent explosive disorder. *Journal of Psychiatric Research, 79*, 34–41.

Meyer-Lindenberg, A., Kolachana, B., Gold, B., Olsh, A., Nicodemus, K. K., Mattay, V., et al. (2009). Genetic variants in AVPR1A linked to autism predict amygdala activation and personality traits in healthy humans. *Molecular Psychiatry, 14*(10), 968–975.

Miczek, K. A., Fish, E. W., De Bold, J. F., & De Almeida, R. M. (2002). Social and neural determinants of aggressive behavior: pharmacotherapeutic targets at serotonin, dopamine and gamma-aminobutyric acid systems. *Psychopharmacology (Berlin), 163*(3–4), 434–458.

Moeller, F. G., Allen, T., Cherek, D. R., Dougherty, D. M., Lane, S., & Swann, A. C. (1998). Ipsapirone neuroendocrine challenge: relationship to aggression as measured in the human laboratory. *Psychiatry Research, 81*(1), 31–38.

Molenberghs, P., Johnson, H., Henry, J. D., & Mattingley, J. B. (2016). Understanding the minds of others: a neuroimaging meta-analysis. *Neuroscience and Biobehavioral Reviews, 65*, 276–291.

New, A. S., Buchsbaum, M. S., Hazlett, E. A., Goodman, M., Koenigsberg, H. W., Lo, J., et al. (2004). Fluoxetine increases relative metabolic rate in prefrontal cortex in impulsive aggression. *Psychopharmacology, 176*, 451–458.

Norden, M. J. (1989). Fluoxetine in borderline personality disorder. *Progress in Neuro-Psychopharmacology & Biological Psychiatry, 13*(6), 885–893.

Northoff, G., Richter, A., Gessner, M., Schlagenhauf, F., Fell, J., Baumgart, F., et al. (2000). Functional dissociation between medial and lateral prefrontal cortical spatiotemporal activation in negative and positive emotions: a combined fMRI/MEG study. *Cerebral Cortex, 10*(1), 93–107.

Ongur, D., & Price, J. L. (2000). The organization of networks within the orbital and medial prefrontal cortex of rats, monkeys and humans. *Cerebral Cortex, 10*, 206–219.

Oquendo, M. A., & Mann, J. J. (2000). The biology of impulsivity and suicidality. *Psychiatric Clinics of North America, 23*, 11–25.

Patel, U. J., Grossman, R. I., Phillips, M. D., Udupa, J. K., McGowan, J. C., Miki, Y., et al. (1999). Serial analysis of magnetization-transfer histograms and Expanded Disability Status Scale scores in patients with relapsing-remitting multiple sclerosis. *AJNR. American Journal of Neuroradiology, 20*(10), 1946–1950.

Perez-Cruet, J., Tagliamonte, A., Tagliamonte, P., & Gessa, G. L. (1971). Stimulation of serotonin synthesis by lithium. *The Journal of Pharmacology and Experimental Therapeutics, 178*(2), 325–330.

Petitto, J. M., Lysle, D. T., Gariepy, J. L., & Lewis, M. H. (1994). Association of genetic differences in social behavior and cellular immune responsiveness: effects of social experience. *Brain, Behavior, and Immunity, 8*(2), 111–122.

Pietrini, P., Guazzelli, M., Basso, G., Jaffe, K., & Grafman, J. (2000). Neural correlates of imaginal aggressive behavior assessed by positron emission tomography in healthy subjects. *The American Journal of Psychiatry, 157*(11), 1772–1781.

Price, J., & Drevets, W. (2012). Neural circuits underlying the pathophysiology of mood disorders. *Trends in Cognitive Sciences, 16*, 61–71.

Rilling, J. K., DeMarco, A. C., Hackett, P. D., Thompson, R., Ditzen, B., Patel, R., et al. (2012). Effects of intranasal oxytocin and vasopressin on cooperative behavior and associated brain activity in men. *Psychoneuroendocrinology, 37*, 447–461.

Ring, R. (2005). The central vasopressinergic system: examining opportunities for CNS drug development. *Current Pharmaceutical Design, 11*, 205–225.

Rolls, E. T. (2004). The functions of the orbitofrontal cortex. *Brain and Cognition, 55*(1), 11–29.

Rolls, E. T., Cahusac, P. M., Feigenbaum, J. D., & Miyashita, Y. (1993). Responses of single neurons in the hippocampus of the macaque related to recognition memory. *Experimental Brain Research, 93*(2), 299–306.

Rutkoski, N. J., Lerant, A. A., Nolte, C. M., Westberry, J., & Levenson, C. W. (2002). Regulation of neuropeptide Y in the rat amygdale following unilateral olfactory bulbectomy. *Brain Research, 951*, 69–76.

Saddoris, M. P., Gallagher, M., & Schoenbaum, G. (2005). Rapid associative encoding in basolateral amygdala depends on connections with orbitofrontal cortex. *Neuron, 46*(2), 321–331.

Saxe, R., & Wexler, A. (2005). Making sense of another mind: the role of the right temporo-parietal junction. *Neuropsychologia, 43*(10), 1391–1399.

Schlüter, T., Winz, O., Henkel, K., Prinz, S., Rademacher, L., Schmaljohann, J., et al. (2013). The impact of dopamine on aggression: an [18F]-FDOPA PET Study in healthy males. *The Journal of Neuroscience, 33*, 16889–16896.

Serradeil-Le Gal, C., Wagnon, J., Valette, G., Garcia, G., Pascal, M., Maffrand, J. P., et al. (2002). Nonpeptide vasopressin receptor antagonists: development of selective and orally active V1a, V2 and V1b receptor ligands. *Progress in Brain Research, 139*, 197–210.

Shaikh, M. B., Steinberg, A., & Siegel, A. (1993). Evidence that substance P is utilized in medial amygdaloid facilitation of defensive rage behavior in the cat. *Brain Research, 625*, 283–294.

Shamay-Tsoory, S. G., Fischer, M., Dvash, J., Harari, H., Perach-Bloom, N., & Levkovitz, Y. (2009). Intranasal administration of oxytocin increases envy and schadenfreude (gloating). *Biological Psychiatry, 66*, 864–870.

Sheard, M. H. (1970). Effect of lithium on foot shock aggression in rats. *Nature, 228*(5268), 284–285.

Sheard, M. H., Marini, J. L., Bridges, C. I., & Wagner, E. (1976). The effect of lithium on impulsive aggressive behavior in man. *The American Journal of Psychiatry, 133*(12), 1409–1413.

Silva, H., Iturra, P., Solari, A., Villarroel, J., Jerez, S., Jimenez, M., et al. (2010). Fluoxetine response in impulsive-aggressive behavior and serotonin transporter polymorphism in personality disorder. *Psychiatric Genetics, 20*(1), 25–30.

Simon, N. G. (2002). Hormonal processes in the development and expression of aggressive behavior. *Hormones, Brain and Behavior, 1*, 339–392.

Simon, N. G., Guillon, C., Fabio, K., Heindel, N. D., Lu, S. F., Miller, M., et al. (2008). Vasopressin antagonists as anxiolytics and antidepressants: recent developments. *Recent Patents on CNS Drug Discovery, 3*, 77–93.

Smith, G. S., Savery, D., Marden, C., Lopez Costa, J. J., Averill, S., Priestley, J. V., et al. (1994). Distribution of messenger RNAs encoding enkephalin, substance P, somatostatin, galanin, vasoactive intestinal polypeptide, neuropeptide Y, and calcitonin gene-related peptide in the midbrain periaqueductal grey in the rat. *Journal of Comparative Neurology, 350*, 23–40.

Spoont, M. R., Kuskowski, M., & Pardo, J. V. (2010). Autobiographical memories of anger in violent and non-violent individuals: a script-driven imagery study. *Psychiatry Research, 183*(3), 225–229.

Suarez, E. C. (2003). Joint effect of hostility and severity of depressive symptoms on plasma interleukin-6 concentration. *Psychosomatic Medicine, 65*(4), 523–527.

Suarez, E. C. (2004). C-reactive protein is associated with psychological risk factors of cardiovascular disease in apparently healthy adults. *Psychosomatic Medicine, 66*(5), 684–691.

Taylor, S. P. (1967). Aggressive behavior and physiological arousal as a function of provocation and the tendency to inhibit aggression. *Journal of Personality, 35*(2), 297–310.

Thomas, D. W., Burns, J., Audette, J., Carroll, A., Dow-Hygelund, C., & Hay, M. (2016). *Clinical development success rates 2006-2015*. Washington, DC: Biotechnology Industry Organization.

Thompson, R., Gupta, S., Miller, K., Mills, S., & Orr, S. (2004). The effects of vasopressin on human facial responses related to social communication. *Psychoneuroendocrinology, 29*, 35–48.

Thompson, R. R., George, K., Walton, J. C., Orr, S. P., & Benson, J. (2006). Sex-specific influences of vasopressin on human social communication. *Proceedings of the National Academy of Sciences of the United States of America, 103*, 7889–7894.

Valzelli, L., & Garattini, S. (1968). Behavioral changes and 5-hydroxytryptamine turnover in animals. *Advances in Pharmacology, 6*(B), 249–260.

Virkkunen, M., De Jong, J., Bartko, J., Goodwin, F. K., & Linnoila, M. (1989). Relationship of psychobiological variables to recidivism in violent offenders and impulsive fire setters. A follow-up study. *Archives of General Psychiatry, 46*(7), 600–603.

Virkkunen, M., Nuutila, A., Goodwin, F. K., & Linnoila, M. (1987). Cerebrospinal fluid monoamine metabolite levels in male arsonists. *Archives of General Psychiatry, 44*(3), 241–247.

Wilcock, G. K., Ballard, C. G., Cooper, J. A., & Loft, H. (2011). Memantine for agitation/ aggression and psychosis in moderately severe to severe Alzheimer's disease: a pooled analysis of 3 studies. *Journal of Clinical Psychiatry, 69*, 341–348.

Wilson, F. A., Scalaidhe, S. P., & Goldman-Rakic, P. S. (1993). Dissociation of object and spatial processing domains in primate prefrontal cortex. *Science, 260*(5116), 1955–1958.

Yanowitch, R., & Coccaro, E. F. (2011). The neurochemistry of human aggression. *Advances in Genetics, 75*, 151–169.

Yip, J., & Chahl, L. A. (2001). Localization of NK1 and NK3 receptors in guinea-pig brain. *Regulatory Peptides, 98*, 55–62.

Young, L. J., Toloczko, D., & Insel, T. R. (1999). Localization of vasopressin (V1a) receptor binding and mRNA in the rhesus monkey brain. *Journal of neuroendocrinology, 11*, 291–297.

Yudofsky, S. C., Silver, J. M., Jackson, W., Endicott, J., & Williams, D. (1986). The Overt Aggression Scale for the objective rating of verbal and physical aggression. *The American Journal of Psychiatry, 143*(1), 35–39.

Zald, D. H., & Sim, S. W. (1996). The anatomy and function of the orbital frontal cortex, II. Function and relevance to obsessive compulsive disorder. *Journal of Neuropsychiatry & Clinical Neurosciences, 8*, 249–261.

Zink, C. F., Kempf, L., Hakimi, S., Rainey, C. A., Stein, J. L., & Meyer-Lindenberg, A. (2011). Vasopressin modulates social recognition-related activity in the left temporoparietal junction in humans. *Translational Psychiatry, 1*, 1–5.

Zink, C. F., Stein, J. L., Kempf, L., Hakimi, S., & Meyer-Lindenberg, A. (2010). Vasopressin modulates medial prefrontal cortex-amygdala circuitry during emotion processing in humans. *Journal of Neuroscience, 30*, 7017–7022.

Hypothesizing Major Depression as a Subset of Reward Deficiency Syndrome (RDS) Linked to Polymorphic Reward Genes: Considerations for Translational Medicine Approaches for Future Drug Development

Kenneth Blum[*,†,‡,§,¶,||,#,**,††,‡‡], Mark S. Gold[‡,§§],
Edward J. Modestino[¶¶], Igor Elman[‡‡], David Baron[*,‡],
Rajendra D. Badgaiyan[||||,##], Abdalla Bowirrat[***]

[*]Western University Health Sciences, Graduate School of Biomedical Sciences, Pomona, CA, United States [†]Division of Addiction Services, Dominion Diagnostics LLC, North Kingstown, RI, United States [‡]Division of Neuroscience and Addiction Research, Pathway Healthcare, Birmingham, AL, United States [§]Division of Neuroscience Research and Addiction Therapy, Shores Treatment and Recovery Center, Port Saint Lucie, FL, United States [¶]Human Integrated Services Unit, University of Vermont Centre for Clinical & Translational Science, College of Medicine, Burlington, VT, United States [||]Eötvös Loránd University, Institute of Psychology, Budapest, Hungary [#]Division of Clinical Neurology, Path Foundation NY, New York, NY, United States [**]Division of Addiction Research & Therapy, Nupathways, Innsbrook, MO, United States [††]Division of Precision Medicine, Geneus Health LLC, San Antonio, TX, United States [‡‡]Department of Psychiatry, Wright State University, Boonshoft School of Medicine, Dayton, OH, United States [§§]Department of Psychiatry, Washington University School of Medicine, St. Louis, MO, United States [¶¶]Department of Psychology, Curry College, Milton, MA, United States [||||]Department of Psychiatry, South Texas Veteran Health Care System, Audie L. Murphy Memorial VA Hospital, San Antonio, TX, United States [##]Long School of Medicine, University of Texas Medical Center, San Antonio, TX, United States [***]Division of Anatomy, Biochemistry and Genetics, Faculty of Medicine and Health Sciences, An-Najah National University, Nablus, Palestine

I INTRODUCTION

Afflicting about 17.6 million Americans (González et al., 2010), major depressive disorder (MDD) has been speculated to become the leading cause of disability in the United States (González et al., 2010; Greenberg, Fournier, Sisitsky, Pike, & Kessler, 2015). MDD now costs an estimated $210 billion annually due to the loss of productivity from medical comorbidities (Knol et al., 2006; Nouwen et al., 2010) and long-term disability (Greenberg et al., 2015), let alone enormous human suffering and suicidal attempts in up to 15% of the depressed

patients (Graßnickel, Illes, Juckel, & Uhl, 2015). Unfortunately, major depression is also an underdiagnosed and undertreated entity, so those who screen positive for depression do not receive adequate therapy, whereas those who are in treatment may not even fulfill the diagnostic criteria (Mental Health America, 2016).

In an attempt to understand reward dependence behaviors and associated neurogenetic impairments leading to brain reward dysphoria or depression and thus social ineptness, more in-depth research is required. Our laboratory coined the term Reward Deficiency Syndrome (RDS) in 1996 to describe a standard genetic rubric of impulsive, compulsive, and addictive behaviors, which is an important emerging psychoneurogenetic concept, now being adopted worldwide, and it is listed as a disorder in SAGE 2017 Encyclopedia of Abnormal Psychology (Blum et al., 1996). To highlight the importance of this topic, a meeting of the World Congress of Psychiatric Genetics was held in New York City, where approximately 60 articles were presented involving genes and depression. Interestingly, out of these research reports, roughly 20% included the comorbidity of depression, and many related RDS behaviors including substance use disorder (SUD), ADHD, obsessive-compulsive disorder, and anxiety. Regarding systematically mapping genes associated with depression alone and in association with RDS behaviors, we are proposing genotyping of diagnosed MDD and RDS probands compared with a highly screened controlled population.

Akin to other neuropsychiatric disorders, MDD could be attributed to a constellation of many factors including developmental, environmental, social, psychological, medical, molecular, demographic, and vocational (Brugha, 2003; Endicott, 1998; Kinney & Tanaka, 2009). Nonetheless the mere fact that the incidence of this disorder has risen over the years (Compton, 2006) despite the possible somatic and psychosocial interventions and intense monitoring efforts calls for novel insights informing better prevention, diagnosis, and management of MDD.

Although MDD is classified as a mood disorder (DSM-5), it is a complex condition with features that extend beyond mood per se. For instance, anhedonia symptoms, that is, the loss of interest and pleasure, may suggest deficits in the brain reward/reinforcement circuits, and such symptoms have been linked with reward function deficits in clinical (Elman et al., 2009) and healthy samples (Chung & Barch, 2015). Indeed, in Freud's seminal paper in 1917, *Mourning and Melancholia*, he highlighted the key role of disturbed social attachments in the course of depression. The function of social attachments is evolutionarily embedded within the reward/motivation neural network, responsible for repetitious actions essential for the existence of individuals and species (Borsook et al., 2016; Elman & Borsook, 2016; Insel, 2003). This may be because human beings were destined for demise from starvation and/or from acts of violence in the absence of affiliation with one another.

Despite supportive animal data (Nestler & Carlezon Jr, 2006) and suggestive clinical literature (Elman et al., 2005), MDD has not yet been proposed for the inclusion within the reward deficiency spectrum; that is to say, a state of depressed striatal dopaminergic activity due to a paucity or aberration of the dopamine D2 dopaminergic receptors (Blum et al., 1995). Such inclusion may highlight potential directions for systematic thinking and for positing novel ideas and questions aimed at understanding the course of MDD and what currently makes many patients partially or wholly resistant to conventional antidepressant modalities. Several lines of evidence support this sort of conceptualization, including the involvement of the reward-related transcription factors (e.g., CREB, ΔFosB, SRF, NFκB, and β-catenin) in the pathophysiology of MDD (Ferguson, 1984; Walters, Kuo, & Blendy, 2003) along with neuroimaging research in MDD implicating the fundamental reward structures including the nucleus accumbens (NAc), amygdala, and medial prefrontal cortex (Szczepanik et al., 2016; Young et al., 2016). The function of this our review is to extend the existing literature by synthesizing the existing genetic evidence in support of the inclusion of MDD within the RDS and their possible resolution with prodopamine regulation.

Most recently, Stein, Rivera, Anderson, and Bailey (2017) reported on many significant findings of the comorbidity of MDD and substance use disorder (SUD). They found the following:

1. Nearly two out of three persons were screened as having MDD, yet only 8.2% were being treated for it before admission.
2. Screening positive for MDD was associated with a 2.95-fold increase in the expected odds of perceived need for depression treatment (PNDT).
3. Nearly half of those MDD (48%) did not PNDT.
4. Approximately 40% of the participants saw that they were not depressed; of these participants, 52% screened positive for MDD.

Stein et al. (2017) apparently encouraged that patients seeking opioid treatment should be screened for MDD and treated accordingly. It is noteworthy that it has been reported earlier that upward of 90% of patients attending treatment for SUD diagnose positive for MDD, however, following residential treatment, only one-third of the patients present with an MDD diagnosis upon leaving treatment (Archer, Oscar-Berman, Blum, & Gold, 2013; Beaulieu et al., 2012; Blum, Wallace, & Hall, 1986; Staiger, Thomas, Ricciardelli, & McCabe, 2011).

Mood disorders are one of the most important causes of disability for human health and the second leading source of disease burden, going beyond cardiovascular diseases, dementia, lung cancer, and diabetes. While genetic risk factors are well established for MDD as a chronic disorder (Bentley, Pagalilauan, & Simpson, 2014), little is known about the putative genomic relationship between impulsive, compulsive, and addictive behaviors and depression. The important fundamental question of understanding the true nature of MDD and resultant comorbidities reside in the genomic expression of polygenes of the mesolimbic pathway and in particular the brain reward cascade (Blum & Braverman, 2000).

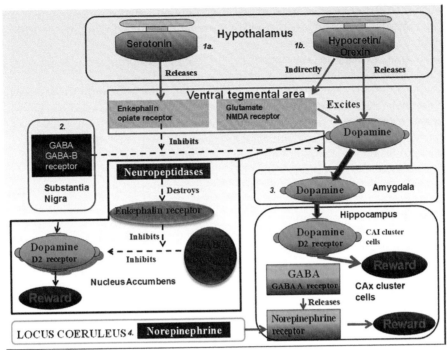

FIG. 30.1 Interactions in brain reward regions associated with Reward Deficiency Syndrome (RDS). (1a) Hypothalamus: Serotonin indirectly activates opiate receptors, causing a release of enkephalins in the ventral tegmental area (VTA). The enkephalins inhibit the firing of GABA, which originates in the substantia nigra. (2) GABA: GABA acts through GABA-B receptors to inhibit/regulate the release of dopamine (DA) at the VTA, projecting to nucleus accumbens (NAcc). The release of DA in NAcc activates DA D2 receptors. This same release is modulated by enkephalins via GABA. Enkephalins are regulated by neuropeptidases. (3) DA also is released in the amygdala. Projecting from the amygdala, DA stimulates the hippocampus (HIPP) where CA cells excite DA D2 receptors. (4) Additionally, norepinephrine (NE) in the locus coeruleus innervates the HIPP around CAx (a cluster of cells yet to be identified). In the HIPP, excitation of GABA-A receptors causes the release of NE. (1b) Hypothalamus: Hypocretin/orexin release from the lateral hypothalamus causes an excitatory modulation of DA in the VTA (directly via receptors on dopaminergic neurons and indirectly by augmenting glutamatergic excitability of dopaminergic neurons via increasing NMDA receptor number). *Reprinted with permission from Modestino, E.J., Blum, K., Oscar-Berman, M., Gold, M.S., Duane, D.D., Sultan, S.G.S., et al. (2015). Reward deficiency syndrome: attentional/arousal subtypes, limitations of current diagnostic nosology, and future research.* Journal of Reward Deficiency Syndrome, 1(1) 6–9.

Related work published in 1990 in JAMA involved the initial discovery of the dopamine D2 receptor gene and alcoholism, with following considerable controversy and over hundreds of published articles. It serves today as the cornerstone of understanding the neuronal and synaptic genetics of many reward dependence behaviors including mood and well-being (Blum et al., 1990; Noble, Blum, Ritchie, Montgomery, & Sheridan, 1991). Between the years 1990 and 1995, polymorphisms (genetic variants) of the DRD2 gene and other brain reward candidate genes (including serotonergic, endocannabinoidergic, enkephalinergic, GABAergic, and dopaminergic; Stice, Yokum, Burger, Epstein, & Smolen, 2012) have been associated with many impulsive, compulsive, and addictive behaviors. In fact the DRD2 gene variant (*TaqI A1 allele*) has been associated with not only alcoholism but also heroin dependence; psychostimulant dependence; nicotine dependence; carbohydrate craving; pathological aggression; pathological gambling; sex addiction; high-risk taking behavior; and certain personality disorders including schizoid avoidance behavior, borderline personality, impaired executive function, inability

to cope with stress in the family, posttraumatic stress disorder, and most recently excessive internet video gaming (Comings & Blum, 2000; Gyollai et al., 2014; Noble, 2003). For a comprehensive overview of the brain reward cascade associated with RDS, please refer to Fig. 30.1 (Modestino et al., 2015).

The possibility that dopamine-related genetic variants may provide information linked to etiology and course of depression has been researched extensively (Pauli et al., 2013). It is well known that depression is highly heritable (Sartor et al., 2012) and that there are several genetic variants that modulate endogenous dopamine neurotransmission (Pauli et al., 2013). While most of the literature favors the role of dopamine in depression (Brown & Gershon, 1993), some association studies failed to find significant associations (Chiesa et al., 2014; Leszczyńska-Rodziewicz et al., 2005).

Importantly, when dopamine-related polymorphisms have been investigated in the context of genome-wide association studies (GWAS), to date, almost none have emerged as significantly associated with depression (Tsang, Mather, Sachdev, & Reppermund, 2017). According to many experts, one likely contributor to these inconsistent findings is that common genetic variants for complex diseases like MDD tend to have small to modest effects. Certainly, tests of association based on a single nucleotide polymorphism (SNP) are unlikely to yield significant effects unless within extensive samples.

One interesting approach, providing an answer to the ongoing controversy of whether or not many polymorphic dopaminergic genes are associated with MDD symptomatology, was published by Pearson-Fuhrhop et al. (2014). They examined the combined sum of five dopamine-related polymorphisms and depressive symptom severity. These included synaptic dopamine availability (*COMT* and *DAT*) and dopamine receptor binding (*DRD1, DRD2, and DRD3*). By utilizing a dopamine genetic risk score based on functional polymorphisms, they found a significant association with the degree of depressive symptoms in healthy individuals and with depression severity in MDD, supporting the original concept of severity as suggested by Blum, Oscar-Berman, Badgaiyan, Braverman, and Gold (2014) and Noble et al. (1991) for alcoholism. Most importantly, this genetic risk score shows stronger associations with the measures of depression than does any single variant. This provides clinicians with a genetic model to determine the best therapeutic intervention concerning dopaminergic therapies for persons with MDD. Furthermore, there has been a significant number of scientific studies suggesting an array of genetic dopaminergic dysfunctions in psychiatric disorders (Blum et al., 1990; Blum et al., 1995, 1996, Blum, Chen, Meshkin, et al., 2006; Blum et al., 2008a; Blum et al., 2008b; Blum, Chen, Williams, et al., 2008; Blum et al., 2012; Blum, Trachtenberg, & Ramsay, 2013; Blum, Oscar-Berman, Badgaiyan, et al., 2014; Blum, Oscar-Berman, Demetrovics, Barh, & Gold, 2014; Blum, Febo, & Badgaiyan, 2016; Blum, 2017; Blum et al., 2017).

While many genes have already been associated with depression, to our knowledge, there has never been a comprehensive candidate approach covering a plethora of polygenes in a single sample to provide information regarding genetic patterns as potential predictors of depression in humans. Certainly, we are cognizant of genomic array scans to delineate chromosomal areas associated and linked to various forms of affective disorders. However, our approach that does not devalue the genomic array approach may provide the basis for a novel personalized medical treatment to combat depression utilizing polymorphic DNA-directed customization of nonpharmacological, antidepressant-like, nutrigenomic solutions (Blum, Chen, et al., 2006).

To date, research in neuropsychiatric genetics, due to limited resources and funding, has resulted in very truncated and in some cases erroneous knowledge. This has

prohibited, to some degree, real progress in both the prevention and treatment of the societal devastation accompanying all forms of mood disorders. In essence, scientific exploration is being hampered by the lack of financial support; and because of this the work has translated into only unlocking a tiny fraction of the neuroscience behind depression. In the future, we hope to be able to make significant strides leading to a candidate genomic map of MDD in humans. This may be the premier step in developing evidence-based, personalized medical solutions not only for MDD but also RDS. This is important as RDS accounts for uncontrollable addictive behaviors affecting over 100 million people (33% of the US population carries the DRD2 gene A1 allele that associates with low dopamine D2 receptors and aberrant substance seeking behavior) in the United States alone (Noble et al., 1991).

To understand the enormous work ahead, the genes involved include, but are not limited to, neurotransmitter synthesis, neuronal vesicle storage, synaptic and neuronal metabolism, and neuronal release. We must mimic the entire brain reward cascade involving serotonergic, cannabinoidergic, enkephalinergic, GABAergic, dopaminergic, adrenergic, and cholinergic pathways (Stice et al., 2012).

Moreover, there are a number of known candidate genes and chromosomal loci that have already been documented to associate with depression and include the following short list: 3Q27.2, 12q24, 4Q35, 13Q14, 3Q35, 13Q14.4, 15Q28, 15Q26, DISC1, CIT, GRIN2A, SIPA1L1, PACAP, AKT1, CCL13 chemokine, POLG, PPARD, CREB1, AANAT, BCL2, 3'UTR, NRID1, MTHFR, CHRM2, cadherin FAT, SCNI8, NPAS3, ST8SIA2, FKBP5, HPA, HTR2A, G72/G30, TPH2, EMID2, PFTKI, TFR2, SMURF1, PBEFI, ACN9, LHFPL3, PILRB, AVPR1B, P2RX7, GLM460ARG, ILORA, DDC, DAT, IL28R, and PAPLN (Gatt, Burton, Williams, & Schofield, 2015), among other genes to be studied. It is our hypothesis that systematic evaluation of the genome via genome-wide scan analysis will be useful for future directions; however, our candidate approach may have a high initial impact, providing DNA targets for possible prevention and treatment of depression.

It is conceivable that one important outcome of this comprehensive systematic genetic evaluation will result in a standard diagnostic gene panel that could be utilized in a clinical setting to provide a generalized genetic pattern to inform patients of their genetic propensity for MDD. Regarding clinical diagnosis of depression and even RDS behaviors, evidence is emerging that brain electric activity mapping (BEAM) may be beneficial in clinical assessment. It is noteworthy that in earlier published work from our laboratory, we have shown that P300 latency was associated with not only substance abuse but also depression (Braverman & Blum,

1996) and ADHD (Braverman et al., 2006; Gold, Blum, Oscar-Berman, & Braverman, 2014).

This should translate to a more targeted and more effective personalized treatment, thereby reducing the burden of recurrent depression, reducing suicidal ideation, reducing Parkinsonian-like symptoms induced by pharmaceutical-based antidepressants (Messiha, 1993), and most importantly supporting a life free of mental anguish and pain and an enhancement of well-being. DNA customization of nutraceutical products is being explored with interesting outcomes. In this regard, "Gene Guided Precision Nutrition" and KB220 variants (a complex mixture of amino acids, trace metals, and herbals) are standard bearers for a state of the art DNA customization. Moreover the technology that is related to KB220 variants may have an impact on reducing or eradicating extreme cravings. This may be accomplished by influencing gene expression as a cornerstone in the pioneering of the practical applications of nutrigenomics (Febo et al., 2017). Continuing discoveries have been an important catalyst for the evolution, expansion, and explicit acknowledgment of the importance of nutrigenomics and its remarkable contributions to human health. Neuro-Nutrigenomics is now a critical field of scientific investigation that offers great promise to improve the human condition. In the beginning is the development of the Genetic Addiction Risk Score (GARS), which has a significant predictive value for severity of alcohol and drug abuse as well as other nonsubstance-related addictive behaviors (Blum, 2015). The customization of such neuro-nutrients has now been commercialized by Geneus Health LLC.

II HAVE WE HATCHED A CUSTOMIZED "PRECISION MEDICINE?"

It is well known that intrinsic reward and drugs of abuse like opioids may converge upon mesolimbic pathways and activate a common mechanism of neural plasticity in the nucleus accumbens (Pitchers et al., 2014). Pitchers et al. showed an endogenous opioid-induced neuroplasticity of dopaminergic neurons in the VTA that influenced natural and opiate (morphine) reward (Pitchers et al., 2014). This finding is of interest because the D-phenylalanine present in the KB220 variant, KB220Z compound, is known to act as an enkephalinase inhibitor (Marcello, Grazia, Sergio, & Federigo, 1986) and may induce recruitment of dopamine-containing neurons especially in carriers of the dopamine receptor (DR)D2 A1 allele. Carriers of the DRD2 A1 allele have 30%–40% fewer D2 receptors than carriers of the DRD2 A2 allele (Noble et al., 1991)

III INITIAL DNA CUSTOMIZED STUDIES

We started out with the mindset to design a study with which to evaluate DNA customization with nutritional solutions for both wellness and especially weight management. In terms of nutrigenomics, we review the results of a number of studies (Blum, Chen, Chen, et al., 2008a; Blum, Chen, Chen, et al., 2008b; Blum, Chen, Williams, et al., 2008; Blum, Downs, et al., 2016; Blum, Febo, Fried, et al., 2016; Blum, Meshkin, & Downs, 2006), whereby Blum's laboratory genotyped 1058 subjects, and these subjects were administered KB220Z (formerly LG9939, Recomposize, Genotrim) (a complex neuroadaptagen nutraceutical that includes dl-phenylalanine, chromium, l-tyrosine other select amino acids, and adaptogens) based on polymorphic outcomes. The resultant customized formulae involved a minimum of 175 single nucleotide polymorphisms (SNPs) covering 16 genes relevant to the brain reward cascade (BRC; Blum et al., 2017) and most importantly involved in "dopamine homeostasis." In this small cohort, using simple t-tests comparing many parameters before and after 80 days of consumption of the nutraceutical, we found positive significant changes in a number of important parameters including reduced weight and lower body mass index (BMI; Blum, Chen, Chen, et al., 2008a; Blum, Chen, Chen, et al., 2008b; Blum, Chen, Williams, et al., 2008). Importantly, of all the outcomes and gene polymorphisms, only the DRD2 gene polymorphism (Al allele) had a significant Pearson correlation using days on treatment ($r = 0.42$, and $P = .045$). This twofold increase is a significant genotype for compliance in treatment (Blum, Chen, Chen, et al., 2008a; Blum, Chen, Chen, et al., 2008b; Blum, Chen, Williams, et al., 2008).

Also, Blum's research team methodically assessed the impingement of polymorphisms from five possible genes and their potential as targets for the development of a DNA-customized nutraceutical KB220Z to combat obesity with particular emphasis on body recomposition as measured by BMI (Blum, Chen, Chen, et al., 2008b). We included specific alleles such as DRD2 Al, MTHFR C 677T, 5HT2a 1438G/A, PPAR-γProl2Ala, and Leptin Ob1875 < 208bp. Pre- and post hoc analyses revealed a significant difference between the starting BMI and the BMI following an average of 41 days (28–70 days) of KB220Z intake in the 21 individuals. Similarly the pretreatment weight in pounds was 183.52 compared with the posttreatment weight of 179, a statistically significant ($P < .047$) change. In this particular group, 53% lost on average over 2.5% of their starting weight (Blum, Chen, et al., 2006). The results of these studies in obesity are presented to provide some idea of the potential success utilizing nutrigenomics-based solutions, which in the future may pave the way to treat and prevent RDS-like behaviors.

IV SUMMARY

We are hypothesizing that major depression, especially anhedonia (excluding bipolar disorder), should be included as a subtype of Reward Deficiency Syndrome (RDS) following more extensive genetic and molecular neurobiological research. RDS, first conceptualized by one of the authors (KB) in 1995, is a failure of the system that usually confers satisfaction, resulting in behavior such as overeating, heavy cigarette smoking, drug and alcohol abuse, gambling, and hyperactivity and is based on hypodopaminergia. Importantly, RDS has been recently designated for inclusion in Wenzel (2017).

We are proposing the potential use of a genetic risk score in our approach. This combines the assumed effects of multiple polymorphisms in the same biological system, as a new tool in addiction medicine that could have therapeutic value in the future for improved diagnostic clarity. In this brief hypothesis-based chapter, we present both neurogenetic and molecular neurobiological studies that support our premise. While this is a paradigm shift from traditional psychiatric practice, its adaptation seems to have heuristic value. The inclusion of depression in RDS will assist in more appropriate treatment in the addiction recovery community, whereby the goal of achieving dopamine stabilization or homeostasis will result in better clinical outcomes. One example of this laudable goal can be achieved by epigenetic induction, involving the administration of the nutraceutical KB220Z (Blum, Chen, et al., 2006) that has shown positive results in food addiction and obesity. We hereby encourage future research into dopaminergic homeostasis in SUD with MDD.

References

Archer, T., Oscar-Berman, M., Blum, K., & Gold, M. (2013). Epigenetic Modulation of Mood Disorders. *Journal of Genetic Syndromes & Gene Therapy, 4*(120), pii: 1000120.

Beaulieu, S., Saury, S., Sareen, J., Tremblay, J., Schütz, C. G., & McIntyre, R. S. (2012). The Canadian network for mood and anxiety treatments (CANMAT) task force recommendations for the management of patients with mood disorders and comorbid substance use disorders. *Annals of Clinical Psychiatry, 24*(1), 38–55.

Bentley, S. M., Pagalilauan, G. L., & Simpson, S. A. (2014). Major depression. *The Medical Clinics of North America, 98*(5), 981–1005. https://doi.org/10.1016/j.mcna.2014.06.013.

Blum, K. (2015). Coupling neurogenetics (GARS™) and a nutrigenomic based dopaminergic agonist to treat reward deficiency syndrome (RDS): targeting polymorphic reward genes for carbohydrate addiction algorithms. *Journal of Reward Deficiency Syndrome, 1*(2). https://doi.org/10.17756/jrds.2015-012.

Blum, K. (2017). Reward deficiency syndrome. In A. Wenzel (Ed.), *The SAGE encyclopedia of abnormal and clinical psychology seven volume set*. New York, NY: Sage Publishing.

Blum, K., & Braverman, E. R. (2000). Reward deficiency syndrome: a biogenetic model for the diagnosis and treatment of impulsive, addictive and compulsive behaviors. *Journal of Psychoactive Drugs, 32*, 1–112.

Blum, K., Chen, A. L., Chen, T. J., Rhoades, P., Prihoda, T. J., et al. (2008b). LG839: anti-obesity effects and polymorphic gene correlates of reward deficiency syndrome. *Advances in Therapy, 25*(9), 894–913. https://doi.org/10.1007/s12325-008-0093-z.

Blum, K., Chen, A. L. C., Chen, T. L. C., Rhoades, P., Prihoda, P., et al. (2008a). Dopamine D2 receptor Taq A1 allele predicts treatment compliance of LG839 in a subset analysis of a pilot study in The Netherlands. *Gene Therapy and Molecular Biology, 12*(1), 129–140.

Blum, K., Chen, T. J., Meshkin, B., et al. (2006). Reward deficiency syndrome in obesity: a preliminary cross-sectional trial with a Genotrim variant. *Advances in Therapy, 23*(6), 1040–1051.

Blum, K., Chen, T. J. H., Williams, L., et al. (2008). A short term pilot open label study to evaluate efficacy and safety of LG839, a customized DNA directed nutraceutical in obesity: exploring nutrigenomics. *Gene Therapy and Molecular Biology, 12*(2), 371–382.

Blum, K., Downs, B. W., Dushaj, K., Li, M., Braverman, E. R., Fried, L., et al. (2016). The benefits of customized DNA directed nutrition to balance the brain reward circuitry and reduce addictive behaviors. *Precision Medicine, 1*(1), 18–33.

Blum, K., Febo, M., & Badgaiyan, R. D. (2016). Fifty years in the development of a glutaminergic-dopaminergic optimization complex (KB220) to balance brain reward circuitry in reward deficiency syndrome: a pictorial. *Austin Addiction Sciences, 1*(2). pii: 1006.

Blum, K., Febo, M., Badgaiyan, R. D., Demetrovics, Z., Simpatico, T., Fahlke, C., et al. (2017). Common neurogenetic diagnosis and meso-limbic manipulation of hypodopaminergic function in reward deficiency syndrome (RDS): changing the recovery landscape. *Current Neuropharmacology, 15*(1), 184–194.

Blum, K., Febo, M., Fried, L., Li, M., Dushaj, K., Braverman, E. R., et al. (2016). Hypothesizing that neuropharmacological and neuroimaging studies of glutaminergic-dopaminergic optimization complex (KB220Z) are associated with "Dopamine Homeostasis" in reward deficiency syndrome (RDS). *Substance Use & Misuse, 52*(4), 535–547. https://doi.org/10.1080/10826084.2016.1244551.

Blum, K., Meshkin, B., & Downs, B. W. (2006). DNA based customized nutraceutical "gene therapy" utilizing a genoscore: a hypothesized paradigm shift of a novel approach to the diagnosis, stratification, prognosis and treatment of inflammatory processes in the human. *Medical Hypotheses, 66*(5), 1008–1018. https://doi.org/10.1016/j.mehy.2005.09.029.

Blum, K., Noble, E. P., Sheridan, P. J., Montgomery, A., Ritchie, T., Jagadeeswaran, P., et al. (1990). Allelic association of human dopamine D2 receptor gene in alcoholism. *JAMA, 263*, 2055–2060.

Blum, K., Oscar-Berman, M., Badgaiyan, R. D., Braverman, E. R., & Gold, M. S. (2014). Hypothesizing darkness induced alcohol intake linked to dopaminergic regulation of brain function. *Psychology, 05*(04), 282–288. https://doi.org/10.4236/psych.2014.54038.

Blum, K., Oscar-Berman, M., Demetrovics, Z., Barh, D., & Gold, M. S. (2014). Genetic addiction risk score (GARS): molecular neurogenetic evidence for predisposition to reward deficiency syndrome (RDS). *Molecular Neurobiology, 50*(3), 765–796. https://doi.org/10.1007/s12035-014-8726-5.

Blum, K., Oscar-Berman, M., Giordano, J., Downs, B., Simpatico, T., Han, D., et al. (2012). Neurogenetic impairments of brain reward circuitry links to reward deficiency syndrome (RDS): potential nutrigenomic induced dopaminergic activation. *Journal of Genetic Syndromes & Gene Therapy, 03*(04). https://doi.org/10.4172/2157-7412.1000e115.

Blum, K., Sheridan, P. J., Wood, R. C., Braverman, E. R., Chen, T. J., & Comings, D. E. (1995). Dopamine D2 receptor gene variants: association and linkage studies in impulsive-addictive-compulsive behaviour. *Pharmacogenetics, 5*(3), 121–141. https://doi.org/10.1097/00008571-199506000-00001.

Blum, K., Sheridan, P. J., Wood, R. C., Braverman, E. R., Chen, T. J., Cull, J. G., et al. (1996). The D2 dopamine receptor gene as a determinant of reward deficiency syndrome. *Journal of the Royal Society of Medicine, 89*(7), 396–400.

Blum, K., Trachtenberg, M. C., & Ramsay, J. C. (2013). Improvement of inpatient treatment of the alcoholic as a function of neurotransmitter restoration: a pilot study. *The International Journal of the Addictions*, 23(9), 991–998.

Blum, K., Wallace, J. E., & Hall, W. C. (1986). A commentary on the pathogenesis and biochemical profile of alcohol-induced depression. *Journal of Psychoactive Drugs*, 18(2), 161–162.

Borsook, D., Linnman, C., Faria, V., Strassman, A., Becerra, L., & Elman, I. (2016). Reward deficiency and anti-reward in pain chronification. *Neuroscience & Biobehavioral Reviews*, 68, 282–297. https://doi.org/10.1016/j.neubiorev.2016.05.033.

Braverman, E. R., & Blum, K. (1996). Substance use disorder exacerbates brain electrophysiological abnormalities in a psychiatrically-III population. *Clinical EEG and Neuroscience*, 27(4 Suppl), 5–28. https://doi.org/10.1177/1550059496027s0402.

Braverman, E. R., Chen, T. J., Schoolfield, J., Martinez-Pons, M., Arcuri, V., Varshavskiy, M., et al. (2006). Delayed P300 latency correlates with abnormal test of variables of attention (TOVA) in adults and predicts early cognitive decline in a clinical setting. *Advances in Therapy*, 23(4), 582–600. https://doi.org/10.1007/bf02850047.

Brown, A. S., & Gershon, S. (1993). Dopamine and depression. *Journal of Neural Transmission. General Section*, 91(2-3), 75–109.

Brugha, T. S. (2003). The effects of life events and social relationships on the course of major depression. *Current Psychiatry Reports*, 5(6), 431–438. https://doi.org/10.1007/s11920-003-0081-6.

Chiesa, A., Lia, L., Alberti, S., Lee, S. J., Han, C., Patkar, A. A., et al. (2014). Lack of influence of rs4680 (COMT) and rs6276 (DRD2) on diagnosis and clinical outcomes in patients with major depression. *International Journal of Psychiatry in Clinical Practice*, 18(2), 97–102. https://doi.org/10.3109/13651501.2014.894073.

Chung, Y. S., & Barch, D. (2015). Anhedonia is associated with reduced incentive cue related activation in the basal ganglia. *Cognitive, Affective, & Behavioral Neuroscience*, 15(4), 749–767. https://doi.org/10.3758/s13415-015-0366-3.

Comings, D. E., & Blum, K. (2000). Reward deficiency syndrome: genetic aspects of behavioral disorders. *Progress in Brain Research*, 126, 325–341.

Compton, W. (2006). Changes in the prevalence of major depression and comorbid substance use disorders in the United States between 1991–1992 and 2001–2002. *American Journal of Psychiatry*, 163(12), 2141. https://doi.org/10.1176/appi.ajp.163.12.2141.

Elman, I., Ariely, D., Mazar, N., Aharon, I., Lasko, N. B., Macklin, M. L., et al. (2005). Probing reward function in post-traumatic stress disorder with beautiful facial images. *Psychiatry Research*, 135(3), 179–183. https://doi.org/10.1016/j.psychres.2005.04.002.

Elman, I., & Borsook, D. (2016). Common brain mechanisms of chronic pain and addiction. *Neuron*, 89(1), 11–36. https://doi.org/10.1016/j.neuron.2015.11.027.

Elman, I., Lowen, S., Frederick, B. B., Chi, W., Becerra, L., & Pitman, R. K. (2009). Functional neuroimaging of reward circuitry responsivity to monetary gains and losses in posttraumatic stress disorder. *Biological Psychiatry*, 66(12), 1083–1090. https://doi.org/10.1016/j.biopsych.2009.06.006.

Endicott, J. (1998). Gender similarities and differences in the course of depression. *The Journal of Gender-Specific Medicine*, 1(3), 40–43.

Febo, M., Blum, K., Badgaiyan, R. D., Perez, P. D., Colon-Perez, L. M., Thanos, P., et al. (2017). Enhanced functional connectivity and volume between cognitive and reward centers of naïve rodent brain produced by pro-dopaminergic agent KB220Z. *PLoS One*, 12(4). https://doi.org/10.1371/journal.pone.0174774.

Ferguson, V. (1984). Nursing care delivery in the US: challenges for the decade. *Taehan Kanho*, 23(3), 42–43.

Gatt, J. M., Burton, K. L., Williams, L. M., & Schofield, P. R. (2015). Specific and common genes implicated across major mental disorders: a review of meta-analysis studies. *Journal of Psychiatric Research*, 60, 1–13. https://doi.org/10.1016/j.jpsychires.2014.09.014.

Gold, M. S., Blum, K., Oscar-Berman, M., & Braverman, E. R. (2014). Low dopamine function in attention deficit/hyperactivity disorder: should genotyping signify early diagnosis in children? *Postgraduate Medicine*, 126(1), 153–177. https://doi.org/10.3810/pgm.2014.01.2735.

González, H. M., Vega, W. A., Williams, D. R., Tarraf, W., West, B. T., & Neighbors, H. W. (2010). Depression care in the United States. *Archives of General Psychiatry*, 67(1), 37. https://doi.org/10.1001/archgenpsychiatry.2009.168.

Graßnickel, V., Illes, F., Juckel, G., & Uhl, I. (2015). Loudness dependence of auditory evoked potentials (LDAEP) in clinical monitoring of suicidal patients with major depression in comparison with non-suicidal depressed patients and healthy volunteers: a follow-up-study. *Journal of Affective Disorders*, 184, 299–304. https://doi.org/10.1016/j.jad.2015.06.007.

Greenberg, P. E., Fournier, A., Sisitsky, T., Pike, C. T., & Kessler, R. C. (2015). The economic burden of adults with major depressive disorder in the United States (2005 and 2010). *The Journal of Clinical Psychiatry*, 76(2), 155–162. https://doi.org/10.4088/jcp.14m09298.

Gyollai, A., Griffiths, M. D., Barta, C., Vereczkei, A., Urbán, R., Kun, B., et al. (2014). The genetics of problem and pathological gambling: a systematic review. *Current Pharmaceutical Design*, 20(25), 3993–3999.

Insel, T. R. (2003). Is social attachment an addictive disorder? *Physiology & Behavior*, 79(3), 351–357. https://doi.org/10.1016/s0031-9384(03)00148-3.

Kinney, D. K., & Tanaka, M. (2009). An evolutionary hypothesis of depression and its symptoms, adaptive value, and risk factors. *The Journal of Nervous and Mental Disease*, 197(8), 561–567. https://doi.org/10.1097/nmd.0b013e3181b05fa8.

Knol, M. J., Twisk, J. W., Beekman, A. T., Heine, R. J., Snoek, F. J., & Pouwer, F. (2006). Depression as a risk factor for the onset of type 2 diabetes mellitus. A meta-analysis. *Diabetologia*, 49(5), 837–845.

Leszczyńska-Rodziewicz, A., Hauser, J., Dmitrzak-Weglarz, M., Skibińka, M., Czerski, P., Zakrzewska, et al. (2005). Lack of association between polymorphisms of dopamine receptors, type D2, and bipolar affective illness in a polish population. *Medical Science Monitor: International Medical Journal of Experimental and Clinical Research*, 11(6), CR289–295.

Marcello, F., Grazia, S. M., Sergio, M., & Federigo, S. (1986). Pharmacological "enkephalinase" inhibition in man. *Advances in Experimental Medicine and Biology*, 198(Pt B), 153–160.

Mental Health America. (2016). *New state rankings shines light on mental health crisis, show differences in blue red states*. Retrieved from: http://www.mentalhealthamerica.net/new-state-rankings-shines-light-mental-health-crisis-show-differences-blue-red-states. Accessed 27 June 2017.

Messiha, F. S. (1993). Fluoxetine: adverse effects and drug-drug interactions. *Journal of Toxicology. Clinical Toxicology*, 31(4), 603–630.

Modestino, E. J., Blum, K., Oscar-Berman, M., Gold, M. S., Duane, D. D., Sultan, S. G. S., et al. (2015). Reward deficiency syndrome: attentional/arousal subtypes, limitations of current diagnostic nosology, and future research. *Journal of Reward Deficiency Syndrome*, 1(1), 6–9.

Nestler, E. J., & Carlezon, W. A., Jr. (2006). Role of the brain's reward circuitry in depression: transcriptional mechanisms. *Biological Psychiatry*, 59(12), 1151–1159.

Noble, E. P. (2003). D2 dopamine receptor gene in psychiatric and neurologic disorders and its phenotypes. *American Journal of Medical Genetics*, 81, 257–267.

Noble, E. P., Blum, K., Ritchie, T., Montgomery, A., & Sheridan, P. J. (1991). Allelic association of the D2 dopamine receptor gene with receptor-binding characteristics in alcoholism. *JAMA Psychiatry*, 48(7), 648–654.

Nouwen, A., Winkley, K., Twisk, J., Lloyd, C. E., Peyrot, M., Ismail, K., et al. (2010). Type 2 diabetes mellitus as a risk factor for the onset of depression: a systematic review and meta-analysis. *Diabetologia*, 53(12), 2480–2486. https://doi.org/10.1007/s00125-010-1874-x.

Pauli, A., Prata, D. P., Mechelli, A., Picchioni, M., Fu, C. H., Chaddock, C. A., et al. (2013). Interaction between effects of genes coding for dopamine and glutamate transmission on striatal and parahippocampal function. *Human Brain Mapping, 34,* 2244–2258. https://doi.org/10.1002/hbm.22061.

Pearson-Fuhrhop, K. M., Dunn, E. C., Mortero, S., Devan, W. J., Falcone, G. J., Lee, et al. (2014). Dopamine genetic risk score predicts depressive symptoms in healthy adults and adults with depression. *PLoS One, 9*(5). https://doi.org/10.1371/journal.pone.0093772.

Pitchers, K. K., Coppens, C. M., Beloate, L. N., Fuller, J., Van, S., Frohmader, KS., et al. (2014). Endogenous opioid-induced neuroplasticity of dopaminergic neurons in the ventral tegmental area influences natural and opiate reward. *The Journal of Neuroscience, 34*(26), 8825–8836. PMID: 24966382. https://doi.org/10.1523/JNEUROSCI.0133-14.2014.

Sartor, C. E., Grant, J. D., Lynskey, M. T., McCutcheon, V. V., Waldron, M., Statham, D. J., et al. (2012). Common heritable contributions to low-risk trauma, high-risk trauma, posttraumatic stress disorder, and major depression. *Archives of General Psychiatry, 69*(3), 293–299. https://doi.org/10.1001/archgenpsychiatry.2011.1385.

Staiger, P. K., Thomas, A. C., Ricciardelli, L. A., & McCabe, M. P. (2011). Identifying depression and anxiety disorders in people presenting for substance use treatment. *The Medical Journal of Australia, 195*(3), S60–S63.

Stein, M. D., Rivera, O. J., Anderson, B. J., & Bailey, G. L. (2017). Perceived need for depression treatment among persons entering inpatient opioid detoxification. *The American Journal on Addictions, 26*(4), 395–399. https://doi.org/10.1111/ajad.12554.

Stice, E., Yokum, S., Burger, K., Epstein, L., & Smolen, A. (2012). Multilocus genetic composite reflecting dopamine signaling capacity predicts reward circuitry responsivity. *Journal of Neuroscience, 32*(29), 10093–10100. https://doi.org/10.1523/jneurosci.1506-12.2012.

Szczepanik, J., Nugent, A. C., Drevets, W. C., Khanna, A., Zarate, C. A., & Furey, M. L. (2016). Amygdala response to explicit sad face stimuli at baseline predicts antidepressant treatment response to scopolamine in major depressive disorder. *Psychiatry Research: Neuroimaging, 254,* 67–73. https://doi.org/10.1016/j.pscychresns.2016.06.005.

Tsang, R. S., Mather, K. A., Sachdev, P. S., & Reppermund, S. (2017). Systematic review and meta-analysis of genetic studies of late-life depression. *Neuroscience and Biobehavioral Reviews, 75,* 129–139. https://doi.org/10.1016/j.neubiorev.2017.01.028.

Walters, C. L., Kuo, Y., & Blendy, J. A. (2003). Differential distribution of CREB in the mesolimbic dopamine reward pathway. *Journal of Neurochemistry, 87*(5), 1237–1244. https://doi.org/10.1046/j.1471-4159.2003.02090.x.

Wenzel, A. (2017). *Sage encyclopedia of abnormal and clinical psychology.*

Young, C. B., Chen, T., Nusslock, R., Keller, J., Schatzberg, A. F., & Menon, V. (2016). Anhedonia and general distress show dissociable ventromedial prefrontal cortex connectivity in major depressive disorder. *Translational Psychiatry, 6*(5). https://doi.org/10.1038/tp.2016.80.

Traveling Through the Storm: Leveraging Virtual Patient Monitoring and Artificial Intelligence to Observe, Predict, and Affect Patient Behavior in CNS Drug Development

Adam Hanina, Laura Shafner

AiCure, New York, NY, United States

Over the next few decades, CNS drug development will become increasingly precise and targeted, moving from a system-level understanding of brain function to a granular approach where the brain is viewed similarly to a microchip and computer software and moving from neural circuitry to information architecture. The hope is that a veritable shift toward the underlying mechanics that contribute to illness onset and progression would yield more effective treatments. This shift will also require the development of more portable and accurate monitoring systems and a better grasp of how information and biological processes move and impact our biological hardware. It is a slow and inevitable march toward a software representation of human awareness, where layers of emotion and consciousness are represented through a complex contour map distorting, managing, and interpreting data.

For the time being, we find ourselves at the system level. New sensors and mobile devices can help us capture more objective and longitudinal data sets to better understand and potentially predict complex behavioral patterns. Virtual monitoring in ambulatory patients may provide greater resolution into symptom and disease progression. Daily insights on a patient level combined with rapid advances in computing and artificial intelligence have the potential to amplify signal detection, moving away from average scores on the primary response endpoint scores to a refined understanding of the response at the patient level. The heterogeneity of this patient population precludes a one-size-fits-all approach.

In this chapter, we will cover methodologies to increase drug exposure and provide greater signal detection and approaches for more objective endpoint detection. In many ways, simple tactics can have a significant difference between trial failure and success. However, the lack of FDA guidance to validate these types of novel instruments presents some difficulty to the industry adapting to new standards quickly. Existing rating assessments, while currently on the gold standard, do not have the specificity or sensitivity to be used as a calibration tool for new technology methods. Strengthened FDA guidance in this area is needed.

Need for future

I DRAWING SYSTEM ANALYSIS FROM WEATHER PATTERNS

In many ways the modeling of weather patterns draws the closest comparison to measuring and understanding the complex CNS. Weather forecasting has a fascinating history, and its evolution has moved over the centuries from observation and subjective assessment to a sophisticated analysis that includes sensors and predictive analytics (Earth Observatory, 2002). The American Meteorological Society recommends system-level analysis to include two primary areas:

1. Build data sets: "Application of statistical techniques to correct systemic biases. Past model forecasts are compared to observations over a long period of time to quantify expected errors."

2. Multimodal approach to developing models: "Using an ensemble of models to explore forecast uncertainty that results from the chaotic nature of the atmosphere and errors in the modeling system."[1]

seen in trials and undermine our ability to see clear efficacy and safety properties of the drug being tested. Missing data causes assumptions and inferences to be made, and a plethora of statistical methodologies are currently

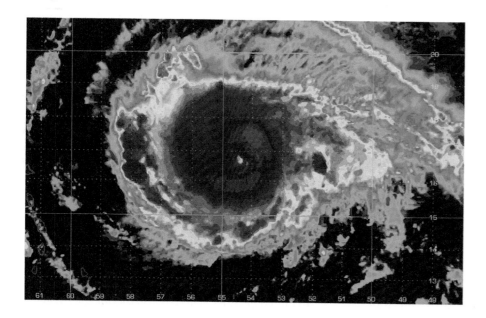

Vast amounts of data are captured to build these temporal models, and baseline calculations are created to ensure variance is identified. An array of sensors, often overlapping, is required to fuel data models for meaningful prediction. Some data parameters captured by these sensors are more closely correlated to weather patterns than others, and naturally a data hierarchy exists. However, the data must be longitudinal, objective, and frequently sampled to provide an insight into the severity and predictability of weather patterns. Furthermore, since each pattern is unique, a baseline for that specific weather pattern must be created.

II COMPLEXITY AND NOISE THROUGH INTERMITTENT DATA CAPTURE AND INTERPRETATION IN CLINICAL TRIALS

Many existing data collection methods or tools are insensitive and uncalibrated to specific endpoints. Most trials depend on brick-and-mortar types of infrequent rater-led assessments. Yet system-level analysis requires longitudinal, objective, and frequent assessments; a lack thereof may contribute to the current small effect size

being used to act as data patches. Unfortunately, these gaps can be quite large. This noise contributes to the high degree of trial failure in CNS.

III INFORMATION HIERARCHY, STANDARDIZATION, AND SCALE

The obvious answer to missing data and infrequent data collection is to fill the gaps through technology. Many groups are investigating the applications of new sensor and digital toolsets. Unfortunately, not all data are created equally, and a bias occurs within the technology community to try to capture as many data inputs as possible under the guise of big data. This runs the risk of distorting the endpoint focus and creating distraction. The need for an information hierarchy in the face of these big, commoditized, and accessible data sets (GPS, accelerometer, and self-reports) is clear. Within this hierarchy, strategies should be developed that prioritize not only data sets that provide insights into a specific indication within a single trial but also data sets that can ensure intertrial comparisons.

On a system level the data that must be captured within a clinical trial include drug exposure and response. A core framework is needed to identify which data will offer sufficient visibility into the disease state being studied. These data run from the very obvious,

---

[1] https://www.ametsoc.org/ams/index.cfm/about-ams/ams-statements/statements-of-the-ams-in-force/weather-analysis-and-forecasting/ (Accessed 28 March 2018).

for example, whether the patient is ingesting the investigational product, to the more nuanced behavioral patterns that could be indicative of a response. Mechanisms to capture these data may be more or less reliable.

Even after a hierarchy of data has been defined and linked to a specific disease state, the process of data capture must be based on a standardized monitoring method. Standardization yields multiple benefits, including quality control (Sweetman & Doig, 2011; Woodcock, 2010), patient safety, reduced variance, efficiency, and regulatory oversight. These are particularly important given the increasingly globalized nature of clinical trials (Glickman et al., 2009). The only tools that are sufficiently scalable and that lend themselves to the capture of vast data sets in a standardized fashion across millions of patients globally are mobile- and web-based technologies.

IV TRAVELING WITH THE STORM: OBJECTIVE ENDPOINT ASSESSMENT

Determining if a patient is improving on therapy is a nontrivial matter. The lack of biomarkers in CNS obliges researchers to rely on mostly subjective measures. While assessments performed by trained raters are convenient to administer during clinic visits, the results obtained depend on interpretation, interrater variability, and a paucity of data points over the course of the trial. The obligatory reliance on average scores across categories of symptoms compounds the problem of interpretation in largely heterogeneous patient populations, potentially letting a signal go undetected in subpopulations. Ecological momentary assessments, while no longer relying on a rater in the clinic and allowing for frequent data capture, do not solve the subjectivity bias due to the nature of patient self-reporting.

In an ideal situation a human rater would travel with and observe the patient every second of every day, relying on an array of active and passive assessments. This is typically the role that care providers take on, and while untrained and usually only privy to a single patient, they frequently are able to observe and predict symptoms more accurately than a treating physician. In other words, they are unconsciously collecting frequent and longitudinal data sets through passive observation and active interactions and most importantly automatically calibrating across a shifting and unique baseline specific to that patient—in essence, they are traveling alongside the weather pattern. The difficulty is that these experiences are limited to a sample size of one and are not scientifically generalizable. Generalizability requires the same type of continuous observation on a population level to replicate the sensitivity and reliability of existing instruments, requiring a multimodal approach to data capture.

Going back to early research by Intel, landlines could provide insight into cognitive decline; for example, the time it took to answer the phone or remember the name of the person on the other end could be tracked over time and form the basis of an assessment. There is significant research into the applications of voice and audio analysis in CNS research due to the accessibility of the data sources through mobile phones and newer interactive voice interfaces. Insights may be drawn from prosodic and audio analysis; however, it is unclear whether a single mode approach to endpoint analysis is generalizable and repeatable. Rater assessments are rarely conducted solely through a telephone.

New methods are being developed for multimodal analysis that synthesize the clinician–patient interaction and combine both passive and active types of data capture mechanisms.

Passive	Longitudinal analysis
	Temporal analysis
	Visual analysis
Active	Audio analysis
	Speech analysis
	Vocal and content response

Once a clinical paradigm has been developed, it can be tested and measured based on known insights of a specific disease. The example in the succeeding text shows data captured by AiCure that combines both active and passive data (Fig. 31.1). Passive longitudinal data sets ensure a baseline analysis is established. Passive visual data are captured from the camera of the smartphone each time the patient doses. Active tests include a combination of evocative visual information and open- and close-ended questions and ecological momentary assessments where the system automatically interviews the patient and asks them how they feel. These active tests are intermittently presented to the subject and are designed to be significantly shorter than in-clinic assessments, taking anywhere between 30 seconds to 10 minutes to administer.

Once the passive and active data are collected, a multimodal analysis combines computer vision, audio analysis, and deep learning neural networks to explore trends in the data (Fig. 31.2). This "complete" picture of the patient is normalized against the patient's baseline captured from the passive data. While exploratory, these multimodal approaches to data capture are being launched by clinical trial sponsors. Results will be compared with existing instruments to determine whether these new tools are equally or perhaps more robust and to test sensitivity and reliability.

Passive	Longitudinal analysis
	Temporal analysis
	Visual analysis
Active	Audio analysis
	Speech analysis
	Vocal & content response

Baseline Shift Analysis

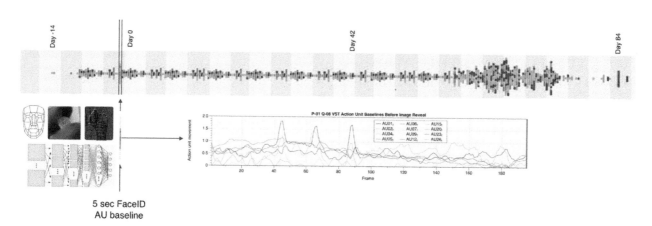

5 sec FaceID
AU baseline

FIG. 31.1 Baseline shift analysis over time. Multimodal computational diagnostics (CDx). *Data retrieved from healthy volunteers, January–July 2018.*

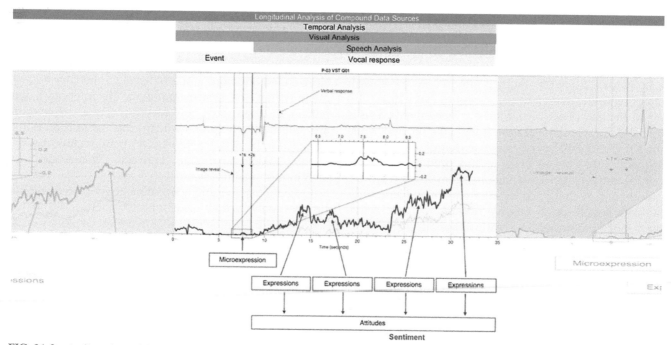

FIG. 31.2 Audio and visual data from single use. Multimodal computational diagnostics (CDx). *Data retrieved from healthy volunteers, January–July 2018.*

V NO DRUG EXPOSURE, NO SIGNAL

Despite exciting developments in continuous virtual patient assessment, a fundamental question still needs to be addressed if clinical trialists are to detect a signal: is the participant taking the investigational medicinal product (IMP)? The tendency is to assume that patients who consent to take part in a clinical trial are looking to advance science and/or may benefit from the treatment and therefore will be adherent to the all aspects of the protocol, including the IMP. Yet the literature on professional subjects (Devine et al., 2013) and high levels of nonadherence in real-world settings suggest that a significant proportion of trial participants are partially adherent or not adherent at all. In some disease states in CNS, up to 39% of participants have been identified as nonadherent based on plasma levels (McCann et al., 2015). Poor drug exposure diminishes the reliability of the data collected, whether these data are based on rater-led assessments or any type of novel multimodal approach.

A lack of drug exposure contributes to a significant statistical power reduction. For example, if 20% of subjects are nonadherent, the statistical power in a clinical trial may be reduced by 30%, potentially jeopardizing the likelihood of seeing a signal (see, e.g., Fig. 31.3). Indeed, small effect sizes of 12%–15% may easily be drowned out. It is estimated that between 30% and 50% of CNS trials fail due to low levels of drug exposure. Although nonadherence is a known problem, current statistical methodologies do not take it into account. While our reliance on intent to treat (ITT) ensures subjects are not cherry-picked out of the primary analysis, it also relies on the assumption that all subjects who enter into a clinical trial contract do so with the intent of receiving treatment. How to accurately measure and maximize drug exposure and account for subjects who may be negatively contributing to the signal by feigning adherence are twin challenges in current clinical trial methodology. Many consider it a scientific lapse in itself not to view adherence as an independent variable, since variation in dose equals variation in drug response (Urquhart, 1991). A reliance on ITT analysis may be responsible for many potentially good drugs not reaching patients and conversely unsafe medications getting into the marketplace, in addition to the decimated valuations following each announcement of a failed drug trial.

The lack of published results and the paucity of accurate dosing data make it unclear exactly how many pharmaceutical drug trials have failed or have had to rely on low drug exposure data when analyzing treatment response.

VI METHODS TO INCREASE DRUG EXPOSURE FOR GREATER STATISTICAL POWER: TRIAL ENRICHMENT STRATEGIES

A plethora of solutions have been developed to fill the missing data gaps caused by nonadherence. However, few have been able to demonstrate increases in drug exposure. Virtual patient monitoring solutions must have a number of attributes to prove effective:

i) Assistive: ability to train and assist the patient
ii) Accurate: validated and accurate
iii) Frequency: easily accessible to ensure frequency of data collection
iv) Scalable: scalable and easy to operationalize by all stakeholders

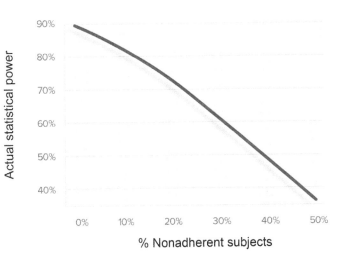

Leveraging Artificial Intelligence and Digital Tools to Observe, Predict, and Characterize Patient Behavior in CNS Drug Development

If 1 in 5 participants does not take the medication, Power will fall to 74%

FIG. 31.3 Power versus nonadherence. From Shiovitz et al. (2016), Fig. 31.1. The impact of noninformative subjects on study power. If noninformative data are not excluded, then a study intended to be powered at 90% would have an actual power between 50% and 87% depending on the percentage of subjects (10%–40%).

Pill counts and patient-reported data are unreliable and overestimate adherence; they do not confirm ingestion and do not offer any information on the time of day a medication is taken or whether the patient has skipped a dose. In a recent phase 2 CNS study, up to 75% of subjects incorrectly self-reported the date and time of last dose taken at the point of PK sampling (Shafner et al., 2017). Other methods such as electronic pill bottles are equally ineffective as they do not confirm ingestion. Smart pills that embed microchips or sensors into the medication are expensive and require a modification to the manufacturing process the need for additional regulatory approval.

Artificial intelligence (AI) platforms that visually confirm the patient, the drug, and correct administration have been shown to accurately measure and increase drug exposure across several therapeutic areas, including CNS. Real-time dosing data allow clinical sites to intervene with subjects during the trial, potentially reducing dropout and ensuring optimal dosing behavior.

VII ACCURATE DOSING PROFILES

The following graphs (Fig. 31.4) highlight four patient behavioral profiles. While all four patients have 90% cumulative average adherence, current ITT methodology would not take into account the very different dosing patterns displayed. These data were captured by an artificial intelligence platform in a 6-month phase 2 CNS study. The blue dots represent a unique dosing event, and the y-axis represents the time of dosing each day. These graphs highlight the importance of capturing dosing variance to understand how forgiving a drug is and variance in response.

Leveraging Artificial Intelligence and Digital Tools to Observe, Predict, and Characterize Patient Behavior in CNS Drug Development

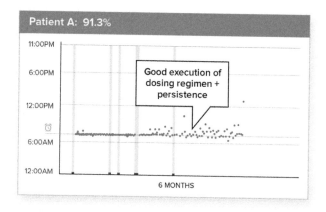

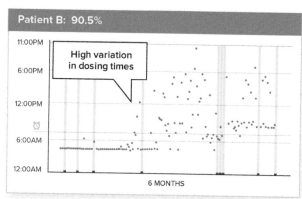

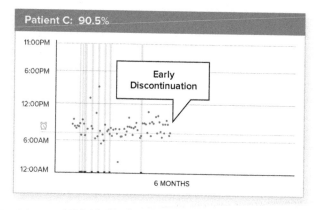

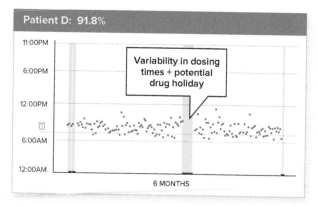

FIG. 31.4   Participant dosing charts. *Bain, E.E., Shafner, L., Walling, D.P., Othman, A.A., Chuang-Stein, C., Hinkle, J., & Hanina, A. (2017). Use of a novel artificial intelligence platform on mobile devices to assess dosing compliance in a phase 2 clinical trial in subjects with schizophrenia. JMIR Mhealth Uhealth, 5(2) e18. doi:10.2196/mhealth.7030. PubMed PMID: 28223265. https://www.ncbi.nlm.nih.gov/pubmed/28223265.*

VIII PREDICTING PATIENT BEHAVIOR

Based on virtual patient monitoring through AI platforms, it appears that early behavior predicts future behavior, that is, a subject's adherence level during the first week or 2 of a trial is highly predictive of adherence throughout the rest of the trial (see figure later). Using a placebo lead-in period to identify nonadherence could help sponsors exclude potentially poor-performing subjects prior to randomization. Different adherence thresholds may be applied to different disease states.

X WHERE IS THIS ALL GOING?

The collection of large visual data sets over time through multimodal approaches based on artificial intelligence offers a new tool in the conduct of clinical trials in CNS. The most fundamental advantage will be to amass high-quality data—dosing data, behavioral patterns, and symptom tracking—that may be used toward a more nuanced, personalized, and objective understanding of patient response. Similar to all scientific advances, the challenge will be in the validation, acceptance, and

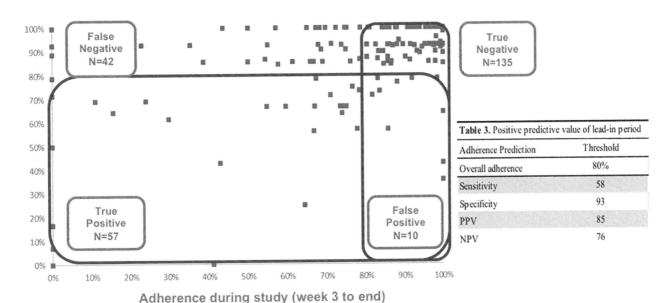

Table 3. Positive predictive value of lead-in period

Adherence Prediction	Threshold
Overall adherence	80%
Sensitivity	58
Specificity	93
PPV	85
NPV	76

Predictive validity analysis: predicting non-adherence in schizophrenia trials (Shafner, Bardsley, Hall, & Hanina, 2018).

IX IDENTIFYING PROFESSIONAL SUBJECTS AND CONCEALMENT NOT TO TAKE THE IMP

Different types of fabrication exist that may result in subjects not following the protocol as intended. A number of groups offer database checks, including biometric databases, to determine if subjects are duplicate enrolling. Concealment of not taking the IMP has been more difficult to detect, particularly in real time. Unblinded data at the end of the study may reveal subjects with little or no drug in their blood, but few approaches allow for the identification of this behavior during the trial. Artificial intelligence platforms that rely on visual data may offer new insights and real-time evidence of this type of behavior. Data acquired by AiCure indicate that intentional nonadherence, while feigning to take the IMP, may range between 5% and 31% in CNS clinical trials.

adoption by the scientific community. Partnerships between the pharmaceutical and technology sectors are rapidly increasing as ubiquitous methods of collecting hitherto unseen behavioral patterns are viewed as potentially promising, in particular in disease states that lack clear biomarkers, such as CNS disorders. Artificial intelligence will allow for the management and parsing of huge data sets that travel from clinical development to clinical practice and back again. Unlike meteorology the hope will be not only to interpret and predict patterns but also to affect clinical outcomes.

References

Devine, E. G., Waters, M. E., Putnam, M., Surprise, C., O'Malley, K., Richambault, C., et al. (2013). Concealment and fabrication by experienced research subjects. *Clinical Trials, 10*(6), 935–948.

Earth Observatory. (2002). https://earthobservatory.nasa.gov/Features/WxForecasting/wx2.php.

Glickman, S. W., McHutchison, J. G., Peterson, E. D., Cairns, C. B., Harrington, R. A., Califf, R. M., et al. (2009). Ethical and scientific implications of the globalization of clinical research. *The New England Journal of Medicine, 360*(8), 816–823.

McCann, D. J., Petry, N., Bresell, A., Isacsson, E., Wilson, E., & Alexander, R. (2015). Medication nonadherence, "professional subjects," and apparent placebo responders: overlapping challenges for medications development. *Journal of Clinical Psychopharmacology, 35*(5), 566–573.

Shafner, L., Bardsley, R., Hall, G., & Hanina, A. (2018). Using artificial intelligence platforms to enhance study design in schizophrenia trials. In *Presented at the schizophrenia international research society meeting, April 4–8, 2018, Florence, Italy.*

Shafner, L., McCue, M., Rubin, A., Dong, X., Hanson, E., Mahableshwarkar, A. R., et al. (2017). Using artificial intelligence platforms to monitor and identify early nonadherence activity based on visual confirmation of medication ingestion. In *Presented during the American Society of Clinical Psychopharmacology (ASCP) annual meeting, May 29–June 2, 2017, Miami Beach, FL, USA.* Study 2002: NCT02477020.

Sweetman, E. A., & Doig, G. S. (2011). Failure to report protocol violations in clinical trials: a threat to internal validity? *Trials, 12,* 214.

Urquhart, J. (1991). Patient compliance as an exploratory variable in four selected cardiovascular studies. In J. Cramer & B. Spilker (Eds.), *Patient compliance in medical practice and clinical trials.* New York: Raven Press.

Woodcock, J. (2010). *Director of the FDA's Center for Drug Evaluation and Research (CDER), speaking at the Institute of Medicine's (US) Forum on drug discovery, development, and translation. Summary.* Washington, DC: National Academies Press (US). http://www.ncbi.nlm.nih.gov/books/NBK50892/.

Index